Praise for
MADAME BOVARY'S OVARIES

"A whimsical and unique analysis of the forces motivating literary characters. . . . This well-researched and humorous narrative will appeal to lovers of evolutionary science as well as of literature." —*Science News*

"With a very readable prose style, David and Nanelle Barash give account of an important new development in literary criticism: the incorporation of the biology of human nature." —Edward O. Wilson, Pulitzer Prize–winning author of *On Human Nature*

"Lies at the crossroads between literary studies and biology, and has much to offer students of either subject." —*Nature*

"Delightful, deliberately provocative and quick-footed . . . they persuasively argue that to understand literature, one should understand evolutionary biology. . . . An English teacher or book club hoping to provoke a spirited discussion should take a look." —*Seattle Times*

"An engrossing, provocative work of literary speculation." —*San Francisco Chronicle*

"An intriguing and groundbreaking book. Barash and Barash take us on a romp through the classics to explain why these characters and plots ring so true. . . . You will never look at Othello, Jay Gatsby, Hester Prynne or Leopold Bloom the same way again." —Helen Fisher, author of *Why We Love*

"Witty and insightful . . . their explicit goal is to apply the basic principles of sociobiology (think Richard Dawkins' *The Selfish Gene*) to the study of literature . . . the result is a surprisingly lighthearted romp through both literature and the animal kingdom, aimed at a casual reader who's interested in either or both." —*Publishers Weekly*

Madame Bovary's Ovaries

A Darwinian Look at Literature

DAVID P. BARASH
and
NANELLE R. BARASH

DELTA TRADE PAPERBACKS

MADAME BOVARY'S OVARIES: A DARWINIAN LOOK AT LITERATURE
A Delta Book

PUBLISHING HISTORY
Delacorte Press hardcover edition published May 2005
Delta Trade Paperback edition / June 2006

Published by
Bantam Dell
A Division of Random House, Inc.
New York, New York

Book design by Ellen Cipriano

Library of Congress Catalog Card Number: 2005298037

Delta is a registered trademark of Random House, Inc.,
and the colophon is a trademark of Random House, Inc.

ISBN-13: 978-0-385-33802-8
ISBN-10: 0-385-33802-3

Printed in the United States of America
Published simultaneously in Canada

www.bantamdell.com

BVG   10   9   8   7   6   5   4   3   2

To the memory of
Nat Barash
1915–2005

CONTENTS

Madame Bovary, c'est moi.

—GUSTAVE FLAUBERT

1

THE HUMAN NATURE OF STORIES

A Quick Hit of Bio-Lit-Crit

———

*O**thello*** isn't just a story about a jealous guy. Huckleberry Finn isn't just a rebellious, headstrong kid. Madame Bovary isn't just a horny married woman. As students, we are told about various ways to understand fiction: that Othello may also teach us about deceit and loyalty (among other things), how Huck will tell us about the American national character, that Madame Bovary will help reveal the meaning of social transgression. In addition, those who get deep and sophisticated enough may be urged to examine what they read from various perspectives: those of Marx, Freud, Jung, or maybe the French literary theorists Derrida or Foucault, not to omit feminist and "queer" studies, socioeconomic analyses, and the historical facts of each author's personal biography. The list is nearly endless: New Criticism, old criticism, new historicism, old historicism, critical theory, and sometimes crackpot theory.

It's all fine, up to a point. No one has a monopoly on how to read what others have written. There is much to be said for examining literature as a reflection of class struggle (Marx), unconscious drives (Freud), power relations (Foucault), social mores, sexual repression, historical forces, or even—as postmodernists often insist—of "texts" that signify nothing more than themselves. But in fact, *Othello is* a story about a jealous guy. Huck Finn *is* a rebellious, headstrong boy. And Madame Bovary *is* a horny married woman.

The reason *Othello* is still being read and performed five hundred years after Shakespeare wrote it is because this play tells us something timeless and universal . . . not so much about a fellow named Othello but about ourselves. It speaks to the Othello within everyone: our shared human nature. *Othello* the play is about a jealous guy, and, as we shall see, jealousy is a particularly potent and widespread human emotion, one to which *men* are especially vulnerable. That's precisely why it's okay to talk about Othello or Madame Bovary or Huckleberry Finn in the present tense: they live on, at least in part, because they have distinctly human characteristics that transcend the artistry by which they were depicted. Their tribulations, responses, loves and hates, fears and delights are in some way recognizable to all readers, to marvel at, agree or disagree with, learn from, or be shocked by.

It may be startling to some—especially those who have not kept up with recent advances in biological science—but the evidence is now undeniable that much of human life is not socially constructed. In short, even though learning and cultural traditions exert a powerful influence, there also exists an underlying human nature, universally valid and characteristic of all *Homo sapiens*. People live in many different places, following many different traditions and cultural trajectories, but beneath this wonderful diversity there is something else that is equally wonderful, and maybe even more so: a common thread of recognizable humanity, woven of human DNA and shared by everyone who reads and writes (as well as those who don't). Othello's jealousy, Huck's rebelliousness, and Emma's urges are just three examples of that common thread.

In his advice to the traveling players, Hamlet suggested that the role of the artist is to hold a mirror up to nature—not, as some theorists would have it, to hold a mirror up to another mirror and thereby reflect only the infinite emptiness of mirrors. The "nature" at issue here isn't wild animals, pretty landscapes, or magnificent wilderness, but *human* nature. And human nature isn't like a unicorn or some other mythical beast. It exists. It does so because human genes exist and have produced a different kind of creature than horse genes or hyacinth genes have. "Read deeply," writes Harold Bloom in *How to Read and Why*, ". . . not to believe, not to accept, not to contradict, but to learn to share in that one nature that writes and reads."

As Bloom intuits, connecting literature and human nature isn't really all that new. Until recently, in fact, our most enduring images of human nature have resided in literature. Where better to find it? Psychology, for instance, didn't even exist until scarcely a hundred years ago, and during most of the twentieth century, it was torn between two equally unhelpful poles: the semimystical mythologizing of Freud and the sterile behaviorism of John Watson and B. F. Skinner. Anyone wanting to get a sense of human nature in, say, the Bronze Age can do no better than to excavate among the words of Homer, or for the Elizabethan Age, Shakespeare.

Universal human nature was perceived thousands of years ago by our greatest storytellers, from the early authors of the Hindu *Mahabharata*, the Babylonian tale of Gilgamesh, and Homer's *Iliad* and *Odyssey* to Virgil's *Aeneid* and the writings of Dante, Cervantes, and Shakespeare. It wasn't until Charles Darwin, however, that the scientific basis for human nature was identified. Actually, some of the most important breakthroughs didn't occur until a century and more *after* Darwin, when the genetic basis of evolution by natural selection was discovered and its implications for human behavior were made clear.

These breakthroughs haven't had much direct effect on the conscious *creation* of literature, but we hope to show that they can be immensely useful in its *interpretation*, since they help the reader to see what was always there, albeit generally unacknowledged by writers and readers alike. Roland Barthes's celebrated essay "The Death of the Author" proclaimed that the intentions of an author do not matter in interpreting his or her text. We agree, to a point. It doesn't matter, for example, whether authors are *intentionally* presenting a biologically accurate view of human beings. In fact, it is even more telling if they have no such aim and yet end up doing just that; nature whispers within their work nonetheless. Just as it did among the ancients, biology continues to flourish inside the best of our modern writers.

The key concept is that human beings, like all other living things, are biological critters, products of evolution by natural selection. As a result, people are strongly inclined to behave in ways that enhance their fitness. Not physical fitness, although being strong, smart, and healthy can certainly help. Rather, fitness is the fundamental evolutionary bottom line: a measure of success in projecting genes into the future. If living things seek food when hungry, sleep when tired, have sex when horny, if they scratch when they itch, do a good job of pumping their blood, and learn to keep their heads down when predators are about, it is because those who do so have been more successful in promoting genes for eating, sleeping, mating, scratching, pumping, hiding, and so forth. Such individuals are, in short, more fit than other individuals whose ancestors were less adroit (or, to be more accurate, whose genes were). This means that whatever else they may be—artful manipulators of language and symbol, composers of symphonies, splitters of atoms no less than of logs—human beings are concatenations of genes that have evolved to do their best at copying themselves and then kicking those copies into the future.

This does not mean that everyone is desperately seeking to have as many children as possible, or even necessarily to survive. But since we have inherited the genes of men and women who *did* reproduce and survive, we unconsciously behave in ways designed to enhance our success in doing so, that is, to benefit what biologists call our fitness. These behaviors are the stubborn, indelible core of human nature. To be sure, human beings have also been blessed, or cursed, with unique self-consciousness and the ability to say no (at least on occasion) to their biological inclinations. Breathing, too, is part of our human, and animal, nature. So is digesting. We cannot say no to them. By contrast, people are not absolutely obliged to have children or even to have sex, not to mention engage in the other, more arcane activities we shall shortly explore. Human beings can and sometimes do say no to many of the fitness-enhancing tendencies that make up human nature. But this doesn't mean that those tendencies aren't there. Indeed, even the occasional decision to say no, and the conscious effort it requires, is testimony to the existence of those deep, internal yearnings in the first place; otherwise, there would be nothing to rebel against. All of this means that people, whether they acknowledge it or not, are fitness-focused creatures. And that isn't all. As we shall explore in chapters ahead, an evolutionary view of human nature goes beyond simply identifying the importance of fitness itself to make specific predictions as to how human beings are likely to behave depending on whether, for example, they are men, women, children, parents, someone's friend, or someone else's enemy.

Although evolutionary biology is so new that it deserves in many ways to be called "revolutionary biology," the idea that great literature reflects certain human universals is actually as old as literary analysis itself, having been foreshadowed in the first organized attempt to make sense of fiction, Aristotle's *Poetics*. "Poetry is something more philosophic and of graver import than history," wrote the great Greek himself, "since its statements are of the nature of universals, whereas those of history are singulars." By "poetry," Aristotle meant creative fiction, not just poetry in the narrow sense but also theater (novels were unknown in his day). His point is that the power of poetry lies in its ability to capture fundamental truths about the human condition, including, most notably, the way people act . . . which, in turn, derives from the nature of what people *are*. And this is where a biological perspective has much to offer.

It won't always be pretty. Indeed, throughout *Madame Bovary's Ovaries*, we'll point to a number of inclinations that are regrettable, sometimes downright despicable, but always, as Nietzsche has written, "human, all too human." Men taking sexual advantage of women; women often doing the same thing, although typically in different ways. Competitiveness, whether violent or more subtle. The selfish underbelly of friendship. Nepotism (favoritism toward relatives) often combined with discrimination against strangers. Abuse and neglect of stepchildren. The catalog is intriguing but not necessarily inspiring. Please note that throughout, we offer descriptions, not prescriptions, in the hope of illuminating what people—and their literature—are like, not necessarily how they ought to be. (In fact, a case can be made that part of the burden of being human is to behave counter to some of our all-too-human inclinations, but that is another story.)

Seeking to understand *Homo sapiens*, not to condone ethically unacceptable behavior, a growing band of scientists has been busily unraveling the nature of human nature. They are known as evolutionary biologists, sociobiologists, behavioral ecologists, Darwinian anthropologists, and, increasingly, evolutionary psychologists, an expanding group in which literary critics are not (yet) included. But at last, the nature of human nature is becoming clear. One of its cardinal principles—reflected in literature—is the gravitational pull exerted by what Richard Dawkins first labeled "selfish genes," a force that influences not only what people do but also the stories they tell about themselves, including what they find interesting, boring, perplexing, and frightening.

Dostoyevsky's Ivan Karamazov worried that "without God, everything is permissible." Without human nature, too, everything is permissible. There could be worlds of the imagination in which people don't eat, sleep, communicate with each other, or reproduce. Or in which there is no sexual identity, no predisposition to care preferentially for one's children or relatives, no predictable patterns of love, anger, competition, or cooperation. The result would be a kind of science fiction or wild fantasy, yet it is noteworthy that such imaginary excursions of extreme inhumanness are rarely undertaken, almost certainly because wholesale departures from the recognizably human are not only very difficult to portray but also genuinely incomprehensible and thus unlikely to be interesting. Even the physically bizarre creatures conjured up in the Harry Potter books, Tolkien's *The Lord of the Rings*, or the various *Star Wars* movies (especially the wonderful bar scene in the first film), for all their imaginative anatomic variety, retain demonstrably human motivations and relationships to each other, just as—centuries after their conception—Hamlet, Don Quixote, and Achilles retain their vitality because they retain their humanity.

It has been said that great writers (Shakespeare, Cervantes, Homer, Tolstoy, Dickens, Austen, James, and Chekhov, among others) peopled the world with characters that seem so real that they appear to have existed even before they were written about. Certainly these characters remain "alive"—almost literally—long after their creators have passed away.

This leads to a key notion in nearly all approaches to fiction, but especially in an evolutionary analysis of literature: believability. "The only difference between fiction and nonfiction," observed Mark Twain, only somewhat ironically, "is that fiction should be completely believable." How, then, is this achieved? Fictional characters are believable when they reveal their human nature, which is to say, when they behave in concert with biological expectation. This is what lies behind Falstaff's expansive humor, Heathcliff's obsessive passion, Jane Eyre's spunkiness, Huck Finn's mixture of naïveté and wisdom, Augie March's antic yearning for self-realization. Harold Bloom once more: "[Shakespeare's] uncanny ability to present consistent and different actual-seeming voices of imaginary beings stems in part from the most abundant sense of reality ever to invade literature."

To be sure, literary characters may sometimes behave counter to reality, and thus in defiance of expectation. After all, they are fictional! But such exceptions are notably rare, and also of particular interest; their impact comes from the drama of seeing patterns contrary to our anticipation of how a biological creature—as opposed to a crystal, say, or a robot—would likely act.

"From the crooked timber of humanity," wrote Kant, "nothing straight was ever fashioned." And from the squishy stuff of humanity, nothing nonbiological was ever fashioned. Even the loftiest products of human imagination are, first of all, emanations of that gooey, breathing, eating, sleeping, defecating, reproducing, evolving, and evolved creature known as *Homo sapiens*. We aren't idealized, ethereal essences but genuine biological beings, shaped by evolution and twisted and gnarled by life itself. This is why the most damning observation that can be made about a character in a novel (or play or movie) is that he or she isn't believable, which is another way of saying that for fiction to make sense, it must accord with a kind of evolutionary reality. Too much artificial straightness won't do.

Interestingly, the exceptions—although not proving this rule—provide paradoxical confirmation. Thus for millennia readers have been fascinated by Achilles, despite his unrealistic immunity to injury; significantly, although he is physically inhuman, when it comes to his psychology Achilles is human indeed. His unbelievable invulnerability is combined with other altogether realistic traits, such as intense competitiveness, a penchant for sulking famously when he feels unappreciated, and a tendency toward anger when deprived of a loved one. Or take the character Pilate in Toni Morrison's *Song of Solomon*, who lacked a navel, thereby magically demonstrating her profound independence. Not only can such suspensions of disbelief be consistent with a biological approach to literature, but they also add to its richness insofar as they help italicize the underlying humanity of the characters in question.

Evolutionary psychology isn't currently part of the standard approach to interpreting literature, but perhaps its time is coming. After all, it is their universality, their human believability, that makes Shakespeare's characters readily comprehensible, despite the fact that they are now five hundred years old. Their language may sometimes be dauntingly archaic, but this only emphasizes the fact that, as the French say, the more things change, the more they remain the same.

There is something instantly recognizable about such basic, such obviously natural traits as Romeo and Juliet's hormonally overheated teenage love, Hamlet's intellectualized indecisiveness, Lady Macbeth's ambition as well as her remorse, Falstaff's drunken cavorting, Viola's resourcefulness, Lear's impotent rage, Othello's jealousy, and Puck's . . . well, his puckishness. And when the last concludes wonderingly in *A Midsummer Night's Dream*, "What fools these mortals be," the reader or theatergoer cannot but agree, because deep down she knows what human beings they—and each of us—all are.

Our point is that yes, Virginia, there is human nature, just as there is hippopotamus nature, halibut nature, even hickory tree nature. And the greatest of storytellers have been those who depict it. Let's grant, with Hamlet, that literature holds that mirror up to nature, including the nature of human beings. And so we are led to the assertion that evolutionary biological insights yield a powerful set of instruments with which to understand literature and, in the process, ourselves.

In *Madame Bovary's Ovaries*, we merge two worlds, literature and science, showing how fiction can be illuminated by the single most important idea in biology (evolution) newly applied to human behavior. We hope that our dissection of Madame Bovary's ovaries, Othello's jealousy, Holden Caulfield's alienation, and the like will reveal a novel way to read and understand. Not the *only* way, mind you—our intent is not to sweep away any current literary theories in favor of science—but a new one, a useful tool to add to each reader's kit. Our basic premise is simple enough, although oddly revolutionary at the same time: that people are biological creatures and that as such they share a universal, evolved human nature. Add to this our second basic principle: that evolutionary psychology, a decidedly nonfiction science, has been discovering why human beings behave as they do, and that it offers a raft of refreshing, rewarding, challenging insights into the world of fiction no less than that of fact. You hold the result in your hands: a new set of spectacles for the reader who is discerning, perplexed, or just plain curious and ready for something new.

At this point, it may be tempting to ask why any particular slant, whether theoretical or scientific, is needed in order to make sense of literature. Why not just read the books and let them speak for themselves? The answer is simple: people always use some sort of interpretive lens, whether they admit it or not. John Maynard Keynes wrote, for example, that when economists propose to do their work "without theory," this merely means that they are in the grip of some other, unacknowledged theory. There is no such thing as a truly naive reader, which is why experts in literature have long searched for new and useful ways to approach their subject.

Several decades ago, geneticist Theodosius Dobzhansky gave this title to a now-famous article he wrote for a technical journal: "Nothing in Biology Makes Sense Except in the Light of Evolution." Dobzhansky's dictum applies to the world of letters just as it does to the world of life, because the former is merely one manifestation of the latter: literature is life written down. Accordingly, literary critics—and, more important, garden-variety readers—should profit by adding the cardinal principle of the life sciences to their armamentarium. According to renowned literary theorist Northrop Frye, "Criticism is badly in need of an organizing principle, a central hypothesis which, like the theory of evolution in biology, will see the phenomena it deals with as parts of a whole." Such an organizing principle already exists, however, needing only to be recognized and developed. Ironically, it is the same one that Frye gestured toward so longingly: evolution. Whereas the various warhorses of traditional literary analysis offer intellectual richness, so does Darwin. Moreover, Darwin has this additional appeal: he was right.

Don't get the wrong impression. We are not proposing scientific veracity as the sole guidepost when it comes to approaching literature. After all, physics is also valid (despite what some postmodernists—many of them, incidentally, literary theorists—often claim), and yet it would be pointless and, indeed, comical to base an approach to literature on quantum mechanics, string theory, or general relativity. Even though, for example, suitable calculations of force, mass, and momentum would doubtless yield insights into Anna Karenina's terminal encounter with a train, we suspect that a strictly Newtonian analysis of Tolstoy's great novel would leave out some important (non-Newtonian) dynamics.

Nonetheless, given the near-universal scientific consensus when it comes to the origins and nature of life in general (including that of *Homo sapiens*), one might think that literary critics would long ago have rushed to embrace biology or at least to explore its potential usefulness. This, to our knowledge, has not happened, although from time to time scholars have at least gestured in that direction, and there is a nascent movement among a tiny minority of humanities professors to take Darwin seriously at last.

This is *not* to claim that biological realism should be the touchstone for literary quality (in fact, we doubt that there should or could be any single measure), although we have already suggested that to some extent, believability may be what logicians refer to as "necessary but not sufficient." There will always be room for the uniquely subjective qualities of literature, with its richly imaginative textures. Fictional accounts of what people *might* be like are in no danger of being supplanted by nonfiction representations of what they really do. Great literature is not rendered great simply by accurately portraying human nature; fiction is not ethnography or photography, nor should it aspire to be.

We find it significant, however, that for all the expressive freedom of literature, there is virtually no written equivalent of abstract expressionism, which is to say, arrays of words seemingly disconnected from any reference to the real world. Or think of the vast difference between literature and what psychiatrists call "word salad," the random, chaotic verbalizations of people suffering from a variant of schizophrenia. The point is that literature deals, however impressionistically and subjectively, with people, and the nature of people (just like the nature of their languages) follows certain consistent patterns, even as that nature is relatively open-ended and malleable. Couldn't there, shouldn't there, be room for readers to take this into account?

Literature, when done well, not only is timeless (think of *The Iliad, The Odyssey, The Aeneid, Hamlet, Don Quixote, War and Peace, Madame Bovary*) but also travels widely and deeply. The Japanese adore Shakespeare. Americans read Tolstoy and Dostoyevsky. *Don Quixote* has been translated into dozens of languages. "Nothing can please many, and please long," wrote Samuel Johnson in the preface to his edition of Shakespeare's plays,

> but just representations of general [human] nature . . . The irregular combinations of fanciful invention may delight awhile . . . but the pleasures of sudden wonder are soon exhausted, and the mind can only repose on the stability of truth.

Dr. Johnson is talking here about our old friend human nature—those universals that Aristotle so admired, that same source of "stability of truth" that serves as a kind of unmoving polestar throughout *Madame Bovary's Ovaries.*

Memorable literature owes its greatness to many things, including the artistry and imagination with which characters and situations are portrayed, not to mention the richness of the language employed, all made possible by the genius of the author. But if human nature isn't somehow in or behind the picture, then literature will have less staying power. For this reason alone, it seems certain that despite all the hand-wringing, the Western canon is solid; its grounding in the biology of human nature is what keeps it from becoming a loose cannon. In this regard, we must take issue with one of critic Harold Bloom's more quotable assertions: Shakespeare didn't "invent the human," evolution did.

Arnold Weinstein's book *A Scream Goes Through the House* is a fine contemporary work of more traditional literary criticism, sub-titled *What Literature Teaches Us About Life*. It delivers generously on this promise. *Madame Bovary's Ovaries* might have had the reverse subtitle: *What Life Teaches Us About Literature*.

Get ready, in short, to get in touch with your "one nature," the one that not only reads and writes but also lives and is therefore abundantly reflected in the books and stories all around us. Let's see what happens when we marry the rich world of creative fiction to the equally rich world of biological insights. Our approach will be simple enough: we'll describe some of the key concepts in modern, Darwinian behavioral biology and try to show how they flourish in literature. In the process, we'll spread our nets widely, and if successful, we'll come up with an interesting and diverse catch.

We believe that the current offering is new,[1] and also, not coincidentally, going to be controversial. We hope that it will also be productive and—most of all—fun.

[1] Well, not entirely new. As already mentioned, some scholars—such as Brian Boyd, Joseph Carroll, Ellen Dissanayake, Nancy Easterlin, Jonathan Gottschall, and Michelle Sugiyama—have begun exploring the potential of "Darwinian literary criticism," but thus far their work has been directed toward a technical audience, and they certainly represent a minority, even among scholars.

2

OTHELLO AND
OTHER ANGRY FELLOWS

Male Sexual Jealousy

O*thello* is a terrific tale of jealousy, murder, revenge, high passion, and low cunning. The tragedy has fascinated audiences
for centuries, to the point that Othello himself has become
emblematic of lethal and credulous jealousy, his wife, Desdemona,
the embodiment of innocent victimhood, and his "friend" Iago the
archfiend and purveyor of deadly disinformation. The play is great
because it is wonderfully written, but its timeless, universal appeal
may well be due at least as much to the fact that it taps into a deep
underlying current of human frailty: sexual jealousy. Indeed, the
depth of *Othello*, the play, derives from the extent to which it
plumbs the depths of Othello, the deeply human, vibrantly organic
creature.

Let's look, therefore, at Othello as a member of the species
Homo sapiens. As a believable human being. And, most important,
as a sexually jealous male.

The story, in brief: Othello is military commander of the armed
forces of Venice. He has just married the young and lovely
Desdemona. He has also just passed over one assistant, Iago, and
named another, Cassio, as his chief deputy. Iago, boiling with
hatred and rage, schemes to convince Othello that Desdemona
has been unfaithful to him with Cassio. Othello eventually comes

under Iago's spell and, driven nearly mad with jealousy, kills Desdemona, then himself. Exit stage right.

Othello wasn't simply jealous, a black African, and a war hero; he was also a *man*. This brings up some important biology. Men are different from women, not simply because of social traditions, the cut of their clothes, or, for that matter, the structure of their private parts. Even the notorious Elizabethan codpiece didn't cover the real meaning of "manhood" (birds, for instance, are unquestionably either male or female, and yet even the most "masculine" feathered fellows, such as hawks or eagles, lack external genitalia).[1] Rather, Othello's maleness stems from something that he and all men share with other male animals: sperm. Males are simply those creatures that produce an amazingly large number of very small sex cells. Whether bird or mammal, Venetian or Moor, this and this alone is how biologists distinguish the two sexes. Females are egg makers; males, sperm squirters. The truly important thing about Othello wasn't the color of his skin, his age, or his war record. Rather, Othello was all about sperm; Desdemona, eggs. So when the evil Iago egged on Othello, he was doing more than one might think. And herein lies a tale.

The most important consequence of the biological difference between the sexes is that one sperm maker can fertilize many egg makers. A female, by contrast, can generally be fertilized by the prompt exertions of merely one male. Therefore, the payoff for an already impregnated female to have additional male lovers is not nearly as great as it is for males to inseminate additional females. A woman's reproductive potential is necessarily limited: she can usually only have one—rarely, two—children every nine months. During that period, more sleeping around won't yield more offspring. On the other hand, consider the man's situation (whether or not his wife is pregnant): if he has sex with another woman, he has given himself the possibility of yet more descendants. If he

---

[1] Nearly all birds have a cloaca, which is a common opening for reproductive and excretory products. There are, however, some interesting exceptions: ducks, geese, and ostriches, for example, have penises. But this really doesn't matter for our purposes.

repeats the process yet again, with someone else, he might conceiv-ably have conceived yet again. And again. And again.

This isn't to recommend such behavior, but to understand it. Moreover, we'll shortly see that women aren't as sexually reticent as just made out, but the general pattern is nonetheless clear: men are more likely to seek multiple sexual partners than are women. Similarly, they are more likely to be stimulated by the prospect of a casual sexual conquest, and also—tragically for both Othello and Desdemona—prone to sexual rivalry with other men as well as sex-ual jealousy of "their" women.

Another way of looking at all this from nature's point of view: eggs are large, expensive, and relatively rare since they only come one at a time and require a lot of energy if they are to be turned into children. Sperm, on the other hand, are small and cheap, and there is an almost inexhaustible supply. As a result, eggs and egg makers are limiting resources for the success of sperm and sperm makers, which is to say that the number of children a male can produce depends less on him than on how many women he can impregnate (and also on whether or not other males succeed in impregnating the woman or women he associates with).

These simple differences have enormous consequences for the behavior of males and females, men and women, Othello and Desdemona. First, males of most species make the most of their evolutionary prospects by having sex with as many females as possi-ble. Each female is a potential target of opportunity and worth competing over because she is an egg maker, uterus bearer, and potential pregnancy maintainer. Although only one male is likely to hit a woman's reproductive bull's-eye at any one time, many are likely to try. The result, all too often, is trouble, especially among those avid, competing spermatic archers.

Under these circumstances, successful males tend to be those who have excluded others. Moreover, *really* successful males are those who have succeeded in monopolizing the sexual favors of more than one female. (This is equivalent to firing successfully at more than one target, a strategy that makes sense if the arrows are cheap and the rules of the game permit it.)

So what's a male to do? If his species is highly social—as in a variety of mammals, including *Homo sapiens*—he will likely aim at as many females as possible. In short, he'll try to obtain a harem, and when that's not possible, he will likely be prone to sleeping around . . . all the while worrying about the fact that his counterparts are trying to do the same thing. Othello's credulity, although excessive and eventually lethal, is exactly what one might expect from a man who, like most men, has an inkling of the lascivious inclinations of his competitors, in no small part because he doubtless feels them in himself.

After all, human beings are perfectly good mammals, and the evidence—from biology, anthropology, and history—is overwhelming that for most of their evolutionary history, men have aspired to be harem keepers and have engaged in sexually oriented male-male competition. It shows in our bodies, our brains, and our behavior—especially in the differences between men and women. Biologists call this sexual dimorphism (*di* = "two," *morph* = "shape"), which refers to the simple but consequential fact that among many living things, males and females are different, not only in their gametes but also in their anatomy as well as their inclinations.

The general pattern is simple and consistent: among harem-keeping species, males are larger, more aggressive, and more sexually assertive than females. Male gorillas, for instance, are considerably larger and more aggressive than females. Generation after generation, the largest, most aggressive silverback males outcompeted their rivals, and as a result they bequeathed their genes—for largeness and aggressiveness, among other things—to future generations.

To be sure, not all animals, or even all primates, have evolved this mating strategy. Creatures such as beavers, coyotes, or gibbons, for example, are mostly monogamous, and they show very few differences between males and females: very little sexual dimorphism. Chimpanzees, on the other hand, are quite close to human beings—measured by their DNA—but their flamboyantly erotic lifestyle has little resemblance to the more staid love life of *Homo sapiens*. Every species is distinct, some more monogamous and some less so, but certain general patterns nonetheless remain, including the fact that sexual dimorphism of the sort found among gorillas (and many other creatures as well, such as deer and seals) bespeaks males competing for the sexual attention of females.

What, then, about our own species? Human beings are pretty much in between the gibbon and the gorilla: more dimorphic than the former but less so than the latter. Although Judeo-Christian morality, not to mention the restraints of civil law, restricts westerners to monogamy, the biological fact remains that inside the most faithful husband there is a fervent philanderer—and, at least in part, a hopeful harem master—just waiting to emerge. And, by the same token, there is a sexually jealous competitor whose anxiety and anger is all too readily evoked. Just ask Othello.

Most of us are (or claim to be) monogamous, just like the ill-starred Moor. It is an open question, hotly contested among biologists and historians, why Western cultural tradition has overridden our natural polygynous inclinations. One possibility is that since under monogamy each man is more likely to get a wife, whereas polygyny necessitates that for every harem keeper there are frustrated, angry, and potentially troublesome bachelors, monogamy serves to diminish sexual discontent and thus promote social harmony. (Women, by contrast, will generally be mated whether the prevailing system is monogamy or polygyny.) In any event, anthropologists know that harem keeping is closer to our natural state than the current regime of culturally imposed monogamy. Thus, prior to the cultural homogenization of the last few centuries, upward of 85 percent of human societies were preferentially polygynous. For yet more evidence of our polygynous patriarchy, look at boy-girl differences at the age of sexual maturation. Girls grow up several years earlier than boys, a pattern that is consistent with polygyny in other animals: when the social life of a species is characterized by intense male-male competition, it's a bad idea to enter the fray when you are too young, too small, too weak, and/or too stupid to prosper, or maybe even to survive. Hence, in harem-keeping species, males don't start breeding—and, more to the point, struggling with other males for the opportunity to breed—until they are older and thus larger, stronger, and cagier. Moreover, the greater the degree of polygyny (that is, the larger the average harem size and hence the greater the intensity of male-male competition), the greater the disparity between the ages at which males and females become sexually mature.

Put this all together and there is simply no question about it: a Martian zoologist visiting the Earth would take one look at *Homo sapiens* and our sexual dimorphism in body size, inclinations toward violence and having multiple sexual partners, and male-female differences in age of maturation and would confidently conclude that at heart—and body and brain—human beings are preferentially polygynous. And one thing is true about a harem-keeping species: the payoff for success is large and the consequences of failure severe. Since there are roughly equal numbers of men and women, the reality is that the more polygynous the society, the greater the number of men left out in the nonreproductive cold, doomed to be frustrated, resentful bachelors. Most men can be grateful that monogamy is legally mandated: harem keeping is no picnic. The result? A high level of sexual competitiveness, even in a species such as ours, in which serious harem keeping is largely a distant biological memory.

Finally, add to this the fact that any male—whether highly dominant harem keeper or run-of-the-mill monogamist—can be cuckolded, and to top it off, the additional fact that (as explored more lasciviously in Chapter 5) females have their own hankering for occasional infidelity. The result is an evolutionary witches' brew in which men are especially subject to a heavy dose of sexual jealousy. Even if, like Othello as well as the rest of us, he inhabits a society in which polygyny is outlawed, and moreover even if he is not a frequent philanderer, the typical human male has been outfitted by a lengthy evolutionary past with a profound distaste for the womanizing of his fellow men as well as a sensitivity to being cuckolded by his fellow women.

To be sure, females are typically less than delighted if their male courts and inseminates other women. That's why Frankie killed Johnny, why harems tend to be difficult places, and why even in ostensibly monogamous households, woman-woman competition can be pretty intense (think of the movie *Fatal Attraction*). Still, men have more to lose than women when their partners are unfaithful. Once again, biology holds the key. The male's investment in mating isn't much to speak of; there is no comparison between a man's "lovin' spoonful" and a woman's pregnancy and lactation. More important, even if a philandering man ends up providing assistance to his extramarital lover and her offspring, a deceived woman is not going to be deceived as to *her* maternity, although if she also fools around she may have her doubts about the father.

A woman whose husband is unfaithful can still bear her own children, no matter who the father. A husband whose wife is unfaithful could end up in the same situation as a sparrow victimized by a cuckoo: manipulated into rearing someone else's offspring (hence the origin of the antique-sounding but very up-to-the-minute word "cuckold").

And so we return to Othello, a dominant bull elk or silverback male if ever there was one, but who—like the rest of us—has been culturally restricted to just one Desdemona at a time. It should be clear that it doesn't really matter whether such a male is ostensibly monogamous or polygynous: either way, it pays for males to be sexually jealous and thus highly protective of their reproductive prerogatives. During the rut, bull elk are notoriously aggressive and intolerant of each other, while cow elk are comparatively placid; even among monogamous songbirds, males regularly patrol their territories, alert for intruders.

Men have evolved to be similar to bull elk, or maybe even nastier, since women are fertile all year round. Not surprisingly, therefore, the

Othellos (and Cassios and Iagos and Roderigos—the latter a minor character who lusts for Desdemona and is manipulated by Iago into attacking Cassio, twice) among us pant after each other's women while simultaneously ready to be roused to jealous rage by any suggestion that someone else might do unto their woman (or women) as they long to do unto others'. To paraphrase Winston Churchill, men are often jealous beasts, with much to be jealous about.[2]

To make matters worse, Shakespeare's Othello-elk has a "buddy" whispering in his ear, introducing the poisonous pretense that a different young buck—Othello's lieutenant, Cassio—has been "enjoying" his Desdemona. Iago taunts, insinuates, and sets up innocent circumstances that appear to the credulous Othello to prove Desdemona's infidelity.

Of course, Iago is himself male, and thus he is both attuned to Othello's anxiously susceptible jealousy and also afflicted by his own, acknowledging at one point a suspicion that his wife, Emilia, might have been unfaithful . . . with Othello. "I hate the Moor," Iago soliloquizes, "and it is thought abroad that 'twixt my sheets, he's done my office." All in all, a nasty business, enmeshed in yet more intimations of male sexual jealousy via some of the peripheral characters.

It is also worth noting that Othello is considerably older than Desdemona, a pattern that lends itself to an additional slate of evolutionary insights. Yet another biological asymmetry between the sexes is that whereas women go through menopause, men remain potentially reproductive into old age. In addition, older, successful men—such as the rich, charismatic Moor—are further likely to be attractive to women because they offer the prospect of what we might call r-cubed: reproductively relevant resources. At the same time, such men are vulnerable to women seeking to profit from their superior wealth and social prestige, who then cuckold them

---

[2] Churchill once described his political rival Clement Atlee as "a modest man, with much to be modest about."

with younger, more physically attractive specimens. In short, there is a female proclivity to get wealth, power, and protection from one male and sperm from another.

As we'll see later, when we confront Madame Bovary and her sisterly soul mates, this is a very ancient pattern, well ensconced in the animal world no less than in our own, and it leads to yet another variant on the theme of male-male sexual competition: older, more powerful men versus others who are younger and less socially successful—but often more physically attractive—over access to sexually appealing, nubile women. No one should be surprised, therefore, when the middle-aged Othello is prone to anxiety about possibly being sexually supplanted by his young lieutenant.

Othello has plenty of company. Think of King Arthur contending, unsuccessfully, with Sir Lancelot over Guinevere. Or the tale of Tristan and Isolde: King Mark is a bit like Othello—somewhat elderly, albeit influential—and sure enough, his bride-to-be, Isolde, falls for Mark's emissary, Tristan, who is youthful and dashing and whose appeal is also, truth to tell, abetted by Isolde's consumption of a magic love potion. (Isn't it possible, moreover, that the love potion part of this myth is an attempt to explain an attraction that biology interprets more directly and plausibly?) According to Homer's account, Menelaus is older and a king but loses Helen to the younger Paris. Euripides' tragic tale *Hippolytus* recounts the erotic infatuation of Phaedra, wife of the older Theseus, with her eponymous stepson. This story had sufficient resonance, moreover, for Racine to update it in his greatest play, *Phaedra*. It is a dynamic that isn't restricted to tragedy, however, having become the subject of many a barbed comedy of manners, from Boccaccio's *Decameron* to Chaucer's *Canterbury Tales* and the plays of Molière, Oscar Wilde, Tom Stoppard, and Edward Albee.

Put all this together, and Othello's jealousy is not an isolated peculiarity; it reflects an enduring, universal theme. To be sure, the plot—like much of Shakespeare—is overwrought, and yet the audience cannot simply dismiss *Othello* the play as absurd or downright preposterous, because men in particular cannot simply dismiss the plight of Othello the person: sexual jealousy is too much a part of everyone's biological and social experience. George Bernard Shaw managed nonetheless to despise *Othello*, complaining of its "police court morality and commonplace thought." But Shaw's greatest weakness as a playwright lies precisely in his tone deafness to male-female dynamics, which includes, crucially, male-male sexual competition à la *Othello*. As to "commonplace thought," that's precisely the point: nothing is more devastatingly commonplace than male sexual jealousy, especially when it takes a violent turn.

Not surprisingly, sexual jealousy isn't unique to *Othello*. It emerges in much of literature, as in life, and when it does—especially if it is violent—it is likely to be a guy thing. People, like elk, are only human.

For example, male-male competition animates "The Knight's Tale," the first and longest of Chaucer's *Canterbury Tales*. It is the story of two cousins and—until sexual rivalry intervenes—best friends, Arcite and Palamon, who have been imprisoned and who both fall in love with the same woman, whom they ogle from their prison window. One is eventually released and the other escapes, whereupon, in a lovely melding of chivalric propriety with Darwinian sexual competitiveness, they agree to a lethal fight to the finish, with the winner to get the girl. Thus doth best friendship give way to murderous violence, once the best friends have—hath?—become sexual competitors. "The Knight's Tale" being a medieval romance, there is, of course, more: before Arcite and Palamon proceed to hack each other to bits, they are met by King Theseus, who had originally imprisoned them, and whose sister, Emilye, is their shared inamorata. Theseus demands that the two contestants put up their swords and meet one year hence, each attended by one hundred armored knights, to battle it out for the hand of the fair damsel. Which they do.

Shakespeare was evidently taken with the story, so much so that he borrowed it for his last play, *Two Noble Kinsmen*, coauthored with John Fletcher. Others have also been taken with the theme of male-male competition—not surprising, since it takes up so much of human endeavor.

F. Scott Fitzgerald, for example, is generally seen as the great chronicler of the decadent Jazz Age, but he was at least as much a bard of male sexual jealousy, which drives *The Great Gatsby* just as much as Jay Gatsby drives himself and his famed automobile. When, in the novel's ironic denouement, the cuckolded George Wilson kills Gatsby, it is in a fit of sexually jealous rage, thinking—wrongly—that his wife, Myrtle Wilson, had been unfaithful with Gatsby, when in fact Tom Buchanan was the culprit, that same philandering Tom who was married to Daisy, Gatsby's secret flame. Most critics and readers focus on the mystery of Jay Gatsby's past and his quintessentially American striving for money and success, all the while missing the role of male-male competition not only in motivating Gatsby's personal trajectory but also in powering its tragic denouement.

Nor are sexual jealousy and male-male competition necessarily limited to duos, such as Othello and Cassio, Arcite and Palamon, or Tom Wilson and Jay Gatsby. For a case of multiple male sexual jealousy, consider *The Sun Also Rises*, by Ernest Hemingway, a great writer whose star has recently been in partial eclipse within the American literary establishment, partly because of his uninhibited sexism. It is likely, however, that the critical pendulum will once again swing in Papa's direction, due in no small measure to the acuity with which this most masculine of all writers perceived the underlying persistence of sexual tension among men. Thus, in traditional interpretations, *The Sun Also Rises* depicts the "lost generation" between World Wars I and II. But it also does a masterly job of depicting male-male competition, in this case for the beautiful, pleasure-seeking Lady Brett Ashley. Or rather, call it male-male-male-male competition, since the rivals are Mike Campbell (her fiancé, a wealthy Scot), Jake Barnes (an impotent, wounded war veteran), Robert Cohn (an aspiring writer and amateur boxer), and Pedro Romero (a handsome, up-and-coming young bullfighter).

Lady Brett can't keep away from men, who buzz around her like besotted flies. She really loves Jake, who really loves her, too, but can't do much about it. Cohn keeps pestering her, but his insecurity is a big-time turnoff. By contrast, the confident, sexy bullfighter Pedro Romero is definitely a turn-on, and indeed, Brett and Pedro have an affair, but it doesn't last and she ends up going back to Mike Campbell, with his alcoholic, wealthy, upper-class ways. Meanwhile, Cohn—who is an accomplished boxer—picks fights with Jake, Mike, and the erotically charged bullfighter before remorsefully leaving town just as the story seems to be heading for an impressive climax of passion and jealousy. In Hemingway's hands, the "lost generation" is so lost that even sexual rivalry can't energize it. Othello might well wish for a bit of the lost generation's ironic indifference. But in fact, the pugilistic Robert Cohn—who beats up Romero, Mike Campbell, and even the nonthreatening Jake Barnes—more accurately reflects the male sexual style.

Recall that sperm are cheap and easy to replace. So it has been biologically advantageous for males to spread their seed widely, or at least to give it a try. As with Robert Cohn, this may involve punching people in the nose if they think it will help, or, in the case of Pedro Romero, engaging in high-risk, flashy activities (e.g., bullfighting) as a way of getting noticed—and, if possible, bedded—by the ladies. Or being rich and upper-class, like Mike Campbell. But not to get shot in the balls, like Jake Barnes! By the same token, a cuckolded husband is widely considered pitiable; he is a loser, not only socially but also biologically. Indeed, he is a double loser, since on top of the injury of being potentially nonreproductive is heaped this heavy material insult: he may well end up expending time, effort, and expense in the rearing of someone else's child. No wonder there are Othellos, Gatsbys, jilted lovers, and jealous husbands of all kinds echoing in our literature: they resound no less in our biology. The real wonder is that there aren't more.

Actually, there are. Threatened adultery pops up as a theme in many other depictions, not least those of Shakespeare. It abounds, for example, not only in *Othello* but also in *Much Ado About Nothing*, in which young Claudio makes much ado about what seems to him something indeed: his perception (entirely incorrect, it turns out) that his bride-to-be, a heroine aptly named Hero, has been "unchaste." Interrupting their wedding, Claudio denounces Hero as promiscuous: "she knows the heat of a luxurious bed," or so he claims, and it requires much of the rest of the play to clarify that she is altogether innocent in every sense of the word.

Claudio is young, inexperienced, and easily confounded.[3] The same applies to another male Shakespearean creation whose credulity falls for yet another false report of female infidelity: Posthumus, husband in *Cymbeline*—which, incidentally, lays claim to flaunting the most preposterous of all Shakespearean plots, the only believable component of which is Posthumus's anguish when he is led to believe (falsely, once again) that his wife, the faithful Imogen, has had sex with the devious Iachimo (shades of Iago). Iachimo bets Posthumus that he can seduce Imogen. He fails—Imogen is as pure as Hero, or Desdemona—but succeeds in smuggling himself into Imogen's bedchamber, after which, by describing its contents (including Imogen, whom he evidently ogled most indecently), he persuades Posthumus that he won the bet:

> *Under her breast—*
> *Worthy the pressing—lies a mole, right proud*
> *Of that most delicate lodging.*
> *By my life, I kissed it, and it gave me present hunger*
> *To feed again, though full.*
> *You do remember*
> *This stain upon her?*

---

[3] Actually, there is a far more interesting couple in *Much Ado About Nothing*: Benedick and Beatrice, who are older, wiser, and wittier. Their byplay speaks less to male sexual jealousy than to matters of female choice in particular, à la Jane Austen, which we take up in the next chapter.

At this, Posthumus is thrown into a paroxysm of grief and rage, proclaiming his wife as suffering from "another stain, as big as hell can hold." Posthumus goes on to torture himself by imagining Iachimo as superstud: "Perchance he spoke not, but, like a full-acorned boar [i.e., one whose "nuts" are intact], . . . cried 'O!' and mounted."

And so, like Claudio denouncing Hero, Posthumus rejects Imogen, even scheming to have her killed. Fear not: everything is straightened out in the end. But Shakespeare knew, as well as Darwin, that it will happen again, that men are not only sexually jealous and often violently so, but astoundingly willing to believe the worst, especially if it involves a threat to what they fancy as their sexual monopoly over a woman.

Such violent sexual credulity is hardly limited to the young and callow, neither in nature nor in Shakespeare. Our exemplar, Othello, was a middle-aged specimen, "declined into the vale of years." And in *The Winter's Tale*, we see another man in the throes of a sexually charged middle-aged crisis. The play begins with King Leontes unaccountably—and again, falsely—accusing his faithful wife of committing adultery with his childhood friend, then visiting Leontes's court. It ends with forgiveness and reconciliation. Ditto for *Much Ado About Nothing* as well as *Cymbeline*. These plays have all been criticized for their bizarre plots (seeming statues that come to life, a headless corpse mistaken by the faithful Imogen—of all people—for her husband), which are literally incredible. Yet even the most carping critic or jaded theater patron is likely to agree that these plays, amid all their absurdity, remain stunningly believable in one respect, and probably only this one: each paints a realistic picture of male sexual jealousy.

Of course, male-male competition has been going on for hundreds of millions—perhaps billions—of years, and Shakespeare wasn't the first to write about it.

Legend has it, for instance, that the Trojan War was set in motion when Paris (son of the Trojan king) ran off with Helen (wife of Menelaus, a major Greek king). At its end, the victorious Greeks kill the male Trojans and divvy up the women—a bloody, inhumane, but all-too-human resolution that occurred often in the Old Testament as well and is powerful testimony to the persistence of the competitive impulse. Over and over, whether in the Bible or *The Iliad*, men fight other men, with women as the prize; indeed, the earliest descriptions of war involve male-male competition far more than faceless armies maneuvering against each other. In such cases, two goals are at stake: direct access to fertile females and achieving social status.

Make no mistake: underlying both is the promise of male competitive success, and especially sexual—and thus reproductive—fulfillment. At one point, when the Greek army is especially demoralized, the wise and aged Nestor cheers them up by reminding them that Zeus has promised victory, which means that no Greek will have to "hurry to return homeward until after he has lain down alongside a wife of some Trojan." But even in such simplistically sexual situations, men are typically also maneuvering for status and prestige, mostly because where status and prestige are found, women are not far behind.

An animal parallel comes to mind: when choosing among possible mates, female red-winged blackbirds elect to nest with males whose territories are especially desirable, which means, for this species, that they contain predator-proof nest sites. Not surprisingly, this motivates male redwings to compete for access to those desirable territories. And who gets such access? Those who are socially dominant. Nor is this pattern unique to blackbirds. In essentially every species that forms a hierarchy of social dominance, the individuals at the top get the lion's share of the ladies. Even among chimpanzees, which don't partake of a simple, linear dominance hierarchy or an easily depicted sex life, high-status males tend to have more offspring than do subordinates.

The famous wrath of Achilles—an undoubtedly dominant male—is kindled by the fact that Agamemnon, brother to Menelaus and chief military honcho among the Greeks, had taken the gorgeous Briseis from Achilles. Not insignificantly, Briseis had been "given" to Achilles as a reward for some of his earlier heroic exploits, all of which involved killing other men. It is made clear, however, that Achilles isn't so much enamored of Briseis as he is of his own reputation. His anger derives from the threat to his social status and prestige, which is inextricably linked to his sex appeal.

Combat isn't simply mano a mano by force of arms alone; it also involves one-on-one verbal jousting, with extensive taunts, threats, and boasts, all of which are attempts to diminish the other and enhance oneself. Moreover, such vigorous competition is only marginally less intense within the armies on each side.

As anthropologist Robin Fox describes it, "The Greek warriors, beached on the sands of Troy, very much resemble giant elephant seals angling for upward mobility in the social and reproductive hierarchy. Those who top the hierarchy are men like Achilles, Ajax, Odysseus, Agamemnon, and Diomedes—they are huge, sleek, bellicose animals."[4] Just as alpha elephant seals get the most females, so do the most dominant warriors. Such dominance is typically established by contests staged among members of the same army. Robin Fox once again:

> The games are designed to award the talents typical of elite warriors: speed, guile, martial art, level-headedness, muscular bulk, and power. As on the battlefield, there are arguments, spear-thrusting and hurtling, punching and grappling, flying feet, clattering chariots, rock hurling, and arrow-shooting. . . . A very similar dynamic plays out in all other intra-army confrontations. They include all varieties of manly posturing, posing, rumbling, bellowing, flexing, chest puffing, and teeth baring.

---

[4] "Sexual Conflict in the Epics," *Human Nature* 6 (1995): 135–44.

All this is not to claim that men are *just like* elephant seals; rather, they are *like* elephant seals. Whether men, or women, like it or not.

At one point in the games of *The Iliad*, Diomedes and Ajax have at each other in a ritualized spear and sword fight that becomes so potentially deadly that the observers have to separate the two. Then there is a boxing match involving someone named Epeios, who proclaims that he will thoroughly destroy his opponent: "Utterly will I break apart his flesh and crush his bones. Let his mourners, who are his kin, wait in a throng so they can carry him away after my hands have broken him."

And these guys are supposed to be on the same side!

Nor is such posturing limited to literary depictions of the heroes of yore. Moreover, it isn't even necessary that women be the immediate prize. Like stallions rearing and snorting in a stable even with no mares nearby, men will often posture and puff themselves up even in an all-male environment. In such cases, women are nonetheless the behind-the-scenes motivators. Here, for example, is Mark Twain's account of a fight witnessed by Huckleberry Finn, when he had stowed away on a barge making its way down the Mississippi. A dispute broke out between two "rough-looking men," and one of them

> jumped up in the air and cracked his heels together again and shouted out: "Whoo-oop! I'm the original iron-jawed, brass-mounted copper-bellied corpse-maker from the wilds of Arkansaw! Look at me! I'm the man they call Sudden Death and General Desolation! . . . I take nineteen alligators and a bar'l of whisky for breakfast when I'm in robust health, and a bushel of rattlesnakes and a dead body when I'm ailing. . . . Stand back and give me room according to my strength! Blood's my natural drink and the wails of the dying is music to my ear. . . ."
>
> Then the man that had started the row . . . jumped up and he began to shout like this: "Whoo-oop! Bow your neck . . . for the massacre of isolated communities is the pastime of my idle

moments, the destruction of nationalities the serious business of my life! The boundless vastness of the great American desert is my enclosed property, and I bury the dead on my own premises! . . . Whoo-oop! . . . the Pet Child of Calamity's a'coming!"[5]

In Twain's hands, male-male competition is larger than life, filled with bluff and bluster, and comically absurd, whereas in Homer's, it is tragic. This may be due in part to the simple fact that in *Huckleberry Finn*, the competition occurs in the absence of women; it is men's entertainment, but also deadly serious and undergirded—as is all male-male competition—by the portentous prospect of who comes out on top. Hence the whooping and hollering, the speaking in exclamation points, and if need be beating each other over the head. Once men have established their dominance hierarchy, the likelihood is that a parallel hierarchy of reproductive access and thus evolutionary success will follow.

It isn't necessary, incidentally, for the participants to understand the connection between male aggressive bravado and reproductive success. It isn't even necessary that immediate sexual triumph follow from competitive social success. The key connection is that in the past, men who succeeded in besting other men were likely, as a result, to succeed in conveying their genes into future generations, either because they won more women directly or because they won prestige and social success, which in turn resulted in their being chosen by more women as well as deferred to by subordinate men.

Reading Twain, one cannot help laughing, but with a rueful recognition of the plausibility that underlies such absurdly excessive posturing. But the posturing isn't merely an arbitrary artifact of frontier America. Reading Homer, it is similarly easy to assume that one is eavesdropping on culture-specific behaviors that have

---

[5] We thank Steven Pinker and his book *The Blank Slate* (New York: Viking, 2002), for making us aware of this quotation.

nothing to do with animals, but dig deeper in either case—beneath the veneer of nineteenth-century America or classical Greece—and you strike underlying roots in biology, a bedrock that all human beings share with elephant seals, elk, gorillas, and much of the animal world.

To an extent only rarely appreciated by social scientists, and essentially not at all by self-proclaimed experts in literature—including classicists, whose contact with the raw ferocity of the blood-soaked Homeric epics should have taught them better—the human and animal estate are now, and have ever been, fundamentally identical.

Sometimes the violent outcome of male-male sexual competition is wildly exaggerated, although the underlying motivation rarely is. Let's turn briefly to *The Odyssey*, whose grand finale features a classic—in more ways than one—case of male-male competition over women, in this case the famously faithful Penelope and the mob of 108 (count 'em!) ill-mannered suitors, whose primary ill-manneredness consists of the fact that they have been lusting after another, higher-ranking man's woman. In the climactic scene, that horny horde of heavy-handed bachelors is challenged to string Odysseus's great bow, a test of physical strength and ancient manliness. Of course, only Odysseus can do it, whereupon, with the help of his son Telemachus, he proceeds to slaughter all the upstarts and reassert his claim to the throne, and to Penelope.

In the animal world, there is no end to the competitive hijinks in which males will engage, struggling to supplant each other in order to get their Penelopes. At the same time, it must be acknowledged that male-male competition, whether animal or human, needn't always generate violence. Sometimes it merely festers, leaving its participants marinating in a hormonal broth of equal parts anxiety, frustration, and barely repressed jealousy.

Take Marcel Proust's *In Search of Lost Time*, a masterpiece of great art distilled from memories and recalled perceptions. It is difficult to imagine two worlds of literature that are more disparate than the blood-soaked, brain-bespattered, lusty epics of Homer— dating from several thousand years ago—and the delicate, mannered, cultivated, and almost painfully fussy recollections of a rather effete and retiring scion of the French upper middle class at the onset of the twentieth century, who spent much of his adulthood carefully crafting his words in a darkened, cork-lined room. And yet they be of one blood, Homer and Proust.

Much of the richness of Proust's oeuvre consists of its evocation of—you guessed it—male sexual jealousy. Thus, *In Search of Lost Time* (more popularly mistranslated as *Remembrance of Things Past*) recounts the obsessive competitive anguish of three major characters regarding the sexual infidelities of their lovers. Saint-Loup, Swann, and Marcel cannot shake off the worries and woes associated with their lady friends. Marcel, for instance, spends much of his early adulthood tracking down the multihued love life of Albertine, the love of his young life, not at all daunted by the fact that by this time she had been killed in a horseback riding accident.

As Arnold Weinstein describes it in *A Scream Goes Through the House* (New York: Random House, 2003),

> the narrator continues to imagine, with ever more frenzy, scenes of betrayal committed by Albertine, acts of sexual independence that can (now) never be corroborated or disproved because the actress is no longer available for interrogation (not that much was ascertained even when she was). The dead Albertine romps through her imagined sexual repertory, thanks to the prodigiously creative jealousy of her grieving lover.

Nor is Marcel's jealousy diminished by his discovery that Albertine was a lesbian. One might expect that insofar as sexual jealousy is reproductively driven, homosexual infidelity to one's heterosexual partner would be of little account, but this fails to reckon with the powerful emotional undertow of motivations driven by evolution. When biology confronts the need to implant a potent response—as is certainly the case for sexual jealousy—it is likely to do so by linking that response to behaviors that are powered by emotion, rather than by reason, and that are quickly and reliably aroused. The fine-tuning of whether infidelity is same-sex (and hence, less likely to have reproductive consequences) is less important than the simple, blunt, and powerful panhuman abhorrence of a partner's infidelity, period. Especially when the "infidel" is female.

By the same token, one might expect that heterosexual infidelity could also be easily ignored, so long as the unfaithful partner used a condom. Of course, it isn't that simple. After all, birth control is very new on the evolutionary horizon; there isn't time for *Homo sapiens* to have evolved a suitably nuanced response to it. And sex with someone else, even avowedly nonreproductive sex, cannot help conveying a threat of restructured affection as well as the prospect of further infidelity in the future.

Sexual competition dies hard, if it ever does. In Proust's huge multivolume novel, as in this brief selection from one of its components, *The Captive*, the dogged persistence of sexual jealousy is the reason jilted lovers insist on "ransacking the past, in search of a clue," a search that may well turn up nothing, but continues nonetheless:

> Always retrospective, it is like the historian who has to write the history of a period from which he has no documents; always belated, it dashed like an enraged bull to the spot where it will not find the dazzling, arrogant creature who is tormenting it and whom the crowd admires for his splendour and cunning. Jealousy thrashes around in the void.

To be sure, even men can get along. And literature isn't shy about depicting it. Significantly, however, in such cases, the two (rarely more) collaborating Y chromosome carriers usually have a clear dominance relationship: Don Quixote and Sancho Panza, Robinson Crusoe and "his man" Friday, the Lone Ranger and Tonto.

Social equals in similar situations are expected to fight, or at least to strut and posture and act like jerks on occasion. They might well display their professed "manliness" like Charles Bukowski, whose writing reflected a persona that was self-consciously hard-drinking, wildly womanizing, heavily gambling, two-fisted, usually unemployed, abused, and abusive. Bukowski's first collection of short stories was given the memorable—and biologically appropriate—title *Erections, Ejaculations, Exhibitions, and General Tales of Ordinary Madness*. Most of its erecting, ejaculating, exhibiting, and otherwise maddening behavior takes place in a male-dominated context, in which, not surprisingly, various aspects of male-male domination figure prominently. For another, even more recent array of male-male violence (typically over one or more women, but sometimes with sex in the background . . . although never very far), pick up Cormac McCarthy's *Blood Meridian* or any of his Border Trilogy: *All the Pretty Horses, The Crossing,* and *Cities of the Plain.*

And of course, there is nothing like competition over a woman to drive a wedge between men who might otherwise be the best of friends. Probably the iconic modern example in this case is *McTeague,* a classic example of literary naturalism written by Frank Norris and published in 1899. The eponymous hero is physically imposing but not especially bright, charming yet also prone to drunkenness and violence. McTeague works as an unlicensed dentist, sometimes extracting teeth with his bare hands. His best friend is Marcus, who is engaged to Trina. But McTeague also falls for Trina, and vice versa, whereupon Marcus obligingly steps aside and the two marry.

All is not well, however. Barely repressed sexual jealousy and competitiveness between the two men bubble just beneath the surface. Their buddyhood inevitably and irrevocably dissolves. While attending a picnic at which wrestling matches are held, McTeague and Marcus are the two winners. When they face each other,

Marcus bites off part of his "friend's" ear, and McTeague responds by breaking Marcus's arm. Shortly after, McTeague is barred from practicing dentistry: someone had informed the San Francisco authorities that McTeague was practicing without a license. It was Marcus.

McTeague becomes increasingly alcoholic and violent, physically abuses Trina, and eventually kills her. He flees to Death Valley, California, where he is eventually confronted by a man with a gun: none other than Marcus, who had volunteered for the manhunt to capture Trina's killer. The two men—friends no more—struggle, during which Marcus is mortally injured, but before dying he succeeds in handcuffing himself to McTeague. Exposed to the broiling desert sun, without water, and lethally linked to his sworn enemy, McTeague, too, will die.

McTeague is in part a critique of American acquisitiveness and materialism; in 1924, director Erich von Stroheim adapted it into a renowned silent film titled Greed. But Frank Norris's novel may be even more powerful as a depiction of the destructive effects of male-male sexual competition. Thus, when the reader first meets McTeague, he is basically gentle and harmless, a "draft horse, immensely strong, stupid, docile, obedient." After being sexually awakened by Trina, however, his brutishness is aroused, tragically revealing itself in his treatment both of her and of his erstwhile friend, Marcus. When Marcus bites McTeague's ear during their wrestling match, this evokes "the hideous yelling of a hurt beast, the squealing of a wounded elephant. . . . It was something no longer human; it was rather an echo from the jungle." Here, author Norris missed a crucial point: echoes from the jungle are often profoundly human, especially when they involve males struggling with each other over females.

Nor are females necessarily passive, as we have already noted. They are more than capable of stimulating male-male competition; in fact, a frequent pattern involves the femme fatale encouraging her lover to kill another man, not uncommonly her current husband. It has become almost a genre unto itself: sexy, somewhat sociopathic drifter dude meets equally sexy, alienated wife of boring or downright beastly husband, whereupon drifter/lover kills husband, with encouragement—either covert or overt—of wife. For example, James Cain's *The Postman Always Rings Twice*, an explicit tale of adultery and murder, and a sensation when it appeared in 1933, employs a kind of tough-guy, Mickey Spillane objectivity to describe the murderous outcome when Frank Chambers, a young drifter, seduces Nora Papadakis, wife of service station owner Nick Papadakis. In this lurid tale, Frank and Nora decide to murder Nick (obese, middle-aged, and no longer appealing to his young wife) and collect the insurance money. Consider how rare, by contrast, is the mirror image: sexy woman meets unhappily married man and then proceeds to murder his wife. It's not that women don't compete, but rather that they generally do so with more subtlety and less violence. What do you think of, for example, when you consider the well-known fact that girls or women commonly engage in "telephone aggression"? Perhaps this: that they are likely to employ catty, undermining gossip. Now think about men or boys making hurtful use of a telephone: likely by hitting one another over the head with it.

At the same time, female-female competition is no less real than its male-male variety; as will be seen in the following chapter, it has its own peculiar flavor and style, less violent than the ill-fated Othello or McTeague, but nonetheless more vigorous than that of mere cheerleaders egging on their male counterparts. Accordingly, we turn next to females, and ask: what does the nature of femaleness, and thus womanhood, teach us about those women we meet within the stories we tell ourselves about ourselves?

3

THE KEY TO JANE AUSTEN'S HEART

What Women Want, and Why

———

Freud famously asked "What do women want?" but didn't even try to answer. Biologists have a pretty good idea, however. So do the world's great novelists.

Take Jane Austen. Her scope was remarkably limited—in geography, in events, in diversity of people: just a few middle- and upper-class families in one or another rural village in England during the late eighteenth and early nineteenth centuries, when the author herself lived. Add to this the fact that Jane Austen is, of all famous novelists, perhaps the *least* interested in nature, animals, and biology; she barely even bothers with basic physical descriptions of her characters. (Were they tall or short, thin or fat, brunette or blond?) And yet Austen captured universal truths, making her one of the English-speaking world's most beloved writers. So here we have the redoubtable Ms. Austen, writing about a very restricted range of human experiences, yet touching on one of the most fundamental and universal of all situations.

That situation is simply this: finding the right mate. All of Jane Austen's novels are love stories; all of them are concerned with getting the major characters suitably married; and in all of them, things work out eventually. Sounds boring—especially, perhaps, to men—but it isn't! Austen understands that the stakes are high, the outcomes weighty. And regardless of gender, the reader can't help sharing in her sense of bemused urgency, since the personalities are handled in such a masterly way, their characters lively and compelling, and, of course, because they are dealing with something that has obsessed our own species (and most other living things) for millions of years. Darwin called it sexual selection, and as we have seen, when it involves competition between individuals of the same sex, and particularly when those individuals are male, things can get aggressive, even violent.

At the same time, there is another side of sexual selection, namely, mate choice. Males and females "hooking up" isn't just a matter of pushy, angry, potentially violent Othellos competing with each other and occasionally even murdering their consorts; it's also a question of Desdemonas carefully and often peacefully choosing their Othellos. We have already described how, in the most dramatic cases, males push and shove and otherwise seek to overmaster each other for the biological bottom line: the reproductive payoff that comes from getting a mate. (Or, if they are especially lucky or well positioned, several mates.)

Jane Austen's domain is less confrontational but, in its way, no less dramatic, and certainly no less important. It is the female-oriented realm of sexual selection: choosing a mate and getting chosen. Just as males have largely cornered the market on violent same-sex competition, females occupy the spotlight when it comes to the choosy component of sexual selection. Why this difference? Essentially, it's because females have something that males want. Since males are comparatively eager and more or less sexually undiscriminating ("easy come, easy go" is a biologically accurate account of male sexuality, on several levels), females are often in the driver's seat: mate choice is largely their province. Of course, mate choice can be a female prerogative only among those species in which mate choice occurs at all. The typical female elephant seal, gorilla, or elk—creatures that for various reasons have thrown in their lot with a large-harem strategy—have woefully little opportunity to exercise sexual preferences. As we saw in the last chapter, species in which males keep big harems are those in which males are, well, particularly big, and generally rather nasty to boot. Cow elephant seals, female elk, and lady gorillas therefore have little choice but to mate with the master.

But then there are those other species, such as human beings, that still possess various stigmata of harem keeping but are also somewhat predisposed to pair bonding. In such cases, although male-male competition is an unavoidable fact of life, females also get to have their say, all the more so because they are the keepers of what biologists call "parental investment." This refers to eggs and everything that follows: a placenta, pregnancy, lactation, a high probability of maternal devotedness, and so forth. By contrast to this rich bounty of female attractions, male mammals have little to offer but their sperm, and so, in a strictly anatomical sense, they are beggars looking for a handout. In fact, a singles ad placed by males might read, "Nearly naked packages of DNA looking for a suitable home." Biologically, females are the wealthy ones; all males, even those with the fattest wallets, are paupers. So in our own species as well as many others, females get to exercise choice, deciding who among the eager suitors is to get their hand . . . and their breasts, vagina, and most crucially uterus, with all that entails.

As a result, women are in a position to drive a hard bargain, saying essentially, "You can have my goodies, but only if you provide assets of your own in return." What assets? Money, social prestige, good genes, a guarantee of parental assistance, what we referred to in the last chapter as r-cubed: reproductively relevant resources.

As part of courtship, men the world over commonly ply their dates with flowers and seek to impress them with fancy cars, expensive clothes, and elegant restaurants, often expecting sex in return. Rare is the man who doesn't see sex as a desired (if often unfulfilled) goal at the end of a date. Women, in turn, send countermessages. By acquiescing sexually, they are signaling that the man has what it takes to win their affection; by refusing, they may be saying either that he does not or that they are simply not sure, or they may be emphasizing that their own investment is so valuable that it cannot be obtained quickly. Women are exercising their choice, toting up the available r-cubed, waiting for the right fellow to come along.

And this is where Jane Austen shines. She is the poet laureate of female choice. Selection of the right marital partner is central to her writing, more so than for any other major novelist. Nearly always, Austen's women are in the driver's seat (and never more so than when they adroitly lead a man to think that *he* is). Darwin suggested that female choice is the motive force behind sexual selection; he pointed especially to the elaborate tail feathers of male Argus pheasants and peacocks, suggesting that such seeming absurdities evolved because generations of females have preferred to mate with the more elaborately bedecked males. (As we'll see, there are two major reasons why this peculiar preference might have evolved. For one, fancy males may be healthier as well as genetically better endowed than their plainer compatriots. For another, by preferring males who are appealingly ornamented, a female makes it likely that she will produce sons whose ornamentation will be similarly appealing to their own future mates. The result? She will have more grandchildren; her choosiness causes more of her genes—including those for choosiness—to be passed on.)

In any event, bright colors, pendulous wattles, shiny plumage, an elaborate song repertoire, or large canines or horns or antlers have come to characterize the males of many species. Whenever fancier males are preferred by females, natural selection will automatically ramp up the fanciness of males, simply because the plainer models are more likely to go unmated, and thus their plainness dies with them.

Modern biologists understand that this applies not only to physical characteristics but also to such traits as intelligence, generosity, and "control of resources" (which is to say, wealth and social standing). There is an old, sarcastic saying that the rich get richer and the poor get children. In the biological world, things are a bit more complex and even less fair: the rich get children, and the poor get next to nothing. Thus, "wealthy" blackbird boys—that is, those who own desirable, predator-free territories—get more black- bird girls and, in turn, more offspring. The "richest" male wood- chucks, whose burrows are deep and safe and who occupy good feeding areas, attract the healthiest, womanliest woodchucks, with whom, once again, they have more offspring. Evolution rewards immediate success (including but not limited to wealth and social prestige) with reproductive success, and one way it does so is by in- stilling a female preference for successful males. Or at least a prefer- ence for males who are more successful than they are.

"If I were a carpenter, and you were a lady," asks the male voice in a popular folk song, "would you marry me anyway, would you have my baby?" To this, the female responds reassuringly (espe- cially for carpenters) in the affirmative. But the truth is that most women, throughout most of history, would have answered with a resounding "No way!"

Jane Austen showed how it actually works among human be- ings. She wrote six novels, each justly famous today. In every one, the reader quickly meets the ideal mate for her heroine, but along with him are introduced many other characters, some of them plausible partners but each flawed in some crucial way.

And so we watch and wait, in a mixture of fascination, delight, and even on occasion a shiver of fear, as her protagonists work their way through their inevitable patches of sexual quicksand or hack through innumerable threatening thickets of social entanglements. Just as in our shared evolutionary past (and present), sexual choice for Austen's characters takes place in a social environment, chock- full of possibilities, and the tension mounts as we wonder who will end up with whom and whether the heroine will recognize that her own best interest lies in connecting with one fellow rather than another, often despite the urgings of society and family. Austen provides us with just enough obstacles and alternative outcomes to leave things uncertain, often until the last few pages.

What do Jane's young ladies look for in a mate? The same traits that female animals generally look for in their swains. Call them the three goods: good genes, good behavior, and good stuff. In other words, looks, personality, and money, although not necessarily in that order. Once again, given that males bring very little, anatomically and physiologically, to the reproductive marketplace, it is only natural—literally—that females choose among males based on what they *do* have to offer: the quality of their genes, their ability and inclination to be a good partner, and the quantity of their r-cubed—the size not of their penis but of their wallets.

It deserves to be repeated: wealth isn't merely a human construct. Among living things generally, parents who have valuable stuff—food-rich or predator-free territories, good nest sites, a den that lends itself to successful birth or hibernation, and so forth—can count on having a larger number of more successful offspring. And in turn, they can count on being preferred by members of the opposite sex for precisely this reason. This approach also suggests why human wealth—especially in its more arbitrary forms, such as gold or fancy trinkets—is considered desirable at all: because in the past at least, those who possessed it were able to exchange it for evolutionary success, even if such a transaction is less reliable in today's world.

Of the three goods, r-cubed loomed especially large in Austen's day and—despite the supposedly liberated twenty-first-century world—in ours. Thus, human beings, no less than blackbirds or woodchucks, find wealth appealing, not only as a goal for themselves but also as a sexual attractant when present in a potential partner. As already mentioned, resource-rich males of nearly every species become remarkably attractive at a level that often goes beyond (or beneath) conscious awareness. And in this regard, *Homo sapiens* may well take the cake.

"It is a truth, universally acknowledged," writes Jane Austen in the famous opening sentence of *Pride and Prejudice*, "that a single man, in possession of a good fortune, must be in want of a wife." She goes on:

However little known the feelings or views of such a man may be on his first entering a neighborhood, this truth is so well fixed in the minds of surrounding families that he is considered the rightful property of some one or other of their daughters.

In *Pride and Prejudice*, Mrs. Bennet—mother of five daughters— is laughably obsessed with "eligible" men, that is, single gentlemen possessing such fortunes. Indeed, part of Austen's genius has been to identify, dissect, and at the same time give life to this "truth," which other writers have more often accepted as simply embedded in the social landscape, so taken for granted as to go unremarked upon.

Pride and Prejudice opens as a country estate has been rented to Mr. Bingley, a wealthy London gentleman, which greatly encourages Mrs. Bennet to hope that her chief goal in life—finding suitably wealthy husbands for her marriageable daughters—will attain fruition. Mr. Bennet has the temerity to suggest that perhaps Bingley hasn't rented Netherfield Park with the express intent of marrying one of the Bennet daughters. (But we all know better.) As it happens, Bingley hits it off splendidly with Jane, the oldest Bennet daughter; moreover, he has a close friend, as rich as he is unmarried, named Darcy. Darcy, however, is very proud, and he initially insults Elizabeth Bennet, the second oldest sister; Darcy's pride evokes corresponding prejudice in Elizabeth, the sprightly, attractive, clever, and altogether admirable heroine. He also evokes something else. (More on this in a moment.)

Other characters appear: Mr. Collins, a ridiculously pompous clergyman who is also a conceited bore, but who will shortly come into possession of the Bennet property because Mr. Bennet has no male offspring; Wickham, a dashing officer who is an immediate hit with the ladies but who, we are led to understand, is something of a bounder, not above misrepresenting his assets as well as his intentions; Charlotte Lucas, a friend of Elizabeth's who is getting a bit long in the tooth (all of twenty-seven years); Elizabeth's three younger, rather ditzy sisters, Mary, Kitty, and Lydia, who are led more by their hormones than their heads; and an array of others. Eventually, Lydia runs off with Wickham, the regrettable Collins marries Charlotte Lucas, and—after many on-again, off-again embarrassments, annoyances, near misses, and misunderstandings—Bingley becomes engaged to Jane Bennet. Darcy's pride is sufficiently humbled and Elizabeth's prejudice adequately mollified so that Darcy and Elizabeth, too, announce their forthcoming nuptials.

A useful term for much of this is *hypergamy*, literally "marrying up." Men try it, too, but given the biology of male-female differences, it makes sense that hypergamy is a female specialty. This is to say that women can—and often do—insist on mating only with the best male if possible, one who represents some sort of improvement over their present condition. After all, since one male can fertilize many females, there is a certain logic to females insisting—insofar as they have choice in the matter—that their mate be an especially good specimen. Compared to many other species, human beings are often able to exercise such choice, which, when the opportunity arises, comes down to women choosing men who are particularly well-off. Dominant, wealthy, well-positioned males therefore not uncommonly mate with subordinate, poorer, less-advantaged females, and not the other way around. In fact, marriages in which women are wealthier or otherwise "above" their husbands generally have a poor long-term prognosis, whereas for women, "marrying up" is a widespread goal, perhaps even the norm.

In fact, this pattern is so well established—in the animal world generally and the human one in particular—that dominant, healthy seals and elk, as well as wealthy, upper-class people, are more likely to favor their sons, while the socially disadvantaged tend to invest more in daughters.

In F. Scott Fitzgerald's *Tender Is the Night*, we are given a masterly depiction of reverse hypergamy (we might call it "hypogamy," although, significantly, such a term doesn't even exist). Here, the beautiful, exceedingly wealthy Nicole Warren is married to up-and-coming psychiatrist Dick Diver. He is bright enough and attractive, but given that she—not he—is the wealthy one, the relationship would likely be biologically unstable, except for one fact: Nicole is mentally ill (the victim, we later discover, of childhood sexual abuse) and Dick is/was her shrink. As time goes on, however, Nicole gets stronger—and even richer when her father dies—while Dr. Dick gets weaker. His comparative poverty is increasingly evident, he has an affair with a much younger woman, and he begins to drink excessively. Dick Diver's downward spiral is matched by Nicole's upward progress, culminating—not surprisingly—in her leaving him for the dashing soldier Tommy Barban. Dick Diver dives indeed, into obscurity.

Another, related take on *Tender Is the Night* relies on the possibility, in twentieth-century America, for women to accumulate resources through inheritance, which in turn facilitates reverse hypergamy (women associating romantically with men who are "below" them). Nicole Warren is the one with the money; Dick's the one with the looks, personality, smarts, and, relatively speaking, good mental health (i.e., likely good genes). But the Warren-Diver duo is still not the norm, a situation of reverse hypergamy that is emphasized when Nicole's condition improves, whereupon her relationship with Dick becomes increasingly untenable.

Untenable relationships are sometimes maintained nonetheless, as memorably portrayed, for example, in Edward Albee's *Who's Afraid of Virginia Woolf?* Here we witness the painfully dissatisfied marriage of Martha, daughter of the local university president, to George, a middle-ranking, middle-aged disappointment (at least to Martha) who, as she likes to point out, is *in* the college's history department, as opposed to *being* the history department. George has failed to live up to both Martha's and her father's expectations; her marriage, like that of Nicole Warren to Dick Diver, is another case of reverse hypergamy, with a predictably rocky outcome. Martha explains her feelings toward her husband as follows:

> "He was the groom . . . he was going to be groomed. He'd take over some day . . . first, he'd take over the History Department,

and then when Daddy retired, he'd take over the college . . . you know? That's the way it was supposed to be. . . . And Daddy seemed to think it was a pretty good idea, too. For a while . . . Until he watched for a couple of years and started thinking maybe it wasn't such a good idea after all . . . that maybe Georgie-boy didn't have the *stuff* . . . that he didn't have it in him! . . . So, here I am, stuck with this flop . . . who's married to the President's daughter, who's expected to *be* somebody, not just some nobody, some bookworm, somebody who's so damn . . . contemplative, he can't make anything out of himself, somebody without the *guts* to make anybody proud of him."

George and Martha seem unlikely to divorce, although in Albee's lacerating depiction of the pain generated by such role reversals, the audience cannot help wishing that they would.

For a more traditional account of hypergamy "as it ought to be"—or at least as evolutionary biology would predict—take one of the most famous depictions of a female adventuress: Becky Sharp in Thackeray's *Vanity Fair*. Published several decades after the works of Jane Austen, *Vanity Fair* presents a much less sympathetic portrayal of Becky's hypergamic efforts. Whereas Jane Austen's heroines are smart, plucky, deserving, and socially sensitive, Becky Sharp is scheming, unethical, and selfish, but nonetheless exceedingly attractive to men. She sets her cap for Rawdon Crawley, son of a very rich family, and marries him secretly, but in the process she ends up alienating his wealthy father and even wealthier aunt, who—horror of horrors—disinherits her nephew.

At this point, a biologically attentive reader is likely to anticipate trouble. Sure enough, growing impatient with her husband's "reduced circumstances," Becky becomes the mistress of rich old Lord Steyne, eventually taking up with the prosperous Jos Sedley, brother of her closest childhood friend. Jos unwisely takes out a large insurance policy with Becky as beneficiary, and then—surprise—dies under unexplained circumstances, leaving Becky Sharp a wealthy woman at last.

Hypergamy triumphs.

But not everyone cheers. It is likely not a mere coincidence that Thackeray, a man, takes a much darker, more critical view of hypergamy than does Austen, a woman. Readers can't help siding with Jane's heroines, so sympathetically portrayed in their efforts to get the right man, whereas Becky Sharp may well represent male distrust of scheming hypergamous females. Especially considering that she ends up with all the stuff without having provided the female side of the bargain: offspring, . . . or even sexual fidelity.

Often, "marrying up" is at least as much a goal for a young woman's family as for the sweet innocent thing herself. Gabriel García Márquez's *Love in the Time of Cholera* is a love story beginning with the death of an elderly doctor, whereupon his equally elderly wife, Fermina Díaz, is the surprised recipient of protestations of eternal love from a comparably aged gentleman, Florentino Ariza. As the story unfolds, we learn that nearly fifty-two years earlier, Fermina and Florentino had been young lovers, but, in a fit of Austenesque bio-logic, she rejected him as unworthy. Shortly after, Fermina's father pressured her to marry the doctor—an especially "good catch" because he was already a prestigious physician—and the two lived together, although with more stability than happiness, for upward of five decades. Fermina's earlier hypergamy sets the stage for the ensuing novel, which follows the renewal of affection between Fermina and Florentino once the doctor (the object of Fermina's father's hypergamic lust by proxy) has died.

We also learn, interestingly, that during the ensuing five decades between Fermina's rejection of Florentino and their late-life reunion, Florentino had numerous affairs, despite vowing a kind of distant fidelity to Fermina, and that he had also become a successful businessman in an effort—ultimately rewarded—to make himself attractive to her. Throughout this marvelously evocative and bittersweet novel, love is equated with illness (including but not limited to the cholera of its title), but it is also presented as a disease that can be ameliorated, if not cured, with enough "medicine" in the form of money and social prestige.

Let's be clear, however. Whereas successful hypergamy can indeed be a winning ticket for a woman (and, of course, for men such as Fermina's physician-husband, who constitute the jackpot), it can be a disaster for guys at the bottom. Had Florentino remained as fiscally unprepossessing as he had been fifty-two years previously, the widowed Fermina never would have considered taking him as her geriatric lover. Someone who is unable to seduce a woman by his physical or intellectual charms alone (this includes most men) and who also lacks money is unlikely to make it with the ladies, regardless of his age. In *Down and Out in Paris and London*, George Orwell wrote perceptively about the situation of the English hobo, or "tramp":

> Any presentable woman can, in the last resort, attach herself to some man. The result, for a tramp, is that he is condemned to perpetual celibacy. For of course it goes without saying that if a tramp finds no women at his own level, those above—even a very little above—are as far out of his reach as the moon. The reasons are not worth discussing [note: we beg to differ on this point!], but there is no doubt that women never, or hardly ever, condescend to men who are much poorer than themselves. A tramp, therefore, is a celibate from the moment when he takes to the road. He is absolutely without hope of getting a wife, a mistress, or any kind of woman except—very rarely, when he can raise a few shillings—a prostitute.

In *Vanity Fair*, the matrimonial desirability of the various male characters varied predictably as their wealth ebbed and flowed, but at least people were pretty much what they seemed. (Except for Becky Sharp, whose male victims are consistently blinded by her coquetry and beauty.) In other cases, people actively misrepresent their resources; when they do, for some reason they are much more likely to err on the side of appearing richer! In the world of most living things, neediness isn't especially attractive; wealthiness is.

Edith Wharton's *The Custom of the Country*, one of her most notable books, introduces one of literature's strongest if most unpleasant characters, Undine Spragg, whose pursuit of money rivals Becky Sharp's. We can do no better than to repeat the following by critic Harold Bloom (*How to Read and Why*, New York, Scribners, 2001), who—all the more appropriately for our purposes—summarizes Ms. Spragg's hypergamic frenzy without knowing that he is doing so:

> The story of Undine Spragg, as created by Edith Wharton, has epic dimensions and thuggish protagonists, a contrast that keeps it lively. Undine is an unstoppable sexual force, almost occult in her destructive drive. . . . [She] boils up out of Kansas into New York City, where she marries the wealthy socialite and would-be artist Ralph Marvell. Later, she gives herself to Peter van Degen for a two-month affair. After rejecting poor Marvell, inducing his suicide, Undine devours a French aristocrat, Raymond de Chelles, and then returns to her first, secret Kansan marriage with Elmer Moffat, now a New York billionaire. That is the gist of Wharton's fable; Elaine Showalter sees Undine as the answer to Freud: "While Freud asks, 'What do women want?,' Wharton replies, 'What have you got?' "

Edith Wharton was a snob, an anti-Semite, and an apostle of wealth and social class who, in *The Custom of the Country*, appropriated Thackeray's Becky Sharp and brought her up to date. Hypergamy, however, needs no updating. It is older than the human species. And its depiction leaves a well-trodden trail.

One of the earliest English novels, *Moll Flanders*, by Daniel Defoe, provides a marvelously picaresque account of attempted hypergamy, with an added double cross. This book, published in 1722, actually bears the following remarkable—and descriptive—title: *The Fortunes and Misfortunes of the famous Moll Flanders, who was born in Newgate, and during a life of continued variety, was twelve years a Whore, five times a Wife (thereof once to her own brother), twelve years a Thief, eight years a transported Felon in Virginia, at last grew rich, lived honest, and died a penitent.* The story of Moll, the world's best-known "picaroon," is one of unrelieved efforts at hypergamy. At one notable point, she leaves a banker with whom she had been flirting to marry Jemmy E., a wealthy Lancashire gentleman. Moll had led Jemmy to believe that she, too, had "means," although she didn't. And sure enough, she soon discovers that Jemmy was similarly leading her on! The two rogues, well matched, turn out to be a congenial couple after all.

Fast-forward nearly three hundred years to a much lesser novel, but one that continues the hearty hypergamous tradition: *Maneater*, by Gigi Levangie Grazer. It tells the story of Clarissa Alpert, a twenty-first-century California girl whose sole purpose in life is to bag a rich husband. She sets her sights on Aaron Mason, checks him out on Google, decides he'll do just fine, and marries him, but soon ends up leaving this frantic phone message: "Mommy, it's Clarissa. There's been an emergency. Aaron is poor."

If the matter of good resources is so obvious as to require little elaboration, the question of good behavior is so complex that it is impossible to do it justice. For one thing, resources are pretty much obvious: it takes money to buy—or even to rent—a gracious country estate, to maintain an elegant team of horses, to afford to do nothing with one's days beyond attending balls, going on hunting parties, or visiting one's equally well-to-do neighbors. Sometimes, though, behavior (whether good or bad) is equally obvious: Collins, the soon-to-be-wealthy minister in Austen's *Pride and Prejudice*, is also a here-and-now pompous jerk, too pompous and too much a jerk to hide it. But on occasion behavior can be faked. For example, Wickham, the military officer, is a liar and a blackguard. And so it behooves Elizabeth Bennet—as with other female mammals whose precious parental investment is at stake—to be a careful comparison shopper: squeezing the Charmin, checking for signs of good or bad behavior beneath the surface.

One especially important trait to assess in a would-be mate is what we identify as parenting potential, notably a man's inclination and ability to take care of kids, either his own, perhaps by a previous relationship, or someone else's. After all, human beings are extreme among mammals in the degree to which their babies are born helpless, as well as in the amount of subsequent parental care they require. Any woman prospecting for a reproductive partner can therefore be forgiven—nay, expected—to evaluate whether he seems suitable in this regard. Speaking biologically, it is a matter of predicting a partner's likely level of parental investment—which requires not only that a presumed investor has the wherewithal to invest, but also the inclination to do so. (Note, by the way, that such assessments are likely to be hardwired into the human psyche, not at all attenuated by a possible commitment to intentional childlessness.)

Significantly, it isn't lost on Elizabeth Bennet that for all his superior airs, Darcy extends protection to Elizabeth's sister, Lydia, when she is vulnerable. Elizabeth also learns to discriminate the realities of character when it may be hidden behind peculiar or idiosyncratic behavior; for example, she learns that Wickham may be agreeable enough, but that he is lacking in conscientiousness. Similarly, she must detach herself from her father (who is lamentably careless, despite his pleasant, cultivated ways) and affiliate with Darcy, who, albeit stiff-necked, is also more responsible. Much of the mating game à la Austen—and also à la Darwin—is carried out by assessing personality structure by way of telling incidents, which help separate likely dads from cads.

To understand what appeals to many women, listen to Czech novelist Milan Kundera as he describes a male conquest that was powered by a young man's expressed interest in a young woman and his solicitude for her:

> On that fateful day, a young man in jeans sat down at the counter. Tamina was all alone in the café at the time. The young

man ordered a Coke, and sipped the liquid slowly. He looked at Tamina. Tamina looked out into space.

Suddenly he said, "Tamina."

If that was meant to impress her, it failed. There was no trick to finding out her name. All the customers in the neighborhood knew it.

"I know you are sad," he went on.

That didn't have the desired effect either. She knew that there were all kinds of ways to make a conquest, and that one of the surest roads to a woman's genitals was through her sadness. All the same she looked at him with greater interest than before.

They began talking. What attracted and held Tamina's attention was his questions. Not what he asked, but the fact that he asked anything at all.

Incidentally, in Kundera's novel it didn't hurt that the young man was also driving an expensive red sports car. Resources, anyone?

Why is it seductive to be a good questioner? Or even a questioner at all? Almost certainly, the mere act of inquiry says something about one's solicitude for the person being interrogated. Consider the joke about the self-obsessed movie star who in the course of a date during which he has spoken continually about himself finally announces: "Let's talk about you now. What did you think of my latest movie?"

It is seductive to be a good listener no less than a questioner, and once again we suspect this is because listening is another sign of reproductively relevant good behavior: taking the time to really listen to someone indicates attentiveness and hence a greater probability of committing oneself to the person being attended to, of being more likely to stick around and help out when things get tough, and so forth. A cartoon in the *New Yorker*, titled "Male Prostitute," featured a well-dressed man on a street corner, speaking with a woman: "Oh yeah, baby, I'll listen to you. I'll listen to you all night long!"

Then there is the question of good genes. When it comes to the courtship antics of animals, genes are typically on display in the size, color, and style of one's body, as well as via the ability to sing an elaborate courtship song, bellow loudly, patrol a hotly contested territory, and so forth. Austen barely notes her characters' physical appearance except for observing whether they are "pleasing" or "plain." At the same time, she is hardly unaware of the consequences. Elizabeth Bennet's friend Charlotte Lucas is described—somewhat offhandedly—as homely. When the lamentable Mr. Collins proposes to Elizabeth (who rejects him, much to her mother's consternation and her father's approbation), Collins proceeds immediately to make the same offer to the less attractive Charlotte, who promptly accepts. Here once more is Jane Austen as a discerning, intuitive biologist who understands what modern researchers have only recently discovered: that animals are capable of remarkably accurate self-assessment. Dominant, desirable specimens are likely to drive a harder reproductive bargain, insisting on comparably desirable mates. Those such as Charlotte Lucas are less discriminating. Beggars can't be choosers. Or at least, they are often obliged to be less choosy.

If Jane Austen's heroines seem overtly unimpressed by their suitors' physical appearance (in the early nineteenth century, it would have been more than a bit unseemly to comment on a guy's nice buns; a "square, manly jaw" was about as far as one could go), they more than compensate by their awareness of the minutest details of each other's behavior and speech. In fact, one of the special delights of *Pride and Prejudice*, as well as Austen's other novels, is the witty repartee, the clever observations, the insightful comments of the major characters, which is to say, the verbal and thus mental adroitness of the heroine and her anointed husband-to-be.

In a now-famous study of sexual attitudes in thirty-seven different societies, evolutionary psychologist David Buss found that men and women differed consistently in their preferences for a partner, with men emphasizing youth and physical attractiveness and women being especially focused on wealth and reliability. Relatively little attention has been paid, however, to what Buss's research found about the shared preference of both men and women: for kindness and intelligence. For kindness, substitute "good behavior," and for intelligence, "good genes." Moreover, in a highly social species such as ours, mental functioning can readily be assessed by competence at conversation. Here again, Jane Austen provides a textbook case of sexual selection in action, as her protagonists reveal their intellects—while stimulating the readers'—via their verbal adroitness.

We've already mentioned the peacock's tail, so intriguing to Darwin (as well as to the peahen). Darwin was looking for an explanation for why animals—especially males—would grow such ridiculously ornate structures when they don't seem to add to success in the travails of daily life; if anything, in fact, they are liabilities, since fancy feathers and the like take time and energy to construct and, in the case of telltale tails, make it more difficult to hide from predators as well as possibly precipitating an awkward or even life-threatening tangle in the bushes.

The key is that while peahens are indeed looking at the peacock's tail, their motivation is more practical than Darwin's: they are on the prowl for signs that whoever they choose as a sperm donor is up to the job. This means that the male in question is strong, healthy, and more or less free of parasites and disease. What better way to ascertain this than to lose your heart only to males who are able to construct a veritable Taj Mahal of a tail? Although it's clear that the best way to a peahen's eggs is by a whale of a peacock's tail, that appendage is really a signifier of something else: male quality. A peacock who is struggling to fight off diseases or internal parasites would have a hard time also devoting the thousands of calories needed to grow such a fancy structure. So a well-ornamented peacock is either comparatively pathogen-free or genetically well endowed (in somewhat old-fashioned human discourse, he must have a "strong constitution").

Moreover, feather-based fussiness on the part of the peahen conveys an extra benefit: a peacock who passes muster would also be likely to bestow comparable anatomical architecture on his (and thus her) offspring. As a result, when the next generation of peacocks and peahens go a-courting, any male offspring of a choosy peahen will probably have a tail that is a chip off the old cock's block, and thus he, too, will attract his share of starry-eyed peahens. This argument—known as the "sexy son hypothesis"—has received strong support from theoretical biologists, who have tested it via computer simulations, as well as fieldworkers, who have observed it among many different species.

In his book *The Mating Mind*, evolutionary psychologist Geoffrey Miller has proposed that the mind of *Homo sapiens*, elaborated as it is far beyond the straightforward necessities of survival and reproduction, represents the human equivalent of a peacock's tail, with mental agility serving as an indicator of competence, whereby the capacity to charm members of the opposite sex—as by constructing elaborate, syntactically accurate, suitably polysyllabic, and intellectually stimulating sentences, perhaps such as this one!—is an evolutionary consequence of a process not dissimilar to what Jane Austen's heroines and heroes engage in so brilliantly. (And which we imitate at our peril.)

Evolutionary geneticists refer to "runaway sexual selection," in which preference for a trait can take on a biological life of its own, whereupon something in the opposite sex will be preferred simply because it is preferred. Thus, like peahens choosing a peacock with a fancy tail because their ancestors did so, and thereby producing sons with fancy tails that were preferred in their turn, Austen's people—and people in general—probably choose others with a fancy vocabulary not only because this indicates intelligence, mental health, and education but also because it suggests that the fancy talkers' offspring will themselves be fancy talkers and thus verbally seductive in their own right when show-off time comes around for the next generation.

In any event, it is nearly impossible not to be charmed by the intellectual alacrity of Austen's courting couples, just as the couples themselves are presumed to have been. If people were peacocks, we expect that Jane would be dilating with enthusiasm upon the glories of visually contrasting spots, especially when combined with the shimmering effulgence of natural iridescence.

For another example of sexual selection and Austen in action, consider her novel *Mansfield Park*. Here is the tale of Fanny Price, a poor relative fostered to the wealthy Bertram family, where she joins her cousins Edmund, Tom, Maria, and Julia, and—of course—their various suitors. These include Mr. Rushworth, as rich as he is stupid, and Henry Crawford, charming but self-centered and unprincipled, as well as Henry's sister, the shallow and frivolous Mary Crawford. Maria Bertram loves Henry but is engaged to Rushworth. Henry falls for Fanny (the novel's heroine), who secretly loves Edmund, who thinks he loves Mary. Fanny turns down Henry, thereby showing her spunk but also angering the elder Mr. Bertram, who can't understand why she should refuse such an "advantageous match." Maria, who eventually gives up on Henry, marries Rushworth, then runs off scandalously with Henry, while sister Julia elopes with another eligible bachelor and man of fashion, Mr. Bates. Got it?

But things work out reasonably well. The Bates dude turns out to be a tolerably good catch after all, since he has fewer debts and more income than Julia's father had supposed, while Maria eventually leaves her disreputable union with Henry. Moreover, Edmund comes to realize that Mary Crawford isn't for him: she didn't appreciate the seriousness of the transgression committed when her brother ran off with the already married Maria, thereby suggesting that Mary herself might well be less than reliable as a potential wife. Most important, it finally dawns on Edmund that he actually loves Fanny, who early on had recognized that stalwart, wealthy, intelligent, and thoughtful Edmund was just the man for her. We all breathe a sigh of relief.

It surely is noteworthy that the first *Mansfield Park* character to marry, Maria Bertram, chooses Rushworth, the richest guy in the county. Maria lacks Fanny's keen judgment and ability to see beyond Rushworth's "good resources." After all, she isn't the heroine. Fanny, on the other hand, has the strength of character and perspicacity of judgment to resist pressures that she accept the caddish Henry Crawford and to recognize Edmund's superior overall value.

Most readers find it easy enough to arrive at a similar recognition. Why shouldn't Austen's characters be equivalently perspicacious? Indeed, they often are. And so, just as in real life it is not unknown for two or more women to choose the same man, in literature, too, we find women facing off over men. Add to this the fact that people are unusual among mammals in that fathers/husbands often provide a fair amount of parental investment (protecting the family from predators, getting food, helping rear children, and so forth). The upshot is that good men may indeed be hard to find, and worth competing over.

The study of female-female competition among animals has long taken a backseat to concern with its male-male counterpart, in large part because the latter is more eye-catching and decibel-raising, as well as ear- and skull-splitting. But increasingly, biologists are discovering a rough edge to the former. For example, when female monkeys "babysit" each other's offspring, their apparent solicitude is often mixed with a substantial dose of child abuse. And among many species, including wolves, dominant females inhibit the ovulation of subordinates. To be sure, there are even cases of female-female violence, as with this discovery: when a male house sparrow dallies with an otherwise unattached female, the male's "lawful" mate will occasionally kill any offspring that result. More often, however, female-female competition is less obvious but perhaps no less effective. Among starlings, for example, if a mated male begins showing sexual interest in any of the local bachelorettes, his "wife" typically ramps up her own sexual receptivity, thereby monopolizing his amorous attention (or at least trying to).

Female-female competition features prominently in Greek mythology, particularly among the goddesses Athena, Aphrodite, and Hera. The last, in addition, seems to spend much of her time getting even with the various nubile earthly women who have been the recipients of her husband Zeus's extracurricular visitations. Nor is the Roman version lacking in comparable competitiveness. Virgil's *Aeneid* is underpinned by a constant struggle between Venus (Aeneas's mother and protector) and Juno (his sworn foe) as the two goddesses vie to undermine, outsmart, and outmaneuver each other, using human beings as proxies.

Goddesses, however, don't kill each other, although we hesitate to present this as evidence that female-female competition is less violent than the male-male variety (after all, one of the perquisites of divinity is immortality, and even the male gods don't typically murder one another). To be sure, there are lethal manifestations of female-female competition in the world of letters, although they are less frequent and less prominent than their male-male counterparts. Regan and Goneril, murderous daughters of Shakespeare's King Lear, both lust after the equally vicious Edmund, as a result of which they alternately poison and stab each other. Mostly, however, the literary imagination pictures rival women as schemers and bad-mouthers who limit their backstabbing to the metaphoric. Consider, for example, the hilarious vitriol heaped upon each other by Hermia and Helena, both of whom have developed a love-potion-induced infatuation with the same young man, in Shakespeare's *A Midsummer Night's Dream*. Or the comic dressing-down served up by Polly Peachum and Lucy Brown in Brecht's *Threepenny Opera* (which was based, in turn, on an eighteenth-century play, *The Beggar's Opera*). Polly and Lucy, rivals for the affection of Mack the Knife, compare waistlines and bustlines and argue over whose ankles are the slenderest.

Jane Austen, not surprisingly, is fully up to showing us how female-female competition works, in the process weaving it carefully into her beautifully realized depictions of female choice. Toward the end of her novel *Emma*, such competition emerges with the suddenness of a literary lightning bolt. Up to this point, our eponymous heroine had been unavailingly meddling in the love lives of everyone around her. Then Emma's protégée, Harriet Smith, reveals that she has lost her heart to George Knightley, a longtime friend of Emma's family, surpassingly rich, woefully above Harriet's social station, but—as every reader discerns early on—just right for Emma herself. No sooner does Harriet announce her "unseemly attachment" than Emma realizes that she herself loves Knightley and that she couldn't possibly imagine him married to anyone else!

Incidentally, studies of mate selection in fish have shown that a male guppy initially rejected by females will suddenly become highly attractive to guppy gals after the experimenter rigs things so that it appears that other females have been attracted to him. It seems to be the sexy-son hypothesis in action, with females clamoring to mate with a male once he is identified as popular. Perhaps it is the piscine equivalent of the George Knightley effect.

Back in the world of human beings and Jane Austen, female-female competition is depicted in *Mansfield Park* as well as in *Emma*. Maria threw in her lot with Rushworth because of his money, then realizes her error when Henry Crawford shows up; in fact, the early phase of the novel is taken up with the courtly catfight between sisters Maria and Julia over Henry, who has both money and brains (although he is a bit underfunded in the scruples department). There is also another—and more significant—case of female-female competition in the works between Fanny Price and Henry's sister, Mary, over Edmund. Initially, Fanny seems too insignificant (too poor, too dependent, too young) to compete with Mary, but eventually she triumphs. When Henry Crawford decamps—he wasn't really interested in either sister—Maria goes ahead and marries Rushworth, so when Henry returns, he turns his charm upon Fanny. Henry is a thoroughgoing rake, but his masculine wiles are not adequate for him to win over our Fanny, who remains sufficiently discerning to hold out for Edmund.

Back at Mansfield Park, once Henry has been rejected by Fanny, he runs into Maria in London; this is when he gets her to run off with him. Maria's tale is thus a mini *Madame Bovary*, but our attentions are elsewhere. Henry's sister, Mary, refuses to be "shocked, *shocked*" at her brother's behavior—behavior that involved Edmund's sister, no less—and so she reveals herself as insufficiently attuned to the merits of female fidelity, not only in the reader's mind but in Edmund's as well.

Why are Mary's attitudes about fidelity so important to Edmund? Or any man, for that matter? And wouldn't any woman be equally concerned about Henry's fidelity? Well, yes and no. Here's the biological angle: women get pregnant, not men, and as a result, men can be cuckolded—induced to rear someone else's offspring—whereas women are at least guaranteed to be genetically related to their own children. So in all human societies, a woman's marital fidelity is considered (by men) as far more important than a man's to a woman. The double standard, which is essentially pancultural, is not due so much to patriarchal churlishness as to the biology of internal fertilization, which is equally universal. If, as is the case, female infidelity is often punishable by death and its male counterpart by a slap on the wrist, this is likely underpinned by the fact that philandering by the two sexes is asymmetric in its genetic consequences. Thus, it is noteworthy that although philandering by a married man is typically not seen as a big deal in most cultures, it is a transgression not only moral but potentially mortal if the woman in question is married, in which case it is seen as a crime against the husband . . . largely because, on an evolutionary plane, it is a crime against his confidence of paternity.

Consistent with these concerns, Edmund listens to his own internal whisperings of biological doubt, which suggest that Mary Crawford might well turn out to be the female equivalent of tomcatting Henry. It is then that he notices Fanny—not just as a pleasant companion but as a prospective mate—and the two are married.

Like all of Austen's best characters, Edmund and Fanny excel at reading the social dynamics around them and are ultimately rewarded for their acute assessments. Biologically sophisticated readers will translate this into a genetic reward not unlike that of a discerning peahen or a gullible guppy. This, in turn, means that Fanny's characteristics—including not only her wise assessment of Edmund but also her success in female-female competition—will likely be projected into the future, abetted by Edmund's resources and good behavior.

Pride and Prejudice and *Mansfield Park* are widely acknowledged among literary critics as prototype Regency romances, in which a proud, rich, sometimes overbearing, and wealthy aristocrat (typically male) eventually marries the comparatively poor, good, intelligent, spunky young girl. *Jane Eyre*, which came several decades later, is closer to the Gothic genre: the plucky heroine, terribly alone, arrives in a gloomy and threatening place and encounters a hero who is equally gloomy and threatening; he is initially distrusted but eventually wins the girl's love, and she his. If the mechanics have changed (admittedly not a lot), the basic biology hasn't. In *Jane Eyre*, the austere Mr. Rochester has a terrible secret: he is married, and to a frightening madwoman, Bertha Mason, who lives in the attic and periodically attacks him with demonic fury. And yet Rochester shows unmistakably good behavior in that he commits resources to caring for his crazy, scary wife.

It is worth noting that Rochester's prior marriage does not itself make him any less suitable as a partner for Jane Eyre; by contrast, however, prior sexual experience on the part of a woman—the impregnated sex—is typically much more troublesome (see Chapter 4). Rochester is obviously not a virgin, but we can all safely infer, as he doubtless does, that Jane is. The novel comes to a close with Bertha safely dead and Rochester—albeit blinded and crippled by the fire she started—finally free. Moreover, in attempting to rescue Bertha, despite all her liabilities, Rochester demonstrated no small amount of devotion as well as courage. *Jane Eyre* ends with this famous statement: "Reader, I married him."

Reader, now you know why.

From Jane Austen to Jane Eyre is a transition from female choice in an environment of delicacy and propriety to one that is considerably darker and more fraught. *Wuthering Heights*, by Emily Brontë (sister of Charlotte, who wrote *Jane Eyre*), continues this trend and if anything accelerates it. We are transported, on one hand, to Thrushcross Grange, a civilized, pleasant, sheltered place in the valley inhabited by the Lintons, who are appropriately civilized, pleasant, and sheltered, and who could have made an appearance in any of Austen's novels. On the other, we encounter Wuthering Heights: bleak, ferocious, subject to violent winds, and occupied by the Earnshaws, who are as crude, violent, passionate, and quintessentially male as the Lintons are sophisticated, soft, gentle, and iconically female. From the perspective of a "choosing" female, one can hardly imagine two more contrasting options. Catherine Earnshaw is hypergamically scheduled to marry the higher-class Edgar Linton. But Heathcliff, the rough-and-ready orphan who had been taken in by the Earnshaws, possesses a passionate, animal nature that cannot forgo Catherine any more than she, in the end, can give him up.

It might seem at first glance that Catherine Earnshaw's enduring embrace of Heathcliff over Linton runs headlong into our argument about what women want, but not so fast! After all, part of the fascination and excitement of literature comes specifically from the tension it reflects between our desires and those preferences instilled by social tradition. Our wants don't always correspond with what is best for us or what society permits. Herein lies tragedy.

It is not uncommon for women to yearn for a guy who is at least a little bad, offering the thrill of an unpredictable, forbidden animal attraction and uncontrollability. Consider this, from a typical Harlequin romance published in 1995:

Faith looked as innocent as an angel in her plain white blouse and midcalf blue chambray skirt. Her hair was drawn back in a loose ponytail at the nape of her neck, tied with a yellow bandanna that left little wisps and tendrils free to frame her face like a halo. He looked like the ragged end of an all-nighter, towering over her, dark and disreputable, with his uncombed hair and screaming eagle tattoo and beard-stubbled jaw. Together, they looked like something out of a fifties movie: the Earth Angel and the Bad-Ass Biker.[1]

Women buy—and read—more novels than men do. Although not all of them are "romance" novels, a remarkable proportion deal with the trials and tribulations of finding the right man. These difficulties often involve a conflict described with admirable succinctness by Richard Dawkins as involving a choice between dads and cads. Dads are those males who promise good resources and good behavior; cads lack staying power but are likely to dazzle with attractive appearance and other stimulating indications of good genes. You might think that women should have evolved to see through the latter and choose the former, and to a large extent they have. But there is a catch: good genes have staying power, too! Remember, those good genes can indicate underlying bodily health, and—through the interposition of sexual selection's runaway process and the production of sexy sons—offer the prospect of ongoing reproductive success in future generations. In other words, if you can get Bill Gates, Hugh Grant, and Jane Eyre's Mr. Rochester all rolled into one, you're in luck. But what do you do if you must choose one and reject the others?

So just as men are said to suffer from a "madonna/whore complex" (more on this in the next chapter), there is a predictable tension within women, deriving from the attractiveness of both dads and cads. Catherine Earnshaw felt it and stuck with Linton, a dad,

---

[1] Candace Schuler, *Lovers and Strangers* (New York: Harlequin Temptation, 1995).

while yearning for her cad, Heathcliff. Jane Austen's heroines consistently feel it, too, but with equal consistency they opt for dads.

Although most women might dispute Erica Jong's infamous endorsement of the "zipless fuck" (a sexual consummation, presumably with a cad, so quick, so spontaneous, and so "in the moment" that unzipping isn't even called for), contemporary writing seems, if anything, more likely to celebrate—or at least to describe—the appeal of cads. According to Jong's heroine, the vivacious Isadora Wing, a zipless fuck is a "platonic ideal," a perfect encounter in which "zippers fell away like rose petals, underwear blew off in one breath like dandelion fluff." It seems unlikely that modern women are more likely to couple with cads—ziplessly or not—than their predecessors were. Maybe they are simply more likely to acknowledge the temptation.

In any event, from Jane Austen to the present day, women—and especially women writers—have been moved to note the dad/cad tension, albeit without the nifty verbal shorthand. Perhaps the optimum (biologically, if not ethically) is to marry a dad but mate, surreptitiously, with a cad. But at the same time, there looms the danger of being seduced and abandoned by a cad; after all, that's why they're called cads!

Consider, for example, *Bridget Jones's Diary*, the 1990s bestseller by journalist Helen Fielding that began as a column in the London *Independent* newspaper and then expanded into a hilarious novel (and a pretty good movie, too). Bridget Jones isn't only wonderfully clever and funny; she also strikes a familiar chord in her obsessive worries about—surprise!—finding a mate. Bridget vows that she will keep away from "misogynists, megalomaniacs, adulterers, workaholics, chauvinists, or perverts," but, interestingly, not paupers. That goes without saying. It is hard to imagine Elizabeth Bennet emulating Bridget Jones and drinking, smoking, or even eating too much. But the pursuit of self-improvement and of a male with *r*-cubed? That's another story.

Generations of XX chromosome carriers (and even a few XYs)

can also sympathize with Bridget's aversion to the intolerable "smug marrieds" professing concern for her and her fellow "single-tons." Bridget/Fielding points out, "We wouldn't rush up to *them* and roar, 'How's your marriage going? Still having sex?' " Neither would Jane Austen. Jane would also no doubt sympathize with Bridget's parents, who want her to marry glamorous lawyer Mark Darcy. (Does that last name ring a bell?) But Bridget compounds her comical ineptitude by being rude to Darcy, the inverse of what transpires in *Pride and Prejudice*. It may be stretching things to picture Mrs. Bennet in the position of Bridget's mom—sorry, Mum—starting a new career as the host of the TV program called *Suddenly Single* and then disappearing with a Portuguese gigolo. But nearly two centuries post-Austen, some things haven't changed. Good genes, good behavior, good resources: Mark Darcy has them all, and finally Bridget gets it . . . and him, and thus them. The benefits of a cad and of a dad, all wrapped up in one sexy guy.

It is noteworthy that the premier scribe of "what women want" was herself a woman. It is also beyond dispute that women typically read books about relationships, while men choose stories of adventure. By contrast to accounts of female choice, modern novels—and movies and magazine articles—that cater to male sensibilities tend to be more directly violent and explicitly sex-laden, often adventures of the search-and-destroy genre, filled with casual erotic conquests. Picture James Bond, pulling out his gun and his penis with about equal frequency. It is also noteworthy that 007 and his ilk feature in numerous sequels, since there is little in the male sexual imagination that limits a desirable love life to just one partner; by contrast, the world of romance novels—which caters almost exclusively to women—necessarily must create new characters with every book, since the typical climax of these tales involves a satisfying and presumably monogamous union between heroine and hero.

Sadly, things don't always end up that way. Natural selection operates by the gradual accumulation of whatever works—it does not guarantee perfection. After all, more than 90 percent of all species that ever lived are now extinct. Not surprisingly, even though evolution has equipped both men and women to be adroit intuitive biologists, neither sex has a perfect track record when it comes to assessing the desires of the other. Women, for instance— especially in modern-day America—consistently overrate the importance of slenderness as a sexual turn-on for men, most of whom prefer their partners to look healthy rather than anorexic.

And men have historically ended up wide of the mark at one extreme or the other when it comes to sizing up the sexuality of women, either seeing them as sexually rapacious and unutterably lascivious and thus responsible for the world's evils (thereby almost certainly projecting their own lust onto the ladies) or seeing women as altogether chaste, albeit chased.

Thus at one time, Talmudic scholars entertained such an overblown estimate of women's sexuality (and society's responsibility to repress it) that widows were forbidden to keep male dogs as pets! At the other extreme, an influential nineteenth-century Victorian tract by one Dr. William Acton announced that "the majority of women (happily for society) are not very much troubled with sexual feelings of any kind. What men are habitually, women are only exceptionally."

The truth, of course, is that women are neither sexually rapacious nor downright asexual. Central to that truth is the role of female as sexual chooser.

But male befuddlement as to what women really want hasn't stopped them from guessing, worrying, and often fantasizing, frequently about their own anatomy. The reality is that nearly always what women want is not what men think—or fear—that they do. *Lady Chatterley's Lover* is a perfect case. D. H. Lawrence thought he knew what turned women on, but he was wrong. The story, in brief:

Lady Chatterley is married to Lord Clifford Chatterley, rendered impotent by a war wound. She finally finds sexual satisfaction with Mellors, the virile gamekeeper. Maybe the fantasy of Lady Chatterley going gaga over her lover's erect penis—which she does with almost boring repetition—was a turn-on for Lawrence, and for many men who mostly read the "dirty parts," but it doesn't do nearly so much for women:

> The sun through the low window sent in a beam that lit up his thighs and slim belly and the erect phallos rising darkish and hot-looking from the little cloud of vivid gold-red hair. She was startled and afraid.
>
> "How strange!" she said slowly. "How strange he stands there! So big! and so dark and cock-sure! Is he like that?"
>
> The man looked down the front of his slender white body, and laughed. Between the slim breasts the hair was dark, almost black. But at the root of the belly, where the phallos rose thick and arching, it was gold-red, vivid in a little cloud.
>
> "So proud!" she murmured, uneasy. "And so lordly! Now I know why men are so overbearing! But he's lovely, *really*. Like another being! A bit terrifying! But lovely really! And he comes to *me*!—" She caught her lower lip between her teeth, in fear and excitement. . . . And he was helpless, as the penis in slow soft undulations filled and surged and rose up, and grew hard, standing there hard and overweening, in its curious towering fashion. The woman too trembled a little as she watched.

We have no problem with D. H. Lawrence as a creative and imaginative writer (although we are more than a little troubled by his protofascism), and we applaud his willingness to challenge prudish conventions as to what constitutes "obscenity." But Jane Austen, for all her nineteenth-century restraint and delicacy, was a much better evolutionary psychologist, and a far sexier writer. And even Bridget Jones—not to mention Charles Darwin—knew more about what makes women, as well as peahens, tick.

4

How to Make Rhett Give a Damn

What Men Want, and Why

L et's start with the "madonna/whore complex." (The madonna of the Bible, not the *Billboard* charts.) According to tradition—and biology—men fool around readily enough with whores but prefer to marry virgins. These days, the expectation of premarital female chastity is more than a bit quaint, and perhaps less realistic than ever. And in an age of AIDS, sex with prostitutes can be downright dangerous. But the caricature stands: what men want for a night's entertainment is one thing, what they want in a spouse is another. In other words, men like bad girls but marry good ones. Which brings us to *Tess of the d'Urbervilles*.

Written by Thomas Hardy in 1891, this modern classic tells the travails of Tess, a young, poor, inexperienced lass who, in a moment of weakness, allowed herself to be "possessed" by a dapper young cad named Alec d'Urberville. As a result, Tess became pregnant; her baby died, but Alec kept pestering her for more "possession." Tess escapes to a dairy farm, where she works as a milkmaid while falling in love with a fellow named Angel Clare, who reciprocates her feelings. Angel is especially attracted to Tess's "innocence," an innocence that is, ironically, quite genuine, and which made her susceptible to Alec's earlier advances in the first place. Angel proposes marriage. Initially, Tess turns him down, feeling in her heart of hearts that she is too impure to marry this earthly angel, but later she accepts. The night before their wedding, Tess writes a letter to Angel, confessing her one-night stand with Alec d'Urberville, feeling glumly confident that with this knowledge, Angel will change his mind. But the next morning, Angel is as loving as ever.

Tess delightedly assumes that, soiled as she is, Angel has nonetheless decided to accept her as his bride. It turns out, however, that Angel never found the letter. He tells her about a night of debauchery in his own past, whereupon she tells him of hers. Tess forgives Angel; Angel, however, cannot bring himself to forgive Tess, saying, essentially, that he couldn't possibly live as husband to someone so besmirched. Angel goes off to Brazil. Tess returns to hard work as an itinerant farm laborer, barely avoiding starvation. At this time, the pesteringly possessive Alec d'Urberville shows up once again, claiming to have reformed his ways and behaving generously toward Tess's indigent parents. Tess writes to Angel, begging him to take her back and save her from her previous possessor and present pursuer. But it takes several months for her letter to reach Brazil, during which time Tess concludes, in misery, that Angel has abandoned her for good. Then, sure enough, Angel returns to England, ready to accept Tess at last . . . only to find that, despairing of his forgiveness and desperate to escape a life of poverty, she has moved in with Alec. Tess completes the tragedy by stabbing Alec and then confessing her deed to Angel, with whom she finally reconciles, enjoying a few days of bliss before she is arrested and taken away to be hanged.

Tess of the d'Urbervilles is a grim, compelling story of class differences, economic desperation, and the helplessness of good people caught in bad circumstances. For many, it is primarily a tragedy of errors, notably Tess's assumption that Angel had read her letter and forgiven her prior sexual encounter, whereas in fact he had not. (And when he found out, he could not.) Tragedy ensued. Seen through modern eyes, Angel admittedly did not live up to his name, although for late Victorians, his stiff-necked adherence to the double standard made perfect sense. For Hardy and his audience—many of whom were appalled by the explicit consideration of sex even though no sexual acts as such were described—the tragedy that bound Tess, Angel, and Alec was powered by economic circumstances and social conventions. Like lumbering prehistoric creatures mired in the La Brea tar pits, this unhappy trio was stuck, constrained from moving freely by forces beyond their ken and greatly exceeding their pitiful strength.

At the same time, we can ask why. Why couldn't Angel accept Tess after knowing that she was no longer a virgin? After all, Tess was more than willing to overlook Angel's parallel behavior. It isn't enough simply to point to the double standard, whereby "boys will be boys"—which includes "sowing their wild oats" on occasion—while "good girls don't do that." Of course, some girls do (or else boys wouldn't have the chance to be boys), at least in part because not all are "good." Maybe these looser young ladies are just weak-willed compared to their "better" sisters? Or is it more a question of strategy? If their aim is to mate with the best male available, then do some females intuitively realize that they have less to offer and therefore their best bet is to be more available than the standoffish good girls with their more desirable traits?

What is clear, however, is that in society after society, throughout most of human history, it is far more acceptable for a man to have premarital sexual experience than it is for a woman. The same applies to extramarital affairs. An evolutionary explanation presents itself, one that fastidious Victorians might have found appalling but to which their DNA was already resonating. Recall Othello's rage and its biological underpinnings: if Othello had really had an affair with Emilia, Iago's wife, that could have been a genetic disaster for Iago, who might have ended up rearing Othello's child instead of his own. For Desdemona, it might have been a *social* problem, but not, strictly speaking, a genetic one. Any little Desdemonas would still be hers. Not so, of course, for Othello, if Desdemona really had been making "the beast with two

backs" with Cassio, as Iago alleged. We've already seen how this
helps make sense of male sexual jealousy generally.

To be sure, most women are not casual about a husband's infidelity. In literature, as in life, women treat the philandering of their
mates as a very serious matter indeed, a response that is surely
steeped in biology; that is, they're concerned that their mates
might divert important resources to the offspring of other females,
or even possibly leave them for their new girlfriend. Yet as sexually
possessive and jealous as women are, men are even more so. And as
much as the divergence between male and female responses to infidelity is undergirded by cultural tradition, both those responses—
and even the cultural expectations themselves—are built upon a
biological foundation.

In large part, it's a reproductive translation of the old real estate
dictum that three things are especially important in determining
the value of a house: location, location, and location. When it
comes to confidence of genetic relatedness, location once again is
key. This time, it's the location of fertilization: internal or external.

Among all mammals, fertilization is internal, taking place deep
within the female's body. And among all mammals, babies are
cared for—almost exclusively—by their mothers. In every mammal
species (including *Homo sapiens*), females do considerably more
mothering than men do fathering. Although most people take this
parental double standard for granted, it is actually profoundly unfair. After all, a female mammal has carried her offspring for a prolonged period, nurturing it from her own bloodstream, only to
undergo the strenuous and sometimes risky process of birth itself.
At this point, wouldn't it be fair, and even fitness-enhancing, for
her reproductive partner to pitch in and take up some of the slack?
In short, why don't men lactate?

It's not that they lack breasts, or even nipples. The sad truth is
that not even the most doting father can nurse his offspring, basically because for the past hundred million or so years of mammalian history, not even the most confident father could know for

certain that he really was the father. And nursing requires a tremendous expenditure of time and energy; it simply would not be fitness-enhancing for them when "their" child might really have been fathered by someone else. Of course, using this logic we would expect that in species employing external fertilization, with, say, eggs and sperm simply discharged into the surrounding water, males and females should be about equally likely to do the child care. And this is precisely what we find in fish and amphibians.

Back to the tribulations of Tess, now revealed to reflect, in large part, the anguish of Angel and of men generally. Deprived as they are of the serene confidence of genetic parenthood, which is the (literal) birthright of every woman, men in societies as diverse as the Amazonian Yanomamo, the Alaskan Inuit, and the late Victorian English have struggled to increase their confidence. Certain stick insects remain *in copula* literally for weeks, keeping other stick insects from inseminating "their" female. Among some species of bees, males actually explode after mating; they die in the process, but only after converting themselves into a postmortem chastity belt by jamming part of their genitalia inside that of their lady love. Certain sharks, on the other hand, precede copulation by giving the female a high-pressure saltwater douche via their double-barreled penis, which evidently serves to wash out any sperm that may have been deposited by a prior suitor. Not to be outdone, many mammals (including most primates) produce semen that coagulates into a rubbery "copulatory plug," which serves as yet another organic chastity belt. Finally, among ringdoves—a species in which males are expected to provide a fair amount of child care after the young have hatched—males actually reject females who reveal, by their overeager receptivity to courtship, that they have already received the amorous attentions of another male. And Angel Clare? He rejected Tess when he learned that she had already had sex with Alec d'Urberville, even though she didn't love Alec and the child hadn't survived.

The biological bottom line is that men, far more than women, have to deal with a heavy dose of sexual anxiety—not so much performance anxiety as the deeper, biologically generated uncertainty of parental confidence. Concern about a potentially wilted penis is nothing compared with concern about one's genetic posterity. Call it "Angel's anxiety," and alleviating it is an important component of what men want. We should pity, therefore, the poor Y chromosome bearers among us. They have fragile biological egos, and for biologically appropriate reasons. What they want, as a result, is reassurance, a neediness that inclines men to crave indications that any prospective mate is likely to be sexually faithful—and, conversely, to be turned off by indications that she isn't, or wasn't, and therefore might not be in the future.

Maybe men want virgins because it gratifies their ego to think that they will not be compared as lovers with anyone else. More likely, Angel's anxiety about Tess was generated by a concern—heightened, to be sure, by nineteenth-century mores—that Tess's earlier dalliance indicated that she might subsequently prove to be a "loose woman," in which case Angel could lose his genetic posterity. It is always possible as well that men have long been attracted to virgin brides because virginity itself generally correlates with youth, and younger women have a longer reproductive future in store. There is little doubt that virgins tend to be, on average, younger than their sexually experienced sisters, but there is even less doubt as to their previous sexual history.

Even Rhett Butler—the epitome of suave, sexually competent manliness and rebellious indifference to traditional values—shows his fragility in *Gone With the Wind*. He desires Scarlett O'Hara and eventually marries her (albeit as husband number three). But even this alluring, dissolute, cynical, unscrupulous war profiteer and blockade runner cannot quite get over his jealousy of Scarlett's nonstop infatuation with Ashley Wilkes. By the story's end, what's really gone with the wind in Margaret Mitchell's celebrated novel (and even more celebrated motion picture) is Rhett's ability to tolerate Scarlett O'Hara's yearning for another man, even though Scarlett and Ashley never do more than kiss. By the novel's end, just as Scarlett is coming to see Ashley for the diminished and weakened figure that he is—and always was—Rhett finally gives up; in his sexual insecurity, Rhett is finally revealed to be as inadequate and insecure as Ashley, although no more so than any male mammal, burdened with the uncertainties of internal fertilization, can be expected to be.

For the evolutionary biologist, men and women are condemned to a delicate mating dance in which the woman must not be too readily available. Sex can be an important way for a woman to demonstrate her love for a man (as well as responding to her own inclinations), but if she is too "free," too sexually avaricious, or even—ironically—too sexually skillful, she runs the risk of turning him off, at least as a potential long-term mate.

In *Humboldt's Gift*, by Saul Bellow, the protagonist, an on-again, off-again successful writer named Charlie Citrine, is alternately fascinated and repelled by his lover, Renata, who is several decades younger than he. Citrine's little monologue reveals a widespread male concern about a prospective mate who may be just a little bit too good in the sack: "As a carnal artist she was disheartening as well as thrilling, because, thinking of her as wife-material, I had to ask myself where she had learned all this." Bellow's hero/narrator is not simply reflecting a stubborn sexist refusal to tolerate as sauce for the goose what he assumes is acceptable in the gander, a puerile inadequacy of the male psyche, or simply a perverse ambivalence. Nor is it merely a refraction of cultural norms. (Because if so, we must ask why it is the cultural norm of virtually every human society; the one thing all human societies have in common is the biology of humanness.)

Recall that in the madonna/whore complex, men are attracted to brief relationships with the latter—mostly because they are available—but prefer to marry someone who is reliable, that is, relatively chaste, who will not squander the husband's resources on another man's offspring. (Recall the ringdoves, among whom the males reject potential mates who may have already been courted, and perhaps inseminated, by another male.)

But what, you might ask, about the saga of the "hooker with the heart of gold," such as in the movie *Pretty Woman*, in which a millionaire falls for a prostitute? First of all, once the viewer accepts Julia Roberts as a prostitute, it is entirely believable that almost any man would lose his heart to her! (Men, as we shall see, are not entirely indifferent to signs of youth and attractiveness in their prospective partners.) In addition, men seem to have a recurrent fantasy of *rescuing* prostitutes through romantic love. This may have something to do with the prostitute—if sufficiently classy—representing the most sexually desirable woman, normally out of reach for the average man. Her previous record of sexual promiscuity is simultaneously a turnoff and turn-on. And significantly, such fantasies always include the expectation that upon being "rescued," the woman's sexual careerism will definitely be turned off, as the "rescuer" manages somehow to appropriate the woman exclusively for himself.

In any event, most viewers would doubtless agree that any story in which a wealthy, high-status man finds marital bliss with a prostitute would be highly unlikely—about as unlikely as a prudish but decent would-be gentleman farmer from the English dairy country finding marital bliss with a young woman who had been courted and inseminated by a rascally smart-Alec.

If sexual fidelity is high on most men's list of female desirables, youth is not far behind. Men want women who are not only young but typically at least several years younger than themselves. As with the saying "I chased her until she caught me," this may simply reflect a wily triumph of *female* choice. This may seem paradoxical since many women bemoan this male tendency. But recall that one of the keys to Jane Austen's heart (Chapter 3) is good resources, which tend to accumulate with age, providing older men with an advantage when it comes to attracting younger women. From the male perspective, at the same time, youthfulness is desirable in a romantic partner because a younger woman offers a longer and thus richer reproductive future. After all, women have a fixed childbearing window that closes abruptly in middle age. Men don't. Should anyone be surprised, therefore, that around the world, husbands are consistently older than their wives? When was the last time you heard someone refer to a May-December marriage in which May was the man and December the woman?

The consequences of this disparity can be as predictable as they are infuriating: middle-aged and even elderly men find themselves attracted to young women, not uncommonly abandoning their middle-aged wives for a newer model. Think, for instance, of *The First Wives Club*, Olivia Goldsmith's 1992 novel in which three middle-aged women (later portrayed on film by Goldie Hawn, Bette Midler, and Diane Keaton), all first wives inhabiting the headier spheres of New York society, were dumped by their husbands in favor of younger, blonder, and more decorative "trophy wives." Not surprisingly, this happened after the men had achieved sufficient wealth and power to be attractive to younger, more fertile women. Incidentally, the three "first wives" eventually find suitable revenge against their heedless husbands, if not against their troublesome biology.

Probably the most dramatic depiction of male preference for youth is Vladimir Nabokov's *Lolita*. Its main character, Humbert Humbert, doesn't leave his aging wife for a younger woman; rather, he marries an older woman in order to be with her preteen daughter and then takes advantage of the fact that his wife dies in an automobile accident to run off with the girl. As hard to believe as it may be, the novel is most compelling simply for its lush language and extraordinary evocation of 1950s America. Yet as a pedophile sexual predator, Humbert is also nothing less than despicable. And while it may seem from their focus on sex that evolutionary biologists are amoral cultural relativists, they, too, recognize something not merely immoral but biologically off base in the hero/villain's obsession with a "nymphet." Nabokov never clearly states whether Lolita has begun menstruating and thus is fertile, and legally this does not matter—Humbert is without doubt guilty of statutory rape. But biologically it does matter.

The unconscious age-related double standard that most people carry within them is italicized in those very rare cases when a young man is portrayed as lusting for an old woman: *Harold and Maude*, for instance. This 1971 film imagined a romance, fully sexual, between a teenager and an octogenarian woman, leaving the audience alternately shocked and amused. Stories in which older men romance younger women capture a chunk of biological reality and hence are not uncommon; *Harold and Maude* captured an audience because it brazenly defied that reality. The truth is that relationships between young men and old women are exceedingly rare precisely because such liaisons do not make reproductive, evolutionary sense, whereas the converse, although occasionally ridiculous, is nonetheless consistent with human nature because it is consistent with nature itself.

Of course, the great majority of men are neither Humbert nor Harold, although more can relate to the former than the latter. Male preference for youthful sexual partners is what anthropologists call a "cross-cultural universal." There is, for instance, no human society in which an eighty-year-old woman is seen as more sexually alluring than a twenty-year-old. The bottom line isn't youth as such, any more than it is virginity per se. Both virginity and youth bespeak the prospect of reproductive success for the man in question: the former as a promise of fidelity, the latter of fertility. Both sexes are turned on by sexual partners who offer them the prospect of the greatest possible reproductive success, even though they typically don't know that this is happening. A man slavering over a *Playboy* Playmate of the Month, ogling an attractive young thing in a strip joint, or admiring the healthy hair and smooth skin of a coworker isn't thinking: "She would probably bear healthy babies." Indeed, the prospect of actual baby making might well be a psychological turn*off* these days, but this doesn't mean that biology isn't driving the system, giving impetus to what men find sexy. In other words, these men need not have reproduction in mind; it is in their genes. Sex is the route, reproduction the destination. Most living things—including human beings—have been programmed to enjoy the trip, giving little or no thought to where it might lead.

Often, they'd rather end up somewhere else. Consider the opening line of *Genesis*, a 2003 novel by Jim Crace: "Every woman he dares to sleep with bears his child." Felix Dern has sex with six different women—voilà, six children! But rather than revel in this evolutionary success, our hero learns that "to be so fertile is a curse." Felix didn't want all these offspring; what he did want, however, was sex with each of the women in question.

How many hungry people of either sex look at a meal and say to themselves: "This looks like an appropriate array of carbohydrates, proteins, and fats, which would optimally satisfy my intracellular metabolic needs"? Instead, a horny man looks at an attractive sexual partner just like a hungry one looks at a well-prepared meal and says to himself: "Yum!"

Not surprisingly, there are certain aspects of the female body that men generally find especially yummy, not least among them breasts. We could fill an entire chapter—perhaps an entire book—with literary references to the attractive female figure, running from "shapely" and "curvaceous" to the Victorian era's "amply provided with womanly charms" to more explicit modern accounts. Any way you write it, men like breasts. It is a near obsession that writers have long acknowledged, although only rarely have they glimpsed the underlying biology as eagerly as they have alluded to the décolletage of their female characters.

What is this all about? Why are breasts so endlessly fascinating—to men, at least? It all has to do with two closely related substances: milk and fat. Add to these a simple fact that we looked at earlier: only women get pregnant. This explains not only why women have a uterus as well as organs of lactation but why men have neither. It also seems to explain why women have a higher proportion of body fat. Extra metabolic reserves, in the form of adipose tissue, almost certainly contributed to the success of our ancestral mothers in carrying healthy babies to term, then providing milk for them afterward. Indeed, it appears that a certain minimum of fat is necessary for women to start menstruating. Serious female athletes even cease having regular periods; when they stop their intense training and, as a result, increase their fat reserves, normal cycling resumes. It is interesting that no similar correlation between body weight, fat levels, and sperm production has ever been demonstrated for males, although the last waxes and wanes according to other things, notably body temperature and frequency of ejaculation. Couples trying unsuccessfully to become pregnant often must be counseled, paradoxically, to *reduce* their frequency of sexual intercourse so as to increase the man's sperm count, but men are not advised to become chubbier so as to become more reproductively competent. The *f*-cubed connection (female, fat, and fertility) makes evolutionary sense because pregnancy and the developing fetus require lots of calories. Moreover, extra body fat represents insurance in the event of lean times. By contrast, the strictly biological demands for a man's reproduction are concluded with an ejaculation.

The likelihood, therefore, is that fat distribution in women has been directly favored by natural selection. At the same time, men's preference for the curvy hourglass figure of sexually mature women almost certainly reflects an unconscious choice of those body types most likely to produce healthy children (for much the same reason that people prefer eating bread over bark, or steak over stones—obviously, the former taste better, but we developed these taste preferences *because* certain things are more nutritious than others). Broad hips, after all, facilitate labor and delivery, and breasts are not just sexual adornment: they provide nourishment for the infant. Not surprisingly, these considerations show up in literature.

Chaucer, in *The Canterbury Tales*, had no doubt what made a woman sexually attractive: "buttokes brode and brestes rounde and huge." And Stephen Dedalus, James Joyce's young fictional hero in *Portrait of the Artist as a Young Man*, came up with this analysis while musing with his friends on the nature of female beauty:

Every physical quality admired by men in women is in direct connection with the manifold functions of women for the propagation of the species. . . . [Y]ou admired the great flanks of Venus because you felt that she would bear you burly offspring and admired her great breasts because you felt that she would give good milk to her children and yours.

To be sure, notions of physical attractiveness reflect social convention and not just biology alone; accordingly, they are far from universal. The full-bodied earth mother goddess or Botticelli beauty has largely been replaced in the United States, for example, by the nearly anorexic fashion model. Among other societies, tattooed faces are de rigueur, or elongated necks, teeth that have been filed to points, and so on. In some places, women typically shave their legs and underarms; in others, they don't.

In a world of scarce food, an ample figure probably did indicate something important about likely success as a parent, while today, in the Western world at least, obstetricians can readily deliver babies by cesarean section, and infant formula can substitute for breast-feeding. Moreover, for those enjoying abundant—often excessively abundant—food, a degree of slimness indicates other components of health, such as sufficient exercise, prudent eating habits, and so forth. But beauty has never really been in the eye of the beholder; rather, it's in the brain, which interprets what the eye sees. And that brain was produced by natural selection. What it interprets as "beautiful"—if "beautiful" means "desirable to mate with"—is whatever contributes to evolutionary success. So it can be predicted that within a spectrum of cultural variation, the basic evolved male desire for large-breasted, slim-waisted women is still at work and is likely to continue. And so those skinny—bordering on scrawny—models have made a booming business of silicone implants.

Human beings are unique among mammals in sporting prominent mammaries when not lactating, and also in making erotic use of them. Why? One intriguing possibility is that for more than 99 percent of our evolutionary past, a primordial battle of the sexes has been under way, with the battlegrounds being the female figure and the male unconscious mind. Let's begin with the assumption that Stephen Dedalus was correct, and well-developed female mammaries have been preferred by men as a promise of subsequent milk for their offspring. This would have led, in turn, to large-breastedness—even when not lactating—on the part of women, because insofar as busty women were especially desirable, they would have been mated by the fittest men, and accordingly would have produced more offspring . . . the daughters of whom were likely to be ample-bosomed themselves.

But the situation is almost certainly more complicated than the simple promise of breasts as milk providers. There is, for example, very little correlation between the size of a woman's nonlactating breasts and her eventual milk production—the greatest proportion of a nonlactating breast is occupied by fat, with virtually no glandular tissue. Milk-producing glands develop only later, during pregnancy, although of course the breasts of a nursing woman are typically full indeed.

Here is one possible explanation: let's assume that long ago, men preferred women with relatively large breasts and hips because the latter indicated room for the baby to be born and the former was at least observed to characterize a successfully nursing mother. This would set the stage for women to have evolved fatty deposits on their breasts and hips, essentially to deceive men. "Choose me," they would have been saying, in the Stephen Dedalus interpretation, "and you will get successful children," although in reality, fatty breasts are no more a guarantee of subsequent milk production than fatty hips truly indicate a wide birth canal. If so, then prominent fatty breast tissue is a case of successful but false advertising. Round one goes to women.

But the sparring need not have ended here. Men could have counterattacked, insisting that their spouses have comparatively small waists as a way of ensuring relatively low fat levels and in the process keeping women honest about what they have to offer. If so, then round two would have gone to the men. Not to be outdone, however, women might have been able to parry their mates' counterattack, arranging their anatomy such that their waists tend to remain slim while fat is deposited elsewhere. The sexually appealing, hourglass figure of women may owe much to deception and counterdeception of this sort.

Devendra Singh, a psychologist at the University of Texas, has explored male preference for the female figure, specifically the ratio of waist to hip measurements. Before puberty, the waist/hip ratios of boys and girls are indistinguishable; then they diverge. Boys lose fat from their thighs and buttocks, while girls deposit fat on their upper thighs and hips. By the time they are adults, women have 40 percent more fat than men in their lower trunk. At the prime of their reproductive years, healthy adult women will have a waist/hip ratio of .67 to .80, while for healthy men, that ratio is between .85 and .95. After menopause, women's waist/hip ratios once again approach that of men. Not surprisingly, hormones seem to be responsible. High levels of estrogen (actually, high estrogen-to-testosterone ratios) stimulate fat deposition around the hips and inhibit it in the abdominal region. Therefore, a low waist/hip ratio signals relatively high reproductive ability, while a high waist/hip ratio can be a sign of illness or a previous or current pregnancy, none of which bodes well for the evolutionary success of men who choose such women.

In a number of studies, Singh has found—as we would expect—that men rated images of women as more attractive the lower this ratio: .80 is preferred to .90, and .70 is preferred to .80. Examining images of *Playboy* centerfolds and of beauty contest winners during the past thirty years, Singh discovered that although preference for overall thinness increased, the ideal waist/hip ratio remained unchanged at precisely .70. (As mentioned, this does not deny that a fondness for total skinniness varies quite a bit: plumpness is valued in societies where resources are scarce, while slenderness is sought after in the affluent West, in which wealthy, healthy people watch their weight and work out.) Regrettably, we cannot apply a tape measure to the female figure as described by the world's creative literary artists, but it's a good bet that if such an assessment could be performed, it would reveal that writers from Ovid to Ian Fleming depicted that trusty, lusty .70.

In any event, it seems clear that female physical traits that men find attractive are heavily weighted toward those that indicate health and, almost certainly, reproductive potential. Regularity of features, skin and hair quality, youth, appropriate physical size, and so forth: all these traits correspond to bodily health. Another feature that is receiving substantial attention is symmetry. Most living things tend to be symmetrical, and human beings are no exception. And yet it is no small trick for a growing body to end up physically balanced. In fact, it is well documented that disturbances during embryology and growth—due to inadequate nutrition, genetic anomalies, and so on—produce disturbances in symmetry. When people are asked to evaluate the attractiveness of human faces that have been computer-altered to reflect differing degrees of symmetry, the results suggest a rewrite of Keats's famous line about the interconnection of beauty and truth. Biologically, we might say, "Beauty is symmetry and symmetry, beauty."

Put it all together and we can at last make sense of the great concluding line to the song "There Is Nothing Like a Dame," in *South Pacific*, when a sailor named Stewpot sings, with deep voice and mounting, robust enthusiasm: "There is absolutely nothing like the frame . . . of . . . a . . . dame."

This story is still told in New Zealand. It seems an Episcopal bishop had been visiting an isolated Maori village. As everyone was about to retire for the night after an evening of high-spirited feasting and dancing, the local headman, wanting to show profound hospitality to his honored guest, called out: "A woman for the bishop." Seeing the scowl of disapproval on the prelate's face, he roared even louder, "Two women for the bishop!"

Clearly, the chieftain was an accomplished evolutionary psychologist. He knew that male sexuality includes a fondness for multiple partners. To some extent, the more the merrier. Lord Byron wondered, "How the devil is it that fresh features / Have such a charm for us poor human creatures?" Biologically, the explanation is obvious: insofar as each prospective partner represents a potential pregnancy—and thus an increase in the reproductive success of the successful inseminator—males generally and human males in particular are partial to "multiple sexual partners." (Not necessarily at the same time; sequentially will do just fine.)

It's that old penchant for polygyny rearing its visage once again. Interestingly, men are often inclined to fantasize about how wonderful it would be if society would only grant them the opportunity to live out their harem-having inclinations, conveniently forgetting that in such a case, other men would also be striving similarly. It is a bit like those people who claim to remember their past lives, in which they were always Cleopatra or Napoleon, never a slave or a peasant. But that doesn't stop men from imagining how it might be if only the world were organized just for them.

Amidst the "oxen of the sun" chapter in James Joyce's *Ulysses*, a horny young man named Buck Mulligan expresses, in comically exaggerated language, a similarly widespread male fantasy: providing sexual services for erotically and reproductively frustrated women, no questions asked and no fee charged.

It grieved him, plaguily, he said, to see the nuptial couch defrauded of its dearest pledges and to reflect upon so many agreeable females . . . who lose their womanly bloom . . . when they might multiply the inlets of happiness, sacrificing the inestimable jewel of their sex when a hundred pretty fellows were at hand to caress; this, he assured them, made his heart weep. To curb this inconvenience . . . he proposed to set up a national fertilising farm . . . with an obelisk hewn and erected after the fashion of Egypt and to offer his dutiful yeoman services for the fecundation of any female of what grade of life soever who should there direct to him with the desire of fulfilling the functions of her natural. Money was no object, he said, nor would he take a penny for his pains. The poorest kitchenwench no less than the opulent lady of fashion . . . would find in him their man.

Our young Buck goes on, describing how he would nourish himself so as to maintain necessary strength for the servicing of his imagined clients, to the great approval of his assembled friends. No mention of love or even affection for his partners, and certainly no second thoughts about child care or the maintenance of long-term relationships. The overriding consideration is sexual variety: lots of women, the more they, the merrier he.

Speaking a bit more delicately, W. S. Gilbert, in *Trial by Jury*, alluded to the flip side of the male fondness for variety with the knowing line "Love unchanged will cloy." It isn't necessarily true that monogamy equals monotony, but the biology of male sexuality ordains that novelty itself is a turn-on for men—not simply new positions or new accoutrements, but literally new partners. A few centuries before Gilbert and Sullivan, Shakespeare had described Cleopatra as follows: "Age cannot wither her, nor custom stale her infinite variety." But then, Cleopatra was supposed to have been remarkable precisely because by contrast, "other women cloy the appetites they feed."

One of the first novels to explicitly and directly depict this male penchant for multiple sexual partners, Henry Miller's *Tropic of Cancer*, has been called the most honest book ever written. It is a confessional, autobiographical account of Miller's years as an expatriate in Paris. No punches are pulled in its sexual descriptions, which include a (perhaps unhealthy) dose of gritty reality: accounts of watching a whore use a bidet before sex, of lice, feces, dirt, poverty, mooching off one's friends, cuckolding one's colleagues, trying to create literature while celebrating (or succumbing to; choose your own interpretation) a yearning for multiple partners that would make a Maori chieftain proud and his Episcopal visitor wince (perhaps at least in part out of envy).

As a species that forms pair bonds and in which females—as well as social and religious tradition—exert strong pressure on males to be monogamous, the achievement of serial promiscuity à la Henry Miller is difficult, to say the least. As a result—and for sound biological reasons—not many men have managed to lead the life that Henry Miller did.

Tropic of Cancer was initially published in 1934 but not allowed into the United States until 1961. (Humanity, wrote T. S. Eliot, cannot stand too much reality; in this regard, prudish America may be exceptionally human.) Miller gives his readers graphic accounts of his own exploits with Elsa, Tania, Irene, Llona, Germaine, Claude, even his mysterious wife, Mona (have we left anyone out?), and that's just in the first forty pages. Throughout, Miller's enthusiasm for female flesh—especially when freshly encountered—is undiminished.

In his preface to *Tropic of Cancer*, poet Karl Shapiro called Henry Miller the "greatest living writer," adding, "It is possible that he is the first writer outside the Orient who has succeeded in writing as naturally about sex on a large scale as novelists ordinarily write about the dinner table or the battlefield."[1] The three Fs— feeding, fighting, and fornication—are about as biological as anything gets.

Equally libidinal, and more current, is the Czech expatriate novelist Milan Kundera, several times nominated for a Nobel Prize. Kundera's notorious sexism is said to have worked against him, at least within the Nobel committee's secret deliberations, but at the same time, it has powered much of the compelling honesty of his writing. *The Unbearable Lightness of Being* introduces us to Tomas, a

---

[1] Karl Shapiro, "The Greatest Living Writer," preface to Henry Miller, *Tropic of Cancer* (New York: Grove Press, 1961).

brilliant Prague surgeon who lives happily as an irresistible sexual adventurer. His signature move is to demand of his potential conquests: "Take off your clothes," whereupon they generally do just that. As with D. H. Lawrence's depiction of Lady Chatterley swooning over her lover's erect penis, we suspect that the acquiescence of Tomas's numerous inamoratas is more a reflection of Kundera's overheated imagination than a believable portrayal of what most women are likely to do, but it certainly demonstrates what men want.

Here is another dose of Kundera, from his other masterpiece, *The Book of Laughter and Forgetting*:

> Every man has two erotic biographies. Usually people talk only about the first: the list of affairs and of one-night stands. The other biography is sometimes more interesting: the parade of women we wanted to have, the women who got away. It is a mournful history of opportunities wasted.

Biologists have a term for the male tendency to be sexually stimulated by relations with new female partners: it is known as the "Coolidge effect." It seems the president and his wife were separately touring a model farm during the 1920s. Mrs. Coolidge, commenting on the many chickens associated with a single rooster, commented, "He must be kept quite busy." She suggested that this be brought to the president's attention. Accordingly, when the presidential party arrived later that same day, the guide announced, "Mrs. Coolidge wished me to point out that our single rooster copulates many times each day." "Always with the same female?" asked the president. "No, sir." "Well," said Mr. Coolidge, "tell *that* to Mrs. Coolidge."

Is it just men? Yes and no. Yes: women as a rule aren't turned on by multiple partners. No: males of other species generally act the same way. A ram, for example, left to copulate with the same ewe, will do so several times, then stop, apparently satiated. If he is then presented with a different ewe, his sexual enthusiasm returns, until once again his interest wanes. But each time he is given an opportunity to copulate with a new female, his enthusiasm—and capacity—comes roaring back. A new ewe makes a new him.

This phenomenon was known long before the modern science of animal behavior and before evolutionary biologists understood the consequences of making sperm versus eggs. "I have put to stud an old horse who could not be controlled at the scent of mares," wrote the sixteenth-century essayist Montaigne. "Facility presently sated him toward his own mares: but toward strange ones, and the first one that passes by his pasture, he returns to his importunate neighings and his furious heats, as before."

As a species, human beings generally stop short of importunate neighings and furious heats, but not by much. One of the characters in the musical *Finian's Rainbow* proclaims in song, "When I'm not near the girl I love, I love the girl I'm near." He goes on: "When I'm not facing the face that I fancy, I fancy the face I face." Russell Banks's novel *Continental Drift* includes several infidelities on the part of Bob Dubois, its primary male character. At one point, while Bob is admiring the face of Marguerite, his most recent mistress, he realizes that he can't even remember the details of his wife's face:

> Now he can't recall it. His memory is only of having paid attention to something that has disappeared, . . . so that now, when he . . . remembers it, all he sees is the center of her eyes, as if her face has somehow gradually become invisible without his ever having noticed until after it was gone, lost to him, he is sure, forever.

It is a fascinating irony as well that although men stand to gain more—in terms of producing offspring—from copulations with multiple partners, women are physiologically capable of having more sex than men. Yet social systems are commonly structured the other way around. In his bitingly satiric *Letters from the Earth*, Mark Twain had great fun with this paradox. Here is Twain's Devil reporting his discoveries, after visiting our planet:

Now there you have a sample of man's "reasoning powers," as he calls them. He observes certain facts. For instance, that in all his life he never sees the day that he can satisfy one woman; also, that no woman ever sees the day that she can't overwork, and defeat, and put out of commission any ten masculine plants that can be put to bed to her. He puts those strikingly suggestive and luminous facts together, and from them draws this astonishing conclusion: The Creator intended the woman to be restricted to one man.

Now if you or any other really intelligent person were arranging the fairnesses, and justices between man and woman, you would give the man a one-fiftieth interest in one woman, and the woman a harem. Now wouldn't you? Necessarily, I give you my word, this creature with the decrepit candle has arranged it exactly the other way.

From an evolutionary perspective, it is more logical for one man to mate with multiple women than to have one woman mated to several men. And in this case, evolutionary logic has won out, not only in the great majority of social systems described by anthropologists but also in most of the domestic dramas described by writers.

Another key difference between the sexes is that men are much more susceptible to sexual stimulation than are women. If it's weird sex—involving animals, unusual objects, even the occasional corpse—it's likely to be male sex. Men buy nearly all the pornography. Men visit prostitutes.

Picture the following scene: a man and woman have just spent a pleasant first date together and are now in the privacy of her apartment. She turns to him and, saying nothing, slowly begins to unbutton her blouse, then starts to unhook her bra. All the while, the man watches: fascinated, excited, and delighted. Then she uncovers one breast and he . . .

What is he likely to do? Slap her across the face? Beg her to stop and put her clothes back on? Scream for help? Dial 911? If you were to write a plausible ending to this steamy encounter, it would almost certainly involve a passionate sexual response by the man. The sort of female behavior described above is a definite turn-on for most men.

Now run through the scene again, this time reversing the roles: without explanation, the man suddenly unzips his fly and takes out his erect penis. The likelihood is that in such circumstances the woman would react with something less than delight, fascination, and sexual enthusiasm. Rather than being excited, she may well be repulsed, almost certainly by his social inappropriateness, and perhaps by the sexual organ itself. In any event, she is not likely to be aroused by the visual image of his genitals. By contrast, nearly every heterosexual man is excited by comparable exposure to the intimate anatomy of an attractive woman.

Simply put, women are more often interested in romance, with its implication of caring and commitment (and the promise of follow-through), while men are most interested in sex and immediate availability. This particular gender gap helps explain why women are indignant about being treated as sex objects and why men—often despite their best intentions—keep doing so. In Colette's novel *The Vagabond*, a woman criticizes (and sees through) her admirer as follows:

> If he pretends, cunning as an animal, to have forgotten that he wants to possess me, neither does he show any eagerness to find out what I am like, to question me or read my character, and I notice that he pays more attention to the play of light on my hair than to what I am saying.

What's love got to do with it? Everything and nothing. "Who can explain it, who can tell you why?" ask Rodgers and Hammerstein in their song "Some Enchanted Evening." They conclude, "Fools give you reasons; wise men never try." Well, we have tried. And so have generations of writers. For the biologist, love is a mechanism whereby individuals bond in the interest of maximizing their fitness. What hunger is to metabolism, love is to reproduction. For all its evident irrationality, it is more logical than most people realize. Most writers prefer to treat love—as distinct from lust—as magical, mysterious, inexplicable, and impenetrable, even as they describe its unfolding in precisely the same way that most biologists would predict, albeit adding a twist of the inchoate.

Here is Jay Gatsby acknowledging to Nick Carraway that Daisy Buchanan represents all his hopes and dreams, and recalling their first kiss:

> He knew that when he kissed this girl, and forever wed his unut-terable visions to her perishable breath, his mind would never romp again like the mind of God. So he waited, listening for a moment longer to the tuning-fork that had been struck upon a star. Then he kissed her. At his lips' touch she blossomed for him like a flower and the incarnation was complete.

We can struggle against our evolutionary heritage, constantly yearning for a new day, a new success: "Gatsby believed in the green light," notes Nick Carraway, the narrator of Fitzgerald's renowned novel, "the orgiastic future that year by year recedes before us. It eluded us then, but that's no matter—tomorrow we will run faster, stretch out our arms farther. . . . And one fine morning—" What? One fine morning will we outrun, outreach, somehow outperform our evolutionary heritage? Don't count on it. When *The Great Gatsby* famously concludes, "So we beat on, boats against the current, borne back ceaselessly into the past," it is a more distant and more clinging past than Fitzgerald, or most of his readers, ever imagined.

5

Madame Bovary's Ovaries

The Biology of Adultery

When Gustave Flaubert's grand and sexy novel *Madame Bovary* first appeared, readers were shocked, *shocked*. Here was a respectable middle-class woman who had a lover—several, in fact. And she was married! One hundred fifty years later, when students of animal behavior discovered that in nature many babies are fathered by someone other than the mother's social partner, they, too, were shocked, *shocked* (or at least surprised). These upsetting findings all point in the same direction: there is more than a little hanky-panky going on in the world, human as well as animal, and it isn't just a "guy thing." Females—even married ones—are active participants.

To be sure, there is much to be said for marital fidelity. But if you agree with John Milton—who in *Paradise Lost* hailed "wedded love" as follows—

> . . . *mysterious Law, true source*
> *Of human offspring . . .*
> *By thee adulterous Lust was driv'n from men*
> *Among the bestial herds to range . . .*

—then we have a large bridge in Brooklyn that you might want to purchase.

"Adulterous lust" has hardly been driven from men. Or women.

Infants have their infancy, and adults . . . ? Adultery. It "ranges," to be sure, among Milton's "bestial herds," but that's not the only place. The world of great literature, just like the great world of life, is filled with philandering. In addition to *Madame Bovary*, the list stretches from *The Iliad* through *Ulysses*, with numerous stops in between and after. Nor is it ignored by the creators and consumers of not-so-great literature, preoccupying, for example, the suburban occupants of John Updike's "marriage novels" as well as nearly every soap opera ever broadcast.

If monogamy itself is something of a myth, many of our most enduring myths are powered by extramarital events, too: think of the ill-fated Mediterranean house of Atreus (featuring, among others, Agamemnon and his concubine, Cassandra, as well as his wife, Clytemnestra, and her extracurricular lover, Aegisthus), the ancient Teutonic travails of Tristan and Isolde, as well as the Anglo-French exploits of Queen Guinevere and Sir Lancelot while King Arthur stood by. Adultery launched not only a thousand ships to Troy but also the earliest piece of organized literature in the Western world (*The Iliad*) and provided much of the motive force behind Malory's *Le Morte d'Arthur*.

Consider that in all of these cases it is sexual transgression on the part of a married *woman* that generates outrage and, nearly always, her punishment. Consider as well that married women, just as married men, are not only *like* mated animals, they *are* mated animals.

Let's look, then, at a few animals first.

Several decades ago, some intrepid wildlife biologists performed vasectomies on male blackbirds, trying to come up with a nonlethal way of controlling their numbers. The operations were a success, but to the researchers' amazement, female blackbirds kept laying fertile eggs! Evidently, someone other than her "husband" was laying Madame Blackbird. It should have been a wake-up call, at least to the biologists (by all accounts, male blackbirds were— and still are—as naive about their mates' extracurricular activities as the unfortunate Charles Bovary, about whom more later).

Then, in the 1990s, DNA fingerprinting became widely available, and with it, incontrovertible evidence that females of nearly every species engage in EPCs, or extrapair copulations.[1] In other words, they screw around. This was unexpected, since, as we have seen, eggs are produced in relatively small numbers and would seem easily fertilized by just one male, presumably the female's "rightful" mate. This is especially true since males make so much sperm. Besides, additional lovers should only bring additional hassles: an outraged husband could behave violently toward the "errant" wife (remember Othello?) or—equally devastating for any species that relies on shared parental care—renege on his child support payments. After all, it is biologically costly to expend resources on behalf of someone else's offspring (more on this in Chapters 6 and 7). For now, the derivation of the word "adultery" is itself revealing: from the Latin *adulterare*, meaning "to alter or change." To adulterate means to "debase by adding inferior materials or elements; making impure by admixture." From a male's perspective, the crucial admixture is someone else's sperm, which is to say, his genes.

Let's also examine the word "cuckold," which, as we've already seen, derives from the European cuckoo, renowned for its behavior as a nest parasite. Cuckoos lay their eggs in the nest of other species, who then become unwitting foster parents. When cuckoo chicks hatch, they add injury to insult by ejecting the host's biological offspring, thus monopolizing their foster parents' resources for themselves. To be cuckolded is to suffer the fate of those unwitting males who fail to see they have been displaced by a lover and end up not only biological failures but social laughingstocks.

In *Love's Labour's Lost*, Shakespeare gives us this cynical song: "The cuckoo then in every tree mocks married men; for thus sings he, 'Cuckoo; cuckoo; cuckoo; O word of fear; unpleasing to a married ear!' " If a man ends up unknowingly rearing another man's

---

[1] This phenomenon is detailed in *The Myth of Monogamy*, by D. P. Barash and J. E. Lipton (New York: Henry Holt, 2002).

children conceived with his spouse, his love's labor is lost indeed. Even during the French Revolution, at a time when enthusiasm for creating a new society was so great that the names for months of the year were tossed out and replaced with new ones, sexual asymmetries of the old regime were retained in one regard: legal sanctions against a *wife's* adultery.

So evolution should have given us faithful female blackbirds, as well as among nearly every species in which females have little to gain and much to lose by "cheating." And yet the facts are undeniable. In species after species, biologists have found that 10, 20, in some cases as many as 70 percent of the offspring reveal "extrapair paternity"—they're not genetic chips off the old paternal block. Rather, some other male has literally chipped in. In the movie *Heartburn*, Nora Ephron's character complains to her father about her philandering husband (journalist Carl Bernstein), whereupon the elder Mr. Ephron responds, not very sympathetically: "You want monogamy? Marry a swan." Well, it turns out that not even swans are monogamous. And not only males stray. She may be married to a sleek and sober swain, with a lovely house on the lake, yet periodically, Madame Swan is likely to sneak off into the cattails with someone else. It's not quite what Marcel Proust meant by *Swann's Way*, but it is the way of the world.

In fact, the only guaranteed monogamous species appears to be a small worm that parasitizes fish intestines. The males and females of this odd creature, *Diplozöon paradoxicum*, meet in adolescence, after which their bodies literally fuse together, till death do they not part. (We like to imagine that they are happily mated, but in any event they are irremediably united and thus faithful.) For almost everyone else, EPCs are the rule, not the exception.

In *Ars amatoria*, the Roman poet Ovid justifies what is perhaps the most notorious, and ruinous, of all cases of adultery: Helen's affair with Paris, which precipitated the Trojan War and caused unimaginable misery. It seems that Helen's husband, Menelaus, was away at the time:

> *Afraid of lonely nights, her spouse away*
> *Safe in her guest's warm bosom, Helen lay.*
> *What folly, Menelaus, forth to wend,*
> *Beneath one rooftree leaving wife and friend? . . .*
> *Blameless is Helen, and her lover too:*
> *They did what you or anyone would do.*

Ovid's lighthearted view of infidelity so irritated the emperor Augustus that the poet was exiled from Rome. We are less sanguine about extracurricular sexuality than Ovid but less judgmental than Augustus (who had some lingering anxiety about his daughter's erotic predilections). In exploring the biology of extrapair copulation and connecting it to humanity and to literature, our intent is not to legitimize adultery by either sex, nor to stigmatize. Rather, it is to open eyes while stimulating that sexiest part of the human anatomy, the one between our ears.

Biologists have a good idea why males are typically interested in sexual variety, but they are more perplexed when it comes to females. The reason, paradoxically, is that there are many possible explanations, as least some of which also help us make sense of Madame Bovary.[2] First, a female may get a leg up on the evolutionary ladder by mating with a "better" partner, obtaining superior genes to go with the material resources and parental assistance provided by her "official" mate. Or, oddly enough, she may find that another male looks good simply because other females feel the same. As noted earlier, according to this "sexy son hypothesis," females

---

[2] Mark Twain once noted that it was easy to stop smoking: he had done it hundreds of times! Similarly, it is easy to explain female sexual infidelity: there are many explanations.

prefer to mate with males whose traits (bright feathers, large antlers, seductive wattles and jowls, broad shoulders, or a resemblance to, say, Pierce Brosnan) are attractive to other females in general and likely to be inherited by their sons.[3] If so, then a swan or antelope— or human being—who swoons and sighs, "I want to have *his* baby," is doing so because her genes are whispering, "He's *so* cute."

Why such whispers? Because of this unconscious correlation: females who mate with Pierce Brosnan are likely to produce sons who are similarly attractive to the next generation of females, and who therefore will enhance the evolutionary success of their mothers, which is probably why those females were so readily seduced to begin with. In short, in the realm of sexual attraction, nothing succeeds like sexual attraction itself. Once set in motion, it can have a self-maintaining dynamic, quite separate from a male's wealth, social position, qualities as a father, and so forth.

Among birds known as bluethroats, the throats of the males are, not surprisingly, bright blue (females are comparatively drab). When researchers enhance a male's blueness by use of iridescent spray paint, these newly anointed beacons of bluethroat beefcake beauty become very attractive to surrounding females; even already mated female bluethroats start fluttering around, seeking EPCs. Similarly, when a male's throat is artificially bleached to be *less* blue, his "wife" starts looking for another lover. As already noted, appearance counts in the animal world, almost certainly because mating with a sexy-looking partner is likely to produce sexy-looking offspring.

At the same time, there is growing evidence that animal sexiness often reflects genuine healthfulness. A bright, shiny blue throat, for example, isn't easy to come by—in the absence of helpful biologists armed with spray paint—and so individuals who succeed in bluing

---

[3] You may have noticed that this leaves unanswered another question: how did such preferences get started in the first place? This is the subject of vigorous research and equally vigorous debate among biologists. It seems probable that a trait is perceived as sexy insofar as it contributes in some way to the evolutionary success of the perceiver, that is, to begin with, it is an "honest signal." Over time, however, such preferences can take on a momentum of their own.

up are likely to have a strong constitution, an efficient metabolism, a body that is comparatively free of parasites, and so on. As a result, even if other females weren't attracted to such secondary sexual characteristics, any correlation between them and likely well-being would probably be enough to make a bluethroat into a genuine heartthrob. Add to this the likelihood that such preferences can contribute to a positive feedback mechanism, and it is not surprising that even a happily married female may have her head turned.

Of course, physical traits are no less important to humans, something that has not been lost on the creators of great literature. Here is Tolstoy's Anna Karenina, contemplating her husband and bemoaning his lack of physical charm:

> "Well, he's a good man; upright, kind, and remarkable in his own line," said Anna to herself when she had returned to her room, as though defending him from attack—from the accusation that he was not lovable. "But why is it his ears stick out so oddly? Or has he had his hair cut too short?"

One can imagine Mrs. Bluethroat—whose mate has just been bleached—engaging in a similar interior monologue.

Physical attractiveness (or its absence) is such an obvious factor in infidelity that it can easily be overlooked. In animal societies that are ostensibly monogamous, females must be "married" in order to reproduce, since a partner's assistance is often necessary in order to defend a territory, construct a nest, and provision the offspring. But this doesn't mean that every female must combine her genes with the male that she has settled for: if she has to have a mate but he turns out to be less desirable than some other, why not set up housekeeping with the former but make babies with the latter? In short, get resources and parental assistance from one and the best possible genes from another.

Guinevere's adultery is said to have undone King Arthur's Round Table, and it should occasion no surprise that Sir Lancelot was hardly a stable boy. Not only was he renowned for his military prowess and his overall goodness, but he was also said to be an extraordinary physical specimen. Thus, according to Malory's account, after iron bars were placed at Guinevere's window to protect her chastity, Lance pulls them apart by sheer strength, in the process wounding his hands and shedding blood that nearly gives him away, after which he spends the night with the queen: "Syr Launcelot wente to bedde with the quene and toke no force of hys hurte honde, but toke hys pleasaunce and his lykynge untyll hit was the dawnyying of the day."

Nor are similar considerations missing from more popular media. In the movie *Unfaithful*, Connie Sumner, a beautiful woman living in what seems to be upper-middle-class bliss—and married to Richard Gere, besides—nonetheless has what turns out to be a disastrous affair with a young, charming, sexy hunk. Why? Because he is a young, charming, sexy hunk, whereupon Mrs. Sumner is swept off her feet and into her lover's bed . . . and couch, and stairway, and the bathroom of a nearby restaurant, and so forth.

Once again, it doesn't matter that this unfaithful woman, just like Queen Guinevere, didn't become pregnant by her EPC partner, nor even that she would presumably engage in strenuous methods of birth control, if necessary, to prevent conception. The point is that human beings, like other creatures, have been endowed with an inclination to have sex with partners who are especially attractive. Delving into that inclination, we see that it is an evolutionary bequeathal from a precontraceptive world when such choices were more directly connected to reproductive consequences. Although the technology has changed, our genes have hardly noticed.

This is not to say that Darwinian urges are necessarily unconscious. Anna and Vronsky know perfectly well they want to have sex. But it's the evolutionary logic behind it—the reproductive payoff and ultimate motivating force—that operates not only below the belt but also below the level of conscious awareness.

By the same token, it is worth noting that Emma Bovary and her beaux—the wealthy rake Rodolphe and the attractive law student Léon—seemed no more eager to reproduce than any of the other renowned lascivious couples of literary or cinematic lore. It would not have been surprising, for example, if Anna Karenina and Count Vronsky had employed the latest advances in nineteenth-century Russian birth control, whatever that may have involved. They wanted each other, not a baby. Biologists understand that a major reason why Emma wanted sex with Rodolphe, Léon, and the marquis (the last unconsummated) was because deep inside (in the DNA of her brain) she heard a subliminal Darwinian whisper that tickled her ovaries, even though she may not have acknowledged it and would likely have even acted consciously against such an outcome. Smart women sometimes really do make foolish choices, and a whiff of Darwin enables us to glimpse some of the reasons why.

Madame Bovary evidently found her various lovers sexually exciting, just as a hungry person—even if she knows nothing of digestive physiology—can be seduced by a tasty meal. Most people eat not in order to stoke their metabolism but rather to slake their hunger. By seeing such urges for what they are, the modern reader can also see how the prospect of enhancing her evolutionary situation undergirds a beleaguered heroine's erotic hunger. Especially, as we shall soon explore, if she is otherwise undernourished.

This inscription, from Thomas Parke D'Invilliers, begins F. Scott Fitzgerald's *The Great Gatsby*:

> *Then wear the gold hat, if that will move her;*
> *If you can bounce high, bounce for her too,*
> *Till she cry "Lover, gold-hatted, high-bouncing lover,*
> *I must have you!"*

But what makes a woman cry "I must have you"? It turns out that being gold-hatted, no less than high-bouncing, counts for quite a lot. Aside from being turned on by physical attractiveness, female animals often look for lovers who are especially wealthy and willing to share that wealth with sexual partners (male animals with a rich territory will often allow their mistresses to forage there in exchange for sexual favors), including when their "official" mate is a lax provider. For example, an osprey female whose husband is a failure at bringing home the salmon is especially liable to copulate with other males while he is gone; in return, these obliging gents help provision her and her offspring.

Other zoological Madame Bovarys appear to use their adulterous unions to establish a possible bridge to a new relationship, a behavior that is assuredly within the human repertoire as well. But one of the most interesting theories of female infidelity is that primates in particular may do the monkey equivalent of Emma's carnal carriage ride as a way of purchasing infanticide and child abuse insurance. This isn't about, as one might think, a maternal response to an abusive simian husband. Rather, it involves males other than the parent: among many primate species it is not uncommon for males to kill the offspring of others and then reinseminate the mothers, a brutally effective case of getting rid of the other guy's genes to make room for yours. Thus, it would make sense for a female to be generous with her sexual favors so as to induce the neighborhood males to "think" that they might have fathered her offspring, which, in turn, makes the males less likely to murder the juveniles. (Presumably, a male chimpanzee who encounters a female with an infant is likely to say to himself: "Isn't that my old flame, with whom I had such a hot time behind the baobab tree? Cute little tyke with her, too. Looks just like me!")

There just isn't a single, simple answer to the question of why females are sexually unfaithful, although accumulating evidence strongly suggests that infidelity by either sex occurs when it increases the likely evolutionary success of the "infidel." This area of investigation is one of the newest, most exciting realms of research in animal behavior and evolutionary biology; the above discussion only skims the surface. But it helps open a door upon one of the oldest, most pervasive themes in literature. Just as evolutionary insights help us understand Othello's jealousy, a dollop of biology sheds light on Madame Bovary and her soul mates. If we are correct, then Emma Bovary is a reflection not only of what goes on "out there" in the animal world but also of what exists "in here," within our own hearts . . . and genes, and gonads. And what has accordingly found its way onto the printed page as well.

And so we come to Emma Bovary herself, literary poster woman for female infidelity. How well does Emma's behavior conform to biological expectations? Very well indeed. Emma's father was a farmer, and her marriage to Charles Bovary is a typical example of "marrying up," or hypergamy, whereby women seek to pair with men who are socioeconomically above them (recall the thrust of many a Jane Austen novel, described in Chapter 3). Flaubert tells us that Emma's father was in debt and "losing money every year." With the family in bad financial shape, Charles Bovary therefore seemed a good catch, at least to Monsieur Rouault, his prospective father-in-law:

> Charles was a bit namby-pamby, not his dream of a son-in-law; but he was said to be reliable, thrifty, very well educated, and he probably wouldn't haggle too much over the dowry. Moreover, Rouault was soon going to have to sell twenty-two of his acres. . . . "If he asks me for her," he said to himself, "I won't refuse."

This is precisely what happens in many species. Given the existence of competition among males, the winners are those who are a bit smarter, stronger, richer, and better connected, who offer a better deal to the females (and, in the case of human beings, their families). Or—another way to look at it—women have been able to insist on mating preferentially with only the most desirable men, which means that women should be more likely to marry up than down. If harem keeping were the current norm, the result would be a relatively small number of socially, financially, and/or physically dominant men monopolizing most of the women. By contrast, in a social system of ostensible monogamy, as currently found in the West, the stage is set for female infidelity, as some women seek to better their situation. As we have already suggested, this pattern also suggests an especially poor prognosis for marriages that go the other way, when the woman is more competent, educated, or otherwise desirable than her mate.

This might seem to augur well for the Charles/Emma union, and initially it did. But twenty-first-century readers need to realize that the situation of a physician in nineteenth-century France was not all that impressive. Moreover, Charles Bovary wasn't even a physician; it is quite clear from the novel that he was an *officier de santé*, a "health officer," and thus one step below an actual doctor: less training, less skill, and less renown, more like a licensed practical nurse. And so Charles Bovary, although initially a good catch for lower-class Emma, is not *that* good. Especially not when her expectations expanded. Plus, Charles is depicted as rather boring and dull-witted, lacking the human equivalent of a bright blue throat or fancy tail feathers:

> By working hard he [Charles] managed to stay about the middle of the class; once he even got an honorable mention in natural history. . . . He understood absolutely nothing of any of it. He listened in vain: he could not grasp it. In the performance of his daily task he was like a mill-horse that treads blindfolded in a circle, utterly ignorant of what he is grinding.

Nor was Charles Bovary especially ambitious:

> Emma looked at him and shrugged her shoulders. Why didn't she at least have for a husband one of those silent, dedicated men who spend their nights immersed in books and who by the time they're sixty and rheumatic have acquired a row of decorations to wear on their ill-fitting black coats? She would have liked the name Bovary—her name—to be famous, on display in all the bookshops, constantly mentioned in the newspapers, known all over France. But Charles had no ambition! . . . "It's pathetic!" she whispered to herself, despair in her heart. "What a booby!"

We cannot help noting, with some irony, that "booby" also refers to several species of oceangoing birds; there are blue-footed, red-footed, pink-footed, and brown-footed boobies, among others. And within these species, females are especially prone to copulating with other males when their mate is, well, a genuine booby. It is also noteworthy that Emma Bovary had a sharp eye for the differences among men. Having seen one booby, she would doubtless conclude, you have not seen them all.

Thus at one point (before Emma's first affair), the Bovarys are invited to a ball at the elegant home of a marquis:

> She thought of Les Bertaux [the farm where she grew up]: . . . and she saw herself as she had been there, skimming cream with her finger from the milk jars in the dairy. But amid the splendors of this night her past life, hitherto so vividly present, was vanishing utterly; indeed she was beginning almost to doubt that she had lived it. She was here: and around the brilliant ball was a shadow that veiled all else.

Here in the marquis's home was something closer to her idealized life of perfection, and both the marquis and another aristocrat, a viscount, came much closer to Emma's ideal than did Charles: "However she imagined him [her perfect husband], he wasn't a bit like Charles. He might have been handsome, witty, distinguished, magnetic—the kind of man her convent schoolmates had doubtless married." Emma didn't enter into an affair with either the marquis or the viscount, but if she could have, she probably would have. After all, in comparison Charles was poor not only in resources but also in intellect (which is to say, probably offering genes that are less than stunning):

> Charles's conversation was flat as a sidewalk, a place of passage for the ideas of everyman. . . . He couldn't swim or fence or fire a pistol; one day he couldn't tell her the meaning of a riding term she had come upon in a novel. Wasn't it a man's role, though, to

know everything? Shouldn't he be expert at all kinds of things, able to initiate you into the intensities of passion, the refinements of life, all the mysteries? *This* man could teach you nothing; he knew nothing, he wished for nothing. He took it for granted that she was content; and she resented his settled calm, his serene dullness, the very happiness she herself brought him.

Sure enough, Emma proceeds to have an affair with Rodolphe, wealthy, witty, debonair, and something of an accomplished sexual predator:

Monsieur Rodolphe Boulanger was thirty-four. He was brutal and shrewd. He was something of a connoisseur: there had been many women in his life. This one [Emma] seemed pretty, so the thought of her and her husband stayed with him.

"I have an idea he's stupid. I'll bet she's tired of him. His fingernails are dirty and he hasn't shaved in three days. He trots off to see his patients and leaves her home to darn his socks. How bored she must be! . . . A compliment or two and she'd adore me, I'm positive. She'd be sweet! But—how would I get rid of her later?" . . . [After thinking about his present mistress:] "Ah, Madame Bovary is much prettier—and what's more, much fresher. Virginie's certainly growing too fat."

When their affair is in danger of being revealed to Charles, Rodolphe reveals himself to be, for all his faults, attractive to Emma because of his swaggering courage:

"Have you got your pistols?"

"What for?"

"Why—to defend yourself," said Emma.

"You mean against your husband? That poor . . . ?"

And Rodolphe ended his sentence with a gesture that meant that he could annihilate Charles with a flick of his finger.

This display of fearlessness dazzled her, even though she sensed in it a crudity and bland vulgarity that shocked her.

But Rodolphe eventually dumps her, at about the same time that Emma's husband experiences a major professional reversal. A treatment that Charles attempted on a patient had a disastrous outcome, causing him humiliation and lowering his stature in the community as well as in his wife's eyes:

> He was sitting quite calmly, utterly oblivious of the fact that the ridicule henceforth inseparable from his name would disgrace her as well. And she had tried to love him! She had wept tears of repentance at having given herself to another!

It isn't just Emma Bovary who is especially likely to be unfaithful when her mate has suffered a decline in status. A recent study of black-capped chickadees, for instance, found that whereas females are generally faithful, they occasionally stray, but only with males who are socially dominant to their husbands. When a male chickadee experiences a reversal in his dominance stature, he had better look to his "wife." Charles Bovary might have been similarly advised.

Meanwhile, things have been going from bad to worse in the Bovary household. Emma predictably begins another affair, this time with young, handsome Léon Dupuis, an aspiring attorney, at the same time that Charles Bovary is going deeply into debt, largely because of Emma's profligacy. Recall that Charles was boring and intellectually dull. By contrast, Léon radiated mental sprightliness, a characteristic that is "romantic," in large part because it suggests the presence of good genes in a prospective mating partner. (Recall also our earlier look at Jane Austen's heroes and heroines and their crucially witty repartee.)

Although Monsieur Bovary was a dutiful husband, he was decidedly unglamorous, smelling of medicines and body fluids instead of fancy perfumes. By contrast, Léon

> was thought to have very gentlemanly manners. He listened respectfully to his elders, and seemed not to get excited about politics—a remarkable trait in a young man. Besides, he was talented. He painted in water colors, could read the key of G,

and when he didn't play cards after dinner he often took up a book. Monsieur Homais esteemed him because he was educated; Madame Homais liked him because he was helpful: he would often spend some time with her children in the garden.

Emma eventually commits suicide as her debts mount and neither Rodolphe nor Léon will pay them. Charles subsequently discovers letters between her and her former lovers, whereupon he, too, dies brokenhearted.

It is said that Flaubert struggled for weeks to craft the perfect phrase for each sentence, and much of *Madame Bovary*'s impact comes from the exquisitely evocative language in which it is written. At the same time, much of the impact of *Madame Bovary* comes from Madame Bovary, the woman whose struggles with monogamy could have been written by a knowledgeable evolutionary biologist (if he or she were also a literary giant). In the novel's pathetic ending, we see the sentimental imprint of nineteenth-century morality: in Flaubert's time, the myth of monogamy was widespread, at least in the case of women. A sexual double standard was firmly ensconced culturally but based equally firmly on the biological asymmetries between men and women. As we've seen earlier, this comes down to who makes sperm and who makes eggs, which sex gets pregnant, and which lacks confidence of parentage. Moreover, it was assumed that although men—even married men—would have multiple lovers, women's libido was essentially nonexistent. Their yearning for quiet, monogamous domesticity was supposed to be the flip side of men's randiness. Now we know the universality of what Flaubert described a century ago: women, too, are sexual creatures, influenced no less than men by their own biology.

"*Madame Bovary, c'est moi*," wrote Flaubert. He might have written, more accurately, "*Madame Bovary, c'est nous*."

Madame Bovary's ovaries weren't really hyperactive (although her romantic imagination may have been regrettably stuck in overdrive). Certainly she isn't the only sexually unfaithful woman in literature.

The American writer Kate Chopin (pronounced like the renowned composer) is often described as a "regionalist," and yet she has of late been acclaimed as nothing less than a universalist, especially in her masterpiece account of female adultery, *The Awakening*. This short novel, published in 1899, was considered so scandalous that it effectively terminated Chopin's career as a publishable author. Today, she is revered as a feminist icon, someone ahead of her time in depicting womanly independence from male-dominated society, and one of the first to probe the hidden secrets of the (sometimes adulterous) female heart.

It is unclear whether Guinevere underwent a particular sexual awakening with Lancelot, beyond whatever she experienced with Arthur. But it is pretty obvious that in Chopin's novella, Edna Pontellier—attractive, young, frustrated spouse of the older, benevolent, but tiresomely traditional Léonce Pontellier—is due for her own awakening:

> The acme of bliss . . . was not for her in this world. As the devoted wife of a man who worshipped her, she felt she would take her place with a certain dignity in the world of reality, closing the portals forever behind her upon the realm of romance and dreams. . . . She grew fond of her husband, realizing with some unaccountable satisfaction that no trace of passion or excessive and fictitious warmth colored her affection.

Edna is needy and unfulfilled: she goes swimming in the heat of the day (itself a shocking violation of turn-of-the-century norms), lies languorously in her hammock refusing her husband's demands that she come inside, defies the rigid social conventions of New Orleans society, becomes increasingly interested in Robert Lebrun—who flirts with married women but can't deal with a serious relationship—and, finally, has an affair with Alcée Arobin, like Emma Bovary's Rodolphe an accomplished Don Juan:

He stayed and dined with Edna. He stayed and sat beside the wood fire. They laughed and talked; and before it was time to go he was telling her how different life might have been if he had known her years before. With ingenuous frankness he spoke of what a wicked, ill-disciplined boy he had been, and impulsively drew up his cuff to exhibit upon his wrist the scar from a saber cut which he had received in a duel outside of Paris when he was nineteen. She touched his hand as she scanned the red cicatrice on the inside of his white wrist. A quick impulse that was somewhat spasmodic impelled her fingers to close in a sort of clutch upon his hand. He felt the pressure of her pointed nails in the flesh of his palm.

Edna doesn't love Alcée. But he satisfies a sexual passion that had been dormant within her and was, ironically, aroused by another man, Lebrun, who returns her feelings but ultimately flees from the prospect of an illicit affair.

Other women in Chopin's novel exemplify alternative routes to personal gratification: one of Edna's friends is the embodiment of maternal virtue; another is a serious pianist, exclusively committed to her art; an old woman (encountered only briefly) signifies religious satisfaction; but Edna's awakening is a matter of social independence combined with pure sexual fulfillment. At one point, the benevolent family doctor, consulted by her worried husband because Edna had been acting so strangely,

noted a subtle change which had transformed her from the listless woman he had known into a being who, for the moment, seemed palpitant with the forces of life. Her speech was warm and energetic. There was no repression in her glance or gesture. She reminded him of some beautiful, sleek animal waking up in the sun.

Although *The Awakening* includes no account of physical intimacy between Edna and her husband, and even its description of Edna Pontellier's "waking up" with Alcée is notably delicate by current standards, there is no question that with her lover, Edna experiences something exciting, compelling, and new: "It was the first kiss of her life to which her nature had really responded. It was a flaming torch that kindled desire." Later, after she and Alcée make love,

> she felt as if a mist had been lifted from her eyes, enabling her to look upon and comprehend the significance of life, that monster made up of beauty and brutality. But among the conflicting sensations which assailed her, there was neither shame nor remorse. There was a dull pang of regret because it was not the kiss of love which had inflamed her, because it was not love which had held this cup of life to her lips.

Edna's affair with Alcée is all about sex, and little else:

> He did not answer, except to continue to caress her. He did not say good night until she had become supple to his gentle, seductive entreaties. . . . [She] unfolded under his delicate sense of her nature's requirements like a torpid, torrid, sensitive blossom.

So let's talk a bit about what Kate Chopin is implying but is too delicate to state explicitly: female orgasm. It had been widely thought until recently that female animals didn't experience orgasm; even now, with clear evidence for female climax in a variety of species, its adaptive significance remains a biological mystery, since clearly it is not necessary in order for females to reproduce. A number of theories have been put forth, suggesting ways in which orgasm might aid a woman's chance of conceiving, but none of them is terribly convincing. There is the "knockout" hypothesis, which suggests that a postorgasmic woman is likely to remain lying down longer, thereby facilitating conception. Another is that female orgasm may be something that is just along for the evolutionary ride, a mere by-product of ejaculation in men. Like nipples, which are nonfunctional in men but biologically useful in women, perhaps orgasm in women persisted simply because it is closely tied to ejaculation in men, although among women it may be physiologically and behaviorally inconsequential.

Another possibility is that female orgasm provides what psychologists call "positive reinforcement," which is to say, a mechanism whereby individuals are internally prodded to continue what they are doing when what they are doing is somehow good for them. Thus, maybe female orgasm, tied as it is to "good sex," is how a woman's body tells her that she is with a desirable sexual partner. Consider, for example, this observation of sex in grizzly bears. When subordinate males copulate, they spend their time looking around, heads swiveling for possible sight of an approaching dominant bear. By contrast, dominant grizzlies are more leisurely when they have mounted a female. It isn't clear whether grizzly sows experience orgasm, but if so, they are probably more likely to do so under the "gentle, seductive entreaties" of a successful, dominant, and unhurried male, whereupon they might well "unfold" "like a torpid, torrid, sensitive blossom," which is to say, like Edna Pontellier with her lover. But, sad to relate, not with her husband.

Among animals, perhaps the major risk that females run when they engage in an EPC is that their mates, if they find out, might abandon them and their offspring. It is interesting that although both Emma Bovary and Edna Pontellier are mothers, the two are presented as lacking any deep maternal inclinations, the absence of which facilitated their adulteries. Edna, unlike her friend Adèle Ratignolle, wasn't a "mother-woman," and Emma Bovary treated her child as a dress-up doll.

By contrast, both Hester Prynne, the key figure in Nathaniel Hawthorne's *The Scarlet Letter* (of which more later), and Anna Karenina, eponymous heroine of Tolstoy's great novel, are devoted mothers, such that the possible loss of their children emerges as a major threat to their happiness—in fact, *the* major threat. Hester struggles with the Puritan authorities to be allowed to keep Pearl, her child by the Reverend Arthur Dimmesdale (and Dimmesdale, in turn, shows a degree of paternal solicitude by arguing success-fully in support of this desperate request). Karenin, on the other hand, is, we might say, paternally challenged, and when he threat-ens Anna with the loss of her child if she persists in demanding a divorce, he is clearly seeking advantage rather than responding out of fatherly solicitude. Indeed, Anna's deceived husband makes it clear that one consequence of Anna's affair is that he has begun to question his paternity itself: "I doubt everything, I cannot bear my son, sometimes I do not believe he is my son."

Anna, meanwhile, has no such doubts. After all, she is a per-fectly good mammal, with all the consequences of internal fertiliza-tion:

> No matter what happened to her, she could not give up her son. Let her husband put her to shame and turn her out, let Vronsky grow cold towards her . . . she could not leave her son. She had an aim in life. And she must act; act to secure this relation to her son, so that he might not be taken away from her.

Anna becomes pregnant by Vronsky, gives birth to a daughter, and the three go off together, while Karenin makes good on his threat to take control of Anna's son, Seriozha. This is an unen-durable blow, and the forced separation contributes greatly to her eventual suicide.

Suicide is transgressive in that it violates the prevailing social code. So, of course, is marital infidelity. Moreover, adultery is also—from the perspective of the cuckolded husband—a transgres-sion against an even deeper, biological code. This is probably why it is universally considered such a serious offense, especially when done by women. "The difference [between the consequences of a man's and a woman's infidelity] is boundless," noted Samuel Johnson. "The man imposes no bastards upon his wife." It is tempt-ing to interpret Anna's fate as the punishment for such a potential imposition, meted out—not coincidentally—by a male writer, except that Kate Chopin exacts similar punishment upon Edna Pontellier. Maybe Ms. Chopin, although remarkably liberated for her time, was nonetheless swayed by the same powerful antiadultery sentiment that prevails even today, under which a sexually unfaithful *woman*, in particular, must suffer for her actions. Or maybe she was simply reflecting the bias of her day, according to which adulterers—especially if female—must come to a bad end.

In *Anna Karenina*, a liberal-minded gentleman named Pestov comments that the real inequality between husband and wife is the fact that infidelity by each is punished differently, to which Karenin responds, "I think the foundations of this attitude are rooted in the very nature of things." Here, at least, Anna's husband is sensitive indeed. Asymmetric punishment for adultery—in effect, the sexual double standard—hardly exhausts the catalog of husband-wife social inequality. Nonetheless, both Pestov and Karenin are correct: adultery by a wife is indeed punished differently than is adultery by a husband. And Karenin is even more correct: this difference is indeed rooted in the "very nature" of things.

Moreover, a man is *expected* to react intolerantly to his wife's adultery. In many countries, men are even legally excused for murdering an unfaithful spouse and her consort. Lest you assume that we're talking only about third world countries or those practicing an extreme form of Islam, until 1974 homicide was fully legal in the state of Texas "when committed by the husband upon the person of anyone taken in the act of adultery with the wife, provided the killing takes place before the parties to the act have separated" (Texas Penal Code 1925, article 1220). What was meant by "in the act" isn't entirely clear, and one has to wonder whether this particular brand of Texas justice required that a homicidal husband slay his rival while sexual intercourse was literally in progress. But the basic idea is clear enough: adultery—especially on the part of a wife—is assumed to be such an immense provocation to a husband that violence is anticipated and even "justified."

Literature gives us many cases of the wronged—"outraged"— husband as avenging angel. In Leo Tolstoy's novella *The Kreutzer Sonata*, a man named Pozdnyshev brags how he killed his wife, thinking her guilty of adultery. She was a pianist who evidently had an affair with a violinist with whom she had played Beethoven's passionate sonata. Pozdnyshev blames the romantic music: "An awakening of energy and feeling unsuited both to the time and place, to which no outlet is given, cannot but act harmfully." We blame biology.

For another icon of literature in which biology looms far larger than most critics have allowed, and which features a paradigmatic "wronged" husband, consider Hawthorne's *The Scarlet Letter*, which isn't so much about adultery as about society's response to it. That response is both vengeful and—albeit indirectly—violent as well, on the part of one Roger Chillingworth. Hester Prynne, young and alone in the Massachusetts Bay Colony, has reason to think that Chillingworth, her husband, has died, whereupon she has an illicit sexual relationship, as evidenced by the incontrovertibly biological fact that she bears a child, and for which she is condemned to wear the famous scarlet letter: an A for adultery.

It turns out, however, that Hester's husband, the not-very-jolly Roger, isn't dead after all. He returns in time to witness Hester's (and, incognito, his own) humiliation, whereupon Chillingworth commits himself to discovering and revenging himself on the man who cuck- olded him. It turns out to have been the community's straitlaced minister, the Reverend Arthur Dimmesdale. Chillingworth proceeds to destroy Dimmesdale physically as well as psychologically, referring to himself along the way as a "fiend." Roger Chillingworth is old, de- formed, and likely wasn't an especially loving or appealing husband even when he and Hester lived as a (presumably faithful) couple: "Hadst thou met earlier with a better love than mine," he admits to

Hester, "this evil [i.e., her adultery] had not been." But following Hester's infidelity—which produced the "love child," Pearl— Chillingworth is chilling indeed in his single-minded pursuit of vengeance:

> There came a glare of red light out of his eyes; as if the old man's soul were on fire. . . . old Roger Chillingworth was a striking evidence of man's faculty of transforming himself into a devil . . . devoting himself, for seven years, to the constant analysis of a heart full of torture, and deriving his enjoyment thence, and adding fuel to those fiery tortures which he analyzed and gloated over. . . . "A mortal man, with once a human heart, has become a fiend for his [Dimmesdale's] especial torment!"

Chillingworth angrily rejects Hester's plea that he leave off persecuting Dimmesdale: "By thy first step awry, thou didst plant the germ of evil; but, since that moment, it has been a dark necessity. . . . Let the black flower blossom as it may!" Chillingworth's pursuit is so violent, persistent, and consuming that he could as well have killed the minister directly. *The Scarlet Letter* is unusual in the annals of literature in that the male participant, rather than the married woman, ends up suffering the most. Another example of male comeuppance—played this time for laughs rather than tragedy—occurs in a more contemporary novel, John Irving's *The World According to Garp*. Here, the transgressing man gets his penis amputated when Garp, the cuckolded husband, accidentally drives his automobile into a parked car within which Mrs. Garp is performing fellatio on her extramarital lover.

Whether lethally violent (as in *The Kreutzer Sonata*), indirectly so (*The Scarlet Letter*), or comically and unintentionally retributive (*The World According to Garp*), biology nonetheless underpins a predictable male response to female infidelity. For most behaviors, evolutionary inclinations whisper within us. But when it comes to certain things—notably, male response to a spouse's adultery—

those whispers fairly rise to a shout. Indeed, the expectation of such a response is so great that when it is *not* met, when a cuckolded husband reacts meekly or not at all, this may be considered evidence for his inadequacy, and even a kind of justification for the wife's adultery in the first place. For example, after Anna Karenina's adultery has been revealed to him, Karenin makes it clear that he values propriety above all else, that his outrage is limited to the embarrassment caused by her behavior, and that he will not grant Anna a divorce because of the social awkwardness it would entail:

> "My decision is as follows. Whatever your conduct may have been, I do not consider myself justified in severing the ties with which a Higher Power has bound us. The family cannot be broken up at the caprice, discretion, or even the sin of one of the partners of that marriage, and our life must continue as before."

Karenin is only concerned with sparing himself the social consequences of acknowledging his wife's affair. His unflappable indifference to the emotional consequences of his wife's behavior, itself so counter to biological expectations, helps illuminate his own character, which in turn sheds light on why Anna was so dissatisfied with Karenin in the first place and why she turned to Vronsky: Karenin's coldness, his valuing of social appearances over the warmth of love and the promptings—sometimes unruly and even violent—of life itself.

> He [Karenin] went on coldly and calmly. . . . "As you know, I look upon jealousy as a humiliating and degrading emotion and I shall never allow myself to be influenced by it; but there are certain laws of propriety which one cannot disregard with impunity."

And here is Anna talking with her lover, Vronsky, about Karenin's response to their adultery:

"In general terms, he [Karenin] will say in his official manner, with all distinctness and precision, that he cannot let me go but will take all measures in his power to prevent a scandal. And he will quietly and punctiliously do what he says. That is what will happen. He's not a human being but a machine, and a cruel machine when he's angry . . . He's not a man, not a human being—he's a . . . puppet! No one knows him but I do. Oh, if I'd been in his place, if anyone else had been in his place, I should long ago have murdered and torn in pieces a wife like me. I shouldn't be calling her '*ma chère*'! He's not a man, he's an automaton."

Throughout this book, we point to cases in which behavior parallels the expectations of biology. Equally revealing, however, are those unusual depictions of people whose actions run *contrary* to prediction. When this happens, it is likely to throw light on what is special about them, whether their abiologia reveals them to be especially cold and thus undesirable (like Karenin) or especially saintly (such as Jesus or the Buddha, whose renunciation of earthly pleasures and selfish pursuits marked them as notable in their own way). In other words, departures from expectation can themselves be meaningful, but only if we begin with an expectation of what human beings "naturally" are like. It is also significant that even Karenin, for all his protestations of propriety, eventually shows some jealousy toward Vronsky and anger toward Anna that go beyond the frozen "disappointment" that he expresses; a cold fish, indeed, but still a fish.

Even when, for whatever reason, the husband doesn't exact physical vengeance as such, a philandering wife is nearly always punished (especially, we might add, in stories written by men). Emma dies. Anna dies. Even the liberated Edna Pontellier dies. In Stendhal's *The Red and the Black*, the social-climbing opportunist Julien Sorel has an ongoing affair with Madame de Rênal, who is married to the local mayor. She dies a few days after Julien is executed (for having tried, earlier, to kill her), embracing her children but still loving

Julien hopelessly and helplessly, driven to religious distraction by her guilt but, by the laws of social and biological propriety, not permitted to recover emotionally from her adulterous affair.

Of course, there is no one-size-fits-all response to adultery. In *The Golden Bowl*, Henry James depicts a situation that in anyone else's hands would have been worthy of Jerry Springer–type fireworks yet is notable for its apparent mildness. At the same time, *The Golden Bowl* shows biology lurking just offstage, fluttering the curtain of social rules and cultural guidelines. Maggie Verver, a wealthy young American living in Europe (most of James's characters, it seems, are wealthy young Americans living in Europe), discovers that her best friend, whom she had encouraged to marry her widower father, previously had an affair with Maggie's current husband, the urbane Prince Amerigo. Moreover, they are still at it! But no suicides in this case, no murder à la Pozdnyshev nor fiendish pursuit à la Chillingworth; rather, sensitivity and delicacy.

Above all, James's sophisticated upper-crust characters are careful not to upset the matrimonial applecart as they strive—successfully—to keep up appearances and the code of upper-crust respectability: make one's life a work of art. Fake it till you make it? Perhaps. At the same time, anyone reading *The Golden Bowl* cannot help but notice the coiled tension generated by the various revelations of past and present infidelity and how this energy suffuses the outwardly unruffled response of the various protagonists. Indeed, it is the power of their evolutionary heritage that drives much of the action, just as the characters' refusal to act upon its dictates highlights the social restraints and sophistication that James so successfully conveys. Like a tornado that races through a museum, miraculously sparing the precious artifacts, adultery slashes through the Verver/Amerigo family, and the reader is left breathless, wondering how so much energy can have done so little apparent damage.

Perhaps the finest literary account of postadultery reconciliation, however, comes from James Joyce's *Ulysses*, widely seen as the greatest novel of the twentieth century. In it, the author parallels the mythic, elevated adventures of the Homeric superhero, Odysseus, with the genuine, down-to-earth, quotidian experiences of a very ordinary guy, Leopold Bloom, during a very ordinary day in early-twentieth-century Dublin. While Bloom goes to work, attends a funeral, eats, and has various quintessentially normal and earthy encounters, his wife, Molly, has a steamy afternoon extramarital dalliance (also all too normal) with the aptly named Blazes Boylan. Our hero knows of this, is greatly disturbed by it, but comes to a kind of peace nonetheless, emphasizing in this reverie the fundamental "naturalness" of Molly's rendezvous by contrasting it with other provocations:

> As natural as any and every natural act of a nature expressed or understood executed in natured nature by natural creatures in accordance with his, her and their natured natures, of dissimilar similarity. As not as calamitous as a cataclysmic annihilation of the planet in consequence of collision with a dark sun. As less reprehensible than theft, highway robbery, cruelty to children and animals, obtaining money under false pretenses, forgery, embezzlement, misappropriation of public money, betrayal of public trust, malingering, mayhem, corruption of minors, criminal libel, blackmail, contempt of court, arson, treason, felony, mutiny on the high seas, trespass, burglary, jailbreaking, practice of unnatural vice, desertion from armed forces in the field, perjury, poaching, usury, intelligence with the king's enemies, impersonation, criminal assault, manslaughter, willful and premeditated murder. As not more abnormal than all other altered processes of adaptation to altered conditions of existence, resulting in a reciprocal equilibrium between the bodily organism and its attendant circumstances, foods, beverages, acquired habits, indulged inclinations, and significant disease.

Molly Bloom is no Penelope, whose legendary faithfulness to Odysseus—despite a slew of suitors—can be contrasted with the wily hero's dalliances with Circe and Calypso. Similarly, Leopold

Bloom is no Odysseus; his sexual conquest during "Bloomsday" consists of masturbating while surreptitiously watching a young woman who, leaning backward, reveals her underwear.

But Molly's adultery doesn't itself demonstrate that Leopold is a loser. Rather, the telling thing would appear to be his lethargic response to her infidelity, verging on nonresponse. But wait! Things are more complex than this; indeed, great literature is made great, in part, by its complexity. To be sure, it would have been possible for Joyce to portray Leopold Bloom as murderously enraged by his wife's adultery, like Tolstoy's Pozdnyshev. But throughout *Ulysses*, Joyce is after bigger game; he portrays Bloom as a modern-day Odysseus, not the legendary, hormonally predictable hero of yore and of straightforward biology. He is, for one thing, a pacifist. But at the same time, Bloom is wonderfully believable as an embodied male animal. Thus, he is hardly indifferent to Molly's affair with Boylan. For much of the novel, in fact, our hero is preoccupied with it, terribly agitated and driven, if not to distraction, then certainly to a high level of guilt, anger, and—especially in the phantasmagorical Circe ("night-town") chapter—to a remarkable and prolonged bout of obsessive erotic imaginings.

Here is yet another layer of complexity: in writing *Ulysses*, Joyce was concerned to show how daily life can rise to its own version of grandeur, with Bloom the pacifist emerging as a memorable counterpart to the homicidal suitor-slaying fury of Homer's Odysseus. In this sense, Leopold's overt tolerance of Molly's behavior elevates him, just as Molly's famous monologue at the novel's end elevates us all, concluding as it does with an affirmation of love and of affirmation itself:

> I asked him with my eyes to ask again yes and then he asked me
> would I yes to say yes my mountain flower and first I put my arms
> around him yes and drew him down to me so he could feel my
> breasts all perfume yes and his heart was going like mad and yes I
> said yes I will Yes.

6

6

WISDOM FROM THE GODFATHER

Kin Selection, or the Enduring Importance of Being Family

———————

The Godfather may not have been an evolutionary geneticist, but he knew a thing or two about life. "How do I love you?" Don Corleone might have written. "Let me count your genes."

Genes, not bodies, are where the evolutionary action is. It may be hard to believe, since superficially the organic world is filled with trees, flowers, birds, fish, mammals, insects, and people—not a speck of DNA to be seen. And yet the genes are there. It is genes, not bodies, that persevere through the trajectory of life, and they do so by hitchhiking (or hijacking) a ride in one body or another. After all, bodies are mortal; not so genes. Bodies come and go; genes can, at least in theory, go on forever. And of course, those that succeed in "going on"—if not forever, at least into those generations that have persevered thus far—are the ones responsible for all of the life around us.

So while the "family" that Mario Puzo wrote about was literally a family—a congregation of distant, individual bodies—shared genes made up the mortar that held the Corleones together; similarly for the Tataglias and the other Mafia clans. The importance of being family, however, isn't limited to criminal enterprises or godfathers. It is the earnest enterprise of life.

In his poem "Heredity," Thomas Hardy had a premonition of modern evolutionary biology and the endurance of genes:

I am the family face
Flesh perishes, I live on
Projecting trait and trace
Through time to times anon,
And leaping from place to place
Over oblivion.

The years-heired feature that can
In curve and voice and eye
Despise the human span
Of durance—that is I;
The eternal thing in man,
That heeds no call to die.

When people refer to evolution with the verbal shorthand "survival of the fittest," only rarely do they understand just what they are talking about. Ever since Herbert Spencer coined this phrase in the nineteenth century, it has been misused to justify one group's domination over another: stronger over weaker, smarter over less intelligent, less scrupulous over more restrained, "pure" races over "mixed," and so forth. Yet, aside from the egregious immorality of such abuse, it is also scientifically wrongheaded. "Fittest" means none of these things, nor does it imply any ethical justification for what is found in nature. The living world simply is what it is. If human beings were to derive ethical guidelines from the natural world, we should probably refrain from walking upright; after all, gravity would dictate crawling on our bellies! And in deference to entropy, rooms should never be tidied. As to "fitness," this biological concept simply indicates a living thing's capacity to pass on genes, in a given environment. Thus, being "fit" may mean being big and strong, or perhaps smart and cunning, or it might be the weak and suitably frightened or the sneaky and unprincipled who reproduce successfully.

The "fittest," as biologists use the term, are, as we've seen, those who outcompete their rivals for whatever reason; more precisely, those most successful in projecting copies *of their genes* into the future. Evolutionary success—that is, fitness—is what genes strive for; bodies are how they do it, and reproducing is the most direct route, but, as we shall see, it isn't the only one. Physical fitness often helps, but the bottom line isn't bodily strength, or even overall health. Rather, what matters to evolution is the ability to stay in the game, for genes to prosper as self-promoters over long stretches of time. And the usual way for genes to benefit themselves is to package themselves in creatures known as offspring. The issue isn't health, happiness, strength, or even longevity (that is, a long-lived body), since even a Methuselah will eventually fall by the evolutionary wayside, to be replaced by genetic lineages that somehow manage to persevere longer than the most elderly individual.

So, survival of the fittest? Certainly. But the fittest *what*?

It can't be the fittest body, since most individuals of most species are actually rather short-lived. In the billions of years that evolution has had to work with, natural selection has perfected all sorts of highly sophisticated, complex adaptations, absolute marvels of effective design: mammals that fly and that use sonar to hunt moths on the wing, caterpillars that look like snakes and vice versa, bacteria that can prosper in superheated seawater, eagles that can make out the face of a dime while hovering a hundred feet in the air. Amid all these marvels, if there had been a significant evolutionary advantage for individuals to live for centuries, it seems very likely that this could have been done, too. After all, in some cases it has. Redwood trees and bristlecone pines survive for literally *thousands* of years. And yet, they are notable exceptions. Most creatures, plant or animal, are ephemeral.

Turning to the vertebrates, where natural selection has shown itself *capable* of producing long-lived bodies (for example, giant tortoises or sturgeon), the sad reality is that most of the time, it doesn't bother. This is because mere survival of the individual is not very high on the evolutionary agenda. In fact, the overwhelming majority of living things have a life span of less than a year. So what *does* matter? What is evolution all about? Not the individual, and not the species (which, after all, is merely the sum total of all individuals), but rather the survival of the "family face." The Corleone line and blood. In short, the gene.

This gene's-eye view of evolution emphasizes the most important quality of DNA: its ability to copy itself. (Second in importance would be its ability to construct bodies and equip those bodies with instructions as to how to grow and develop, as well as how to behave under various circumstances.) In his book *The Selfish Gene*, Richard Dawkins emphasized that what are known as bodies or organisms—that is, individual plants or animals—are really "survival machines" manufactured by genes for their own selfish benefit. This captures not only what life is at present but how it started.

In the beginning was the gene. Or rather, an array of organic molecules, some of which were capable of copying themselves. Most of their fellow molecules likely broke apart over time, but some were self-replicating, perhaps like crystals mechanically repeating their structure. In any event, even copycatting genes wouldn't have left an indelible imprint on the natural world, especially once they found themselves competing with other replicators that happened to blunder into the world's second greatest discovery (after replication itself): building bodies. Here is Dawkins's account, which we quote at length since not only is it suitably dramatic, but the story it tells is laden with meaning and even beauty.

> Replicators began not merely to exist, but to construct for themselves containers, vehicles for their continued existence. The replicators that survived were the ones that built survival machines for them to live in. The first survival machines probably consisted of nothing more than a protective coat. But making a living got steadily harder as new rivals arose with better and more effective survival machines. Survival machines got bigger and more elaborate, and the process was cumulative and progressive.

Was there to be any end to the gradual improvement in the techniques and artifices used by the replicators to ensure their own continuation in the world? There would be plenty of time for improvement. What weird engines of self-preservation would the millennia bring forth? Four thousand million years on, what was to be the fate of the ancient replicators? They did not die out, for they are past masters of the survival arts.

But do not look for them floating loose in the sea; they gave up that cavalier freedom long ago. Now they swarm in huge colonies, safe inside gigantic lumbering robots, sealed off from the outside world, communicating with it by tortuous indirect routes, manipulating it by remote control. They are in you and in me; they created us, body and mind; and their preservation is the ultimate rationale for our existence. They have come a long way, those replicators. Now they go by the name of genes, and we are their survival machines.

Genes are the ultimate survivors.

In his novel *The Sound and the Fury*, William Faulkner pronounces this judgment upon his seemingly doomed yet stubbornly persistent characters: "They endured." The same can be said of genes. As Faulkner relates it, the Compson family of Mississippi endured through lust and violence, madness and incest, thievery and the heavy hand of history. Ditto for genes. Genes also endured through drought and flood, predators and parasites, friends and foes, good times and bad. And unlike Faulkner's fictional inhabitants of Yoknapatawpha County, genes are real.

How, then, did they survive?

Certainly not by giving up on themselves, nor by dropping the ball. Those genetic chunks of DNA that constitute every human—or hippo, halibut, or hyacinth—are directly continuous with similar genes in their ancestors, and theirs before that, and so on, back to the slimy, sloshy, organic soup from which we all have sprung (or swum, or crawled). Going back through evolutionary time, not a single one of your antecedents has failed to reproduce. Not one ever missed a beat. Otherwise, you wouldn't be reading this.

Those genes that "made it" did so by creating bodies that have proved successful when it comes to making it in their own lives and—most important—in helping project copies of themselves into the future. After all, imagine a gene that did a great job of creating a superb body but failed utterly at getting itself passed along to the next generation; it would perish eventually, along with that superb body. Bodies are merely slingshots, designed by their genes to shoot copies of themselves into the future. From a strictly biological perspective, in fact, this is *all* that bodies are. Of course, it isn't adaptive for naked genes to be catapulted about; they'd quickly be destroyed and devoured by the array of rapacious bodies, generated by other genes, that people our planet. And so genes are projected into the future enclosed in bodies of their own.

In Herman Melville's great novel, Captain Ahab strove mightily to "strike through the mask" beyond the outer appearance of his nemesis, Moby-Dick, to its deeper reality (evil, enmity, God?). It may require a similarly mighty act of the imagination to see through the superficial reality of organisms and acknowledge the deeper reality of genes that lies beneath the universal mask of fleshy bodies.

After all, looking around, we see bodies, not naked genes. Bodies, bodies everywhere: eating, sleeping, being eaten, growing, reproducing, laughing, lounging, walking, running, swimming, hopping, and slithering their way to . . . what? To either success or failure, as measured by how well they project their component genes into the future.

This gene's-eye view was first brought to the attention of biologists by William D. Hamilton, a British geneticist and perhaps the most innovative evolutionary thinker since Darwin. In a series of papers in the 1960s, Hamilton pointed out that the business of genes is to look out for identical copies of themselves, packaged in different bodies. In the most obvious cases, we call these particular bodies "children," and those individuals who share an above-average amount of genetic relatedness "families." A family—whether Compson, Corleone, yours, or ours—is thus a constellation of bodies, created by genes, providing the prospect of their own perpetuation.

Hamilton went further, showing mathematically as well as logically that even reproduction is a special case of the more general phenomenon of genetic self-promotion via promulgation. Seen through this gene-ial lens, parents care for their offspring "because" those offspring are the most direct route for parental genes to advance themselves. Parental love is, accordingly, an evolutionary mechanism assuring that all this caretaking actually takes place. Moreover, there is no reason why genes should limit themselves to merely caring for offspring, since children, after all, are merely a special case of the more general phenomenon of genes looking out for copies of themselves tucked into other bodies. Accordingly, these bodies, as particular "persons of interest," needn't be limited to one's children. To varying degrees, people also look after nieces and nephews, grandchildren, cousins, and so forth. And now we know why: because each genetic relative is biologically important in precisely the same sense that a child is, only somewhat less so because the relative is more distantly related. Hamilton was inspired to make a marvelous conceptual leap, one that allowed him to perceive what evolution had been up to all along: he was able, at last, to make sense of the previously puzzling phenomenon of altruism and, to some extent, of social behavior generally.

Here, in brief, is—or was—the conundrum: why does altruism exist? Technically, it shouldn't, since any gene or combination of genes that induces its body to do something that helps another at the expense of itself would by definition benefit other genes, while hurting itself. Why should any of the Corleones *ever* look out for each other? Dysfunctional, you say? Actually, the Godfather's clan differs from others only in the intensity of their self-sacrificial benevolence (not to mention their violent way of expressing it). Altruism, or benevolence toward others generally, seems to fly in the face of evolution's quintessential selfishness. "Mercy did not exist in the primordial life," according to Jack London in his best-known novel, *The Call of the Wild*. "It was misunderstood for fear, and such misunderstanding made for death. Kill or be killed, eat or be eaten, was the law." We are very fond of *The Call of the Wild* and, like generations of readers, have thrilled to the adventures of Buck, the great dog who, at the story's end, answers the wild's call.

And we have been equally entranced by the protagonist of Jack London's other great animal story, *White Fang*, who "became quicker of movement than the other dogs, swifter of foot, craftier, deadlier, more lithe, more lean with iron-like muscle and sinew, more enduring, more cruel, more ferocious, and more intelligent." According to the ostensibly omniscient narrator, Mr. Fang "had to become all these things, else he would not have held his own nor survived the hostile environment in which he found himself."

But Jack London's sense of the "primordial life" was more than a little overheated. Here is a corrective dose of biological reality: cooperation is as genuine, and often as necessary, as competition. It is also perplexing. Thus, animals frequently give alarm calls when a predator shows up, thereby helping alert others to the danger . . . but in the process exposing themselves to greater risk. Similarly, they frequently share food, and Vito Corleone, among others, always seems to be urging his family members to sacrifice for each other. How can evolution arrange for the perpetuation of genes that promote this sort of thing, that appear to be self-sacrificing, if their effect is to *reduce* their own abundance?

Here is Hamilton's in-gene-ious solution: altruism is actually selfishness in disguise. Genes can enhance—not reduce—their fitness by prodding their bodies to help other bodies within which there reside identical copies of themselves. So when an altruist helps a family member—whether an offspring or some other genetic relative—even at some cost to himself, the benefactor can actually be benefiting himself at the gene level. (More precisely, genes for such altruism are helping themselves.) This holds if there is a sufficient probability that genes underlying the altruism are present in the beneficiary, which is to say, so long as the beneficiary is a genetic relative. The "paradox" of altruism, then, is resolved by realizing that what appear to be altruistic bodies are, at the same time, the manifestation of selfish genes. Looking out for a relative, even at personal cost, can be, accordingly, a case of selfish genes looking after themselves. The closer the relative, the greater the probability of shared, identical genes, and hence the more we should expect to see such altruism.

Inspired by Hamilton's insight, students of animal behavior have found that the alarm calling of prairie dogs, for example, is more likely when the alarmist is genetically related to the beneficiaries. Similarly, whenever food is being shared, genes are often shared as well. Here is another seeming incongruity, resolved by a dose of gene-oriented thinking: when zebras are attacked by lions, other zebras often come to their aid; not so, however, for wildebeest. The resolution? Zebras generally live in family groups; accordingly, aid rendered to a zebra in distress is more parsimoniously viewed as gene selfishness than as inexplicable altruism. And as for the selfish disregard of one wildebeest for another, it is significant that wildebeest live in anonymous crowds of nonrelatives, so the evolutionary geneticist expects them to be indifferent to each other's fate.

The list of examples goes on, but probably the most dramatic was recognized and explained by Hamilton himself. This is the case of the social insects (bees, wasps, and ants), in which workers typically do not even attempt to reproduce. Rather, they labor selflessly for the success of someone else, namely, the queen. Such altruistic restraint puzzled biologists since Darwin. Hamilton helpfully pointed out that those insect species with a uniquely intense form of altruism have an equally unique genetic system, known as haplodiploidy. As a result, worker bees, wasps, and ants are actually more closely related to their sisters (by a factor of three-quarters) than they would be to their own offspring (half), were they to reproduce. In short, genes within each worker do more to promote their own success by staying home and caring for their siblings, who are also offspring of the queen, than by seeking to rear their own offspring.

The point is not that because something is true of ants (or bees, or zebras, or prairie dogs) it is also true of people. Rather, it is that evolutionary genetics has revealed a powerful general rule, something to which a remarkably wide range of living things adhere. It takes extraordinary hubris—not to mention a willful denial of the basic facts of life—to claim that we are qualitatively different from the rest of the living world. To be sure, human beings are many things to themselves and others. But, no less than our fellow creatures, people are *also* the way their genes act out their goals.

Think, for example, of this universal human trait: nepotism. It means favoritism toward relatives or, more specifically, treating relatives more kindly than nonrelatives and closer relatives better than more distant ones. And think as well of this secret passcode learned by the feral child, Mowgli, in Rudyard Kipling's *The Jungle Book*: "We be of one blood, thee and I." By it, Mowgli gained the allegiance of all creatures of the tropical Indian jungle. He is literally correct. We really *are* of one blood: frogs, fishes, fowl, and folks. Birds of a feather, all of us, sharing an evolutionary history. More to the point, we share genes right here and now, those who are genetic relatives more so than others. And so, not surprisingly, those who are genetic relatives are more likely than others to share acts of kindness.

Back in the Corleone clan, we have Michael (memorably played in the movie by a youthful Al Pacino) telling his girlfriend, somewhat apologetically, how his father (played even more memorably by Marlon Brando) had once gotten his way by suggesting that a particular gentleman would put either his signature or his brains on a disputed contract. "That's my family," explained Michael. "It's not me." Seeing—or reading—this exchange, we all know that reality will prove otherwise, and not just because such is the foreshadowed narrative arc of the story. Rather, because of the nature of biological reality, it *is* the story, a tale of genes looking out for themselves: our families, our genes, ourselves, how it is that being in the same family means that we are of one blood.

He was not a biologist, but in his letters, nearly two hundred years ago, the English romantic poet John Keats anticipated what would become one of the cornerstones of the newfangled "revolutionary biology." It is a prescient recognition of what life is all about. "I go amongst the fields and catch a glimpse of a stoat or a fieldmouse peeping out of the withered grass," wrote Keats. "The creature hath a purpose and its eyes are bright with it. I go amongst the buildings of a city and I see a man hurrying along—to what? The creature hath a purpose and its eyes are bright with it."

Fifty years before Darwin, Keats realized that animals have a purpose of some sort. And furthermore, he sensed that human beings, too, have a purpose. Although Keats didn't state unequivocally that human bodies and animal bodies share the same purpose, (r)evolutionary biologists are now prepared to do just that. Furthermore, we can even identify that purpose: achieving the greatest possible success of their genes. (Whether it brightens their eyes is another question.)

People can, and do, go about their lives in pursuit of their biological "purpose," even if they don't know what that purpose *is*, or if—like those stoats, field mice, or the first gene molecules of the early soup—they don't realize they are doing so. If we eat when hungry, sleep when tired, engage in sex when the right opportunity presents itself, run from enemies (or overpower them or hide from them, depending on what is called for), we are behaving purposefully, with that purpose going beyond the immediate and obvious motivation of filling our bellies, slaking our desires, saving our skins, and so forth. Behind the superficial facade of satisfying one's needs or responding to one's fears lies the deeper purpose of satisfying the ever-present, bottom-line requirement of evolutionary success.

When this purpose is thwarted, something is awry. This is why Don Vito Corleone, Michael's father, responds as follows to a feature in *Life* magazine describing the exploits of his son, a war hero in World War II: "A friend had shown Don Corleone the magazine (his family did not dare), and the Don had grunted disdainfully and said, 'He performs those miracles for *strangers*.'" Genes are supposed to look after each other, not strangers.

Of course, the Corleone clan is not unique in this regard. The Mafia, fictional as well as real, is constructed around nepotism: in a family, the various members are more firmly committed to each other than to the outside world of laws and customs. Speaking of a competing clan, we learn from author Mario Puzo that "the Bocchichios' one asset was a closely knit structure of blood relationships, a family loyalty severe even for a society where family loyalty came before loyalty to a wife." (Ironic, isn't it, that an enterprise synonymous with violence and brutality is also founded on altruism.)

The Corleone family gets its power by appealing to friendship, duty, and—most crucially—family loyalty, rather than mere threats and extortions; that's why it is so successful. Don Vito Corleone will go out of his way to assist a friend, appealing to his honor and pride, getting him to trust the family, and then expecting similar favors in return. At one point, when singer Johnny Fontane, whom Don Vito Corleone considers his godson, requests assistance in getting a movie contract, the Godfather responds: "Friendship is everything. Friendship is more than talent. It is more than government. *It is almost the equal of family* [our emphasis]. Never forget that." (The theme of reciprocity, essentially the exchange of favors—what biologists call "reciprocal altruism"—looms large in both the animal and human worlds, as we'll see in Chapter 9; for now, we emphasize that despite the special place of friendship in the don's world, it is clear from his advice to Johnny Fontane that even this takes a backseat to family.)

Within the immediate family, the expectation of such loyalty is even more important and even more evident. It goes without saying that the family is very close: Sonny Corleone—Michael's brother—is enraged when his sister Connie is beaten up by Carlo, her husband; members of the family are expected to rely on each other for their lives and mutual protection, and so, Sonny rushes over to "get" Carlo . . . and is then killed en route by a rival clan. And not surprisingly, what eventually lures Michael Corleone back into the family is a series of attempts on the life of his father. Not only is Don Vito the symbol of the family, he is integral to its flesh and blood, progenitor of (half) of its genes. Michael may have claimed otherwise, but to say that it is his family is also to say that it is him.

One of the most powerful incentives that Don Corleone offers to prospective "friends" is that he will always look out for their family in the event of their death; the Godfather guarantees that the children and wife of anyone who gets injured during his service will always be looked after. (A chilling analogy can be found among those indirect sponsors of Middle East terror who reimburse the families of suicidal "martyrs.")

The potency of shared genes (biologists call it "kin selection") is so great that even nonkin can readily be sucked in. People have developed ways of tapping into its gravitational force by alleging kinship even when it doesn't occur. Sociologists and anthropologists use the term "fictive kinship," and we see it operating whenever appeals are made to "brothers" and "sisters" who are neither, in the myriad nationalistic claims about "fatherland" or "motherland," and in supranationalistic appeals to the human "family," and even in the Black Power movement of the 1960s and 1970s.

Interestingly, the very notion of a godparent—even without the Mafia implication—suggests just such a fictive relationship, a pseudo-blood connection between individuals (typically, recipients of favors, the kind of altruism that one expects from a blood relative). The expectation of powerful bonding between Don Vito Corleone and numerous others is precisely why the Godfather became such a powerful man. He successfully expanded his family, fictive as well as genuine, and then reaped the benefit of equally immense family loyalty (underscored, of course, by fear).

An especially interesting case of fictive kinship is played out in the story of Tom Hagen, brought up by the Corleone family since he was orphaned at about the age of ten. Distrusted by other families because he is of Irish descent, rather than Sicilian, Tom's situation emphasizes a further expansion of blood loyalty, since only Sicilians are to be trusted and respected. However, the Don knows how clever and wily Tom is, and therefore welcomes him despite his biological liabilities, eventually even making him *consigliore*, second in command. Tom is pleased, and the narrator is in no doubt as to the conflation of family with loyalty: " 'I would work for you like your sons,' Hagen said, meaning with complete loyalty, with complete acceptance of the Don's parental divinity."

Later, after an attack on the Godfather's life, Tom is discussing with Don Corleone's biological offspring, Sonny and Michael, what to do. Tom advises caution; Sonny, the impatient son, is upset and wants to do something right away. Furthermore, when push comes to shove, all three recognize the compelling power of genetic connection:

> "If your father dies, make the deal. Then wait and see." [said Tom Hagen] Sonny was white-faced with anger. "That's easy for you to say, it's not your father they killed." Hagen said quickly and proudly, "I was as good a son to him as you or Mike, maybe better. I'm giving you a professional opinion. Personally I want to kill those bastards." The emotion in his voice shamed Sonny, who said, "Oh, Christ, Tom, I didn't mean it that way." But he had, really. Blood was blood and nothing else was its equal.

It is all too easy, incidentally, to take it for granted that people love and care for family members, and to ask "What's new or surprising about that?" We would bet that Isaac Newton may well have been met with a similar response when he announced the existence of gravity: "Of course things fall. What's new or surprising about that?" To be sure, objects fell long before Newton recognized that there was something at work, gravity, that was worth exploring and explaining. Similarly, pointing out that relatives look out for each other seems too obvious to mention. But because something is so much a part of our daily existence that we take it for granted doesn't mean that we necessarily understand it. Everyone knows blood is thicker than water; now, thanks to William Hamilton, we know why.

Ever true to life, literature has long assumed that genetic ties are genuine ones, almost gravitational in their force. Sophocles' tragic play *Antigone* gives us the affecting story of how Antigone (daughter of Oedipus) ultimately goes to her death for insisting that her brother Polyneices receive a decent burial. (The tyrant Creon had decreed that Polyneices's body must remain unburied and thus defiled, because he had led a revolt against the duly constituted authorities.) In *The Charterhouse of Parma*, by Stendhal, we get to observe the numerous machinations of Gina Pietranera, aunt of the adventurer Fabrizio del Dongo, during the young man's exploits in wild-and-wooly Napoleonic Italy and France. Aunt Gina gets her nephew released from prison, sets him up as a student in Naples, sponsors a jailbreak when he is imprisoned once again, and even arranges for the poisoning of Fabrizio's mortal enemy. Families, we learn from literature—as well as from life—are likely to stick together.

Fabrizio's father was a domineering, miserly fanatic to whom his mother was altogether subservient. And so Aunt Gina stood up for him. In another wartime tale, from yet another generation, Louis Begley's *Wartime Lies* recounts the experiences of Maciek, a twelve-year-old Polish Jew, and his aunt Tania as they (barely) survive the German invasion of Poland during World War II. In this case, Maciek's mother died in childbirth, and Aunt Tania has been raising him ever since. Having no children of her own, she "settled for being the perfect aunt."

To Antigone, the dutiful sister, or Gina and Tania, both perfect aunts, we could add any number of brothers, sons, daughters,

fathers, mothers, grandparents, nieces and nephews, cousins, and
so forth, nearly always with their nepotistic beneficence varying inversely with the probability of genes held in common . . . that is,
with the degree of genetic closeness. Blood is blood and nothing
else is its equal.

In a perceptive discussion of Ralph Waldo Emerson, Mark Van
Doren once wrote that "he was at his best . . . not when he was
trying to understand the man he was, but when he was being
that man." Literature, similarly, is at its best not when it is self-
consciously trying to understand the human condition but when it
is demonstrating that condition. There may be nothing more emblematic of human nature than those primordial family-based
patterns whereby our bodies/genes interact with other bodies composed, in part, of copies of our own genes.

   If you were to interview an intelligent fish and ask her to describe her environment, probably the last thing you would hear
from your hypothetical piscine interlocutor is "It's mighty wet
down here!" Some things—especially those all around us—are
taken for granted. They constitute the ocean in which we swim.
Kin selection and its corollaries altruism and nepotism are like
that. Every story in the human repertoire takes place within the
context of shared genes (with our relatives) cooperating and competing with not-so-shared genes (everyone else). This is because
every human life takes place in this same context, immersed in that
familiar, familial ocean.

   We have all been swimming in it for a very long time, even before we literally crawled onto the land up into the trees and then
down again to become recognizably human, and doubtless ever
since. It is nearly impossible to cite, or even to imagine, a literary
creation in which kin selection of one sort or another does *not* feature prominently. A revealing exception: William Golding's dark
tale *Pincher Martin*, which recounts the solitary exploits of a pilot

shot down during World War II, who crawls up on a tiny oceanic outcropping and survives by eating sea urchins, sea cucumbers, and starfish . . . until he dies, alone. Even such tales—including those featuring shipwrecked survivors—rarely lack sociality altogether; *The Swiss Family Robinson*, of course, involved a *family*.

The assumptions of kin selection run so deep that they are more likely to be noticed when breached than when followed, just as we take the normal functioning of our bodies for granted and pay attention only when something goes wrong. Take when Cain killed Abel—brothers share 50 percent of their genes and aren't supposed to do this sort of thing! That's precisely the point. Biologically, we *are* expected to be our brother's keeper, or at least not his murderer.

Thus, when Shakespeare's Richard III arranges to have his brother Clarence imprisoned in the Tower of London and then killed, and similarly when Richard contrives to murder his two young nephews (both of whom stand between him and the throne of England), it simply italicizes his monstrous, inhuman nature. It might, of course, be pointed out that human beings are by definition incapable of behaving inhumanly: "Nothing human is foreign to me," wrote the Roman poet Terence. Nevertheless, it also remains that some acts—because they seem unnaturally cruel as well as unexpected—are simply beyond the pale, thereby conveying a powerful message about the perpetrator. The killing of a close relative ranks high among these.

Accordingly, there is something riveting and revealing about fratricide because it so thoroughly defies the expectations of kin selection. By the same token, the most memorable scene in William Styron's novel *Sophie's Choice* occurs when Sophie must choose which of her children is to die in a Nazi concentration camp. And there may be no more powerful indictment of slavery than the shocking realization that Sethe, central figure of Toni Morrison's *Beloved*, has killed her child rather than allow her to grow up in bondage.

Few things, however, are as simple as we might wish, and as we shall see in Chapter 8, there are even exceptions to the seemingly straightforward genetic logic that tells close kin not to harm each other. Cain had one-half of his genes in common with Abel, just as Richard did with Clarence. But at the same time, neither Cain nor Abel was genetically identical to his sibling/victim. They had, in addition to their shared evolutionary interest, their own biological fish to fry, emphasized, perhaps, by high-stakes competition for limited resources (God's approval, the crown of England). Nonetheless, in most cases kin selection generates benevolence toward relatives, such that when close kin do serious harm to one another, our attention is immediately engaged.

Although fratricide and infanticide (and, for that matter, matricide and patricide) are especially chilling crimes, the killing of any family member is generally perceived as highly agitating to the survivors and is therefore especially likely to generate murderous tit-for-tat feuds. Huck Finn, for example, finds himself involved in precisely such a nightmare of lethal retaliation between the Grangerfords and the Shepherdsons, and it may well be noteworthy that shortly after Romeo and Juliet were secretly married—leading to a brief hope that their union might finally dissolve the enmity between Montague and Capulet—Romeo gets embroiled in a fight and ends up killing Juliet's cousin, Tybalt. After this, there is no hope for reconciliation between the two clans. Just as genes promote altruism among relatives, they aren't very forgiving when it comes to its opposite. Fortunately, however, we are most likely to see and to celebrate the beneficent side of shared genes.

For decades, biologists were intrigued by the extent to which certain individuals would run risks, sometimes even laying down their lives, for others. Armed at last with the insights of kin-based altruism, these actions now make sense: whether it is the self-sacrificial stinging of a worker bee or the willingness of adult zebras—being "perfect aunts"—to defend a colt attacked by lions, we now see genes caring for themselves. Armed with those same insights, it is also possible to make sense of those fictional tales in which people reveal desires and connections that had previously seemed incongruous until later, when it is shown that shared genes have been involved after all, which is to say that kin selection has been pulling the strings backstage.

For example, in Henry James's *Portrait of a Lady*, we encounter the machinations of one Madame Merle, who, we eventually discover, is the former mistress of the effete Gilbert Osmond. All along, Madame Merle had been contriving to get the very wealthy Isabel Archer (the lady of the book's title) to marry Gilbert. Why? To promote the success of Pansy, ultimately revealed to be Madame Merle's child by Gilbert. "Aha!" says the reader: a mother helping out her own offspring. ("Aha!" says the evolutionary biologist: genes helping themselves.) The novels of Charles Dickens are similarly filled with cases in which individuals behave with incomprehensible concern for another's welfare until it is revealed that the beneficiary is actually a relative—son, daughter, brother, sister, and so forth—of the "altruist," whereupon the altruism is instantly comprehensible, even if the plot itself defies realism.

Like gravity, kin selection surrounds us. Granted, kin selection is not often made manifest in the many novels in which it figures. But by the same token, neither does imaginative writing (this side of the *Annals of Physics*) bother to discuss why apples fall or why Marcel Proust's teacup, once placed on the table, remained there! At the same time, it is clear that kin selection exerts a kind of gravitational pull upon family members, often bringing them satisfyingly together after circumstances have cruelly pulled them apart. And the closer the kin, the greater the efforts at mutual assistance; moreover, the greater the satisfaction when bonding is achieved. We are thinking, for instance, of Alice Walker's *The Color Purple*, which concludes with the gratifying reunion of separated sisters Celie and Nettie (an important theme in the novel, which, by the way, received short shrift in the otherwise excellent motion picture based on the book).

We don't want to push the gravity metaphor too far, but let's stick with it a bit longer: just as earthly gravity isn't universal (objects float in space), so parental love isn't universal in nature. Some animals simply squirt sperm or pop out their eggs, after which they let success or failure take its course. But just as gravity is stronger when the distance between two objects decreases, altruism among living things is stronger when those relatives are genetically closer.

In one of his least satisfying attempts at scientific explanation, Aristotle once claimed that objects accelerate as they fall because they become increasingly "jubilant" as they approach the earth. Readers can relate to the visceral satisfaction of separated kin who are finally reunited. Thus, the reunion of separated sisters emerges as a cheery bookend theme at the beginning and end of Amy Tan's *The Joy Luck Club*. In this case, the gradually revealed story of Jing-mei Woo provides the thread that binds the book together. At the novel's onset, Jing-mei's mother, Suyuan, has just died. Gradually, we learn her story: how Suyuan, daughter of a Chinese officer in the early days of World War II, had been forced to abandon her twin daughters while fleeing the advancing Japanese. (Another version of *Sophie's Choice*, that is, an agonizing kin selection conundrum.) As we progress through the numerous personal narratives of *The Joy Luck Club*—all, incidentally, involving family ties and the sacrifices various women had been forced to make on their behalf—the novel concludes with a return to the story of Jing-mei. Fulfilling her mother's long-cherished wish, Jing-mei returns to China to meet her two half-sisters, so reluctantly abandoned by their mother decades earlier and now adult, who welcome her warmly.

A warm welcome for wayward relatives: the theme recurs regularly in the literary mind, and although one needn't be familiar with evolutionary genetics to share the feeling of satisfaction it evokes, there is additional satisfaction (if not jubilation) to be had in understanding *why*. There is nothing like a sudden—even if implausible—revelation that so-and-so is actually the son or daughter of such-and-such to neatly tie up the loose ends of a sentimental narrative. The implausibility in such cases resides in the coincidence that people who have encountered each other outside a recognized family context are—typically unknown to themselves—actually genetic relatives as well. There is nothing at all implausible, however, about the warmth and benevolence that results; it is pure biology.

For example, Esther Summerson, the admirable young heroine of Dickens's *Bleak House*, turns out to be the illegitimate daughter of the formidable Lady Dedlock. The two are joyously united and their relationship acknowledged when Esther is deathly ill with smallpox (don't worry: she recovers; Lady Dedlock, however, is unable to survive her shame). Henry Fielding's *Tom Jones*, one of the earliest English novels—published in 1749—recounts the bawdy adventures of the tale's namesake, thought to have been the bastard child of one Jenny Jones, a serving girl in the wealthy Allworthy household. Tom was accordingly brought up as a foundling alongside a spoiled and rather dastardly young man, known as Master Blifil, who in turn was the acknowledged son of Bridget Allworthy, sister of the all-worthy squire. As it happens, Tom became the apple of Allworthy's eye (Bridget having died and young Blifil being a spoiled brat), although because of his low birth, Tom always played second fiddle. When things look bleakest for our hero, however, Tom is revealed to be not the son of Jenny Jones after all—which is just as well, since he had previously enjoyed a sportive one-night stand with Ms. Jenny—but yet another son of the secretly prolific Bridget. Thus, Tom isn't merely a foundling but Squire Allworthy's altogether genetically worthy, fully related nephew after all. The squire gets to be a "perfect uncle."

What a relief!

And why is it such a relief? Quite possibly because the affirmation of a genetic bond reassures nervous males that in fact they may well be the biological progenitors they assume themselves to be . . . but sometimes aren't. (More of this in Chapter 7.)

When biologists speak of the "literature on kin selection," they are referring to various technical accounts of how the process works in nature. And there is in fact an immense literature of this sort. At the same time, it should by now be apparent that there is an even larger body of literature on kin selection within literature itself. Much of it examines the agony of how genes respond when their own best interests are thwarted. Here, for obvious reasons, the death of a child looms especially large, as in Alice Sebold's *The Lovely Bones*, Barbara Kingsolver's *The Poisonwood Bible*, or Louisa May Alcott's *Little Women*. Let's take a closer look at this perennial favorite, with its bittersweet mix of desolation and gratification.

Little Women tells of the March family: a mother and her four daughters while the husband/father is away dealing with lesser matters than attending to the demands of kin selection (i.e., fighting the Civil War). Back in Massachusetts, Beth—who eventually dies—is ever the paragon of family devotion. When her sisters complain about one thing or another, Beth points out, "We've got Father and Mother and each other." Moreover, she says this "contentedly." Discussing their future plans at a later point, Beth reveals, "Mine is to stay at home safe with Father and Mother, and help take care of the family. . . . I only wish we may all keep well and be together; nothing else."

Beth may be exceptional but not unique. The Marches are a tightly bound genetic unit, unflaggingly devoted to each other, even when this means they must stay poor as a result. At one point, when the family is particularly destitute, a rich aunt offers to adopt one of the sisters. The response is immediate: "We can't give up our girls for a dozen fortunes. Rich or poor, we will keep together and be happy in one another." If there is something saccharine about all this mutual devotion—and there is—there have also been hundreds of thousands of readers who have basked in its genuine, biological sweetness. It takes a hard heart indeed, one inured to suffering, perhaps, to remain unwarmed by the benevolent caretaking shown by the March sisters toward each other:

> Meg was Amy's confidant and monitor, and, by some strange attraction of opposites, Jo was gentle Beth's. . . . The two older

girls were a great deal to one another, but each took one of the younger into her keeping, and watched over her in her own way; "playing mother" they called it, . . . with the maternal instinct of little women.

At one point, after a teacher has been cruel to young Amy, the entire family offers support:

Mrs. March did not say much, but looked disturbed, and comforted her afflicted little daughter in her tenderest manner. Meg bathed the insulted hand with glycerine and tears; Beth felt that even her beloved kittens would fail as a balm for griefs like this; Jo wrathfully proposed that Mr. David be arrested without delay; and Hannah shook her fist at the "villain" and pounded potatoes for dinner as if she had him under her pestle.

When Jo finally receives some payment for her writing, her immediate inclination is to give it to her family. Similarly, toward the end—of the book, and of Beth's life—Jo and Beth remain devoted to each other, even at the cost of becoming nearly oblivious to others:

It was not a fashionable place, but, even among the pleasant people there, the girls made few friends, preferring to live for one another. Beth was too shy to enjoy society, and Jo too wrapped up in her to care for anyone else; so they were all in all to each other, and came and went, quite unconscious of the interest they excited in those about them, who watched with sympathetic eyes the strong sister and the feeble one, always together.

It is said that there is no such thing as a single ant. Similarly, there is no such thing as a solitary human being. Even when people aren't doling out benefits as a function of their genetic relatedness, they are looking ahead to their lineage.

Take *The Aeneid*, a classic example—in both senses of the word—of kin selection. This epic, two-thousand-year-old poem has been interpreted as many things: a history/allegory of Rome's founding, an effort by its author to ingratiate himself with the emperor Augustus, the greatest surviving example of Latin verse, a prefiguration of Christianity, even a work of divine inspiration. Whatever one's take on *The Aeneid*, Virgil's masterpiece also reflects our shared human biology.

Homer's earlier tale, *The Iliad*, concludes with the coming end of the Trojan War. According to legend, the task of founding Rome as a glorious successor to Troy fell to Aeneas, son of Venus, whose divine parentage doubtless bolstered Roman pride. (It can only be an added evolutionary plus if those shared genes are also divine.)

The Aeneid is divided into twelve books, the first six of which describe the tribulations of Aeneas and his men at sea as they recount the fall of Troy, travel to their promised land, and pause for a time in Carthage, where Aeneas famously dallies with Queen Dido. The second half of the poem details the war between Aeneas and his rival, Turnus, for control of Italy.

Throughout, Aeneas has a mission: to establish a new city, modeled after Troy, and—not coincidentally—to initiate a genetic line that will come to rule much of the known world. Aeneas has been promised that his children, grandchildren, and great-grandchildren will become great rulers. For the pre-Romans, as for most people today, posterity was a prime objective. So even though the genetics of kin selection and fitness maximization were not consciously acknowledged, the iconic image of Aeneas fleeing Troy has the renowned progenitor carrying Anchises, his father, on his shoulders, and leading his son, Ascanius, by the hand.

But what of Aeneas's sojourn with Dido in Carthage? It seems a perfect deal, at least for many contemporary men: to have the resources of an entire city and the love and affection of a beautiful queen, no strings attached. Why, then, does he leave for uncharted waters? What, furthermore, can one say about Aeneas's personal qualities if he abandons his lover, thereby causing her suicide? Such coldheartedness seems counter to Aeneas's heroic reputation, and indeed, it has divided students of the poem for ages: was he right to leave, or an inexcusable bounder? Whatever else he was doing, Aeneas was following human, biological impulses, conveniently projected onto the gods, who scold him for impeding his posterity.

Informed of Aeneas's initial dalliance with Dido, Jupiter sends his messenger Mercury to remonstrate with the errant hero. Mercury asks Aeneas, "Are you forgetful of what is your own kingdom, your own fate?" The message is clear: he must depart and settle eventually in Italy. The final blow comes with Jupiter's insinuation that were he to remain in Carthage, Aeneas would be not only reneging on his duty to found Rome but begrudging his son, and subsequent descendants-to-be, their rightful inheritance.

If Aeneas's genes could spell out their reckoning, it would go somewhat like this: although staying with Dido is great fun, you— and, more important, your genes—have bigger fish to fry. When the alternative is founding a dynasty to rule the world and establish an eternal city and way of life, a sterile dalliance with a middle-aged woman just doesn't cut it. Given the option of maximizing your inclusive fitness—the sum total of your success in projecting your genes into the future[1]—by founding a dynasty, the calculation is self-evident, even for nonbiologists. So Aeneas sets sail once again, revealing, as he departs, an intuitive comprehension of his actions. (Of course, it probably helps to know that the success of your endeavors is divinely ensured.) As he pleads for Dido's understanding, Aeneas explains, "It is not my own free will that leads me to Italy." In his conscious mind, it is the gods who drive Aeneas's actions, but deep down, it is his biological impulses that compel him to leave. Indeed, it is biology—the yet-to-be established lineage of Rome—that lies behind the gods' dictate. It would have

---

[1] Or, more accurately, the success of one's genes in projecting copies of themselves into the future.

been more than a bit unliterary, as well as ahead of his time scientifically, for Virgil to have Aeneas announce, "My genes made me do it," but in fact, for much Greek and Roman literature, the gods serve as a personification of Darwinian desires and needs, the embodiment of otherwise shadowy genetic imperatives.

Genes—or the Roman equivalent, "blood"—are at work throughout The Aeneid. Toward the end of Book VIII, Venus, Aeneas's mother, gives him a great shield upon which is engraved the fate of Rome and, thus, of Aeneas's descendants. In wonder, Aeneas surveys this future and then, not entirely understanding what he sees, he puts the shield to his shoulder, "lifting up the fame and fate of his sons' sons." The biological metaphor should be clear: Aeneas carries the burden of starting this new line. No one but he can father those descendants engraved on his shield. The pressure is enormous, yet he must shoulder the burden—risk his and his friends' lives—to provide for his progeny and ensure success for the generations to come. His genes work like a Trojan.

The ancient Romans evidently understood evolutionary genetics nearly as well as the Godfather, since additional manifestations of kin selection abound throughout The Aeneid. For example, after Aeneas and his men leave the burning Troy (and before they arrive at Carthage), they come upon an island, where our stalwart hero encounters a large dogwood that drips blood. Understandably, Aeneas is less than delighted at this discovery, especially when an unearthly moan rises from the plant. But then it beseeches, "Spare my body. . . . I am no stranger to you; I am Trojan." It turns out that Polydorus, a Trojan, was murdered here by the Greeks. Aeneas's fear is immediately quenched. Trojans—even when turned into a tree—aren't a threat to other Trojans, which also explains why Polydorus first refers to their common ancestry when speaking to Aeneas. Polydorus expects that Aeneas will honor their shared genetic tie and follow his wishes. In fact, the Trojans hold a proper funeral for poor Polydorus, an act that takes considerable time and resources, but one deemed appropriate for a blood relative.

The connection between kinship and altruism also appears notably at the beginning of Book V, after Aeneas has left Carthage. While at sea, with a great storm brewing, the helmsman, Palinurus, suggests that the Trojans take shelter on land. He recalls that "the faithful shores of Eryx, your [Aeneas's] brother, are not far off," to which Aeneas agrees: "Can any country . . . offer me more welcome harbor than the land that holds my Dardan [Trojan] friend?" When in doubt, go to a friend and fellow countryman who is, moreover, also one's brother. Sure enough, when the Trojans reach land, Aeneas's kin responds as expected to this internal genetic reckoning: "not forgetting his old parentage, he welcomes their return with joy."

Consider two initial greetings between Aeneas and Evander, a petty king in Italy. In theory, Aeneas should be wary of Evander, who came from Arcadia, a district of Greece. But Aeneas points to an ancient biological connection between their two lineages, explaining, "I was not afraid because you were a Danaan [Greek] . . . our related ancestors join me to you." He then launches into a detailed description of how the two chieftains are genetically related: the mother of Dardanus, the ancestor of the Trojans, was Electra, whose father was Atlas. Evander's father is Mercury, whose mother was Maia. And by lucky chance, Maia was Atlas's daughter. After rattling off their shared genealogy, Aeneas finally concludes that "both our races branch out of one blood. Trusting to this, I shunned ambassadors or sly approaches."

When people struggle—or simply work—on behalf of their children and other relatives, when they concern themselves with the "family name" in any of a variety of manifestations, they are also, whether consciously or not, carrying out a deep-seated biological mandate. Not uncommonly, this involves competing with other, parallel lineages.

Shakespeare's history plays give us a panorama of two such squabbling genetic lines, the Yorks and the Lancasters, each struggling to bestow benefits on their own kin, and *not* someone else's.

Beginning with *Richard II*, we witness the replacement of a York with a Lancaster (Henry IV), whose offspring—Prince Hal—and associated cohorts struggle against those affiliated with the competing lineage. On goes the tumult, through *Henry IV* parts I and II, *Henry V*, *Henry VI* parts I, II, and III, and concluding with *Richard III*, which recounts the return to power, and subsequent overthrow, of the Yorks. In evolutionary terms: York genes are supplanted by Lancaster genes, to be defeated for a time by Yorks, after which the Lancaster lineage emerges once more on top. It's a very old story, but a startlingly new perspective, in which people are proxies for their DNA.

Such tales of dynastic succession, whether from ancient Rome or late medieval England, are tales of kin selection, writ in equal parts kindness and blood. Why is blood thicker than water? Not simply because of its various formed elements (red blood cells, white blood cells, and so forth), but because it is chock-full of human genes, which stands as a literal fact no less than a metaphor. Insofar as our greatest storytellers take this thickness of blood and blood relations for granted, they are using their personal genius to build upon the thick, firm foundations of life itself.

Who better to close a chapter on family ties than William Faulkner, the master of southern gothic relations in all their wayward intricacies?

Faulkner is the great American portraitist of pain, specifically the South's anguish over its bitter legacy of slavery and the Civil War, of hatred, fear of miscegenation, the rancidity of racism, and the aftermath of lust, typically focused through multiple points of view. But nearly always, the different viewpoints and even the oft-confusing shifts of time and narrative style that Faulkner serves up are united in their concern with various members of a family: the Bundrens in *As I Lay Dying*, the Compsons in *The Sound and the Fury*, the Sutpens in *Absalom, Absalom!*

There isn't a whole lot of overt altruism to be found among Faulkner's people, but no one has ever said that this writer—probably the greatest American novelist—was straightforward in any respect! Yet it is the family ties, however pained and stressed, that literally tie his narratives together, providing coherence to a body of work that is often devilishly fractured in all other respects. Kin are the glue of Faulkner's greatest masterpieces: confused kin, incestuous kin, angry and agitated kin, rebellious and resentful kin, morose and murderous kin . . . but kin nonetheless. And so, when the Bundrens, Compsons, and Sutpens begin to fall apart, we know that things are bad indeed.

Even as Faulkner's folks undergo their familial decline, they also struggle and scratch and strain against each other, closely bound by their blood ties while also undergoing the push and pull of history, racism, and plain old-fashioned individual idiosyncrasy. They are like escaped convicts, shackled together by their genetic legacy while simultaneously yearning for personal freedom. To some extent, whereas the March family, depicted by Louisa May Alcott, shows us the glass of kin selection half full, the denizens of Faulkner's Yoknapatawpha County show us that glass half empty. After all—as we'll investigate more fully in Chapter 8—although the genetic interests of relatives are similar, they are not identical. So if you find *Little Women* too treacly and sweet, concerning itself so unrelentingly with the positive kin-selected ties among the March sisters, you might well find in Faulkner's families either a bracing corrective or a painful minefield. His people struggle against each other, yet they also suffer *together*, within the context—both bitter and better—of shared genes.

As I Lay Dying gives us the tragicomic tale of how the Bundren family endeavors to bury Addie, wife and mother, dragging her corpse through biblical travails of flood and fire, losing many of their most treasured possessions on the way, suffering pain, fracture, infection, and sexual betrayal. They finally bury Addie a full nine days after her death, by which time the coffin is followed by ominous circling vultures and resented by anyone with the ill luck to be downwind . . . except for her kinfolk.

In the only chapter narrated from Addie's viewpoint, the dying woman recalls her father's advice: "the reason for living was to get ready to stay dead a long time." Bleak words, to be sure. Yet the Darwinian reader cannot help seeing the natural wisdom in them: you won't be around for long, so do well by your progeny, the only entities that can persevere. And the same reader can't help but wonder about the role of kin: to assist in the process and, along the way, to assist each other, sometimes in spite of themselves and the awful burdens those Bundrens bear.

Another Faulkner masterpiece, *The Sound and the Fury*, recalls Macbeth's famed soliloquy, the one that describes life as "a tale told by an idiot, full of sound and fury, signifying nothing." Sure enough, *The Sound and the Fury* begins as just that: a tale told by a genuine idiot, thirty-three-year-old Benjy Compson, who is mentally retarded. But it signifies plenty.

The Compson family, once a part of the southern aristocracy, is—like so many of Faulkner's families—in serious decline. Daughter Candace ("Caddy") is loving, and even loyal in her own way, but also libidinous; her promiscuity was an intolerable burden to her brother, Quentin, who, we learn, committed suicide several decades earlier, shortly after his sister married someone other than the father of the child she was then carrying. Quentin had somehow managed to convince himself that he had committed incest with Caddy—although he hadn't—and he "identified family with his sister's membrane." Presumably, a rupture of one was therefore a rupture of the other.

Quentin Compson clearly is in bad straits, as his suicide makes clear, but strangely, it seems to have been missed by most critics that Quentin's despair—like the struggles of the Bundrens—is almost entirely family-based, the consequence of ill-satisfied kin concerns. Quentin despairs over his brother Benjy's feeblemindedness and over the Compson clan's deterioration due to the weight of pride, history, and a seemingly ineluctable biological degeneration.

Quentin is also contorted with guilt that the last of his family's land was sold to finance his Harvard education. But most of all, he is obsessed with Caddy: the loss of her honor as well as the fact that it was a stranger—and not himself—who impregnated her.

Brother Benjy, meanwhile, also harbors a persistent yearning for sister Caddy that, unlike Quentin's, isn't sexual but filial: she was the only one who offered him love and protection. With her gone, brother Benjy hangs out at the former Compson pasture (now a golf course, since being sold), and in his uncomprehending way is emotionally lacerated whenever he hears the golfers shout "Caddy!" Once more, it is a special strength of Faulkner's stories that his characters mix anguish about family members with an unspoken, but pervasive and often obsessive, involvement with them.

Jason Compson, another Compson brother, is no great shakes, either: mean-spirited, petulant, bitter, and deceitful. His most creative act is to have his feebleminded brother, Benjy, castrated, and to vow eternal bachelorhood. It is an understatement to note that the Compson family is fraying. Quentin is dead, Benjy is castrated, Jason is sworn to bachelorhood, and Caddy's offspring is illegitimate (and no less prone to promiscuity than her mother). At the same time, it is also an understatement to note that family, dilapidated as it may be, is all that the Compsons can cling to.

The most admirable figure in *The Sound and the Fury* is Dilsey, the black "mammy" who is the only consistently present, loving, and reliable character. A modern-day Greek chorus, she intones, "I've seed de first en de last. . . . I seed de beginnin, en now I sees de endin." Still, Faulkner's focus on family—even decaying and fractured families such as the Compsons—implies that the end isn't yet in sight. It was Faulkner, after all, who proudly noted that whatever else can be said about the families he described, "they endured." We, in turn, point out that the endurance of families means the endurance of genes, just as it was Faulkner who demonstrated more clearly than any other writer that the network of shared genes can be both supportive and ensnaring.

The Cinderella Syndrome

Regarding the Struggles of Stepchildren

─────────

These days, more Americans have seen the musical *Les Miz* than have read Victor Hugo's novel *Les Misérables*. In both versions, however, Jean Valjean adopts the young and vulnerable Cosette, who had been living as the fosterling of Monsieur and Madame Thénardier. In both versions, Cosette's situation under the roof of the innkeeper is less than enviable, although unlike the musical's presentation of the Thénardiers as lovable scoundrels, Hugo's account is much grimmer and their treatment of Cosette sinister and downright abusive:

> Cosette was in her usual place, seated on the cross-bar under the kitchen table near the hearth. Clad in rags, her bare feet in wooden clogs, she was knitting woolen stockings for the Thénardier children by the light of the fire. . . . Two fresh child-ish voices could be heard laughing and chattering in the next room, those of Éponine and Azelma [the Thénardiers' biological offspring; in the musical version, Azelma was deleted]. A leather strap hung from a nail in the wall near the hearth.

As to the condition of Cosette herself:

> She was thin and pale, and so small that although she was eight years old she looked no more than six. Her big eyes in their

shadowed sockets seemed almost extinguished by the many tears they had shed. Her lips were drawn in the curve of habitual suffering that is to be seen on the faces of the condemned and the incurably sick. Her hands . . . were smothered with chilblains . . . she was always shivering. . . . Her clothes were a collection of rags which would have been lamentable in summer and in winter were disgraceful—torn garments of cotton, with no wool anywhere. Here and there her skin was visible, and her many bruises bore witness to her mistress's [Madame Thénardier's] attentions. Her bare legs were rough and red, and the hollow between her shoulder-blades was pathetic.

By contrast, note the state of the Thénardiers' own children, Éponine and Azelma:

They were two very pretty little girls with a look of the town rather than of the country, very charming, the one with glossy chestnut curls and the other with long dark plaits down her back, both of them lively and plump and clean with a glow of freshness and health that was pleasant to see. They were warmly clad but with a maternal skill which ensured that the thickness of the materials did not detract from their elegance. Winter was provided for but spring was not forgotten. They brought brightness with them, and they entered like reigning beauties. There was assurance in their looks and gaiety, and in the noise they made.

Next, jump ahead from a nineteenth-century French bestseller to a publishing phenomenon of the late twentieth and early twenty-first century: Harry Potter. Any child can confirm that Harry's early days, growing up as a stepchild within—but not part of—the Dursley family, were more like the experience of Cosette than like that of Éponine and Azelma. The Dursleys' own son, the despicable Dudley, is ugly, stupid, and spoiled beyond comprehension. Photos of the young brat adorn the

Dursley walls, which reveal "no sign at all that another boy lived in the house, too."[1]

Contrasted with the overindulged Dudley, Harry never gets enough to eat, never gets new clothes, never gets toys, never gets a birthday celebration, sleeps in a cupboard under the stairs, does menial labor, and is yelled at constantly. Moreover—and for many children, perhaps the unkindest cut of all—Dudley gets to have birthday parties and presents, but not Harry:

> Harry got slowly out of bed and started looking for socks. He found a pair under his bed and, after pulling a spider off one of them, put them on. Harry was used to spiders, because the cupboard under the stairs was full of them, and that was where he slept. When he was dressed he went down the hall into the kitchen. The table was almost hidden beneath all Dudley's birthday presents. . . . Every year on Dudley's birthday, his parents took him and a friend out for the day, to adventure parks, hamburger restaurants, or the movies. Every year, Harry was left behind with Mrs. Figg, a mad old lady who lived two streets away. Harry hated it there. The whole house smelled of cabbage and Mrs. Figg made him look at photographs of all the cats she'd ever owned.

Harry is also expected to wear the equivalent of Cosette's rags:

> One day in July, Aunt Petunia took Dudley to London to buy his Smeltings uniform, leaving Harry at Mrs. Figg's. . . . There was a horrible smell in the kitchen the next morning when Harry went in for breakfast. It seemed to be coming from a large metal tub in the sink. He went to have a look. The tub was full of what looked like dirty rags swimming in gray water. "What's this?" he asked Aunt Petunia. Her lips tightened as they always did if he dared to

---

[1] Granted, Harry is nephew to Dudley's mother, so presumably a bit of kin selection ameliorates his rotten treatment, but our point is that compared to the overindulged Dudley, Harry is a stepchild indeed.

ask a question. "Your new school uniform . . . I'm dyeing some of Dudley's old things gray for you."

Shades of Cinderella? Indeed. But Cosette and Harry aren't the only cases of abused stepchildren depicted via story. In fact, Cinderella herself, unlike Mickey Mouse or Donald Duck, did not spring fully formed out of the Walt Disney studios, or even out of the fertile forehead of Uncle Walt himself. She exists, in various forms, in many cultures, but always with the same recognizable tale of woe.

In an ancient Japanese folktale, a kind, gentle, and honest young lady named Benizara was much put-upon by her stepmother. Benizara was so virtuous (and also—a lovely Japanese touch—so good at extemporizing a poem) that she wins the heart of a nobleman. Her wicked stepmother, however, tries to substitute her own daughter—Cinderella's (sorry, Benizara's) stepsister, Kakezara—at the wedding, but stepmom screws up and Kakezara ends up dead. Which is probably just as well for Benizara: there is a broad-leaved and fiendishly sharp-spined plant, similar to the devil's club found in the Pacific Northwest, known in Japan as *mamako-no-shiri-nugui*, or stepchild's bottom wiper!

And that's not all. According to evolutionary psychologists Martin Daly and Margo Wilson, the Russian folktale of Baba Yaga begins as follows:

> Once upon a time there was an old couple. The husband lost his wife and married again. But he had a daughter by his first mar-riage, a young girl, and she found no favour in the eyes of her evil stepmother, who used to beat her, and consider how she could get her killed outright.[2]

The stepmother urges the girl to go to her "aunt," sister of the stepmother, who is a witch, a cannibal, and, moreover, not very

---

[2] M. Daly and M. Wilson, *The Truth About Cinderella* (New Haven: Yale University Press, 1998).

nice, but our Slavic Cindy cleverly consults her real aunt first and thereby learns how to elude the snare. "As soon as her father heard all about it, he became wroth with his wife, and shot her. But he and his daughter lived on and flourished." Living and flourishing is precisely what Cinderellas around the world have had a hard time doing, especially if their welfare is left to the not-so-tender mercies of stepparents. In this regard, the Grimm brothers' fairy tales are especially grim as well as consistent: in addition to Cindy herself, there is also Snow White, Hansel and Gretel, and The Juniper Tree. Whenever a deserving child is mistreated, cherchez la stepmother.

From Cinderella and her fellow sufferers to Les Misérables and Harry Potter, and lots more in between: what in Darwin's name is going on?

Biology, that's what. Evolution frowns on taking care of someone else's kids because there is no payoff in offering parental assistance when you aren't really the parent, when the genes thus promoted are not your own. From the dawn of human existence to the present day, people have lived in an amazing variety of different social systems, from capitalism to communism, feudalism to democracy, hunting and gathering to high-tech lifestyles of the rich, famous, and forgettable; we can be industrialists, serfs, monarchists, democrats, subsistence farmers, computer programmers. Yet for all this diversity, there is not now and has never been a single society in which people routinely give up their reproduction to someone else. Sure, we may delegate child care (typically for pay), but actual reproduction? No way. People indulge in all sorts of specialization and division of labor, but propelling genes into the future is something nearly everyone chooses to do for him- or herself. As we saw in Chapter 6, this can be achieved by promoting the success of relatives; the most obvious and direct route to genetic advancement, however, is to package your own genes in your own child. (More accurately, for genes to package copies of themselves in bodies known as children.)

The clear-cut biological significance of reproduction is why altruism is so interesting: when we first encounter it, altruism appears to go against this most basic principle of the living world. Being a parent, by contrast, is so obvious, so appropriate, so biologically de rigueur that up until recently it has largely escaped the scrutiny it deserves.

The obviousness of reproduction and parenting derives from the simple fact that it is the most straightforward way for genes to enhance their evolutionary success. If you want something done right, goes the saying, do it yourself. And nowhere in the living world is this more true than when it comes to parenting. Whenever offspring are being produced, fed, trained, kept warm or cool or wet or dry, protected from enemies or introduced to friends, something fundamental is going on: genes are nurturing copies of themselves.

"He that hath wife and children," wrote the sixteenth-century English philosopher Francis Bacon, "hath given hostages to fortune, for they are impediments to great enterprise, either of virtue or mischief." Bacon, one of the great architects of modern science and philosophy, lived too early to understand this important finding of evolution: children may be impediments to some things, but they are also passports to the most pressing enterprise—indeed, the only persistent enterprise—of life itself. In this respect, all living things are hostages, not to fortune but to natural selection.

"We had lots of kids, and trouble and pain," goes the folk song "Kisses Sweeter than Wine," "but oh Lord, we'd do it again." Why would they do it again, given that having kids involves so much trouble and pain? And why did they do it the first time? Hint: not simply because of those oh-so-sweet kisses. In fact, natural selection has only contrived to make love and sex and kisses sweet in the first place because this is how biology gets us to do it.

Try asking spiders of the African species *Stegodyphus mimosarum* about trouble and pain. Comfortably housed within silken nest chambers woven by the mother just for this purpose, baby spider-lings perch on her arachnid abdomen and cheerfully fill their own bellies with their mother's flesh (or whatever passes for flesh among spiders), eventually killing her in the process.

Why does Mommy *Stegodyphus mimosarum* permit such an out-rage? Because they are so irresistibly cute. Or because it feels so wonderfully good, perhaps like having an itch scratched. Or maybe because she simply feels no alternative, like the mammalian need to breathe or circulate blood. Whatever the immediate mecha-nism, the end result is that her spiderlings are sent off into the world with a full stomach at their mother's expense. This is some-what more extreme than the suburban parent making sure Junior goes off to school with a freshly made peanut butter and jelly sandwich, but at the most basic biological level it's not altogether different.

By contrast, picture a sleek, plump female elephant seal, mater-nally nursing her baby. Along comes an interloper, someone else's pup, who attempts to sneak-suckle. This youngster already has a mother, of course, but is trying to cadge an additional meal. (In the immortal phrase coined by elephant seal guru Burney Le Boeuf of the University of California at Santa Cruz, it is seeking to become a "double mother sucker.") What happens? Sometimes the sneaky little tyke succeeds, but most often the seal cow is outraged, and sometimes murderously so: she bites the little thief, occasionally killing him for his larcenous presumption.

Undoubtedly much of the difference between a self-sacrificial spider mom and a milk-withholding momma seal can be chalked up to differences between spiders and elephant seals. But you can still be sure that Spiderwoman wouldn't cheerfully offer herself for dinner to the arachnid equivalent of a swarm of double mother suckers, hatchlings of some *other* reproducing female, just as even the most puppicidal pinniped behaves quite benevolently toward her own offspring. Parenthood matters. Parental benevolence in the natural world is not broadcast indiscriminately.

And so we come to a tiny, abundant, and intriguing animal, the Mexican free-tailed bat. Although each female produces only one young at a time, these creatures congregate in immense gatherings, hundreds of thousands and more in a single cave. While the adults are out cruising for insects, the young crowd together in "crèches," with population densities as high as two pups per square inch. Utter chaos appears to reign, especially when females return from foraging and the nurslings swarm all over them. Consistent with good-of-the-species thinking, instead of its good-of-the-gene alternative, batologists had long thought that the babies were fed indiscriminately and rather communistically: from each lactating female according to her ability, to each according to his or her need. Looking at the melee, it is difficult even for a modern biologist to imagine how parent-offspring pairs are ever sorted out.

But when Gary F. McCracken of the University of Tennessee studied these nurseries, he found that female Mexican free-tailed bats were not at all free when it came to dispensing milk. Mothers recognized their own young 83 percent of the time, apparently by sound and smell. Whenever a female suckled young not her own, it was evidently a result of error (on her part) and/or milk stealing (by the little batling). Stepbats need not apply.

Next, a bird, specifically the mountain bluebird. Birds are especially interesting when it comes to parental care because unlike mammals, whose females are uniquely adapted to nurse their offspring, both males and females contribute about equally in the avian world. Rutgers University ornithologist Harry Power asked whether male mountain bluebirds who were manipulated into being stepparents rather than biological parents would behave differently as a result. In one experiment, mountain bluebirds were provided with nest boxes designed to be especially attractive to them. Bluebird pairs quickly moved in and started families, after which the males were removed, leaving the females single parents. They did not remain single for long, however. In what passes for bluebird society, they were "wealthy widows," since good nests are hard to find, and, thanks to the researchers, each of these females now owned a valuable piece of property. They and their dependent offspring were soon joined by fortune-hunting males. Significantly, these new arrivals—stepfathers of the nestlings—did *not* participate in feeding the youngsters, and only one in twenty-five gave alarm calls in response to possible predators (biological fathers sound an alarm nearly 100 percent of the time).

Similar examples can be multiplied almost indefinitely. Indeed, it is now a commonplace among biologists that parenting in any species means caring for one's own children. Not for other living things the blithe assumption that parental solicitude is a mere social convention, as in this dialog from Bernard Malamud's novel *The Fixer* between Yakov Bok, falsely imprisoned in a Russian jail, and his estranged wife:

> "I've come to say I've given birth to a child."
>
> "So what do you want from me? . . ."
>
> ". . . it might make things easier if you wouldn't mind saying you are my son's father. . . ."
>
> "Who's the father . . . ?"
>
> ". . . He came, he went, I forgot him. . . . Whoever acts the father is the father."

It may seem perverse to cite the above, which points *away* from biology; after all, our basic argument is that most literature makes sense in the light of biology rather than contradicting it. But those few cases that go against evolutionary wisdom stand out because of their rarity. When it comes to parenting, the truth is more often the precise opposite of Mrs. Bok's wishful thinking: whoever *is* the father acts the father. (The same applies, of course, to mothers, although they are less likely to be deceived.)

Whether bat or bird, seal or spider, living things reproduce because this is the major way their genes propagate themselves. It is also the major reason for love, including love of adults for each other and of parents for children. And it goes a long way toward telling us why stepparenting is so often a conflicted and difficult business.

In the extreme case, stepparenting among animals leads to outright murder. The paradigmatic example was first reported by anthropologist Sarah Blaffer Hrdy, who studied langur monkeys in India. In this harem-forming species, one male monopolizes a number of adult females and breeds with them, while a corresponding bunch of langur bachelors languish resentfully in the background. Every now and then, a revolution takes place in langurland, whereupon the dominant male is ousted and one of the bachelors takes over the troop of females. In such cases, the newly ascendant male is likely to methodically pursue and kill the nursing infants, who are offspring of the previous male. Without suckling infants, the newly bereaved mothers stop lactating, their ovaries begin cycling once again, and they mate with their offspring's murderer. So, because nursing females are less likely to ovulate, by killing their infants a male not only eliminates the offspring of his predecessor but also improves his own reproductive prospects.

When Hrdy first presented her findings in the late 1970s, anthropologists and even some biologists were disbelieving. How could a behavior that is so hurtful to the species have evolved? (At that time, many scientists were still in thrall to species-level benefit.) It must be some sort of pathology, they insisted, or perhaps a result of overcrowding or malnutrition, or maybe just some weird anecdotal rarity. But subsequent decades have supported Hrdy's interpretation; moreover, a similar pattern of infanticide on the part of animal stepparents has been documented for lions, chimpanzees, and various species of rodents, almost wherever biologists have looked. The evidence is overwhelming: infanticide—and nearly always by nonbiological "parents"—is distressingly commonplace. It may well be downright terrible for the species, but so long as it's a net plus for the infanticidal individual, there it is.

What about people?

The Canadian husband-and-wife team of psychologists Martin Daly and Margo Wilson, professors at McMaster University in Hamilton, Ontario, have pioneered the evolutionary underpinnings of stepparenting as a risk factor for child neglect and abuse. Since child rearing is difficult, costly, and prolonged in our species, they reasoned, natural selection is unlikely to have produced indiscriminate parenting. (If even the Mexican free-tailed bat can be fussy about dispensing parental care, so can human beings.) In fact, parental feelings are expected to vary with the evolutionary interest that children hold for the adults in question: the greater the genetic return, the greater the inclination to invest time, energy, and love. And similarly, the greater the disinclination to mistreat them.

To summarize two decades of research on human beings: youngsters living with a stepparent are from forty to sixty times more at risk of neglect, abuse, and infanticide than are comparable children living with their biological parents. This is true even when other factors such as income, education level, and ethnicity are taken into account. Like it or not—and, given the high frequency of stepparenting and blended families, many people don't like it—the step relationship is by far the highest predictor of a child's maltreatment.

These numbers are staggering, and once again we must ask, what is going on here? Does this mean that all stepparents are abusive? Of course not. Neither does it imply that biological parents are necessarily doting. Nor that biologists have it in for nonbiological parents (one of the present authors, as a stepparent as well as a biologist, can testify on both accounts). But these startling findings do mean that in daily life, stepfamilies—because they are out of step with biology—are liable to be stressful places, demanding the best within us and sometimes bringing out the worst, leading in some cases to reduced caretaking, increased intolerance, and even, in extreme situations, violence. This is where langurs, lions, and chimpanzees come in, contributing to an understanding of child abuse, neglect, and even, on occasion, murder in human beings.

Despite its many rewards, child rearing can, after all, be stressful, even for the most well-balanced and devoted parents. It is understandable—if not pardonable—that without genetic connection to ameliorate the rough edges, there would be a lower threshold for adults' ability to tolerate infant crying, children's interrupting, and the normal demands of even the most well-behaved youngsters, not to mention the predictable requirements of food, clothing, education, and so forth, which often can only be satisfied at some cost to the stepparents' own biological children. Moreover, for unstable adults already teetering on the edge of self-control, stepparenthood could well make a tragic difference.

Earlier, we visited with *Jane Eyre* as an example of the girl-chooses-boy theme among Gothic novels. This romantic classic also tells several stepparent tales. Start with Jane and her aunt Reed, who, it must be noted, is her aunt through marriage and not blood. As the story begins, Mrs. Reed has assumed responsibility for the orphaned Jane, though not out of any altruistic feelings on her part, but rather because while on his deathbed, her husband (who had a genetic tie to Jane) admonished her to do so. Jane describes the case as follows:

> I knew that he was my own uncle—my mother's brother—that he had taken me when a parentless infant to his house; and that in his last moments he had required a promise of Mrs. Reed that she would rear and maintain me as one of her own children. Mrs. Reed probably considered she had kept this promise; and so she had, I dare say, as well as her nature would permit her; but how could she really like an interloper not of her race [i.e., a nonrelative], and unconnected with her, after her husband's death, by any tie? It must have been irksome to find herself bound by a hard-wrung pledge to stand in the stead of a parent to a strange child she could not love, and to see an uncongenial alien permanently intruded on her own family group.

Although Aunt Reed kept her promise, she did so only half-heartedly, after having first "entreated him [her husband, Jane's uncle] rather to put it [Jane] out to nurse and pay for its maintenance." And so Aunt Reed proceeds to abuse our young heroine—psychologically for the most part—including a famous incident in which the girl is locked in the spooky room where her uncle died. Jane is then sent to an even more abusive school (in loco stepparentis) before eventually encountering Mr. Rochester of Thornfield Hall. It is not that Aunt Reed didn't understand her role as stepparent or that, as sociologists like to claim these days, she lacked role models, but rather that she didn't want to lavish the same solicitude on Jane that she made readily available to her own offspring.

As Daly and Wilson point out,

> There is a commonsense alternative hypothesis about why some "roles" seem easy and "well-defined" while others are difficult and "ambiguous." It is simply that the former match our inclinations while the latter defy them. Stepparents do not find their roles less satisfying and more conflictual than natural parents because they don't *know* what they are supposed to do. Their problem is that they don't *want* to do what they feel obliged to do, namely to make a substantial investment of "parental" effort without receiving the usual emotional rewards. The "ambiguity" of the stepparent's situation does not reside in society's failure to define his role, but in genuine conflicts of interest within the stepfamily.

To this, we add that those "genuine conflicts of interest" are genuine because they are, at heart, genetic.

Consider now another classic English novel, *Oliver Twist*. We first meet young Oliver when he is a famously abused orphan who outrages the establishment by asking, among other things, for more porridge. As punishment, he is sent away to an abusive stepfamily of undertakers and thence circuitously to the underworld of London crime, where his newest "stepfather," Fagin, is interested only in how Oliver's small, deft hands can contribute, by picking pockets, to Fagin's own material advancement. Obviously, Oliver isn't nurtured in this environment; rather, he is provided with just enough food to keep him alive and functioning, albeit unwillingly, as a participant in the gang's nefarious activities. Things end well, however, for our young waif precisely when his waifhood ends, that is, when he connects with his own biological relatives, the Maylies, and with a nonbiological protector, Mr. Brownlow (who eventually recognizes Oliver as the offspring of a dear friend's child).

Oliver Twist is unusual not only in turning out well but also in ending up remarkably sweet and good-natured, given his very difficult upbringing. More commonly, the stepchild—often a bastard child as well—is portrayed as angry at being dispossessed and ill-treated. To some extent, bastardy has thus been equated with nastiness (just think of the epithet), which provides a way for the biologically unsophisticated to understand what might otherwise be inexplicable. Consider Mordred, bastard child of King Arthur, who is ultimately responsible for nothing less than the fall of Camelot; the murderous Smerdyakov in Dostoyevsky's *The Brothers Karamazov*; or the violent and vengeful Edmund, the illegitimate child of Gloucester in *King Lear*, whose machinations result in his father's blinding and the deaths of all three of Lear's (biological) daughters.

As we've seen already, the evolutionarily optimal male strategy is to make as many children as possible but to invest preferentially only in those that are legitimate. Nonetheless, there are many stories of bastard children inheriting large amounts of money upon the father's death, especially if there are no natural children with a competing claim. (Despite his illegitimacy, Pierre Bezuhov, one of the central characters of Tolstoy's *War and Peace*, inherits a large fortune when his father dies; notably, Pierre has no siblings.)

Denial of the bastard seems counter-Darwinian, since parents should presumably want the best for their children, regardless of whether those children derived from a legally consecrated marriage.

But a parent's solicitude toward his or her illegitimate children is often complicated by conflict with a current spouse, particularly if other children have been produced legitimately (Smerdyakov had to deal with the three acknowledged brothers Karamazov, Dmitri, Ivan, and Alyosha, just as Edmund competed with Edgar, Gloucester's acknowledged son). Moreover, it is typically a *father* who has to deal with a bastard child, and fathers, as we have already seen, cannot be entirely confident that they are in fact fathers.

Uncertain paternity cuts both ways, from father to child and back again. Thus, referring to Odysseus, young Telemachus remonstrates, "My mother saith he is my father. Yet for myself I know it not. For no man knoweth who hath begotten him." If this is true of the offspring of the famously faithful Penelope, how much more true must it be of everyone else!

Whether coincidentally or not, this issue—who is whose issue?—reappears in James Joyce's *Ulysses*, developed by none other than Stephen Dedalus, who in fact represents a twentieth-century Telemachus, (re)united at the novel's end with his pseudofather, Leopold Bloom. In the "Scylla and Charybdis" chapter of *Ulysses*, Stephen expounds on the unknowability of fatherhood:

> Fatherhood, in the sense of conscious begetting, is unknown to man. It is a mystical estate, an apostolic succession, from only begetter to only begotten. On that mystery and not on the madonna which the cunning Italian intellect flung to the mob of Europe the church is founded and founded irremovably because founded, like the world, macro- and microcosm, upon the void. Upon the incertitude, upon unlikelihood. Amor matris, subjective and objective genitive, may be the only true thing in life. Paternity may be a legal fiction. Who is the father of any son that any son should love him or he any son?

As Stephen notes, Telemachus's lament also works the other way: no man knoweth for certain whom he hath begot. And this, in turn, leads to a painful but prominent literary theme. In his mordant play *The Father*, the Swedish writer August Strindberg describes the dilemma of a husband tormented by whether he is the biological parent of his child:

> I know of nothing so ludicrous as to see a father talking about his children. "My wife's children," he should say. Did you never feel the falseness of your position, had you never had any pinpricks of doubt?

It is a good guess that by and large, fathers also find it easier to let go than do mothers, not necessarily because of any conscious doubt, but rather because of a biologically inspired diminution in confidence. In "Walking Away," the poet C. Day Lewis gives a predictably male view when he describes his eldest son going away to school, and the poet's awareness of "how selfhood begins with a walking away / And love is proved in the letting go."

Letting go is especially likely when there is doubt as to paternity. And not surprisingly, the phenomenon is cross-cultural. Take, for example, *The Tale of Genji*, written one thousand years ago by Murasaki Shikibu, and believed to be the first novel written in Asia, perhaps the first of all time. It tells of the picaresque adventures of Genji, offspring of the emperor and a concubine. Genji is exceptionally handsome and aristocratic, possessing a definite "way with the ladies." He has many lovers, including his own stepmother, Fujitsubo, with whom he has a child. He is also cuckolded: his favorite wife, the Third Princess, dawdles with Kashiwagi, the son of Genji's longtime political and sexual competitor, To no Chujo. Accordingly, when the Third Princess gave birth, Genji was deeply distressed: "How vast and unconditional his joy would be, he thought, were it not for his doubts about the child." And later, "It would not be easy to guard the secret [that the Third Princess had dallied with Kashiwagi] if the resemblance to the father was strong." The princess eventually abdicates and becomes a nun.

Anyone who still needs persuading that paternity matters should take a good look at Thomas Hardy's dark masterpiece *The Mayor of*

Casterbridge. A significant part of this novel examines how the relationship of the protagonist, Michael Henchard, with his purported daughter, Elizabeth-Jane, changes as Michael realizes that the girl he had thought to be his child turns out to be someone else's.

The book begins with a much younger Michael arriving at the English equivalent of a county fair with his wife and baby daughter, Elizabeth-Jane; in a drunken stupor, he sells them both—for five guineas—to a passing sailor. Eighteen years later, after Michael Henchard has reformed and made a name for himself as the highly respected mayor of Casterbridge, who should show up but his long-abandoned wife, with Elizabeth-Jane in tow, claiming that the sailor died and left them penniless. Elizabeth-Jane believes the sailor to be her father, and Henchard, seeking to avoid the obloquy of owning up to his despicable behavior eighteen years previously, but also wanting to make good on his obligations, suggests that he (re)marry his wife, after which their daughter will consider Michael her stepfather.

Henchard describes the scheme as follows:

> "I meet you, court you, and marry you, Elizabeth-Jane coming to my house as my step-daughter. . . . the secret would be yours and mine only; and I should have the pleasure of seeing my own only child under my roof, as well as my wife."

He believes Elizabeth-Jane to be the daughter he sold eighteen years ago, an assumption that is shown by his benevolent concern for her upbringing: "The freedom [Elizabeth-Jane] experienced, the indulgence with which she was treated, went beyond her expectations." Although hints abound, it never crosses Michael Henchard's conscious mind that this child might not be his after all, even when the answer is literally staring him in the face, as in the following discourse:

> "I thought Elizabeth-Jane's hair promised to be black when she was a baby?" [Henchard] said to his wife.

"It did; but they alter so," replied Susan.

"Their hair gets darker, I know—but I wasn't aware it light-ened ever?"

"O yes." And the same uneasy expression came out on her face, to which the future held the key.

Later, after his wife's death, Michael discovers that Elizabeth-Jane is not in fact his child, the original Elizabeth-Jane having died soon after the sale to the sailor and been replaced by the present girl, who is of course the sailor's daughter. When this becomes clear, Michael rethinks the troubling matter of Elizabeth-Jane's lack of resemblance to himself, acknowledging that he and his now-identified stepdaughter look nothing alike:

> He steadfastly regarded her features. . . . They were fair: his were dark. . . . In the present statuesque repose of the young girl's countenance [the sailor] Richard Newson's was unmistakably re-flected.

And how did Michael Henchard respond to that newly per-ceived reflection? As follows: "He could not endure the sight of her."

From this point on, Michael changes from being a secretly dot-ing daddy merely pretending to be a stepparent into a predictable, withholding, and resentful stepparent trying unsuccessfully to act like a parent. Ironically, just as he tells Elizabeth-Jane the truth of their situation, Henchard notes that it has changed drastically: "The mockery was, that he should have no sooner taught a girl to claim the shelter of his paternity than he discovered her to have no kinship with him." And so he disowns and rejects her, further com-pounding the tragedy all around.

T. S. Eliot once noted that there was so much in Shakespeare—and, we would add, so much in Shakespearean criticism—that the best one could hope for is to be wrong about Shakespeare in a new way. Well, here is a new way to look at *Hamlet:* as a stepparent story. It is clear that for all Hamlet's complexity and depth, he wasn't happy with Claudius, his stepfather, even before he got the unwelcome news from the ghost of dear old departed Dad that Claudius had done him in. In any event, Hamlet isn't shy about berating his mother, Gertrude, for her "o'er-hasty marriage" and urging her to refrain from sex with her new husband. Forget about Freud: should Gertrude become pregnant, this would further cloud Hamlet's future and give him a likely unwelcome competitor.

There is a French proverb, dating from about the same era as Shakespeare, that speaks not only to Hamlet and his uncle/stepfather Claudius, but also to many others:

> *The mother of babes who decides to rewed*
> *Has taken their enemy into her bed.*

David Copperfield would have to agree. As Dickens presents it, young David is a male version of Cinderella. Cindy's life went rapidly downhill after her father remarried. David's took a dive when his father died and his mother took Mr. Murdstone—who quickly revealed himself to be David's enemy—into her bed. Earlier, when courting Clara Copperfield, Murdstone had gone out of his way to seem well disposed toward young David, bringing him gifts, taking him for pleasant daily outings, and allowing him to ride on the saddle in front of him, never seeming the least offended when David was predictably cold toward him.

Almost immediately after the marriage, however, Mr. Murdstone begins to reveal himself for who he really is: cold, domineering, and cruelly indifferent to David's needs. He moves quickly to separate David from Clara, first emotionally, by, as David puts it, "preventing [Clara] from ever being alone with [him] or talking lovingly to [him]," and then physically, by packing him off to boarding school. By the time David Copperfield returns for the holidays, his little brother—Murdstone's child by Clara—has been born. Mr. Murdstone is not literally a male langur monkey, killing an infant so as to breed with the mother, but the parallel is strikingly close.

Once Clara dies, Murdstone's true relationship with David becomes clear: young David is sent to work in Murdstone's factory at age ten, where he is ill-fed and overworked. When the boy runs away to the safety of his great-aunt, Murdstone is quick to give up on him altogether. Here is that great-aunt, Betsy Trotwood, confronting the stepfather with his behavior:

"Do you think I don't know what kind of life you must have led that poor unworldly, misdirected baby? First you come along smirking and making great eyes at her [David's mother], all soft and silky. . . . Yes, you worshipped her. You doted on her boy, too. You would be another father to him. And the poor deluded innocent believed you. She had never seen such a man. Yes, you were all to live together in a garden of roses. . . . And when you had caught the poor little fool you must begin to make a caged bird of her and try to teach her your own ugly notes. So it was, Mr. Murdstone, that you eventually succeeded in breaking her heart. And if that were not enough you have done your best to break her boy's spirit."

For her part, Clara Copperfield had been clueless and perhaps somewhat desperate, not unlike Hamlet's mother, Gertrude, who evidently assumed that Claudius would treat young Hamlet in a benevolent, fully paternal manner and that the prince, similarly, would promptly accept Claudius as a replacement father. Instead, Hamlet persisted in his view that something was rotten, not only in Denmark generally but in his personal situation as well. But at least Hamlet was a young adult; David Copperfield, by contrast, is langurlike in his helplessness.

Parentless waifs tug at our heartstrings; we cannot help recognizing their need for family. But Oliver Twist and David

Copperfield are exceptions, at least in retaining their own good humor: stories often suggest that once they become someone's stepchild, waifs who have been sinned against ere long become sinners in their own right. Joe Christmas, protagonist of William Faulkner's *Light in August*, starts life in an orphanage, where he has the bad fortune to accidentally eavesdrop on a sexual encounter involving two of the adult employees. In punishment—and fear that he will expose the couple—he is sent to become the foster child of Mr. McEachern, a willful, violent Christian fundamentalist who regularly beats young Joe for stubbornly refusing to memorize the catechism. As his life unfolds, Joe grows up tough, resentful, and violent, eventually killing a woman who sought to befriend and help him (and who also became his lover). Joe Christmas—everybody's stepchild, forced by a violent and rejecting world to become violent and rejecting in turn—is soon captured, castrated, and killed by a pursuing mob of vigilantes.

A century earlier, Heathcliff grew up the disfavored stepchild of *Wuthering Heights*. Wild (as bespeaks the gusty implications of the book's title), abandoned (in both senses of the word), and fiercely in love with Catherine Earnshaw but deemed unsuitable to marry her because of his unknown background and lowly step status, Heathcliff wreaks revenge on both the Earnshaw family, which denied his legitimacy, and the Linton family, into which his beloved Catherine eventually marries. Heathcliff's retaliation involves destroying the offspring of both lineages in a passionate quest to compensate, somehow, for his intolerable outsider situation.

The stepchild or fosterling as violent outsider isn't a characterization limited to Joe Christmas and Heathcliff. In fact, it is difficult to identify an imaginative literary depiction of the stepchild as a happy, well-adjusted, wholly accepted member of either family or society.

When it comes to stepparenting stories, one need only consult the immense, six-volume compendium *Motif-Index of Folk Literature* to see just how universal is the idea of the wicked stepparent. Tale after tale features evil stepmothers, but nary a case in which stepmothers are benevolent or well-intended. As for stepfathers, the *Motif-Index* identifies two categories: "cruel" and "lustful."

It is fruitless to deny the ubiquity of the negative stepparent image. But maybe the very myth itself of the malevolent stepparent is the cause of the problem. Maybe it is because stepparents are widely seen to be so difficult that they in fact are difficult. However, this begs the question of *why* stepparents are so widely viewed in a negative light. Most likely, stepparents are perceived as potentially dangerous and liable to treat their stepchildren badly because—all over the world—they occasionally do so, and, more to the point, they are on average more prone to do so than are biological parents. Whereas most stepparents are decent, humane, and loving—and some genetic parents are truly despicable—the fact remains that on balance, the former are significantly more liable to be "bad parents" than are their genetically connected counterparts. And nonbiological children—although perhaps less dramatically than Heathcliff, Joe Christmas, or even Cinderella—are liable to suffer the consequences.

Parenting is an extreme form of kin selection. To be a parent is to participate in a one-directional flow of benefits, a disparity that is necessitated by the fact that parents are so much older, larger, wiser, and more powerful than their children. Parents are biologically (and therefore psychologically as well as socially) expected to conform to these expectations and to accept an asymmetrical relationship with their children. With some notable exceptions, discussed in the next chapter, it is an arrangement that works well for all parties.

Stepparents, on the other hand, often struggle to mimic genetic parents, in their behavior as well as their feelings. And yet it is a difficult undertaking, even for the best of them. The most well-intended advice, anecdotes, and pop psychology—even when utterly non-Darwinian—acknowledge that stepparenting is difficult, as is being a stepchild. By contrast with its widespread if unintentional depiction in literature, the evolutionary biology of stepparenting has not made impressive headway into the traditional wisdom of social science, which, as we have said, still tends to attribute its near-universal stress to problems of "role definition" and "social expectations." It is far more likely that the fault lies not in our stars and not in society but in ourselves, that is, in the fact that we are biological creatures, carrying on a long-standing tradition by which genes struggle with other genes and favor copies of themselves.

Such struggles aren't hopeless, however. Indeed, we would argue that they are much of what being human is all about. Think of the scene in *The African Queen* (based, incidentally, on a novel by C. S. Forester), in which Katharine Hepburn's character sternly points out to a grimy, boozy Humphrey Bogart, who has sought to excuse his alcoholic excess with the claim that somehow his nature made him do it: "Nature, Mr. Allnut, is what we are put on earth to rise above."

When it comes to rising above nature, what about adoption? Doesn't its success show the inadequacy of biology as an interpreter of family function and dysfunction? Quite the opposite. What adoption really demonstrates is how easy it is, in certain cases, for us to fool Mother Nature.

To understand how this can happen, we must first emphasize an important rule: natural selection doesn't do more than is needed. This is because it takes "selection pressure" to create something out of nothing. No pressure, no adaptation. This is relevant in the case of adoption because would-be adopting parents must somehow fool themselves into feeling that their adopted child is "theirs." As it happens, evolution strongly promotes mechanisms that enable parents to recognize their children—and which therefore work against adoption—but only if there is a threat that otherwise parents would waste their precious care on someone else's offspring.

In other words, parent-offspring recognition should be acutely developed when mixups are likely and absent when they aren't. People aren't biologically prone to such errors, and so we lack mechanisms to prevent them. For a good case of this notion, Mike Beecher of the University of Washington turned to a pair of bird species, the bank swallow and the rough-winged swallow. Bank swallows nest in burrows dug in clay banks and are colonial, with many nesting pairs closely associated. Hence, breeding members of this species run the risk that their parental care might be misdirected to someone else's young. Rough-winged swallows, on the other hand, although closely related, are essentially solitary, each pair maintaining a nest that is isolated from other rough-winged swallows. So there is very little chance that a rough-wing will accidentally proffer food to nestlings other than its own.

Beecher found that the vocalizations of young bank swallows (the colonial species, vulnerable to mixups) are much more distinctive than those of rough-wings (the go-it-alone guys). Having a unique vocal fingerprint makes it easy for bank swallow parents to learn the distinctive vocal traits of their offspring. As a result, when bank swallow youngsters land at the wrong nest—which they often do, since in this colonial species, nest entrances are typically close to each other—the adults shoo them away, reserving food and protection for their own offspring. By contrast, rough-winged swallows can afford to be undiscriminating, since under normal conditions they run no risk of being importuned by strangers. There are no rough-winged swallow equivalents of double mother suckers, as among elephant seals. Interestingly, rough-winged swallows can be fooled by an experimenter, duped to accept strangers introduced into their nest, something that bank swallows never do. Rough-wing parents will even feed bank swallow babies; bank swallow parents reject any babies not their own.

To recapitulate: rough-winged swallows (the species that, because of its solitary lifestyle, doesn't possess offspring-recognition mechanisms) can essentially be induced to adopt, whereas bank swallows (whose social tendencies put them at risk of misdirecting their parental efforts) are equipped with a built-in tendency to recognize their offspring, and to reject nestlings that aren't their own.

Although people are more social than solitary, when it comes to offspring recognition, we are rough-wings rather than bank swallows. After a woman gives birth, there is simply no question whether the baby is hers. Switching newborns may take place in Gilbert and Sullivan operettas or—very rarely—in a modern, crowded metropolitan hospital, but not among a small band of early hominids trudging around the Pleistocene savannah. There is simply no way an African Eve could find herself *accidentally* nursing someone else's child. Lacking the threat of misidentifying our babies, our ancestors would almost certainly have also lacked any automatic, lock-and-key recognition mechanisms. (Recall the minimalism of evolutionary adaptations.) As a result, we have a wonderfully open program when it comes to identifying children as our own.

Granted that our evolutionary past gives us leeway to fool ourselves, in a sense, into responding as though someone else's offspring is actually our own, but why do so? After all, to adopt is to expend time and resources on behalf of someone *unrelated* to the adopter. Accordingly, it would appear to be an evolutionary blunder, comparable to genuine altruism (that is, beneficence toward another without any genetic compensation). Yet human beings can be quite insistent upon adopting, often struggling against heavy odds and bureaucratic red tape to do so. Adopted children, moreover, are typically well loved and cared for, and about as successful as biological children.

To start with, let's point out the obvious: adoption, overwhelmingly, is *not* most people's first choice. When it comes to children, the vast majority prefer to make their own. Only if this is not an option are most people inclined to satisfy their desire to be a parent—a desire that is almost certainly a highly adaptive legacy of evolution—by parenting someone else's kids. In addition, bear in mind that for perhaps 99.99 percent of its evolutionary past, *Homo sapiens* lived in small hunter-gatherer bands that almost certainly numbered fewer than a hundred. Within such groups, most individuals were related. As a result, anyone who adopted a child was likely to be caring for a genetic relative (say hello once again to our old friend kin selection). Even individuals who cared for an unrelated child may well have positioned themselves to receive a return benefit from the child's genetic relatives as well as becoming a possible recipient of social approval, and hence biological benefit, from within the local group.

Note once more that stepparenting and adoption are not the same, the former being a much darker and more troubled phenomenon. To be sure, stepparenting is similar to adoption in that non-genetic "parents" end up taking care of someone else's children. But whereas adoption involves a specific commitment to the adopted child (typically on the part of *both* adopting adults), stepparenting nearly always comes about as a side effect of two adults' commitment *toward each other*. Stepchildren, if any, are generally thrown in as an unavoidable—and, if the truth be acknowledged, often unwanted—part of the deal.[3] Adoption, moreover, typically takes place when the child is an infant, thereby enhancing the prospects that adopting parents can fool their unconscious selves into responding as though the adopted child is genetically their own. On the other hand, it usually isn't until they are older that stepchildren enter the stepparent's life, which is a further obstacle to parental devotion.

Put it all together, and whereas stepparenting and adoption are both clearly part of the human behavioral repertoire, the latter is likely to be much less conflictual and—at the biological level, at least—downright easy. Not surprisingly, literary depictions of adop-

---

[3] Technically, fostering is yet another category, which applies when a child is taken into the house of adults, neither of whom is the biological parent. It pertains to the situations of Cosette and Oliver Twist, and especially in the past meant virtual slavery. For our purposes, it is essentially equivalent to stepparenting, but with two stepparents and no biological parent to leaven the child's plight.

tion tend to be correspondingly comfortable as well as comforting. Barbara Kingsolver's first novel, *The Bean Trees*, was a heartwarming, thought-provoking, yet thoroughly genuine depiction of the adventures of Taylor Greer, a delightful, headstrong, and impulsive young woman who adopts Turtle when the young child is deposited into the front seat of her car. The connection between adoptive parent and child constitutes the core of the book, and it is one that the reader never doubts, despite the fact that Taylor and Turtle are not genetic relatives. Although they don't share genes, they do share needs, and the affiliation is one that Taylor enters into of her own free will, not carried along in the slipstream of a higher-priority adult relationship.

Earlier, we looked at how Monsieur and Madame Thénardier treated Cosette very differently than their own daughters. Now it's time to consider another relationship depicted in *Les Misérables*, that between Jean Valjean and Cosette. The middle-aged Valjean didn't find himself stuck with the young girl because of a sought-for union with her mother, Fantine; rather, he adopted Cosette, because, as Hugo makes clear, she met his need for a child just as he met her need for a parent:

> The gulf that nature had created between Valjean and Cosette, the gap of fifty years, was bridged by circumstance. The overriding force of destiny united these two beings so sundered by the years and so akin in what they lacked. Each fulfilled the other, Cosette with her instinctive need of a father, Valjean with his instinctive need of a child. For them to meet was to find, and in the moment when their hands first touched, they joined. Seeing the other, each perceived the other's need. In the deepest sense of the words it may be said that in their isolation Jean Valjean had been a widower, as Cosette was an orphan; and in this sense he became her father.

Another example of adoption, and to our mind the most suitable and heartwarming account, is George Eliot's *Silas Marner*. A lonely, reclusive miser, Silas is a painfully nearsighted weaver whose only pleasure comes from counting his accumulated gold pieces. One day his trove is plundered, plunging him into despair.

> Formerly, his heart had been as a locked casket with its treasure inside; but now the casket was empty, and the lock was broken. Left groping in darkness, with his prop utterly gone, Silas had inevitably a sense, though a dull and half-despairing one, that if any help came to him it must come from without.

Shortly thereafter, upon entering his isolated cottage, Silas sees a golden gleam in front of his fireplace; it isn't his gold, but a tiny, yellow-haired girl who somehow managed to reach safety after her destitute mother died in the snow nearby.

Silas adopts Eppie, and she transforms him in return, helping the old man regain his life just as he saved hers. The once-bare windows of chez Marner are soon festooned with lacy curtains, and Silas, for his part, finds himself opened as never before to joy and fulfillment. Eppie is a replacement, and more, for the former miser's lost gold:

> He could only have said that the child was come instead of the gold—that the gold had turned into the child. . . . The gold had kept his thoughts in an ever-repeated circle, leading to nothing beyond itself; but Eppie was an object compacted of changes and hopes that forced his thoughts onward, and carried them far away from their old eager pacing towards the same blank limit— carried them away to the new things that would come with the coming years, when Eppie would have learned to understand how her father Silas cared for her.

But there is trouble in paradise, arriving in the person of Godfrey Cass, the biological father of Eppie and the sole surviving son of Squire Cass, the town's wealthiest man. It seems that eighteen years previously, in a fit of drunken foolishness, Godfrey had failed to rise above "nature" and had impulsively and secretly married a coarse and common woman, who became Eppie's mother. Mortified by his earlier error, Godfrey had kept secret not only his marital indiscretion but also his fatherhood until, many years into a childless marriage, he revealed the truth to his wife, Nancy. (Godfrey's truth-telling was also stimulated by the discovery of the body of his good-for-nothing brother, along with Silas's gold, which he had stolen.)

Godfrey and Nancy Cass go to Silas and Eppie, demanding custody of the now budding young lady, pointing out that they can offer her wealth and a "suitable upbringing" far beyond anything available from the rustic weaver. Eppie, however, elects to remain with her adoptive father, and also marries a fine young chap, thereby concluding our tale.

There is nothing demeaning about adoption being a win-win proposition, benefiting Jean Valjean as well as Cosette, Silas Marner as well as Eppie. And it is neither surprising nor disreputable that the childless Casses cast a longing parental eye on Eppie as well. Nor is it irrelevant that by their action, adopters are often seen to demonstrate their good character and even, on occasion, their marriageability: don't forget the importance of "good behavior" (Chapter 3) in demonstrating one's suitability as a mate.

Also, keep in mind Jane Eyre and her eventual encounter with the lordly, and also secretly married, Rochester. Jane was initially hired as governess to Rochester's young ward, a girl named Adèle who was the offspring of a French prostitute and maybe, just maybe, Rochester's natural child as well. He denies it, however, claiming that he adopted Adèle out of disinterested altruism:

> "I see no proofs of such grim paternity written in her countenance. . . . I acknowledged no natural claim on Adèle's part to be supported by me; nor do I now acknowledge any, for I am not her father; but hearing that she was quite destitute, I e'en took the poor thing out of the slime and mud of Paris, and transplanted her here, to grow up clean in the wholesome soil of an English country garden."

If you were Jane Eyre, wouldn't you, too, be moved by the kindliness of such a man, however forbidding and distant he appears in other respects? And wouldn't his benevolent adoption of Adèle go a long way toward modulating any sense you might have of him as cold and unfeeling? Isn't it interesting as well that the villagewide reputation of Silas Marner, who spent decades trying unsuccessfully to live down an unjust accusation of thievery in his youth, was finally rehabilitated when his devotion to baby Eppie, his adopted child, became public?

"All happy families are happy in the same way," wrote Leo Tolstoy in the famous opening sentence of *Anna Karenina*. "Every unhappy family is unhappy in its own way." As to happy families, there is room for debate, but when it comes to unhappiness, Tolstoy was certainly onto something: people have devised—or blundered into—innumerable ways of being unhappy. Even this diversity, however, resolves itself into some recognizable patterns, many of which involve the struggles of stepparenting.

8

On the Complaints of Portnoy, Caulfield, Huck Finn, and the Brothers Karamazov Everywhere

Parent-Offspring Conflict

Portnoy has a complaint. Holden Caulfield is pissed off. And let's face it, even that all-American boy Huck Finn has a hard time with adults. Nor are these three alone. In literature, as in life, children and parents don't always get along. In fact, one might say that they never do—at least, not without genuine friction—even though in the long run things have a habit of working out.

No one is shocked when men struggle over status, dominance, or money, not to mention direct one-on-one duels over mates. Similarly for comparatively toned-down conflict among women. There is even a powerful biological rationale to battles *between* the sexes, insofar as men and women are different genetically as well as in their reproductive tactics and strategies, despite the fact that their interests converge in reproduction. (As George Burns once pointed out, no wonder there's trouble between the sexes. After all, they want different things: men want women and women want men!) Yet it may seem surprising that parents and offspring are so often at odds, especially when we look through evolutionary spectacles.

The genetic interests of parents and their offspring would seem to coincide perfectly, if only because parents want their children to succeed, as do the children themselves. And so, as Portnoy's put-upon parents would no doubt observe, what's to complain about? Even a hard-eyed biological perspective, by which children are merely vehicles enabling parental genes to replicate themselves, leads to the suggestion that the latter achieve their goals via the former (think Little League). So little or no conflict might be expected (think *Little Women*).

Years of Walt Disney's *True Life Adventures*, combined with animated films from *Dumbo* to *The Lion King*, have both reflected and generated the presumption that parent and child—especially mother and child—are the epitome of shared goals and perfect amiability. The image among human beings is, if anything, even more clearly established: madonna and child convey a sense of peace and contentment that transcends the merely theological.

When rough spots emerge in the parent-child nexus, the traditional view has long been that the culprit is mutual misunderstanding, with its attendant failures of communication: everyone is supposed to mean well. The conventional wisdom is that in the course of conveying heartfelt parental assistance, advice, and information to the child, sometimes there are problems, largely because the child is necessarily inexperienced, and also liable to be headstrong and uninformed as to where its true interests lie. According to this view, the well-socialized and gradually more mature youngster eventually recognizes that it is best served by going along with parental inclinations, at which point conflict ceases and socialization has been achieved. Thus when conflict arises between parent and offspring, the traditional explanation is that children are primitive, even barbaric little creatures who need time to become incorporated into the society of responsible adults. And so we indulge Huck and his buddy Tom Sawyer, their boys-will-be-boyhood, in the comfortable assurance that eventually, they will grow up and shape up.

Then there is the psychoanalytic tradition, which focuses on sexual rivalry, especially between sons and fathers. As a result of presumed Oedipal conflict, boys are supposed to be terrified that their fathers will castrate them, while girls resent not having a penis; trouble therefore ensues until eventually things quiet down when the child gets older and less obstreperous, reconciling its primitive conflicts by assuming the social role appropriate to its gender.

The view from evolutionary biology is quite different, rather darker . . . and much more persuasive. It is closer to that expressed in Shakespeare's *A Midsummer Night's Dream*: "The course of true love never did run smooth." Which is not to deny the love of parent and offspring, but simply to point out their interests. The point at issue, now known to biologists as "parent-offspring conflict," was developed in a brilliant paper published in 1974 by the evolutionary theorist Robert Trivers. The following discussion owes much to his insights (many of which now seem obvious, but only after Trivers first pointed them out).

One of Trivers's most basic insights is that there is no genetic identity between parent and offspring; rather, there is simply a 50 percent probability that any gene present in a parent is also present in the child. Nonetheless, prior to recognizing the consequences of this simple fact, biologists—and to an even greater extent most psychologists and sociologists—tended to treat the child as an appendage to the parent, rather than a separate being with its own strengths as well as weaknesses and, even more important, its own agenda. It needs to be emphasized that there is a parent-offspring genetic glass of DNA that is half empty: the 50 percent of parental genome *not* shared between parent and child.

When we looked at the biology and literature of kin selection (Chapter 6), the point was that when genes look out for themselves, the result, ironically, is altruism. But just as shared genes result in shared interests, unshared genes result in conflicting agendas and even outright conflict.

So it isn't really surprising after all that Holden Caulfield, in *The Catcher in the Rye*, becomes disgusted with his parents and with adults in general, mortified at the "phoniness" that he sees around him, suspicious of the grown-up world. After all, the biological interests of a parent are limited to making the most of the *parent's* prospects, not *necessarily* those of the parent's offspring. Remember, while the parent does have a strong genetic interest in his or her child, the only biological reason for creating offspring in the first place is as a means of advancing the *parent's* fitness and not someone else's! Although it's in each parent's interest—or rather that of parental genes—to invest in offspring, they may not be well served by putting all their genetic eggs in one basket (just to mix the metaphor up even further). As a result, investing in any individual offspring must be balanced against maintaining the ability to invest in others, later on or simultaneously. Hence a potential clash of interests.

Among species with a short life span and not much to contribute to their offspring, such as most insects and many fish, adults typically reproduce in a single, gung-ho burst of enthusiasm, known rather indelicately to biologists as a "big bang." On the other hand, in the case of birds and mammals—including *Homo sapiens*—parents are selected to hold back a bit, rather than spend all their time and energy on behalf of any one offspring.

Accordingly, parents make unconscious calculations about how best to invest their reproductive resources, including the need to hedge their bets and keep enough in reserve for additional children. Insofar as parents hold back, it isn't simply because they have a 50 percent genetic overlap with their children, but because they need to remain capable of rearing other offspring, as well as to assist with additional kin such as nieces, nephews, grandchildren, and so forth.

At the same time, every child is ultimately interested in making the most of *its* fitness, not necessarily that of its parent. Another way of looking at it: a child is 100 percent related to itself but only 50 percent related to any parent. As a result, the child devalues its parents' interests by a factor of one-half, as parents do for each child. Children and parents can, in a sense, feel only one-half of each other's pain, just as they enjoy only one-half of each other's genetic gain. No wonder there's conflict. (As George Burns might have added: parents and children each want different things, namely, their own fitness, not each other's.)

The upshot has been an important new understanding in evolutionary psychology, one that makes sense of precisely those parent-offspring rough spots that literature has recognized for centuries, although without benefit of current genetic wisdom: despite their profound bonds, parent and child are also profoundly disconnected from each other. Thus, they are locked in a battle of evolutionary wills, with each party biologically primed to demand more than the other is inclined to give. Parents want grandchildren; once they reach a certain age, children want sex but not necessarily their own children. Parents want their offspring to be self-sufficient; children want their parents to be financially and emotionally generous. Parents want the chores to be done; children want freedom to live their own lives. Parents want to see themselves in their children; children want to see themselves in themselves. King Lear wants his daughters to vie with each other in expressing undying love for him; two of the three are brutally interested only in themselves.

At last, transgenerational conflict—so much a fixture of human life, and providing so much grist for the literary mill—is open to be examined, predicted, and understood.

Among animals, including human beings, some of the conflict between parents and offspring reveals itself during weaning, with offspring often seeking to obtain more milk than parents are inclined to provide. Lactating mothers who had been the epitome of maternal devotion become increasingly short-tempered with their nurslings. One can almost hear them complaining, like Portnoy's mother, "Enough already!" It is almost unheard of for a nursing cat, dog, lion, or chimpanzee to snarl at one of her babies—until, that is, those babies have gotten big enough and able to make it on their own, while at the same time remaining so demanding that they are beginning to impinge on the mother's future options.

How much milk should a mommy mammal provide? It depends who you ask. From the offspring's perspective, enough to guarantee the child's maximum growth and eventual success, with a small adjustment to make sure that Mom doesn't overdo it and so deplete herself that she cannot breed again; after all, given the exigencies of kin selection, it is often in the interest of each offspring to be provided with siblings. And Mom is the one to do so. From the mother's perspective, she should endow each child with enough milk to give it a good start, but with a much larger allowance for her own well-being and especially, granting her sufficient strength to breed again, or at least, to provide milk for any other siblings. Again, as Trivers emphasized, the mother and the offspring are each entirely related to itself and only one-half related to the other. The stage is set for trouble.

Something similar to the mammal pattern of weaning conflict even takes place among birds. Large nestlings—big enough to fly, hence known as fledglings—can often be found pursuing their harried parents, importuning them for food. In late spring throughout North America, it is common to see fledglings of many different species quivering their wings and uttering incessant begging calls while the parents back away, look far into the distance (as though trying to ignore what is in front of them), and often literally take wing, pursued by their nearly grown but indefatigably demanding offspring.

Conflict over weaning, or its equivalent in birds, does not exhaust the potential for parents and offspring to disagree. In this new evolutionary way of looking at parent-offspring relations, the root of all evils is conflict over parental investment, anything that parents provide to their offspring that contributes to the offspring's success but which carries with it a cost as measured by the parent's ability to produce and invest similarly in additional offspring. Among human beings, parental investment includes money but is not limited to it. It also goes on longer, and is more intense, than in any other species. Many a harried parent, struggling to provide for even the most loved and rewarding child, will answer the question "What do you want your child to be?" with an immediate reply:

"Self-supporting!" And many a put-upon child yearns for its parents to back off . . . but keep sending money.

And so we return to *Portnoy's Complaint*. How great a leap is it from a demanding fledgling to a complaining Alexander Portnoy? Not that far. The evolutionary theory of parent-offspring conflict tells us that parents seek to manipulate children in their own ways and for their own ends, whereas children can be expected to resist. And not surprisingly, it also predicts that many of these conflicts revolve around sex and reproduction, in which a youngster's developing sexual urges are more disruptive than reproductive.

In one notable scene from that antic, comic, iconic tale of generational conflict, Portnoy's father waits in constipated agony outside the bathroom door while young Alexander Portnoy masturbates within, after which our temporarily depleted hero must endure his mother's insistence on searching his nonexistent feces (after all, he was supposed to be pooping) for evidence of nonkosher hamburgers. It is more than a little challenging, as Portnoy complains, "being a nice Jewish boy, publicly pleasing my parents while privately pulling my putz!"

Nor is Alexander Portnoy's incessant masturbation the only indication that he and his parents aren't quite on the same wavelength. With all its hilarious, ribald, offbeat humor, *Portnoy's Complaint* is filled with parent-offspring conflict, notably his mother's "ubiquity," his father's blockheaded insistence as to how young Alexander should behave, and constant demands that he "capitulate," punctuated by his own occasional, if irresolute, efforts at self-assertion:

> "I'm sorry," I mumble . . . "but just because it's your religion doesn't mean it's mine."
>
> "What did you say? Turn around, Mister. I want the courtesy of a reply from your mouth."
>
> "I don't have a religion," I say. . . .

"You don't, eh?"

"I can't."

"And why not? You're something special? Look at me! You're somebody too special?"

"I don't believe in God."

"Get out of those dungarees, Alex, and put on some decent clothes."

"They're not dungarees, they're Levi's."

"It's Rosh Hashanah, Alex, and to me you're wearing overalls! Get in there and put a tie on and a jacket and a pair of trousers and a clean shirt, and come out looking like a human being. And shoes, Mister, hard shoes."

"My shirt is clean—"

"Oh, you're riding for a fall, Mr. Big. You're fourteen years old, and believe me, you don't know everything there is to know. Get out of those moccasins! What the hell are you supposed to be, some kind of Indian?"

What, then, is Portnoy's complaint? Basically, this: how hard it is to be yourself when your parents are constantly breathing down your neck, unendingly exhorting, complaining, warning, demanding, insisting, and—of course, this being a Jewish tale—guilt-mongering, all the while loving, to be sure:

> "Call, Alex. Visit, Alex. Alex, keep us informed. Don't go away without telling us, please, not again. Last time you went away you didn't tell us, your father was ready to phone the police. You know how many times a day he called and got no answer? Take a guess, how many?"

It may be significant that the traditional view of parent-offspring relations, which assumes that father and mother know best, and which has never taken the perspective of offspring very seriously, is one that has been promulgated by adults. Similarly, we can expect that parents would be likely to present their teachings, manipulations, guilt-mongering, and arm-twisting as "for your own good," or even accompanied by protestations that "this hurts me more than it hurts you," emphasizing, in short, the value of their opinions and advice as well as the degree to which the parental perspective is solely in the best interest of the child. (Nonetheless, sometimes parents do have something worthwhile to offer their children, and not all interactions are hurtfully self-serving. Mark Twain once noted that when he was a teenager, his parents knew nothing; as a young man in his early twenties, he was astounded how much they had learned in just a few years!)

Philip Roth's novel is an extended comic monologue, with the adult Alexander (the not-so-great) telling his tale of developmental woe to an analyst. Included is this memorable recitation of Momma Portnoy's appeal to her son, followed by an added prod from Poppa Portnoy:

"Do you remember Seymour Schmuck, Alex?" she asks me. . . . "Well, I met his mother on the street today, and she told me that Seymour is now the biggest brain surgeon in the entire Western Hemisphere. He owns six different split-level ranch-type houses . . . and last year [he took] his wife and his two little daughters . . . to Europe for an eighty-million-dollar tour of seven thousand countries, some of them you never even heard of . . . and that's how big your friend Seymour is today! *And how happy he makes his parents!*"

And you, the implication is, when are you going to get married already? In Newark and the surrounding suburbs this apparently is the question on everybody's lips: WHEN IS ALEXANDER PORTNOY GOING TO STOP BEING SELFISH AND GIVE HIS PARENTS, WHO ARE SUCH WONDERFUL PEOPLE, GRANDCHILDREN? "Well," says my father, the tears brimming up in his eyes, "well," he asks, *every single time I see him*, "is there a serious girl in the picture, Big Shot? Excuse me for asking, I'm only your father, but since I'm not going to be alive forever, and you in case you forgot carry the family name, I wonder if maybe you could let me in on the secret."

Classic parental guilt-mongering (whether Jewish or otherwise) involves urging a child to make its *parents* happy, never mind him. Ironically, as in Alexander Portnoy's case, this can even involve parental desires for grandchildren, resisted by the child, now a young adult. The younger Portnoy is plenty interested in girls, lots of them; his parents, on the other hand, want to know if there is a "serious girl" in his future. They want grandchildren, not Alexander Portnoy's personal sexual satisfaction. There is abundant biological research—much of it involving sophisticated mathematical models—on the optimal timing of reproduction in animals; suffice it to note that what is optimal for a would-be grandparent isn't necessarily the best option for a future parent. A postadolescent oyster has few opportunities to argue with its parents over the perfect time to start making little oysterlings. Human beings aren't so lucky.

Which leads us to Caulfield's complaint. In J. D. Salinger's modern classic, the hero isn't especially hung up on sex (at least not like Portnoy) or on his "parental unit." Nonetheless, there is plenty of parent-offspring conflict: over Holden's failure in school, his yearning for a lost innocence, and the fact that he is at great pains to avoid encountering the elder Caulfields, who, he is convinced, couldn't possibly understand him. He's probably right. From the story's outset, Holden Caulfield is vigorously dismissive of his parents:

> If you really want to hear about it, the first thing you'll probably want to know is where I was born, and what my lousy childhood was like, and how my parents were occupied and all before they had me, and all that David Copperfield kind of crap, but I don't feel like going into it, if you want to know the truth.

Insofar as parents are often predisposed to push, pull, and prod their children to meet the parents' needs, it makes perfect sense that adolescents in particular are prone to fight back, and also that they are likely to respond favorably to depictions of their peers doing just that. (This helps explain the spectacular popularity of *The Catcher in the Rye*, which almost certainly exceeds its literary merit.) Furthermore, given that children are smaller, weaker, poorer, and less experienced than their parents, it also makes sense that in their case, parent-offspring conflict often involves a series of strategic retreats and more of a guerrilla conflict than out-and-out warfare.

Holden Caulfield impulsively runs away from the boarding school from which he is about to be dismissed—the fourth one—after failing in his halfhearted efforts to explain his dissatisfactions to a fellow student, and resenting his roommate's selfish (but successful) relationship with girls. Significantly, the only people to whom he relates positively are his two siblings: a brother, now dead, and his younger sister, Phoebe (whom he hopes to save from adults).

As to the adult world, it is a continuing source of conflict for young Master Caulfield, leaving him alienated, misunderstood, and profoundly alone: keeping away from his parents, he wanders lonely and alienated through deserted city streets, an empty and brooding hotel lobby, a silent dormitory. Here are his final words as he leaves school:

> When I was all set to go, when I had my bags and all, I stood for a while next to the stairs and took a last look down the goddam corridor. I was sort of crying. I don't know why. I put my red hunting hat on, and turned the peak around to the back, the way I liked it, and then I yelled at the top of my goddam voice, *"Sleep tight, ya morons!"*

No one understands him and, to be fair, he doesn't understand anyone else, either. Holden's attempts to connect with the world are all pathetic failures: he goes to crowded bars where he feels painfully isolated; he looks up an old girlfriend only to find her unable to relate to his concerns; he tries to make a date with someone he had heard described as "loose" but doesn't follow through; he spends time with one of his former teachers only to flee in terror when the man appears to make a homosexual advance; he invites an unresponsive taxicab driver to have a drink with him; he even, at the instigation of a hotel elevator operator, has a prostitute to his room, only to spend the time talking and then being beaten by the hotel man.

In some ways, the biology of parent-offspring conflict is really about a wider kind of conflict: between every young individual and the adult world that he or she must learn to negotiate. We have seen that doing so often requires performing a delicate dance with/against one's own parents, who have their own agendas. It also involves an even more dangerous dance with/against the rest of the adult world, which, after all, is often primed to view the newly emerging individual as a competitor and/or someone to be exploited. No wonder youngsters, being vulnerable, are often both alienated and wary.

Holden is appalled when the adult world intrudes on his sister's life in the form of obscene graffiti scrawled on the walls of her school. The demands and preoccupations of adult reality fall short of his hopes, leaving him convinced that everyone is a "jerk," "phony," a "moron," "corny," "crummy," "lousy," or "dopey." He idealizes himself as a protector of children, a defender of offspring in their conflict with the grown-up world, dreaming idealistically that he might some-day be the "catcher in the rye" who rescues little children as they play in a field of rye and keeps them from falling off a cliff:

> Anyway, I keep picturing all these little kids playing some game in this big field of rye and all. Thousands of little kids, and no-body's around—nobody big, I mean—except me. And I'm stand-ing on the edge of some crazy cliff. What I have to do, I have to catch everybody if they start to go over the cliff—I mean if they're running and they don't look where they're going I have to come out from somewhere and *catch* them. That's all I'd do all day. I'd just be the catcher in the rye and all.

Holden Caulfield probably wants to be caught and protected himself, hence his fantasy about catching and protecting younger children. Yet, ironically, he is misquoting the Robert Burns song, which asks whether there is anything wrong "if a body *meet* [not *catch*] a body comin' thro' the rye," and if that meeting results in casual sex between them. Holden's fantasy involves catching children before they plunge over a cliff into the adult world, with its demands, its insistent "phoniness," and of course its knowledge of sex. Meanwhile, Holden falls off his own cliff, leading to a "victory" of sorts by adult society, since he tells his story while incarcerated in a mental institution.

For a half century, *The Catcher in the Rye* has had a firm grip on the imaginations of millions of readers, most of them teenage, despite the fact that Holden Caulfield is an absolute jerk: he lies to everyone, cannot complete anything or connect with anyone, and is unrelentingly critical of those around him. Most likely, this is why Holden Caulfield speaks so convincingly to adolescents: as he says to one of his teachers, he feels trapped on "the other side" of life, and his story is one of nearly frantic and continually unsuccessful attempts to find companionship, to satisfy E. M. Forster's dictum, "Only connect." He has, of course, his red hunting hat (worn backward long before this became popular) as an enduring symbol of his individuality. It's about all that he has.

Holden Caulfield sought desperately to freeze childhood innocence, thereby protecting it from the polluting effects of the adult world, while at the same time, not coincidentally, shielding children—including himself—from the conflicts that the adult world unavoidably generates. Nonetheless, parent-offspring conflict, like shit, happens. Nearly every story of growing up therefore involves some degree of rebellion, dissatisfaction, and disputation with parents and other "authority figures." Earlier, we wrote about kin relationships as an ocean in which we swim. There are other natural bodies of water so pervasive that they are largely taken for granted. Conflict between parent and offspring are prominent among them.

Long-lasting tales of young people owe much of their enduring appeal to varying degrees of immersion in this enduring theme. *Huckleberry Finn* is a perfect example. Admittedly, Huck's mother is dead and his biological father is of no great consequence in his story except as an abusive alcoholic. But significantly, Huck has numerous substitute parents: Judge Thatcher, the Widow Douglas, Tom Sawyer's Aunt Polly, Aunt Sally, Miss Watson, the Grangerfield family, Jim, the rather sinister Duke and King. Some are helpful, some hurtful, some admirable, some despicable, but in all cases they constitute the adult background against which our young hero must struggle to find himself.

By contrast to the wealthy Holden Caulfield, Huckleberry Finn—that quintessential, mythic American boy—is utterly impoverished when it comes to money but infinitely richer in relationships with adults, with whom he has no dearth of conflicts. As Mark Twain's novel begins, the Widow Douglas and Miss Watson, her sister, are determined to civilize Huck; they want him to stop swearing and smoking, to wear clean clothes, go to school and sleep in a bed. Listen to Huck's account:

> The widow rang a bell for supper, and you had to come to time. When you got to the table you couldn't go right to eating, but you had to wait for the widow to tuck down her head and brumble a little over the victuals, though there warn't really anything the matter with them—that is nothing only everything was cooked by itself. In a barrel of odds and ends it is different; things get mixed up, and the juice kind of swaps around, and the things go better.
>
> After supper she got out her book and learned me about Moses and the Bulrushes, and I was in a sweat to find out all about him; but by she let it out that Moses had been dead a considerable long time; so then I didn't care no more about him, because I don't take no stock in dead people.

The old ladies are well-meaning but annoyingly self-righteous, and Huck certainly notices:

Pretty soon I wanted to smoke, and asked the widow to let me. But she wouldn't. She said it was a mean practice and wasn't clean, and I must try to not do it any more. That is just the way with some people. They get down on a thing when they don't know nothing about it. Here she was a-bothering about Moses, which was no kin to her, and no use to anybody, being gone, you see, yet finding a power of fault with me for doing a thing that had some good in it. And she took snuff, too; of course that was all right, because she done it herself.

Children are notoriously sensitive to adult hypocrisy, especially when it emanates from parents or—in Huck's case—parent substitutes. Perhaps this is part of our specieswide legacy of such conflict: an acute awareness of who is doing what, and who is trying to get whom to do what, and why.

Huck rises above the mean, low people he encounters, and even above the expectations of a society and of his elders, by eventually repudiating the slave-owning culture in which he had been raised. Thus Huck initially decides to write a letter telling where Jim, an escaped slave, can be found and returned to slavery. But then, in his signal crisis of conscience, Huck reconsiders:

> [G]ot to thinking over our trip down the river; and I see Jim before me, all the time, in the day, and in the night-time, sometimes moonlight, sometimes storms, and we a floating along, talking, and singing, and laughing. But somehow I couldn't seem to strike no places to harden me against him, but only the other kind. I'd see him standing my watch on top of his'n, stead of calling me, so I could go on sleeping, and see him how glad he was when I come back out of the fog; and when I come to him again in the swamp, up there where the feud was, and such-like times . . . and how good he always was; and at last I struck time I saved him by telling the men we had small-pox aboard, and he was so grateful, and said I was the best friend old Jim ever had in the world, and the only one he's got now; and then I happened to look around, and see that paper.

It was a close place. I took it up, and held it in my hand. I was a-trembling, because I'd got to decide, forever, betwixt two things, and I knowed it. I studied a minute, sort of holding my breath, and then says to myself: "All right, then, I'll go to hell"— and tore it up.

Finally, at the book's end, Huck rejects civilized, adult, faux-parental society once and for all, and we cheer: "But I reckon I got to light out for the Territory ahead of the rest, because Aunt Sally she's going to adopt me and sivilize me, and I can't stand it. I been there before."

It seems like a one-sided fight. After all, parents are bigger, older, stronger, wealthier, and presumably wiser than their children. The little tykes shouldn't stand a chance, at least so long as the parents in question are still young enough to remain vigorous. At the same time, however, offspring aren't altogether helpless: as Trivers emphasized, they can be expected especially to resort to psychological techniques. August Wilson's play *Fences* describes the struggles of family patriarch Troy Maxson as he confronts his children. Maxson's son Lyons, although thirty-four, keeps showing up and requesting money; as Trivers predicted, Lyons Maxson has ways of getting his way, in particular playing upon his father's guilt over the fact that the elder Maxson had not been substantially involved in his son's upbringing:

> "You can't change me, Pop. I'm thirty-four years old. If you wanted to change me, you should have been there when I was growing up. I come by to see you . . . ask for ten dollars and you want to talk about how I was raised. You don't know nothing about how I was raised."

Troy is also in conflict with his other son, Cory, who wants to go to college and play football instead of taking a more "reliable" job.

At the play's end, Troy ruthlessly and unsympathetically pushes Cory out of his house, in part to make room for Troy's infant daughter (by another woman), who is coming to live with them, and in part as a reenactment of Troy's own childhood experience. Thus at the age of fourteen, he was disowned by *his* father after it became clear that the two were fighting over the same woman:

> "Now I thought he was mad cause I ain't done my work. But I see where he was chasing me off so he could have the gal for himself. When I see what the matter of it was, I lost all fear of my daddy. Right there is where I become a man . . . at fourteen years of age. . . . The only thing I knew was the time had come for me to leave my daddy's house. And right there the world suddenly got big."

To be sure, a dose of tough love can sometimes be a worthwhile component of good parenting for human beings who—at some point—insist that their children pay their own way, or at least accept responsibility for their own actions. And there are things parents can and should do to help their children's world "get big." At the same time, a dose of evolutionary theory helps clarify how it is that in many cases such actions are more than a little self-serving.

King Lear is doubtless correct when he says that it is "sharper than a serpent's tooth to have a thankless child." The converse also holds, as biology helps make plain: it is a genuine pain in the ass to have a selfish, demanding parent!

Although parent-offspring conflict seems tailor-made for adolescence, no one is ever too old to participate. Parents and offspring can be expected to disagree over many things beyond the relatively simple question of how much parents should provide for a child or whether they should do so at all. They will predictably disagree, for instance, over the child's behavior toward a third party—in the most obvious and probably most frequent case, a brother or sister.

Here's how it works. For a parent, who is equally related to each child, every one is equally important. Parents come out ahead any-time one of their children acts altruistically toward a brother or sis-ter, so long as the benefit to the recipient is greater than the cost to the altruist. But the child can be expected to see things differently, since he or she is related by only 50 percent to a sibling (as to a par-ent), but 100 percent to itself. The result is conflict, not only be-tween siblings—sibling rivalry indeed—but also between parent and offspring.

Think of parents urging a child to play nicely with its brother or sister, typically more nicely than the child wants to do. Or pressur-ing a child to share when the youngster is inclined to be more self-ish. Part of the power of the evolutionary approach comes from its recognition that such cases are not simply due to stubbornness or sheer perversity on the part of a rivalrous or selfish child. Rather, the key is that children are inclined to act in response to *their own* biological interests rather than those of their siblings, their parents, society, or anyone else, for that matter. In most such cases, the bot-tom line is that parents are expected to exhort, extort, or otherwise try to induce their offspring to act differently—toward each other as well as toward the parent—than the offspring would choose if left to themselves.

When extreme, sibling rivalry can become siblicide, and it is more frequent in the natural world than biologists used to think. Embryonic sharks, for example, begin their predatory lives with an early burst of sibling competition, devouring each other as they swim about in utero before they are born. And pronghorn antelope fetuses kill each other in the womb, presumably thereby increasing the amount of mother's milk that they will eventually obtain. Siblicide has even been reported for plants. The Dalbergia tree of India disperses its seeds via pods that float in the wind; lighter pods travel farther. The first Dalbergia seed to develop produces chemi-cals that kill its pod siblings before it leaves the tree, thereby giving this particular "bad seed" sole occupancy of the vehicle, and thus, the advantage of greater dispersal distance.

Fortunately, there is no evidence—as yet—that human beings partake of siblicide as an evolutionary strategy. But the story of Cain killing Abel (which has resonance, in one form or another, in a variety of cultures) suggests that brothers and sisters may be less mutually supportive than many, especially the parents, might wish. Such an insufficiency of brotherly love is exactly what John Steinbeck recognized in *East of Eden*, his dark tale of fratricidal competition, and its sisterly counterpart inspired Shakespeare when he wrote *King Lear*.

Ivan Turgenev's classic *Fathers and Sons* has long been seen as social criticism, representing the conflict between old and new, between traditional and "scientific" conceptions of society, and a novel in which the younger generation's fascination with nihilism foreshadows the forthcoming Russian Revolution. And so it is. Arkady and Bazarov are recent graduates of the university in St. Petersburg, filled with new and "nihilistic" ideas, critical of their families but also assuming that they will be taken care of by them. At the same time, it is worth noting that Turgenev chose parents and offspring to exemplify these issues. Indeed, *Fathers and Sons* is a textbook depiction of parent-offspring conflict, in which two sons (Arkady and Bazarov) struggle with their fathers (Nikolai and Vassily), to establish their own independent identities, complete with sexual pursuits and predictable concerns about inheritance.

There is even some overt sexual conflict across the generations, when Bazarov makes an advance on his friend's father's mistress and must fight a duel as a result. Parent-offspring conflict is real and troublesome enough, but add a hefty dose of plain old-fashioned male-male sexual competition and the dynamic becomes dynamite.

This explosive mix is dramatically portrayed in Dostoyevsky's *The Brothers Karamazov*. Here, we are given the panoramic tale of genetic and sexual conflict—ultimately lethal—between a father and his offspring.

Fyodor Karamazov is a philanderer, a brutal buffoon, a man of means, a drunkard, a "sensualist" (which is to say, a very horny and unrestrained SOB), and, least of all, a father. Old man Karamazov had abandoned all of his offspring, one by one, leaving each in turn to be cared for by Grigory, a devoted servant, and eventually placed with other relatives while the old ogre indulged in his orgies.

Dmitri, the eldest Karamazov son, and offspring of Fyodor's first wife, grew up believing that he was entitled to a substantial inheritance via his mother. Dmitri became a soldier and a sensualist like his father, a prodigal son whose life is wild, passionate, and violent. Fyodor's two other legitimate children, both via his second wife, are Ivan, an atheist and intellectual, and Alyosha, the youngest, a paragon of moral and religious probity and would-be man of God, follower of the saintly Father Zossima. Then there is Smerdyakov, Fyodor's illegitimate son by the town idiot, "stinking Lizaveta." (Fyodor Karamazov's previous wives are now dead, so in Dostoyevsky's novel, parent-offspring conflict is reduced to conflict between father and offspring.) Smerdyakov is kept on as a servant in the old rake's household, and it is he who murders Fyodor, a crime for which Dmitri is convicted.

Dmitri had plenty of motive. In addition to conflict over his inheritance—wanting more resources than the parent wants to provide—there is direct competition over the same woman: the lovely and seductive Grushenka, a luscious lady of ill repute. Imagine the struggles of McTeague and his erstwhile friend Marcus (see Chapter 2), this time taking place between father and son. Although shared genes may help soften the (literal) blows, such outright conflict within the context of kin is likely to be especially painful if only because of the ambivalence involved. Dmitri acknowledges the "hideous horror" of his struggle with "the old voluptuary" over Grushenka, a battle in which the hotheaded son had become increasingly enraged since his father and sexual competitor was also the holder of the purse strings, access to which he was unfairly—but understandably—denying his son. Things are more complicated yet. Dmitri is engaged, not to Grushenka but to Katerina, who in turn is loved by his brother Ivan! (Now sibling rivalry rears its unlovely head; moreover, as evolutionary theory would predict, it is especially intense among these two since they are half-brothers.) Ivan wants Dmitri to go ahead and marry Grushenka so that he can marry Katerina. But old man Karamazov also has hot pants for Grushenka, which is why he refuses to give any money to Dmitri. Fyodor's bankbook is all that he has with which to lure the gorgeous Gru.

Enter the saintly Father Zossima, patron and idealized father figure of the third brother, Alyosha, who tries to mediate an end to the family feuding, especially between Dmitri and his father, Fyodor, over Dmitri's inheritance. But to the embarrassment of all, the old rake simply acts like a buffoon, later insisting that he will hoard all the family money to use on young women. Specifically, Fyodor plans to offer the seductive Grushenka three thousand rubles in return for her sexual favors. Thus, the two men, father and son, literally do battle in competition over both resources and sex. Dmitri later beats his father, threatening to kill him if he in fact seduces Grushenka, so when Fyodor is murdered, the blame falls naturally enough on Dmitri. But actually Smerdyakov did it.

Freud claimed that *The Brothers Karamazov* was the greatest novel of all time, evidently because it "confirmed" his theory, developed in *Totem and Taboo*, that humankind began with a primal act of patricide, in which sons gathered together and killed the father (who, Fyodor Karamazov–like, had appropriated all the nubile women). Dostoyevsky's book is extraordinary, but its quality— or indeed that of any work of imaginative fiction—should emphatically not be judged by its consistency with any particular theory, even evolution. If biological accuracy made for great writing, then the next Nobel Prizes for literature should come from field and laboratory notebooks or the output of DNA sequencing machines. Indeed, writing that sets literary achievement second to any other goal is invariably second-rate. T. S. Eliot once noted, for example, that the problem with most religious poetry is that it is more religion than poetry.

Themes of violent sexual competition between fathers and sons are more widespread than one might think; moreover, the underlying biological reality of so-called Oedipal competition likely resides not at all in sexual competition for the mother, but for other women, with parent-offspring conflict thrown in for good measure. Take, for example, Eugene O'Neill's *Desire Under the Elms*, in which a son goes so far as to have sex with his stepmother, thereby cuckolding his father. In O'Neill's dark masterpiece of parent-offspring conflict set among stony people in equally stony, rural New England, old Ephraim Cabot had gained control of his second wife's farm, worked her to death, and brutalized his three sons. The youngest, Eben, stubbornly insists that the property is his birthright—shades of Dmitri Karamazov—and he proceeds to pay off his two older brothers, getting them to leave for California and forgo any claim to the farm in return. But flinty old Ephraim then shows up with his third wife, attractive Abbie Putnam, another tool in the contest between parent and child.

Not only does Eben resent the arrival of his young stepmother, he also recognizes that she threatens his inheritance. Abbie, in turn, wants to solidify her situation, and with unerring biological wisdom, she recognizes that producing a child is just the ticket. She proceeds to do so by seducing Eben, bearing his child, and announcing it as Ephraim's. It is as though Grushenka had married old Fyodor, then had a child by Dmitri so as to ensure herself the Karamazov kopeks. Good revenge for Dmitri—that is, Eben—as well, except for the fact that whoever cuckolds his father in this way also assures that he will be disinherited in favor of his biological son, the alleged child of the old man.

In O'Neill's version, Abbie comes to love Eben, who, in turn, is infuriated when he realizes how he has been used by her. Abbie then smothers her child, hoping thereby to prove that she really wanted Eben as a man and not just a sperm donor with whom to deceive her elderly husband into thinking she had borne him an heir.

The theme of parent-offspring conflict via sexual competition is not merely one for the nineteenth or twentieth century. It appears in Western literature as early as Euripides' ancient tragedy *Hippolytus*, which details the doomed passion of Phaedra, wife of Theseus, for her stepson, Hippolytus. Racine's seventeenth-century play continued the story into the beginning of modern times. In this tale, Hippolytus revered his father, the renowned monster slayer Theseus, but also had to deal with Theseus's darker side: his career as a womanizer who readily abandoned his conquests. Theseus's wife, Phaedra, was the sister of Ariadne, who had obligingly provided Theseus with the thread by which he found his way back out of the labyrinth after killing the notorious Minotaur. After this, Theseus ran off with both Ariadne *and* Phaedra, only to dump the previously helpful Ariadne in favor of her more nubile sister. Neither of these ladies, incidentally, gave birth to Hippolytus; that was a matter of Theseus hooking up—briefly, once again—with the queen of the Amazons. Theseus certainly got around.

Hippolytus is a perfect example of the lopsided playing field for children when battling with their parents, not only because of the obvious asymmetry in age, power, and psychological influence but also because of the latter's accumulated past. In Racine's play, considered one of the great cultural milestones of French literature, Hippolytus doesn't even make a pass at his stepmother; rather, Phaedra lusts unavailingly for him, then falsely accuses the youngster of rape, whereupon Theseus has him killed. And Phaedra kills herself.

The theme of cross-generational conflict, often with violent and sexual overtones, is alive and well in recent and contemporary literature, and we can be grateful that sometimes, at least, it stops short of murder. On occasion, it emerges as humorous, ironic, and even touching, as in John Millington Synge's finest work, *The Playboy of the Western World*. A young man named Christopher Mahon arrives one evening at a tavern on the wild Mayo coast, announcing that he has run away from home because he killed his father, Old Mahon, during a fight. This indication of manly independence greatly impresses the locals, especially Pegeen, the tavernkeeper's a-bit-too-easily-impressed daughter. (It is tempting to suggest that apparent victors in parent-offspring conflict are especially appealing to the opposite sex, because such victors are likely to be successful in turn; shades of the "sexy son hypothesis," described earlier. There is an added and ironic complication, however: insofar as success in parent-offspring conflict is inheritable, then offspring who emerge victorious over their parents are presumably prone to produce offspring who are, in turn, liable to best *them*.)

In any event, Christopher Mahon is prevailed upon to repeat, over and over, his tale: how he had always been a meek, obedient, put-upon son until Old Mahon insisted that he marry an elderly rich widow. Christopher's disobedience precipitated a battle between father and son, in the course of which the old man was struck lethally on the head. Young Chris becomes quite the local hero, admired by all the ladies, and as his fame spreads, he comes to believe his own tale and to relish his renown. This newly crowned playboy of the Western world is eventually accosted, however, by Old Mahon, who is not in the least dead after all and who humiliates Christopher in front of the village and demands that he return home. But at this point the son isn't so meek as before, and, energized by his status as young lord of filial defiance, he proceeds, much to the old man's amazement, to strike his father over the head once again; this time, moreover, it appears that he has finally, really, and truly killed his father. The townspeople—even Christopher's sweetheart, the pretty and impressionable Pegeen—now turn against him, being sensible of the difference between a rumored patricide in a distant county and the real thing, up close and personal. (As per our earlier suggestion, maybe Pegeen has second thoughts about possibly raising sons who are inclined to bash parental skulls.) But as Christopher Mahon, undutiful son in the extreme, is about to be hanged for his lethal lack of filial piety, his father comes to, for the second time—Old Mahon, we must conclude, has a hard head indeed—staggers over to his son, and unties him, whereupon the two leave arm in arm, reconciled at last and commenting contemptuously about how naive some country bumpkins can be.

Such cheerful foolishness aside, parent-offspring conflict generally assumes a more somber tone, one in which, as we have noted, the parents hold most of the cards. It dominates, for example, at least two of Tennessee Williams's plays: *Cat on a Hot Tin Roof* and *The Glass Menagerie*. In the former, generational conflict is played out most notably between Big Daddy Pollitt and his sons, once again, as in the Karamazov clan, over an inheritance. Big Daddy is the richest landowner in the Mississippi Delta, with two sons: Brick—his favorite—a bitter, alcoholic former football star (played in one film version by Paul Newman), and Gooper, a hypocritical, greedy father of five, with number six on the way. By contrast, Brick and his wife, Maggie (in which role Elizabeth Taylor sizzled opposite Newman), are childless. The stakes are huge, since Big Daddy is dying of cancer and, true to evolutionary expectations, no one expects the Pollitt plantation to go to a childless couple.

As the drama unfolds, we discover that Maggie is sexually (and financially) frustrated, describing herself as being "nervous as a cat on a hot tin roof," because Brick has no carnal interest in her, which makes it unlikely that either Brick or Maggie will produce the child necessary for them to get their share of Big Daddy's millions. The family's dirty laundry is revealed: Maggie had seduced Skipper, Brick's closest friend, in a peculiar effort to snap Brick out of his homosexual longings, and Skipper had subsequently committed suicide. As this lacerating play comes to an end, Maggie makes the lying announcement that she is, at last, carrying Brick's child, and then sets about trying to make it so.

While *Cat on a Hot Tin Roof* tells of parent-offspring estrangement, with parental expectations exceeding filial inclinations, in *The Glass Menagerie* Tennessee Williams presented his audience with a different side of parent-offspring conflict, this one proceeding via pathological enmeshment. It is the story of middle-aged Amanda, by turns infuriating and pathetic, who lives via the illusions of her youth, in the process stifling her children, one of whom eventually escapes while the other remains entrapped, surrounded by a menagerie of miniature glass statues. Parent-offspring conflict all right, in which a neurotic, controlling parent—whose pathology presumably precludes an accurate assessment of where her ultimate biological interest really lies—is unable to let go of her daughter; her son, meanwhile, must pry his way out of this particular menagerie.

Jonathan Franzen's *The Corrections* brings us yet another vision of conflicted parent-offspring interactions, a take that once again proves especially rewarding when subjected to an evolutionary perspective. In *The Corrections* we see a moderately dysfunctional but altogether recognizable family, the Lamberts, whose patriarch is slipping into dementia while the matriarch yearns for the perfect Midwest Christmas, which would meet parental needs but not those of her grown children. The younger Lamberts—Chip, Denise, and Gary—are pretty much making a mess of things, each failing to meet parental expectations in how they ply their own lives; at the same time, the siblings are unable or unwilling to support each other as the disappointed elder Lamberts would prefer. (Recall that the theory of parent-offspring conflict predicts conflict between parents and offspring not only over parental resources but also with regard to offspring behavior toward each other.) In Franzen's prize-winning novel, numerous "corrections" are attempted—and some even succeed, for a time—but underlying it all is a family dynamic that cannot be altogether corrected (because it derives from our human nature) and of which even the author was apparently unaware, even as he depicted it so faithfully.

Although growing up isn't a disease requiring correction, it evidently feels that way to many people. In the process, it's hard not to notice that when serious parent-offspring conflict is portrayed in literature, it is overwhelmingly a son (Holden Caulfield, Huck Finn,

Alexander Portnoy, and so forth) who does most of the conflicting. Why not daughters? There is nothing in the evolutionary origin of parent-offspring conflict that is sex-specific; certainly, daughters are no less endowed than sons with a 50 percent probability of *not* sharing genes with their parents. Perhaps the relative scarcity of depicted parent-daughter conflict is a result of the fact that, as we've seen, females are more likely to employ subtle styles of conflict, whereas males are more overt. But on the other hand, good literature is more than capable of revealing subtle interactions.

A notable—and notably gentle—portrayal of mother-daughter conflict comes from Laura Esquivel's magical re-creation of life in turn-of-the-century Mexico, *Like Water for Chocolate*. Tita, youngest daughter of the tyrannical Mama Elena, is required by her mother to remain unmarried, so as to be available to care for the aging matriarch. (It's a common practice in a number of cultures, analogous perhaps to those bird species in which the youngest offspring is often undersized and undernourished, essentially sacrificed for the benefit of other family members.) Tita must watch helplessly and miserably as her older sister gets to marry Pedro, the man of her dreams. Tita, however, succeeds via cooking, by which she imbues magnificent meals with her anger, grief, and longing. And, this being magical realism, Tita eventually triumphs.

In the less magical world of realistic biology, such triumphs are, sad to say, rare.

For an ominous and complex perspective on parent-offspring conflict (and another one in which sons are the centerpiece of conflict between parents and offspring), let's turn to William Faulkner's *Absalom, Absalom!*—the novel's title echoing the cry of grief and loss uttered by King David at the death of the son who had engaged in some no-nonsense parent-offspring obstreperousness. (Absalom had rebelled against his father and in the process killed one of his brothers.) In Faulkner's fictional Yoknapatawpha County, the reader gradually comes to understand the saga of Thomas Sutpen—a violent and dangerous Extra-Big Daddy—who is an even harder act to follow than is Theseus, less foolish than Fyodor Karamazov, less pathetic than Jack Portnoy, and a more potent presence than the parents of Huck Finn and Holden Caulfield put together. Sutpen is a larger-than-life figure (indeed, a larger than larger-than-life figure) who carves "Sutpen's hundred"—a hundred square miles of land—and a mansion out of the Mississippi wilderness by brutality and force of will. He is a genuine, genetic force of nature, a man whose vitality both defies and demands conflict from offspring and strangers alike.

From the onset, Sutpen has his "design," his plan for his own life and that of his posterity; all others are left to struggle in his wake. Sutpen, in Faulkner's hands, is history made flesh. His story and that of his descendants is reconstructed, piecemeal, by a southerner, Quentin Compson (who also figures prominently in *The Sound and the Fury*). The account is complex and multidimensional, and—as bespeaks true conflict—there is no single "correct" story line. Just as the "best" course of action varies depending on whether one is, for example, parent or offspring, and each gene and genetic lineage can rightly see itself as central to the process of life, everyone in *Absalom, Absalom!* has a different version of events. But despite the fractured and competing narratives, some things are clear: Thomas Sutpen was a fierce and driven progenitor who reveled in conflict and provoked more than his share. When not working on his land or building his mansion, he liked to relax by fighting hand-to-hand with his most powerful slaves. Once when his two children, Henry and Judith, observed this lurid scene, Henry fainted while Judith watched with fascination and delight. It doesn't take a literary genius to see that young Henry was somewhat unlike his father, whereas Judith was a chip off the old genetic block, yet Thomas Sutpen demands a *male* heir worthy, in his mind, of himself.

Henry attends college, where he befriends Charles Bon, an immensely seductive young man who is only later revealed to be Henry's half-brother, offspring of Thomas Sutpen by a Haitian woman many years ago, and who, because of his mixed blood, has been rejected by his father. Charles desperately seeks the elder Sutpen's acknowledgment but is continually denied. So he woos and wins his half-sister, Judith, without divulging their genetic relationship; when he arrives to marry her, he is murdered on the outskirts of the Sutpen plantation by Henry, his friend and half-brother, who evidently had learned of Charles's background and could not countenance the forthcoming incest. Having acted out his lethal share of half-sibling rivalry, Henry becomes a fugitive. Desperate to carry on his line, the aging but still "demonic" Sutpen proceeds to impregnate Milly Jones, young enough to be his granddaughter.

Milly is the granddaughter of Wash Jones, a squatter on Thomas Sutpen's land and Sutpen's occasional drinking companion. When Milly gives birth to a girl child instead of a boy, Sutpen brutally rejects her, which drives Wash to behead Sutpen with a scythe. As the curtain closes on several more Sutpen generations, each bedeviled by conflict, even Thomas Sutpen's brute strength and ferocious will are unable to carry the day against the destructive, divisive consequences of race hatred, brutality, and incest. Parent-offspring conflict pervades this story, just as the brooding and demanding presence of Sutpen himself weighs like a suffocating blanket on his offspring and, by extension, all children of the South. Their filial and sororal love eventually give way to fratricide as patterns of generational and individual conflict work their malignant and obsessive ends. Faulkner's plots are marinated in conflict—notably within as well as between families—and so is life.

Just as there is no single point of view that offers a unitary truth about Faulkner's South or its people, biological reality dictates that there is no unitary perspective when it comes to the process of living. Natural selection acts—not always successfully—to "maximize the fitness" of each individual, which means that even in the case of parents and offspring, personal agendas often clash and in the process, victory may be elusive, even impossible. Sutpen's grand design isn't necessarily shared by others, and this is why he fails. Here is Quentin Compson, meditating (in poetic stream of consciousness) on Thomas Sutpen in particular, but also on the demands of pushy parents more generally:

> Mad impotent old man who realized at last that there must be some limit even to the capabilities of a demon for doing harm, who . . . can deliver just one more fierce shot and crumble to dust in its own furious blast and recoil, who looked about upon the scene which was still within his scope and compass and saw son gone, vanished, more insuperable to him now than if the son were dead since now (if the son still lived) his name would be different and those to call him by it strangers and whatever dragon's outcropping of Sutpen blood the son might sow on the body of whatever strange woman would therefore carry on the tradition, accomplish the hereditary evil and harm under another name and upon and among people who will never have heard the right one.

A similar disconnect is central to the evolutionary concept of parent-offspring conflict: individuals, even close relatives such as parent and child, have their own distinct interests, since everyone occupies center stage in his or her own personal drama. Sutpen's design, and on a smaller scale the designs of all parents, are readily upended by historical realities, current situations, bad luck, and most of all by the stubborn but altogether understandable insistence of others—notably including, but not limited to, their own offspring—to pursue their own goals, their own designs, be they grand or petty. Whether the stuff of comedy, tragedy, or quotidian reality, such conflicts are part of evolution's bequeathal to us all.

Finally, for readers who crave yet more violence as well as those who may decry the lack of women in these accounts of parent-offspring conflict thus far, there is *The Oresteia*, the fall of the house of Atreus. If you think the Sutpen saga reveals that a house divided against itself cannot stand, the ancient Greek tragedy of Agamemnon and company shows just how divided and destructive such conflict can become. No one, it seems, outdid the Greeks.

Ten years before the action depicted in Aeschylus's classic tragedy, Agamemnon, commander of the Greek armies sailing for Troy, had sacrificed his own daughter, Iphigenia, in order to propitiate the gods and obtain a good wind for his fleet. He had lured Iphigenia to her doom, by the way, by falsely promising that she was there to be married to the Greek hero Achilles. Iphigenia wanted Achilles; Agamemnon wanted a favorable wind by which to pursue his own goals. How's that for conflicting interest between father and daughter? News of Agamemnon's deed generated no small amount of husband-wife conflict when it was reported to Mrs. Agamemnon, aka Clytemnestra. (It was not coincidental, incidentally, that Faulkner used this name—Clytie for short—in *Absalom, Absalom!*)

Generations before the killing of Iphigenia, the House of Atreus was not unfamiliar with lethal deeds. Indeed, it started out with a few skeletons in the family closet, notably the corpses of the children of Thyestes, who had been murdered by his competitor, Atreus, tyrant of Argos, and then served up as dinner to their father. (Parent-offspring consumption?) Actually, things are genetically more sinister yet: Atreus and Thyestes were brothers, whose sibling rivalry was such that the latter seduced the former's wife. This, in turn, is what had led Atreus to seek the culinary comeuppance of Thyestes, but by doing so via the shedding of kindred blood, Atreus contributed to a bad precedent.

Atreus, in turn, had two sons, Menelaus and Agamemnon, the former the cuckolded husband of Helen (she of Troy) and the latter chosen to lead the avenging Greek army against that city, and who, in the process, willingly presided over the killing of his own daughter. When Agamemnon returns to Argos after the Trojan War, he is murdered by the infuriated Clytemnestra—Helen's sister, by the way; the two brothers had married two sisters—who had been plotting her bloody retaliation while shacking up with a fellow named Aegisthus. All quite complicated, but it is about to get more so. Aegisthus is the sole surviving son of Thyestes, the one who had seduced Atreus's wife and was then unwittingly deceived into eating his own murdered children. Accordingly, Aegisthus has a bone or two to pick with the family of Atreus.

Agamemnon's ill-treatment of Iphigenia thus stands in a long tradition of the elders of the House of Atreus looking out for their own interests at the expense of their offspring. "It is bitter, bitter being the chief," complains Agamemnon in Aeschylus's play. "To slay my own little girl? With my hand to pour her virgin's blood on an altar to go to war? And yet, if I fail we never shall sail to Troy, as we have pledged to each other to do, and I shall dishonor myself and each of you." Murdering Iphigenia may well have been bitter, but as we suggested earlier, the Greek gods have often served as external excuses for the unpleasant but biologically influenced behavior of human beings. In Agamemnon's case, it is at least convenient that he is able to repackage the nastiness of especially intense parent-offspring conflict as divinely mandated: the gods made me do it.

True it is that by sacrificing Iphy, Aggy lost a daughter, but he gained immense prestige among his fellow Greeks and, with the ultimately successful war against Troy, his pick of their women. One of these was a daughter of King Priam, Cassandra, the one who famously warned in vain of Troy's coming fall and whose very presence added to Clytemnestra's fury . . . and whom Clytemnestra also slew. Part of the biology of parent-offspring conflict is that whereas each parent may be inclined to take advantage of his or her offspring, the other parent is likely to disagree vigorously, especially if, as in Agamemnon's case, the spouse's actions are asymmetric. Agamemnon's lopsided "conflict" with Iphigenia benefited only Agamemnon; not only was there no compensating benefit for Clytemnestra, there was even an additional loss beyond the death of her daughter, since Agamemnon ended up with a concubine whose success would in no way help Clytemnestra. No wonder she was annoyed.

So much for *Agamemnon*, the first of Aeschylus's three monumental tragedies. In part two, *The Libation Bearers*, Orestes—son of the murderess Clytemnestra and the murdered murderer Agamemnon—teams up with his sister, Electra, to kill Clytemnestra and Aegisthus. Killing Aegisthus is no big deal, but for a son to kill his mother is parent-offspring conflict with a vengeance, even when done with a substantial dose of vengeance in mind. And so, in part three, *The Furies*, Orestes is pursued by none other than the Furies themselves (representing his anguished guilt and society's demands for justice, or perhaps the agonized cries of overwrought genes whose interests have ultimately been betrayed because their possessors have overplayed their hand). In the course of the extended legal proceeding that follows, Apollo testifies on Orestes's behalf, making the biologically novel argument that matricide isn't as serious as it might seem, since mothers aren't really related to their children: only the father, who plants his seed in the mother's womb, is the true parent. Orestes beats the rap!

More likely, some things do matter, and one of them, whatever its biological origins, is the pain that comes from conflicts that evolutionary theory tells us may be modulated but not avoided altogether. Perhaps, however, they can at least be understood, and thereby rendered less hurtful, since unlike a lower mammal caught in the grip of weaning conflict or an adult bird bedeviled by a crew of insatiable fledglings, people are capable of wisdom (a capacity that is also part of our biology). Here is some of the wisdom of *Agamemnon*, reflecting the agony of parent-offspring conflict turned to murder, chanted by a Greek chorus and filtered through poetry that, although twenty-five-hundred years old, is transfixingly human:

> Even in our sleep, pain that cannot forget falls drop by drop upon the heart, and in our own despair, against our will, comes wisdom to us by the awful grace of God.

9

OF MUSKETEERS AND MICE AND MEN
AND WRATH AND RECIPROCITY
AND FRIENDSHIP

In Steinbeck Country and Elsewhere

———————

A ll for one," as the Three Musketeers were wont to say, "and one for all." Sounds good, but the biological reality is a bit more selfish and sobering, and so, for the most part, are its literary depictions. Even though virtue is reputed to be its own reward, the evolutionary process has a hard time rewarding virtuous behavior unless it is directed toward genetic relatives, either offspring (Chapter 8) or other kin (Chapter 6). In either case the reward comes from direct payoff to the genes in question, so altruism is revealed to be selfishness in disguise.

Yet altruism doesn't absolutely require shared genes; shared favors will do. The key is positive payback, or reciprocity: you scratch my back, I'll scratch yours. As a result, people—and other animals—can stick together through thick and thin, better or worse, so long as no one individual ends up with all the benefit and another with all the cost.

Buddy stories, which abound in literature as in life, offer models of seemingly selfless devotion. Scratch the surface, however, and there are inevitably payoffs working both ways. Otherwise, either the relationship isn't friendship (rather, compulsion and/or manipulation) or it isn't destined to last very long. Perhaps the oldest text of the Western written tradition is the enduring Babylonian tale of friendship and mutual assistance exchanged between Gilgamesh and

Enkidu. Also Homer's recounting of Achilles and Patrocles, the stir-
ring medieval romances featuring Roland and Oliver, the Knights of
the Round Table, Don Quixote and Sancho Panza, and more recent
if lighter fare, such as the Lone Ranger and Tonto or Butch Cassidy
and the Sundance Kid. In all these cases, favors are *exchanged*; never
is one friend only a giver and the other only a receiver.

As to those fabled Musketeers, even they weren't all that self-
abnegating. Throughout Dumas's swashbuckling novel, the implication was clear: buckles aren't swashed simply for the fun of it, nor just
to do good. Either they are payback for assistance rendered in the
past or they are done in the hope of generating a comparable behavior to be directed toward the current do-gooder sometime in the future. When our four heroes (Athos, Aramis, and Porthos—the
original Three Musketeers—plus D'Artagnan, the fourth) are needy
and poor, "the hungry friends, followed by their lackeys, are seen
haunting the quays and guard-rooms, picking up among their friends
abroad all the dinners they could meet with." This generosity on the
part of the Musketeers' friends and acquaintances is encouraged by
Aramis, who points out that "it was prudent to sow repasts left and
right in prosperity in order to reap a few in times of need." Biologists
call it "reciprocal altruism." Regular people call it friendship.

Here is Aramis, once again (who seemed to have an especially
acute sense of the role of reciprocity in mediating relationships
among nonrelatives). Aramis has just received some money from
his girlfriend, whereas D'Artagnan has been poor and hungry for
some time. "My dear D'Artagnan," says he, "if you please, we will
join our friends; as I am rich, we will to-day begin to dine together
again, expecting that you will be rich in return." The rules are
clear, even among these established buddies. They all share money
and food; in fact whenever they pawn a ring—which they do regularly, seeming to have an infinite supply of precious jewelry—they
share the money evenly between them, no matter who originally
owned it, and even if it was a family heirloom or lover's keepsake.
The Musketeers constitute a kind of inchoate Marxist co-op,

everyone making sure that the others can eat, drink, and be merry in safety and comfort.

This friendship among the Musketeers is especially intense because of the situations of great risk in which they find themselves. It's a familiar story, reflected, for example, in Shakespeare's famous wartime rallying speech, entrusted to Henry V: "We few, we happy few; we band of brothers; for he today that sheds his blood with me shall be my brother." Unsaid but clearly implied: he who chickens out isn't.

Camaraderie among the Musketeers is thus based as well on a kind of "I've got your back" theory. Dumas's heroes live in a very violent, dangerous time in which people kill each other merely for being disrespectful and in which traitors and double-crossers abound. They are the "inseparables" in part because there is safety in numbers, and they have made a commitment to help each other as much as possible. Thus, when D'Artagnan goes on a dangerous mission to England, they all go with him even though there is nothing in it for them (at least, not on the surface). They all risk their lives, with only D'Artagnan getting to his destination; the rest are left on the road in various fights and other encounters.

Because *The Three Musketeers* is really about D'Artagnan, it might seem that they assist him without any expectation of positive payback, but after D'Artagnan returns from England he undertakes the highly risky course of going out by himself, leading a bunch of fancy, valuable horses given to him by Lord Buckingham and which he means to give, most generously, to his buddies. In the course of events, D'Artagnan attempts to find and rescue the other Musketeers; after all, they went with him and sacrificed for him, so he will sacrifice for them by looking out for their safety. (Isn't that what friends are for?)

While it warms our hearts, friendship based on reciprocal benevolence is strangely delicate, always teetering on the uncertainty that whoever gives might not get back. Kin-based altruism, by contrast, is more secure: after one individual does something to benefit a relative, it doesn't matter whether the recipient is especially grateful, whether he or she reciprocates sometime down the road, or if the good deed is even identified as such. Part of the charm of selfish genetics is that the altruist's payoff inheres directly in his action, regardless of what the recipient does or doesn't do. Any benefit to the recipient automatically bounces back to the donor as a result of their shared genes.

This is a lesson that Blanche DuBois could have learned. "I have always relied," she famously acknowledges in Tennessee Williams's *A Streetcar Named Desire*, "on the kindness of strangers." Not necessarily a bad policy, but only if those others are likely to play ball, and that, in turn, depends in large part on whether you are in a position, and of an inclination, to reciprocate when called to do so. Otherwise, there is little reason for those strangers to continue to help you, and much reason for them to act less benevolently. In Williams's play, Blanche also relied on the kindness of her kin—notably her sister Stella, with whom she moved in when she needed a place to live. But Stella turned out to be sexually in thrall to the earthy Stanley Kowalski (played in the movie version, torn T-shirt, rippling muscles, and all, by a young and sexy, "Stella!"-screaming Marlon Brando). And Blanche turns out to be helpless and hapless, sexually violated by Stanley and left to the not-so-tender mercies of a notably unkind, late-arriving crew of operatives from the Louisiana state mental hospital.

Altruism based on reciprocity involves a short-term asymmetry: the altruist gives while the beneficiary gets. What's to keep the beneficiary from gathering in his benefits but refusing to reciprocate when and if the opportunity arises? This is why relying on the kindness of strangers is so risky. What if, once D'Artagnan becomes wealthy, he snubs Aramis and doesn't pay him back after all? We can assume that if this were to happen, their friendship would end pretty quickly, just as it does in our own twenty-first-century world when a couple repeatedly has dinner at another's house but doesn't invite them over "in turn." What, for instance, is likely to happen to those who keep receiving Christmas cards from others while not sending out any of their own?

At the same time, reputation counts, and people often restrain themselves or sacrifice on behalf of a friendship, so long as it is one that has endured, which is to say, so long as the friends have already shown themselves to be reliable reciprocators. At one point in *The Three Musketeers*, the evil Cardinal Richelieu has just offered D'Artagnan a high office in his own guards—a real honor, and very tempting—but he refuses, at which point the cardinal warns that given D'Artagnan's penchant for getting in trouble, and considering the enemies he has made and the schemes he has been involved in, our fourth Musketeer is in serious danger. To be safe, the cardinal insinuates, he really should take up the cardinal's offer. D'Artagnan again refuses. After this,

> D'Artagnan went out, but at the door his heart almost failed him, and he felt inclined to return. But the noble and severe countenance of Athos crossed his mind: if he made the compact with the Cardinal which he required, Athos would no more give him his hand, Athos would renounce him. It was this fear that restrained him, so powerful is the influence of a truly great character on all that surrounds it.

It isn't always easy, but, as Ringo Starr once sang—and Athos, Aramis, Porthos, and D'Artagnan would doubtless agree—it is often at least possible to "get by with a little help from my friends."

In a sense, reciprocal altruism is plain common sense: one good turn deserves another. Its echo resounds whenever we assess some-one as "reliable," "trustworthy," or "somebody you can (or can't) count on," as well as in such laments as "What did I do to deserve that?" or the satisfying "She had it coming." But sometimes it takes a while for common sense to become scientifically legitimate. In this case, the problem is as follows: when genes aren't shared and thus nepotism (aka kin selection) isn't at issue, is there some way that natural selection can still promote altruism?

The answer is, it can.

Here is the idea. Imagine you have some food and are about to eat it. Along comes someone else who begs for a portion. If you give in and share, there is less for yourself. Of course, if your hungry importuner is a genetic relative, you should be generous—that is, altruistic—and all the more so if you are closely related. You can also be expected to be more generous if he or she needs the food more than you do, less so if you can really use the extra bite yourself. But there is an additional set of circumstances in which it would pay your genes to donate food to the beggar, *even if the two of you are not related at all.*

The key provision is that in the long run, the ultimate benefit of being altruistic has to overcome the cost. In other words, there must be a good chance that sometime in the future the tables will be turned—in our example, literally. If the recipient of your generosity, for example, will someday have food when you don't, and if your good deed will be remembered and repaid in kind, then reciprocity may be a tasty feast for everyone involved. (And in this case, once again, the act isn't really altruistic at all! Thus are the Musketeers—for all their friendly collegiality—revealed as, at heart, the three must-get-theirs.)

It sounds simple enough, but actually these are demanding, rigorous conditions, which probably don't occur very often. The problem is that as soon as you have given some food to a beggar, unrelated to you, your evolutionary bank account has gone down by one notch, while the beggar's has gone up. This wouldn't be a problem in the long run *if* you will eventually gain more than you lost by donating the food in the first place. Not impossible, for example, if you have just killed a large animal, since you probably wouldn't require every bit of the meat. For a hungry beggar, by contrast, just one slice might mean the difference between survival and starvation. So you lose a little; he gains a lot. If sometime later *you* are starving and your beggar buddy has hit the jackpot, it could be a good trade all around: give a little when you are fat and happy in return for a likely assist when you're really in need. In some ways, the transaction is equivalent to a healthy person banking some of her bone marrow to be used later, if necessary, when the situation is dire. But when it comes to bone marrow, there is no question that the donor will be able to retrieve the investment. In the case of reciprocity, on the other hand, there is a huge and lingering doubt: the beggar/beneficiary might *not* pay you back, or might not be around even if he is inclined to do so. It's a risk, sometimes a big one. Perhaps this is what Polonius, in Shakespeare's *Hamlet*, was driving at when he advised his son, "Neither a borrower nor a lender be."[1]

Remember, after the first act of altruism, you (the altruist) have lost something—in our example, some food, which translates into some fitness, the basic biological currency—while the recipient has benefited. What is to stop him or her from profiting from your generosity, then snubbing you when payback time rolls around? It is more blessed, we are told, to give than to receive. But it is awfully tempting to take and not to repay. (Maybe that's why giving is so universally blessed!)

In any event, this is the major difficulty that any system of reciprocal altruism must confront. Economists refer to it as the "free rider problem," referring to the ever-present temptation of taking advantage of another's benevolence. "Good people," of course, don't do it. Take, for example, one of those high-minded souls encountered by Jack Kerouac in *On the Road*:

---

[1] If Polonius were only worried that his son might not be paid back, he presumably would have left it "Don't be a lender." The problem with being a borrower is that then *you* are expected to pay back or lose your friend.

I was just about giving up and planning to sit over coffee when a fairly new car stopped, driven by a young guy. I ran like mad.

"Where you going?"

"Denver."

"Well, I can take you a hundred miles up the line."

"Grand, grand, you saved my life."

"I used to hitchhike myself, that's why I always pick up a fellow."

Besides what economists call the free rider problem, there are other difficulties as well, all of which must be resolved for reciprocity to work. For example, altruist and beneficiary must have a good chance of meeting again later. Otherwise, a debt incurred cannot be paid back. Hence, reciprocity isn't anticipated among total strangers, who are likely to melt away into an anonymous crowd. Part of the early hippie ethos—celebrated and to some extent created by Kerouac in his book—is that crowds need not be anonymous and that treating people as reliable reciprocators might become a self-fulfilling prophecy. There must also be enough likelihood that at some time the tables really will be turned, that the beneficiary will be in a position to reciprocate and repay the debt, and that he will in fact do so when the opportunity arises. This notoriously didn't happen among most of Kerouac's generation, so the benevolent, prosocial, indiscriminately altruistic hippie ideal has largely fallen by the wayside, victimized in part by the fact that crowds are too large and too anonymous to promote reciprocity. (To be sure, there are still bumper stickers that urge us all to "perform random acts of kindness," but relatively few actually do so.)

If a youthful hippie never grows up to own a car or if D'Artagnan were just an average schlemiel, then picking up a hitchhiker or buying D'Artagnan a meal wouldn't be terribly wise. Given D'Artagnan's demonstrated competence with a sword, however, Aramis's "altruism" shows not just generosity but some cagey evolutionary wisdom. (We're not so sure about the hitchhiker.) The beneficiary must also be able to recognize the altruist and not dispense repayment to others who don't deserve it. But above all looms the temptation to cheat.

Most people assume that cheating is nasty, unfair, maybe even despicable. And there is probably a good reason why this is so, why cheaters generally evoke such intolerance. In fact, it may be precisely *because* reciprocity is so important to people and at the same time so vulnerable to cheaters that the nonreciprocator is singled out for such criticism. "There is no duty more indispensable," wrote Cicero, "than that of returning a kindness. All men distrust one forgetful of a benefit."

Pity the poor cheater, however, for she is doubly tempted. First she gains by receiving something for nothing; then she gains yet again by ducking out when it is her turn to reciprocate (if she *does* reciprocate, then she would instantaneously be losing something, however small, just as the initial altruist did). Of course, the cheater may really be forgetful, as Cicero suggests, but it seems more likely that just as there can be Freudian slips, there are "Freudian forgets," or rather, biologically mediated inclinations to deny one's obligation. Maybe beggars can't be choosers; they can, however, be cheaters.

Emile Zola's huge multivolume panorama of human foible and folly, *The Human Comedy*, includes many examples of such moral defection, notably including *Nana*, the tale of a beautiful and heartless woman who takes love and money from many, promising (and sometimes giving) sex in return, but no love. Nana is vain, shallow, stupid, voluptuous, and a genuine, 100 percent cheater, not only in the customary sense of contemporary country ballads ("Oh, your cheatin' heart") but also in the deeper, biological sense of someone who fails the basic criterion of reciprocity, leaving havoc in her wake.

The situation, however, is not hopeless. Just as there are costs to reciprocating (hence temptations to cheat), there are also some good reasons to do one's part and play by the rules. Most important, if cheaters can be identified and then punished, this in itself would make cheating costly. Such punishment could be administered directly, as by physical punishment or some form of reprimand, or indirectly, by excluding the cheater from subsequent exchanges of favors. Biologist Robert Trivers first opened the topic of reciprocal altruism to scientific scrutiny in a landmark paper published in 1971, when he was a graduate student at Harvard, just as he was to open biologists' eyes to parent-offspring conflict three years later. Although Trivers pointed to a few possible animal examples, he also suggested that the requirements for reciprocal altruism are so stringent, and by the same token the temptations to cheat are so great, that full-blown reciprocity might well be a human specialty. Trivers even coined the phrase "moralistic aggression" for the peculiarly intense pressure that people exert upon suspected cheaters in an effort to bring them into line and keep them there. Hell hath no fury like an altruist whose generosity is not reciprocated.

That, at least, is the theory. The practice—at least among animals—is best demonstrated by that favorite of horror stories, the vampire. Not Bram Stoker's novelistic version, nor the much mythologized Romanian nobleman, Vlad the Impaler, but *Desmodus rotundus*, a species of bat that exists only in Central and South America, not Transylvania. These creatures don't actually suck blood; rather, they lick it after making a tiny incision with their needle-like incisors. And their preferred prey is livestock—cattle and especially horses—not people. (There are two other species of vampire bat, white-winged and hairy-legged, which feed mostly on bird blood and about which very little is known.) In Central America, where they have been most carefully studied, vampire bats are reported to land on the mane or tail of their victims, although one of the authors has seen them in the cattle country of Colombia, landing on the ground nearby, and then tiptoeing incongruously, shoulders hunched high, apparently trying not to awaken their prey.

Belying their nasty reputations, vampires use a buddy system. They are, if you please, reciprocating altruists.

Vampire bats roost in caves or hollow trees, in groups of eight to twelve adult females, including their female offspring (young males leave home early). Gerald Wilkinson of the University of Maryland, who has studied these creatures for many years, finds that vampire bats' daytime roosts are not made up of close relatives, however. Instead, female vampire bats within these roosts establish long-term associations, often lasting several years, and sometimes more than ten. These creatures are female musketeers. (It may also be noteworthy that despite their small size, most bats have a long life span, which in itself seems to be favorable for reciprocity, since it would increase the chances of eventual payback.)

What do vampires share? Food. To put it frankly, they regurgitate blood to each other. A well-fed bat has a distended belly, easily visible to the human observer and presumably even more obvious—and attractive—to a hungry bat. The solicitor licks under the potential donor's wing, then her lips; if successful, she gets a vomited meal of blood in return. This may sound less than delightful, but for a vampire bat, it is the very stuff of life. Such sharing is important, since one-third of all vampire bats under two years of age come back from their nocturnal flights empty-stomached, generally because the cows or horses on which they attempted to feed woke up and brushed them away. This wariness by their victims poses a serious problem for the bats, since they need to consume 50 to 100 percent of their weight in blood every twenty-four hours, and if they fail to obtain a meal just two nights running, they can starve to death. All vampire bats come back hungry on occasion, and it appears that individuals who are successful one night are liable to fail, unpredictably, on another. So, the life of a vampire bat is one of constant alternation between success and failure, which satisfies that key requirement for reciprocal altruism: recipients and donors often trade places.

This helps them avoid what has been called the banker's paradox: when we most need help, we are typically least able to reciprocate. Thus a bank is most eager to lend money to those who need it *least*, that is, those who are the best credit risk. As it happens, a starving vampire bat is about as good a credit risk as one who is fat and happy.

Wilkinson found that bats establish relationships of reciprocating regurgitation, and furthermore, that such pairs are composed only of those that have been together at the roost at least 60 percent of the time. In addition—and as expected by reciprocity theory—the cost of donating blood is less than the benefit of receiving it: a recently fed bat can save the life of its roostmate at relatively little cost to itself. Wilkinson also found that well-fed vampires direct their bounty toward others who are especially needy and close to starvation. And finally, bats that have been starving and received mouth-to-mouth transfusions are likely to reciprocate the next time around, when their hunting has been good and the previous donors are in need. It is not clear whether there are any cheaters among vampire bats, but these animals evidently can recognize each other as individuals, so it seems likely that any cheaters could readily be made to suffer a dire penalty for nonreciprocation: denial of food when they need it.

Whether human beings are natural-born killers is still up for debate, but we are definitely natural-born reciprocators, despite the glum warning of Welsh writer Alice Thomas Ellis: "There is no reciprocity. Men love women. Women love children. Children love hamsters." It is probably significant as well that whenever it comes to such exchanges, our emotions and sense of fairness are immediately engaged. One good turn deserves another—we all give at least lip service to this bromide, partly because it is hammered into us by our teachers, our families, our codes of morality and ethics, the constant prodding and insistent concern of society at large. Might there not also be a deeper, biologically based inclination as well, one that probably involves a complicated mix of tendencies, including—but not limited to—"reciprocate when necessary" and "cheat when possible"? Maybe social rules wouldn't push so hard for reciprocity if people were naturally inclined to play fair most of the time, to return good for good without the urging of ethical principles and moral teachings.

Or perhaps this is a false dichotomy if ethical principles are in fact externalized instinct, reflecting our intuitive sense that moral behavior is in our long-term self-interest. But it's not likely: even though natural selection does not operate importantly at higher levels such as groups or species, social groups nonetheless exist, and not surprisingly, they seek to indoctrinate their members to behave in ways that contribute to their own success.

Nonreciprocators, those who take and do not give, are quickly singled out as selfish undesirables, people one had better not associate with. It probably takes a fair amount of intellect to achieve this: you have to remember who got what in the past, and whether he or she gave back. It is even possible that our braininess is due in part to precisely these demands. And it may not be coincidental that vampires, notable reciprocators that they are, also have the distinction of being the brainiest of all bats.

Essentially, reciprocity is a social contract, of the sort that operates in our most intimate, day-to-day lives. It is so powerful, in fact, that most people are discomfited to receive a handout without some prospect of repayment. Charity is devoid of reciprocity, to be sure, and thus it is notably hard to take. Consider this account in *The Grapes of Wrath*, by John Steinbeck (of which more later). Here, a woman is remembering how she once took charity:

> "We was hungry—they made us crawl for our dinner. They took our dignity. They—I hate 'em! . . . Mis' Joad, we don't allow nobody in this camp to build theirself up that-a-way. We don't allow nobody to give nothing to another person. They can give it to the camp, and the camp can pass it out. We won't have no charity!" Her voice was fierce and hoarse. "I hate 'em," she said. "I ain't never seen my man beat before, but them—them Salvation Army done it to 'im."

Even so, most people are uncomfortable when expectations of reciprocity are made explicit, just as friendship itself is somehow supposed to transcend "mere" considerations of payback and the troubling question "What have you done for me lately?"

As with so many aspects of human behavior, it seems that the dignity of an act varies inversely with the extent to which it can be explained; for many, to understand the underpinnings of our behavior is to deprive it of legitimacy. To these readers, we apologize, but we do not retract! Like language, reciprocal altruism is clearly learned, and it follows rules that are handed down by local tradition. Just as all human beings use language, and all languages have certain deep structures in common, all people engage in some form of reciprocity, patterns that are shared worldwide and that everyone can understand at some level. Although generosity is praised, reciprocity is expected.

"There is some benevolence, however small," wrote David Hume in 1750, ". . . some particle of the dove kneaded into our frame, along with the elements of the wolf and serpent."[2] That "particle of the dove," responsible for our most benevolent, altruistic inclinations, is the same part of human beings—as well as wolves and serpents—that responds to the adaptive call of either kin selection or reciprocity or both. And that writes about it.

In theory, reciprocal altruism is quite different from kin selection. It could even operate between individuals who are members of different species, since at the genetic level, reciprocity is promoted by paybacks that benefit genes within the body of the individual doing the behaving rather than through relatives. For this reason, "reciprocity" is a better term than "reciprocal altruism," since it refers to a process that is actually selfish, working at the level of the individual as well as his or her genes.

A large proportion of human behavior—both real and depicted in literature—might well be encompassed and in a sense explained

---

[2] David Hume, An Enquiry Concerning the Principles of Morals (Oxford: Clarendon Press, 1975).

by adding together the combined effects of kin selection and reciprocity. With whom, for example, do people interact? Offspring, mates, friends, and those with whom we do business or exchange information. Of these, the first relates to nepotism and thus kin selection, whereas marriage represents a mixed case by which people typically marry nonrelatives and create genetic relatives. The last two—friends and business associates—are ripe for reciprocity. (And cheating.)

Pure kin-selected altruism is probably rare: most people generally expect some kind of reciprocity even from their relatives. And although pure kin selection without reciprocity may be infrequent, pure reciprocity without kin selection is comparatively common. It is, we suggest, what friendship is all about. "The only reward of virtue is virtue," wrote Ralph Waldo Emerson. "The only way to have a friend is to be one." We count on our friends. And they count on us. Put more strongly, we identify someone as a friend if, and often only if, we *can* count on him or her. What do we count on friends *for*? For reciprocity: for the reliable, useful, and pleasurable exchange of favors, assistance, valued company, sympathy, shared interests, and so forth. And, if you are a musketeer, for saving one's life when needed.

But that's not all.

"A friend," we learn in Shakespeare's *Julius Caesar*, "should bear his friend's infirmities." Think of how often the phrase "That's what friends are for" pops up when someone does a good turn for another. At the same time, doesn't "One good turn deserves another" also imply that if you fail to do someone a good turn, you might well be ignored in the future? Or that if you have received a good turn, failing to pay it back would be a wrong turn indeed?

It may be cynical to point out that friendship ultimately rests on the expectation of reciprocity. But cynicism and validity are not mutually exclusive. When it comes to business associates and others with whom we interact but do not strictly consider friends, relationships are governed even more objectively and impersonally by the expectation of getting fair value in return for our offerings. We exchange goods or favors, for example, with the shared expectation that each side is being fairly recompensed. Even a simple request for information—such as the time of day—is repaid with a "thank you," which acknowledges the small debt and also, in a sense, confers a degree of status upon the donor.

Among chimpanzees, interestingly, being nice is often a matter of selfish tactics in disguise. There is a method to the beneficent madness of chimps who give food away. As primatologist Frans de Waal puts it, "sharing is no free-for-all." Or, as Marcus Aurelius noted in the second century, "The art of living is more like wrestling than dancing."

Examining more than five thousand food transfers among chimpanzees housed at the Yerkes Regional Primate Research Center, in Georgia, de Waal and his associates found that "the number of transfers in each direction was related to the number in the opposite direction; that is, if A shared a lot with B, B generally shared a lot with A, and if A shared little with C, C also shared little with A." De Waal also reports, interestingly, that chimpanzees don't insist that reciprocation always take place in the same currency. For example, "A's chances for getting food from B improved if A had groomed B earlier that day."[3]

Robert Trivers, the guru of reciprocal altruism, has pointed out that systems of reciprocity are likely to be inherently unstable, relying on a variety of psychological and social mechanisms in order to keep them going. Diverse psychological traits and social teachings may in fact owe their existence to the pressures of reciprocal altruism. To take one example, the Golden Rule is a statement of idealized reciprocity: Do unto others as you would have others do unto you. And while it's true that there exist such injunctions as

---

[3] Frans de Waal, *Good Natured* (Cambridge, Mass.: Harvard University Press, 1996), 153.

"Whomsoever shall smite thee on thy right cheek, turn to him the other also," lessons similar to the Golden Rule have been identified for most of the world's religions and ethical systems. They are especially important since, as we've seen, when those others are truly "other"—that is, unrelated—there is a powerful yet subtle pressure to behave more selfishly.

When cheating is obvious, retaliation can be an equally obvious response, at least among creatures intelligent enough to recognize the transgression. Human beings are experts at blaming one another for what they have or have not done, whether their behavior was warranted given previous events, and so forth. "Friendship is friendship," goes an old Chinese proverb, "but accounts must be kept." Punishment for nonreciprocation—although implied among vampire bats—has thus far been reported for only one non-human animal species, and it should surprise no one that this animal is the highly intelligent chimpanzee. Here is an account by de Waal, describing a complex encounter among captive chimps at the Arnhem Zoo in Holland:

> A high-ranking female, Puist, took the trouble and risk to help her male friend, Luit, chase off a rival, Nikkie. Nikkie, however, had a habit after major confrontations of singling out and cornering allies of his rivals, to punish them. This time Nikkie displayed at Puist shortly after he had been attacked. Puist turned to Luit, stretching out her hand in search of support. But Luit did not lift a finger to protect her. Immediately after Nikkie had left the scene, Puist turned on Luit, barking furiously. She chased him across the enclosure and even pummeled him.

It is easy to imagine Puist saying to Luit: "I helped you. Why didn't you help me?" It seems that the wisdom of reciprocity is no more lost on chimpanzees than it is on humans. And much of this wisdom seems to come from the penalty of being caught refusing to pay back the good turn that one has received.

For another example, de Waal compares the behavior of two adult female chimpanzees, Gwinnie and Mai:

> If Gwinnie obtained [food] . . . she would take it to the top of a climbing frame, where it could easily be monopolized. Except for her offspring, few others managed to get anything. Mai, in contrast, shared readily and was typically surrounded by a cluster of beggars.

Later, Gwinnie, with her stingy personality, encountered more resistance and threats when she begged food from the others. De Waal concludes, "It is as if the other apes are telling Gwinnie, 'You never share with us, why should we share with you!' "

In short, reputation matters. As a result, people scramble to associate themselves with the ideals of kindness, goodness, and benevolent altruism, although actually practicing them is often a different matter. In Shakespeare's *Richard III*, there is a hilarious scene in which Richard—one of the most despicable, altogether unredeemable villains in literature—is trying to convince the citizens of London that he is ethical, altruistic, and trustworthy. So he arranges to be seen reading the Bible, something he assuredly does not do when left to his own devices.

Friendship is not just a guy thing, nor is reciprocity. In fact, given that males are biologically primed to be if anything more competitive than females, we might expect female friendships to be more pronounced than their male counterparts.

Indeed, "sisterhood" is reputed to be powerful, and never more so than when it is backed by reciprocity. In Rebecca Wells's *Divine Secrets of the Ya-Ya Sisterhood*, the mother-daughter team of Vivi (mother) and Siddalee (daughter) are going through a rough patch when the Ya-Yas—Vivi's band of stubborn and endearing girlfriends, who go back more than sixty years together and are now "bucking seventy"—sashay in and conspire to bring everyone back together. They get Vivi to send Sidda "The Divine Secrets of the Ya-Ya Sisterhood," a scrapbook of their early years together.

As Sidda enters into the early lives of this tribe of Louisiana wild women, she comes to appreciate the power of relationships in their lives, and her own. We meet Vivi, Necie, Teensy, and Caro, growing up in central Louisiana during the 1930s and 1940s, and through Sidda's eyes we learn how her mother suffered and survived, leaning all the while on her Ya-Ya friends and being leaned on in turn.

The Ya-Yas' most divine secret is their reciprocity. When Necie is unable to accompany the others to see the premiere of *Gone With the Wind*, Vivi writes her daily letters:

> You asked me to write you about every little thing, and that is just what I'm going to do. I'll save everything and we can paste it into my Divine Secrets album when we get home! . . . So that is our day, Countess Singing Cloud. And every single word is true. And when we get home, we will act things out for you. . . . Well, we all slept real late, especially me who was a real lazyhead after staying up all night writing to you. . . . You are our blood sister, remember, and blood sisters can never really go away from each other, no matter how lonesome train whistles sound in the night air.

Vivi's letters are detailed enough to tell an entire story that goes on for about thirty pages. Writing so much is itself a sacrifice on her part, one of many that each of the Ya-Yas makes for each other. Then when Vivi has a breakdown and runs away, the Ya-Yas look after Vivi's kids. Here are excerpts from Vivi's letters to each of her friends talking about her children and her gratitude:

> Teensy Baby . . . That beautiful white gown you gave her [Sidda]. How did you find something so perfect? . . .
>
> Caro Dahlin . . . My dearest friend—I am . . . at a loss for

words to thank you for all you have done for me and my gang. Taking care of my boys for almost three solid months. . . .

Dear dear Necie . . . You were the one who kept my world running while I was gone. . . . The ten thousand basketball games and altar-boy practices and Girl Scout and Brownie meetings and dentist appointments and God knows what else. Baby doll, you must have *lived* in your station wagon between taxiing your kids *and* mine.

Rebecca Wells's gem of a novel is filled with poignant little reciprocation moments, and also telling cases of nonreciprocation, as when Lyle Johnson refuses to do Vivi a favor, but she eventually gets around him and then adds, " 'Lyle, I'm looking forward to the day when you have to ask *me* for a favor.' Mama winked at me, and I winked back." There is a lot of winking in *The Divine Secrets of the Ya-Ya Sisterhood*, a paean to the widely acknowledged power of female friendship, as well as to the typically unacknowledged impact of reciprocity.

Nor are the Ya-Yas alone. The world abounds in beautifully realized accounts of women's reciprocity: old, young, and in-between. Terry McMillan's *Waiting to Exhale* gives us the earthy, hilarious tribulations of Bernadine, Gloria, Robin, and Savannah, four thirtyish African American women who struggle with men in modern Phoenix, supporting each other all the while, sometimes literally holding one another up (depending on the vagaries of alcohol and other events). And Amy Tan's phenomenally successful *The Joy Luck Club* owes much of that success to its meticulous description of how four aged Chinese women, seated around a mah-jongg table in San Francisco over a span of forty years, offer friendly sustenance to each other's past as well as to their present and future via one another's daughters.

Interestingly, even though the biological theory of mutual exchange is not sex-specific, there seem to be relatively few accounts of male-female friendships based on pure reciprocity; perhaps this is because such relationships, uncontaminated by sexual overtones, are themselves rare.

When it comes to depicting bonding and friendship among men, John Steinbeck is our top candidate. If not the most profound, his portraits are at least the most generous, warmhearted, and sympathetic of any in literature. Steinbeck country is the realm of reciprocity.

Probably the finest buddy novel in modern times is Steinbeck's *Of Mice and Men*, the tragic, beautiful, bittersweet tale of Lennie Small, a dim-witted but well-meaning giant, and George Milton, "small and quick, with restless eyes and sharp features." The two travel the dusty roads of rural California, dreaming of someday owning their own land. As George explains to Lennie,

> "Guys like us, that work on ranches, are the loneliest guys in the world. They got no family. They don't belong no place . . . [but] with us it ain't like that. We got a future. We got somebody to talk to that gives a damn about us. . . ." Lennie broke in, "But not us! An' why? Because . . . because I got you to look after me, and you got me to look after you, and that's why." He laughed delightedly.

George is smart, a bit of a schemer, and—when necessary—a liar, but with a good heart. Lennie could never survive without him, and indeed, his uncontrolled strength constantly gets the two of them into trouble. On the other hand, George is able to benefit from Lennie's physical capacity as a worker. Theirs is a reciprocating system: "We kinda look after each other," acknowledges George. "He ain't bright. Hell of a good worker, though. Hell of a nice fella, but he ain't bright. I've knew him for a long time."

At one point, when Crooks, one of the ranch hands, teases Lennie that maybe George had gotten hurt and won't be coming back from a night on the town, we get a hint of Lennie's protectiveness:

Suddenly Lennie's eyes centered and grew quiet, and mad. He stood up and walked dangerously toward Crooks. "Who hurt George?" he demanded. Crooks saw the danger as it approached him. . . . "I was just supposin'," he said. . . . Lennie stood over him. "What you supposin' for? Ain't nobody goin' to suppose no hurt to George."

Later, when Lennie has accidentally killed a young woman and is about to be hunted by an angry posse, we hear George announcing reciprocally, "I ain't gonna let 'em hurt Lennie." At the novel's end, it is ironically George who kills Lennie, but he does it humanely, with immense sadness, in a final, desperate act of friendship and kindness, the best that down-and-out people—committed to helping each other—can come up with in a cruel and brutal world.

In another of his great stories of reciprocity, *Tortilla Flat*, Steinbeck chronicles the denizens of a particular stretch of Monterey, California, populated by the "paisanos." This varied band of merry men share Mexican, Indian, Spanish, and Caucasian backgrounds as well as a cheerful fondness for booze, women, each other's company, and the avoidance of hard work. Danny, Pilon, Jesus Maria, Big Joe Portagee, and the Pirate consume wine by the gallon and whores by the brothelful. There is, however, a moral code—in a word: reciprocity.

Big Joe Portagee was happy to be with Pilon. "Here is one who takes care of his friends," he thought. "Even when they sleep he is alert to see that no harm comes to them." He resolved to do something nice for Pilon sometime.

Everyone in *Tortilla Flat*, it appears, resolves to do something nice for someone else sometime. And often they do. Mostly they exchange favors, and although they do not explicitly keep track, implicitly everyone maintains careful accounts:

It is impossible to say whether Danny expected any rent, or whether Pilon expected to pay any. If they did, both were disappointed. Danny never asked for it, and Pilon never offered it. The two friends were often together. Let Pilon come by a jug of wine or a piece of meat and Danny was sure to drop in and visit. And, if Danny were lucky or astute in the same way, Pilon spent a riotous night with him. Poor Pilon would have paid the money if he ever had any, but he never did have.

The dust jacket of our copy of *Tortilla Flat* describes the book as follows: "A story of love, laughter, and larceny in an uninhibited private world." Uninhibited by many standards, but rigid when it comes to the unwritten demands of friendship. In fact, the expectations of reciprocity are so ingrained that our bighearted band of rascals even convince themselves that they had behaved according to "code" when they hadn't, as in this exchange, in which Pablo and Pilon complain that one of them—Danny—has had the bad grace to hint at money that he is due:

> "We have been his friends for years. When he was in need, we fed him. When he was cold, we clothed him."
> "When was that?" Pablo asked.
> "Well, we would have, if he needed anything and we had it. That is the kind of friends we were to him."

One of the vagabonds—the Pirate—used to live in a chicken coop with his five dogs, earning twenty-five cents a day for hauling kindling into town. The others contrive to get hold of his money, which they estimate at about a hundred dollars, but are stymied when he relies on *their* friendship, asking them to help him keep it safe. It seems that the Pirate owes a gold candlestick to St. Francis, in thanks for having cured one of his dogs. When they learn that the Pirate is planning to use the money to repay this debt, they— who had been planning to steal it—now guard it diligently. The responsibilities of reciprocity are instantly appreciated and respected, even if the debt holder is a plaster saint.

Danny is the central character in *Tortilla Flat* and the social glue that holds everything together. He also owns the house where everyone lives but for which no one, it seems, pays rent. This introduces a constant, mild undercurrent of good-natured guilt, since, as we have seen, favors are supposed to be repaid. And so there recurs throughout the book a heartwarming, almost childlike eagerness on the part of Danny's friends to "do the right thing" and "make it up" to him. Thus, when the house they had previously been occupying burns down and all are invited to come and live in Danny's place, the paisanos celebrate Danny's generosity, also immediately offering to repay him:

> Although no one had mentioned it, each of the four knew they were all going to live in Danny's house. Pilon sighed with pleasure. Gone was the worry of rent; gone the responsibility of owing money. No longer was he a tenant, but a guest. In his mind he gave thanks for the burning of the other house.
>
> "We will all be happy here, Danny," he said. "In the evenings we will sit by the fire and our friends will come in to visit. And sometimes maybe we will have a glass of wine to drink for friendship's sake."
>
> Then Jesus Maria, in a frenzy of gratefulness, made a rash promise. It was the grappa that did it, and the night of the fire, and all the deviled eggs. He felt that he had received great gifts, and he wanted to distribute a gift in return. "It shall be our burden and our duty to see that there is always food in the house of Danny," he declaimed. "Never shall our friend go hungry."

Part of the humor of *Tortilla Flat* derives from the fact that these high-minded hopes are never quite realized, but at the same time, the friends keep trying to meet their responsibilities. Toward the novel's end, they attempt to find a treasure, committing themselves to using it to repay Danny:

All the idealism in Pilon came out then. He told Big Joe how good Danny was to his friends. "And we do nothing for him," he said. "We pay no rent. Sometimes we get drunk and break the furniture. We fight with Danny when we are angry with him, and we call him names. Oh, we are very bad, Big Joe. And so all of us, Pablo and Jesus Maria and the Pirate and I, talked and planned. We are all in the woods, tonight, looking for treasure. And this treasure is to be for Danny. He is so good, Big Joe. He is so kind; and we are so bad. But if we take a great sack of treasure to him, then he will be glad. It is because my heart is clean of selfishness that I can find this treasure."

Of course he doesn't. The important thing, however, is that he tries.

People also try to behave decently, and even altruistically, throughout Steinbeck's acknowledged masterpiece, *The Grapes of Wrath*. Their accomplishment is to do so reciprocally. As the novel opens, Tom Joad agrees to disclose his past to a truck driver who gave him a ride to his family farm. Even though he would rather not divulge this information, Tom recognizes it as a kind of exchange: "You been a good guy. You give me a lift. Well, hell! I done time."

Reciprocity means more than sharing information, though. At one point, Tom Joad and his friend Jim Casy have come across Muley Graves, who has killed a rabbit. Tom and Casy are hungry, and ask to share:

> Muley fidgeted in embarrassment. "I ain't got no choice in the matter." He stopped on the ungracious sound of his words. "That ain't like I mean it. That ain't. I mean"—he stumbled—"what I mean, if a fella's got somepin to eat, and another fella's hungry— why, the first fella ain't got no choice. I mean, s'pose I pick up my rabbits an' go off somewhere an' eat 'em. See?"

Later, when the Joad family is debating whether they can afford to take Casy with them on their trip, Ma says they must:

"It's a long time our folks been here and east before, an' I never heerd tell of no Joads or no Hazletts, neither, ever refusin' food an' shelter or a lift on the road to anybody that asked. They's been mean Joads, but never that mean. . . . One more ain't gonna hurt; an' a man, strong an' healthy, ain't never no burden."

The last part is redolent with reciprocity. The Joads will do Casy a favor by taking him along, but he will be expected to pull his weight and contribute what he can. A man, strong and healthy, ain't never no burden because he'll help out, as friends are expected to do.

Such exchanges permeate *The Grapes of Wrath*, a novel that could have served as a sourcebook for Trivers's now-classic article that opened the eyes of biologists to the world of reciprocity. For example, when Grampa dies in the Wilsons' tent,

"Fine friendly folks," Pa said softly. . . . "We're thankful to you folks."

"We're proud to help," said Wilson.

"We're beholden to you," said Pa.

"There's no beholden in a time of dying," said Wilson, and Sairy echoed him, "Never no beholden."

Al said, "I'll fix your car—me an' Tom will." And Al looked proud that he could return the family's obligation.

A bit later, Ma Joad is once again thanking Sairy Wilson for being so kind, whereupon Sairy's response hinges on reciprocity. She feels safe because she knows that others will help her, in large part because she has helped them: "Sairy said, 'You shouldn' talk like that. We're proud to help. I ain't felt so—safe in a long time. People needs—to help.'"

Sure enough, the Joads end up helping the Wilsons, combining forces in order to keep their two trucks running. The families cooperate by lightening the load on the Joads' truck, while Al and Tom constantly monitor the Wilsons':

Ma said, "You won't be no burden. Each'll help each, an' we'll all git to California. Sairy Wilson he'ped lay Grampa out," and she stopped. The relationship was plain.

Steinbeck's people are notable for recognizing and acting upon the bond of friendship and solidarity, which, although heartwarming, also possesses a consistent logic, born of necessity: "A man with food fed a hungry man, and thus ensured himself against hunger." Toward the book's end, Ma Joad is thanking the neighbors for helping when Rose of Sharon's baby was stillborn:

> "You been frien'ly," she said. "We thank you."
> The stout woman smiled. "No need to thank. Ever'body's in the same wagon. S'pose we was down. You'd a give us a han'."
> "Yes," Ma said, "we would."
> "Or anybody."
> "Or anybody. Use' ta be the fambly was fust. It ain't so now. It's anybody. Worse off we get, the more we got to do."

Maybe John Steinbeck went too far in idealizing the poor and the downtrodden. But he also captured some deep truths, paralleling this observation by Albert Camus in the closing lines of his great novel *The Plague*: "What we learn in a time of pestilence: that there are more things to admire in men than to despise." Steinbeck also recognized—without need of evolutionary theory—the important principle that kin-based altruism is (in a manner of speaking) closely related to reciprocity-based altruism. Consider, for example, this account of the camps of the Dust Bowl migrants:

> And because they were lonely and perplexed, because they had all come from a place of sadness and worry and defeat, and because they were all going to a new mysterious place, they huddled together; they talked together; they shared their lives, their food, and the things they hoped for in the new country. . . . In the evening a strange thing happened: the twenty families became one family, the children were the children of all.

The Grapes of Wrath goes on to make a momentous claim, one that is widely acknowledged today but was new in the 1930s, a time that hadn't yet absorbed either an ecological conscience or an evolutionary sensibility. That claim—although not stated in so many words—is that reciprocity extends beyond interpersonal friendship to encompass the connection of people to the land. For that relationship to be healthy, it, too, must be reciprocal. If treated well, the land will reciprocate; if not, it won't. In Steinbeck's world, "the bank"—representative of uncaring industrial capitalism—supplants the genuine, interdependent, interpersonal relationships that make for a caring, just, and healthy society. The relationships among Steinbeck's poor folks are all rich and warm by contrast with the "inhuman" sterility of the bank. Here, in an intercalary chapter, the bank's representatives are trying to explain why the "real owners" must leave, even though they have worked the same land for years:

> Sure, cried the tenant men, but it's our land. We measured it and broke it up. We were born on it, and we got killed on it, died on it. Even if it's no good, it's still ours. That's what makes it ours—being born on it, working it, dying on it. That makes ownership, not a paper with numbers on it.
>
> We're sorry. It's not us. It's the monster. The bank isn't like a man.
>
> Yes, but the bank is only made of men.
>
> No, you're wrong there—quite wrong there. The bank is something else than men. It happens that every man in a bank hates what the bank does, and yet the bank does it. The bank is something more than men, I tell you. It's the monster. Men made it, but they can't control it.

Sure enough, a few pages later, we are given this description of a bulldozer and its driver, once the bank got its way and took control of the land: "The man sitting in the iron seat did not look like a man: gloved, goggled, rubber dust mask over the nose and mouth, he was part of the monster, a robot in the seat."

A healthy man-land relationship is reciprocal and productive. By contrast, that between machine and land is likely to be neither:

> The man who is more than his chemistry, walking on the earth, turning his plow point for a stone, dropping his handles to slide over an outcropping, kneeling in the earth to eat his lunch; that man who is more than his elements knows the land that is more than its analysis. But the machine man, driving a dead tractor on land that he does not know and love, understands only chemistry; and he is contemptuous of the land and of himself. When the corrugated iron doors are shut, he goes home, and his home is not the land.

Finally, any consideration of friendship and reciprocity among Steinbeck's souls must revisit the famous final scene of *The Grapes of Wrath*, in which Rose of Sharon grows from being a whiny girl into a real, caring woman, wise in the ways of the world and the need to give if you are to get. Her baby having died, she gives her milk to a starving man.

> For a minute Rose of Sharon sat still in the whispering barn. Then she hoisted her tired body up and drew the comforter about her. She moved slowly to the corner and stood looking down at the wasted face, into the wide, frightened eyes. Then slowly she lay down beside him. He shook his head slowly from side to side. Rose of Sharon loosened one side of the blanket and bared her breast. "You got to," she said. She squirmed closer and pulled his head close. "There!" she said. "There." Her hand moved behind his head and supported it. Her fingers moved gently in his hair. She looked up and across the barn, and her lips came together and smiled mysteriously.

10

Epilogue

Foxes, Hedgehogs, Science, and Literature

―――――――

In one of his most influential essays, "The Hedgehog and the Fox," Isaiah Berlin expanded on a well-known observation by the ancient Greek poet Archilochus: "The fox knows many things, but the hedgehog knows one big thing." Archilochus had been contrasting Athens, a multifaceted, foxy city-state if ever there was one, with Sparta, which, like a hedgehog, was single-minded. But Berlin, a philosopher, saw in these zoological contrasts a metaphor for comparing intellectual styles (Shakespeare was a fox, Dante a hedgehog, Tolstoy a fox who desperately wanted to be a hedgehog). In *Madame Bovary's Ovaries*, we have looked a fox squarely in the eyes and called it a hedgehog.

Human beings, while basically animals, are also something more—animals *plus*. Many believe this plus to be a soul; most biologists maintain that specialness resides in humanity's plus-sized brain and that language, imagination, culture, symbolism, and self-consciousness all result from big-braininess. But however you slice it, no objective observer worth his salt, upon studying the life-forms of planet Earth, would conclude that human beings are in any sense nonbiological.

"What a strange scene you describe and what strange prisoners. They are just like us." This comes from Plato's *Republic* (and is also used as an epigraph in José Saramago's *The Cave*). One consequence of being the same species is that everyone is astoundingly like everyone else. What a strange scene indeed! And yet how much stranger, how much more astounding, if literature were to represent people who in any deep sense were *not* "just like us."

According to Joseph Conrad in *'Twixt Land and Sea*, the creative artist speaks to

> the subtle but invincible conviction of solidarity that knits together the loneliness of innumerable hearts; to the solidarity in dreams, in joy, in sorrow, in aspirations, in illusions, in hope, in fear, which binds men to each other, which binds together all humanity—the dead to the living and the living to the unborn.

Throughout this book, we have sought to point out that the solidarity of which Conrad speaks so movingly is born of a shared biology, and that we are bound together in dreams, joy, sorrow, and so forth precisely because we are bound together by an evolutionary past formed of a continuous thread: the double helix of DNA.

"But, but, but," some might protest, "there is so much more to literature than *that*." Of course there is. Biology isn't *the* key to understanding and appreciating literature. It is *a* key. A skeleton key of sorts, which opens more doors, permits more access, and sheds more light than any of its more narrowly designed alternatives.

At the same time, even the most devoted biology buffs should beware hardening of the categories. One characteristic of human nature, it seems, is a yearning for simplicity, often achieved by setting up fundamental dichotomies: yes/no, black/white, either/or, humanities/science, culture/biology. The world, however, is rarely composed of such straightforward alternatives: shades of gray abound, and culture *and* biology, the humanities *and* the sciences interpenetrate and illuminate rather than oppose each other.

"I am large," bragged Walt Whitman. "I contain multitudes." All of literature—and not just Whitman's contribution—is large.

It, too, contains multitudes, and among these, biology has its rightful and hitherto neglected place.

Within biology, evolution, too, is "large." Indeed, according to philosopher Daniel Dennett, "evolution is the greatest idea, ever." Certainly it isn't the only idea, or even the only idea worth entertaining . . . but equally certainly, it deserves attention not only from biologists but from serious readers as well. And evolution, the fundamental driving force of the living world, also contains multitudes. If, as Henry James wrote, "it takes a great deal of history to produce a little literature," it also takes a great deal of biology to make up something as large as the human literary imagination.

Regrettably, however, there is a long-standing dispute between science and the humanities when it comes to tapping the wellsprings of human nature. More than seventy years ago, Max Eastman, in a book whose title does a fine job of summarizing his quest (*The Literary Mind: Its Place in an Age of Science*), concluded that science was on the verge of answering "every problem that arises" and that literature, therefore, "has no place in such a world." He was wrong on both counts. First, science seems unlikely to have all the answers (ever), and second, literature has its place and always will—all the more when thoughtfully combined with science. This has been our lofty goal in *Madame Bovary's Ovaries*, although we have admittedly sought to bring it down to earth by leavening both literature and science, here and there, with a bit of levity.

Eastman's perspective has nonetheless been repeatedly rediscovered and rebroadcast (more often, interestingly, by humanists than by scientists). Consider this, from playwright Eugène Ionesco, in 1970:

> I wonder if art hasn't reached a dead-end, if indeed in its present form, it hasn't already reached its end. Once, writers and poets were venerated as seers and prophets. They had a certain intuition, a sharper sensitivity than their contemporaries, better still, they discovered things and their imaginations went beyond the discoveries even of science itself, to things science would only

establish twenty-five or fifty years later. . . . But for some time now, science has been making enormous progress, whereas the empirical revelations of writers have been making very little . . . can literature still be considered a means to knowledge?

As though to balance Ionesco—the humanist distrustful of the humanities—Noam Chomsky is a scientist who is radically distrustful of science: "It is quite possible—overwhelmingly probable, one might guess—that we will always learn more about human life and human personality from novels than from scientific psychology."

The truth, almost certainly, is in between. Literature is unquestionably a means to knowledge, and so, without doubt, is science (for our purposes, biological science). The good news is that we needn't discard either. In fact, it would be foolish to do so, just as it would be foolish to define oneself as exclusively a fox or a hedgehog.

It is now more than half a century since C. P. Snow warned about the existence of "two cultures"—the humanities and the sciences—which rarely meet. Snow's observation was valid then and, by and large, has remained true ever since. But these days, as John Brockman has pointed out, a kind of "third culture" has been emerging, generating a new natural philosophy that examines our place in the universe by combining insights largely from the natural sciences. Shouldn't our perceptions be as wide as the subject we seek to perceive?

"Every man, wherever he goes," according to philosopher Bertrand Russell, "is encompassed by a cloud of comforting convictions which move with him like flies on a summer day." This is true even of the most open-minded people; indeed, maybe one of the most comforting convictions is of being open-minded! It is also true of serious practitioners of literary criticism, too many of whom are comfortably convinced that theirs is the only way to read. At the same time, Russell's warning doubtless applies to the notions set forth in this book as well: literature is more than just a manifestation of human DNA, and yet how could biology *not* be relevant to literature?

In his magisterial *History of English Literature*, written in the late nineteenth century between the appearance of Darwin's two masterpieces, *On the Origin of Species by Means of Natural Selection* and *The Descent of Man and Selection in Relation to Sex*, the French scholar Hippolyte Taine struggled to glimpse an evolutionary conception of literature, reveling in the fact that modern science finally "approaches man." More than a century ago, Taine suggested that science "has gone beyond the visible and palpable world of stars, stones, plants, amongst which man disdainfully confined her. It reaches the heart, provided with exact and penetrating implements." How much more penetrating are the analytic implements of today's biology! Taine's approach was relatively primitive, emphasizing the importance of a writer's inherited tendencies as well as his or her cultural and historical circumstances. Modern biological science has come a long way since then, although by and large, literary theorists have not noticed.

Maybe they still won't. That's okay with us, since our concern is not with the official, scholarly establishment of theorists and critics but with readers, those people who—almost literally—consume novels and plays, seeking sustenance along with pleasure.

We have based our approach on the conviction that literature, after all, is by human beings, about human beings, and for human beings. And whatever else they are, human beings are *beings*, biological creatures through and through, from beginning to end, ashes to ashes and dust to dust, hydrogen and oxygen and carbon and a sprinkling of sulfur, potassium, calcium, sodium, iron, and phosphrous, shaken and stirred and winnowed and selected by millions of years of evolutionary history.

If there is a single take-home lesson to be derived from the progress of biological science, from Darwin through the outset of the twenty-first century, it is continuity. People are natural, organic, biological, evolved critters, just like every other natural, organic, biological, evolved critter on this planet. When the great biologist Julian Huxley warned against what we called "nothing butism," the tempting oversimplification that because we are animals we are nothing but animals, his point was well taken, but it can also be usefully turned around: just because human beings are capable of great leaps of imagination and intellect and are also possessors of complex culture, this doesn't mean that we are nothing but creatures of imagination, intellect, and culture. We are also animals.

Evolutionary psychologists John Tooby and Leda Cosmides have argued that people "should be interested in evolutionary biology for the same reason that hikers should be interested in an aerial map of an unfamiliar territory that they plan to explore on foot. If they look at the map, they are much less likely to lose their way." Or as William Faulkner put it, "The past isn't dead. It isn't even past." It still persists, in our hands, our hearts, our heads, and in the most revered and beloved creations of our intellect and imagination. Human beings can no more walk away from their biology—or from their creative, artistic imaginations—than we can outrun our shadows or stand apart from our own thoughts. And when we read, seeking perhaps to gain deeper insight into ourselves, or maybe just to enjoy a good story, we can be all the more dazzled by human creativity and all the more grateful for our shared humanity in direct proportion as that reading is enriched by the insights that modern biology has to offer.

Thus, in calling attention to Madame Bovary's ovaries, we have no wish to ignore the rest of her anatomy, nor to diminish her as a person, as a creation, or even as a metaphor. Rather, throughout this book we have suggested that reading makes more sense and is also more fun when informed by modern science's current knowledge of biology and of human nature. We come to expand the appreciation of literature, not to limit it. Human nature pulses inside every writer and, when artfully communicated, is understood by every reader, because it is so deeply shared. It is the breath and beat of living organisms embodied in an organic world of sex, blood, food, fear, anger, love, hopes, trees, animals, air, water, sky, rocks, and dirt. Now that biologists have begun clarifying their perspective on what it means to be human, it is time to look for it—for ourselves, in the deepest sense—where it has always been: in our greatest, most resonant stories.

Moreover, the path is open to all. You needn't be a trained biologist to proceed. In fact, as a Darwinian reader you have an advantage over traditional scientists, who, after all, don't typically learn about discoveries firsthand but rather by reading other people's accounts. By contrast, literature—the original data for anyone wanting to practice Darwinian lit-crit—is there for anyone to encounter directly. So try spicing up your reading with a bit of biology. Go ahead, dive in, swim about, look around, and see what you can see. Most likely, it will be yourself.

ACKNOWLEDGMENTS

We thank Malcolm Scully for warmly embracing the first incarnation of this project as an article in *The Chronicle of Higher Education*. Also our agent, John Michel, for telling us there was a book "here," for finding it a happy home, and then, going beyond the call of agently duty and improving it. Editor Bill Massey at Bantam Dell was doubly astute: pointing out just where our arguments needed sharpening, and also encouraging and supporting us when they stood on their own. Micahlyn Whitt was wonderfully effective at shepherding this project through production, making it easy, at least for us! Also helpful: the students of Honors Seminar 397 at the University of Washington, and particularly Kevin Comartin and Rachel Liebman, who came up with more than their share of good suggestions.

We also want to thank and congratulate . . . each other, for being such compatible coauthors and showing that—theory notwithstanding (see Chapter 8)—parent-offspring conflict needn't be inevitable. We are delighted, in addition, to thank those writers, from Homer to Rebecca Wells, who provded us with the tasty material upon which we have feasted, and to Charles Darwin and his intellectual descendants, for giving us the utensils. Most of all, we express our deepest gratitude to the many millions of creatures, going back in an unbroken line to the primordial Precambrian slime, who gave rise to evolutionary biologists as well as writers of fiction, including us, you, dear reader, and everyone else.

—David P. & Nanelle R. Barash

About the Authors

DAVID BARASH is currently Professor of Psychology at the University of Washington and the author of two dozen books, including *The Myth of Monogamy*, written with his wife, psychiatrist Judith Lipton.

NANELLE BARASH is currently studying biology and literature at Swarthmore College. This is her first book . . . although probably not her last.

INDEX

Additional copies of *Private Pilot and Recreational Pilot FAA Knowledge Test* are available from

Gleim Publications, Inc.
P.O. Box 12848 • University Station
Gainesville, Florida 32604

(352) 375-0772
(800) 87-GLEIM or (800) 874-5346
Fax: (352) 375-6940

www.gleim.com | avmarketing@gleim.com

The price is $19.95 (subject to change without notice). Orders must be prepaid. Call us, order online, or use the order form on page 371. Shipping and handling charges apply to all orders. Add applicable sales tax to shipments within Florida.

Gleim Publications, Inc., guarantees the immediate refund of all resalable texts, unopened and un-downloaded Test Prep Software, and unopened and un-downloaded audios returned within 30 days of purchase. Aviation Test Prep Online may be canceled within 30 days of purchase if no more than the first study unit has been accessed. Other Aviation online courses may be canceled within 30 days of purchase if no more than two study units have been accessed. This policy applies only to products that are purchased directly from Gleim Publications, Inc. No refunds will be provided on opened or downloaded Test Prep Software or audios, partial returns of package sets, or shipping and handling charges. Any freight charges incurred for returned or refused packages will be the purchaser's responsibility. Returns of books purchased from bookstores and other resellers should be made to the respective bookstore or reseller.

REVIEWERS AND CONTRIBUTORS

Jamie Beckett, CMEL, CFII, MEI, AGI, A&P, is one of our aviation editors. Mr. Beckett researched questions, wrote and edited answer explanations, and incorporated revisions into the text.

Eric L. Crump, CMEL, CFII, AGI, B.S., Middle Tennessee State University, is one of our aviation editors. Mr. Crump researched questions, wrote and edited answer explanations, and incorporated revisions into the text.

Scott Krogh, CMEL, CFII, MEI, is the Gleim 141 Chief Flight Instructor and one of our aviation editors. Mr. Krogh researched questions, wrote and edited answer explanations, and incorporated revisions into the text.

The CFIs who have worked with us throughout the years to develop and improve our pilot training materials.

The many FAA employees who helped, in person or by telephone, primarily in Gainesville, Orlando, Oklahoma City, and Washington, DC.

The many pilots who have provided comments and suggestions about *Private Pilot and Recreational Pilot FAA Knowledge Test* during the past several decades.

A PERSONAL THANKS

This manual would not have been possible without the extraordinary effort and dedication of Jacob Brunny, Julie Cutlip, Eileen Nickl, Teresa Soard, Justin Stephenson, Joanne Strong, and Candace Van Doren, who typed the entire manuscript and all revisions and drafted and laid out the diagrams and illustrations in this book.

The authors also appreciate the production and editorial assistance of Ellen Buhl, Jessica Felkins, Chris Hawley, Katie Larson, Cary Marcous, Jean Marzullo, Shane Rapp, Drew Sheppard, and Martha Willis.

Finally, we appreciate the encouragement, support, and tolerance of our families throughout this project.

2014 EDITION

PRIVATE PILOT

AND RECREATIONAL PILOT

FAA KNOWLEDGE TEST

for the FAA Computer-Based Pilot Knowledge Test

Private Pilot - Airplane　　*Recreational Pilot - Airplane*
Private Pilot - Airplane Transition

by

Irvin N. Gleim, Ph.D., CFII

and

Garrett W. Gleim, CFII

ABOUT THE AUTHORS

Irvin N. Gleim earned his private pilot certificate in 1965 from the Institute of Aviation at the University of Illinois, where he subsequently received his Ph.D. He is a commercial pilot and flight instructor (instrument) with multi-engine and seaplane ratings and is a member of the Aircraft Owners and Pilots Association, American Bonanza Society, Civil Air Patrol, Experimental Aircraft Association, National Association of Flight Instructors, and Seaplane Pilots Association. He is the author of flight maneuvers and practical test prep books for the sport, private, instrument, commercial, and flight instructor certificates/ratings and the author of study guides for the sport, private/recreational, instrument, commercial, flight/ground instructor, fundamentals of instructing, airline transport pilot, and flight engineer FAA knowledge tests. Three additional pilot training books are *Pilot Handbook*, *Aviation Weather and Weather Services*, and *FAR/AIM*.

Dr. Gleim has also written articles for professional accounting and business law journals and is the author of widely used review manuals for the CIA (Certified Internal Auditor) exam, the CMA (Certified Management Accountant) exam, the CPA (Certified Public Accountant) exam, and the EA (IRS Enrolled Agent) exam. He is Professor Emeritus, Fisher School of Accounting, University of Florida, and is a CFM, CIA, CMA, and CPA.

Garrett W. Gleim earned his private pilot certificate in 1997 in a Piper Super Cub. He is a commercial pilot (single- and multi-engine), ground instructor (advanced and instrument), and flight instructor (instrument and multi-engine), and he is a member of the Aircraft Owners and Pilots Association and the National Association of Flight Instructors. He is the author of study guides for the sport, private/recreational, instrument, commercial, flight/ground instructor, fundamentals of instructing, and airline transport pilot FAA knowledge tests. He received a Bachelor of Science in Economics from The Wharton School, University of Pennsylvania.

Gleim Publications, Inc.
P.O. Box 12848 · University Station
Gainesville, Florida 32604

(352) 375-0772
(800) 87-GLEIM or (800) 874-5346
Fax: (352) 375-6940

Internet: www.gleim.com
Email: admin@gleim.com

For updates to the first printing of the 2014 edition of
**Private Pilot and Recreational Pilot
FAA Knowledge Test**

Go To: www.gleim.com/updates

Or: Email update@gleim.com with **PPKT 2014-1** in the subject line. You will receive our current update as a reply.

Updates are available until the next edition is published.

ISSN 1080-4900
ISBN 978-1-58194-382-5

First Printing: July 2013

Copyright © 2013 by Gleim Publications, Inc.

SOURCES USED IN *PRIVATE PILOT AND RECREATIONAL PILOT FAA KNOWLEDGE TEST*

The first lines of our answer explanations contain citations to authoritative sources of the answers. These publications can be obtained from the FAA, the Government Printing Office, and aviation bookstores. These citations are abbreviated as provided below:

A/FD	Airport/Facility Directory	AWS	Aviation Weather Services
AAH	Advanced Avionics Handbook	FAR	Federal Aviation Regulations
AC	Advisory Circular	FI Comp	Flight Computer
ACL	Aeronautical Chart Legend	IFH	Instrument Flying Handbook
AFH	Airplane Flying Handbook	NTSB	National Transportation Safety Board
AIM	Aeronautical Information Manual		Regulations
AvW	Aviation Weather	PHAK	Pilot's Handbook of Aeronautical Knowledge
AWBH	Aircraft Weight and Balance Handbook		

HELP!!

This 2014 edition is designed specifically for private pilots. Please send any corrections and suggestions for subsequent editions to the authors, c/o Gleim Publications, Inc. The last two pages in this book have been reserved for you to make comments and suggestions. They can be torn out and mailed to Gleim Publications, Inc.

Two other volumes are also available. *Private Pilot Flight Maneuvers and Practical Test Prep* focuses on your flight training and the FAA practical test, just as this book focuses on the FAA knowledge test. *Pilot Handbook* is a complete pilot ground school text in outline format with many diagrams for ease in understanding.

Save time, money, and frustration -- see the order form on page 371. Please bring Gleim books to the attention of flight instructors, fixed base operators, and others with a potential interest in flying. Wide distribution of these books and increased interest in flying depend on your assistance, good word, etc. Thank you.

NOTE: ANSWER DISCREPANCIES and UPDATES

Our answers have been carefully researched and reviewed. Inevitably, there will be differences with competitors' books and even the FAA. If necessary, we will develop an UPDATE for *Private Pilot and Recreational Pilot FAA Knowledge Test*. Send an email to update@gleim.com as described above, and visit our website for the latest updates and information on all of our products. Updates for this 2014 edition will be available until the next edition is published. To continue providing our customers with first-rate service, we request that questions about our materials be sent to us via email to aviation@gleim.com. The appropriate staff member will give each question thorough consideration and a prompt response. Questions concerning orders, prices, shipments, or payments will be handled via telephone by our competent and courteous customer service staff.

TABLE OF CONTENTS

NOTE: The FAA no longer releases the complete database of test questions to the public. Instead, sample questions are released on the Airmen Testing page of the FAA website on a quarterly basis. These questions are similar to the actual test questions, but they are not exact matches.

Gleim utilizes customer feedback and FAA publications to create additional sample questions that closely represent the topical coverage of each FAA knowledge test. In order to do well on the knowledge test, you must study the Gleim outlines in this book, answer all the questions under exam conditions (i.e., without looking at the answers first), and develop an understanding of the topics addressed. You should not simply memorize questions and answers. This will not prepare you for your FAA knowledge test, and it will not help you develop the knowledge you need to safely operate an aircraft.

Always refer to the Gleim update service (www.gleim.com/updates) to ensure you have the latest information that is available. If you see topics covered on your FAA knowledge test that are not contained in this book, please contact us at aviation@gleim.com to report your experience and help us fine-tune our test preparation materials.

Thank you!

PREFACE

The primary purpose of this book is to provide you with the easiest, fastest, and least expensive means of passing the FAA knowledge tests for the private pilot (airplane) and/or the recreational pilot certificates. The publicly released FAA knowledge test bank does **not** have questions grouped together by topic. We have organized them for you. We have

1. Reproduced all previously released knowledge test questions published by the FAA. We have also included many additional similar test questions, which we believe may appear in some form on your knowledge test.

2. Reordered the questions into 118 logical topics.

3. Organized the 118 topics into 11 study units.

4. Explained the answer immediately to the right of each question.

5. Provided an easy-to-study outline of exactly what you need to know (and no more) at the beginning of each study unit.

Accordingly, you can thoroughly prepare for the FAA pilot knowledge test by

1. Studying the brief outlines at the beginning of each study unit.

2. Answering the question on the left side of each page while covering up the answer explanations on the right side of each page.

3. Reading the answer explanation for each question that you answer incorrectly or have difficulty answering.

4. Facilitating this Gleim process with our **FAA Test Prep Online**. Our software allows you to emulate the FAA test (CATS or PSI/LaserGrade). By practicing answering questions on a computer, you will become at ease with the computer testing process and have the confidence to PASS. See pages 17 through 19.

5. Using our **Online Ground School**, which provides you with our outlines, practice problems, and sample tests. This course is easily accessible through the Internet. Also, we give you a money-back guarantee with our **Online Ground School**. If you are unsuccessful, you get your money back!

Additionally, this book will introduce our entire series of pilot training texts, which use the same presentation method: outlines, illustrations, questions, and answer explanations. For example, *Pilot Handbook* is a textbook of aeronautical knowledge presented in easy-to-use outline format, with many charts, diagrams, figures, etc., included. While this book contains only the material needed to pass the FAA pilot knowledge test, *Pilot Handbook* contains the textbook knowledge required to be a safe and proficient pilot.

Many books create additional work for the user. In contrast, this book and its companion, *Private Pilot Flight Maneuvers and Practical Test Prep*, facilitate your effort. They are easy to use. The outline/illustration format, type styles, and spacing are designed to improve readability. Concepts are often presented as phrases rather than as complete sentences – similar to notes that you would take in a class lecture.

Also, recognize that this study manual is concerned with **airplane** flight training, not balloon, glider, or helicopter training. We are confident this book, **FAA Test Prep Online**, and/or **Online Ground School** will facilitate speedy completion of your knowledge test. We also wish you the very best as you complete your private pilot and/or recreational pilot certification, in related flying, and in obtaining additional ratings and certificates.

Enjoy Flying Safely!

Irvin N. Gleim
Garrett W. Gleim
July 2013

INTRODUCTION: THE FAA PILOT KNOWLEDGE TEST

The beginning of this Introduction explains how to obtain a private pilot certificate, and it explains the content and procedures of the Federal Aviation Administration (FAA) knowledge test, including how to take the test at a computer testing center. The remainder of this Introduction discusses and illustrates the Gleim **Online Ground School** and **FAA Test Prep Online**. Achieving a private pilot certificate is fun. Begin today!

Private Pilot and Recreational Pilot FAA Knowledge Test is one of five books contained in the Gleim Private Pilot Kit. The other four books are

1. ***Private Pilot Flight Maneuvers and Practical Test Prep***
2. ***Private Pilot Syllabus***
3. ***Pilot Handbook***
4. ***FAR/AIM***

Private Pilot Flight Maneuvers and Practical Test Prep presents each flight maneuver you will perform in outline/illustration format so you will know what to expect and what to do before each flight lesson. This book will thoroughly prepare you to complete your FAA practical (flight) test confidently and successfully.

Private Pilot Syllabus is a step-by-step syllabus of ground and flight training lesson plans for your private pilot training.

Pilot Handbook is a complete pilot reference book that combines over 100 FAA books and documents, including *AIM*, FARs, ACs, and much more. Aerodynamics, airplane systems, airspace, and navigation are among the topics explained in *Pilot Handbook*. This book, more than any other, will help make you a better and more proficient pilot.

FAR/AIM is an essential part of every pilot's library. The Gleim *FAR/AIM* is an easy-to-read reference book containing all of the Federal Aviation Regulations (FARs) applicable to general aviation flying, plus the full text of the FAA's *Aeronautical Information Manual (AIM)*.

The Gleim *Aviation Weather and Weather Services* is not included in the Private Pilot Kit, but you may want to purchase it if you do not already have it. This book combines all of the information from the FAA's *Aviation Weather* (AC 00-6), *Aviation Weather Services* (AC 00-45), and numerous FAA publications into one easy-to-understand book. It will help you study all aspects of aviation weather and provide you with a single reference book.

WHAT IS A PRIVATE PILOT CERTIFICATE?

A private pilot certificate is much like a driver's license. A private pilot certificate will allow you to fly an airplane and carry passengers and baggage, although not for compensation or hire. However, operating expenses may be shared with your passengers. The certificate, which is plastic (similar to a driver's license), is sent to you by the FAA upon satisfactory completion of your training program, a pilot knowledge test, and a practical test. A sample private pilot certificate is reproduced below. The recreational pilot certificate is the same except it says "recreational."

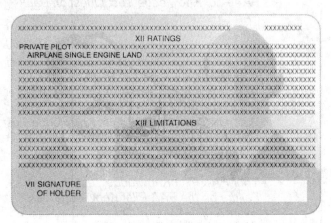

WHAT IS A RECREATIONAL PILOT CERTIFICATE?

The FAA added a recreational pilot certificate in 1989 for those who want to fly locally for fun, i.e., recreation. The objective is to provide only the flight training required for those **NOT** aspiring to fly on trips, at night, with more than one passenger, or to airports with an operating control tower or other airspace requiring air traffic control (ATC) communication. A recreational pilot certificate will take you less time and money to obtain, but your flying privileges will be restricted. The recreational pilot certificate is, however, upgradable to a private pilot certificate.

The recreational pilot knowledge test has 50 questions and includes most of the topics on the private pilot knowledge test. There are 27 questions specific to recreational pilots in Study Unit 4. For the recreational pilot knowledge test, we suggest you study the entire book except the following:

Subunit	Subunit Name		Subunit	Subunit Name
3.3	Beacons and Taxiway Lights		3.14	ATC Traffic Advisories
3.7	Collision Avoidance (Qs on night visual scanning)		3.15	ATC Light Signals
3.8	ATIS and Ground Control		Study Unit 4 FARs	61.31, 61.113, 91.123, 91.130,
3.11	Terminal Radar Programs			91.131, 91.135, 91.157

REQUIREMENTS TO OBTAIN A PRIVATE PILOT CERTIFICATE

1. Be at least 17 years of age.
2. Be able to read, write, and understand English (certificates with operating limitations may be available for medically related deficiencies).
3. Obtain at least a third-class FAA medical certificate (see the sample on the next page).
   a. You must undergo a routine medical examination, which may be administered only by FAA-designated doctors called aviation medical examiners (AME).
      1) For operations requiring a private, recreational, or student pilot certificate, a first-, second-, or third-class medical certificate expires at the end of the last day of the month either
         a) 5 years after the date of examination shown on the certificate, if you have not reached your 40th birthday on or before the date of examination, or
         b) 2 years after the date of examination shown on the certificate, if you have reached your 40th birthday on or before the date of examination.

b. Even if you have a physical handicap, medical certificates can be issued in many cases. Operating limitations may be imposed depending upon the nature of the disability.

c. Your certificated flight instructor (CFI) or fixed-base operator (FBO) will be able to recommend an AME.

    1) An FBO is an airport business that gives flight lessons, sells aviation fuel, repairs airplanes, etc.

    2) Also, the FAA publishes a directory that lists all authorized AMEs by name and address. Copies of this directory are kept at most FAA offices, ATC facilities, and Flight Service Stations (FSSs). Alternatively, go to the Gleim website at www.gleim.com/aviation/amesearch.php.

d. As a student pilot, your medical certificate will also function as your student pilot certificate once it is signed by you and your AME.

    1) Alternatively, a separate student pilot certificate can be obtained from an FAA Flight Standards District Office (FSDO) or any designated pilot examiner.

        a) This would be necessary if your AME issued you a third-class medical certificate instead of a student pilot certificate (you should request a student pilot certificate).

Front

Back

UNITED STATES OF AMERICA
Department of Transportation
Federal Aviation Administration
BB-5342031

MEDICAL CERTIFICATE ___3rd___ **CLASS AND STUDENT PILOT CERTIFICATE**

This certifies that *(Full name and address):*

Richmond, Kane Everett
7771 Coral Way
N. Ft. Myers, Fl. 33903

Date of Birth	Height	Weight	Hair	Eyes	Sex
1/30/67	5'9"	140	Brn.	Brn.	M

has met the medical standards prescribed in Part 67, Federal Aviation Regulations, for this class of Medical Certificate.

Limitations

None

Date of Examination	Examiner's Serial No.
06/26/13	11967-1

Examiner | Signature *E.W. Williams, II, D.O.*
Typed Name | E.W. Williams II, D.O.

AIRMAN'S SIGNATURE *Kane Everett Richmond*

FAA Form 8420-2 (3-99) Supersedes Previous Edition

PASSENGER-CARRYING PROHIBITED

CONDITIONS OF ISSUE: This certificate shall be in the personal possession of the airman at all times while exercising the privileges of his or her airman certificate. The issuance of a medical certificate by an Aviation Medical Examiner may be reversed by the FAA within 60 days. Section 61.19 of the Federal Aviation Regulations (FAR) sets forth the duration of a student pilot certificate. Unless otherwise limited, the duration of a medical certificate is set forth in §61.23 of the FAR. The holder of this certificate is governed by the provisions of FAR §§ 61.53, 63.19, and 65.49(d) relating to physical deficiency (14 CFR Parts 61, 63, and 65).

CERTIFICATED INSTRUCTOR'S ENDORSEMENT FOR STUDENT PILOTS

I certify that the holder of this certificate has met the requirements of the regulations and is competent for the following:

		DATE	MAKE AND MODEL OF AIRCRAFT	INSTRUCTOR'S SIGNATURE	INSTRUCTOR'S CERT. No.	Exp. Date
A. To Solo The Following Aircraft		7-3-13	C-152	*Tracey Lin Law*	264750091	8/2014
B. To Make Solo Cross-Country Flights	AIRCRAFT CATEGORY Airplane / Glider / Rotorcraft					

    2) The only substantive difference between a regular medical certificate and a medical certificate/student pilot certificate is that the back of the medical certificate/student pilot certificate provides for flight instructor signature.

        a) Also, the combined medical certificate/student pilot certificate is on slightly heavier paper and is yellow instead of white.

    3) Note that the back of the student pilot certificate must be signed by your CFI prior to solo flight (flying by yourself).

    4) You must be at least 16 years of age to receive a student pilot certificate.

4.   Receive and log ground training from an authorized instructor or complete a home-study course (such as studying this book, *Private Pilot Flight Maneuvers and Practical Test Prep*, and *Pilot Handbook* or using the Gleim **Online Ground School**) to learn

    a.   *Applicable Federal Aviation Regulations ... that relate to private pilot privileges, limitations, and flight operations*

    b.   *Accident reporting requirements of the National Transportation Safety Board*

    c.   *Use of the applicable portions of the Aeronautical Information Manual and FAA ACs (advisory circulars)*

    d.   *Use of aeronautical charts for VFR navigation using pilotage, dead reckoning, and navigation systems*

    e.   *Radio communication procedures*

    f.   *Recognition of critical weather situations from the ground and in flight, windshear avoidance, and the procurement and use of aeronautical weather reports and forecasts*

    g.   *Safe and efficient operation of aircraft, including collision avoidance, and recognition and avoidance of wake turbulence*

    h.   *Effects of density altitude on takeoff and climb performance*

    i.   *Weight and balance computations*

    j.   *Principles of aerodynamics, powerplants, and aircraft systems*

    k.   *Stall awareness, spin entry, spins, and spin recovery techniques for the airplane...category ratings*

    l.   *Aeronautical decision making and judgment*

    m.   *Preflight action that includes*

        1)   *How to obtain information on runway lengths at airports of intended use, data on takeoff and landing distances, weather reports and forecasts, and fuel requirements*

        2)   *How to plan for alternatives if the flight cannot be completed or delays are encountered*

5.   Pass a knowledge test with a score of 70% or better. All FAA tests are administered at FAA-designated computer testing centers. The private pilot knowledge test consists of 60 multiple-choice questions selected from the airplane-related questions in the FAA's private pilot test bank; the remaining questions are for balloons, helicopters, etc. The FAA's published airplane-related questions, along with our own similar questions, are reproduced in this book with complete explanations.

6.   Accumulate flight experience (FAR 61.109). Receive a total of 40 hr. of flight instruction and solo flight time, including

    a.   20 hr. of flight training from an authorized flight instructor, including at least

        1)   3 hr. of cross-country, i.e., to other airports

        2)   3 hr. at night, including

            a)   One cross-country flight of over 100 NM total distance
            b)   10 takeoffs and 10 landings to a full stop at an airport

        3)   3 hr. of instrument flight training in an airplane

        4)   3 hr. in airplanes in preparation for the private pilot practical test within 2 calendar months prior to that test

    NOTE: A maximum of 2.5 hr. of instruction may be accomplished in an FAA-approved flight simulator or flight training device representing an airplane.

    b.   10 hr. of solo flight time in an airplane, including at least

        1)   5 hr. of cross-country flights

        2)   One solo cross-country flight of at least 150 NM total distance, with full-stop landings at a minimum of three points and with one segment of the flight consisting of a straight-line distance of more than 50 NM between the takeoff and landing locations

        3)   Three solo takeoffs and landings to a full stop at an airport with an operating control tower

7.   Receive flight instruction and demonstrate skill (FAR 61.107).

    a.   Obtain a logbook sign-off by your CFI on the following areas of operations:

        1)   *Preflight preparation*
        2)   *Preflight procedures*
        3)   *Airport and seaplane base operations*
        4)   *Takeoffs, landings, and go-arounds*
        5)   *Performance maneuvers*
        6)   *Ground reference maneuvers*
        7)   *Navigation*
        8)   *Slow flight and stalls*
        9)   *Basic instrument maneuvers*
        10)   *Emergency operations*
        11)   *Night operations*
        12)   *Postflight procedures*
        13)   *Multi-engine operations (for only multi-engine airplanes)*

    b.   Alternatively, enroll in an FAA-certificated pilot school that has an approved private pilot certification course (airplane).

        1)   These are known as Part 141 schools or Part 142 training centers because they are authorized by Part 141 or Part 142 of the FARs.

            a)   All other regulations concerning the certification of pilots are found in Part 61 of the FARs.

8.   Successfully complete a practical (flight) test, which will be given as a final exam by an FAA inspector or designated pilot examiner and conducted as specified in the FAA's Private Pilot Practical Test Standards (FAA-S-8081-14).

    a.   FAA inspectors are FAA employees and do not charge for their services.

    b.   FAA-designated pilot examiners are proficient, experienced flight instructors and pilots who are authorized by the FAA to conduct practical tests. They do charge a fee.

    c.   The FAA's Private Pilot Practical Test Standards are outlined and reprinted in the Gleim ***Private Pilot Flight Maneuvers and Practical Test Prep*** book.

FAA PILOT KNOWLEDGE TEST

1.   This book is designed to help you prepare for and pass the following FAA knowledge tests:

   a.   Private Pilot-Airplane (PAR) consists of 60 questions. Time limit is 2.5 hours.
   b.   Recreational Pilot-Airplane (RPA) consists of 50 questions. Time limit is 2 hours.
   c.   Private Pilot-Airplane Transition (PAT) consists of 30 questions (to upgrade from recreational to private). Time limit is 1.5 hours.

NON-AIRPLANE TESTS

If you are using this book to study for a non-airplane pilot knowledge test, you should skip all of the obviously airplane-related questions. Then, go to www.gleim.com/aviation/naqas/ for the appropriate non-airplane questions.

---

2.   The FAA legends and figures are in a book titled *Computer Testing Supplement for Recreational Pilot and Private Pilot*, which you will be given to use at the time of your test.

   a.   For the purpose of test preparation, the appropriate legends and figures are reproduced in this book.

3.   In an effort to develop better questions, the FAA frequently **pretests** questions on knowledge tests by adding up to five "pretest" questions. The pretest questions will not be graded.

   a.   You will NOT know which questions are real and which are pretest, so you must attempt to answer all questions correctly.
   b.   When you notice a question NOT covered by Gleim, it might be a pretest question.

      1)   We want to know about each pretest question you see.
      2)   Please email (aviation@gleim.com) or call (800-874-5346) with your recollection of any possible pretest questions so we may improve our efforts to prepare future pilots.

FAA PILOT KNOWLEDGE TEST QUESTION BANK

In an effort to keep applicants from simply memorizing test questions, the FAA does not currently disclose all the questions you might see on your FAA knowledge test. We encourage you to take the time to fully learn and understand the concepts explained in the knowledge transfer outlines contained in this book. Using this book or other Gleim test preparation material to merely memorize the questions and answers is unwise, unproductive, and will not ensure your success on your FAA knowledge test. Memorization also greatly reduces the amount of information you will actually learn during your study.

The questions and answers provided in this book include all previously released FAA questions in addition to questions developed from current FAA reference materials that closely approximate the types of questions you should see on your knowledge test. We are confident that by studying our knowledge transfer outlines, answering our questions under exam conditions, and not relying on rote memorization, you will be able to successfully pass your FAA knowledge test and begin learning to become a safe and competent pilot.

FAA QUESTIONS WITH TYPOGRAPHICAL ERRORS

Occasionally, FAA test questions contain typographical errors such that there is no correct answer. The FAA test development process involves many steps and people and, as you would expect, glitches occur in the system that are beyond the control of any one person. We indicate "best" rather than correct answers for some questions. Use these best answers for the indicated questions.

Note that the FAA corrects (rewrites) defective questions as they are discovered; these changes are explained in our updates--see page iv. However, problems due to faulty or out-of-date figures printed in the FAA Computer Testing Supplements are expensive to correct. Thus, it is important to carefully study questions that are noted to have a best answer in this book. Even though the best answer may not be completely correct, you should select it when taking your test.

REORGANIZATION OF FAA QUESTIONS

1. In the public FAA knowledge test question bank releases, the questions are **not** grouped together by topic; i.e., they appear to be presented randomly.

    a. We have reorganized and renumbered the questions into study units and subunits.

2. Pages 349 through 356 contain a list of all the questions in FAA learning statement code order, with cross-references to the study units and question numbers in this book.

    a. For example, question 1-34 is assigned the code PLT003, which means it is found in Study Unit 1 as question 34 in this book and is covered under the FAA learning statement, "Calculate aircraft performance - center of gravity."

    b. Questions relating to helicopters, gliders, balloons, etc., are excluded.

HOW TO PREPARE FOR THE FAA PILOT KNOWLEDGE TEST

1. Begin by carefully reading the rest of this Introduction. You need to have a complete understanding of the examination process prior to initiating your study. This knowledge will make your studying more efficient.

2. After you have spent an hour analyzing this Introduction, set up a study schedule, including a target date for taking your knowledge test.

    a. Do not let the study process drag on and become discouraging; i.e., the quicker the better.

    b. Consider enrolling in an organized ground school course, like the Gleim **Online Ground School**, or one held at your local FBO, community college, etc.

    c. Determine where and when you are going to take your knowledge test.

3. Work through each of Study Units 1 through 11.

    a. All previously released questions in the FAA's private pilot knowledge test question bank that are applicable to airplanes have been grouped into the following 11 categories, which are the titles of Study Units 1 through 11:

    Study Unit 1: Airplanes and Aerodynamics
    Study Unit 2: Airplane Instruments, Engines, and Systems
    Study Unit 3: Airports, Air Traffic Control, and Airspace
    Study Unit 4: Federal Aviation Regulations
    Study Unit 5: Airplane Performance and Weight and Balance
    Study Unit 6: Aeromedical Factors and Aeronautical Decision Making (ADM)
    Study Unit 7: Aviation Weather
    Study Unit 8: Aviation Weather Services
    Study Unit 9: Navigation: Charts and Publications
    Study Unit 10: Navigation Systems
    Study Unit 11: Cross-Country Flight Planning

    b. Within each of the study units listed, questions relating to the same subtopic (e.g., thunderstorms, airplane stability, sectional charts, etc.) are grouped together to facilitate your study program. Each subtopic is called a subunit.

c. To the right of each question, we present

1) The correct answer

2) The appropriate source document for the answer explanation

A/FD	*Airport/Facility Directory*	*AWS*	*Aviation Weather Services*
AAH	*Advanced Avionics Handbook*	FAR	Federal Aviation Regulations
AC	Advisory Circular	FI Comp	Flight Computer
ACL	Aeronautical Chart Legend	*IFH*	*Instrument Flying Handbook*
AFH	*Airplane Flying Handbook*	NTSB	National Transportation Safety Board
AIM	*Aeronautical Information Manual*		Regulations
AvW	*Aviation Weather*	*PHAK*	*Pilot's Handbook of Aeronautical Knowledge*
AWBH	*Aircraft Weight and Balance Handbook*		

a) The codes may refer to an entire document, such as an advisory circular, or to a particular chapter or subsection of a larger document.

i) See page 361 for a complete list of abbreviations and acronyms used in this book.

3) A comprehensive answer explanation, including

a) A discussion of the correct answer or concept

b) An explanation of why the other two answer choices are incorrect

4. Each study unit begins with a list of its subunit titles. The number after each title is the number of questions that cover the information in that subunit. The two numbers following the number of questions are the page numbers on which the outline and the questions for that particular subunit begin, respectively.

5. Begin by studying the outlines slowly and carefully. The outlines in this part of the book are very brief and have only one purpose: to help you pass the FAA knowledge test.

a. **CAUTION:** The **sole purpose** of this book is to expedite your passing the FAA knowledge test for the recreational pilot and/or private pilot certificate. Accordingly, all extraneous material (i.e., not directly tested on the FAA knowledge test) is omitted, even though much more information and knowledge are necessary to be proficient and fly safely. This additional material is presented in two related Gleim books: *Private Pilot Flight Maneuvers and Practical Test Prep* and *Pilot Handbook*.

6. Next, answer the questions under exam conditions. Cover the answer explanations on the right side of each page with a piece of paper while you answer the questions.

a. Remember, it is very important to the learning (and understanding) process that you honestly commit yourself to an answer. If you are wrong, your memory will be reinforced by having discovered your error. Therefore, it is crucial to cover up the answer and make an honest attempt to answer the question before reading the answer.

b. Study the answer explanation for each question that you answer incorrectly, do not understand, or have difficulty with.

c. Use our **Online Ground School** or **FAA Test Prep Online** to ensure that you do not refer to answers before committing to one AND to simulate actual computer testing center exam conditions.

d. Go to www.gleim.com/OGS to view our **Online Ground School**. It is a structured course to assist those who have trouble sitting down to books and software.

7. Note that this test book contains questions grouped by topic. Thus, some questions may appear repetitive, while others may be duplicates or near-duplicates. Accordingly, do not work question after question (i.e., waste time and effort) if you are already conversant with a topic and the type of questions asked.

8. As you move through study units, you may need further explanation or clarification of certain topics. You may wish to obtain and use the following Gleim books described on page 1:

   a. ***Private Pilot Flight Maneuvers and Practical Test Prep***
   b. ***Pilot Handbook***
   c. ***Aviation Weather and Weather Services***

9. Keep track of your work. As you complete a subunit, grade yourself with an A, B, C, or ? (use a ? if you need help on the subject) next to the subunit title at the front of the respective study unit.

   a. The A, B, C, or ? is your self-evaluation of your comprehension of the material in that subunit and your ability to answer the questions.

      A   means a good understanding.
      B   means a fair understanding.
      C   means a shaky understanding.
      ?   means to ask your CFI or others about the material and/or questions, and read the pertinent sections in ***Private Pilot Flight Maneuvers and Practical Test Prep*** and/or ***Pilot Handbook***.

   b. This procedure will provide you with the ability to quickly see (by looking at the first page of each study unit) how much studying you have done (and how much remains) and how well you have done.

   c. This procedure will also facilitate review. You can spend more time on the subunits that were more difficult for you.

   d. **FAA Test Prep Online** provides you with your historical performance data.

Follow the suggestions given throughout this Introduction and you will have no trouble passing the FAA knowledge test the first time you take it.

With this overview of exam requirements, you are ready to begin the easy-to-study outlines and rearranged questions with answers to build your knowledge and confidence and PASS THE FAA's RECREATIONAL PILOT OR PRIVATE PILOT KNOWLEDGE TEST.

The feedback we receive from users indicates that our materials reduce anxiety, improve FAA test scores, and build knowledge. Studying for each test becomes a useful step toward advanced certificates and ratings.

MULTIPLE-CHOICE QUESTION-ANSWERING TECHNIQUE

Because the private pilot knowledge test has a set number of questions (60) and a set time limit (2.5 hours), you can plan your test-taking session to ensure that you leave yourself enough time to answer each question with relative certainty. The following steps will help you move through the knowledge test efficiently and produce better test results.

1. **Budget your time.** We make this point with emphasis. Just as you would fill up your gas tank prior to reaching empty, so too should you finish your exam before time expires.

   a. If you utilize the entire 2.5-hour time limit, that allows you 2.5 minutes per question.

   b. If you are adequately prepared for the test, you should finish it within 45-60 minutes.

      1) Use any extra time you have to review questions that you are not sure about, cross-country planning questions with multiple steps and calculations, and similar questions in your exam that may help you answer other questions.

   c. Time yourself when completing study sessions in this book and/or review your time investment reports from the Gleim **FAA Test Prep Online** to track your progress and adherence to the time limit and your own personal time allocation budget.

2. **Answer the questions in consecutive order.**

   a. Do **not** agonize over any one item. Stay within your time budget.

      1) We suggest that you skip cross-country planning questions and other similarly involved computational questions on your first pass through the exam. Come back to them after you have been through the entire test once.

   b. Mark any questions you are unsure of and return to them later as time allows.

      1) Once you initiate test grading, you will no longer be able to review/change any answers.

   c. Never leave a multiple-choice question unanswered. Make your best educated guess in the time allowed.

      1) Your score is based on the number of correct responses. You will not be penalized for guessing incorrectly.

3. **For each multiple-choice question,**

   a. **Try to ignore the answer choices.** Do not allow the answer choices to affect your reading of the question.

      1) If three answer choices are presented, two of them are incorrect. These choices are called **distractors** for good reason. Often, distractors are written to appear correct at first glance until further analysis.

      2) In computational items, the distractors are carefully calculated such that they are the result of making common mistakes. Be careful, and double-check your computations to the extent that time permits.

   b. **Read the question carefully** to determine the precise requirement.

      1) Focusing on what is required enables you to ignore extraneous information, to focus on the relevant facts, and to proceed directly to determining the correct answer.

         a) Be especially careful to note when the requirement is an **exception**; e.g., "Which of the following is **not** a type of hypoxia?"

   c. **Determine the correct answer** before looking at the answer choices.

      1) Mentally note what you believe the correct response is before ever glancing at the available answer choices.

   d. **Read the answer choices carefully.**

      1) Even if the first answer appears to be the correct choice, do **not** skip the remaining answer choices. Questions often require the "best" answer of the choices provided. Thus, each choice requires your consideration.

      2) Treat each answer choice as a true/false question as you analyze it. Is the statement asserted in the answer choice true or false?

   e. **Click on the best answer.**

      1) If you are uncertain, guess intelligently. Improve on your 33% chance of getting the correct answer with blind guessing.

      2) For many multiple-choice questions, at least one answer choice can be eliminated with minimal effort, thereby increasing your educated guess to a 50-50 proposition.

4. After you have been through all the questions in the test, consult the question status list to determine which questions are unanswered and which are marked for review.

   a. Go back to the marked questions and finalize your answer choices.

   b. Verify that all questions have been answered.

5. **If you don't know the answer,**

    a. Again, guess; but make it an educated guess by selecting the best possible answer.

        1) Rule out answers that you think are incorrect.

        2) Speculate on what the FAA is looking for and/or the rationale behind the question.

        3) Select the best answer or guess between equally appealing answers. Your first guess is usually the most intuitive. If you cannot make an educated guess, re-read the stem and each answer choice and pick the best or most intuitive answer. It's just a guess!

    b. Avoid lingering on any question for too long. Remember your time budget and the overall test time limit.

SIMULATED FAA PRACTICE TEST

Appendix A, "Private Pilot Practice Test," beginning on page 335, allows you to practice taking the FAA knowledge test without the answers next to the questions. This test has 60 questions, randomly selected from the airplane-related questions in our private pilot knowledge test bank. Topical coverage in this practice test is similar to that of the FAA knowledge test.

It is very important that you answer all 60 questions in one sitting. You should not consult the answers, especially when being referred to figures (charts, tables, etc.) throughout this book where the questions are answered and explained. Analyze your performance based on the answer key that follows the practice test.

Also rely on the Gleim **FAA Test Prep Online** to simulate actual computer testing conditions, including the screen layouts, instructions, etc., for CATS and PSI/LaserGrade.

For more information on the Gleim **FAA Test Prep Online**, see pages 17 through 19.

PART 141 SCHOOLS WITH FAA PILOT KNOWLEDGE TEST EXAMINING AUTHORITY

The FAA permits some FAR Part 141 schools to develop, administer, and grade their own knowledge tests as long as they use questions from the FAA test bank, similar to the questions in this book. The FAA does not provide the correct answers to the Part 141 schools, and the FAA only reviews the Part 141 school test question selection sheets. Thus, some of the answers used by Part 141 test examiners may not agree with the FAA or with those in this book. The latter is not a problem but may explain why you may miss a question on a Part 141 pilot knowledge test using an answer presented in this book.

AUTHORIZATION TO TAKE THE FAA PILOT KNOWLEDGE TEST

Before taking the private pilot knowledge test, you must receive an endorsement from an authorized instructor who conducted the ground training or reviewed your home-study in the areas listed in item 4. on page 4, certifying that you are prepared to pass the knowledge test.

Recreational pilots have a similarly worded requirement. For your convenience, standard authorization forms for both the private and recreational pilot knowledge tests are reproduced on page 359, which can be easily completed, signed by a flight or ground instructor, torn out, and taken to the test site.

Note that if you use the Gleim **FAA Test Prep Online** or **Online Ground School**, the program will generate an authorization signed in facsimile by Dr. Gleim that is accepted at all CATS and PSI/LaserGrade locations.

WHEN TO TAKE THE FAA PILOT KNOWLEDGE TEST

1. You must be at least 15 years of age to take the private pilot knowledge test.
2. You must prepare for the test by successfully completing a ground instruction course, or by using this book as your self-developed home study course.
   a. See "Authorization to Take the FAA Pilot Knowledge Test" on the previous page.
3. Take the FAA knowledge test within 30 days of beginning your study.
   a. Get the knowledge test behind you.
4. Your practical test must follow within 24 months.
   a. Otherwise, you will have to retake your knowledge test.

WHAT TO TAKE TO THE FAA PILOT KNOWLEDGE TEST

1. The same flight computer that you use to solve the test questions in this book, i.e., one you are familiar with and have used before
2. Navigational plotter
3. A pocket calculator you are familiar with and have used before (no instructional material for the calculator is allowed)
4. Authorization to take the knowledge test (see previous page and page 359)
5. Proper identification that contains your
   a. Photograph
   b. Signature
   c. Date of birth
   d. Actual residential address, if different from your mailing address

NOTE: Paper and pencils are supplied at the examination site.

COMPUTER TESTING CENTERS

The FAA has contracted with two computer testing services to administer FAA knowledge tests. Both of these computer testing services have testing centers throughout the country. To register for the knowledge test, call one of the computer testing services listed below or call one of their testing centers. You can find a location most convenient to you, get information regarding the cost to take the knowledge test, and confirm the time allowed for the test. When you register, you will pay the fee with a credit card. Information about these testing services and telephone numbers can be found at www.gleim.com/testing_centers.

CATS   (800) 947-4228   PSI/LaserGrade   (800) 211-2754

COMPUTER TESTING PROCEDURES

When you arrive at the testing center, you will be required to provide positive proof of identification and documentary evidence of your age. The identification must include your photograph, signature, and actual residential address if different from the mailing address. This information may be presented in more than one form of identification. Next, you will sign in on the testing center's daily log. Your signature on the logsheet certifies that, if this is a retest, you meet the applicable requirements (see "Failure on the FAA Pilot Knowledge Test" on page 14) and that you have not passed this test in the past 2 years. Finally, you will present your logbook endorsement or authorization form from your instructor, which authorizes you to take the test. A standard authorization form is provided on page 359 for your use. Both **FAA Test Prep Online** and **Online Ground School** generate an authorization signed in facsimile by Dr. Gleim that is accepted at all CATS and PSI/LaserGrade locations.

Next, you will be taken into the testing room and seated at a computer terminal. A person from the testing center will assist you in logging onto the system, and you will be asked to confirm your personal data (e.g., name, Social Security number, etc.). Then you will be prompted and given an online introduction to the computer testing system, and you will take a sample test. If you have used our **FAA Test Prep Online**, you will be conversant with the computer testing methodology and environment, and you will breeze through the sample test and begin the actual test soon after. You will be allowed 2.5 hours to complete the actual test, which equates to 2.5 minutes per question. Confirm the time permitted when you call the testing center to register. When you have completed your test, an Airman Computer Test Report will be printed out, validated (usually with an embossed seal), and given to you by a person from the testing center. Before you leave, you will be required to sign out on the testing center's daily log.

Each testing service has certain idiosyncrasies in its paperwork, scheduling, and telephone procedures as well as in its software. It is for this reason that our **FAA Test Prep Online** emulates both of the FAA-approved computer testing companies.

YOUR FAA PILOT KNOWLEDGE TEST REPORT

1. You will receive your FAA Pilot Knowledge Test Report upon completion of the test. An example test report is reproduced below.

    a. Note that you will receive only one grade as illustrated.
    b. The expiration date is the date by which you must take your FAA practical test.
    c. The report lists the FAA learning statement codes of the questions you missed so you can review the topics you missed prior to your practical test.

Federal Aviation Administration
Airman Computer Test Report

EXAM TITLE: Private Pilot Airplane

NAME: Jones David John

ID NUMBER: 123456789   TAKE: 1

DATE: 07/14/13   SCORE: 82   GRADE: Pass

....

Knowledge area codes in which questions were answered incorrectly. See appropriate FAA knowledge test study guide. A code may represent more than one incorrect response.

PLT015 PLT192 PLT226

EXPIRATION DATE: 07/31/15

DO NOT LOSE THIS REPORT

....

Authorized instructor's statement (if applicable).

I have given Mr./Ms. _____ additional instruction in each subject area shown to be deficient and consider the applicant competent to pass the test.

Last _____ Initial _____ Cert. No. _____ Type _____

Signature _____

CTD's Embossed Seal

2. Use the FAA Listing of Learning Statement Codes on pages 345 through 347 to determine which topics you had difficulty with.

    a. Look them over and review them with your CFI so (s)he can certify that (s)he reviewed the deficient areas and found you competent in them when you take your practical test. Have your CFI sign off your deficiencies on the FAA Pilot Knowledge Test Report.

3. Keep your FAA Pilot Knowledge Test Report in a safe place because you must submit it to the FAA inspector/examiner when you take your practical test.

FAILURE ON THE FAA PILOT KNOWLEDGE TEST

1. If you fail (score less than 70%) the knowledge test (which is virtually impossible if you follow the Gleim system), you may retake it after your instructor endorses the bottom of your FAA Pilot Knowledge Test Report certifying that you have received the necessary ground training to retake the test.

2. Upon retaking the test, you will find that the procedure is the same except that you must also submit your FAA Pilot Knowledge Test Report indicating the previous failure to the computer testing center.

3. Note that the pass rate on the private pilot knowledge test is about 92%; i.e., fewer than 1 out of 10 fail the test initially. Reasons for failure include

    a. Failure to study the material tested and mere memorization of correct answers. (Relevant study material is contained in the outlines at the beginning of Study Units 1 through 11 of this book.)

    b. Failure to practice working through the questions under test conditions. (All of the previously released FAA questions on airplanes appear in Study Units 1 through 11 of this book.)

    c. Poor examination technique, such as misreading questions and not understanding the requirements.

This Gleim Knowledge Test book will prepare you to pass the FAA private pilot knowledge test on your first attempt! In addition, the Gleim *Private Pilot Flight Maneuvers and Practical Test Prep* book will save you time and frustration as you prepare for the FAA practical test.

Just as this book organizes and explains the knowledge needed to pass your FAA knowledge test, *Private Pilot Flight Maneuvers and Practical Test Prep* will assist you in developing the competence and confidence to pass your FAA practical test.

Also, flight maneuvers are quickly perfected when you understand exactly what to expect before you get into an airplane to practice the flight maneuvers. You must be ahead of (not behind) your CFI and your airplane. Our flight maneuvers books explain and illustrate all flight maneuvers so the maneuvers and their execution are intuitively appealing to you. Call (800) 874-5346 or visit www.gleim.com/aviation and order your books today!

GLEIM ONLINE GROUND SCHOOL

1. Gleim **Online Ground School (OGS)** course content is based on the Gleim Knowledge Test books, **FAA Test Prep Online**, FAA publications, and Gleim reference books. The delivery system is modeled on the Gleim FAA-approved online **Flight Instructor Refresher Course**.

   a. Online Ground School courses are available for

      1) Private Pilot
      2) Sport Pilot
      3) CFI/CGI
      4) FOI
      5) Instrument Pilot
      6) Commercial Pilot
      7) ATP
      8) Flight Engineer
      9) Canadian Certificate Conversion

   b. OGS courses are airplane-only and have lessons that correspond to the study units in the Gleim FAA Knowledge Test books.

   c. Each course contains study outlines that automatically reference current FAA publications, the appropriate knowledge test questions, FAA figures, and Gleim answer explanations.

   d. OGS is always up to date.

   e. Users achieve very high knowledge test scores and a near-100% pass rate.

   f. **Gleim Online Ground School is the most flexible course available!** Access your OGS personal classroom from any computer with Internet access--24 hours a day, 7 days a week. Your virtual classroom is never closed!

Number	Study Unit	Status	Score	Time Started	Time Completed	A/V	Outline	Action
1	Airplanes and Aerodynamics	In Progress	N/A	03-22-2012, 1:21 pm	N/A	View	View	Continue
2	Airplane Instruments, Engines, and Systems	Not Started	N/A	N/A	N/A			
3	Airports, Air Traffic Control, and Airspace	Not Started	N/A	N/A	N/A			
4	Federal Aviation Regulations	Not Started	N/A	N/A	N/A			
5	Airplane Performance and Weight and Balance	Not Started	N/A	N/A	N/A			
	Stage Test #1	Not Started	N/A	N/A	N/A	N/A	N/A	
6	Aeromedical Factors and Aeronautical Decision Making (ADM)	Not Started	N/A	N/A	N/A			
7	Aviation Weather	Not Started	N/A	N/A	N/A			
8	Aviation Weather Services	Not Started	N/A	N/A	N/A			
9	Navigation: Charts and Publications	Not Started	N/A	N/A	N/A			
10	Navigation Systems	Not Started	N/A	N/A	N/A			
11	Cross-Country Flight Planning	Not Started	N/A	N/A	N/A			
	Stage Test #2	Not Started	N/A	N/A	N/A	N/A	N/A	
	End-Of-Course Test	Not Started	N/A	N/A	N/A	N/A	N/A	
	Practice Test #1	Not Started	N/A	N/A	N/A	N/A	N/A	
	Practice Test #2	Not Started	N/A	N/A	N/A	N/A	N/A	
	Practice Test #3	Not Started	N/A	N/A	N/A	N/A	N/A	
	Practice Test #4	Not Started	N/A	N/A	N/A	N/A	N/A	
	Practice Test #5	Not Started	N/A	N/A	N/A	N/A	N/A	

g. **Save time and study only the material you need to know!** Gleim **Online Ground School** Certificate Selection will provide you with a customized study plan. You save time because unnecessary questions will be automatically eliminated.

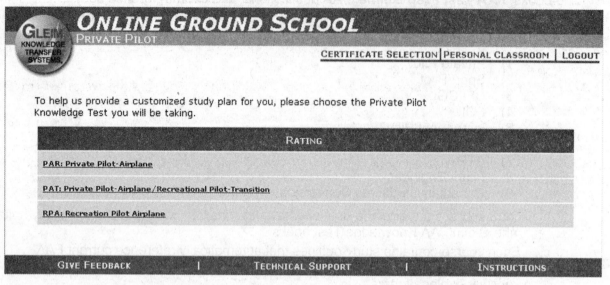

h. **We are truly interactive. We help you focus on any weaker areas.** Answer explanations for wrong choices help you learn from your mistakes.

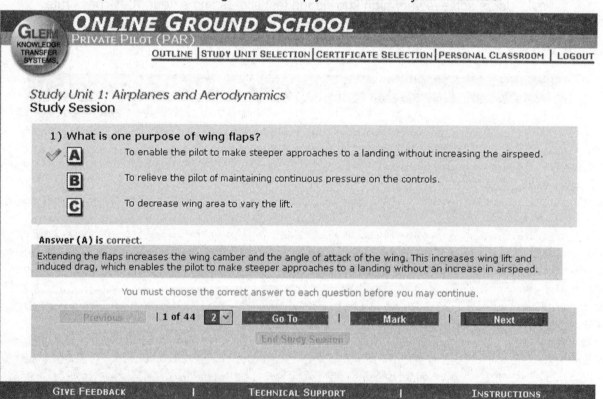

Register for Gleim Online Ground School today:
www.gleim.com/OGS

GLEIM FAA TEST PREP ONLINE

Computer testing is consistent with aviation's use of computers (e.g., DUATS, flight simulators, computerized cockpits, etc.). All FAA knowledge tests are administered by computer.

Computer testing is natural after computer study and computer-assisted instruction is a very efficient and effective method of study. The Gleim **FAA Test Prep Online** is designed to prepare you for computer testing because our software can simulate both CATS and PSI/LaserGrade. We make you comfortable with computer testing!

FAA Test Prep Online contains all of the questions in this book, context-sensitive outline material, and on-screen charts and figures. It allows you to choose either Study Mode or Test Mode.

In Study Mode, the software provides you with an explanation of each answer you choose (correct or incorrect). You design each Study Session:

Topic(s) and/or FAA learning statement codes
  you wish to cover
Number of questions
Order of questions -- FAA, Gleim, or random
Order of answers to each question --
  Gleim or random

Questions marked and/or missed from last session --
  test, study, or both
Questions marked and/or missed from all sessions --
  test, study, or both
Questions never seen, answered, or answered correctly

In Test Mode, you decide the format -- CATS or PSI/LaserGrade. When you finish your test, you can and should study the questions missed and access answer explanations. The software emulates the operation of FAA-approved computer testing companies. Thus, you have a complete understanding of how to take an FAA knowledge test and know exactly what to expect before you go to a computer testing center.

The Gleim **FAA Test Prep Online** is an all-in-one program designed to help anyone with a computer, Internet access, and an interest in flying pass the FAA knowledge tests.

Study Sessions and Test Sessions

Study Sessions give you immediate feedback on why your answer selection for a particular question is correct or incorrect and allow you to access the context-sensitive outline material that helps to explain concepts related to the question. Choose from several different question sources: all questions available for that library; questions from a certain topic (Gleim study units and subunits); questions that you missed or marked in the last sessions you created; questions that you have never seen, answered, or answered correctly; questions from certain FAA learning statement codes; etc. You can mix up the questions by selecting to randomize the question and/or answer order so that you do not memorize answer letters.

You may then grade your study sessions and track your study progress using the performance analysis charts and graphs. The Performance Analysis information helps you to focus on areas where you need the most improvement, saving you time in the overall study process. You may then want to go back and study questions that you missed in a previous session, or you may want to create a Study Session of questions that you marked in the previous session. All of these options are made easy with **FAA Test Prep Online**'s Study Sessions.

After studying the outlines and questions in a Study Session, you can further test your skills with a Test Session. These sessions allow you to answer questions under actual testing conditions using one of the simulations of the major testing services. In a Test Session, you will not know which questions you have answered correctly until the session is graded.

Recommended Study Program

1. Start with Study Unit 1 and proceed through study units in chronological order. Follow the three-step process below.

    a. First, carefully study the Gleim Outline.

    b. Second, create a Study Session of all questions in the study unit. Answer and study all questions in the Study Session.

    c. Third, create a Test Session of all questions in the study unit. Answer all questions in the Test Session.

2. After each Study Session and Test Session, create a new Study Session from questions answered incorrectly. This is of critical importance to allow you to learn from your mistakes.

Example Question Screen

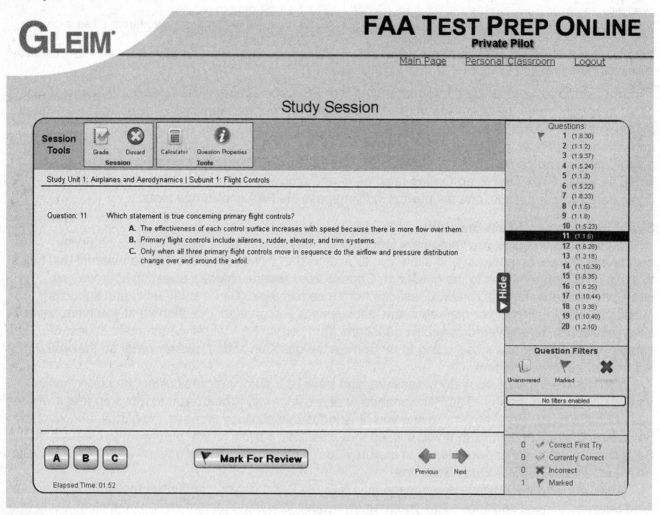

Practice Test

Take an exam in the actual testing environment of either of the major testing centers: CATS or PSI/LaserGrade. **FAA Test Prep Online** simulates the testing formats of these testing centers, making it easy for you to study questions under actual exam conditions. After studying with **FAA Test Prep Online**, you will know exactly what to expect when you go in to take your pilot knowledge test.

On-Screen Charts and Figures

One of the most convenient features of **FAA Test Prep Online** is the easily accessible on-screen charts and figures. Many of the questions refer to drawings, maps, charts, and other pictures that provide information to help answer the question. In **FAA Test Prep Online**, you can pull up any of these figures with the click of a button. You can increase or decrease the size of the images, and you may also use our drawing feature to calculate the true course between two given points (required only on the private pilot knowledge test).

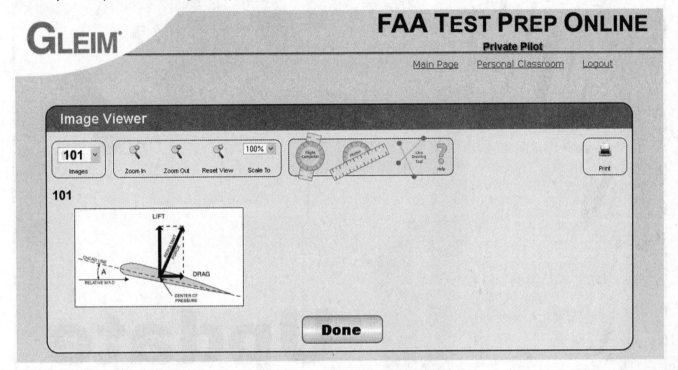

Instructor Sign-Off Sheets

FAA Test Prep Online is capable of generating an instructor sign-off for FAA knowledge tests that require one. This sign-off has been approved by the FAA and can be presented at the computer testing center as authorization to take your test--you do NOT need an additional endorsement from your instructor.

In order to obtain the instructor sign-off sheet for your test, you must first answer all relevant questions in **FAA Test Prep Online** correctly. Then, select "Sign-Off Sheets" under the "Additional Features" area on the Main page. If you have answered all of the required questions, the instructor sign-off sheet will appear for you to print. If you have not yet answered all required questions, a list of the unanswered questions, along with their location, will appear.

Order your copy of FAA Test Prep Online today
(800) 874-5346 • gleim.com

Free Updates and Technical Support

Gleim offers FREE technical support to all users of the current versions. Call (800) 874-5346, send an email to support@gleim.com, or fill out the technical support request form online (www.gleim.com/support/form.php). Additionally, Gleim **FAA Test Prep Online** is always up to date. The program is automatically updated when any changes (e.g., FAA question release) are made, so you can be confident that Gleim will prepare you for your knowledge test. For more information on our email update service for books, turn to page iv.

20

STUDY UNIT ONE
AIRPLANES AND AERODYNAMICS

(5 pages of outline)

This study unit contains outlines of major concepts tested, sample test questions and answers regarding airplanes and aerodynamics, and an explanation of each answer. The table of contents above lists each subunit within this study unit, the number of questions pertaining to that particular subunit, and the pages on which the outline and questions begin, respectively.

CAUTION: Recall that the **sole purpose** of this book is to expedite your passing the FAA pilot knowledge test for the private pilot certificate. Accordingly, all extraneous material (i.e., topics or regulations not directly tested on the FAA pilot knowledge test) is omitted, even though much more information and knowledge are necessary to fly safely. This additional material is presented in *Pilot Handbook* and *Private Pilot Flight Maneuvers and Practical Test Prep*, available from Gleim Publications, Inc. See the order form on page 371.

1.1 FLIGHT CONTROLS

1. The three primary flight controls of an airplane are the ailerons, the elevator (or stabilator), and the rudder.

   a. Movement of any of these primary flight control surfaces changes the airflow and pressure distribution over and around the airfoil.

      1) These changes affect the lift and drag produced and allow a pilot to control the aircraft about its three axes of rotation.

   b. **Ailerons** are control surfaces attached to each wing that move in the opposite direction from one another to control roll about the longitudinal axis.

      1) EXAMPLE: Moving the yoke or stick to the right causes the right aileron to deflect upward, resulting in decreased lift on the right wing. The left aileron moves in the opposite direction and increases the lift on the left wing. Thus, the increased lift on the left wing and and the decreased lift on the right wing cause the airplane to roll to the right.

   c. The **elevator** is the primary control device for changing the pitch attitude of an airplane, changing the pitch about the lateral axis. It is usually located on the fixed horizontal stabilizer on the tail of the airplane.

      1) EXAMPLE: Pulling back on the yoke or stick deflects the trailing edge of the elevator up. This position creates a downward aerodynamic force, causing the tail of the aircraft to move down and the nose to pitch up.

      2) A **stabilator** is a one-piece horizontal stabilizer and elevator that pivots from a central hinge point.

3) A **canard** is similar to the horizontal stabilizer but is located in front of the main wings. An elevator is attached to the trailing edge of the canard to control pitch.

a) The canard, however, actually creates lift and holds the nose up rather than the aft-tail design that prevents the nose from rotating downward.

d. The **rudder** controls movement of the aircraft about its vertical axis.

1) When deflecting the rudder into the airflow, a horizontal force is exerted in the opposite direction. This motion is called yaw.

2) Rudder effectiveness increases with speed because there is more airflow over the surface of the control device.

2. Secondary flight controls may consist of wing flaps, leading edge devices, spoilers, and trim systems.

a. **Flaps** are attached to the trailing edge of the wing and are used during approach and landing to increase wing lift. This allows an increase in the angle of descent without increasing airspeed.

1) The most common flap used on general aviation aircraft today is the slotted flap.

2) When the slotted flap is lowered, high-pressure air from the lower surface of the wing is ducted to the upper surface of the flap, delaying airflow separation.

b. **Spoilers** are high-drag devices deployed from the wings to reduce lift and increase drag. They are found on gliders and some high-speed aircraft.

c. **Trim systems** are used to relieve the pilot of the need to maintain constant back pressure on the flight controls. They include trim tabs, antiservo tabs, and ground adjustable tabs.

1) Trim tabs are attached to the trailing edge of the elevator.

a) EXAMPLE: If the trim tab is set to the full nose-up position, the tab moves full down. This causes the tail of the airplane to pitch down and the nose to pitch up.

1.2 AERODYNAMIC FORCES

1. The four aerodynamic forces acting on an airplane during flight are

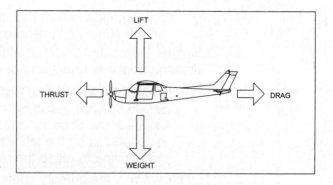

a. Lift – the upward-acting force
b. Weight – the downward-acting force
c. Thrust – the forward-acting force
d. Drag – the rearward-acting force

2. These forces are at equilibrium when the airplane is in unaccelerated flight:

Lift = Weight
Thrust = Drag

3. **Bernoulli's Principle** states in part that "the internal pressure of a fluid (liquid or gas) decreases at points where the speed of the fluid increases." In other words, high speed flow is associated with low pressure, and low speed flow is associated with high pressure.

   a. This principle is applicable to an airplane wing because it is designed and constructed with a curve or camber. When air flows along the upper wing surface, it travels a greater distance in the same period of time (i.e., faster) than the airflow along the lower wing surface.

   b. Therefore, the pressure above the wing is less than it is below the wing. This generates a lift force over the upper curved surface of the wing.

1.3 ANGLE OF ATTACK

1. The angle of attack is the angle between the wing chord line and the direction of the relative wind.

   a. The wing chord line is an imaginary straight line from the leading edge to the trailing edge of the wing.

   b. The relative wind is the direction of airflow relative to the wing when the wing is moving through the air.

2. The angle of attack at which a wing stalls remains constant regardless of weight, airplane loading, etc.

1.4 STALLS AND SPINS

1. An airplane can be stalled at any airspeed in any flight attitude. A stall results whenever the critical angle of attack is exceeded.

2. An airplane in a given configuration will stall at the same indicated airspeed regardless of altitude because the airspeed indicator is directly related to air density.

3. An airplane spins when one wing is less stalled than the other wing.

   a. To enter a spin, an airplane must always be stalled first.

1.5 FROST

1. Frost forms when the temperature of the collecting surface is at or below the dew point of the adjacent air and the dew point is below freezing.

   a. The water vapor changes its physical state through deposition and immediately forms as ice crystals on the wing surface.

2. Frost on wings disrupts the smooth airflow over the airfoil by causing early airflow separation from the wing. This

   a. Decreases lift and
   b. Causes friction and increases drag.

3. Frost may make it difficult or impossible for an airplane to take off.

4. Frost should be removed before attempting to take off.

1.6 GROUND EFFECT

1. Ground effect is the result of the interference of the ground (or water) surface with the airflow patterns about an airplane.

2. The vertical component of the airflow around the wing is restricted, which alters the wing's upwash, downwash, and wingtip vortices.

3. The reduction of the wingtip vortices alters the spanwise lift distribution and reduces the induced angle of attack and induced drag.

    a. Thus, the wing will require a lower angle of attack in ground effect to produce the same lift coefficient, or, if a constant angle of attack is maintained, an increase in the lift coefficient will result.

4. An airplane is affected by ground effect when it is within the length of the airplane's wingspan above the ground. The ground effect is most often recognized when the airplane is less than one-half the wingspan's length above the ground.

5. Ground effect may cause an airplane to float on landings or permit it to become airborne with insufficient airspeed to stay in flight above the area of ground effect.

    a. An airplane may settle back to the surface abruptly after flying through the ground effect if the pilot has not attained recommended takeoff airspeed.

1.7 AIRPLANE TURN

1. The horizontal component of lift makes an airplane turn.

    a. To attain this horizontal component of lift, the pilot coordinates rudder, aileron, and elevator.

2. The rudder on an airplane controls the yaw, i.e., rotation about the vertical axis, but does not cause the airplane to turn.

1.8 AIRPLANE STABILITY

1. An inherently stable airplane returns to its original condition (position or attitude) after being disturbed. It requires less effort to control.

2. The location of the center of gravity (CG) with respect to the center of lift (or center of pressure) determines the longitudinal stability of an airplane.

    a. Changes in the center of pressure in a wing affects the aircraft's aerodynamic balance and control.

3. Airplanes (except a T-tail) normally pitch down when power is reduced (and the controls are not adjusted) because the downwash on the elevators from the propeller slipstream is reduced and elevator effectiveness is reduced. This allows the nose to drop.

4. When the CG in an airplane is located at or rear of the aft CG limit, the airplane

    a. Develops an inability to recover from stall conditions and
    b. Becomes less stable at all airspeeds.

1.9 TORQUE AND P-FACTOR

1. The torque effect (left-turning tendency) is greatest at low airspeed, high angles of attack, and high power, e.g., on takeoff.

2. P-factor (asymmetric propeller loading) causes the airplane to yaw to the left when at high angles of attack because the descending right side of the propeller (as seen from the rear) has a higher angle of attack (than the upward-moving blade on the left side) and provides more thrust.

1.10 LOAD FACTOR

1. Load factor refers to the additional weight carried by the wings due to the airplane's weight plus the centrifugal force.

    a. The amount of excess load that can be imposed on an airplane's wings varies directly with the airplane's speed and the excess lift available.

        1) At low speeds, very little excess lift is available, so very little excess load can be imposed.

        2) At high speeds, the wings' lifting capacity is so great that the load factor can quickly exceed safety limits.

    b. An increased load factor will result in an airplane stalling at a higher airspeed.

    c. As bank angle increases, the load factor increases. The wings have to carry not only the airplane's weight but the centrifugal force as well.

2. On the exam, a load factor chart is given with the amount of bank on the horizontal axis (along the bottom of the graph), and the load factor on the vertical axis (up the left side of the graph). Additionally, a table that provides the load factor corresponding to specific bank angles is found on the left side of the chart. Use this table to answer load factor questions.

    a. Compute the load factor by multiplying the airplane's weight by the load factor that corresponds to the given angle of bank. For example, the wings of a 2,000-lb. airplane in a 60° bank must support 4,000 lb. (2,000 lb. × 2.000).

    b. Example load factor chart:

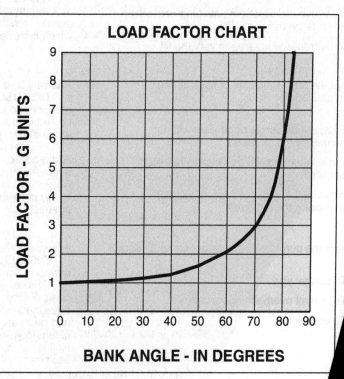

ANGLE of BANK ϕ	LOAD FACTOR n
0°	1.0
10°	1.015
30°	1.154
45°	1.414
60°	2.000
70°	2.923
80°	5.747
85°	11.473
90°	∞

Figure 2. – Load Factor Chart.

3. Load factor (or G units) is a multiple of the regular weight or, alternatively, a multiple of the force of gravity.

    a. Straight-and-level flight has a load factor at 1.0. (Verify on the chart above.)

    b. A 60° level bank has a load factor of 2.0. Due to centrifugal force, the wings hold up twice the amount of weight.

    c. A 50° level bank has a load factor of about 1.5.

QUESTIONS AND ANSWER EXPLANATIONS

All of the private pilot knowledge test questions chosen by the FAA for release as well as additional questions selected by Gleim relating to the material in the previous outlines are reproduced on the following pages. These questions have been organized into the same subunits as the outlines. To the immediate right of each question are the correct answer and answer explanation. You should cover these answers and answer explanations while responding to the questions. Refer to the general discussion in the Introduction on how to take the FAA pilot knowledge test.

Remember that the questions from the FAA pilot knowledge test bank have been reordered by topic and organized into a meaningful sequence. Also, the first line of the answer explanation gives the citation of the authoritative source for the answer.

QUESTIONS

1.1 Flight Controls

1. What is one purpose of wing flaps?

A. To enable the pilot to make steeper approaches to a landing without increasing the airspeed.

B. To relieve the pilot of maintaining continuous pressure on the controls.

C. To decrease wing area to vary the lift.

Answer (A) is correct. *(PHAK Chap 5)*
DISCUSSION: Extending the flaps increases the wing camber and the angle of attack of the wing. This increases wing lift and induced drag, which enables the pilot to make steeper approaches to a landing without an increase in airspeed.
Answer (B) is incorrect. Trim tabs (not wing flaps) help relieve control pressures. Answer (C) is incorrect. Wing area usually remains the same, except for certain specialized flaps which increase (not decrease) the wing area.

One of the main functions of flaps during
approach and landing is to

decrease the angle of descent without
increasing the airspeed.

permit a touchdown at a higher indicated
speed.

increase the angle of descent without
increasing the airspeed.

Answer (C) is correct. *(PHAK Chap 5)*
DISCUSSION: Extending the flaps increases the wing camber and the angle of attack of the wing. This increases wing lift and induced drag, which enables the pilot to increase the angle of descent without increasing the airspeed.
Answer (A) is incorrect. Extending the flaps increases lift and induced drag, which enables the pilot to increase (not decrease) the angle of descent without increasing the airspeed. Answer (B) is incorrect. Flaps increase lift at slow airspeed, which permits touchdown at a lower (not higher) indicated airspeed.

purpose of the rudder on an airplane?

banking tendency.

Answer (A) is correct. *(PHAK Chap 5)*
DISCUSSION: The rudder is used to control yaw, which is rotation about the airplane's vertical axis.
Answer (B) is incorrect. The ailerons (not the rudder) control overbanking. Overbanking tendency refers to the outside wing traveling significantly faster than the inside wing in a steep turn and generating incremental lift to raise the outside wing higher unless corrected by aileron pressure. Answer (C) is incorrect. Roll is movement about the longitudinal axis and is controlled by ailerons.

of the   control surface?

s must hold

Answer (A) is correct. *(PHAK Chap 5)*
DISCUSSION: The three primary flight controls of an airplane are the ailerons, the elevator (or stabilator), and the rudder.
Answer (B) is incorrect. The stabilator, or elevator, is a primary flight control surface. Answer (C) is incorrect. Ailerons are a primary flight control surface.

5. The elevator controls movement around which axis?

    A. Longitudinal.

    B. Lateral.

    C. Vertical.

Answer (B) is correct. *(PHAK Chap 5)*
    DISCUSSION: The elevator is the primary control device for changing the pitch attitude of an airplane about the lateral axis.
    Answer (A) is incorrect. Ailerons are control surfaces attached to each wing that move in the opposite direction from one another to control roll about the longitudinal axis.
Answer (C) is incorrect. The rudder controls movement of the aircraft about its vertical axis.

6. Which statement is true concerning primary flight controls?

    A. The effectiveness of each control surface increases with speed because there is more flow over them.

    B. Only when all three primary flight controls move in sequence do the airflow and pressure distribution change over and around the airfoil.

    C. Primary flight controls include ailerons, rudder, elevator, and trim systems.

Answer (A) is correct. *(PHAK Chap 5)*
    DISCUSSION: Rudder, aileron, and elevator effectiveness increase with speed because there is more airflow over the surface of the control device.
    Answer (B) is incorrect. Movement of any primary flight control surface changes the airflow and pressure distribution over and around the airfoil. Answer (C) is incorrect. The primary flight controls do not include trim systems; these are considered secondary flight controls.

7. Which of the following is true concerning flaps?

    A. Flaps are attached to the leading edge of the wing and are used to increase wing lift.

    B. Flaps allow an increase in the angle of descent without increasing airspeed.

    C. Flaps are high drag devices deployed from the wings to reduce lift.

Answer (B) is correct. *(PHAK Chap 5)*
    DISCUSSION: Flaps are attached to the trailing edge of the wing and are used during approach and landing to increase wing lift. This allows an increase in the angle of descent without increasing airspeed.
    Answer (A) is incorrect. Flaps are attached to the trailing edge, not the leading edge, of the wing. Answer (C) is incorrect. Spoilers, not flaps, are high-drag devices deployed from the wings to reduce lift and increase drag.

8. Which device is a secondary flight control?

    A. Spoilers.

    B. Ailerons.

    C. Stabilators.

Answer (A) is correct. *(PHAK Chap 5)*
    DISCUSSION: Spoilers are high-drag devices that assist an aircraft in slowing down and losing altitude without gaining extra speed. They are common on gliders and some high-speed aircraft.
    Answer (B) is incorrect. Ailerons control the roll of the aircraft and are a primary flight control surface. Answer (C) is incorrect. Stabilators function as both a horizontal stabilizer and an elevator, which makes them a primary control surface.

9. Trim systems are designed to do what?

    A. They relieve the pilot of the need to maintain constant back pressure on the flight controls.

    B. They are used during approach and landing to increase wing lift.

    C. They move in the opposite direction from one another to control roll.

Answer (A) is correct. *(PHAK Chap 5)*
    DISCUSSION: Trim systems are used to relieve the pilot of the need to maintain constant back pressure on the flight controls. They include trim tabs, antiservo tabs, and ground adjustable tabs.
    Answer (B) is incorrect. Flaps, not trim systems, are used during approach and landing to increase lift. This allows an increase in the angle of descent without increasing airspeed. Answer (C) is incorrect. Ailerons are control surfaces attached to each wing that move in the opposite direction from one another to control roll about the longitudinal axis.

1.2 Aerodynamic Forces

10. The four forces acting on an airplane in flight are

A. lift, weight, thrust, and drag.

B. lift, weight, gravity, and thrust.

C. lift, gravity, power, and friction.

Answer (A) is correct. *(PHAK Chap 4)*
   DISCUSSION: Lift is produced by the wings and opposes weight, which is the result of gravity. Thrust is produced by the engine/propeller and opposes drag, which is the resistance of the air as the airplane moves through it.
   Answer (B) is incorrect. Gravity reacts with the airplane's mass, thus producing weight, which opposes lift. Answer (C) is incorrect. Gravity results in weight, power produces thrust, and friction is a cause of drag. Power, gravity, velocity, and friction are not aerodynamic forces in themselves.

11. When are the four forces that act on an airplane in equilibrium?

A. During unaccelerated flight.

B. When the aircraft is accelerating.

C. When the aircraft is at rest on the ground.

Answer (A) is correct. *(PHAK Chap 4)*
   DISCUSSION: The four forces (lift, weight, thrust, and drag) that act on an airplane are in equilibrium during unaccelerated flight.
   Answer (B) is incorrect. Thrust must exceed drag in order for the airplane to accelerate. Answer (C) is incorrect. When the airplane is at rest on the ground, there are no aerodynamic forces acting on it other than weight (gravity).

12. What is the relationship of lift, drag, thrust, and weight when the airplane is in straight-and-level flight?

A. Lift equals weight and thrust equals drag.

B. Lift, drag, and weight equal thrust.

C. Lift and weight equal thrust and drag.

Answer (A) is correct. *(PHAK Chap 4)*
   DISCUSSION: When the airplane is in straight-and-level flight (assuming no change of airspeed), it is not accelerating, and therefore lift equals weight and thrust equals drag.
   Answer (B) is incorrect. Lift equals weight and drag equals thrust. Answer (C) is incorrect. Lift and weight are equal and thrust and drag are equal, but the four are not equal to each other.

13. Which statement relates to Bernoulli's principle?

A. For every action, there is an equal and opposite reaction.

B. An additional upward force is generated as the lower surface of the wing deflects air downward.

C. Air traveling faster over the curved upper surface of an airfoil causes lower pressure on the top surface.

Answer (C) is correct. *(PHAK Chap 3)*
   DISCUSSION: Bernoulli's principle states in part that the internal pressure of a fluid (liquid or gas) decreases at points where the speed of the fluid increases. This same principle applies to air flowing over the curved upper surface of a wing.
   Answer (A) is incorrect. Newton's Third Law of Motion states that, for every action, there is an equal and opposite reaction. Answer (B) is incorrect. The additional upward force that is generated as the lower surface of the wing deflects air downward is related to Newton's Third Law of Motion.

1.3 Angle of Attack

14. The term "angle of attack" is defined as the angle between the

A. chord line of the wing and the relative wind.

B. airplane's longitudinal axis and that of the air striking the airfoil.

C. the airplane's center line and the relative wind.

Answer (A) is correct. *(PHAK Chap 3)*
   DISCUSSION: The angle of attack is the angle between the wing chord line and the direction of the relative wind. The wing chord line is a straight line from the leading edge to the trailing edge of the wing. The relative wind is the direction of the airflow relative to the wing when the wing is moving through the air.
   Answer (B) is incorrect. Angle of attack is the angle between the wing chord line and the relative wind, not the airplane's longitudinal axis. Answer (C) is incorrect. The centerline of the airplane and its relationship to the relative wind is not a factor in defining angle of attack. Angle of attack is the relationship between the wing chord line and the relative wind.

15. The term "angle of attack" is defined as the angle

   A. between the wing chord line and the relative wind.

   B. between the airplane's climb angle and the horizon.

   C. formed by the longitudinal axis of the airplane and the chord line of the wing.

Answer (A) is correct. *(PHAK Chap 3)*
   DISCUSSION: The angle of attack is the angle between the wing chord line and the direction of the relative wind. The wing chord line is a straight line from the leading edge to the trailing edge of the wing. The relative wind is the direction of airflow relative to the wing when the wing is moving through the air.
   Answer (B) is incorrect. The angle between the airplane's climb angle and the horizon does not describe any term. Answer (C) is incorrect. The angle formed by the longitudinal axis of the airplane and the chord line of the wing is the angle of incidence (not attack).

16. Angle of attack is defined as the angle between the chord line of an airfoil and the

   A. direction of the relative wind.

   B. pitch angle of an airfoil.

   C. rotor plane of rotation.

Answer (A) is correct. *(PHAK Chap 3)*
   DISCUSSION: The angle of attack is the angle between the wing chord line and the direction of the relative wind. The wing chord line is a straight line from the leading edge to the trailing edge of the wing. The relative wind is the direction of airflow relative to the wing when the wing is moving through the air.
   Answer (B) is incorrect. Pitch is used in conjunction with the aircraft or longitudinal axis, not the chord line of the airfoil. Answer (C) is incorrect. Rotor plane of rotation deals with helicopters, not fixed-wing aircraft.

17. (Refer to Figure 1 below.) The acute angle A is the angle of

   A. incidence.

   B. attack.

   C. dihedral.

Answer (B) is correct. *(PHAK Chap 3)*
   DISCUSSION: The angle between the relative wind and the wing chord line is the angle of attack. The wing chord line is a straight line from the leading edge to the trailing edge of the wing.
   Answer (A) is incorrect. The angle of incidence is the acute angle formed by the chord line of the wing and the longitudinal axis of the airplane. Answer (C) is incorrect. The dihedral is the angle at which the wings are slanted upward from the wing root to the wingtip.

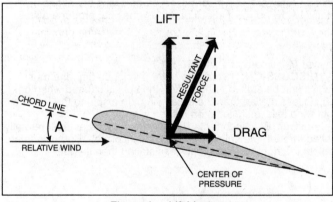

Figure 1. – Lift Vector.

18. The angle between the chord line of an airfoil and the relative wind is known as the angle of

   A. lift.

   B. attack.

   C. incidence.

Answer (B) is correct. *(PHAK Chap 3)*
   DISCUSSION: The angle of attack is the acute angle between the chord line of the wing and the direction of the relative wind.
   Answer (A) is incorrect. The angle of lift is a nonsense term. Answer (C) is incorrect. The angle of incidence is the acute angle formed by the chord line of the wing and the longitudinal axis of the airplane.

19. The angle of attack at which an airplane wing stalls will

    A. increase if the CG is moved forward.

    B. change with an increase in gross weight.

    C. remain the same regardless of gross weight.

Answer (C) is correct. *(PHAK Chap 4)*
    DISCUSSION: A given airplane wing will always stall at the same angle of attack regardless of airspeed, weight, load factor, or density altitude. Each wing has a particular angle of attack (the critical angle of attack) at which the airflow separates from the upper surface of the wing and the stall occurs.
    Answer (A) is incorrect. A change in CG will not change the wing's critical angle of attack. Answer (B) is incorrect. The critical angle of attack does not change when gross weight changes.

1.4 Stalls and Spins

20. As altitude increases, the indicated airspeed at which a given airplane stalls in a particular configuration will

    A. decrease as the true airspeed decreases.

    B. decrease as the true airspeed increases.

    C. remain the same regardless of altitude.

Answer (C) is correct. *(AC 61-67C)*
    DISCUSSION: All the performance factors of an airplane are dependent upon air density. As air density decreases, the airplane stalls at a higher true airspeed. However, you cannot detect the effect of high density altitude on your airspeed indicator. Accordingly, an airplane will stall in a particular configuration at the same indicated airspeed regardless of altitude.
    Answer (A) is incorrect. True airspeed increases, not decreases, with increased altitude, and indicated airspeed at which an airplane stalls remains the same (does not decrease). Answer (B) is incorrect. The indicated airspeed of the stall does not change with increased altitude.

21. In what flight condition must an aircraft be placed in order to spin?

    A. Partially stalled with one wing low.

    B. In a steep diving spiral.

    C. Stalled.

Answer (C) is correct. *(AC 61-67C)*
    DISCUSSION: In order to enter a spin, an airplane must always first be stalled. Thereafter, the spin is caused when one wing becomes less stalled than the other wing.
    Answer (A) is incorrect. The aircraft must first be fully stalled. Answer (B) is incorrect. A steep diving spiral has a relatively low angle of attack and thus does not produce a stall.

22. During a spin to the left, which wing(s) is/are stalled?

    A. Both wings are stalled.

    B. Neither wing is stalled.

    C. Only the left wing is stalled.

Answer (A) is correct. *(AC 61-67C)*
    DISCUSSION: In order to enter a spin, an airplane must always first be stalled. Thereafter, the spin is caused when one wing is less stalled than the other wing. In a spin to the left, the right wing is less stalled than the left wing.
    Answer (B) is incorrect. Both wings must be at least partially stalled through the spin. Answer (C) is incorrect. Both wings are stalled; the right wing is simply less stalled than the left.

1.5 Frost

23. How will frost on the wings of an airplane affect takeoff performance?

    A. Frost will disrupt the smooth flow of air over the wing, adversely affecting its lifting capability.

    B. Frost will change the camber of the wing, increasing its lifting capability.

    C. Frost will cause the airplane to become airborne with a higher angle of attack, decreasing the stall speed.

Answer (A) is correct. *(PHAK Chap 11)*
    DISCUSSION: Frost does not change the basic aerodynamic shape of the wing, but the roughness of its surface spoils the smooth flow of air, thus causing an increase in drag and an early airflow separation over the wing, resulting in a loss of lift.
    Answer (B) is incorrect. Frost will decrease (not increase) lift during takeoff and has no effect on the wing camber. Answer (C) is incorrect. A layer of frost on an airplane will increase drag, which increases (not decreases) the stall speed.

24. Why is frost considered hazardous to flight?

A. Frost changes the basic aerodynamic shape of the airfoils, thereby decreasing lift.

B. Frost slows the airflow over the airfoils, thereby increasing control effectiveness.

C. Frost spoils the smooth flow of air over the wings, thereby decreasing lifting capability.

Answer (C) is correct. *(AvW Chap 10)*
   DISCUSSION: Frost does not change the basic aerodynamic shape of the wing, but the roughness of its surface spoils the smooth flow of air, thus causing an increase in drag and an early airflow separation over the wing, resulting in a loss of lift.
   Answer (A) is incorrect. Frost is thin and does not change the basic aerodynamic shape of the airfoil. Answer (B) is incorrect. The smooth flow of air over the airfoil is affected, not control effectiveness.

25. How does frost affect the lifting surfaces of an airplane on takeoff?

A. Frost may prevent the airplane from becoming airborne at normal takeoff speed.

B. Frost will change the camber of the wing, increasing lift during takeoff.

C. Frost may cause the airplane to become airborne with a lower angle of attack at a lower indicated airspeed.

Answer (A) is correct. *(AvW Chap 10)*
   DISCUSSION: Frost that is not removed from the surface of an airplane prior to takeoff may make it difficult to get the airplane airborne at normal takeoff speed. The frost disrupts the airflow over the wing, which increases drag.
   Answer (B) is incorrect. The smoothness of the wing, not its curvature, is affected and lift is decreased (not increased). Answer (C) is incorrect. Ground effect (not frost) may cause an airplane to become airborne with a lower angle of attack at a lower indicated airspeed.

1.6 Ground Effect

26. What is ground effect?

A. The result of the interference of the surface of the Earth with the airflow patterns about an airplane.

B. The result of an alteration in airflow patterns increasing induced drag about the wings of an airplane.

C. The result of the disruption of the airflow patterns about the wings of an airplane to the point where the wings will no longer support the airplane in flight.

Answer (A) is correct. *(PHAK Chap 4)*
   DISCUSSION: Ground effect is due to the interference of the ground (or water) surface with the airflow patterns about the airplane in flight. As the wing encounters ground effect, there is a reduction in the upwash, downwash, and the wingtip vortices. The result is a reduction in induced drag. Thus, for a given angle of attack, the wing will produce more lift in ground effect than it does out of ground effect.
   Answer (B) is incorrect. The result of the alteration in airflow patterns about the wing decreases, not increases, the induced drag. Answer (C) is incorrect. The disruption of the airflow patterns about the wing decreases induced drag, which causes an increase, not decrease, in lift at a given angle of attack.

27. Floating caused by the phenomenon of ground effect will be most realized during an approach to land when at

A. less than the length of the wingspan above the surface.

B. twice the length of the wingspan above the surface.

C. a higher-than-normal angle of attack.

Answer (A) is correct. *(PHAK Chap 4)*
   DISCUSSION: Ground effect is most usually recognized when the airplane is within one-half of the length of its wingspan above the surface. It may extend as high as a full wingspan length above the surface. Due to an alteration of the airflow about the wings, induced drag decreases, which reduces the thrust required at low airspeeds. Thus, any excess speed during the landing flare may result in considerable floating.
   Answer (B) is incorrect. Ground effect generally extends up to only one wingspan length, not two. Answer (C) is incorrect. Floating will occur with excess airspeed, which results in a lower-than-normal, not higher-than-normal, angle of attack.

28. What must a pilot be aware of as a result of ground effect?

A. Wingtip vortices increase creating wake turbulence problems for arriving and departing aircraft.

B. Induced drag decreases; therefore, any excess speed at the point of flare may cause considerable floating.

C. A full stall landing will require less up elevator deflection than would a full stall when done free of ground effect.

Answer (B) is correct. *(PHAK Chap 4)*
   DISCUSSION: Ground effect reduces the upwash, downwash, and vortices caused by the wings, resulting in a decrease in induced drag. Thus, thrust required at low airspeeds will be reduced, and any excess speed at the point of flare may cause considerable floating.
   Answer (A) is incorrect. Wingtip vortices are decreased, not increased. Answer (C) is incorrect. A full stall landing will require more, not less, up elevator deflection since the wing will require a lower angle of attack in ground effect to produce the same amount of lift.

29. Ground effect is most likely to result in which problem?

    A. Settling to the surface abruptly during landing.

    B. Becoming airborne before reaching recommended takeoff speed.

    C. Inability to get airborne even though airspeed is sufficient for normal takeoff needs.

Answer (B) is correct. *(PHAK Chap 4)*
    DISCUSSION: Due to the reduction of induced drag in ground effect, the airplane may seem capable of becoming airborne well below the recommended takeoff speed. However, as the airplane rises out of ground effect (a height greater than the wingspan) with a deficiency of speed, the increase in induced drag may result in very marginal initial climb performance. In extreme cases, the airplane may become airborne initially, with a deficiency of airspeed, only to settle back on the runway when attempting to fly out of the ground effect area.
    Answer (A) is incorrect. The airplane will experience a little extra lift on landing due to the reduction in induced drag, causing it to float rather than settle abruptly. Answer (C) is incorrect. Ground effect would not hamper the airplane from becoming airborne if the airspeed were sufficient for normal takeoff. Ground effect may allow the airplane to become airborne before reaching the recommended takeoff speed.

1.7 Airplane Turn

30. What force makes an airplane turn?

    A. The horizontal component of lift.

    B. The vertical component of lift.

    C. Centrifugal force.

Answer (A) is correct. *(AFH Chap 3)*
    DISCUSSION: When the wings of an airplane are not level, the lift is not entirely vertical and tends to pull the airplane toward the direction of the lower wing. An airplane is turned when the pilot coordinates rudder, aileron, and elevator to bank in order to attain a horizontal component of lift.
    Answer (B) is incorrect. The vertical component of lift opposes weight and controls vertical, not horizontal, movement. Answer (C) is incorrect. The horizontal component of lift opposes centrifugal force, which acts toward the outside of the turn.

1.8 Airplane Stability

31. An airplane said to be inherently stable will

    A. be difficult to stall.

    B. require less effort to control.

    C. not spin.

Answer (B) is correct. *(PHAK Chap 4)*
    DISCUSSION: An inherently stable airplane will usually return to the original condition of flight (except when in a bank) if disturbed by a force such as air turbulence. Thus, an inherently stable airplane will require less effort to control than an inherently unstable one.
    Answer (A) is incorrect. Stability of an airplane has an effect on stall characteristic, not on the difficulty level of entering a stall. Answer (C) is incorrect. An inherently stable aircraft will spin.

32. What determines the longitudinal stability of an airplane?

    A. The location of the CG with respect to the center of lift.

    B. The effectiveness of the horizontal stabilizer, rudder, and rudder trim tab.

    C. The relationship of thrust and lift to weight and drag.

Answer (A) is correct. *(PHAK Chap 4)*
    DISCUSSION: The location of the center of gravity with respect to the center of lift determines, to a great extent, the longitudinal stability of the airplane. Positive stability is attained by having the center of lift behind the center of gravity. Then the tail provides negative lift, creating a downward tail force, which counteracts the nose's tendency to pitch down.
    Answer (B) is incorrect. The rudder and rudder trim tab control the yaw, not the pitch. Answer (C) is incorrect. The relationship of thrust and lift to weight and drag affects speed and altitude, not longitudinal stability.

33. Changes in the center of pressure of a wing affect the aircraft's

    A. lift/drag ratio.

    B. lifting capacity.

    C. aerodynamic balance and controllability.

Answer (C) is correct. *(PHAK Chap 3)*
    DISCUSSION: Center of pressure (CP) is the imaginary but determinable point at which all of the upward lift forces on the wing are concentrated. In general, at high angles of attack the CP moves forward, while at low angles of attack the CP moves aft. The relationship of the CP to center of gravity (CG) affects both aerodynamic balance and controllability.
    Answer (A) is incorrect. The lift/drag ratio is determined by angle of attack. Answer (B) is incorrect. Lifting capacity is affected by angle of attack, airspeed, and wing planform.

34. An airplane has been loaded in such a manner that the CG is located aft of the aft CG limit. One undesirable flight characteristic a pilot might experience with this airplane would be

    A. a longer takeoff run.

    B. difficulty in recovering from a stalled condition.

    C. stalling at higher-than-normal airspeed.

Answer (B) is correct. *(PHAK Chap 9)*
    DISCUSSION: The recovery from a stall in any airplane becomes progressively more difficult as its center of gravity moves backward. Generally, airplanes become less controllable, especially at slow flight speeds, as the center of gravity is moved backward.
    Answer (A) is incorrect. An airplane with an aft CG has less drag, resulting in a shorter, not longer, takeoff run. Answer (C) is incorrect. An airplane with an aft CG flies at a lower angle of attack, resulting in a lower, not higher, stall speed.

35. What causes an airplane (except a T-tail) to pitch nosedown when power is reduced and controls are not adjusted?

    A. The CG shifts forward when thrust and drag are reduced.

    B. The downwash on the elevators from the propeller slipstream is reduced and elevator effectiveness is reduced.

    C. When thrust is reduced to less than weight, lift is also reduced and the wings can no longer support the weight.

Answer (B) is correct. *(PHAK Chap 5)*
    DISCUSSION: The relative wind on the tail is the result of the airplane's movement through the air and the propeller slipstream. When that slipstream is reduced, the horizontal stabilizer (except a T-tail) will produce less negative lift and the nose will pitch down.
    Answer (A) is incorrect. The CG is not affected by changes in thrust or drag. Answer (C) is incorrect. Thrust and weight have no relationship to each other.

36. Loading an airplane to the most aft CG will cause the airplane to be

    A. less stable at all speeds.

    B. less stable at slow speeds, but more stable at high speeds.

    C. less stable at high speeds, but more stable at low speeds.

Answer (A) is correct. *(PHAK Chap 9)*
    DISCUSSION: Airplanes become less stable at all speeds as the center of gravity is moved backward. The rearward center of gravity limit is determined largely by considerations of stability.
    Answer (B) is incorrect. An aft CG will cause the airplane to be less stable at all speeds. Answer (C) is incorrect. An aft CG will cause the airplane to be less stable at all speeds.

1.9 Torque and P-Factor

37. In what flight condition is torque effect the greatest in a single-engine airplane?

    A. Low airspeed, high power, high angle of attack.

    B. Low airspeed, low power, low angle of attack.

    C. High airspeed, high power, high angle of attack.

Answer (A) is correct. *(PHAK Chap 4)*
    DISCUSSION: The effect of torque increases in direct proportion to engine power and inversely to airspeed. Thus, at low airspeeds, high angles of attack, and high power settings, torque is the greatest.
    Answer (B) is incorrect. Torque effect is the greatest at high (not low) power settings, and high (not low) angle of attack. Answer (C) is incorrect. Torque effect is the greatest at low (not high) airspeeds.

38. The left turning tendency of an airplane caused by P-factor is the result of the

    A. clockwise rotation of the engine and the propeller turning the airplane counterclockwise.

    B. propeller blade descending on the right, producing more thrust than the ascending blade on the left.

    C. gyroscopic forces applied to the rotating propeller blades acting 90° in advance of the point the force was applied.

Answer (B) is correct. *(PHAK Chap 4)*
    DISCUSSION: Asymmetric propeller loading (P-factor) occurs when the airplane is flown at a high angle of attack. The downward-moving blade on the right side of the propeller (as seen from the rear) has a higher angle of attack, which creates higher thrust than the upward-moving blade on the left. Thus, the airplane yaws around the vertical axis to the left.
    Answer (A) is incorrect. Torque reaction (not P-factor) is a result of the clockwise rotation of the engine and the propeller turning the airplane counterclockwise. Answer (C) is incorrect. Gyroscopic precession (not P-factor) is a result of the gyroscopic forces applied to the rotating propeller blades acting 90° in advance of the point the force was applied.

39. When does P-factor cause the airplane to yaw to the left?

A. When at low angles of attack.

B. When at high angles of attack.

C. When at high airspeeds.

Answer (B) is correct. *(PHAK Chap 4)*
DISCUSSION: P-factor or asymmetric propeller loading occurs when an airplane is flown at a high angle of attack because the downward-moving blade on the right side of the propeller (as seen from the rear) has a higher angle of attack, which creates higher thrust than the upward-moving blade on the left. Thus, the airplane yaws around the vertical axis to the left.
Answer (A) is incorrect. At low angles of attack, both sides of the propeller have similar angles of attack and "pull" the airplane straight ahead. Answer (C) is incorrect. At high speeds, an airplane is not at a high angle of attack.

1.10 Load Factor

40. The amount of excess load that can be imposed on the wing of an airplane depends upon the

A. position of the CG.

B. speed of the airplane.

C. abruptness at which the load is applied.

Answer (B) is correct. *(PHAK Chap 4)*
DISCUSSION: The amount of excess load that can be imposed on the wing depends upon how fast the airplane is flying. At low speeds, the maximum available lifting force of the wing is only slightly greater than the amount necessary to support the weight of the airplane. Thus, any excess load would simply cause the airplane to stall. At high speeds, the lifting capacity of the wing is so great (as a result of the greater flow of air over the wings) that a sudden movement of the elevator controls (strong gust of wind) may increase the load factor beyond safe limits. This is why maximum speeds are established by airplane manufacturers.
Answer (A) is incorrect. The position of the CG affects the stability of the airplane but not the total load the wings can support. Answer (C) is incorrect. It is the amount of load, not the abruptness of the load, that is limited. However, the abruptness of the maneuver can affect the amount of the load.

41. Which basic flight maneuver increases the load factor on an airplane as compared to straight-and-level flight?

A. Climbs.

B. Turns.

C. Stalls.

Answer (B) is correct. *(PHAK Chap 4)*
DISCUSSION: Turns increase the load factor because the lift from the wings is used to pull the airplane around a corner as well as to offset the force of gravity. The wings must carry the airplane's weight plus offset centrifugal force during the turn. For example, a 60° bank results in a load factor of 2; i.e., the wings must support twice the weight they do in level flight.
Answer (A) is incorrect. The wings only have to carry the weight of the airplane once the airplane is established in a climb. Answer (C) is incorrect. In a stall, the wings are not producing lift.

42. During an approach to a stall, an increased load factor will cause the aircraft to

A. stall at a higher airspeed.

B. have a tendency to spin.

C. be more difficult to control.

Answer (A) is correct. *(PHAK Chap 4)*
DISCUSSION: The greater the load (whether from gross weight or from centrifugal force), the more lift is required. Therefore, an aircraft will stall at higher airspeeds when the load and/or load factor is increased.
Answer (B) is incorrect. An aircraft's tendency to spin is not related to an increase in load factors. Answer (C) is incorrect. An aircraft's stability (not load factor) determines its controllability.

43. (Refer to Figure 2 on page 35.) If an airplane weighs 2,300 pounds, what approximate weight would the airplane structure be required to support during a 60° banked turn while maintaining altitude?

A. 2,300 pounds.

B. 3,400 pounds.

C. 4,600 pounds.

Answer (C) is correct. *(PHAK Chap 4)*
DISCUSSION: Note on Fig. 2 that, at a 60° bank angle, the load factor is 2. Thus, a 2,300-lb. airplane in a 60° bank would require its wings to support 4,600 lb. (2,300 lb. × 2).
Answer (A) is incorrect. An airplane supporting a load of 2,300 lb. in a 60° banked turn would weigh 1,150 lb., not 2,300 lb. Note on Fig. 2 that, at a 60° bank angle, the load factor is 2. Answer (B) is incorrect. An airplane supporting a load of 3,400 lb. in a 60° banked turn would weigh 1,700 lb., not 2,300 lb. Note on Fig. 2 that, at a 60° bank angle, the load factor is 2.

44. (Refer to Figure 2 below.) If an airplane weighs 3,300 pounds, what approximate weight would the airplane structure be required to support during a 30° banked turn while maintaining altitude?

   A. 1,200 pounds.

   B. 3,100 pounds.

   C. 3,960 pounds.

Answer (C) is correct. *(PHAK Chap 4)*
DISCUSSION: Look on the left side of the chart in Fig. 2 to see that, at a 30° bank angle, the load factor is 1.154. Thus, a 3,300-lb. airplane in a 30° bank would require its wings to support 3,808.2 lb. (3,300 lb. × 1.154). The closest answer choice to this value is 3,960 lb.
Answer (A) is incorrect. An airplane supporting a load of 1,200 lb. in a 30° banked turn would weigh 1,000 lb., not 3,300 lb. Look on the left side of the chart in Fig. 2 to see that, at a 30° bank angle, the load factor is 1.154. Answer (B) is incorrect. An airplane supporting a load of 3,100 lb. in a 30° banked turn would weigh 2,583 lb., not 3,300 lb. Look on the left side of the chart in Fig. 2 to see that, at a 30° bank angle, the load factor is 1.154.

45. (Refer to Figure 2 below.) If an airplane weighs 4,500 pounds, what approximate weight would the airplane structure be required to support during a 45° banked turn while maintaining altitude?

   A. 4,500 pounds.

   B. 6,750 pounds.

   C. 7,200 pounds.

Answer (B) is correct. *(PHAK Chap 4)*
DISCUSSION: Look on the left side of the chart under 45° and note that the load factor curve is 1.414. Thus, a 4,500-lb. airplane in a 45° bank would require its wings to support 6,363 lb. (4,500 lb. × 1.414). The closest answer choice to this value is 6,750 lb.
Answer (A) is incorrect. An airplane supporting a load of 4,500 lb. in a 45° banked turn would weigh 3,000 lb., not 4,500 lb. Look on the left side of the chart under 45° and note that the load factor curve is 1.414. Answer (C) is incorrect. An airplane supporting a load of 7,200 lb. in a 45° banked turn would weigh 4,800 lb., not 4,500 lb. Look on the left side of the chart under 45° and note that the load factor curve is 1.414.

ANGLE of BANK ϕ	LOAD FACTOR n
0°	1.0
10°	1.015
30°	1.154
45°	1.414
60°	2.000
70°	2.923
80°	5.747
85°	11.473
90°	∞

LOAD FACTOR CHART

Figure 2. – Load Factor Chart.

END OF STUDY UNIT

STUDY UNIT TWO
AIRPLANE INSTRUMENTS, ENGINES, AND SYSTEMS

(7 pages of outline)

This study unit contains outlines of major concepts tested, sample test questions and answers regarding the major mechanical and instrument systems in an airplane, and an explanation of each answer. The table of contents above lists each subunit within this study unit, the number of questions pertaining to that particular subunit, and the pages on which the outlines and questions begin, respectively.

CAUTION: Recall that the **sole purpose** of this book is to expedite your passing the FAA pilot knowledge test for the private pilot certificate. Accordingly, all extraneous material (i.e., topics or regulations not directly tested on the FAA pilot knowledge test) is omitted, even though much more information and knowledge are necessary to fly safely. This additional material is presented in *Pilot Handbook* and *Private Pilot Flight Maneuvers and Practical Test Prep*, available from Gleim Publications, Inc. See the order form on page 371.

2.1 COMPASS TURNING ERROR

1. During flight, magnetic compasses can be considered accurate only during straight-and-level flight at constant airspeed.

2. The difference between direction indicated by a magnetic compass not installed in an airplane and one installed in an airplane is called deviation.

   a. Magnetic fields produced by metals and electrical accessories in an airplane disturb the compass needles.

3. In the Northern Hemisphere, acceleration/deceleration error occurs when on an east or west heading. Remember ANDS: Accelerate North, Decelerate South.

   a. A magnetic compass will indicate a turn toward the north during acceleration when on an east or west heading.

   b. A magnetic compass will indicate a turn toward the south during deceleration when on an east or west heading.

   c. Acceleration/deceleration error does not occur when on a north or south heading.

4.   In the Northern Hemisphere, compass turning error occurs when turning from a north or south heading.

   a.   A magnetic compass will lag (and, at the start of a turn, indicate a turn in the opposite direction) when turning from a north heading.

      1)   If turning to the east (right), the compass will initially indicate a turn to the west and then lag behind the actual heading until your airplane is headed east (at which point there is no error).

      2)   If turning to the west (left), the compass will initially indicate a turn to the east and then lag behind the actual heading until your airplane is headed west (at which point there is no error).

   b.   A magnetic compass will lead or precede the turn when turning from a south heading.

   c.   Turning errors do not occur when turning from an east or west heading.

5.   These errors diminish as the acceleration/deceleration or turns are completed.

2.2   PITOT-STATIC SYSTEM

1.   The pitot-static system is a source of pressure for the

   a.   Altimeter
   b.   Vertical-speed indicator
   c.   Airspeed indicator

2.   The pitot tube provides impact (or ram) pressure for the airspeed indicator only.

3.   When the pitot tube and the outside static vents or just the static vents are clogged, all three instruments mentioned above will provide inaccurate readings.

   a.   If only the pitot tube is clogged, only the airspeed indicator will be inoperative.

2.3   AIRSPEED INDICATOR

1.   Airspeed indicators have several color-coded markings (Figure 4 on page 49). (Figure 4 can also be seen in color on page 294.)

   a.   The white arc is the full flap operating range.

      1)   The lower limit is the power-off stalling speed with wing flaps and landing gear in the landing position (V_{S0}).

      2)   The upper limit is the maximum full flaps-extended speed (V_{FE}).

   b.   The green arc is the normal operating range.

      1)   The lower limit is the power-off stalling speed in a specified configuration (V_{S1}). This is normally wing flaps up and landing gear retracted.

      2)   The upper limit is the maximum structural cruising speed (V_{NO}) for normal operation.

   c.   The yellow arc is airspeed that is safe in smooth air only.

      1)   It is known as the caution range.

   d.   The red radial line is the speed that should never be exceeded (V_{NE}).

      1)   This is the maximum speed at which the airplane may be operated in smooth air (or under any circumstances).

2.   The most important airspeed limitation that is **not** color-coded is the maneuvering speed (V_A).

   a.   The maneuvering speed is the maximum speed at which full deflection of aircraft controls can be made without causing structural damage.

   b.   It is usually the maximum speed for flight in turbulent air.

2.4 ALTIMETER

1. Altimeters have three hands (e.g., as a clock has the hour, minute, and second hands; Figure 3, page 50).

2. The three hands on the altimeter are the

   a. 10,000-ft. interval (short needle)
   b. 1,000-ft. interval (medium needle)
   c. 100-ft. interval (long needle)

3. Altimeters are numbered 0-9.

4. To read an altimeter,

   a. First, determine whether the short needle points between 0 and 1 (1-10,000), 1 and 2 (10,000-20,000), or 2 and 3 (20,000-30,000).
   b. Second, determine whether the medium needle is between 0 and 1 (0-1,000), 1 and 2 (1,000-2,000), etc.
   c. Third, determine at which number the long needle is pointing, e.g., 1 for 100 ft., 2 for 200 ft., etc.

2.5 TYPES OF ALTITUDE

1. Absolute altitude is the altitude above the surface, i.e., AGL.

2. True altitude is the actual distance above mean sea level, i.e., MSL. It is not susceptible to variation with atmospheric conditions.

3. Density altitude is pressure altitude corrected for nonstandard temperatures.

4. Pressure altitude is the height above the standard datum plane of 29.92 in. of mercury. Thus, it is the indicated altitude when the altimeter setting is adjusted to 29.92 in. of mercury (also written 29.92" Hg).

5. Pressure altitude and density altitude are the same at standard temperature.

6. Indicated altitude is the same as true altitude when standard conditions exist and the altimeter is calibrated properly.

7. Pressure altitude and true altitude are the same when standard atmospheric conditions (29.92" Hg and 15°C at sea level) exist.

8. When the altimeter is adjusted on the ground so that indicated altitude equals true altitude at airport elevation, the altimeter setting is that for your location, i.e., approximately the setting you would get from the control tower.

2.6 SETTING THE ALTIMETER

1. The indicated altitude on the altimeter increases when you change the altimeter setting to a higher pressure and decreases when you change the setting to a lower pressure.

   a. This is opposite to the altimeter's reaction due to changes in air pressure.

2. The indicated altitude will change at a rate of approximately 1,000 ft. for 1 in. of pressure change in the altimeter setting.

   a. EXAMPLE: When changing the altimeter setting from 29.15 to 29.85, there is a 0.70 in. change in pressure (29.85 − 29.15). The indicated altitude would increase (due to a higher altimeter setting) by 700 ft. (0.70 × 1,000).

2.7 ALTIMETER ERRORS

1.   Since altimeter readings are adjusted for changes in barometric pressure but not for temperature changes, an airplane will be at lower than indicated altitude when flying in colder than standard temperature air when maintaining a constant indicated altitude.

   a.   On warm days, the altimeter indicates lower than actual altitude.

2.   Likewise, when pressure lowers en route at a constant indicated altitude, your altimeter will indicate higher than actual altitude until you adjust it.

3.   Remember, when flying from high to low (temperature or pressure), look out below.

   a.   Low to high, clear the sky.

2.8 GYROSCOPIC INSTRUMENTS

1.   The attitude indicator, with its miniature aircraft and horizon bar, displays a picture of the attitude of the airplane (Figure 7, page 55).

   a.   The relationship of the miniature airplane (labeled "C" on Figure 7) to the horizon bar (labeled "B" on Figure 7) is the same as the relationship of the real aircraft to the actual horizon.

      1)   The banking scale (labeled "A" on Figure 7) shows degrees of bank from level flight.

   b.   The relationship of the miniature airplane to the horizon bar should be used for an indication of pitch and bank attitude, i.e., nose high, nose low, left bank, right bank.

   c.   The gyro in the attitude indicator rotates in a horizontal plane and depends upon rigidity in space for its operation.

   d.   An adjustment knob is provided with which the pilot may move the miniature airplane up or down to align the miniature airplane with the horizon bar to suit the pilot's line of vision.

2.   The turn coordinator shows the roll and yaw movement of the airplane (Figure 5, page 55).

   a.   It displays a miniature airplane, which moves proportionally to the roll rate of the airplane. When the bank is held constant, the turn coordinator indicates the rate of turn.

   b.   The ball indicates whether the angle of bank is coordinated with the rate of turn.

3.   The heading indicator is a gyro instrument that depends on the principle of rigidity in space for its operation (Figure 6, page 55).

   a.   Due to gyro precession, it must be periodically realigned with a magnetic compass.

2.9 GLASS COCKPITS

1.   Glass cockpits, or systems of advanced avionics, are replacing the older round-dial gauges common in many training aircraft.

   a.   These systems vary widely but generally provide flight information such as flight progress, engine monitoring, navigation, terrain, traffic, and weather.

   b.   These systems are designed to decrease pilot workload, enhance situational awareness, and increase the safety margin.

2.   A **primary flight display** (PFD) integrates all flight instruments critical to safe flight in one screen.

   a.   Some PFDs incorporate or overlay navigation instruments on top of primary flight instruments.

      1)   EXAMPLE: An ILS or VOR may be integrated with the heading indicator.

3.   A **multi-function display** (MFD) not only shows primary instrumentation but can combine information from multiple systems on one page or screen.

   a.   Moving maps provide a pictorial view of the aircraft's location, route, airspace, and geographical features.

   NOTE: A moving map should not be used as the primary navigation instrument; it should be a supplement, not a substitute, in the navigational process.

   b.   Onboard weather systems, including radar, may provide real-time weather.
   c.   Other information that could be included on MFDs include terrain and traffic avoidance, checklists, and fuel management systems.

4.   Care should be taken that reliance on glass cockpits does not negate safety. A regular scan, both visually outside and inside on backup gauges, should be combined with other means of navigation and checklists to ensure safe flight.

2.10 ENGINE TEMPERATURE

1.   Excessively high engine temperature either in the air or on the ground will cause loss of power, excessive oil consumption, and excessive wear on the internal engine.

2.   An engine is cooled, in part, by circulating oil through the system to reduce friction and absorb heat from internal engine parts.

3.   Engine oil and cylinder head temperatures can exceed their normal operating range because of (among other causes)

   a.   Operating with too much power
   b.   Climbing too steeply (i.e., at too low an airspeed) in hot weather
   c.   Using fuel that has a lower-than-specified octane rating
   d.   Operating with too lean a mixture
   e.   The oil level being too low

4.   Excessively high engine temperatures can be reduced by reversing any of the previous situations, e.g., reducing power, climbing less steeply (increasing airspeed), using higher octane fuel, enriching the mixture, etc.

2.11 CONSTANT-SPEED PROPELLER

1.   The advantage of a constant-speed propeller (also known as controllable-pitch) is that it permits the pilot to select the blade angle for the most efficient performance.

2.   Constant-speed propeller airplanes have both throttle and propeller controls.

   a.   The throttle controls power output, which is registered on the manifold pressure gauge.
   b.   The propeller control regulates engine revolutions per minute (RPM), which are registered on the tachometer.

3.   To avoid overstressing cylinders, excessively high manifold pressure should not be used with low RPM settings.

2.12 ENGINE IGNITION SYSTEMS

1.   One purpose of the dual-ignition system is to provide for improved engine performance.

   a.   The other is increased safety.

2. Loose or broken wires in the ignition system can cause problems.

   a. EXAMPLE: If the ignition switch is OFF, the magneto may continue to fire if the ignition switch ground wire is disconnected.

   b. If this occurs, the only way to stop the engine is to move the mixture lever to the idle cut-off position, then have the system checked by a qualified aviation maintenance technician.

2.13 CARBURETOR ICING

1. Carburetor-equipped engines are more susceptible to icing than fuel-injected engines.

   a. The operating principle of float-type carburetors is the difference in air pressure between the venturi throat and the air inlet.

   b. Fuel-injected engines do not have a carburetor.

2. The first indication of carburetor ice on airplanes with fixed-pitch propellers and float-type carburetors is a loss of RPM.

3. Carburetor ice is likely to form when outside air temperature is between 20°F and 70°F and there is visible moisture or high humidity.

4. When carburetor heat is applied to eliminate carburetor ice in an airplane equipped with a fixed-pitch propeller, there will be a further decrease in RPM (due to the less dense hot air entering the engine) followed by a gradual increase in RPM as the ice melts.

2.14 CARBURETOR HEAT

1. Carburetor heat enriches the fuel/air mixture,

   a. Because warm air is less dense than cold air.

   b. When the air density decreases (because the air is warm), the fuel/air mixture (ratio) becomes richer since there is less air for the same amount of fuel.

2. Applying carburetor heat decreases engine output and increases operating temperature.

2.15 FUEL/AIR MIXTURE

1. At higher altitudes, the fuel/air mixture must be leaned to decrease the fuel flow in order to compensate for the decreased air density, i.e., to keep the fuel/air mixture constant.

   a. If you descend from high altitudes to lower altitudes without enriching the mixture, the mixture will become leaner because the air is denser at lower altitudes.

2. If you are running up your engine at a high-altitude airport, you may eliminate engine roughness by leaning the mixture,

   a. Particularly if the engine runs even worse with carburetor heat, since warm air further enriches the mixture.

2.16 ABNORMAL COMBUSTION

1. Detonation occurs when the fuel/air mixture explodes instead of burning evenly.

2. Detonation is usually caused by using a lower-than-specified grade (octane) of aviation fuel or by excessive engine temperature.

   a. This causes many engine problems including excessive wear and higher than normal operating temperatures.

3. Lower the nose slightly if you suspect that an engine (with a fixed-pitch propeller) is detonating during climbout after takeoff. This will increase cooling and decrease the engine's workload.

4. Pre-ignition is the uncontrolled firing of the fuel/air charge in advance of the normal spark ignition.

2.17 AVIATION FUEL PRACTICES

1. Use of the next-higher-than-specified (octane) grade of fuel is better than using the next-lower-than-specified grade of fuel. This will prevent the possibility of detonation, or running the engine too hot.

2. Filling the fuel tanks at the end of the day prevents moisture condensation by eliminating the airspace in the tanks.

3. All fuel strainer drains and fuel tank sumps should be drained before each flight to make sure there is no water in the fuel system.

4. In an airplane equipped with fuel pumps, the auxiliary electric fuel pump is used in the event the engine-driven fuel pump fails.

2.18 STARTING THE ENGINE

1. After the engine starts, the throttle should be adjusted for proper RPM and the engine gauges, especially the oil pressure, checked.

2. When starting an airplane engine by hand, it is extremely important that a competent pilot be at the controls in the cockpit.

2.19 ELECTRICAL SYSTEM

1. Most aircraft have either a 14- or 28-volt direct current electrical system.

2. Engine-driven alternators (or generators) supply electrical current to the electrical system and maintain an electrical charge on the battery.

   a. The alternator voltage output should be slightly higher than the battery voltage to keep the battery charged.

      1) EXAMPLE: A 14-volt alternator system would keep a positive charge on a 12-volt battery.

3. The electrical system is turned on by the master switch, providing electrical current to all electrical systems except the ignition system.

   a. Lights, radios, and electric fuel pumps are examples of equipment that commonly use the electrical system.

4. An ammeter shows if the alternator is producing an adequate supply of electrical power and indicates whether the battery is receiving an electrical charge.

   a. A positive indication on the ammeter shows the rate of charge on the battery, while a negative indication means more current is being drawn from the battery than is being replaced.

5. Alternators provide more electrical power at lower engine RPM than generators do.

6. Electrical system failure (battery and alternator) usually results in avionics system failure.

QUESTIONS AND ANSWER EXPLANATIONS

All of the private pilot knowledge test questions chosen by the FAA for release as well as additional questions selected by Gleim relating to the material in the previous outlines are reproduced on the following pages. These questions have been organized into the same subunits as the outlines. To the immediate right of each question are the correct answer and answer explanation. You should cover these answers and answer explanations while responding to the questions. Refer to the general discussion in the Introduction on how to take the FAA pilot knowledge test.

Remember that the questions from the FAA pilot knowledge test bank have been reordered by topic and organized into a meaningful sequence. Also, the first line of the answer explanation gives the citation of the authoritative source for the answer.

QUESTIONS

2.1 Compass Turning Error

1. In the Northern Hemisphere, a magnetic compass will normally indicate a turn toward the north if

A. an aircraft is decelerated while on an east or west heading.

B. a left turn is entered from a west heading.

C. an aircraft is accelerated while on an east or west heading.

Answer (C) is correct. *(PHAK Chap 7)*
    DISCUSSION: In the Northern Hemisphere, a magnetic compass will normally indicate a turn toward the north if an airplane is accelerated while on an east or west heading.
    Answer (A) is incorrect. In the Northern Hemisphere, a magnetic compass will normally indicate a turn toward the north if an airplane is accelerated, not decelerated, while on an east or west heading. Answer (B) is incorrect. There is no compass turning error on turns from a west heading.

2. During flight, when are the indications of a magnetic compass accurate?

A. Only in straight-and-level unaccelerated flight.

B. As long as the airspeed is constant.

C. During turns if the bank does not exceed 18°.

Answer (A) is correct. *(PHAK Chap 7)*
    DISCUSSION: During flight, the magnetic compass indications can be considered accurate only when in straight-and-level, unaccelerated flight. During acceleration, deceleration, or turns, the compass card will dip and cause false readings.
    Answer (B) is incorrect. Even with a constant airspeed, the magnetic compass may not be accurate during a turn.
Answer (C) is incorrect. Due to the compass card dip, the compass may not be accurate even during shallow turns.

3. Deviation in a magnetic compass is caused by the

A. presence of flaws in the permanent magnets of the compass.

B. difference in the location between true north and magnetic north.

C. magnetic fields within the aircraft distorting the lines of magnetic force.

Answer (C) is correct. *(PHAK Chap 7)*
    DISCUSSION: Magnetic fields produced by metals and electrical accessories in the airplane disturb the compass needle and produce errors. These errors are referred to as compass deviation.
    Answer (A) is incorrect. A properly functioning magnetic compass is still subject to deviation. Answer (B) is incorrect. The difference in the location between true and magnetic north refers to magnetic variation, not deviation.

4. Deviation in a magnetic compass is caused by the

A. northerly turning error.

B. certain metals and electrical systems within the aircraft.

C. the difference in location of true north and magnetic north.

Answer (B) is correct. *(PHAK Chap 7)*
    DISCUSSION: The compass in an airplane will align with any magnetic field. Magnetic fields created by metals and the electrical system of the aircraft will hinder the ability of the compass to align with the earth's magnetic field. This phenomenon is known as deviation. Since deviation error varies by heading, a compass correction card is fitted, providing the pilot with the deviation for a given heading.
    Answer (A) is incorrect. Northerly turning error is a product of the pulling-vertical component of the earth's magnetic field. Answer (C) is incorrect. Variation is the error associated with the difference in the location of true and magnetic north.

5. In the Northern Hemisphere, if an aircraft is accelerated or decelerated, the magnetic compass will normally indicate

    A. a turn momentarily.

    B. correctly when on a north or south heading.

    C. a turn toward the south.

Answer (B) is correct. *(PHAK Chap 7)*
    DISCUSSION: Acceleration and deceleration errors on magnetic compasses do not occur when on a north or south heading in the Northern Hemisphere. They occur on east and west headings.
    Answer (A) is incorrect. Acceleration and deceleration errors occur only on easterly and westerly headings. Answer (C) is incorrect. A turn to the north is indicated upon acceleration and a turn to the south is indicated on deceleration when on east or west headings.

6. In the Northern Hemisphere, a magnetic compass will normally indicate initially a turn toward the west if

    A. a left turn is entered from a north heading.

    B. a right turn is entered from a north heading.

    C. an aircraft is accelerated while on a north heading.

Answer (B) is correct. *(PHAK Chap 7)*
    DISCUSSION: Due to the northerly turn error in the Northern Hemisphere, a magnetic compass will initially indicate a turn toward the west if a right (east) turn is entered from a north heading.
    Answer (A) is incorrect. If a left (west) turn were made from a north heading, the compass would initially indicate a turn toward the east. Answer (C) is incorrect. Acceleration/ deceleration error does not occur on a north heading.

7. In the Northern Hemisphere, the magnetic compass will normally indicate a turn toward the south when

    A. a left turn is entered from an east heading.

    B. a right turn is entered from a west heading.

    C. the aircraft is decelerated while on a west heading.

Answer (C) is correct. *(PHAK Chap 7)*
    DISCUSSION: In the Northern Hemisphere, a magnetic compass will normally indicate a turn toward the south if an airplane is decelerated while on an east or west heading.
    Answer (A) is incorrect. Turning errors do not occur from an east heading. Answer (B) is incorrect. Turning errors do not occur from a west heading.

8. In the Northern Hemisphere, a magnetic compass will normally indicate initially a turn toward the east if

    A. an aircraft is decelerated while on a south heading.

    B. an aircraft is accelerated while on a north heading.

    C. a left turn is entered from a north heading.

Answer (C) is correct. *(PHAK Chap 7)*
    DISCUSSION: In the Northern Hemisphere, a magnetic compass normally initially indicates a turn toward the east if a left (west) turn is entered from a north heading.
    Answer (A) is incorrect. Acceleration/deceleration errors do not occur while on a south heading, only on an east or west heading. Answer (B) is incorrect. Acceleration/deceleration errors do not occur while on a north heading, only on an east or west heading.

9. What should be the indication on the magnetic compass as you roll into a standard rate turn to the right from a south heading in the Northern Hemisphere?

    A. The compass will initially indicate a turn to the left.

    B. The compass will indicate a turn to the right, but at a faster rate than is actually occurring.

    C. The compass will remain on south for a short time, then gradually catch up to the magnetic heading of the airplane.

Answer (B) is correct. *(PHAK Chap 7)*
    DISCUSSION: When on a southerly heading in the Northern Hemisphere and you roll into a standard rate turn to the right, the magnetic compass indication precedes the turn, showing a greater amount of turn than is actually occurring.
    Answer (A) is incorrect. The magnetic compass will initially indicate a turn to the left when you roll into a standard rate turn from a north, not south, heading in the Northern Hemisphere. Answer (C) is incorrect. The magnetic compass indication will precede the turn, not remain constant, when you roll into a standard rate turn from a south heading in the Northern Hemisphere.

2.2 Pitot-Static System

10. The pitot system provides impact pressure for which instrument?

    A. Altimeter.

    B. Vertical-speed indicator.

    C. Airspeed indicator.

Answer (C) is correct. *(PHAK Chap 7)*
    DISCUSSION: The pitot system provides impact pressure, or ram pressure, for only the airspeed indicator.
    Answer (A) is incorrect. The altimeter operates off the static (not pitot) system. Answer (B) is incorrect. The vertical-speed indicator operates off the static (not pitot) system.

11. Which instrument will become inoperative if the pitot tube becomes clogged?

  A. Altimeter.

  B. Vertical speed.

  C. Airspeed.

Answer (C) is correct. *(PHAK Chap 7)*
  DISCUSSION: The pitot-static system is a source of pressure for the altimeter, vertical-speed indicator, and airspeed indicator. The pitot tube is connected directly to the airspeed indicator and provides impact pressure for it alone. Thus, if the pitot tube becomes clogged, only the airspeed indicator will become inoperative.
  Answer (A) is incorrect. The altimeter operates off the static system and is not affected by a clogged pitot tube. Answer (B) is incorrect. The vertical speed indicator operates off the static system and is not affected by a clogged pitot tube.

12. If the pitot tube and outside static vents become clogged, which instruments would be affected?

  A. The altimeter, airspeed indicator, and turn-and-slip indicator.

  B. The altimeter, airspeed indicator, and vertical speed indicator.

  C. The altimeter, attitude indicator, and turn-and-slip indicator.

Answer (B) is correct. *(PHAK Chap 7)*
  DISCUSSION: The pitot-static system is a source of air pressure for the operation of the altimeter, airspeed indicator, and vertical speed indicator. Thus, if the pitot and outside static vents become clogged, all of these instruments will be affected.
  Answer (A) is incorrect. The turn-and-slip indicator is a gyroscopic instrument and does not operate on the pitot-static system. Answer (C) is incorrect. The attitude indicator and turn-and-slip indicator are both gyroscopic instruments and do not operate on the pitot-static system.

13. Which instrument(s) will become inoperative if the static vents become clogged?

  A. Airspeed only.

  B. Altimeter only.

  C. Airspeed, altimeter, and vertical speed.

Answer (C) is correct. *(PHAK Chap 7)*
  DISCUSSION: The pitot-static system is a source of air pressure for the operation of the airspeed indicator, altimeter, and vertical speed indicator. Thus, if the static vents become clogged, all three instruments will become inoperative.
  Answer (A) is incorrect. Not only will the airspeed indicator become inoperative, but also the altimeter and vertical speed indicator. Answer (B) is incorrect. Not only will the altimeter become inoperative, but also the airspeed and vertical speed indicators.

2.3 Airspeed Indicator

14. What does the red line on an airspeed indicator represent?

  A. Maneuvering speed.

  B. Turbulent or rough-air speed.

  C. Never-exceed speed.

Answer (C) is correct. *(PHAK Chap 7)*
  DISCUSSION: The red line on an airspeed indicator indicates the maximum speed at which the airplane can be operated in smooth air, which should never be exceeded intentionally. This speed is known as the never-exceed speed.
  Answer (A) is incorrect. Maneuvering speed is not indicated on the airspeed indicator. Answer (B) is incorrect. Turbulent or rough-air speed is not indicated on the airspeed indicator.

15. What is an important airspeed limitation that is not color coded on airspeed indicators?

  A. Never-exceed speed.

  B. Maximum structural cruising speed.

  C. Maneuvering speed.

Answer (C) is correct. *(PHAK Chap 7)*
  DISCUSSION: The maneuvering speed of an airplane is an important airspeed limitation not color-coded on the airspeed indicator. It is found in the airplane manual (Pilot's Operating Handbook) or placarded in the cockpit. Maneuvering speed is the maximum speed at which full deflection of the airplane controls can be made without incurring structural damage. Maneuvering speed or less should be held in turbulent air to prevent structural damage due to excessive loads.
  Answer (A) is incorrect. The never-exceed speed is indicated on the airspeed indicator by a red radial line. Answer (B) is incorrect. The maximum structural cruising speed is indicated by the upper limit of the green arc on the airspeed indicator.

Page Intentionally Left Blank

16. (Refer to Figure 4 on page 49 and in color on page 294.) What is the caution range of the airplane?

    A. 0 to 60 MPH.

    B. 100 to 165 MPH.

    C. 165 to 208 MPH.

Answer (C) is correct. *(PHAK Chap 7)*
    DISCUSSION: The caution range is indicated by the yellow arc on the airspeed indicator. Operation within this range is safe only in smooth air. The airspeed indicator in Fig. 4 indicates the caution range from 165 to 208 MPH.
    Answer (A) is incorrect. The range of 0-60 MPH is less than stall speed. Answer (B) is incorrect. The range of 100-165 MPH is within the normal operating airspeed range, which extends from minimum flap extension speed to maximum structural cruising speed.

17. (Refer to Figure 4 on page 49 and in color on page 294.) The maximum speed at which the airplane can be operated in smooth air is

    A. 100 MPH.

    B. 165 MPH.

    C. 208 MPH.

Answer (C) is correct. *(PHAK Chap 7)*
    DISCUSSION: The maximum speed at which the airplane can be operated in smooth air is indicated by the red radial line. The airspeed indicator in Fig. 4 indicates the red line is at 208 MPH.
    Answer (A) is incorrect. This is the maximum flaps-extended speed, the upper limit of the white arc. Answer (B) is incorrect. This is the maximum structural cruising speed, the upper limit of the green arc.

18. (Refer to Figure 4 on page 49 and in color on page 294.) What is the full flap operating range for the airplane?

    A. 60 to 100 MPH.

    B. 60 to 208 MPH.

    C. 65 to 165 MPH.

Answer (A) is correct. *(PHAK Chap 7)*
    DISCUSSION: The full flap operating range is indicated by the white arc on the airspeed indicator. The airspeed indicator in Fig. 4 indicates the full flap operating range is from 60 to 100 MPH.
    Answer (B) is incorrect. This is the entire operating range of this airplane. Answer (C) is incorrect. This is the normal operating range for this airplane (green arc).

19. (Refer to Figure 4 on page 49 and in color on page 294.) Which marking identifies the never-exceed speed?

    A. Upper limit of the green arc.

    B. Upper limit of the white arc.

    C. The red radial line.

Answer (C) is correct. *(PHAK Chap 7)*
    DISCUSSION: The red radial line represents the never-exceed speed (V_{NE}). Operating an aircraft beyond V_{NE} may result in severe structural damage.
    Answer (A) is incorrect. The upper limit of the green arc represents normal operating speed (V_{NO}). Answer (B) is incorrect. The upper limit of the white arc is the maximum flaps-extended speed (V_{FE}).

20. (Refer to Figure 4 on page 49 and in color on page 294.) Which color identifies the power-off stalling speed in a specified configuration?

    A. Upper limit of the green arc.

    B. Upper limit of the white arc.

    C. Lower limit of the green arc.

Answer (C) is correct. *(PHAK Chap 7)*
    DISCUSSION: The lower airspeed limit of the green arc indicates the power-off stalling speed in a specified configuration. "Specified configuration" refers to flaps up and landing gear retracted.
    Answer (A) is incorrect. The upper limit of the green arc is the maximum structural cruising speed. Answer (B) is incorrect. The upper airspeed limit of the white arc is the maximum flaps-extended speed. Structural damage to the flaps could occur if the flaps are extended above this airspeed.

21. (Refer to Figure 4 on page 49 and in color on page 294.) What is the maximum flaps-extended speed?

    A. 65 MPH.

    B. 100 MPH.

    C. 165 MPH.

Answer (B) is correct. *(PHAK Chap 7)*
    DISCUSSION: The maximum flaps-extended speed is indicated by the upper limit of the white arc. This is the highest airspeed at which a pilot should extend full flaps. At higher airspeeds, severe strain or structural failure could result. The upper limit of the white arc on the airspeed indicator shown in Fig. 4 indicates 100 MPH.
    Answer (A) is incorrect. This is the lower limit of the green arc, which is the power-off stall speed in a specified configuration. Answer (C) is incorrect. This is the upper limit of the green arc, which is the maximum structural cruising speed.

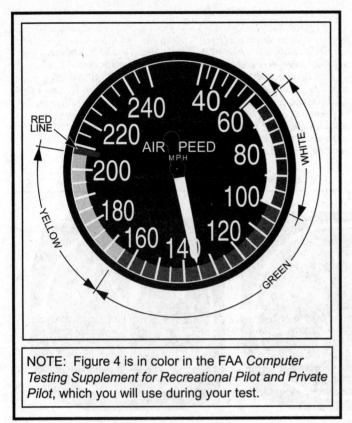

NOTE: Figure 4 is in color in the FAA *Computer Testing Supplement for Recreational Pilot and Private Pilot*, which you will use during your test.

Figure 4. – Airspeed Indicator.

22. (Refer to Figure 4 above and in color on page 294.) Which color identifies the normal flap operating range?

    A. The lower limit of the white arc to the upper limit of the green arc.

    B. The green arc.

    C. The white arc.

Answer (C) is correct. *(PHAK Chap 7)*
    DISCUSSION: The normal flap operating range is indicated by the white arc. The power-off stall speed with flaps extended is at the lower limit of the arc, and the maximum speed at which flaps can be extended without damage to them is the upper limit of the arc.
    Answer (A) is incorrect. The upper limit of the green arc well exceeds the upper limit of the white arc, which is the maximum flap extended speed. Answer (B) is incorrect. The green arc represents the normal operating range.

23. (Refer to Figure 4 above and in color on page 294.) Which color identifies the power-off stalling speed with wing flaps and landing gear in the landing configuration?

    A. Upper limit of the green arc.

    B. Upper limit of the white arc.

    C. Lower limit of the white arc.

Answer (C) is correct. *(PHAK Chap 7)*
    DISCUSSION: The lower limit of the white arc indicates the power-off stalling speed with wing flaps and landing gear in the landing position.
    Answer (A) is incorrect. The upper limit of the green arc is the maximum structural cruising speed. Answer (B) is incorrect. The upper limit of the white arc is the maximum flaps-extended speed.

24. (Refer to Figure 4 above and in color on page 294.) What is the maximum structural cruising speed?

    A. 100 MPH.

    B. 165 MPH.

    C. 208 MPH.

Answer (B) is correct. *(PHAK Chap 7)*
    DISCUSSION: The maximum structural cruising speed is the maximum speed for normal operation and is indicated as the upper limit of the green arc on an airspeed indicator. The upper limit of the green arc on the airspeed indicator shown in Fig. 4 indicates 165 MPH.
    Answer (A) is incorrect. This is the upper limit of the white arc and is the maximum speed at which the flaps can be extended. Answer (C) is incorrect. This is the speed that should never be exceeded. Beyond this speed, structural damage to the airplane may occur.

2.4 Altimeter

25. (Refer to Figure 3 below.) Altimeter 2 indicates

A. 1,500 feet.

B. 4,500 feet.

C. 14,500 feet.

Answer (C) is correct. *(PHAK Chap 7)*

DISCUSSION: Altimeter 2 indicates 14,500 ft. because the shortest needle is between the 1 and the 2, indicating about 15,000 ft.; the middle needle is between 4 and 5, indicating 4,500 ft.; and the long needle is on 5, indicating 500 ft., i.e., 14,500 feet.

Answer (A) is incorrect. For 1,500 ft., the middle needle would have to be between 1 and 2, and the shortest needle between 0 and 1. Answer (B) is incorrect. For 4,500 ft., the shortest needle would have to be between 0 and 1.

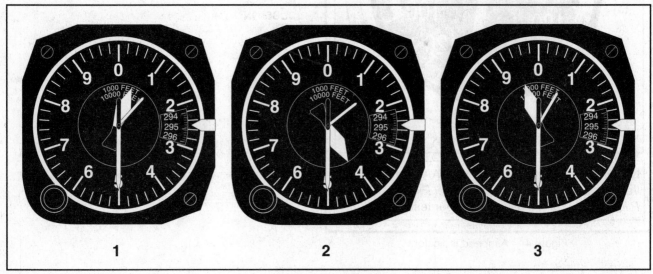

Figure 3. – Altimeter.

26. (Refer to Figure 3 above.) Altimeter 1 indicates

A. 500 feet.

B. 1,500 feet.

C. 10,500 feet.

Answer (C) is correct. *(PHAK Chap 7)*

DISCUSSION: The altimeter has three needles. The short needle indicates 10,000-ft. intervals, the middle-length needle indicates 1,000-ft. intervals, and the long needle indicates 100-ft. intervals. In altimeter 1, the shortest needle is on 1, which indicates about 10,000 feet. The middle-length needle indicates half-way between zero and 1, which is 500 feet. This is confirmed by the longest needle on 5, indicating 500 ft., i.e., 10,500 feet.

Answer (A) is incorrect. If it were indicating just 500 ft., the short and medium needles would have to be on or near zero. Answer (B) is incorrect. If it were 1,500 ft., the shortest needle would be near zero and the middle needle would be between the 1 and the 2.

27. (Refer to Figure 3 above.) Altimeter 3 indicates

A. 9,500 feet.

B. 10,950 feet.

C. 15,940 feet.

Answer (A) is correct. *(PHAK Chap 7)*

DISCUSSION: Altimeter 3 indicates 9,500 ft. because the shortest needle is near 1 (i.e., about 10,000 ft.), the middle needle is between 9 and the 0, indicating between 9,000 and 10,000 ft., and the long needle is on 5, indicating 500 feet.

Answer (B) is incorrect. For 10,950 ft., the middle needle would have to be near the 1 and the long needle would have to be between the 9 and 0. Answer (C) is incorrect. For 15,940 ft., the short needle would have to be between 1 and 2, the middle needle near the 6, and the large needle between the 9 and 0.

28. (Refer to Figure 3 on page 50.) Which altimeter(s) indicate(s) more than 10,000 feet?

A. 1, 2, and 3.

B. 1 and 2 only.

C. 1 only.

Answer (B) is correct. *(PHAK Chap 7)*
  DISCUSSION: Altimeters 1 and 2 indicate over 10,000 ft. because 1 indicates 10,500 ft. and 2 indicates 14,500 feet. The short needle on 3 points just below 1, i.e., below 10,000 feet.
  Answer (A) is incorrect. Altimeter 3 is indicating 9,500 ft., which is less than 10,000 feet. Answer (C) is incorrect. Altimeter 2 is indicating 14,500 ft., which is also more than 10,000 feet.

2.5 Types of Altitude

29. What is absolute altitude?

A. The altitude read directly from the altimeter.

B. The vertical distance of the aircraft above the surface.

C. The height above the standard datum plane.

Answer (B) is correct. *(PHAK Chap 7)*
  DISCUSSION: Absolute altitude is altitude above the surface, i.e., AGL.
  Answer (A) is incorrect. It is indicated altitude. Answer (C) is incorrect. It is pressure altitude.

30. What is true altitude?

A. The vertical distance of the aircraft above sea level.

B. The vertical distance of the aircraft above the surface.

C. The height above the standard datum plane.

Answer (A) is correct. *(PHAK Chap 7)*
  DISCUSSION: True altitude is the actual altitude above mean sea level, i.e., MSL.
  Answer (B) is incorrect. It represents absolute altitude. Answer (C) is incorrect. It is pressure altitude.

31. What is density altitude?

A. The height above the standard datum plane.

B. The pressure altitude corrected for nonstandard temperature.

C. The altitude read directly from the altimeter.

Answer (B) is correct. *(PHAK Chap 7)*
  DISCUSSION: Density altitude is the pressure altitude corrected for nonstandard temperature.
  Answer (A) is incorrect. It defines pressure altitude. Answer (C) is incorrect. It is indicated altitude.

32. Under what condition is pressure altitude and density altitude the same value?

A. At sea level, when the temperature is 0°F.

B. When the altimeter has no installation error.

C. At standard temperature.

Answer (C) is correct. *(PHAK Chap 7)*
  DISCUSSION: Pressure altitude and density altitude are the same when temperature is standard.
  Answer (A) is incorrect. Standard temperature at sea level is 59°F, not 0°F. Answer (B) is incorrect. Installation error refers to pitot tubes and airspeed, not altimeter and altitude.

33. Under what condition is indicated altitude the same as true altitude?

A. If the altimeter has no mechanical error.

B. When at sea level under standard conditions.

C. When at 18,000 feet MSL with the altimeter set at 29.92.

Answer (B) is correct. *(PHAK Chap 7)*
  DISCUSSION: Indicated altitude (what you read on your altimeter) approximates the true altitude (distance above mean sea level) when standard conditions exist and your altimeter is properly calibrated.
  Answer (A) is incorrect. The indicated altitude must be adjusted for nonstandard temperature for true altitude. Answer (C) is incorrect. The altimeter reads pressure altitude when set to 29.92, and that is only true altitude under standard conditions.

34. Under which condition will pressure altitude be equal to true altitude?

A. When the atmospheric pressure is 29.92" Hg.

B. When standard atmospheric conditions exist.

C. When indicated altitude is equal to the pressure altitude.

Answer (B) is correct.   *(AvW Chap 3)*
DISCUSSION: Pressure altitude equals true altitude when standard atmospheric conditions (29.92" Hg and 15°C at sea level) exist.
Answer (A) is incorrect. Standard temperature must also exist. Answer (C) is incorrect. Indicated altitude does not necessarily relate to true or pressure altitudes.

35. What is pressure altitude?

A. The indicated altitude corrected for position and installation error.

B. The altitude indicated when the barometric pressure scale is set to 29.92.

C. The indicated altitude corrected for nonstandard temperature and pressure.

Answer (B) is correct.   *(PHAK Chap 7)*
DISCUSSION: Pressure altitude is the airplane's height above the standard datum plane of 29.92" Hg. If the altimeter is set to 29.92" Hg, the indicated altitude is the pressure altitude.
Answer (A) is incorrect. "Corrected for position and installation error" is used to define calibrated airspeed, not a type of altitude. Answer (C) is incorrect. It describes density altitude.

36. Altimeter setting is the value to which the barometric pressure scale of the altimeter is set so the altimeter indicates

A. calibrated altitude at field elevation.

B. absolute altitude at field elevation.

C. true altitude at field elevation.

Answer (C) is correct.   *(PHAK Chap 7)*
DISCUSSION: Altimeter setting is the value to which the scale of the pressure altimeter is set so that the altimeter indicates true altitude at field elevation.
Answer (A) is incorrect. "Calibrated" refers to airspeed and airspeed indicators, not altitude and altimeters. Answer (B) is incorrect. Absolute altitude is the altitude above the surface, not above MSL.

2.6 Setting the Altimeter

37. If it is necessary to set the altimeter from 29.15 to 29.85, what change occurs?

A. 70-foot increase in indicated altitude.

B. 70-foot increase in density altitude.

C. 700-foot increase in indicated altitude.

Answer (C) is correct.   *(PHAK Chap 7)*
DISCUSSION: When increasing the altimeter setting from 29.15 to 29.85, the indicated altitude increases by 700 feet. The altimeter-indicated altitude moves in the same direction as the altimeter setting and changes about 1,000 ft. for every change of 1" Hg in the altimeter setting.
Answer (A) is incorrect. A change in pressure of .7" Hg is equal to 700 ft., not 70 ft., of altitude. Answer (B) is incorrect. Density altitude is not affected by changing the altimeter setting.

38. If a pilot changes the altimeter setting from 30.11 to 29.96, what is the approximate change in indication?

A. Altimeter will indicate .15" Hg higher.

B. Altimeter will indicate 150 feet higher.

C. Altimeter will indicate 150 feet lower.

Answer (C) is correct.   *(PHAK Chap 7)*
DISCUSSION: Atmospheric pressure decreases approximately 1" of Hg (mercury) for every 1,000 ft. of altitude gained. As an altimeter setting is changed, the change in altitude indication changes the same way (i.e., approximately 1,000 ft. for every 1" change in altimeter setting) and in the same direction (i.e., lowering the altimeter setting lowers the altitude reading). Thus, changing from 30.11 to 29.96 is a decrease of .15 in., or 150 ft. (.15 × 1,000 ft.) lower.
Answer (A) is incorrect. The altimeter indicates feet, not inches, of mercury. Answer (B) is incorrect. The altimeter will show 150 ft. lower, not higher.

2.7 Altimeter Errors

39. If a flight is made from an area of low pressure into an area of high pressure without the altimeter setting being adjusted, the altimeter will indicate

    A. the actual altitude above sea level.

    B. higher than the actual altitude above sea level.

    C. lower than the actual altitude above sea level.

Answer (C) is correct. *(AvW Chap 3)*
    DISCUSSION: When an altimeter setting is at a lower value than the correct setting, the altimeter is indicating less than it should and thus would be showing lower than the actual altitude above sea level.
    Answer (A) is incorrect. The altimeter will show actual altitude only when it is set correctly. Answer (B) is incorrect. The increase in pressure causes the altimeter to read lower, not higher, than actual altitude.

40. If a flight is made from an area of high pressure into an area of lower pressure without the altimeter setting being adjusted, the altimeter will indicate

    A. lower than the actual altitude above sea level.

    B. higher than the actual altitude above sea level.

    C. the actual altitude above sea level.

Answer (B) is correct. *(AvW Chap 3)*
    DISCUSSION: When flying from higher pressure to lower pressure without adjusting your altimeter, the altimeter will indicate a higher than actual altitude. As you adjust an altimeter barometric setting lower, the altimeter indicates lower.
    Answer (A) is incorrect. The decrease in pressure causes the altimeter to read higher, not lower, than actual altitude. Answer (C) is incorrect. The altimeter will show actual altitude only when it is set correctly.

41. Which condition would cause the altimeter to indicate a lower altitude than true altitude?

    A. Air temperature lower than standard.

    B. Atmospheric pressure lower than standard.

    C. Air temperature warmer than standard.

Answer (C) is correct. *(AvW Chap 3)*
    DISCUSSION: In air that is warmer than standard temperature, the airplane will be higher than the altimeter indicates. Said another way, the altimeter will indicate a lower altitude than actually flown.
    Answer (A) is incorrect. When flying in air that is colder than standard temperature, the airplane will be lower than the altimeter indicates ("high to low, look out below"). Answer (B) is incorrect. The altimeter setting corrects the altimeter for nonstandard pressure.

42. Under what condition will true altitude be lower than indicated altitude?

    A. In colder than standard air temperature.

    B. In warmer than standard air temperature.

    C. When density altitude is higher than indicated altitude.

Answer (A) is correct. *(AvW Chap 3)*
    DISCUSSION: The airplane will be lower than the altimeter indicates when flying in air that is colder than standard temperature. Remember that altimeter readings are adjusted for changes in barometric pressure but not for changes in temperature. When one flies from warmer to cold air and keeps a constant indicated altitude at a constant altimeter setting, the plane has actually descended.
    Answer (B) is incorrect. The altimeter indicates lower than actual altitude in warmer than standard temperature. Answer (C) is incorrect. A higher density altitude is usually the result of warmer, not colder, than standard temperature.

43. How do variations in temperature affect the altimeter?

    A. Pressure levels are raised on warm days and the indicated altitude is lower than true altitude.

    B. Higher temperatures expand the pressure levels and the indicated altitude is higher than true altitude.

    C. Lower temperatures lower the pressure levels and the indicated altitude is lower than true altitude.

Answer (A) is correct. *(PHAK Chap 7)*
    DISCUSSION: On warm days, the atmospheric pressure levels are higher than on cold days. Your altimeter will indicate a lower than true altitude. Remember, "low to high, clear the sky."
    Answer (B) is incorrect. Expanding (or raising) the pressure levels will cause indicated altitude to be lower (not higher) than true altitude. Answer (C) is incorrect. Lower pressure levels will cause indicated altitude to be higher (not lower) than true altitude.

2.8 Gyroscopic Instruments

44. (Refer to Figure 7 on page 55.) The proper adjustment to make on the attitude indicator during level flight is to align the

A. horizon bar to the level-flight indication.

B. horizon bar to the miniature airplane.

C. miniature airplane to the horizon bar.

Answer (C) is correct. *(PHAK Chap 7)*
DISCUSSION: The horizon bar (marked as B) on Fig. 7 represents the true horizon. This bar is fixed to the gyro and remains on a horizontal plane as the airplane is pitched or banked about its lateral or longitudinal axis, indicating the attitude of the airplane relative to the true horizon. An adjustment knob is provided, with which the pilot may move the miniature airplane (marked as C) up or down to align the miniature airplane with the horizontal bar to suit the pilot's line of vision.
Answer (A) is incorrect. Aligning the miniature airplane to the horizon bar provides a level-flight indication. Answer (B) is incorrect. The miniature airplane is adjustable, not the horizon bar.

45. (Refer to Figure 7 on page 55.) How should a pilot determine the direction of bank from an attitude indicator such as the one illustrated?

A. By the direction of deflection of the banking scale (A).

B. By the direction of deflection of the horizon bar (B).

C. By the relationship of the miniature airplane (C) to the deflected horizon bar (B).

Answer (C) is correct. *(PHAK Chap 7)*
DISCUSSION: The direction of bank on the attitude indicator (AI) is indicated by the relationship of the miniature airplane to the deflecting horizon bar. The miniature airplane's relative position to the horizon indicates its attitude: nose high, nose low, left bank, right bank. As you look at the attitude indicator, you see your airplane as it is positioned with respect to the actual horizon. The attitude indicator in Fig. 7 indicates a level right turn.
Answer (A) is incorrect. The banking scale (marked as A) may move in the opposite direction, which is confusing. Answer (B) is incorrect. The horizon bar (marked as B) moves in the direction opposite the turn.

46. (Refer to Figure 5 on page 55.) A turn coordinator provides an indication of the

A. movement of the aircraft about the yaw and roll axes.

B. angle of bank up to but not exceeding 30°.

C. attitude of the aircraft with reference to the longitudinal axis.

Answer (A) is correct. *(PHAK Chap 7)*
DISCUSSION: There really are no yaw and roll axes; i.e., an airplane yaws about its vertical axis and rolls about its longitudinal axis. However, this is the best answer since the turn coordinator does indicate the roll and yaw movement of the airplane. The movement of the miniature airplane is proportional to the roll rate of the airplane. When the roll rate is reduced to zero (i.e., when the bank is held constant), the instrument provides an indication of the rate of turn.
Answer (B) is incorrect. The turn coordinator shows the rate of turn rather than angle of bank. Answer (C) is incorrect. The turn coordinator does not show the attitude of the airplane (as does the attitude indicator); it shows the rate of the roll and turn.

47. (Refer to Figure 5 on page 55, Figure 6 on page 55, and Figure 7 on page 55.) To receive accurate indications during flight from a heading indicator, the instrument must be

A. set prior to flight on a known heading.

B. calibrated on a compass rose at regular intervals.

C. periodically realigned with the magnetic compass as the gyro precesses.

Answer (C) is correct. *(PHAK Chap 7)*
DISCUSSION: Due to gyroscopic precession, directional gyros must be periodically realigned with a magnetic compass. Friction is the major cause of its drifting from the correct heading.
Answer (A) is incorrect. The instrument must be periodically reset, not just set initially. Answer (B) is incorrect. There is no calibration of the heading indicator; rather, it is reset.

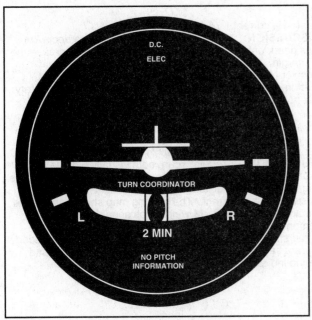

Figure 5. – Turn Coordinator.

Figure 6. – Heading Indicator.

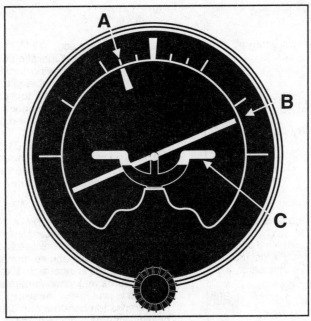

Figure 7. – Attitude Indicator.

2.9 Glass Cockpits

48. What is a benefit of flying with a glass cockpit?

A. There is no longer a need to carry paper charts in flight.

B. Situational awareness is increased.

C. Terrain avoidance is guaranteed.

Answer (B) is correct. *(AAH Chap 5)*
DISCUSSION: Glass cockpits are designed to decrease pilot workload, enhance situational awareness, and increase the safety margin.
Answer (A) is incorrect. Pilots should still have current information and backup electronic navigation to enhance safety. Answer (C) is incorrect. Terrain avoidance is not guaranteed solely by means of relying on advanced avionics.

49. What steps must be taken when flying with glass cockpits to ensure safe flight?

A. Use the moving map for primary means of navigation, use the MFD to check engine systems and weather, back up with supplementary forms of information

B. Regularly scan each item on the PFD, confirm on the MFD.

C. Regularly scan both inside and outside, use all appropriate checklists, and cross-check with other forms of information.

Answer (C) is correct. *(AAH Chap 5)*
DISCUSSION: A regular scan, both visually outside and inside on backup gauges, should be combined with other means of navigation and checklists to ensure safe flight.
Answer (A) is incorrect. The moving map should not be the sole means of navigation. Moving maps should be used as a supplement, not as a replacement. Answer (B) is incorrect. While you should scan both the PFD and MFD, more is needed to ensure a safe flight, such as visually scanning outside and confirming indications from other sources.

2.10 Engine Temperature

50. An abnormally high engine oil temperature indication may be caused by

A. the oil level being too low.

B. operating with a too high viscosity oil.

C. operating with an excessively rich mixture.

Answer (A) is correct. *(PHAK Chap 6)*
DISCUSSION: Operating with an excessively low oil level prevents the oil from being cooled adequately; i.e., an inadequate supply of oil will not be able to transfer engine heat to the engine's oil cooler (similar to a car engine's water radiator). Insufficient oil may also damage an engine from excessive friction within the cylinders and on other metal-to-metal contact parts.
Answer (B) is incorrect. The higher the viscosity, the better the lubricating and cooling capability of the oil. Answer (C) is incorrect. A rich fuel/air mixture usually decreases (not increases) engine temperature.

51. Excessively high engine temperatures will

A. cause damage to heat-conducting hoses and warping of the cylinder cooling fins.

B. cause loss of power, excessive oil consumption, and possible permanent internal engine damage.

C. not appreciably affect an aircraft engine.

Answer (B) is correct. *(PHAK Chap 6)*
DISCUSSION: Excessively high engine temperatures will result in loss of power, excessive oil consumption, and possible permanent internal engine damage.
Answer (A) is incorrect. Excessively high engine temperatures may cause internal engine damage, but external damage is less likely. Answer (C) is incorrect. An excessively high engine temperature can cause a loss of performance and possibly internal engine damage.

52. Excessively high engine temperatures, either in the air or on the ground, will

A. increase fuel consumption and may increase power due to the increased heat.

B. result in damage to heat-conducting hoses and warping of cylinder cooling fans.

C. cause loss of power, excessive oil consumption, and possible permanent internal engine damage.

Answer (C) is correct. *(PHAK Chap 6)*
DISCUSSION: Operating the engine at excessively high temperatures will cause loss of power and excessive oil consumption, and can permanently damage engines.
Answer (A) is incorrect. Overheating can cause excessive oil, not fuel, consumption and a loss, not increase, of power. Answer (B) is incorrect. Hoses are not used to transfer heat in airplane engines. Also, it is extremely unlikely one could overheat an engine to an extent to warp the cylinder cooling fans.

53. For internal cooling, air cooled engines are especially dependent on

    A. a properly functioning thermostat.

    B. air flowing over the exhaust manifold.

    C. the circulation of lubricating oil.

Answer (C) is correct. *(PHAK Chap 6)*
    DISCUSSION: An engine accomplishes much of its cooling by the flow of oil through the lubrication system. The lubrication system aids in cooling by reducing friction and absorbing heat from internal engine parts. Many airplane engines use an oil cooler, a small radiator device that will cool the oil before it is recirculated through the engine.
    Answer (A) is incorrect. Airplanes with air-cooled engines do not use thermostats. Answer (B) is incorrect. Air flowing over the exhaust manifold would have little effect on internal engine parts cooling.

54. If the engine oil temperature and cylinder head temperature gauges have exceeded their normal operating range, the pilot may have been operating with

    A. the mixture set too rich.

    B. higher-than-normal oil pressure.

    C. too much power and with the mixture set too lean.

Answer (C) is correct. *(PHAK Chap 6)*
    DISCUSSION: If the engine oil temperature and cylinder head temperature gauges exceed their normal operating range, it is possible that the power setting is too high and the fuel/air mixture is set excessively lean. These conditions may cause engine overheating.
    Answer (A) is incorrect. A rich mixture setting normally causes lower (not higher-than-normal) engine temperature. Answer (B) is incorrect. A higher-than-normal oil pressure does not normally increase the engine temperature.

55. What action can a pilot take to aid in cooling an engine that is overheating during a climb?

    A. Reduce rate of climb and increase airspeed.

    B. Reduce climb speed and increase RPM.

    C. Increase climb speed and increase RPM.

Answer (A) is correct. *(PHAK Chap 6)*
    DISCUSSION: If an airplane is overheating during a climb, the engine temperature will be decreased if the airspeed is increased. Airspeed will increase if the rate of climb is reduced.
    Answer (B) is incorrect. Reducing airspeed hinders cooling and increasing RPM will further increase engine temperature. Answer (C) is incorrect. Increasing RPM will increase (not decrease) engine temperature.

56. What is one procedure to aid in cooling an engine that is overheating?

    A. Enrich the fuel mixture.

    B. Increase the RPM.

    C. Reduce the airspeed.

Answer (A) is correct. *(PHAK Chap 6)*
    DISCUSSION: Enriched fuel mixtures have a cooling effect on an engine.
    Answer (B) is incorrect. Increasing the RPM increases the engine's internal heat. Answer (C) is incorrect. Reducing the airspeed decreases the airflow needed for cooling, thus increasing the engine's temperature.

2.11 Constant-Speed Propeller

57. How is engine operation controlled on an engine equipped with a constant-speed propeller?

    A. The throttle controls power output as registered on the manifold pressure gauge and the propeller control regulates engine RPM.

    B. The throttle controls power output as registered on the manifold pressure gauge and the propeller control regulates a constant blade angle.

    C. The throttle controls engine RPM as registered on the tachometer and the mixture control regulates the power output.

Answer (A) is correct. *(PHAK Chap 6)*
    DISCUSSION: Airplanes equipped with controllable-pitch propellers have both a throttle control and a propeller control. The throttle controls the power output of the engine, which is registered on the manifold pressure gauge. This is a simple barometer that measures the air pressure in the engine intake manifold in inches of mercury. The propeller control regulates the engine RPM, which is registered on a tachometer.
    Answer (B) is incorrect. The propeller blade angle changes to control the RPM. Answer (C) is incorrect. The throttle controls power output (not RPM), and the mixture controls the fuel to air ratio (not power output).

58. A precaution for the operation of an engine equipped with a constant-speed propeller is to

A. avoid high RPM settings with high manifold pressure.

B. avoid high manifold pressure settings with low RPM.

C. always use a rich mixture with high RPM settings.

Answer (B) is correct. *(PHAK Chap 6)*
   DISCUSSION: For any given RPM, there is a manifold pressure that should not be exceeded. Manifold pressure is excessive for a given RPM when the cylinder design pressure is exceeded, placing undue stress on them. If repeated or extended, the stress would weaken the cylinder components and eventually cause engine failure.
   Answer (A) is incorrect. It is the relationship of high manifold pressure with low RPM that is dangerous (not high RPM with high manifold pressure). Answer (C) is incorrect. The mixture control is related to engine cylinder temperature, not to RPM.

59. What is an advantage of a constant-speed propeller?

A. Permits the pilot to select and maintain a desired cruising speed.

B. Permits the pilot to select the blade angle for the most efficient performance.

C. Provides a smoother operation with stable RPM and eliminates vibrations.

Answer (B) is correct. *(PHAK Chap 6)*
   DISCUSSION: A controllable-pitch propeller (constant-speed) permits the pilot to select the blade angle that will result in the most efficient performance given the flight conditions. A low blade angle and a decreased pitch reduces the propeller drag and allows more engine RPM (power) for takeoffs. After airspeed is attained during cruising flight, the propeller blade is changed to a higher angle to increase pitch. The blade takes a larger bite of air at a lower RPM and consequently increases the efficiency of the flight. This process is similar to shifting gears in an automobile from low to high gear.
   Answer (A) is incorrect. A desired cruising speed is possible with any airplane. Answer (C) is incorrect. Vibrations are eliminated through propeller balancing, not a constant-speed propeller.

2.12 Engine Ignition Systems

60. One purpose of the dual ignition system on an aircraft engine is to provide for

A. improved engine performance.

B. uniform heat distribution.

C. balanced cylinder head pressure.

Answer (A) is correct. *(PHAK Chap 6)*
   DISCUSSION: Most airplane engines are equipped with dual ignition systems, which have two magnetos to supply the electrical current to two spark plugs for each combustion chamber. The main advantages of the dual system are increased safety and improved burning and combustion of the mixture, which results in improved performance.
   Answer (B) is incorrect. The heat distribution within a cylinder is usually not uniform, even with dual ignition. Answer (C) is incorrect. Balanced cylinder-head pressure is a nonsense phrase.

61. If the ignition switch ground wire becomes disconnected, the magneto

A. will not operate because the battery is disconnected from the circuit.

B. may continue to fire.

C. will not operate.

Answer (B) is correct. *(PHAK Chap 6)*
   DISCUSSION: Loose or broken wires in the ignition system can cause problems. For example, if the ignition switch is OFF, the magneto may continue to fire if the ignition switch ground wire is disconnected. If this occurs, the only way to stop the engine is to move the mixture lever to the idle cut-off position, then have the system checked by a qualified aviation maintenance technician.
   Answer (A) is incorrect. The magneto may continue to fire if the ignition switch ground wire is disconnected. Answer (C) is incorrect. The magneto may continue to fire if the ignition switch ground wire is disconnected.

2.13 Carburetor Icing

62. With regard to carburetor ice, float-type carburetor systems in comparison to fuel injection systems are generally considered to be

    A. more susceptible to icing.

    B. equally susceptible to icing.

    C. susceptible to icing only when visible moisture is present.

Answer (A) is correct. *(PHAK Chap 6)*
    DISCUSSION: Float-type carburetor systems are generally more susceptible to icing than fuel-injected engines. When there is visible moisture or high humidity and the temperature is between 20°F and 70°F, icing is possible, particularly at low power settings.
    Answer (B) is incorrect. Fuel injection systems are less susceptible to internal icing than a carburetor system, although air intake icing is equally possible in both systems. Answer (C) is incorrect. Carburetor icing may occur in high humidity with no visible moisture.

63. Which condition is most favorable to the development of carburetor icing?

    A. Any temperature below freezing and a relative humidity of less than 50 percent.

    B. Temperature between 32°F and 50°F and low humidity.

    C. Temperature between 20°F and 70°F and high humidity.

Answer (C) is correct. *(PHAK Chap 6)*
    DISCUSSION: When the temperature is between 20°F and 70°F with visible moisture or high humidity, one should be on the alert for carburetor ice. During low or closed throttle settings, an engine is particularly susceptible to carburetor icing.
    Answer (A) is incorrect. Icing is possible at temperatures up to 70°F and only in high humidity or visible moisture. Answer (B) is incorrect. Low humidity will generally preclude icing and the correct temperature range is 20°F to 70°F.

64. The possibility of carburetor icing exists even when the ambient air temperature is as

    A. high as 70°F and the relative humidity is high.

    B. high as 95°F and there is visible moisture.

    C. low as 0°F and the relative humidity is high.

Answer (A) is correct. *(PHAK Chap 6)*
    DISCUSSION: When the temperature is between 20°F and 70°F with visible moisture or high humidity, one should be on the alert for carburetor ice. During low or closed throttle settings, an engine is particularly susceptible to carburetor icing.
    Answer (B) is incorrect. Icing is usually not a problem above 70°F. Answer (C) is incorrect. Icing is usually not a problem below 20°F.

65. If an aircraft is equipped with a fixed-pitch propeller and a float-type carburetor, the first indication of carburetor ice would most likely be

    A. a drop in oil temperature and cylinder head temperature.

    B. engine roughness.

    C. loss of RPM.

Answer (C) is correct. *(PHAK Chap 6)*
    DISCUSSION: In an airplane equipped with a fixed-pitch propeller and float-type carburetor, the first indication of carburetor ice would be a loss in RPM.
    Answer (A) is incorrect. A carburetor icing condition does not cause a drop in oil temperature or cylinder head temperature. Answer (B) is incorrect. A loss in engine RPM should be evident before engine roughness became noticeable.

66. The operating principle of float-type carburetors is based on the

    A. automatic metering of air at the venturi as the aircraft gains altitude.

    B. difference in air pressure at the venturi throat and the air inlet.

    C. increase in air velocity in the throat of a venturi causing an increase in air pressure.

Answer (B) is correct. *(PHAK Chap 6)*
    DISCUSSION: In a float-type carburetor, air flows into the carburetor and through a venturi tube (a narrow throat in the carburetor). As the air flows more rapidly through the venturi, a low pressure area is created that draws the fuel from a main fuel jet located at the throat of the carburetor and into the airstream, where it is mixed with flowing air. It is called a float-type carburetor in that a ready supply of gasoline is kept in the float bowl by a float, which activates a fuel inlet valve.
    Answer (A) is incorrect. The metering at the venturi is fuel, not air, and this is done manually with a mixture control. Answer (C) is incorrect. The increase in air velocity in the throat of a venturi causes a decrease (not increase) in air pressure (which draws the gas from the main fuel jet into the low-pressure air).

67. The presence of carburetor ice in an aircraft equipped with a fixed-pitch propeller can be verified by applying carburetor heat and noting

    A. an increase in RPM and then a gradual decrease in RPM.

    B. a decrease in RPM and then a constant RPM indication.

    C. a decrease in RPM and then a gradual increase in RPM.

Answer (C) is correct. *(PHAK Chap 6)*
    DISCUSSION: The presence of carburetor ice in an airplane equipped with a fixed-pitch propeller can be verified by applying carburetor heat and noting a decrease in RPM and then a gradual increase. The decrease in RPM as heat is applied is caused by less dense hot air entering the engine and reducing power output. Also, if ice is present, melting water entering the engine may also cause a loss in performance. As the carburetor ice melts, however, the RPM gradually increases until it stabilizes when the ice is completely removed.
    Answer (A) is incorrect. The warm air decreases engine power output and RPM. Ice melting further decreases RPM and then RPM increases slightly after the ice melts. Answer (B) is incorrect. After the ice melts, the RPM will increase gradually (not remain constant).

2.14 Carburetor Heat

68. Generally speaking, the use of carburetor heat tends to

    A. decrease engine performance.

    B. increase engine performance.

    C. have no effect on engine performance.

Answer (A) is correct. *(PHAK Chap 6)*
    DISCUSSION: Use of carburetor heat tends to decrease the engine performance and also to increase the operating temperature. Warmer air is less dense, and engine performance decreases with density. Thus, carburetor heat should not be used when full power is required (as during takeoff) or during normal engine operation except as a check for the presence or removal of carburetor ice.
    Answer (B) is incorrect. Carburetor heat decreases (not increases) engine performance. Answer (C) is incorrect. Carburetor heat does have an effect on performance.

69. Applying carburetor heat will

    A. result in more air going through the carburetor.

    B. enrich the fuel/air mixture.

    C. not affect the fuel/air mixture.

Answer (B) is correct. *(PHAK Chap 6)*
    DISCUSSION: Applying carburetor heat will enrich the fuel/air mixture. Warm air is less dense than cold air, hence the application of heat increases the fuel-to-air ratio.
    Answer (A) is incorrect. Applying carburetor heat will not result in more air going into the carburetor. Answer (C) is incorrect. Applying carburetor heat will enrich the fuel/air mixture.

70. What change occurs in the fuel/air mixture when carburetor heat is applied?

    A. A decrease in RPM results from the lean mixture.

    B. The fuel/air mixture becomes richer.

    C. The fuel/air mixture becomes leaner.

Answer (B) is correct. *(PHAK Chap 6)*
    DISCUSSION: When carburetor heat is applied, hot air is introduced into the carburetor. Hot air is less dense than cold air; therefore, the decrease in air density with a constant amount of fuel makes a richer mixture.
    Answer (A) is incorrect. A drop in RPM as carburetor heat is applied is due to the less dense air and melting ice, not a lean mixture. Answer (C) is incorrect. When carburetor heat is applied, the fuel/air mixture becomes richer, not leaner.

2.15 Fuel/Air Mixture

71. During the run-up at a high-elevation airport, a pilot notes a slight engine roughness that is not affected by the magneto check but grows worse during the carburetor heat check. Under these circumstances, what would be the most logical initial action?

    A. Check the results obtained with a leaner setting of the mixture.

    B. Taxi back to the flight line for a maintenance check.

    C. Reduce manifold pressure to control detonation.

Answer (A) is correct. *(PHAK Chap 6)*
    DISCUSSION: If, during a run-up at a high-elevation airport, you notice a slight roughness that is not affected by a magneto check but grows worse during the carburetor heat check, you should check the results obtained with a leaner setting of the mixture control. At a high-elevation field, the air is less dense and the application of carburetor heat increases the already too rich fuel-to-air mixture. By leaning the mixture during the run-up, the condition should improve.
    Answer (B) is incorrect. This mixture condition is normal at a high-elevation field. However, if after leaning the mixture a satisfactory run-up cannot be obtained, the pilot should taxi back to the flight line for a maintenance check. Answer (C) is incorrect. The question describes a symptom of an excessively rich mixture, not detonation.

72. The basic purpose of adjusting the fuel/air mixture at altitude is to

A. decrease the amount of fuel in the mixture in order to compensate for increased air density.

B. decrease the fuel flow in order to compensate for decreased air density.

C. increase the amount of fuel in the mixture to compensate for the decrease in pressure and density of the air.

Answer (B) is correct. *(PHAK Chap 6)*
DISCUSSION: At higher altitudes, the air density is decreased. Thus, the mixture control must be adjusted to decrease the fuel flow in order to maintain a constant fuel/air ratio.
Answer (A) is incorrect. Air density decreases (not increases) at altitude. Answer (C) is incorrect. The mixture is decreased (not increased) in order to compensate for decreased air density.

73. While cruising at 9,500 feet MSL, the fuel/air mixture is properly adjusted. What will occur if a descent to 4,500 feet MSL is made without readjusting the mixture?

A. The fuel/air mixture may become excessively lean.

B. There will be more fuel in the cylinders than is needed for normal combustion, and the excess fuel will absorb heat and cool the engine.

C. The excessively rich mixture will create higher cylinder head temperatures and may cause detonation.

Answer (A) is correct. *(PHAK Chap 6)*
DISCUSSION: At 9,500 ft., the mixture control is adjusted to provide the proper fuel/air ratio. As the airplane descends, the density of the air increases and there will be less fuel to air in the ratio, causing a leaner running engine. This excessively lean mixture will create higher cylinder temperature and may cause detonation.
Answer (B) is incorrect. As air becomes more dense during the descent, there will be less (not more) fuel in the cylinders than is needed. Answer (C) is incorrect. The mixture will be excessively lean (not rich). Also, a rich mixture would create lower (not higher) cylinder head temperatures.

2.16 Abnormal Combustion

74. Detonation occurs in a reciprocating aircraft engine when

A. the spark plugs are fouled or shorted out or the wiring is defective.

B. hot spots in the combustion chamber ignite the fuel/air mixture in advance of normal ignition.

C. the unburned charge in the cylinders explodes instead of burning normally.

Answer (C) is correct. *(PHAK Chap 6)*
DISCUSSION: Detonation occurs when the fuel/air mixture in the cylinders explodes instead of burning normally. This more rapid force slams the piston down instead of pushing it.
Answer (A) is incorrect. If the spark plugs are "fouled" or the wiring is defective, the cylinders would not be firing; i.e., there would be no combustion. Answer (B) is incorrect. Hot spots in the combustion chamber igniting the fuel/air mixture in advance of normal ignition is pre-ignition.

75. Detonation may occur at high-power settings when

A. the fuel mixture ignites instantaneously instead of burning progressively and evenly.

B. an excessively rich fuel mixture causes an explosive gain in power.

C. the fuel mixture is ignited too early by hot carbon deposits in the cylinder.

Answer (A) is correct. *(PHAK Chap 6)*
DISCUSSION: Detonation occurs when the fuel/air mixture in the cylinders explodes instead of burning progressively and evenly. This more rapid force slams the piston down instead of pushing it.
Answer (B) is incorrect. An excessively rich fuel mixture lowers the temperature inside the cylinder, thus inhibiting the complete combustion of the fuel and producing an appreciable lack of power. Answer (C) is incorrect. Hot carbon deposits in the combustion chamber igniting the fuel/air mixture too early, or in advance of normal ignition, is termed pre-ignition.

76. If a pilot suspects that the engine (with a fixed-pitch propeller) is detonating during climb-out after takeoff, the initial corrective action to take would be to

A. lean the mixture.

B. lower the nose slightly to increase airspeed.

C. apply carburetor heat.

Answer (B) is correct. *(PHAK Chap 6)*
DISCUSSION: If you suspect engine detonation during climb-out after takeoff, you would normally decrease the pitch to increase airspeed (more cooling) and decrease the load on the engine. Detonation is usually caused by a poor grade of fuel or an excessive engine temperature.
Answer (A) is incorrect. Leaning the mixture will increase engine temperature and increase detonation. Answer (C) is incorrect. While carburetor heat will increase the fuel-to-air ratio, hot air flowing into the carburetor will not lower engine temperature. Also, the less dense air will decrease the engine power for climb-out.

77. If the grade of fuel used in an aircraft engine is lower than specified for the engine, it will most likely cause

    A. a mixture of fuel and air that is not uniform in all cylinders.

    B. lower cylinder head temperatures.

    C. detonation.

Answer (C) is correct. *(PHAK Chap 6)*
    DISCUSSION: If the grade of fuel used in an airplane engine is lower than specified for the engine, it will probably cause detonation. Lower grades of fuel ignite at lower temperatures. A higher temperature engine (which should use a higher grade of fuel) may cause lower grade fuel to explode (detonate) rather than burn evenly.
    Answer (A) is incorrect. The carburetor meters the lower-grade fuel quantity in the same manner as a higher grade of fuel. Answer (B) is incorrect. A lower grade of fuel will cause higher (not lower) cylinder head temperatures.

78. The uncontrolled firing of the fuel/air charge in advance of normal spark ignition is known as

    A. combustion.

    B. pre-ignition.

    C. detonation.

Answer (B) is correct. *(PHAK Chap 6)*
    DISCUSSION: Pre-ignition is the ignition of the fuel prior to normal ignition or ignition before the electrical arcing occurs at the spark plug. Pre-ignition may be caused by excessively hot exhaust valves, carbon particles, or spark plugs and electrodes heated to an incandescent, or glowing, state. These hot spots are usually caused by high temperatures encountered during detonation. A significant difference between pre-ignition and detonation is that, if the conditions for detonation exist in one cylinder, they usually exist in all cylinders, but pre-ignition often takes place in only one or two cylinders.
    Answer (A) is incorrect. Combustion is the normal process that takes place inside the cylinders. Answer (C) is incorrect. Detonation is an uncontrolled, explosive ignition of the fuel/air mixture within the cylinder's combustion chamber caused by a combination of excessively high temperature and pressure in the cylinder.

2.17 Aviation Fuel Practices

79. What type fuel can be substituted for an aircraft if the recommended octane is not available?

    A. The next higher octane aviation gas.

    B. The next lower octane aviation gas.

    C. Unleaded automotive gas of the same octane rating.

Answer (A) is correct. *(PHAK Chap 6)*
    DISCUSSION: If the recommended octane is not available for an airplane, the next higher octane aviation gas should be used.
    Answer (B) is incorrect. If the grade of fuel used in an airplane engine is lower than specified for the engine, it will probably cause detonation. Answer (C) is incorrect. Except for very special situations, only aviation gas should be used.

80. Filling the fuel tanks after the last flight of the day is considered a good operating procedure because this will

    A. force any existing water to the top of the tank away from the fuel lines to the engine.

    B. prevent expansion of the fuel by eliminating airspace in the tanks.

    C. prevent moisture condensation by eliminating airspace in the tanks.

Answer (C) is correct. *(PHAK Chap 6)*
    DISCUSSION: Filling the fuel tanks after the last flight of the day is considered good operating practice because it prevents moisture condensation by eliminating airspace in the tanks. Humid air may result in condensation at night when the airplane cools.
    Answer (A) is incorrect. Water is heavier than fuel and will always settle to the bottom of the tank. Answer (B) is incorrect. Filling the fuel tank will not prevent expansion of the fuel.

81. To properly purge water from the fuel system of an aircraft equipped with fuel tank sumps and a fuel strainer quick drain, it is necessary to drain fuel from the

    A. fuel strainer drain.

    B. lowest point in the fuel system.

    C. fuel strainer drain and the fuel tank sumps.

Answer (C) is correct. *(PHAK Chap 6)*
    DISCUSSION: One should purge water from both the fuel strainer drain and all the fuel tank sumps on an airplane. This is the purpose of such drains. They are placed at low areas of the fuel system and should be drained prior to each flight.
    Answer (A) is incorrect. All drains, not just the fuel strainer, should be checked for water. Answer (B) is incorrect. All fuel drains and sumps, not just the lowest point in the system, should be checked for water.

82. On aircraft equipped with fuel pumps, when is the auxiliary electric driven pump used?

   A. All the time to aid the engine-driven fuel pump.

   B. In the event engine-driven fuel pump fails.

   C. Constantly except in starting the engine.

Answer (B) is correct. *(PHAK Chap 6)*
   DISCUSSION: In a fuel pump system, two fuel pumps are used on most airplanes. The main fuel pump is engine-driven, and an auxiliary electric-driven pump is provided for use in the event the engine pump fails.
   Answer (A) is incorrect. An auxiliary fuel pump is a backup system to the engine-driven fuel pump; it is not intended to aid the engine-driven fuel pump. Answer (C) is incorrect. The auxiliary electric fuel pump is normally used in starting the engine.

83. Which would most likely cause the cylinder head temperature and engine oil temperature gauges to exceed their normal operating ranges?

   A. Using fuel that has a lower-than-specified fuel rating.

   B. Using fuel that has a higher-than-specified fuel rating.

   C. Operating with higher-than-normal oil pressure.

Answer (A) is correct. *(PHAK Chap 6)*
   DISCUSSION: Use of fuel with lower-than-specified fuel ratings, e.g., 80 octane instead of 100, can cause many problems, including higher operating temperatures, detonation, etc.
   Answer (B) is incorrect. Higher octane fuels usually result in lower cylinder head temperatures. Answer (C) is incorrect. Higher-than-normal oil pressure provides better lubrication and cooling (although too high an oil pressure can break parts, lines, etc.).

2.18 Starting the Engine

84. What should be the first action after starting an aircraft engine?

   A. Adjust for proper RPM and check for desired indications on the engine gauges.

   B. Place the magneto or ignition switch momentarily in the OFF position to check for proper grounding.

   C. Test each brake and the parking brake.

Answer (A) is correct. *(PHAK Chap 6)*
   DISCUSSION: After the engine starts, the engine speed should be adjusted to the proper RPM. Then the engine gauges should be reviewed, with the oil pressure being the most important gauge initially.
   Answer (B) is incorrect. This check is normally done just prior to engine shutdown. Answer (C) is incorrect. This check is done during taxi.

85. Should it become necessary to handprop an airplane engine, it is extremely important that a competent pilot

   A. call "contact" before touching the propeller.

   B. be at the controls in the cockpit.

   C. be in the cockpit and call out all commands.

Answer (B) is correct. *(PHAK Chap 6)*
   DISCUSSION: Because of the hazards involved in handstarting airplane engines, every precaution should be exercised. It is extremely important that a competent pilot be at the controls in the cockpit. Also, the person turning the propeller should be thoroughly familiar with the technique.
   Answer (A) is incorrect. The person handpropping the airplane yells "gas off, switch off, throttle closed, brakes set" before touching the propeller initially. Contact means the magnetos are on, i.e., "hot." This is not done until starting is attempted. Answer (C) is incorrect. The person handpropping the airplane (not the person in the cockpit) calls out the commands.

2.19 Electrical System

86. An electrical system failure (battery and alternator) occurs during flight. In this situation, you would

A. experience avionics equipment failure.

B. probably experience failure of the engine ignition system, fuel gauges, aircraft lighting system, and avionics equipment.

C. probably experience engine failure due to the loss of the engine-driven fuel pump and also experience failure of the radio equipment, lights, and all instruments that require alternating current.

Answer (A) is correct. *(PHAK Chap 6)*
DISCUSSION: A battery and alternator failure during flight inevitably results in avionics equipment failure due to the lack of electricity.
Answer (B) is incorrect. The engine ignition systems are based on magnetos, which generate their own electricity to operate the spark plugs. Answer (C) is incorrect. Engine-driven fuel pumps are mechanical and not dependent upon electricity.

87. A positive indication on an ammeter

A. indicates the aircraft's battery will soon lose its charge.

B. shows the rate of charge on the battery.

C. means more current is being drawn from the battery than is being replaced.

Answer (B) is correct. *(PHAK Chap 6)*
DISCUSSION: A positive indication on the ammeter shows the rate of charge on the battery.
Answer (A) is incorrect. A battery will not lose its charge while being charged, which is what a positive indication on an ammeter indicates. Answer (C) is incorrect. A negative indication on an ammeter means more current is being drawn from the battery than is being replaced.

88. To keep a battery charged, the alternator voltage output should be

A. less than the battery voltage.

B. equal to the battery voltage.

C. higher than the battery voltage.

Answer (C) is correct. *(PHAK Chap 6)*
DISCUSSION: The alternator voltage output should be slightly higher than the battery voltage to keep the battery charged. For example, a 14-volt alternator system would keep a positive charge on a 12-volt battery.
Answer (A) is incorrect. If the alternator voltage output were less than the battery voltage, the battery would quickly lose its charge. Answer (B) is incorrect. If there were no difference in voltage, the battery would not have or keep a full charge.

89. Which of the following is a true statement concerning electrical systems?

A. The master switch provides current to the electrical system.

B. The airspeed indicator is driven by the electrical system.

C. Lights and radios use the electrical system for power.

Answer (C) is correct. *(PHAK Chap 6)*
DISCUSSION: Lights, radios, and electrical fuel pumps are examples of equipment that commonly use the electrical system.
Answer (A) is incorrect. The master switch provides electrical current to all electrical systems except the ignition system. Answer (B) is incorrect. The airspeed indicator operates on the pitot-static system, not the electrical system.

END OF STUDY UNIT

STUDY UNIT THREE
AIRPORTS, AIR TRAFFIC CONTROL, AND AIRSPACE

(12 pages of outline)

This study unit contains outlines of major concepts tested; sample test questions and answers regarding airports, Air Traffic Control, and airspace; and an explanation of each answer. The table of contents above lists each subunit within this study unit, the number of questions pertaining to that particular subunit, and the pages on which the outlines and questions begin, respectively.

CAUTION: Recall that the **sole purpose** of this book is to expedite your passing the FAA pilot knowledge test for the private pilot certificate. Accordingly, all extraneous material (i.e., topics or regulations not directly tested on the FAA pilot knowledge test) is omitted, even though much more information and knowledge are necessary to fly safely. This additional material is presented in *Pilot Handbook* and *Private Pilot Flight Maneuvers and Practical Test Prep*, available from Gleim Publications, Inc. See the order form on page 371.

3.1 RUNWAY MARKINGS

1. The number at the start of each runway indicates its magnetic alignment divided by 10°; e.g., Runway 26 indicates 260° magnetic; Runway 9 indicates 090° magnetic.

    a. Runways are numbered by the direction in which they point.

2. A displaced threshold is a threshold (marked as a broad solid line across the runway) that is not at the beginning of the full strength runway pavement. The remainder of the runway, following the displaced threshold, is the landing portion of the runway.

    a. See Figure A below.

    b. The paved area before the displaced threshold (marked by arrows) is available for taxiing, the landing rollout, and takeoff of aircraft.

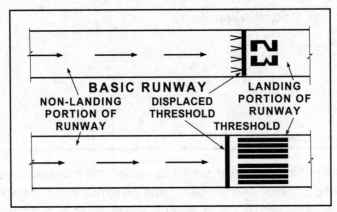

Figure A

3. Chevrons mark any surface or area extending beyond the usable runway that appears usable but that, due to the nature of its structure, is unusable runway.

    a. This area is not available for any use, not even taxiing.

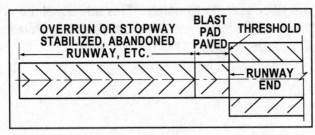

Figure B

4. Closed runways are marked by an "X" on each runway end that is closed.

5. Runway holding position markings indicate where an aircraft is supposed to stop. They consist of four yellow lines, two continuous and two dashed, extending across the width of the taxiway or runway. The solid (continuous) lines are always on the side where the aircraft is to hold.

3.2 TAXIWAY AND DESTINATION SIGNS

1. Destination signs have black characters on a yellow background with an arrow showing the direction of the taxiing route to the destination listed. Outbound destinations commonly show directions to the take-off runways.

    a. Examples of destination signs are shown in Figure 66 on page 293.

        1) They are signs I, J, and K.
        2) In that figure, Sign K designates the direction of taxiway bravo.

2. Taxiway location signs identify the taxiway on which an aircraft is currently located.

    a. Location signs feature a black background with yellow lettering and do not have directional arrows.

3. Taxiway directional signs indicate the designation and direction of a taxiway.

    a. When turning from one taxiway to another, a taxiway directional sign indicates the designation and direction of a taxiway leading out of the intersection.

    b. Taxiway directional signs feature a yellow background with black lettering and directional arrows.

4. When approaching taxiway holding lines from the side with continuous lines, the pilot should not cross the lines without an ATC clearance.

    a. Taxiway holding lines are painted across the width of the taxiway and are yellow.

5. A runway holding position sign is a mandatory instruction sign with white characters on a red background. It is located at the holding position on taxiways that intersect a runway or on runways that intersect other runways.

6. Each of the letters below corresponds to the type of sign or marking in Figure 66.

    A. Runway Holding Position Sign
    B. Holding Position Sign for a Runway Approach Area
    C. Holding Position Sign for ILS Critical Area
    D. Sign Prohibiting Aircraft Entry into an Area
    E. Taxiway Location Sign
    F. Runway Location Sign
    G. Runway Boundary Sign
    H. ILS Critical Area Boundary Sign
    I. Direction Sign for Terminal
    J. Direction Sign for Common Taxiing Route to Runway
    K. Direction Sign for Runway Exit
    L. Runway Distance Remaining Sign
    M. Hold Short-1
    N. Taxiway Ending Sign

7.   Vehicle Roadway Markings

   a.   Vehicle roadway markings define pathways for vehicles to cross areas of the airport used by aircraft.

      1)   Vehicle roadway markings exist in two forms, as indicated by letter C in Figure 65 below and in color on page 292).

         a)   The edge of vehicle roadway markings may be defined by a solid white line or white zipper markings.

      2)   A dashed white line separates opposite-direction vehicle traffic inside the roadway.

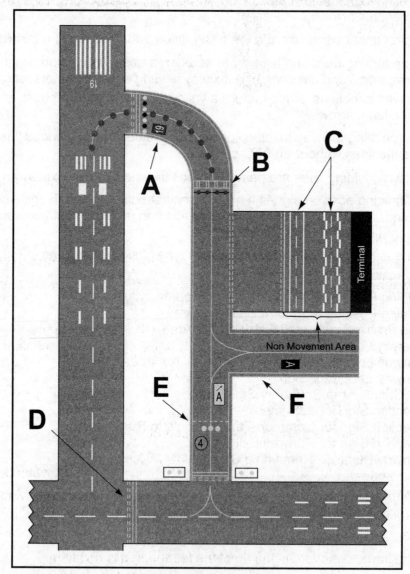

Figure 65. – Airport Markings.

8.   Each of the letters below corresponds to the type of airport markings in Figure 65.

   A.   Holding Position Markings at Beginning of Takeoff Runway 19
   B.   ILS Critical Area Boundary Marking
   C.   Roadway Edge Stripes
   D.   Runway Holding Position Marking
   E.   Taxiway Holding Position Marker
   F.   Taxiway Boundary

9. Yellow Demarcation Bar

   a. The yellow demarcation bar is a 3-ft.-wide, painted yellow bar that separates a displaced threshold from a blast pad, stopway, or taxiway that precedes the runway.

Yellow Demarcation Bar

DEMARCATION BAR
YELLOW, 3′ (1 m) WIDE

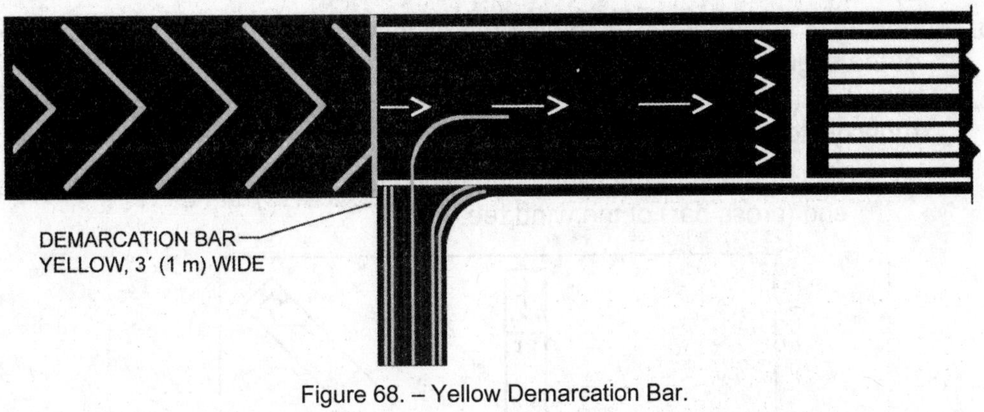

Figure 68. – Yellow Demarcation Bar.

3.3 BEACONS AND TAXIWAY LIGHTS

1. Operation of the green and white rotating beacon at an airport located in Class D airspace during the day indicates that the weather is not VFR; i.e.,

   a. The visibility is less than 3 SM or
   b. The ceiling is less than 1,000 feet.

2. A lighted heliport may be identified by a green, yellow, and white rotating beacon.

3. Military airports are indicated by beacons with two white flashes between each green flash.

4. Airport taxiways are lighted with blue edge lights.

5. To operate pilot-controlled lighting (PCL), you should first click the mic seven times, which turns everything on. For high-intensity lights, leave it alone. For medium-intensity lights, click it five times. For low-intensity lights, click it three times.

3.4 AIRPORT TRAFFIC PATTERNS

1. If you are approaching an airport without an operating control tower,

   a. Left turns are standard, unless otherwise specified.
   b. You must comply with any FAA traffic pattern for that airport when departing.

2. The recommended entry to an airport traffic pattern is 45° to the downwind leg, at the approximate midpoint, at traffic pattern altitude (1,000 ft. AGL).

3. Remember, you land

   a. In the same direction as the tip of the tetrahedron is pointing,
   b. As if you were flying out of the large (open) end of the wind cone, or
   c. Toward the cross-bar end of a wind "T" (visualize the "T" as an airplane with no nose, with the top of the "T" being the wings).

4. If there is no segmented circle installed at the airport, traffic pattern indicators may be installed on or near the end of the runway.

5. The segmented circle system provides traffic pattern information at airports without operating control towers. It consists of the

    a. Segmented circle – located in a position affording maximum visibility to pilots in the air and on the ground, and providing a centralized point for the other elements of the system

    b. Landing strip indicators – showing the alignment of landing runways (legs sticking out of the segmented circle)

    c. Wind direction indicator – a wind cone, wind sock, or wind tee installed near the runways to indicate wind direction

        1) The large end of the wind cone/wind sock points into the wind as does the large end (cross bar) of the wind tee.

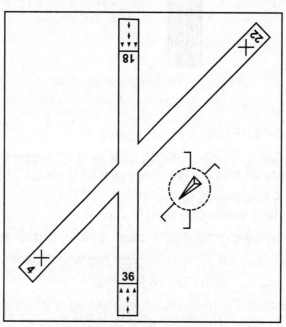

Figure 50. – Airport Diagram.

    d. Landing direction indicator – a tetrahedron on a swivel installed when conditions at the airport warrant its use. It is used to indicate the direction of takeoffs and landings. It should be located at the center of a segmented circle and may be lighted for night operations.

        1) The small end points toward the direction in which a takeoff or landing should be made; i.e., the small end points into the wind.

    e. Traffic pattern indicators – indicators at right angles to the landing strip indicator showing the direction of turn from base to final and upwind to crosswind

        1) In Figure 50 above, Runways 22 and 36 use left traffic, while Runways 4 and 18 use right traffic.

        2) The "X" indicates that Runways 4 and 22 are closed.

        3) The area behind the displaced thresholds of Runways 18 and 36 (marked by arrows) can be used for taxiing and takeoff, but not for landing.

3.5 VISUAL APPROACH SLOPE INDICATORS (VASI)

1. Visual approach slope indicators (VASI) are a system of lights to provide visual descent information during an approach to landing.

2. The standard VASI consists of a two-barred tier of lights. You are

    a. Below the glide path if both light bars are red; i.e., "red means dead."
    b. On the glide path if the far (on top visually) lights are red and the near (on bottom visually) lights are white.
    c. Above the glide path if both light bars are white.

3. Remember, red over white (i.e., R before W alphabetically) is the desired sequence.

    a. White over red is impossible.

4. A tri-color VASI is a single light unit projecting three colors:

    a. The below glide path indicator is red.
    b. The above glide path indicator is amber.
    c. The on glide path indicator is green.

5. VASI only projects a glide path. It has no bearing on runway alignment.

6. On a precision approach path indicator (PAPI),

    a. Low is four red lights (less than 2.5°).
    b. Slightly low is one white and three reds (2.8°).
    c. On glide path is two whites and two reds (3.0°).
    d. Slightly high is three whites and one red (3.2°).
    e. High is four whites (more than 3.5°).

7. On a pulsating approach slope indicator (a VASI with flashing/pulsating signals),

    a. Low is a pulsating red.
    b. On glide path is a steady white or alternating red/white (depending on model).
    c. High is a pulsating white.

8. Each pilot of an airplane approaching to land on a runway served by a visual approach slope indicator shall maintain an altitude at or above the glide slope until a lower altitude is necessary for landing (FAR 91.129).

3.6 WAKE TURBULENCE

1. Wingtip vortices (wake turbulence) are only created when airplanes develop lift.

2. The greatest vortex strength occurs when the generating aircraft is heavy, clean, and slow.

3. The circulation of the vortex is outward, upward, and around each wingtip.

4. Wingtip vortex turbulence tends to sink into the flight path of airplanes operating below the airplane generating the turbulence.

    a. Thus, you should fly above the flight path of a large jet rather than below.
    b. You should also fly upwind rather than downwind of the flight path, since the vortices will drift with the wind.

5. The most dangerous wind, when taking off or landing behind a heavy aircraft, is the light quartering tailwind. It will push the vortices into your touchdown zone, even if you are executing proper procedures.

3.7 COLLISION AVOIDANCE

1. Navigation lights on the aircraft consist of a red light on the left wing, a green light on the right wing, and a white light on the tail. In night flight,

    a. When an airplane is crossing in front of you from your right to left, you will observe a red light.

    b. When an airplane is crossing in front of you from your left to right, you will observe a green light.

    c. When an airplane is flying away from you, you will observe a steady white light.

    d. When an airplane is approaching you head-on, you will observe a red and green light but no white light.

    e. Note that the navigation lights on the wings cannot be seen from the rear.

2. A flashing red light on an aircraft is a rotating beacon and may be seen from any angle.

3. In daylight, the most effective way to scan for other aircraft is to use a series of short, regularly spaced eye movements that bring successive areas of the sky into your central visual field.

    a. Each movement should not exceed 10°, and each area should be observed for at least 1 second to enable detection.

    b. Only a very small center area of the eye has the ability to send clear, sharply focused messages to the brain.

4. At night, collision avoidance scanning must use the off-center portions of the eyes. These portions are most effective at seeing objects at night.

    a. Accordingly, peripheral vision should be used, scanning small sectors and using off-center viewing.

5. Any aircraft that appears to have no relative motion with respect to your aircraft and stays in one scan quadrant is likely to be on a collision course.

    a. If it increases in size, you should take immediate evasive action.

6. Prior to each maneuver, a pilot should visually scan the entire area for collision avoidance.

    a. When climbing or descending VFR on an airway, you should execute gentle banks left and right to facilitate scanning for other aircraft.

7. All pilots are responsible for collision avoidance when operating in an alert area.

8. Most midair collision accidents occur during clear days.

9. Pilots are encouraged to turn on their landing lights when operating below 10,000 feet, day or night, especially when operating in conditions of reduced visibility.

3.8 ATIS AND GROUND CONTROL

1. Automatic Terminal Information Service (ATIS) is a continuous broadcast of recorded noncontrol information in selected high activity terminal areas (i.e., busy airports).

    a. The information is essential but routine.

2. The information included is the latest weather sequence, active runways, and other pertinent remarks.

    a. Ceilings are usually not broadcast if they are above 5,000 ft., and visibility is usually not mentioned if it is more than 5 statute miles.

3. After landing, you should contact ground control only when so instructed by the tower.

4.   A clearance to taxi to the active runway is a clearance to taxi via taxiways to the active runway. You may not cross any runway along your taxi route unless specifically cleared by ATC to do so.

    a.   When cleared to a runway, you are cleared to that runway's runup area, but not onto the active runway itself.

    b.   "Line up and wait" is the instruction to taxi onto the active runway and prepare for takeoff, but not to take off.

3.9 CLASS D AIRSPACE AND AIRPORT ADVISORY AREA

1.   Class D airspace is an area of controlled airspace surrounding an airport with an operating control tower, not associated with Class B or Class C airspace areas.

    a.   Airspace at an airport with a part-time control tower is classified as Class D airspace only when the control tower is operating.

    b.   When the control tower ceases operation for the day, the airspace reverts to Class E, or a combination of Class E and G airspace during the hours the tower is not in operation.

2.   Class D airspace is depicted by a blue segmented (dashed) circle on a sectional chart.

3.   When departing a non-tower satellite airport within Class D airspace, you must establish and maintain two-way radio communication with the primary airport's control tower.

    a.   The primary airport is the airport for which the Class D airspace is designated.

    b.   A satellite airport is any other airport within the Class D airspace area.

4.   Class D airspace is normally the airspace up to 2,500 ft. above the surface of the airport.

    a.   The actual lateral dimensions of Class D airspace are based on the instrument procedures for which the controlled airspace is established.

5.   Two-way radio communication with the control tower is required for landings and takeoffs at all tower-controlled airports, regardless of weather conditions.

6.   Airport Advisory Areas exist at noncontrolled airports that have a Flight Service Station (FSS) physically located on that airport. The FSS provides advisory (not control) information on traffic, weather, etc., to requesting aircraft.

3.10 CLASS C AIRSPACE

1.   Class C airspace consists of a surface area (formerly called the inner circle) and a shelf area (formerly called the outer circle).

    a.   The surface area has a 5-NM radius from the primary airport extending from the surface to 4,000 ft. above the airport elevation.

    b.   The shelf area is an area from 5 to 10 NM from the primary airport extending from 1,200 ft. to 4,000 ft. above the airport elevation.

2.   Surrounding the Class C airspace is the outer area. The outer area is not classified as Class C airspace.

    a.   ATC provides the same radar services as provided in Class C airspace.

    b.   The normal radius of the outer area of Class C airspace is 20 NM from the primary airport.

3.   The minimum equipment needed to operate in Class C airspace

    a.   4096 code transponder,

    b.   Mode C (altitude encoding) capability, and

    c.   Two-way radio communication capability.

4. You must establish and maintain two-way radio communication with ATC prior to entering Class C airspace. A clearance is not required because a clearance relates to IFR operations.

5. When departing from a satellite airport without an operating control tower, you must contact ATC as soon as practicable after takeoff.

3.11 TERMINAL RADAR PROGRAMS

1. Terminal radar programs for VFR aircraft are classified as basic, TRSA, Class C, and Class B service.

    a. Basic radar service provides safety alerts, traffic advisories, and limited vectoring on a workload-permitting basis.

    b. TRSA service provides sequencing and separation for all participating VFR aircraft operating within a Terminal Radar Service Area (TRSA).

2. Terminal radar program participation is voluntary for VFR traffic.

    a. Contact approach control when inbound.

    b. When departing, you should request radar traffic information from ground control on initial contact, along with your direction of flight.

3.12 TRANSPONDER CODES

1. Code 1200 is the standard VFR transponder code.

2. The ident feature should not be engaged unless instructed by ATC.

3. Certain special codes should never be engaged (except in an emergency), as they may cause problems at ATC centers:

    a. 7500 is the hijacking code.
    b. 7600 is the lost radio communication code.
    c. 7700 is the general emergency code.
    d. 7777 is the military interceptor code.

3.13 RADIO PHRASEOLOGY

1. When contacting a flight service station, the proper call sign is the name of the FSS followed by "radio" (e.g., McAlester Radio).

2. When contacting an En Route Flight Advisory Service (EFAS), the proper call sign is the name of the Air Route Traffic Control Center facility serving your area followed by "flight watch" (e.g., "Seattle Flight Watch").

3. Civilian aircraft should start their aircraft call sign with the make or model aircraft (e.g., Cessna 44WH or Baron 2DF).

    a. When a make or model is used, the initial November is dropped from the call sign.

4. Pilots should state each digit of the call sign individually (e.g., 6449U = six, four, four, niner, uniform).

5. When calling out altitudes up to but not including 18,000 ft., state the separate digits of the thousands, plus the hundreds, if appropriate (e.g., 4,500 ft. = four thousand five hundred).

    a. Unless otherwise noted, the altitudes are MSL.

3.14 ATC TRAFFIC ADVISORIES

1. Radar traffic information services provide pilots with traffic advisories of nearby aircraft.

2. Traffic advisories provide information based on the position of other aircraft from your airplane in terms of clock direction in a no-wind condition (i.e., it is based on your ground track, not heading).

    a. 12 o'clock is straight ahead.
    b. 3 o'clock is directly off your right wing.
    c. 6 o'clock is directly behind you.
    d. 9 o'clock is directly off your left wing.
    e. Other positions are described accordingly, e.g., 2 o'clock, 10 o'clock.

3. Traffic advisories usually also include

    a. Distance away in miles
    b. Direction of flight of other aircraft
    c. Altitude of other aircraft

3.15 ATC LIGHT SIGNALS

1. In the absence of radio communications, the tower can communicate with you by light signals.

2. Light signal meanings depend on whether you are on the ground or in the air.

3. Acknowledge light signals in the air by rocking wings in daylight and blinking lights at night.

4. If your radio fails and you wish to land at a tower-controlled airport, remain outside or above the airport's traffic pattern until the direction and flow of traffic has been determined, then join the traffic pattern and maintain visual contact with the tower to receive light signals.

Light Signal	On the Ground	In the Air
Steady Green	Cleared for takeoff	Cleared to land
Flashing Green	Cleared to taxi	Return for landing *(to be followed by steady green at proper time)*
Steady Red	Stop	Give way to other aircraft and continue circling
Flashing Red	Taxi clear of landing area (runway) in use	Airport unsafe -- Do not land
Flashing White	Return to starting point on airport	Not applicable
Alternating Red and Green	General warning signal -- Exercise extreme caution	General warning signal -- Exercise extreme caution

3.16   ELTs AND VHF/DF

1.   ELTs transmit simultaneously on 121.5 and 243.0 megahertz.

   a.   You can monitor either frequency during flight and before shut down (after landing) to ensure your ELT has not been activated.

2.   The VHF/Direction Finder facility is a ground operation that displays the magnetic direction of the airplane from the station each time the airplane transmits a signal to it.

3.   In order to take advantage of VHF/DF radio reception for assistance in locating a position, an airplane must have both a VHF transmitter and a receiver. The transmitter and receiver are necessary to converse with a ground station having VHF/DF facilities.

   a.   The transmitter is also needed to send the signal that the Direction Finder identifies in terms of magnetic heading from the facility.

3.17   LAND AND HOLD SHORT OPERATIONS (LAHSO)

1.   Land and hold short operations (LAHSO) take place at some airports with an operating control tower in order to increase airport capacity and improve the flow of traffic.

   a.   LAHSO requires that you land and hold short of an intersecting runway, an intersecting taxiway, or some other designated point on a runway.

2.   Before accepting a clearance to land and hold short, you must determine that you can safely land and stop within the available landing distance (ALD).

   a.   ALD data are published in the special notices section of the *Airport/Facility Directory*.
   b.   ATC will provide ALD data upon your request.

3.   Student pilots should not participate in the LAHSO program.

4.   The pilot in command has the final authority to accept or decline any land and hold short (LAHSO) clearance.

   a.   Decline a LAHSO clearance if you determine it will compromise safety.

5.   You should receive a LAHSO clearance only when there is a minimum ceiling of 1,000 ft. and visibility of 3 statute miles.

   a.   The intent of having basic VFR weather conditions is to allow pilots to maintain visual contact with other aircraft and ground vehicle operations.

QUESTIONS

3.1 Runway Markings

1. The numbers 9 and 27 on a runway indicate that the runway is oriented approximately

A. 009° and 027° true.

B. 090° and 270° true.

C. 090° and 270° magnetic.

Answer (C) is correct. *(AIM Para 2-3-3)*
DISCUSSION: Runway numbers are determined from the approach direction. The runway number is the whole number nearest one-tenth the magnetic direction of the centerline. Thus, the numbers 9 and 27 on a runway indicate that the runway is oriented approximately 090° and 270° magnetic.
Answer (A) is incorrect. The ending digit, not a leading zero, is dropped. Answer (B) is incorrect. Runways are numbered based on magnetic (not true) direction.

2. The numbers 8 and 26 on the approach ends of the runway indicate that the runway is orientated approximately

A. 008° and 026° true.

B. 080° and 260° true.

C. 080° and 260° magnetic.

Answer (C) is correct. *(AIM Para 2-3-3)*
DISCUSSION: Runway numbers are determined from the approach direction. The runway number is the whole number nearest one-tenth the magnetic direction of the centerline. Thus, the numbers 8 and 26 on a runway indicate that the runway is oriented approximately 080° and 260° magnetic.
Answer (A) is incorrect. The ending digit, not a leading zero, is dropped. Answer (B) is incorrect. Runways are numbered based on magnetic, not true, direction.

3. (Refer to Figure 60 on page 291.) The radius of the procedural outer area of Class C airspace is normally

A. 10 NM.

B. 20 NM.

C. 30 NM.

Answer (B) is correct. *(AIM Chap 3)*
DISCUSSION: A 20-NM radius procedural outer area surrounds the primary airport in Class C airspace. This area is not charted and generally does not require action from the pilot.
Answer (A) is incorrect. Each Class C airspace is individually tailored to the specific area; however, most Class C airspace consists of a charted 5-NM radius core area that extends from the surface to 4,000 ft. AGL and a charted 10-NM radius shelf that extends from 1,200 ft. AGL to 4,000 ft. AGL. Answer (C) is incorrect. A 30-NM outer area does not surround Class C airspace; however, a 30-NM Mode C veil does surround Class B airspace.

4. (Refer to Figure 49 on page 79.) According to the airport diagram, which statement is true?

A. Runway 30 is equipped at position E with emergency arresting gear to provide a means of stopping military aircraft.

B. Takeoffs may be started at position A on Runway 12, and the landing portion of this runway begins at position B.

C. The takeoff and landing portion of Runway 12 begins at position B.

Answer (B) is correct. *(AIM Para 2-3-3)*
DISCUSSION: In Fig. 49, Runway 12 takeoffs may be started at position A, and the landing portion of this runway begins at position B. In this example, a displaced threshold exists at the beginning of Runway 12. The threshold is a heavy line across the runway, designating the beginning portion of a runway usable for landing. The paved area behind the displaced runway threshold is available for taxiing, the landing rollout, and the takeoff of aircraft.
Answer (A) is incorrect. Arresting cables across the operational area of a runway are indicated by yellow circles 10 ft. in diameter painted across the runway at positions of the arresting cables. Area E has chevron markings, which indicates an overrun area. Answer (C) is incorrect. Only the landing portion of RWY 12 begins at position B. The takeoff may be started in the paved area behind the displaced runway threshold (i.e., position A).

5. (Refer to Figure 49 on page 79.) That portion of the runway identified by the letter A may be used for

A. landing.

B. taxiing and takeoff.

C. taxiing and landing.

Answer (B) is correct. *(AIM Para 2-3-3)*
DISCUSSION: The portion of the runway identified by the letter A in Fig. 49 is a displaced threshold, as marked by arrows from the beginning of the runway pointing to the displaced threshold, which means it may be used for taxiing or takeoffs but not for landings.
Answer (A) is incorrect. Area A may be used for the landing rollout but not the actual landing. Answer (C) is incorrect. Area A may be used for the landing rollout but not the actual landing.

6. (Refer to Figure 49 on page 79.) What is the difference between area A and area E on the airport depicted?

A. "A" may be used for taxi and takeoff; "E" may be used only as an overrun.

B. "A" may be used for all operations except heavy aircraft landings; "E" may be used only as an overrun.

C. "A" may be used only for taxiing; "E" may be used for all operations except landings.

Answer (A) is correct. *(AIM Para 2-3-3)*
DISCUSSION: Area A in Fig. 49 is the paved area behind a displaced runway threshold, as identified by the arrows painted on the pavement. This area may be used for taxiing, the landing rollout, and the takeoff of aircraft. Area E is a stopway area, as identified by the chevrons. This area, due to the nature of its structure, is unusable except as an overrun.
Answer (B) is incorrect. Area A cannot be used by any aircraft for landing. Answer (C) is incorrect. Area A can also be used for takeoff and landing rollout. Area E cannot be used for any type of operation, except as an overrun.

7. (Refer to Figure 49 on page 79.) Area C on the airport depicted is classified as a

A. stabilized area.

B. multiple heliport.

C. closed runway.

Answer (C) is correct. *(AIM Para 2-3-6)*
DISCUSSION: The runway marked by the arrow C in Fig. 49 has Xs on the runway, indicating it is closed.
Answer (A) is incorrect. Stabilized areas are designed to be load bearing but may be limited to emergency use only. Area E on the airport indicates a stabilized area. Answer (B) is incorrect. Heliports are marked by Hs, not Xs.

3.2 Taxiway and Destination Signs

8. When approaching taxiway holding lines from the side with the continuous lines, the pilot

A. may continue taxiing.

B. should not cross the lines without ATC clearance.

C. should continue taxiing until all parts of the aircraft have crossed the lines.

Answer (B) is correct. *(AIM Para 2-3-5)*
DISCUSSION: When approaching taxiway holding lines, the solid (continuous) lines are always on the side where the aircraft is to hold. Therefore, do not cross the hold line without ATC clearance.
Answer (A) is incorrect. You cannot cross the hold line without ATC clearance. Answer (C) is incorrect. No part of the aircraft can cross the hold line without ATC clearance.

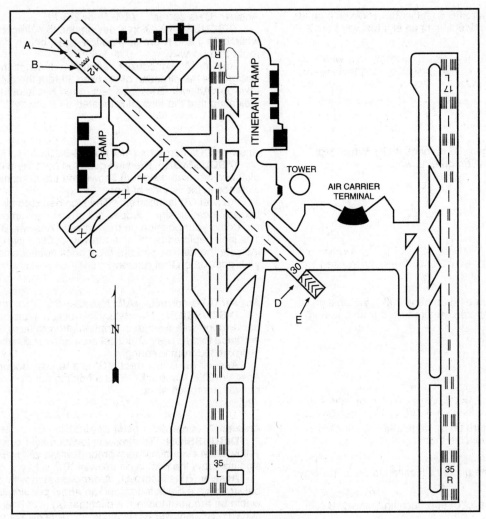

Figure 49. – Airport Diagram.

9. What is the purpose of the runway/runway hold position sign?

    A. Denotes entrance to runway from a taxiway.

    B. Denotes area protected for an aircraft approaching or departing a runway.

    C. Denotes intersecting runways.

Answer (C) is correct. *(AIM Para 2-3-5)*
    DISCUSSION: Runway/runway hold position signs are a type of mandatory instruction sign used to denote intersecting runways. These are runways that intersect and are being used for "Land, Hold Short" operations or are normally used for taxiing. These signs have a red background with white lettering. Runway/runway hold position signs are identical to the signs used for taxiway/runway intersections.
    Answer (A) is incorrect. A runway/runway hold position sign is located on a runway and denotes an intersecting runway, not the entrance to a runway from a taxiway. Answer (B) is incorrect. A runway approach area holding position sign protects an area from approaching or departing aircraft.

10. What does the outbound destination sign identify?

    A. Identifies entrance to the runway from a taxiway.

    B. Identifies runway on which an aircraft is located.

    C. Identifies direction to take-off runways.

Answer (C) is correct. *(AIM Para 2-3-11)*
    DISCUSSION: Outbound destination signs define taxiing directions to takeoff runways. Destination signs have a yellow background with a black inscription. Outbound destination signs always have an arrow showing the direction of the taxiing route to the takeoff runway.
    Answer (A) is incorrect. A runway holding position sign, not an outbound destination sign, identifies the entrance to a runway from a taxiway. Runway holding position signs consist of a red background with white inscription. Answer (B) is incorrect. A runway location sign, not an outbound destination sign, identifies the runway on which the aircraft is currently located. Runway location signs consist of a black background with a yellow inscription and a yellow border.

11. (Refer to Figure 66 on page 293.) Which sign is a designation and direction of an exit taxiway from a runway?

A. J.

B. F.

C. K.

Answer (C) is correct. *(AIM Para 2-3-11)*
  DISCUSSION: Sign K designates the direction of taxiway B; while both J and K are destination signs, only K designates the route to a taxiway.
  Answer (A) is incorrect. Though a destination sign, Sign J designates the direction of Runway 22, not the direction of a taxiway. Answer (B) is incorrect. Sign F is a location sign indicating that the aircraft is located on Runway 22.

12. (Refer to Figure 66 on page 293.) Which sign identifies your position on a runway?

A. E.

B. F.

C. L.

Answer (C) is correct. *(AIM Para 2-3-13)*
  DISCUSSION: The sign depicted by "L" is a runway distance remaining sign. It shows you the remaining runway distance in thousands of feet.
  Answer (A) is incorrect. The sign depicted by "E" is a taxiway location sign. It does not provide any information regarding your position on the runway. Answer (B) is incorrect. The sign depicted by "F" is a runway location sign. While it does indicate the runway you are on, it does not specifically indicate your position on that runway.

13. (Refer to Figure 66 on page 293.) Which sign identifies where aircraft are prohibited from entering?

A. D.

B. G.

C. B.

Answer (A) is correct. *(AIM Para 2-3-8)*
  DISCUSSION: Mandatory instruction signs have a red background with a white inscription and are used to denote an entrance to a runway or critical area and areas where an aircraft is prohibited from entering.
  Answer (B) is incorrect. "G" is a runway boundary sign. Answer (C) is incorrect. "B" is a holding position sign for a runway approach area.

14. (Refer to Figure 66 on page 293.) (Refer to E.) This sign is a visual clue that

A. confirms the aircraft's location to be on taxiway "B."

B. warns the pilot of approaching taxiway "B."

C. indicates "B" holding area is ahead.

Answer (A) is correct. *(AIM Chap 3)*
  DISCUSSION: The taxiway location sign consists of a yellow letter on a black background with a yellow border. This sign confirms the pilot is on taxiway "B."
  Answer (B) is incorrect. A direction sign with a yellow background, a black letter, and an arrow pointing to taxiway "B" would be required to warn a pilot that (s)he is approaching taxiway "B." Answer (C) is incorrect. A taxiway location sign defines a position on a taxiway, not a holding area.

15. (Refer to Figure 66 on page 293.) (Refer to F.) This sign confirms your position on

A. runway 22.

B. routing to runway 22.

C. taxiway 22.

Answer (A) is correct. *(AIM Chap 2)*
  DISCUSSION: A runway position sign has a black background with a yellow inscription and a yellow border. The inscription on the sign informs the pilot (s)he is located on Runway 22.
  Answer (B) is incorrect. A direction sign with a yellow background and black inscription would be required to inform a pilot (s)he is routing to Runway 22. Answer (C) is incorrect. Only runways are numbered. Taxiways are always identified by a letter.

16. (Refer to Figure 66 on page 293.) (Refer to G.) From the cockpit, this marking confirms the aircraft to be

A. on a taxiway, about to enter runway zone.

B. on a runway, about to clear.

C. near an instrument approach clearance zone.

Answer (B) is correct. *(AIM Chap 2)*
  DISCUSSION: When the runway holding position line is viewed from the runway side, the pilot is presented with two dashed bars. The PIC must ensure the entire aircraft has cleared the runway holding position line prior to coming to a stop.
  Answer (A) is incorrect. A pilot entering a runway from a taxiway is presented with the two solid bars on the runway holding position marking, not the dashed lines. Answer (C) is incorrect. The marking depicted is a runway holding position marking and is not related to any form of clearance zone.

17. (Refer to Figure 65 below and in color on page 292.) Which marking indicates a vehicle lane?

    A. A.

    B. C.

    C. E.

Answer (B) is correct. *(AIM Para 2-3-6)*
    DISCUSSION: Vehicle roadway markings define a route of travel for vehicles to cross areas intended for use by aircraft. The roadway is defined by solid white lines, with a dashed line in the middle to separate traffic traveling in opposite directions. White zipper markings may be used instead of solid white lines to define the edge of the roadway at some airports.
    Answer (A) is incorrect. This marking represents a surface painted holding position sign, not a vehicle lane. In this instance, the marking indicates the aircraft is holding short of Runway 19. Answer (C) is incorrect. This marking represents a standard taxiway holding position and is used by ATC to hold aircraft short of an intersecting taxiway.

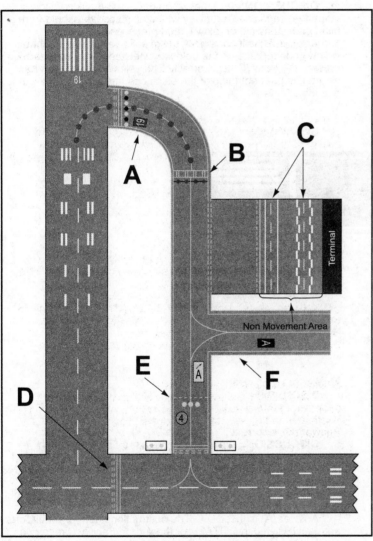

Figure 65. – Airport Markings.

18. When turning onto a taxiway from another taxiway, what is the purpose of the taxiway directional sign?

    A. Indicates direction to take-off runway.

    B. Indicates designation and direction of exit taxiway from runway.

    C. Indicates designation and direction of taxiway leading out of an intersection.

Answer (C) is correct. *(AIM Para 2-3-10)*
    DISCUSSION: Direction signs consist of black lettering on a yellow background. These signs identify the designations of taxiways leading out of an intersection. An arrow next to each taxiway designation indicates the direction that an aircraft must turn in order to taxi onto that taxiway.
    Answer (A) is incorrect. Outbound destination signs, not direction signs, indicate the direction that must be taken out of an intersection in order to follow the preferred taxi route to a runway. Answer (B) is incorrect. The question specifies that you are turning onto a taxiway from another taxiway, not from a runway.

19. What purpose does the taxiway location sign serve?

A. Provides general taxiing direction to named runway.

B. Denotes entrance to runway from a taxiway.

C. Identifies taxiway on which an aircraft is located.

Answer (C) is correct.   *(AIM Para 2-3-9)*
DISCUSSION: Taxiway location signs are used to identify a taxiway on which the aircraft is currently located. Taxiway location signs consist of a black background with a yellow inscription and yellow border.
Answer (A) is incorrect. A runway destination sign, not a taxiway location sign, provides general taxiing information to a named runway. Answer (B) is incorrect. A runway holding position sign, not a taxiway location sign, identifies the entrance to a runway from a taxiway. Runway holding position signs consist of a red background with white inscription.

20. (Refer to Figure 68 below and in color on page 294.) The 'yellow demarcation bar' marking indicates

A. runway with a displaced threshold that precedes the runway.

B. a hold line from a taxiway to a runway.

C. the beginning of available runway for landing on the approach side.

Answer (A) is correct.   *(AIM Para 2-3-6)*
DISCUSSION: A demarcation bar is a 3-ft.-wide yellow stripe that separates a runway with a displaced threshold from a blast pad, stopway, or taxiway that precedes the runway.
Answer (B) is incorrect. A set of solid yellow and dashed yellow lines represents the hold lines between a taxiway and runway. Answer (C) is incorrect. The yellow demarcation bar delineates the beginning of the displaced threshold, which is not a landing surface.

Yellow Demarcation Bar

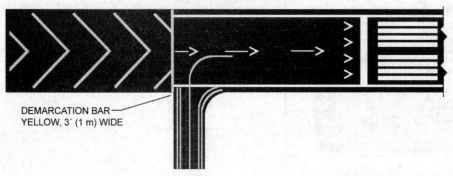

Figure 68. – Yellow Demarcation Bar.

3.3 Beacons and Taxiway Lights

21. An airport's rotating beacon operated during daylight hours indicates

A. there are obstructions on the airport.

B. that weather at the airport located in Class D airspace is below basic VFR weather minimums.

C. the Air Traffic Control tower is not in operation.

Answer (B) is correct.   *(AIM Para 2-1-8)*
DISCUSSION: Operation of the airport beacon during daylight hours often indicates that weather at the airport located in controlled airspace (e.g., Class D airspace) is below basic VFR weather minimums, i.e., less than 1,000 ft. ceiling or 3 SM visibility. Note that there is no regulatory requirement for daylight operation of an airport's rotating beacon.
Answer (A) is incorrect. The obstructions near or on airports are usually listed in NOTAMs or the *Airport/Facility Directory* as appropriate to their hazard. Answer (C) is incorrect. There is no visual signal of tower operation/non-operation.

22. A lighted heliport may be identified by a

A. green, yellow, and white rotating beacon.

B. flashing yellow light.

C. blue lighted square landing area.

Answer (A) is correct.   *(AIM Para 2-1-8)*
DISCUSSION: A lighted heliport may be identified by a green, yellow, and white rotating beacon.
Answer (B) is incorrect. A flashing yellow light is sometimes used to help a pilot locate a lighted water airport. It is used in conjunction with the lighted water airport's white and yellow rotating beacon. Answer (C) is incorrect. A lighted heliport may be identified by a green, yellow, and white rotating beacon, not a blue lighted square landing area.

23. A military air station can be identified by a rotating beacon that emits

    A. white and green alternating flashes.

    B. two quick, white flashes between green flashes.

    C. green, yellow, and white flashes.

Answer (B) is correct. *(AIM Para 2-1-8)*
    DISCUSSION: Lighted land airports are distinguished by white and green airport beacons. To further distinguish it as a military airport, there are two quick white flashes between each green.
    Answer (A) is incorrect. White and green alternating flashes designate a lighted civilian land airport. Answer (C) is incorrect. Green, yellow, and white flashes designate a lighted heliport.

24. How can a military airport be identified at night?

    A. Alternate white and green light flashes.

    B. Dual peaked (two quick) white flashes between green flashes.

    C. White flashing lights with steady green at the same location.

Answer (B) is correct. *(AIM Para 2-1-8)*
    DISCUSSION: Military airport beacons flash alternately white and green but are differentiated from civil beacons by two quick white flashes between the green flashes.
    Answer (A) is incorrect. Alternating white and green beacon light flashes indicate lighted civil land airports. Answer (C) is incorrect. There is no such airport signal.

25. Airport taxiway edge lights are identified at night by

    A. white directional lights.

    B. blue omnidirectional lights.

    C. alternate red and green lights.

Answer (B) is correct. *(AIM Para 2-1-9)*
    DISCUSSION: Taxiway edge lights are used to outline the edges of taxiways during periods of darkness or restricted visibility conditions. These lights are identified at night by blue omnidirectional lights.
    Answer (A) is incorrect. White lights are standard runway edge lights. Answer (C) is incorrect. Alternate red and green lights are a light gun signal, which means exercise extreme caution to all aircraft.

26. To set the high intensity runway lights on medium intensity, the pilot should click the microphone seven times, and then click it

    A. one time within 4 seconds.

    B. three times within 3 seconds.

    C. five times within 5 seconds.

Answer (C) is correct. *(AIM Para 2-1-7)*
    DISCUSSION: To turn on and set the runway lights on medium intensity, the recommended procedure is to key the mic seven times; this ensures that all the lights are on and at high intensity. Next, key the mic five times to get the medium-intensity setting. Lighting systems are activated by keying the mic within a 5-second interval.
    Answer (A) is incorrect. Keying only one time will not adjust or turn the lights on at all. Answer (B) is incorrect. Three additional microphone clicks will give the low-intensity setting.

3.4 Airport Traffic Patterns

27. Which is the correct traffic pattern departure procedure to use at a noncontrolled airport?

    A. Depart in any direction consistent with safety, after crossing the airport boundary.

    B. Make all turns to the left.

    C. Comply with any FAA traffic pattern established for the airport.

Answer (C) is correct. *(FAR 91.127)*
    DISCUSSION: Each person operating an airplane to or from an airport without an operating control tower shall (1) in the case of an airplane approaching to land, make all turns of that airplane to the left unless the airport displays approved light signals or visual markings indicating that turns should be made to the right, in which case the pilot shall make all turns to the right, and (2) in the case of an airplane departing the airport, comply with any FAA traffic pattern for that airport.
    Answer (A) is incorrect. The correct traffic pattern departure procedure at a noncontrolled airport is to comply with any FAA established traffic pattern, not to depart in any direction after crossing the airport boundary. Answer (B) is incorrect. The FAA may establish right- or left-hand traffic patterns, not only left-hand traffic.

28. (Refer to Figure 51 below.) The segmented circle indicates that the airport traffic is

    A. left-hand for Runway 36 and right-hand for Runway 18.

    B. left-hand for Runway 18 and right-hand for Runway 36.

    C. right-hand for Runway 9 and left-hand for Runway 27.

Answer (A) is correct. *(AIM Para 4-3-4)*
    DISCUSSION: A segmented circle (see Fig. 51) is installed at uncontrolled airports to provide traffic pattern information. The landing runway indicators are shown coming out of the segmented circle to show the alignment of landing runways. In Fig. 51 (given the answer choices), the available runways are 18-36 and 9-27.
    The traffic pattern indicators are at the end of the landing runway indicators and are angled out at 90°. These indicate the direction of turn from base to final. Thus, the airport traffic is left-hand for Runway 36 and right-hand for Runway 18. It is also left-hand for Runway 9 and right-hand for Runway 27.
    Answer (B) is incorrect. Runway 18 is right, not left, and Runway 36 is left, not right. Answer (C) is incorrect. Runway 9 is left, not right, and Runway 27 is right, not left.

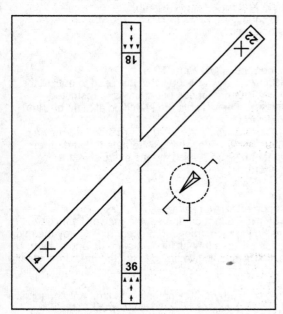

Figure 50. – Airport Diagram.

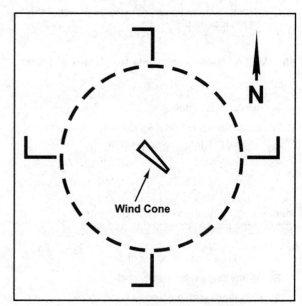

Figure 51. – Airport Landing Indicator.

29. (Refer to Figure 51 above.) The traffic patterns indicated in the segmented circle have been arranged to avoid flights over an area to the

    A. south of the airport.

    B. north of the airport.

    C. southeast of the airport.

Answer (C) is correct. *(AIM Para 4-3-4)*
    DISCUSSION: The traffic patterns indicated in the segmented circle depicted in Fig. 51 have been arranged to avoid flights over an area to the southeast of the airport. All departures from the runways are to the north or west. All approaches to the airport indicate a pattern of arrival from 180° clockwise to 90°, leaving the southeastern quadrant free of flight.
    Answer (A) is incorrect. Arrivals on Runway 36 and departures on Runway 18 result in traffic to the south. Answer (B) is incorrect. Runway 9-27 produces traffic to the north in addition to Runway 36 departures and Runway 18 arrivals.

30. (Refer to Figure 51 above.) The segmented circle indicates that a landing on Runway 26 will be with a

    A. right-quartering headwind.

    B. left-quartering headwind.

    C. right-quartering tailwind.

Answer (A) is correct. *(AIM Para 4-3-4)*
    DISCUSSION: The wind cone at the center of the segmented circle depicted in Fig. 51 indicates that a landing on Runway 26 will be with a right-quartering headwind. The large end of the wind cone is pointing to the direction from which the wind is coming, i.e., a northwest headwind on the right quarter of an airplane landing from the east to the west.
    Answer (B) is incorrect. A left-quartering headwind would be encountered landing on Runway 35. Answer (C) is incorrect. A right-quartering tailwind would be encountered landing on Runway 17.

31. (Refer to Figure 51 on page 84.) Which runway and traffic pattern should be used as indicated by the wind cone in the segmented circle?

    A. Right-hand traffic on Runway 9.

    B. Right-hand traffic on Runway 18.

    C. Left-hand traffic on Runway 36.

Answer (C) is correct. *(AIM Para 4-3-4)*
    DISCUSSION: The appropriate traffic pattern and runway, given a wind from the northwest (Fig. 51), is left-hand traffic on Runway 36, which would have a quartering headwind.
    Answer (A) is incorrect. Runway 9 uses a left-hand pattern. Also, this would be a tailwind landing. Answer (B) is incorrect. Even though there is right traffic on Runway 18, this would be a tailwind landing.

32. (Refer to Figure 50 on page 84.) If the wind is as shown by the landing direction indicator, the pilot should land on

    A. Runway 18 and expect a crosswind from the right.

    B. Runway 22 directly into the wind.

    C. Runway 36 and expect a crosswind from the right.

Answer (A) is correct. *(AIM Para 4-3-4)*
    DISCUSSION: Given a wind as shown by the landing direction indicator in Fig. 50, the pilot should land to the south on Runway 18 and expect a crosswind from the right. The tetrahedron points to the wind that is from the southwest.
    Answer (B) is incorrect. Runways 4 and 22 are closed, as indicated by the X at each end of the runway. Answer (C) is incorrect. The wind is from the southwest (not the northeast). The landing should be into the wind.

33. (Refer to Figure 50 on page 84.) The arrows that appear on the end of the north/south runway indicate that the area

    A. may be used only for taxiing.

    B. is usable for taxiing, takeoff, and landing.

    C. cannot be used for landing, but may be used for taxiing and takeoff.

Answer (C) is correct. *(AIM Para 2-3-3)*
    DISCUSSION: The arrows that appear on the end of the north/south runway (displaced thresholds) as shown in Fig. 50 indicate that the area cannot be used for landing but may be used for taxiing, takeoff, and the landing rollout.
    Answer (A) is incorrect. Takeoffs as well as taxiing are permitted. Answer (B) is incorrect. Landings are not permitted on the area before the displaced threshold.

34. (Refer to Figure 50 on page 84.) Select the proper traffic pattern and runway for landing.

    A. Left-hand traffic and Runway 18.

    B. Right-hand traffic and Runway 18.

    C. Left-hand traffic and Runway 22.

Answer (B) is correct. *(AIM Para 4-3-4)*
    DISCUSSION: The tetrahedron indicates wind direction by pointing into the wind. On Fig. 50, Runways 4 and 22 are closed, as indicated by the X at each end of the runway. Accordingly, with the wind from the southwest, the landing should be made on Runway 18. Runway 18 has right-hand traffic, as indicated by the traffic pattern indicator at a 90° angle to the landing runway indicator in the segmented circle.
    Answer (A) is incorrect. Runway 18 uses a right-hand (not left-hand) pattern. Answer (C) is incorrect. The X markings indicate that Runways 4 and 22 are closed.

35. The recommended entry position to an airport traffic pattern is

    A. 45° to the base leg just below traffic pattern altitude.

    B. to enter 45° at the midpoint of the downwind leg at traffic pattern altitude.

    C. to cross directly over the airport at traffic pattern altitude and join the downwind leg.

Answer (B) is correct. *(AIM Para 4-3-3)*
    DISCUSSION: The recommended entry position to an airport traffic pattern is to enter 45° at the midpoint of the downwind leg at traffic pattern altitude.
    Answer (A) is incorrect. The recommended entry to an airport traffic pattern is to enter 45° at the midpoint of the downwind, not base, leg and at traffic pattern altitude, not below. Answer (C) is incorrect. The recommended entry to an airport traffic pattern is to enter 45° at the midpoint of the downwind, not to cross directly over the airport and join the downwind leg. Also, flying at traffic pattern altitude directly over an airport is an example of poor judgment in collision avoidance precautions.

3.5 Visual Approach Slope Indicators (VASI)

36. An on glide slope indication from a tri-color VASI is

A.   a white light signal.

B.   a green light signal.

C.   an amber light signal.

Answer (B) is correct.   *(AIM Para 2-1-2)*
   DISCUSSION: Tri-color visual approach slope indicators normally consist of a single light unit projecting a 3-color visual approach path into the final approach area of the runway, upon which the indicator is installed. The below glide path indicator is red. The above glide path indicator is amber. The on glide path indicator is green. This type of indicator has a useful range of approximately 1/2 to 1 mi. in daytime and up to 5 mi. at night.
   Answer (A) is incorrect. Tri-color VASI does not emit a white light. Answer (C) is incorrect. Amber indicates above (not on) the glide slope.

37. An above glide slope indication from a tri-color VASI is

A.   a white light signal.

B.   a green light signal.

C.   an amber light signal.

Answer (C) is correct.   *(AIM Para 2-1-2)*
   DISCUSSION: The tri-color VASI has three lights: amber for above the glide slope, green for on the glide slope, and red for below the glide slope.
   Answer (A) is incorrect. Tri-color VASI does not emit a white light. Answer (B) is incorrect. A green light means on (not above) the glide slope.

38. A below glide slope indication from a tri-color VASI is a

A.   red light signal.

B.   pink light signal.

C.   green light signal.

Answer (A) is correct.   *(AIM Para 2-1-2)*
   DISCUSSION: The tri-color VASI has three lights: amber for above the glide slope, green for on the glide slope, and red for below the glide slope.
   Answer (B) is incorrect. A pink light may be seen on a pulsating (not tri-color) VASI when in the area of on, to slightly below, the glide slope. Answer (C) is incorrect. A green light means you are on the glide slope.

39. A below glide slope indication from a pulsating approach slope indicator is a

A.   pulsating white light.

B.   steady white light.

C.   pulsating red light.

Answer (C) is correct.   *(AIM Para 2-1-2)*
   DISCUSSION: A pulsating VASI indicator normally consists of a single light unit projecting a two-color visual approach path into the final approach area of the runway upon which the indicator is installed. The below glide slope indication is a pulsating red, the above glide slope is pulsating white, and the on glide slope is a steady white light. The useful range of this system is about 4 mi. during the day and up to 10 mi. at night.
   Answer (A) is incorrect. A pulsating white light is an above glide slope indication. Answer (B) is incorrect. Steady white is the on glide slope indication.

40. When approaching to land on a runway served by a visual approach slope indicator (VASI), the pilot shall

A.   maintain an altitude that captures the glide slope at least 2 miles downwind from the runway threshold.

B.   maintain an altitude at or above the glide slope.

C.   remain on the glide slope and land between the two-light bar.

Answer (B) is correct.   *(FAR 91.129)*
   DISCUSSION: An airplane approaching to land on a runway served by a VASI shall maintain an altitude at or above the glide slope until a lower altitude is necessary for a safe landing.
   Answer (A) is incorrect. A VASI should not be used for descent until the airplane is visually lined up with the runway. Answer (C) is incorrect. It is unsafe to concentrate on the VASI after nearing the approach end of the runway; i.e., turn your attention to landing the airplane.

41. (Refer to Figure 48 below.) While on final approach to a runway equipped with a standard 2-bar VASI, the lights appear as shown by illustration D. This means that the aircraft is

    A. above the glide slope.

    B. below the glide slope.

    C. on the glide slope.

Answer (B) is correct. *(AIM Para 2-1-2)*
    DISCUSSION: In illustration D of Fig. 48, both rows of lights are red. Thus, the aircraft is below the glide path. Remember, "red means dead."
    Answer (A) is incorrect. If the airplane is above the glide path, the lights would both show white, as indicated by illustration C. Answer (C) is incorrect. If the airplane is on the glide path, the lights would be red over white, as indicated by illustration A.

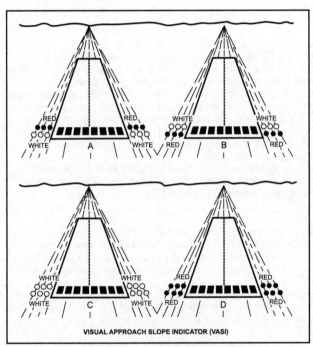

Figure 48. – VASI Illustrations.

42. (Refer to Figure 48 above.) VASI lights as shown by illustration C indicate that the airplane is

    A. off course to the left.

    B. above the glide slope.

    C. below the glide slope.

Answer (B) is correct. *(AIM Para 2-1-2)*
    DISCUSSION: In illustration C of Fig. 48, both rows of lights are white, which means the airplane is above the glide path.
    Answer (A) is incorrect. The VASI does not alert a pilot as to runway alignment, but a pilot who is excessively to the left or right may not be able to see the VASI lights at all. Answer (C) is incorrect. If the airplane is below the glide path, both rows of lights would show red, as indicated by illustration D.

43. (Refer to Figure 48 above.) Illustration A indicates that the aircraft is

    A. below the glide slope.

    B. on the glide slope.

    C. above the glide slope.

Answer (B) is correct. *(AIM Para 2-1-2)*
    DISCUSSION: Illustration A indicates that the airplane is on the glide path (glide slope). The basic principle of the VASI is that of color differentiation between red and white. Each light unit projects a beam of light having a white segment in the upper part and a red segment in the lower part of the beam. Thus, to be on the glide slope you need to be on the lower part of the far light (red) and on the upper part of the near light (white).
    Answer (A) is incorrect. If the airplane is below the glide path, both rows of lights will be red, as indicated in D. Answer (C) is incorrect. If the aircraft is above the glide path, both lights will be white, as indicated in C.

44. While operating in class D airspace, each pilot of an aircraft approaching to land on a runway served by a visual approach slope indicator (VASI) shall

A. maintain a 3° glide until approximately 1/2 mile to the runway before going below the VASI.

B. maintain an altitude at or above the glide slope until a lower altitude is necessary for a safe landing.

C. stay high until the runway can be reached in a power-off landing.

Answer (B) is correct.   *(FAR 91.129)*
DISCUSSION: When approaching to land on a runway served by a VASI, each pilot of an airplane must fly at or above the VASI glide path until a lower altitude is necessary for a safe landing.
Answer (A) is incorrect. A VASI may be adjusted to provide a glide slope more or less than 3°. Answer (C) is incorrect. Higher than the VASI glide path is not required.

45. A slightly high glide slope indication from a precision approach path indicator is

A. four white lights.

B. three white lights and one red light.

C. two white lights and two red lights.

Answer (B) is correct.   *(AIM Para 2-1-2)*
DISCUSSION: A precision approach path indicator (PAPI) has a row of four lights, each of which is similar to a VASI in that they emit a red or white light. Above the glide slope (more than 3.5°) is indicated by four white lights, a slightly above glide slope (3.2°) is indicated by three white lights and one red light, on glide slope (3°) is indicated by two white and two red lights, slightly below glide slope (2.8°) is indicated by one white and three red lights, and below (too low) the glide slope (less than 2.5°) is indicated by four red lights.
Answer (A) is incorrect. Four white lights is a high or more than 3.5° glide slope. Answer (C) is incorrect. Two white and two red lights is an on glide slope (3°).

46. Each pilot of an aircraft approaching to land on a runway served by a visual approach slope indicator (VASI) shall

A. maintain a 3° glide to the runway.

B. maintain an altitude at or above the glide slope.

C. stay high until the runway can be reached in a power-off landing.

Answer (B) is correct.   *(FAR 91.129)*
DISCUSSION: When approaching to land on a runway served by a VASI, each pilot of an airplane must fly at or above the VASI glide path until a lower altitude is necessary for a safe landing.
Answer (A) is incorrect. A VASI may be adjusted to provide a glide slope more or less than 3°. Answer (C) is incorrect. Higher than the VASI glide path is not required.

47. Which approach and landing objective is assured when the pilot remains on the proper glidepath of the VASI?

A. Continuation of course guidance after transition to VFR.

B. Safe obstruction clearance in the approach area.

C. Course guidance from the visual descent point to touchdown.

Answer (B) is correct.   *(AIM Para 2-1-2)*
DISCUSSION: The visual approach slope indicator (VASI) provides safe obstruction clearance within ± 10° of the extended runway centerline out to 4 NM. Pilots are advised to remain on the VASI-directed glide path throughout the entire approach to ensure obstruction clearance.
Answer (A) is incorrect. The VASI provides visual descent guidance, not course guidance. Course guidance implies lateral as well as vertical guidance. Answer (C) is incorrect. The VASI provides visual descent guidance, not course guidance. Course guidance implies lateral as well as vertical guidance.

3.6 Wake Turbulence

48. Wingtip vortices are created only when an aircraft is

A. operating at high airspeeds.

B. heavily loaded.

C. developing lift.

Answer (C) is correct.   *(AIM Para 7-3-4)*
DISCUSSION: Wingtip vortices are the result of the pressure differential over and under a wing when that wing is producing lift. Wingtip vortices do not develop when an airplane is taxiing, although prop blast or jet thrust turbulence can be experienced near the rear of a large airplane that is taxiing.
Answer (A) is incorrect. The greatest turbulence is produced from an airplane operating at a slow airspeed. Answer (B) is incorrect. Even though a heavily loaded airplane may produce greater turbulence, an airplane does not have to be heavily loaded in order to produce wingtip vortices. Wingtip vortices are produced only when an airplane is developing lift.

49. Wingtip vortices created by large aircraft tend to

    A. sink below the aircraft generating turbulence.

    B. rise into the traffic pattern.

    C. rise into the takeoff or landing path of a crossing runway.

Answer (A) is correct. *(AIM Para 7-3-4)*
    DISCUSSION: Wingtip vortices created by large airplanes tend to sink below the airplane generating the turbulence.
    Answer (B) is incorrect. Wingtip vortices sink, not rise. Answer (C) is incorrect. Wingtip vortices do not rise or gain altitude, but sink toward the ground. However, they may move horizontally left or right depending on crosswind conditions.

50. How does the wake turbulence vortex circulate around each wingtip?

    A. Inward, upward, and around each tip.

    B. Inward, upward, and counterclockwise.

    C. Outward, upward, and around each tip.

Answer (C) is correct. *(AIM Para 7-3-4)*
    DISCUSSION: Since the pressure differential is caused by a lower pressure above the wing and a higher pressure below the wing, the air from the bottom moves out, up, and around each wingtip.
    Answer (A) is incorrect. The air moves out around the edge of the wing, not in underneath the wing. Answer (B) is incorrect. The air moves out around the edge of the wing. From behind, the left wingtip vortex is clockwise and the right wingtip vortex is counterclockwise.

51. When taking off or landing at an airport where heavy aircraft are operating, one should be particularly alert to the hazards of wingtip vortices because this turbulence tends to

    A. rise from a crossing runway into the takeoff or landing path.

    B. rise into the traffic pattern area surrounding the airport.

    C. sink into the flightpath of aircraft operating below the aircraft generating the turbulence.

Answer (C) is correct. *(AIM Para 7-3-4)*
    DISCUSSION: When taking off or landing at a busy airport where large, heavy airplanes are operating, you should be particularly alert to the hazards of wingtip vortices because this turbulence tends to sink into the flight paths of airplanes operating below the airplane generating the turbulence. Wingtip vortices are caused by a differential in high and low pressure at the wingtip of an airplane, creating a spiraling effect trailing behind the wingtip, similar to a horizontal tornado.
    Answer (A) is incorrect. Wingtip vortices always trail behind an airplane and descend toward the ground. However, they do drift with the wind and will not stay directly behind an airplane if there is a crosswind. Answer (B) is incorrect. Wingtip vortices sink, not rise.

52. The greatest vortex strength occurs when the generating aircraft is

    A. light, dirty, and fast.

    B. heavy, dirty, and fast.

    C. heavy, clean, and slow.

Answer (C) is correct. *(AIM Para 7-3-3)*
    DISCUSSION: Vortices are the greatest when the wingtips are at high angles of attack. This occurs at high gross weight, flaps up, and low airspeed (heavy, clean, and slow).
    Answer (A) is incorrect. Light aircraft produce less vortex turbulence than heavy aircraft. The use of flaps, spoilers, etc., (i.e., dirty) diminishes vortex turbulence. Answer (B) is incorrect. Being dirty and/or fast causes the wingtip to be at a lower angle of attack, presenting less of a danger than when clean and/or slow.

53. The wind condition that requires maximum caution when avoiding wake turbulence on landing is a

    A. light, quartering headwind.

    B. light, quartering tailwind.

    C. strong headwind.

Answer (B) is correct. *(AIM Para 7-3-4)*
    DISCUSSION: The most dangerous wind condition when avoiding wake turbulence on landing is a light, quartering tailwind. The tailwind can push the vortices forward, which could put it in the touchdown zone of your aircraft even if you used proper procedures and landed beyond the touchdown point of the preceding aircraft. Also, the quartering wind may push the upwind vortices to the middle of the runway.
    Answer (A) is incorrect. Headwinds push the vortices out of your touchdown zone if you land beyond the touchdown point of the preceding aircraft. Answer (C) is incorrect. Strong winds help diffuse wake turbulence vortices.

54. When departing behind a heavy aircraft, the pilot should avoid wake turbulence by maneuvering the aircraft

    A. below and downwind from the heavy aircraft.

    B. above and upwind from the heavy aircraft.

    C. below and upwind from the heavy aircraft.

Answer (B) is correct. *(AIM Para 7-3-6)*
    DISCUSSION: The proper procedure for departing behind a large aircraft is to rotate prior to the large aircraft's rotation point, then fly above and upwind of the large aircraft. Since vortices sink and drift downwind, this should keep you clear.
    Answer (A) is incorrect. You should remain above and upwind from the heavy aircraft. Answer (C) is incorrect. You should fly above the flight path of the large aircraft to avoid the sinking vortices.

55. When landing behind a large aircraft, the pilot should avoid wake turbulence by staying

    A.   above the large aircraft's final approach path and landing beyond the large aircraft's touchdown point.

    B.   below the large aircraft's final approach path and landing before the large aircraft's touchdown point.

    C.   above the large aircraft's final approach path and landing before the large aircraft's touchdown point.

Answer (A) is correct. *(AIM Para 7-3-6)*
    DISCUSSION: When landing behind a large aircraft, your flight path should be above the other aircraft's flight path since the vortices sink. When the aircraft touches down, the vortices will stop, so you should thus touch down beyond where the large aircraft did.
    Answer (B) is incorrect. Below the flight path, you will fly through the sinking vortices generated by the large aircraft. Answer (C) is incorrect. By landing before the large aircraft's touchdown point, you will have to fly below the preceding aircraft's flight path.

56. When landing behind a large aircraft, which procedure should be followed for vortex avoidance?

    A.   Stay above its final approach flightpath all the way to touchdown.

    B.   Stay below and to one side of its final approach flightpath.

    C.   Stay well below its final approach flightpath and land at least 2,000 feet behind.

Answer (A) is correct. *(AIM Para 7-3-6)*
    DISCUSSION: When landing behind a large aircraft, stay above its final approach flight path all the way to touchdown; i.e., touch down beyond the touchdown point of the large aircraft.
    Answer (B) is incorrect. You should stay at or above, not below, its flight path. Answer (C) is incorrect. You should stay at or above, not below, its flight path, and land beyond, not behind, its touchdown point.

3.7 Collision Avoidance

57. During a night flight, you observe a steady red light and a flashing red light ahead and at the same altitude. What is the general direction of movement of the other aircraft?

    A.   The other aircraft is crossing to the left.

    B.   The other aircraft is crossing to the right.

    C.   The other aircraft is approaching head-on.

Answer (A) is correct. *(AFH Chap 10)*
    DISCUSSION: Airplane position lights consist of a steady red light on the left wing (looking forward), a green light on the right wing, and a white light on the tail. Accordingly, if you observe a steady red light, you are looking at the tip of a left wing, which means the other plane is traveling from your right to left (crossing to the left). The red flashing light is the beacon.
    Answer (B) is incorrect. If the airplane were crossing to the right, you would see a steady green light. Answer (C) is incorrect. If the airplane were approaching head-on, you would see both the red and the green lights.

58. During a night flight, you observe a steady white light and a flashing red light ahead and at the same altitude. What is the general direction of movement of the other aircraft?

    A.   The other aircraft is flying away from you.

    B.   The other aircraft is crossing to the left.

    C.   The other aircraft is crossing to the right.

Answer (A) is correct. *(AFH Chap 10)*
    DISCUSSION: A steady white light (the tail light) indicates the other airplane is moving away from you. The flashing red light is the beacon light.
    Answer (B) is incorrect. You would observe a red light if another plane were crossing to your left. Answer (C) is incorrect. You would observe a green light if another airplane were crossing to your right.

59. During a night flight, you observe steady red and green lights ahead and at the same altitude. What is the general direction of movement of the other aircraft?

    A.   The other aircraft is crossing to the left.

    B.   The other aircraft is flying away from you.

    C.   The other aircraft is approaching head-on.

Answer (C) is correct. *(AFH Chap 10)*
    DISCUSSION: If you observe steady red and green lights at the same altitude, the other airplane is approaching head-on. You should take evasive action to the right.
    Answer (A) is incorrect. If the airplane were crossing to the left, you would observe only a red light. Answer (B) is incorrect. If the other airplane were headed away from you, you would observe a white (tail) light.

60. The most effective method of scanning for other aircraft for collision avoidance during daylight hours is to use

A. regularly spaced concentration on the 3-, 9-, and 12-o'clock positions.

B. a series of short, regularly spaced eye movements to search each 10-degree sector.

C. peripheral vision by scanning small sectors and utilizing offcenter viewing.

Answer (B) is correct. *(AC 90-48C)*
DISCUSSION: The most effective way to scan for other aircraft during daylight hours is to use a series of short, regularly spaced eye movements that bring successive areas of the sky into your central visual field. Each movement should not exceed 10°, and each area should be observed for at least one second to enable detection. Only a very small center area of the eye has the ability to send clear, sharply focused messages to the brain. All other areas provide less detail.
Answer (A) is incorrect. The spacing between the positions should be 10°, not 90°. Answer (C) is incorrect. This is the recommended nighttime scanning procedure.

61. The most effective method of scanning for other aircraft for collision avoidance during nighttime hours is to use

A. regularly spaced concentration on the 3-, 9-, and 12-o'clock positions.

B. a series of short, regularly spaced eye movements to search each 30-degree sector.

C. peripheral vision by scanning small sectors and utilizing offcenter viewing.

Answer (C) is correct. *(AC 90-48C)*
DISCUSSION: At night, collision avoidance scanning must use the off-center portions of the eyes. These portions are most effective at seeing objects at night. Accordingly, peripheral vision should be used, scanning small sectors and using off-center viewing. This is in contrast to daytime searching for air traffic, when center viewing should be used.
Answer (A) is incorrect. All areas (up, below, and on all sides) should be scanned for other air traffic. Answer (B) is incorrect. Smaller than 30° sectors should be scanned.

62. How can you determine if another aircraft is on a collision course with your aircraft?

A. The other aircraft will always appear to get larger and closer at a rapid rate.

B. The nose of each aircraft is pointed at the same point in space.

C. There will be no apparent relative motion between your aircraft and the other aircraft.

Answer (C) is correct. *(AIM Para 8-1-8)*
DISCUSSION: Any aircraft that appears to have no relative motion and stays in one scan quadrant is likely to be on a collision course. Also, if a target shows no lateral or vertical motion but increases in size, take evasive action.
Answer (A) is incorrect. Aircraft on collision courses may not always appear to grow larger and/or to close at a rapid rate. Frequently, the degree of proximity cannot be detected. Answer (B) is incorrect. You may not be able to tell in exactly which direction the other airplane is pointed. Even if you could determine the direction of the other airplane, you may not be able to accurately project the flight paths of the two airplanes to determine if they indeed point to the same point in space and will arrive there at the same time (i.e., collide).

63. Prior to starting each maneuver, pilots should

A. check altitude, airspeed, and heading indications.

B. visually scan the entire area for collision avoidance.

C. announce their intentions on the nearest CTAF.

Answer (B) is correct. *(AIM Para 4-4-14)*
DISCUSSION: Prior to each maneuver, a pilot should visually scan the entire area for collision avoidance. Many maneuvers require a clearing turn, which should be used for this purpose.
Answer (A) is incorrect. Altitude, speed, and heading may not all be critical to every maneuver. Collision avoidance is! Answer (C) is incorrect. CTAF is used for operations at an uncontrolled airport, not for pilots doing maneuvers away from an airport.

64. What procedure is recommended when climbing or descending VFR on an airway?

A. Execute gentle banks left and right for continuous visual scanning of the airspace.

B. Advise the nearest FSS of the altitude changes.

C. Fly away from the centerline of the airway before changing altitude.

Answer (A) is correct. *(AC 90-48C)*
DISCUSSION: When climbing (descending) VFR on an airway, you should execute gentle banks left and right to facilitate scanning for other aircraft. Collision avoidance is a constant priority and especially pertinent to climbs and descents on airways where other traffic is expected.
Answer (B) is incorrect. An FSS provides no en route traffic service. Answer (C) is incorrect. It is not necessary to leave the center of the airway, only to scan for other aircraft.

65. Responsibility for collision avoidance in an alert area rests with

A. the controlling agency.

B. all pilots.

C. Air Traffic Control.

Answer (B) is correct. *(AIM Para 3-4-6)*
DISCUSSION: Alert areas may contain a high volume of pilot training or other unusual activity. Pilots using the area as well as pilots crossing the area are equally responsible for collision avoidance.
Answer (A) is incorrect. Pilots are responsible for collision avoidance, not controlling agencies. Answer (C) is incorrect. Pilots are responsible for collision avoidance, not ATC.

66. Most midair collision accidents occur during

A. hazy days.

B. clear days.

C. cloudy nights.

Answer (B) is correct. *(AC 90-48C)*
DISCUSSION: Most midair collision accidents and reported near midair collision incidents occur during good VFR weather conditions (i.e., clear days) and during the hours of daylight. This is when more aircraft are likely to be flying.
Answer (A) is incorrect. During hazy days, fewer pilots will be flying, and those who are will be more vigilant in their scanning for other traffic. Answer (C) is incorrect. During cloudy nights, fewer pilots will be flying, and those who are will be more vigilant in their scanning for other traffic.

67. Pilots are encouraged to turn on their landing lights when operating below 10,000 feet, day or night, and especially when operating

A. within 5 miles of a controlled airport.

B. in conditions of reduced visibility.

C. in Class B airspace.

Answer (B) is correct. *(AIM Para 4-3-23)*
DISCUSSION: The FAA has a voluntary pilot safety program, Operation Lights On, to enhance the see-and-avoid concept. Pilots are encouraged to turn on their landing lights during takeoff and when operating below 10,000 feet, day or night, especially when operating within 10 miles of any airport, or in conditions of reduced visibility and in areas where flocks of birds may be expected.
Answer (A) is incorrect. Pilots are encouraged to turn on their landing lights when operating within 10 miles of any airport, not within 5 miles of a controlled airport. Answer (C) is incorrect. Pilots are encouraged to turn on their landing lights during takeoff and when operating below 10,000 feet, day or night, especially when operating within 10 miles of any airport, or in conditions of reduced visibility and in areas where flocks of birds may be expected.

3.8 ATIS and Ground Control

68. After landing at a tower-controlled airport, when should the pilot contact ground control?

A. When advised by the tower to do so.

B. Prior to turning off the runway.

C. After reaching a taxiway that leads directly to the parking area.

Answer (A) is correct. *(AIM Para 4-3-20)*
DISCUSSION: After landing at a tower-controlled airport, you should contact ground control on the appropriate frequency only when instructed by the tower.
Answer (B) is incorrect. A pilot should not change frequencies unless instructed to do so by the tower. Sometimes the tower controller will be handling both tower and ground frequencies. Switching without permission may be confusing to ATC. Answer (C) is incorrect. A pilot should not change frequencies unless instructed to do so by the tower. Sometimes the tower controller will be handling both tower and ground frequencies. Switching without permission may be confusing to ATC.

69. If instructed by ground control to taxi to Runway 9, the pilot may proceed

A. via taxiways and across runways to, but not onto, Runway 9.

B. to the next intersecting runway where further clearance is required.

C. via taxiways and across runways to Runway 9, where an immediate takeoff may be made.

Answer (B) is correct. *(AIM Para 4-3-18)*
DISCUSSION: A taxi clearance from ATC authorizes the pilot to utilize taxiways along the taxi route, but a specific crossing clearance must be issued for all runways along the route.
Answer (A) is incorrect. A clearance to taxi to the active runway means a pilot has been given permission to taxi via taxiways to, but not onto, the active runway. ATC must issue a specific clearance to cross any runway along the taxi route. Answer (C) is incorrect. The clearance to taxi to a runway does not permit taxiing onto the active runway.

70. Automatic Terminal Information Service (ATIS) is the continuous broadcast of recorded information concerning

    A. pilots of radar-identified aircraft whose aircraft is in dangerous proximity to terrain or to an obstruction.

    B. nonessential information to reduce frequency congestion.

    C. noncontrol information in selected high-activity terminal areas.

Answer (C) is correct. *(AIM Para 4-1-13)*
    DISCUSSION: The continuous broadcast of recorded noncontrol information is known as the Automatic Terminal Information Service (ATIS). ATIS includes weather, active runway, and other information that arriving and departing pilots need to know.
    Answer (A) is incorrect. A controller who has a radar-identified aircraft under his/her control will issue a terrain or obstruction alert to an aircraft that is in dangerous proximity to terrain or to an obstruction. Answer (B) is incorrect. ATIS is considered essential (not nonessential) information, but routine, i.e., noncontrol.

71. Absence of the sky condition and visibility on an ATIS broadcast indicates that

    A. weather conditions are at or above VFR minimums.

    B. the sky condition is clear and visibility is unrestricted.

    C. the ceiling is at least 5,000 feet and visibility is 5 miles or more.

Answer (C) is correct. *(AIM Para 4-1-13)*
    DISCUSSION: The ceiling/sky condition, visibility, and obstructions to vision may be omitted from the ATIS broadcast if the ceiling is above 5,000 ft. with visibility more than 5 statute miles.
    Answer (A) is incorrect. The absence of the sky condition and visibility on an ATIS broadcast implies that the ceiling is above 5,000 ft. and the visibility is more than 5 statute miles. Answer (B) is incorrect. The absence of the sky condition and visibility on an ATIS broadcast implies that the ceiling is above 5,000 ft., not clear, and the visibility is more than 5 SM, not unrestricted.

3.9 Class D Airspace and Airport Advisory Area

72. A blue segmented circle on a Sectional Chart depicts which class airspace?

    A. Class B.

    B. Class C.

    C. Class D.

Answer (C) is correct. *(AIM Para 3-2-5)*
    DISCUSSION: A blue segmented circle on a sectional chart depicts Class D airspace.
    Answer (A) is incorrect. Class B airspace is depicted on a sectional chart by a solid, not segmented, blue circle. Answer (B) is incorrect. Class C airspace is depicted on a sectional chart by a solid magenta, not a blue segmented, circle.

73. Airspace at an airport with a part-time control tower is classified as Class D airspace only

    A. when the weather minimums are below basic VFR.

    B. when the associated control tower is in operation.

    C. when the associated Flight Service Station is in operation.

Answer (B) is correct. *(AIM Para 3-2-5)*
    DISCUSSION: A Class D airspace area is automatically in effect when and only when the associated part-time control tower is in operation regardless of weather conditions, availability of radar services, or time of day. Airports with part-time operating towers only have a part-time Class D airspace area.
    Answer (A) is incorrect. A Class D airspace area is automatically in effect when the tower is in operation, regardless of the weather conditions. Answer (C) is incorrect. A Class D airspace area is in effect when the associated control tower, not FSS, is in operation.

74. When a control tower located on an airport within Class D airspace ceases operation for the day, what happens to the airspace designation?

    A. The airspace designation normally will not change.

    B. The airspace remains Class D airspace as long as a weather observer or automated weather system is available.

    C. The airspace reverts to Class E or a combination of Class E and G airspace during the hours the tower is not in operation.

Answer (C) is correct. *(AIM Para 3-2-5)*
    DISCUSSION: When a tower ceases operation, the Class D airspace reverts to Class E or a combination of Class G and Class E.
    Answer (A) is incorrect. Class D airspace is designated when there is an operating control tower. When the tower ceases operation for the day, the airspace reverts to Class E or a combination of Class G and Class E airspace. Answer (B) is incorrect. The airspace reverts to Class E, not Class D, when the tower ceases operation for the day and an approved weather observer or automated weather system is available.

75. A non-tower satellite airport, within the same Class D airspace as that designated for the primary airport, requires radio communications be established and maintained with the

A. satellite airport's UNICOM.

B. associated Flight Service Station.

C. primary airport's control tower.

Answer (C) is correct. *(AIM Para 3-2-5)*
DISCUSSION: Each pilot departing a non-tower satellite airport, within Class D airspace, must establish and maintain two-way radio communications with the primary airport's control tower as soon as practicable after departing.
Answer (A) is incorrect. When departing a satellite airport without an operating control tower in Class D airspace, you must establish and maintain two-way radio communications with the primary airport's control tower, not the satellite airport's UNICOM. Answer (B) is incorrect. When departing a satellite airport without an operating control tower in Class D airspace, you must establish and maintain two-way radio communications with the primary airport's control tower, not the associated FSS.

76. Unless otherwise authorized, two-way radio communications with Air Traffic Control are required for landings or takeoffs

A. at all tower controlled airports regardless of weather conditions.

B. at all tower controlled airports only when weather conditions are less than VFR.

C. at all tower controlled airports within Class D airspace only when weather conditions are less than VFR.

Answer (A) is correct. *(FAR 91.129)*
DISCUSSION: Two-way radio communications with air traffic control (ATC) are required for landing and taking off at all tower controlled airports, regardless of weather conditions. However, light signals from the tower may be used during radio failure.
Answer (B) is incorrect. Radio communication is also required in VFR weather as well as IFR weather at all tower-controlled airports. Answer (C) is incorrect. Radio communication is required in both VFR and IFR weather when landing at or taking off at all tower-controlled airports within Class D airspace.

77. The lateral dimensions of Class D airspace are based on

A. the number of airports that lie within the Class D airspace.

B. 5 statute miles from the geographical center of the primary airport.

C. the instrument procedures for which the controlled airspace is established.

Answer (C) is correct. *(AIM Para 3-2-5)*
DISCUSSION: The lateral dimensions of Class D airspace are based upon the instrument procedures for which the controlled airspace is established.
Answer (A) is incorrect. While the FAA will attempt to exclude satellite airports as much as possible from Class D airspace, the major criteria for the lateral dimension will be based on the instrument procedures for which the controlled airspace is established. Answer (B) is incorrect. The lateral dimensions of Class D airspace are based on the instrument procedures for which the Class D airspace is established, not a specified radius from the primary airport.

78. If a control tower and an FSS are located on the same airport, which function is provided by the FSS during those periods when the tower is closed?

A. Automatic closing of the IFR flight plan.

B. Approach control services.

C. Airport Advisory Service.

Answer (C) is correct. *(AIM Para 4-1-9)*
DISCUSSION: Flight Service Stations (FSS) co-located on airport provide Local Airport Advisory (LAA) service. When located at an airport with a control tower, the FSS provides this service when the control tower is not in operation.
Answer (A) is incorrect. Flight Service Stations do not automatically close flight plans. Rather, you as the pilot must request this service. Answer (B) is incorrect. Flight Service Stations do not provide approach control service. Rather, they provide pilot advisory services.

79. Prior to entering an Airport Advisory Area, a pilot should

A. monitor ATIS for weather and traffic advisories.

B. contact approach control for vectors to the traffic pattern.

C. contact the local FSS for airport and traffic advisories.

Answer (C) is correct. *(AIM Para 4-1-9)*
DISCUSSION: Airport Advisory Areas exist at noncontrolled airports that have a Flight Service Station (FSS) located on that airport. The FSS provides advisory (not control) information on traffic, weather, etc., to requesting aircraft. Accordingly, pilots should (not must) contact FSSs for advisory services.
Answer (A) is incorrect. ATIS (automatic terminal information service) provides prerecorded weather and airport data but not traffic advisories. Answer (B) is incorrect. Approach control may provide vectors to the airport, but the controller will instruct you to switch to the advisory frequency for airport and traffic advisories.

80. When should pilots state their position on the airport when calling the tower for takeoff?

    A. When visibility is less than 1 mile.

    B. When parallel runways are in use.

    C. When departing from a runway intersection.

Answer (C) is correct. *(AIM Para 4-3-10)*
    DISCUSSION: Intersection departures are often performed at busy, tower-controlled airports. When notifying the tower that you are ready for departure, you must inform the controller of your location so that (s)he can positively identify you before clearing you for takeoff.
    Answer (A) is incorrect. When visibility is less than 1 mile, the field will be operating under instrument flight rules. As a private pilot without an instrument rating, you will not be operating in these conditions. Answer (B) is incorrect. Controllers only require that you notify them of your position when departing from a runway intersection.

3.10 Class C Airspace

81. The normal radius of the outer area of Class C airspace is

    A. 5 nautical miles.

    B. 15 nautical miles.

    C. 20 nautical miles.

Answer (C) is correct. *(AIM Para 3-2-4)*
    DISCUSSION: Do not confuse the outer area (20-NM radius) with the shelf area (10-NM radius) of Class C airspace. Communication with ATC in the outer area of Class C airspace is recommended but not required to create an efficient flow of traffic through the area. Communication with ATC in the shelf area as well as in the inner area of Class C airspace is mandatory.
    Answer (A) is incorrect. This is the radius of the surface area of the Class C airspace, not the outer area. Answer (B) is incorrect. This is not a dimension used for any component of Class C airspace.

82. All operations within Class C airspace must be

    A. on a flight plan filed prior to arrival or departure.

    B. in communications with the responsible ATC facility.

    C. in an aircraft equipped with a transponder with automatic altitude reporting capability.

Answer (B) is correct. *(AIM Para 3-2-4)*
    DISCUSSION: All aircraft operating in Class C airspace are required to contact and establish communications with approach control on the appropriate frequency.
    Answer (A) is incorrect. Flight plans are not required in Class C airspace. Answer (C) is incorrect. ATC may allow an aircraft without a functioning transponder to enter Class C airspace.

83. The vertical limit of Class C airspace above the primary airport is normally

    A. 1,200 feet AGL.

    B. 3,000 feet AGL.

    C. 4,000 feet AGL.

Answer (C) is correct. *(AIM Para 3-2-4)*
    DISCUSSION: The vertical limit (ceiling) of Class C airspace is normally 4,000 ft. above the primary airport elevation.
    Answer (A) is incorrect. This is the floor, not the vertical limit, of the Class C airspace shelf area (5 to 10 NM from primary airport). Answer (B) is incorrect. The vertical limit of Class C airspace is normally 4,000 ft. AGL, not 3,000 ft. AGL, above the elevation of the primary airport.

84. Under what condition may an aircraft operate from a satellite airport within Class C airspace?

    A. The pilot must file a flight plan prior to departure.

    B. The pilot must monitor ATC until clear of the Class C airspace.

    C. The pilot must contact ATC as soon as practicable after takeoff.

Answer (C) is correct. *(AIM Para 3-2-4)*
    DISCUSSION: Aircraft departing from a satellite airport within Class C airspace with an operating control tower must establish and maintain two-way radio communication with the control tower and thereafter as instructed by ATC. When departing a satellite airport without an operating control tower, the pilot must contact and maintain two-way radio communication with ATC as soon as practicable after takeoff.
    Answer (A) is incorrect. Flight plans are not required in Class C airspace. Answer (B) is incorrect. The pilot must maintain communication with ATC, not just monitor ATC, in Class C airspace.

85. Which initial action should a pilot take prior to entering Class C airspace?

    A. Contact approach control on the appropriate frequency.

    B. Contact the tower and request permission to enter.

    C. Contact the FSS for traffic advisories.

Answer (A) is correct. *(AIM Para 3-2-4)*
    DISCUSSION: Prior to entering Class C airspace, a pilot must contact and establish communication with approach control on the appropriate frequency.
    Answer (B) is incorrect. The tower normally controls the air traffic in the traffic pattern, not the aircraft entering the Class C airspace area. Answer (C) is incorrect. The pilot should contact approach control, not FSS, prior to entering Class C airspace.

3.11 Terminal Radar Programs

86. TRSA Service in the terminal radar program provides

    A. IFR separation (1,000 feet vertical and 3 miles lateral) between all aircraft.

    B. warning to pilots when their aircraft are in unsafe proximity to terrain, obstructions, or other aircraft.

    C. sequencing and separation for participating VFR aircraft.

Answer (C) is correct. *(AIM Para 4-1-17)*
    DISCUSSION: TRSA service in the terminal radar program provides sequencing and separation for all participating VFR aircraft within the airspace defined as a Terminal Radar Service Area (TRSA). Pilot participation is urged but is not mandatory.
    Answer (A) is incorrect. TRSA service provides VFR aircraft with a 500-ft., not 1,000-ft., vertical clearance from other aircraft. Answer (B) is incorrect. TRSA service is for traffic advisories, separation between aircraft, and vectoring, not for obstruction clearance.

87. From whom should a departing VFR aircraft request radar traffic information during ground operations?

    A. Clearance delivery.

    B. Tower, just before takeoff.

    C. Ground control, on initial contact.

Answer (C) is correct. *(AIM Para 4-1-17)*
    DISCUSSION: Pilots of departing VFR aircraft are encouraged to request radar traffic information by notifying ground control on initial contact with their request and proposed direction of flight.
    Answer (A) is incorrect. Clearance delivery is usually used at busier airports where radar traffic information may be provided without request. Answer (B) is incorrect. Ground control rather than tower control is the appropriate place to make the request (giving ATC more time to coordinate your request).

88. Basic radar service in the terminal radar program is best described as

    A. safety alerts, traffic advisories, and limited vectoring to VFR aircraft.

    B. mandatory radar service provided by the Automated Radar Terminal System (ARTS) program.

    C. wind-shear warning at participating airports.

Answer (A) is correct. *(AIM Para 4-1-17)*
    DISCUSSION: Basic radar service in the terminal radar program provides safety alerts, traffic advisories, and limited vectoring (on a workload-permitting basis) to VFR aircraft.
    Answer (B) is incorrect. Mandatory radar service is required only in Class B and Class C airspace. Answer (C) is incorrect. The Low-Level Wind Shear Alert System (LLWAS) is based on information gathered from various wind (speed and direction) sensors on and around the airport, not on radar.

3.12 Transponder Codes

89. If Air Traffic Control advises that radar service is terminated when the pilot is departing Class C airspace, the transponder should be set to code

    A. 0000.

    B. 1200.

    C. 4096.

Answer (B) is correct. *(AIM Para 4-1-19)*
    DISCUSSION: The code 1200 designates VFR operations when another number is not assigned by ATC.
    Answer (A) is incorrect. This is not a transponder code. Answer (C) is incorrect. The numbers only go up to 7, so a 9 is not possible.

90. When making routine transponder code changes, pilots should avoid inadvertent selection of which codes?

A. 0700, 1700, 7000.

B. 1200, 1500, 7000.

C. 7500, 7600, 7700.

Answer (C) is correct. *(AIM Para 4-1-19)*
DISCUSSION: Some special codes set aside for emergencies should be avoided during routine VFR flights. They are 7500 for hijacking, 7600 for lost radio communications, and 7700 for a general emergency. Additionally, you should know that code 7777 is reserved for military interceptors.
Answer (A) is incorrect. Any of these may be assigned by ATC. Answer (B) is incorrect. The standard VFR code is 1200.

91. When making routine transponder code changes, pilots should avoid inadvertent selection of which code?

A. 1200.

B. 7600.

C. 4096.

Answer (B) is correct. *(AIM Para 4-1-19)*
DISCUSSION: Some special codes set aside for emergencies should be avoided during routine VFR flights. They are 7500 for hijacking, 7600 for lost radio communications, and 7700 for a general emergency. Additionally, you should know that code 7777 is reserved for military interceptors.
Answer (A) is incorrect. This is the standard VFR code. Answer (C) is incorrect. This is the number of possible codes, not an actual code itself. Additionally, the transponder digits only go up to 7, so a 9 is not possible.

92. When operating under VFR below 18,000 feet MSL, unless otherwise authorized, what transponder code should be selected?

A. 1200.

B. 7600.

C. 7700.

Answer (A) is correct. *(AIM Para 4-1-19)*
DISCUSSION: The standard VFR transponder code is 1200. Since all flight operations above 18,000 ft. MSL are to be IFR, code 1200 is not used above that height.
Answer (B) is incorrect. This is the lost radio communications code. Answer (C) is incorrect. This is the general emergency code.

93. At an altitude below 18,000 feet MSL, which altimeter code should be selected?

A. Mode 3/A, Code 1200.

B. Mode F, Code 1200.

C. Mode C, Code 4096.

Answer (A) is correct. *(AIM Para 4-1-19)*
DISCUSSION: The standard VFR transponder code is 1200. Since all flight operations above 18,000 ft. MSL are to be IFR, code 1200 is not used above that height.
Answer (B) is incorrect. The standard VFR transponder code is 1200, but Mode F is not a valid transponder type. Answer (C) is incorrect. Mode C is a valid transponder type, but 4096 is the number of possible codes, not an actual code itself. Additionally, the transponder digits only go up to 7, so a 9 is not possible.

94. Which of the following codes should be set for VFR flight in Class E airspace?

A. 1200, Mode 3/A.

B. 1200, Mode F.

C. 4600, Mode S.

Answer (A) is correct. *(AIM Para 4-1-19)*
DISCUSSION: The standard VFR transponder code is 1200. Since all flight operations above 18,000 ft. MSL are to be IFR, code 1200 is not used above that height.
Answer (B) is incorrect. The standard VFR transponder code is 1200, but Mode F is not a valid transponder type. Answer (C) is incorrect. Mode S is a valid transponder type, but 4600 is not the correct standard VFR transponder code.

3.13 Radio Phraseology

95. When flying HAWK N666CB, the proper phraseology for initial contact with McAlester FSS is

A. "MC ALESTER RADIO, HAWK SIX SIX SIX CHARLIE BRAVO, RECEIVING ARDMORE VORTAC, OVER."

B. "MC ALESTER STATION, HAWK SIX SIX SIX CEE BEE, RECEIVING ARDMORE VORTAC, OVER."

C. "MC ALESTER FLIGHT SERVICE STATION, HAWK NOVEMBER SIX CHARLIE BRAVO, RECEIVING ARDMORE VORTAC, OVER."

Answer (A) is correct. *(AIM Para 4-2-3)*
DISCUSSION: When calling a ground station, pilots should begin with the name of the facility and the type of facility. Any FSS is referred to as "Radio." When the aircraft manufacturer's name or model is stated, the prefix "N" is dropped. When transmitting and receiving on different frequencies, indicate the name of the VOR or frequency on which a reply is expected. Thus, the proper phraseology on initial contact with McAlester FSS is McAlester Radio, Hawk Six Six Six Charlie Bravo, Receiving Ardmore VORTAC, Over. (NOTE: The word "over" has been dropped from common usage.)
Answer (B) is incorrect. It is McAlester radio, not station, and CB is Charlie Bravo, not cee bee. Answer (C) is incorrect. It is radio, not flight service station. November is dropped in favor of Hawk; also, it is six, six, six Charlie Bravo (not six Charlie Bravo).

96. The correct method of stating 4,500 feet MSL to ATC is

A. "FOUR THOUSAND FIVE HUNDRED."

B. "FOUR POINT FIVE."

C. "FORTY-FIVE HUNDRED FEET MSL."

Answer (A) is correct. *(AIM Para 4-2-9)*
  DISCUSSION: The proper phraseology for altitudes up to but not including 18,000 ft. MSL is to state the separate digits of the thousands, plus the hundreds, if appropriate. It would be "four thousand, five hundred."
  Answer (B) is incorrect. Four point five is slang (not correct) phraseology. Answer (C) is incorrect. The thousand is spoken separately from the hundreds and not together. A stated altitude is understood to be MSL, unless otherwise stated.

97. The correct method of stating 10,500 feet MSL to ATC is

A. "TEN THOUSAND, FIVE HUNDRED FEET."

B. "TEN POINT FIVE."

C. "ONE ZERO THOUSAND, FIVE HUNDRED."

Answer (C) is correct. *(AIM Para 4-2-9)*
  DISCUSSION: The proper phraseology for altitudes up to but not including 18,000 ft. MSL is to state the separate digits of the thousands, plus the hundreds, if appropriate. It would be one zero thousand, five hundred.
  Answer (A) is incorrect. It is one zero, not ten. Answer (B) is incorrect. Ten point five is slang (not correct) phraseology.

3.14 ATC Traffic Advisories

98. When an air traffic controller issues radar traffic information in relation to the 12-hour clock, the reference the controller uses is the aircraft's

A. true course.

B. ground track.

C. magnetic heading.

Answer (B) is correct. *(AIM Para 4-1-14)*
  DISCUSSION: When issuing radar traffic information, the controller will provide the direction of the traffic from your airplane in relation to the 12-hr. clock based on your ground track or magnetic course.
  Answer (A) is incorrect. The controller will issue radar traffic information in relation to the 12-hr. clock based on your airplane's magnetic, not true, course. Answer (C) is incorrect. The controller will issue radar traffic information in relation to the 12-hr. clock based on your airplane's ground track, not magnetic heading.

99. An ATC radar facility issues the following advisory to a pilot flying on a heading of 090°:

"TRAFFIC 3 O'CLOCK, 2 MILES, WESTBOUND..."

Where should the pilot look for this traffic?

A. East.

B. South.

C. West.

Answer (B) is correct. *(AIM Para 4-1-14)*
  DISCUSSION: If you receive traffic information service from radar and are told you have traffic at the 3 o'clock position, traffic is in the direction of the right wingtip, or to the south.
  Answer (A) is incorrect. East is the 12 o'clock position. Answer (C) is incorrect. West is the 6 o'clock position.

100. An ATC radar facility issues the following advisory to a pilot flying on a heading of 360°:

"TRAFFIC 10 O'CLOCK, 2 MILES, SOUTHBOUND..."

Where should the pilot look for this traffic?

A. Northwest.

B. Northeast.

C. Southwest.

Answer (A) is correct. *(AIM Para 4-1-14)*
  DISCUSSION: The controller is telling you that traffic is at 10 o'clock and 2 mi. 9 o'clock is the left wingtip, and 10 o'clock is 2/3 of the way from the nose of the airplane (12 o'clock) to the left wingtip. Thus, you are looking northwest.
  Answer (B) is incorrect. Northeast would be in the 1 to 2 o'clock position. Answer (C) is incorrect. Southwest would be in the 7 to 8 o'clock position.

101. An ATC radar facility issues the following advisory to a pilot during a local flight:

"TRAFFIC 2 O'CLOCK, 5 MILES, NORTHBOUND..."

Where should the pilot look for this traffic?

A. Between directly ahead and 90° to the left.

B. Between directly behind and 90° to the right.

C. Between directly ahead and 90° to the right.

Answer (C) is correct. *(AIM Para 4-1-14)*
 DISCUSSION: The right wingtip is 3 o'clock, and the nose is 12 o'clock. A controller report of traffic 2 o'clock, 5 mi., northbound indicates that the traffic is to the right of the airplane's nose, just ahead of the right wingtip.
 Answer (A) is incorrect. The area directly ahead to 90° left is the area from 12 o'clock to 9 o'clock. Answer (B) is incorrect. The area directly behind to 90° right is the area from 6 o'clock to 3 o'clock.

102. An ATC radar facility issues the following advisory to a pilot flying north in a calm wind:

"TRAFFIC 9 O'CLOCK, 2 MILES, SOUTHBOUND..."

Where should the pilot look for this traffic?

A. South.

B. North.

C. West.

Answer (C) is correct. *(AIM Para 4-1-14)*
 DISCUSSION: Traffic at 9 o'clock is off the left wingtip. The nose of the airplane is 12 o'clock, the left wingtip is 9 o'clock, the tail is 6 o'clock, and the right wingtip is 3 o'clock. With a north heading, the aircraft at 9 o'clock would be west of you.
 Answer (A) is incorrect. South would be the 6 o'clock position. Answer (B) is incorrect. North would be the 12 o'clock position.

3.15 ATC Light Signals

103. While on final approach for landing, an alternating green and red light followed by a flashing red light is received from the control tower. Under these circumstances, the pilot should

A. discontinue the approach, fly the same traffic pattern and approach again, and land.

B. exercise extreme caution and abandon the approach, realizing the airport is unsafe for landing.

C. abandon the approach, circle the airport to the right, and expect a flashing white light when the airport is safe for landing.

Answer (B) is correct. *(FAR 91.125)*
 DISCUSSION: An alternating red and green light signaled from a control tower means "exercise extreme caution" whether to an airplane on the ground or in the air. The flashing red light received while in the air indicates the airport is not safe, and the pilot should not land.
 Answer (A) is incorrect. A flashing green (not red) light means to return for a landing. Answer (C) is incorrect. A flashing green (not red) light means to return for a landing, and a flashing white light does not have a meaning to aircraft in flight.

104. A steady green light signal directed from the control tower to an aircraft in flight is a signal that the pilot

A. is cleared to land.

B. should give way to other aircraft and continue circling.

C. should return for landing.

Answer (A) is correct. *(FAR 91.125)*
 DISCUSSION: A steady green light signal from the tower to an airplane in flight means cleared to land.
 Answer (B) is incorrect. Give way to other aircraft and continue circling is signaled by a steady red light to an airplane in the air. Answer (C) is incorrect. Return for landing is signaled by a flashing green light to an airplane in the air.

105. A flashing white light signal from the control tower to a taxiing aircraft is an indication to

A. taxi at a faster speed.

B. taxi only on taxiways and not cross runways.

C. return to the starting point on the airport.

Answer (C) is correct. *(FAR 91.125)*
 DISCUSSION: A flashing white light given to an aircraft taxiing along the ground means to return to the aircraft's starting point.
 Answer (A) is incorrect. There is no light signal that means to taxi at a faster speed. Answer (B) is incorrect. There is no light signal (by itself) that means to taxi only on taxiways and not cross runways.

106. If the control tower uses a light signal to direct a pilot to give way to other aircraft and continue circling, the light will be

A. flashing red.

B. steady red.

C. alternating red and green.

Answer (B) is correct. *(FAR 91.125)*
   DISCUSSION: A steady red light signal given to an aircraft in the air means to give way to other aircraft and continue circling.
   Answer (A) is incorrect. When in the air, a flashing red light means airport unsafe, do not land. Answer (C) is incorrect. Alternating red and green light always means exercise extreme caution.

107. Which light signal from the control tower clears a pilot to taxi?

A. Flashing green.

B. Steady green.

C. Flashing white.

Answer (A) is correct. *(FAR 91.125)*
   DISCUSSION: A flashing green gives the pilot permission to taxi.
   Answer (B) is incorrect. A steady green light means cleared to take off if on the ground or to land if in the air. Answer (C) is incorrect. A flashing white light means to return to the starting point on the airport for aircraft only on the ground.

108. An alternating red and green light signal directed from the control tower to an aircraft in flight is a signal to

A. hold position.

B. exercise extreme caution.

C. not land; the airport is unsafe.

Answer (B) is correct. *(FAR 91.125)*
   DISCUSSION: A flashing red and green light given anytime means exercise extreme caution.
   Answer (A) is incorrect. A steady red when taxiing means hold your position. There is no light signal to tell you to hold your position when in flight, only to give way to other aircraft and continue circling. Answer (C) is incorrect. A flashing red light means do not land; airport unsafe.

109. If the aircraft's radio fails, what is the recommended procedure when landing at a controlled airport?

A. Observe the traffic flow, enter the pattern, and look for a light signal from the tower.

B. Enter a crosswind leg and rock the wings.

C. Flash the landing lights and cycle the landing gear while circling the airport.

Answer (A) is correct. *(AIM Para 4-2-13)*
   DISCUSSION: If your radio fails and you wish to land at a tower controlled airport, remain outside or above the airport's traffic pattern until the direction and flow of traffic has been determined, then join the airport traffic pattern and maintain visual contact with the tower to receive light signals.
   Answer (B) is incorrect. Crosswind entry is not required; also, you rock the wings to acknowledge light signals during daylight hours. Answer (C) is incorrect. Flashing the landing light is a method of acknowledging light signals at night, and cycling the landing gear is not an option available to fixed-gear aircraft.

3.16 ELTs and VHF/DF

110. When activated, an emergency locator transmitter (ELT) transmits on

A. 118.0 and 118.8 MHz.

B. 121.5 and 243.0 MHz.

C. 123.0 and 119.0 MHz.

Answer (B) is correct. *(AIM Para 6-2-5)*
   DISCUSSION: When activated, an emergency locator transmitter (ELT) transmits simultaneously on the international distress frequencies of 121.5 and 243.0 megahertz.
   Answer (A) is incorrect. These are not emergency frequencies. Answer (C) is incorrect. These are not emergency frequencies.

111. Which procedure is recommended to ensure that the emergency locator transmitter (ELT) has not been activated?

A. Turn off the aircraft ELT after landing.

B. Ask the airport tower if they are receiving an ELT signal.

C. Monitor 121.5 before engine shutdown.

Answer (C) is correct. *(AIM Para 6-2-5)*
   DISCUSSION: To ensure that your ELT has not been activated, you can monitor 121.5 MHz or 243.0 MHz in flight when a receiver is available and prior to engine shutdown at the end of each flight.
   Answer (A) is incorrect. If you turn off the ELT, there is no way of telling whether it has been activated. Answer (B) is incorrect. The tower or ATC should not be bothered by questions about your ELT transmissions. If, however, you do receive signals on 121.5, IMMEDIATELY report it to ATC and/or FSS.

112. The letters VHF/DF appearing in the *Airport/Facility Directory* for a certain airport indicate that

A. this airport is designated as an airport of entry.

B. the Flight Service Station has equipment with which to determine your direction from the station.

C. this airport has a direct-line phone to the Flight Service Station.

Answer (B) is correct. *(AIM Para 1-1-16)*
DISCUSSION: The VHF/Direction Finder (DF) facility is a ground operation that displays the magnetic direction of the airplane from the station each time the airplane communication radio transmits a signal to it. It is used by ATC and FSS to assist lost pilots by telling them which direction they are from the receiving station.
Answer (A) is incorrect. An airport of entry is indicated by the letters AOE. Answer (C) is incorrect. A direct phone line to FSS is indicated by the letters DL following the FSS identifier.

113. To use VHF/DF facilities for assistance in locating an aircraft's position, the aircraft must have a

A. VHF transmitter and receiver.

B. 4096-code transponder.

C. VOR receiver and DME.

Answer (A) is correct. *(AIM Para 1-1-16)*
DISCUSSION: The VHF/Direction Finder (DF) facility is a ground operation that displays the magnetic direction of the airplane from the station each time the airplane communication (VHF) radio transmits a signal to it. Thus, to use such facilities for assistance in locating an airplane position, the airplane must have both a VHF transmitter (to send the signal) and a receiver (to communicate with the operator, who reads out the displayed magnetic direction).
Answer (B) is incorrect. Transponders are received by radar, not VHF/DF. Answer (C) is incorrect. VORs and DMEs relate to (rely on) VORTACs.

3.17 Land and Hold Short Operations (LAHSO)

114. Who should not participate in the Land and Hold Short Operations (LAHSO) program?

A. Recreational pilots only.

B. Military pilots.

C. Student pilots.

Answer (C) is correct. *(AIM Para 4-3-11)*
DISCUSSION: Land and hold short operations (LAHSO) take place at some airports with an operating control tower in order to increase the total capacity and improve the flow of traffic. LAHSO requires that a pilot not use the full length of the runway, but rather that (s)he stop and hold short before reaching an intersecting runway, taxiway, or other specified point on the landing runway. Student pilots or pilots who are not familiar with LAHSO should not participate in the program.
Answer (A) is incorrect. A recreational pilot cannot operate at an airport with an operating control tower (unless working on obtaining his/her private pilot certificate under the supervision of a CFI) and would not have a choice as to whether or not to participate. Answer (B) is incorrect. Student pilots or pilots unfamiliar with LAHSO, not military pilots, should not participate in the program.

115. Who has final authority to accept or decline any land and hold short (LAHSO) clearance?

A. Pilot in command.

B. Owner/operator.

C. Second-in-command.

Answer (A) is correct. *(AIM Para 4-3-11)*
DISCUSSION: Land and hold short operations (LAHSO) take place at some airports with an operating control tower in order to increase the total capacity and improve the flow of traffic. LAHSO requires that a pilot not use the full length of the runway, but rather that (s)he stop and hold short before reaching an intersecting runway, taxiway, or other specified point on the landing runway. LAHSO requires familiarity with the available landing distance (ALD) for given LAHSO combinations and with the landing performance of the aircraft. The pilot in command has the final authority to accept or decline any land and hold short clearance.
Answer (B) is incorrect. The pilot in command, regardless of whether or not (s)he is the owner, has the final authority to accept or decline a LAHSO clearance. Answer (C) is incorrect. The pilot in command, not the second in command, has final authority to accept or decline a LAHSO clearance.

116. Who has final authority to accept or decline any land and hold short (LAHSO) clearance?

- A. Pilot in command.
- B. Air Traffic Controller.
- C. Second in command.

Answer (A) is correct.   *(AIM Para 4-3-11)*
DISCUSSION: Land and hold short operations (LAHSO) take place at some airports with an operating control tower in order to increase and improve the flow of traffic. LAHSO requires that a pilot not use the length of the runway, but rather that (s)he stop and hold short before reaching an intersecting runway, taxiway, or other specified point on the landing runway. LAHSO requires familiarity with the available landing distance (ALD) for given LAHSO combinations and with the landing performance of the aircraft. The pilot in command has the final authority to accept or decline any land and hold short clearance.
Answer (B) is incorrect. An air traffic controller cannot accept, nor can (s)he force a pilot in command to accept, a LAHSO clearance. Answer (C) is incorrect. The pilot in command, not the second in command, has the final authority to accept or decline a LAHSO clearance.

117. When should pilots decline a land and hold short (LAHSO) clearance?

- A. When it will compromise safety.
- B. Only when the tower operator concurs.
- C. Pilots cannot decline clearance.

Answer (A) is correct.   *(AIM Para 4-3-11)*
DISCUSSION: Land and hold short operations (LAHSO) take place at some airports with an operating control tower in order to increase the total capacity and improve the flow of traffic. LAHSO requires that a pilot not use the full length of the runway, but rather that (s)he stop and hold short before reaching an intersecting runway, taxiway, or other specified point on the landing runway. LAHSO requires familiarity with the available landing distance (ALD) for given LAHSO combinations and with the landing performance of the aircraft. Pilots are expected to decline a land and hold short clearance if they determine that it will compromise safety.
Answer (B) is incorrect. The pilot in command has the final authority to accept or decline a land and hold short clearance; agreement from the tower operator is not required. Answer (C) is incorrect. The pilot in command has the authority to decline a land and hold short clearance.

118. Where is the "Available Landing Distance" (ALD) data published for an airport that utilizes Land and Hold Short Operations (LAHSO)?

- A. Special Notices section of the *Airport/Facility Directory (A/FD)*.
- B. *Aeronautical Information Manual (AIM)*.
- C. 14 CFR Part 91, *General Operating and Flight Rules*.

Answer (A) is correct.   *(AIM Para 4-3-11)*
DISCUSSION: Land and hold short operations (LAHSO) take place at some airports with an operating control tower in order to increase the total capacity and improve the flow of traffic. LAHSO requires that a pilot not use the full length of the runway, but rather that (s)he stop and hold short before reaching an intersecting runway, taxiway, or other specified point on the landing runway. LAHSO requires familiarity with the available landing distance (ALD) for given LAHSO combinations and with the landing performance of the aircraft. ALD data are published in the special notices section of the *Airport/Facility Directory*.
Answer (B) is incorrect. The ALD data are published in the *A/FD*, not in the *AIM*, which contains information on how LAHSO are to be conducted. Answer (C) is incorrect. ALD data are published in the *A/FD*, not in 14 CFR Part 91.

119. What is the minimum visibility for a pilot to receive a land and hold short (LAHSO) clearance?

- A. 3 nautical miles.
- B. 3 statute miles.
- C. 1 statute mile.

Answer (B) is correct.   *(AIM Para 4-3-11)*
DISCUSSION: You should receive a land and hold short (LAHSO) clearance only when there is a minimum ceiling of 1,000 ft. and visibility of 3 statute miles. The intent of having basic VFR weather conditions is to allow pilots to maintain visual contact with other aircraft and ground vehicle operations.
Answer (A) is incorrect. The minimum visibility for a pilot to receive a land and hold short (LAHSO) clearance is 3 SM, not 3 nautical miles. Remember, visibility is reported in statute miles, not nautical miles. Answer (C) is incorrect. The minimum visibility for a pilot to receive a special VFR clearance, not a land and hold short clearance, is 1 statute mile.

END OF STUDY UNIT

STUDY UNIT FOUR
FEDERAL AVIATION REGULATIONS

(16 pages of outline)

This study unit contains outlines of major concepts tested, sample test questions and answers regarding Federal Aviation Regulations (FARs), and an explanation of each answer. The table of contents above and on the previous page lists each subunit within this study unit, the number of questions pertaining to that particular subunit, and the pages on which the outlines and questions begin, respectively.

CAUTION: Recall that the **sole purpose** of this book is to expedite your passing the FAA pilot knowledge test for the private pilot certificate. Accordingly, all extraneous material (i.e., topics or regulations not directly tested on the FAA pilot knowledge test) is omitted, even though much more information and knowledge are necessary to fly safely. This additional material is presented in *Pilot Handbook* and *Private Pilot Flight Maneuvers and Practical Test Prep*, available from Gleim Publications, Inc. See the order form on page 371.

The 27 questions beginning on page 131 specifically refer to recreational pilots. Recreational pilots are responsible for all applicable FARs. Subpart D of FAR Part 61, "Recreational Pilots," specifically applies. It is numbered 61.96 through 61.101.

61.96   Applicability and Eligibility Requirements: General
61.97   Aeronautical Knowledge
61.98   Flight Proficiency
61.99   Aeronautical Experience
61.100   Pilots Based on Small Islands
61.101   Recreational Pilot Privileges and Limitations

All of the 27 recreational pilot questions test FAR 61.101.

4.1 FAR PART 1

1.1 General Definitions

1. **Night** means the time between the end of evening civil twilight and the beginning of morning civil twilight, as published in the American Air Almanac converted to local time.

    a. Note that for "recency of experience" (FAR 61.57), night is defined as from 1 hr. after sunset to 1 hr. before sunrise.

    b. Be careful; there are questions on both definitions.

2. **Aircraft categories** (for certification of airmen); broad classifications of aircraft

    a. Airplane
    b. Rotorcraft
    c. Glider
    d. Lighter-than-air

3. **Airplane classes** (for certification of airmen)

    a. Single-engine land
    b. Multi-engine land
    c. Single-engine sea
    d. Multi-engine sea

4. **Rotorcraft classes** (for certification of airmen)

    a. Helicopter
    b. Gyrocopter

5. **Lighter-than-air classes** (for certification of airmen)

    a. Airship
    b. Free balloon
    c. Hot air balloon
    d. Gas balloon

6. Note the previous category and class definitions are for certification of airmen purposes. For certification of aircraft there are different definitions:

    a. **Category** (for certification of aircraft purposes) is based on intended use or operating limitations.

        1) Transport
        2) Normal
        3) Utility
        4) Limited
        5) Restricted
        6) Acrobatic
        7) Provisional

    b. **Classes** as used for certification of aircraft are the same as, or very similar to, categories for certification of airmen, e.g., airplane, rotorcraft, glider, lighter-than-air.

7. **Air traffic control (ATC) clearance** means an authorization to proceed under specific traffic conditions in controlled airspace.

1.2 Abbreviations and Symbols

1. V_{FE} means maximum flap extended speed.

2. V_{LE} means maximum landing gear extended speed.

3. V_{NO} means maximum structural cruising speed.

4. V_A means design maneuvering speed.

5. V_{S0} means the stalling speed or the minimum steady flight speed in the landing configuration.

6. V_X means speed for best angle of climb.

7. V_Y means speed for best rate of climb.

4.2 FAR PART 21

21.181 Duration of Airworthiness Certificates

1. Airworthiness certificates remain in force as long as maintenance and alteration of the aircraft are performed per FARs.

4.3 FAR PART 39

39.1 Applicability

1. Airworthiness Directives (ADs) are issued under FAR Part 39 by the FAA to require correction of unsafe conditions found in an airplane, an airplane engine, a propeller, or an appliance when such conditions exist and are likely to exist or develop in other products of the same design.

   a. Since ADs are issued under FAR Part 39, they are regulatory and must be complied with, unless a specific exemption is granted.

39.3 General

1. No person may operate a product to which an AD applies except in accordance with the requirements of that AD.

   a. Thus, you may operate an airplane that is not in compliance with an AD, if such operation is allowed by the AD.

4.4 FAR PART 43

43.3 Persons Authorized to Perform Maintenance, Preventive Maintenance, Rebuilding, and Alterations

1. A person who holds a pilot certificate (e.g., private pilot) may perform preventive maintenance on any airplane owned or operated by that pilot that is not used in air carrier services.

43.7 Persons Authorized to Approve Aircraft Airframes, Aircraft Engines, Propellers, Appliances, or Component Parts for Return to Service after Maintenance, Preventive Maintenance, Rebuilding, or Alteration

1. To approve the airplane for return to service, after preventive maintenance was done by a pilot, the pilot must hold at least a private pilot certificate.

43.9 Maintenance Records

1.   After preventive maintenance has been performed, the signature, certificate number, and kind of certificate held by the person approving the work, the date, and a description of the work must be entered in the aircraft maintenance records.

43 Appendix A. Major Alterations and Repairs and Preventive Maintenance

1.   Preventive maintenance means simple or minor preservation operations and the replacement of small standard parts not involving complex assembly operations. Examples include replenishing hydraulic fluid and servicing landing gear wheel bearings.

4.5 FAR PART 61

61.3 Requirements for Certificates, Ratings, and Authorizations

1.   When acting as a pilot in command or as a required pilot flight crewmember, you must have a valid pilot certificate and a current and appropriate medical certificate in your personal possession or readily accessible in the airplane.

2.   You must present your pilot certificate or medical certificate upon the request of the Administrator of the FAA or his/her representative; the NTSB; or any federal, state, or local law enforcement officer.

61.15 Offenses Involving Alcohol or Drugs

1.   Each person holding a certificate under Part 61 shall provide a written report of each motor vehicle action involving alcohol or drugs to the FAA, Civil Aviation Security Division, no later than 60 days after the motor vehicle action.

61.23 Medical Certificates: Requirement and Duration

1.   For operations requiring a private, recreational, or student pilot certificate, a first-, second-, or third-class medical certificate

   a.   At the end of the 60th month after the date of examination shown on the certificate, if you have not reached your 40th birthday on or before the date of examination or

   b.   At the end of the 24th month after the date of examination shown on the certificate, if you have reached your 40th birthday on or before the date of examination.

61.31 Type Rating Requirements, Additional Training, and Authorization Requirements

1.   To act as pilot in command of a complex airplane, you must receive and log ground and flight training and receive a logbook endorsement.

   a.   A complex airplane is defined as an airplane with retractable landing gear, flaps, and a controllable pitch propeller.

2.   To act as pilot in command of a high-performance airplane, you must receive and log ground and flight training and receive a logbook endorsement.

   a.   A high-performance airplane is defined as an airplane with an engine of more than 200 horsepower.

3.   A person may not act as pilot in command of any of the following aircraft unless (s)he holds a type rating for that aircraft:

   a.   A large aircraft (i.e., over 12,500 lb. gross weight)
   b.   A turbojet-powered airplane
   c.   Other aircraft specified by the FAA through aircraft type certification procedures

61.56 Flight Review

1. A flight review must have been satisfactorily completed within the previous 24 calendar months to act as pilot in command of an aircraft for which that pilot is rated. A flight review consists of a minimum of 1 hour of flight training by an authorized instructor and 1 hour of ground training.

2. The expiration of the 24-month period for the flight review falls on the last day of the 24th month after the month of the examination date (i.e., 24 calendar months).

61.57 Recent Flight Experience: Pilot in Command

1. To carry passengers, you must have made three landings and three takeoffs within the preceding 90 days.

   a. All three landings must be made in aircraft of the same category, class, and, if a type rating is required, type as the one in which passengers are to be carried.

      1) The categories are airplane, rotorcraft, glider, and lighter-than-air.
      2) The classes are single-engine land, single-engine sea, multi-engine land, and multi-engine sea.

   b. The landings must be to a full stop if the airplane is tailwheel (conventional) rather than nosewheel.

2. To carry passengers at night, you must, within the last 90 days, have made three takeoffs and three landings to a full stop at night in an aircraft of the same category, class, and type, if required.

   a. Night in this case is defined as the period beginning 1 hr. after sunset and ending 1 hr. before sunrise.

61.60 Change of Address

1. You must notify the FAA Airmen Certification Branch in writing of any change in your permanent mailing address.

2. You may not exercise the privileges of your pilot certificate after 30 days from moving unless you make this notification.

61.69 Glider Towing: Experience and Training Requirements

1. Any person may tow a glider if that person has

   a. At least a private pilot certificate

   b. 100 hr. of pilot in command time in the aircraft category, class, and type, if required, that the pilot is using to tow a glider

   c. A logbook endorsement from an authorized instructor certifying that the person has received ground and flight training in gliders

   d. Within the preceding 24 months

      1) Made at least three actual or simulated glider tows while accompanied by a qualified pilot or

      2) Made at least three flights as pilot in command of a glider towed by an aircraft

61.113 Private Pilot Privileges and Limitations: Pilot in Command

1. Private pilots may not pay less than an equal (pro rata) share of the operating expenses of a flight with the passengers.

   a. These operating expenses may involve only fuel, oil, airport expenditures, or rental fees.

2.   Private pilots may operate an aircraft carrying passengers on business only if the flight is incidental to that business or employment and the pilot is not paid as a pilot.

    a.   For example, a CPA who is a private pilot might fly an aircraft carrying CPAs to a client. Such flight is incidental to the CPA's professional duties or business.

3.   A pilot may act as a pilot in command of an aircraft used in a passenger-carrying airlift sponsored by a charitable organization for which passengers make donations to the organization if

    a.   The local FSDO (FAA Flight Standards District Office) is notified at least 7 days before the flight;

    b.   The flight is conducted from an adequate public airport;

    c.   The pilot has logged at least 500 hours;

    d.   No acrobatic or formation flights are performed;

    e.   The aircraft holds a standard airworthiness certificate and is airworthy;

    f.   The flight is day-VFR; and

    g.   The flight is non-stop, begins and ends at the same airport, and is conducted within a 25 NM radius of that airport.

4.6   RECREATIONAL PILOT RELATED FARS

61.101   Recreational Pilot Privileges and Limitations

NOTE: This section is not tested on the Private Pilot Airplane (PAR) or Private Pilot-Airplane Transition (PAT) knowledge test.

1.   A recreational pilot may carry only one passenger.

    a.   A recreational pilot may not pay less than the pro rata (equal) share of the operating expenses of a flight with a passenger, provided the expenses involve only fuel, oil, airport expenses, or aircraft rental fees.

2.   A recreational pilot may act as pilot in command of an airplane

    a.   Only when the flight is within 50 NM of an airport at which the pilot has received ground and flight training from an authorized flight instructor. The pilot must have in his/her possession a logbook endorsement that permits flight within 50 NM from the departure airport.

    b.   When the flight exceeds 50 NM if (s)he receives ground and flight training on the cross-country training requirements for a private pilot and has his/her logbook endorsed certifying proficiency in cross-country flight by an authorized instructor

3.   A recreational pilot may not act as pilot in command of an aircraft

    a.   Certificated for more than four occupants, with more than one engine, with an engine of more than 180 horsepower, or with retractable landing gear

    b.   Classified as a multi-engine airplane, powered-lift, glider, airship, or balloon

    c.   Carrying a passenger or property for compensation or hire

    d.   In furtherance of a business

    e.   Between sunset and sunrise (e.g., night time)

    f.   In airspace in which communication with ATC is required

    g.   At an altitude of more than 10,000 ft. MSL or 2,000 ft. AGL, whichever is higher

    h.   With flight or surface visibility of less than 3 SM

        1)   In Class G airspace, the cloud clearance requirements are clear of clouds when 1,200 ft. AGL or less and 1,000 ft. above, 500 ft. below, and 2,000 ft. horizontally from clouds when more than 1,200 ft. AGL but less than 10,000 ft. MSL.

   i. Without visual reference to the surface

   j. On a flight outside the U.S. unless authorized by the country in which the flight is conducted

   k. For demonstration of that aircraft in flight to a prospective buyer

   l. Used in a passenger-carrying airlift and sponsored by a charitable organization

   m. Towing any object

4. A recreational pilot may not act as a required pilot flight crewmember on any aircraft for which more than one pilot is required.

5. A recreational pilot who has logged fewer than 400 flight hr. and who has not logged pilot-in-command time in an aircraft within the preceding 180 days may not act as pilot in command of an aircraft until the pilot has received flight training from an authorized flight instructor who certifies in the pilot's logbook that the pilot is competent to act as pilot in command.

6. The recreational pilot certificate states, "Holder does not meet ICAO requirements."

7. For the purpose of obtaining additional certificates or ratings, while under the supervision of an authorized flight instructor, a recreational pilot may fly as sole occupant of an aircraft

   a. For which the pilot does not hold an appropriate category or class rating

   b. Within airspace that requires communication with air traffic control

   c. Between sunset and sunrise, provided the flight or surface visibility is at least 5 SM

   d. In excess of 50 NM from an airport at which flight instruction is received

NOTE: For any of these situations, the recreational pilot shall carry the logbook that has been properly endorsed for each flight by an authorized flight instructor.

8. When flying a transponder-equipped aircraft, a recreational pilot should set that transponder on code (squawk) 1200, which is the VFR code.

4.7 FAR PART 71

71.71 Extent of Federal Airways

1. Federal airways include that Class E airspace

   a. Extending upward from 1,200 ft. AGL to and including 17,999 ft. MSL

   b. Within parallel boundary lines 4 NM each side of the airway's centerline

4.8 FAR PART 91: 91.3 – 91.151

91.3 Responsibility and Authority of the Pilot in Command

1. In emergencies, a pilot may deviate from the FARs to the extent needed to maintain the safety of the airplane and passengers.

2. The pilot in command of an aircraft is directly responsible for, and is the final authority as to, the operation of that aircraft.

3. A written report of any deviations from FARs should be filed with the FAA upon request.

91.7 Civil Aircraft Airworthiness

1. The pilot in command is responsible for determining that the airplane is airworthy prior to every flight.

   a. The pilot in command shall discontinue the flight when unairworthy conditions (whether electrical, mechanical, or structural) occur.

91.9 Civil Aircraft Flight Manual, Marking, and Placard Requirements

1. The airworthiness certificate, the FAA registration certificate, and the aircraft flight manual or operating limitations must be aboard.

2. The acronym ARROW can be used as a memory aid. The FCC (Federal Communications Commission), not the FAA, requires the radio station license. As of January 1, 1997, the radio station license is required only for international flights.

    A irworthiness certificate
    R egistration certificate
    R adio station license (FCC requirement for international flight)
    O perating limitations, including
    W eight and balance data

3. The operating limitations of an airplane may be found in the current FAA-approved flight manual, approved manual material, markings, and placards, or any combination thereof.

91.15 Dropping Objects

1. No pilot in command of a civil aircraft may allow any object to be dropped from that aircraft in flight that creates a hazard to persons or property.

    a. However, this section does not prohibit the dropping of any object if reasonable precautions are taken to avoid injury or damage to persons or property.

91.17 Alcohol or Drugs

1. No person may act as a crewmember of a civil airplane while having .04 percent by weight or more alcohol in the blood or if any alcoholic beverages have been consumed within the preceding 8 hours.

2. No person may act as a crewmember of a civil airplane if using any drug that affects the person's faculties in any way contrary to safety.

3. Operating or attempting to operate an aircraft as a crewmember while under the influence of drugs or alcohol is grounds for the denial of an application for a certificate, rating, or authorization issued under 14 CFR Part 91.

91.103 Preflight Action

1. Pilots are required to familiarize themselves with all available information concerning the flight prior to every flight, and specifically to determine,

    a. For any flight, runway lengths at airports of intended use and the airplane's takeoff and landing requirements, and

    b. For IFR flights or those not in the vicinity of an airport,

        1) Weather reports and forecasts,
        2) Fuel requirements,
        3) Alternatives available if the planned flight cannot be completed, and
        4) Any known traffic delays.

91.105 Flight Crewmembers at Stations

1. During takeoff and landing, and while en route, each required flight crewmember shall keep his/her safety belt fastened while at his/her station.

    a. If shoulder harnesses are available, they must be used for takeoff and landing.

91.107 Use of Safety Belts, Shoulder Harnesses, and Child Restraint Systems

1. Pilots must ensure that each occupant is briefed on how to use the safety belts and, if installed, shoulder harnesses.

2. Pilots must notify all occupants to fasten their safety belts before taxiing, taking off, or landing.

3. All passengers of airplanes must wear their safety belts during taxi, takeoffs, and landings.

   a. A passenger who has not reached his/her second birthday may be held by an adult.
   b. Sport parachutists may use the floor of the aircraft as a seat (but still must use safety belts).

91.111 Operating near Other Aircraft

1. No person may operate an aircraft in formation flight except by prior arrangement with the pilot in command of each aircraft in the formation.

91.113 Right-of-Way Rules: Except Water Operations

1. Aircraft in distress have the right-of-way over all other aircraft.

2. When two aircraft are approaching head on or nearly so, the pilot of each aircraft should turn to his/her right, regardless of category.

3. When two aircraft of different categories are converging, the right-of-way depends upon who has the least maneuverability. Thus, the right-of-way belongs to

   a. Balloons over
   b. Gliders over
   c. Airships over
   d. Airplanes or rotorcraft

4. When aircraft of the same category are converging at approximately the same altitude, except head on or nearly so, the aircraft to the other's right has the right-of-way.

   a. If an airplane of the same category as yours is approaching from your right side, it has the right-of-way.

5. When two or more aircraft are approaching an airport for the purpose of landing, the aircraft at the lower altitude has the right-of-way.

   a. This rule shall not be abused by cutting in front of or overtaking another aircraft.

6. An aircraft towing or refueling another aircraft has the right-of-way over all engine-driven aircraft.

91.115 Right-of-Way Rules: Water Operations

1. When aircraft, or an aircraft and a vessel, are on crossing courses, the aircraft or vessel to the other's right has the right-of-way.

91.117 Aircraft Speed

1. The speed limit is 250 kt. (288 MPH) when flying below 10,000 ft. MSL and in Class B airspace.

2. When flying under Class B airspace or in VFR corridors through Class B airspace, the speed limit is 200 kt. (230 MPH).

3. When at or below 2,500 ft. AGL and within 4 NM of the primary airport of Class C or Class D airspace, the speed limit is 200 kt. (230 MPH).

91.119 Minimum Safe Altitudes: General

1. Over congested areas (cities, towns, settlements, or open-air assemblies), a pilot must maintain an altitude of 1,000 ft. above the highest obstacle within a horizontal radius of 2,000 ft. of the airplane.

2. The minimum altitude over other than congested areas is 500 ft. AGL.

    a. Over open water or sparsely populated areas, an airplane may not be operated closer than 500 ft. to any person, vessel, vehicle, or structure.

3. Altitude in all areas must be sufficient to permit an emergency landing without undue hazard to persons or property on the surface if a power unit fails.

91.121 Altimeter Settings

1. Prior to takeoff, the altimeter should be set to the current local altimeter setting. If the current local altimeter setting is not available, use the departure airport elevation.

2. The altimeter of an airplane is required to be set to 29.92 at or above 18,000 ft. MSL to guarantee vertical separation of airplanes above 18,000 ft. MSL.

91.123 Compliance with ATC Clearances and Instructions

1. When an ATC clearance is obtained, no pilot may deviate from that clearance, except in an emergency, unless an amended clearance is obtained, or the deviation is in response to a traffic alert and collision avoidance system resolution advisory. If you feel a rule deviation will occur, you should immediately advise ATC.

2. If you receive priority from ATC in an emergency, you must, upon request, file a detailed report within 48 hr. to the chief of that ATC facility even if no rule has been violated.

3. During an in-flight emergency, the pilot in command may deviate from the FARs to the extent necessary to handle the emergency.

    a. The pilot should notify ATC about the deviation as soon as possible.
    b. If priority is given, a written report (if requested) must be submitted in 48 hours.

91.130 Operations in Class C Airspace

1. Class C airspace is controlled airspace that requires radio communication with ATC.

    a. A pilot must establish two-way radio communication prior to entering Class C airspace and maintain it while within Class C airspace, regardless of weather conditions.

91.131 Operations in Class B Airspace

1. Class B airspace is controlled airspace found at larger airports with high volumes of traffic.

2. Requirements for operating within Class B airspace:

    a. A pilot must hold at least a private pilot certificate or a student pilot certificate with the appropriate logbook endorsements.
    b. Authorization must be received from ATC, regardless of weather conditions.
    c. The airplane must have a two-way communications radio and a transponder equipped with Mode C permits ATC to obtain an altitude readout on its radar screen.

        1) A VOR receiver is required only when operating IFR.

3. Student pilot operations in Class B airspace are only permitted with appropriate logbook endorsements.

    a. For flight through Class B airspace, the student pilot must

        1) Receive ground and flight instructions pertaining to that specific Class B airspace area

        2) Have a CFI logbook endorsement within 90 days for solo flight in that specific Class B airspace area

    b. For takeoffs and landings at an airport within Class B airspace, the student pilot must

        1) Receive ground and flight instructions pertaining to that specific Class B airspace area

        2) Have a CFI logbook endorsement within 90 days for solo flight at that specific airport

    c. No student pilot may take off or land at the following airports:

Andrews AFB	Miami International
Atlanta Hartsfield	Newark International
Boston Logan	New York Kennedy
Chicago O'Hare International	New York La Guardia
Dallas/Fort Worth International	San Francisco International
Los Angeles International	Washington National

4. With certain exceptions, all aircraft within a 30-NM radius of a Class B primary airport and from the surface up to 10,000 ft. MSL must have an operable transponder with Mode C.

91.133 Restricted and Prohibited Areas

1. Restricted areas are a type of special use airspace within which your right to fly is limited.

    a. Restricted areas have unusual and often invisible hazards to aircraft (e.g., balloons, military operations, etc.).

    b. Although restricted areas are not always in use during the times posted in the legend of sectional charts, permission to fly in that airspace must be obtained from the controlling agency. The controlling agency is listed for each restricted area at the bottom of sectional charts.

91.135 Operations in Class A Airspace

1. Since Class A airspace requires operation under IFR at specific flight levels assigned by ATC, VFR flights are prohibited.

91.151 Fuel Requirements for Flight in VFR Conditions

1. During the day, FARs require fuel sufficient to fly to the first point of intended landing and then for an additional 30 min., assuming normal cruise speed.

2. At night, sufficient fuel to fly an additional 45 min. is required.

4.9   FAR PART 91: 91.155 – 91.519

91.155 Basic VFR Weather Minimums

Airspace	Flight Visibility	Distance from Clouds
Class A	Not Applicable	Not applicable
Class B	3 SM	Clear of Clouds
Class C	3 SM	500 ft. below 1,000 ft. above 2,000 ft. horiz.
Class D	3 SM	500 ft. below 1,000 ft. above 2,000 ft. horiz.
Class E		
Less than 10,000 ft. MSL	3 SM	500 ft. below 1,000 ft. above 2,000 ft. horiz.
At or above 10,000 ft. MSL	5 SM	1,000 ft. below 1,000 ft. above 1 SM horiz.

Airspace	Flight Visibility	Distance from Clouds
Class G:		
1,200 ft. or less above the surface (regardless of MSL altitude)		
Day	1 SM	Clear of clouds
Night, except as provided in 1. below	3 SM	500 ft. below 1,000 ft. above 2,000 ft. horiz.
More than 1,200 ft. above the surface but less than 10,000 ft. MSL		
Day	1 SM	500 ft. below 1,000 ft. above 2,000 ft. horiz.
Night	3 SM	500 ft. below 1,000 ft. above 2,000 ft. horiz.
More than 1,200 ft. above the surface and at or above 10,000 ft. MSL	5 SM	1,000 ft. below 1,000 ft. above 1 SM horiz.

1. An airplane may be operated clear of clouds in Class G airspace at night below 1,200 ft. AGL when the visibility is less than 3 SM but more than 1 SM in an airport traffic pattern and within 1/2 NM of the runway.

2. Except when operating under a special VFR clearance,

   a. You may not operate your airplane beneath the ceiling under VFR within the lateral boundaries of the surface areas of Class B, Class C, Class D, or Class E airspace designated for an airport when the ceiling is less than 1,000 feet.

   b. You may not take off, land, or enter the traffic pattern of an airport in Class B, Class C, Class D, or Class E airspace unless the ground visibility is at least 3 SM. If ground visibility is not reported, flight visibility must be at least 3 statute miles.

91.157 Special VFR Weather Minimums

1. With some exceptions, special VFR clearances can be requested in Class B, Class C, Class D, or Class E airspace areas. You must remain clear of clouds and have visibility of at least 1 statute mile.

2. Flight under special VFR clearance at night is only permitted if the pilot has an instrument rating and the aircraft is IFR equipped.

3. Special VFR is an ATC clearance obtained from the control tower. If there is no control tower, obtain the clearance from the appropriate air traffic control facility.

91.159 VFR Cruising Altitude or Flight Level

1. Specified altitudes are required for VFR cruising flight at more than 3,000 ft. AGL and below 18,000 ft. MSL.

    a. The altitude prescribed is based upon the magnetic course (not magnetic heading).

    b. The altitude is prescribed in ft. above mean sea level (MSL).

    c. Use an odd thousand-foot MSL altitude plus 500 ft. for magnetic courses of 0° to 179°, e.g., 3,500, 5,500, 7,500 feet.

    d. Use an even thousand-foot MSL altitude plus 500 ft. for magnetic courses of 180° to 359°, e.g., 4,500, 6,500, or 8,500 feet.

    e. As a memory aid, "East is odd; west is even odder."

91.203 Civil Aircraft: Certifications Required

1. No person may operate a civil aircraft unless the aircraft has a U.S. airworthiness certificate displayed in a manner that makes it legible to passengers and crew.

2. To operate a civil aircraft, a valid U.S. registration issued to the owner of the aircraft must be on board.

91.207 Emergency Locator Transmitters

1. ELT batteries must be replaced (or recharged, if rechargeable) after 1 cumulative hr. of use or after 50% of their useful life expires.

2. ELTs may only be tested on the ground during the first 5 min. after the hour. No airborne checks are allowed.

91.209 Aircraft Lights

1. Airplanes operating (on the ground or in the air) between sunset and sunrise must display lighted position (navigation) lights, except in Alaska.

91.211 Supplemental Oxygen

1. All occupants must be provided with oxygen in an airplane operated at cabin pressure altitudes above 15,000 ft. MSL.

    a. Pilots and crewmembers may not operate an airplane at cabin pressure altitudes above 12,500 ft. MSL up to and including 14,000 ft. MSL for more than 30 min. without supplemental oxygen.

    b. Pilots and crewmembers must use supplemental oxygen at cabin pressure altitudes above 14,000 ft. MSL.

91.215 ATC Transponder and Altitude Reporting Equipment and Use

1. All aircraft must have and use an altitude encoding transponder when operating

    a. Within Class A airspace
    b. Within Class B airspace
    c. Within 30 NM of the Class B airspace primary airport
    d. Within and above Class C airspace
    e. Above 10,000 ft. MSL except at and below 2,500 ft. AGL

91.303 Aerobatic Flight

1. Aerobatic flight includes all intentional maneuvers that

    a. Are not necessary for normal flight and
    b. Involve an abrupt change in the airplane's attitude.

    2.   Aerobatic flight is prohibited

        a.   When flight visibility is less than 3 SM;

        b.   When altitude is less than 1,500 ft. above the ground;

        c.   Within the lateral boundaries of the surface areas of Class B, Class C, Class D, or Class E airspace designated for an airport;

        d.   Within 4 NM of the centerline of any federal airway; or

        e.   Over any congested area or over an open-air assembly of people.

91.307 Parachutes and Parachuting

    1.   With certain exceptions, each occupant of an aircraft must wear an approved parachute during any intentional maneuver exceeding

        a.   60° bank or

        b.   A nose-up or nose-down attitude of 30°.

    2.   Parachutes that are available for emergency use must be packed within a specific time period, based on the materials from which they are constructed.

        a.   Parachutes that include a canopy, shrouds, and harness that are composed exclusively of nylon, rayon, or other similar synthetic fibers must have been repacked by a certificated and appropriately rated parachute rigger within the preceding 180 days.

        b.   Parachutes that include any part that is composed of silk, pongee, or other natural fiber or materials must be repacked by a certificated and appropriately rated parachute rigger within the preceding 60 days.

91.313 Restricted Category Civil Aircraft: Operating Limitations

    1.   Restricted category civil aircraft may not normally be operated

        a.   Over densely populated areas,

        b.   In congested airways, or

        c.   Near a busy airport where passenger transport is conducted.

91.319 Aircraft Having Experimental Certificates: Operating Limitations

    1.   No person may operate an aircraft that has an experimental or restricted certificate over a densely populated area or in a congested airway unless authorized by the FAA.

91.403 General

    1.   The owner or operator of an aircraft is primarily responsible for maintaining that aircraft in an airworthy condition and for complying with all Airworthiness Directives (ADs).

    2.   An operator is a person who uses, or causes to use or authorizes to use, an aircraft for the purpose of air navigation, including the piloting of an aircraft, with or without the right of legal control (i.e., owner, lessee, or otherwise).

        a.   Thus, the pilot in command is also responsible for ensuring that the aircraft is maintained in an airworthy condition and that there is compliance with all ADs.

91.405 Maintenance Required

    1.   Each owner or operator of an aircraft shall ensure that maintenance personnel make the appropriate entries in the aircraft maintenance records indicating the aircraft has been approved for return to service.

91.407 Operation after Maintenance, Preventive Maintenance, Rebuilding, or Alteration

1. When aircraft alterations or repairs change the flight characteristics, the aircraft must be test flown and approved for return to service prior to carrying passengers.

    a. The pilot test flying the aircraft must be at least a private pilot and rated for the type of aircraft being tested.

91.409 Inspections

1. Annual inspections expire on the last day of the 12th calendar month after the previous annual inspection.

2. All aircraft that are used for compensation or hire including flight instruction must be inspected on a 100-hr. basis in addition to the annual inspection.

    a. 100-hr. inspections are due every 100 hr. from the prior due time, regardless of when the inspection was actually performed.

91.413 ATC Transponder Tests and Inspections

1. No person may use an ATC transponder unless it has been tested and inspected within the preceding 24 calendar months.

91.417 Maintenance Records

1. An airplane may not be flown unless it has been given an annual inspection within the preceding 12 calendar months.

    a. The annual inspection expires after 1 year, on the last day of the month of issuance.

2. The completion of the annual inspection and the airplane's return to service should be appropriately documented in the airplane maintenance records.

    a. The documentation should include the current status of airworthiness directives and the method of compliance.

3. The airworthiness of an airplane can be determined by a preflight inspection and a review of the maintenance records.

91.519 Passenger Briefings

1. The pilot in command is responsible for ensuring that all passengers have been orally briefed prior to takeoff. The areas that should constitute this briefing are

    a. Smoking,
    b. Use of safety belts and shoulder harnesses,
    c. Location and means of opening the passenger entry door and emergency exits,
    d. Location of survival equipment,
    e. Ditching procedures and the use of flotation equipment, and
    f. Normal and emergency use of oxygen equipment if installed in the airplane.

4.10 NTSB PART 830

830.5 Immediate Notification

1. Even when no injuries occur to occupants, an airplane accident resulting in substantial damage must be reported to the nearest National Transportation Safety Board (NTSB) field office immediately.

2. The following incidents must also be reported immediately to the NTSB:

   a. Inability of any required crewmember to perform normal flight duties because of in-flight injury or illness
   b. In-flight fire
   c. Flight control system malfunction or failure
   d. An overdue airplane that is believed to be involved in an accident
   e. An airplane collision in flight
   f. Turbine (jet) engine failures

830.10 Preservation of Aircraft Wreckage, Mail, Cargo, and Records

1. Prior to the time the Board or its authorized representative takes custody of aircraft wreckage, mail, or cargo, such wreckage, mail, or cargo may not be disturbed or moved except to

   a. Remove persons injured or trapped,
   b. Protect the wreckage from further damage, or
   c. Protect the public from injury.

830.15 Reports and Statements to Be Filed

1. The operator of an aircraft shall file a report on Board Form 6120.1/2 within 10 days after an accident.

   a. A report must be filed within 7 days if an overdue aircraft is still missing.

2. A report on an incident for which immediate notification is required (830.5) shall be filed only when requested by an authorized representative of the Board.

QUESTIONS AND ANSWER EXPLANATIONS

All of the private pilot knowledge test questions chosen by the FAA for release as well as additional questions selected by Gleim relating to the material in the previous outlines are reproduced on the following pages. These questions have been organized into the same subunits as the outlines. To the immediate right of each question are the correct answer and answer explanation. You should cover these answers and answer explanations while responding to the questions. Refer to the general discussion in the Introduction on how to take the FAA pilot knowledge test.

Remember that the questions from the FAA pilot knowledge test bank have been reordered by topic and organized into a meaningful sequence. Also, the first line of the answer explanation gives the citation of the authoritative source for the answer.

QUESTIONS

4.1 FAR Part 1

<u>1.1 General Definitions</u>

1. With respect to the certification of airmen, which is a category of aircraft?

    A. Gyroplane, helicopter, airship, free balloon.

    B. Airplane, rotorcraft, glider, lighter-than-air.

    C. Single-engine land and sea, multiengine land and sea.

Answer (B) is correct. *(FAR 1.1)*
    DISCUSSION: Category of aircraft, as used with respect to the certification, ratings, privileges, and limitations of airmen, means a broad classification of aircraft. Examples include airplane, rotorcraft, glider, and lighter-than-air.
    Answer (A) is incorrect. Gyroplane, helicopter, airship, and free balloon are classes (not categories) used with respect to the certification of airmen. Answer (C) is incorrect. Single-engine land and sea and multiengine land and sea are classes (not categories) used with respect to the certification of airmen.

2. With respect to the certification of airmen, which is a class of aircraft?

    A. Airplane, rotorcraft, glider, lighter-than-air.

    B. Single-engine land and sea, multiengine land and sea.

    C. Lighter-than-air, airship, hot air balloon, gas balloon.

Answer (B) is correct. *(FAR 1.1)*
    DISCUSSION: Class of aircraft, as used with respect to the certification, ratings, privileges, and limitations of airmen, means a classification of aircraft within a category having similar operating characteristics. Examples include single engine, multiengine, land, water, gyroplane, helicopter, airship, and free balloon.
    Answer (A) is incorrect. Airplane, rotorcraft, glider, and lighter-than-air are categories, not classes, used with respect to the certification of airmen. Answer (C) is incorrect. Lighter-than-air is a category, not class, of aircraft used with respect to the certification of airmen.

3. The definition of nighttime is

    A. sunset to sunrise.

    B. 1 hour after sunset to 1 hour before sunrise.

    C. the time between the end of evening civil twilight and the beginning of morning civil twilight.

Answer (C) is correct. *(FAR 1.1)*
    DISCUSSION: "Night" means the time between the end of evening civil twilight and the beginning of morning civil twilight, as published in the American Air Almanac, converted to local time.
    Answer (A) is incorrect. "Sunset to sunrise" is the time during which navigation lights must be used. Answer (B) is incorrect. From 1 hr. after sunset to 1 hr. before sunrise is the definition for nighttime recency of experience requirements.

4. With respect to the certification of aircraft, which is a category of aircraft?

    A. Normal, utility, acrobatic.

    B. Airplane, rotorcraft, glider.

    C. Landplane, seaplane.

Answer (A) is correct. *(FAR 1.1)*
    DISCUSSION: Category of aircraft, as used with respect to the certification of aircraft, means a grouping of aircraft based upon intended use or operating limitations. Examples include transport, normal, utility, acrobatic, limited, restricted, and provisional.
    Answer (B) is incorrect. Airplane, rotorcraft, and glider are categories of aircraft used with respect to the certification of airmen, not aircraft. Answer (C) is incorrect. Landplane and seaplane are classes, not categories, of aircraft used with respect to the certification of aircraft.

5. With respect to the certification of aircraft, which is a class of aircraft?

    A. Airplane, rotorcraft, glider, balloon.

    B. Normal, utility, acrobatic, limited.

    C. Transport, restricted, provisional.

Answer (A) is correct. *(FAR 1.1)*
    DISCUSSION: Class of aircraft, as used with respect to the certification of aircraft, means a broad grouping of aircraft having similar characteristics of propulsion, flight, or landing. Examples include airplane, rotorcraft, glider, balloon, landplane, and seaplane.
    Answer (B) is incorrect. Normal, utility, acrobatic, and limited are categories, not classes, of aircraft used with respect to the certification of aircraft. Answer (C) is incorrect. Transport, restricted, and provisional are categories, not classes, of aircraft used with respect to the certification of aircraft.

6. An ATC clearance provides

    A. priority over all other traffic.

    B. adequate separation from all traffic.

    C. authorization to proceed under specified traffic conditions in controlled airspace.

Answer (C) is correct. *(FAR 1.1)*
    DISCUSSION: A clearance issued by ATC is predicated on known traffic and known physical airport conditions. An ATC clearance means an authorization by ATC, for the purpose of preventing collision between known airplanes, for an airplane to proceed under specified conditions within controlled airspace.
    Answer (A) is incorrect. An ATC clearance does not necessarily give priority over other traffic (although it might in some instances). Answer (B) is incorrect. An ATC clearance only provides separation from other participating traffic.

1.2 Abbreviations and Symbols

7. Which V-speed represents maximum flap extended speed?

    A. V_{FE}.

    B. V_{LOF}.

    C. V_{FC}.

Answer (A) is correct. *(FAR 1.2)*
    DISCUSSION: V_{FE} means the maximum flap extended speed.
    Answer (B) is incorrect. V_{LOF} means liftoff (not maximum flap extended) speed. Answer (C) is incorrect. V_{FC} means maximum speed for stability characteristics, not maximum flap extended speed.

8. Which V-speed represents maximum landing gear extended speed?

    A. V_{LE}.

    B. V_{LO}.

    C. V_{FE}.

Answer (A) is correct. *(FAR 1.2)*
    DISCUSSION: V_{LE} means the maximum landing gear extended speed.
    Answer (B) is incorrect. V_{LO} is the maximum landing gear operating (not extended) speed. Answer (C) is incorrect. V_{FE} is the maximum flap (not landing gear) extended speed.

9. V_{NO} is defined as the

    A. normal operating range.

    B. never-exceed speed.

    C. maximum structural cruising speed.

Answer (C) is correct. *(FAR 1.2)*
    DISCUSSION: V_{NO} is defined as the maximum structural cruising speed.
    Answer (A) is incorrect. The normal airspeed operating range is indicated by the green arc on the airspeed indicator. There is no V-speed for this range. Answer (B) is incorrect. V_{NE} (not V_{NO}) is the never-exceed speed.

10. Which V-speed represents maneuvering speed?

    A. V_A.

    B. V_{LO}.

    C. V_{NE}.

Answer (A) is correct. *(FAR 1.2)*
    DISCUSSION: V_A means design maneuvering speed.
    Answer (B) is incorrect. V_{LO} is the maximum landing gear operating, not the maneuvering, speed. Answer (C) is incorrect. V_{NE} is the never-exceed, not the maneuvering, speed.

11. V_{S0} is defined as the

    A. stalling speed or minimum steady flight speed in the landing configuration.

    B. stalling speed or minimum steady flight speed in a specified configuration.

    C. stalling speed or minimum takeoff safety speed.

Answer (A) is correct. *(FAR 1.2)*
    DISCUSSION: V_{S0} is defined as the stalling speed or minimum steady flight speed in the landing configuration.
    Answer (B) is incorrect. V_{S1} (not V_{S0}) is the stalling speed or minimum steady flight speed in a specified configuration. Answer (C) is incorrect. V_S (not V_{S0}) is the stalling speed, and V_2 min. (not V_{S0}) is the minimum takeoff safety speed.

12. Which would provide the greatest gain in altitude in the shortest distance during climb after takeoff?

    A. V_Y.

    B. V_A.

    C. V_X.

Answer (C) is correct. *(FAR 1.2 and AFH Chap 3)*
    DISCUSSION: V_X means the best angle of climb airspeed (i.e., the airspeed which will provide the greatest gain in altitude in the shortest distance).
    Answer (A) is incorrect. V_Y is the airspeed for the best rate (not angle) of climb. Answer (B) is incorrect. V_A is the design maneuvering airspeed, not the best angle of climb airspeed.

13. After takeoff, which airspeed would the pilot use to gain the most altitude in a given period of time?

    A. V_Y.

    B. V_X.

    C. V_A.

Answer (A) is correct. *(FAR 1.2 and AFH Chap 3)*
    DISCUSSION: V_Y means the airspeed for the best rate of climb (i.e., the airspeed that you use to gain the most altitude in a given period of time).
    Answer (B) is incorrect. V_X is the airspeed for the best angle (not rate) of climb. Answer (C) is incorrect. V_A is the design maneuvering airspeed, not the best rate of climb airspeed.

4.2 FAR Part 21

21.181 Duration of Airworthiness Certificates

14. How long does the Airworthiness Certificate of an aircraft remain valid?

    A. As long as the aircraft has a current Registration Certificate.

    B. Indefinitely, unless the aircraft suffers major damage.

    C. As long as the aircraft is maintained and operated as required by Federal Aviation Regulations.

Answer (C) is correct. *(FAR 21.181)*
    DISCUSSION: The airworthiness certificate of an airplane remains valid as long as the airplane is in an airworthy condition, i.e., operated and maintained as required by the FARs.
    Answer (A) is incorrect. The registration certificate is the document evidencing ownership. A changed registration has no effect on the airworthiness certificate. Answer (B) is incorrect. The airplane must be maintained and operated according to the FARs, not indefinitely. Even if the aircraft suffers major damage, as long as all required repairs are made, the Airworthiness Certificate remains valid.

4.3 FAR Part 39

39.1 Applicability

15. What should an owner or operator know about Airworthiness Directives (AD's)?

    A. For informational purposes only.

    B. They are mandatory.

    C. They are voluntary.

Answer (B) is correct. *(FAR 39.1)*
    DISCUSSION: Airworthiness Directives (ADs) are issued under FAR Part 39 by the FAA to require correction of unsafe conditions found in an airplane, an airplane engine, a propeller, or an appliance when such conditions exist and are likely to exist or develop in other products of the same design. Since ADs are issued under FAR Part 39, they are regulatory and must be complied with, unless a specific exemption is granted.
    Answer (A) is incorrect. ADs outline required maintenance; they are not for informational purposes only. Answer (C) is incorrect. ADs are mandatory, not voluntary.

39.3 General

16. May a pilot operate an aircraft that is not in compliance with an Airworthiness Directive (AD)?

    A. Yes, under VFR conditions only.

    B. Yes, AD's are only voluntary.

    C. Yes, if allowed by the AD.

Answer (C) is correct. *(FAR 39.3)*
    DISCUSSION: An AD is used to notify aircraft owners and other interested persons of unsafe conditions and prescribe the conditions under which the product (e.g., an aircraft) may continue to be operated. An AD may be one of an emergency nature requiring immediate compliance upon receipt or one of a less urgent nature requiring compliance within a relatively longer period of time. You may operate an airplane that is not in compliance with an AD, if such operation is allowed by the AD.
    Answer (A) is incorrect. An AD, not the operating conditions, may allow an aircraft to be operated before compliance with the AD. Answer (B) is incorrect. ADs are mandatory, not voluntary.

4.4 FAR Part 43

43.3 Persons Authorized to Perform Maintenance, Preventive Maintenance, Rebuilding, and Alterations

17. What regulation allows a private pilot to perform preventive maintenance?

   A. 14 CFR Part 91.403.

   B. 14 CFR Part 43.3.

   C. 14 CFR Part 61.113.

Answer (B) is correct. *(FAR 43.3)*
   DISCUSSION: Preventive maintenance means simple or minor preservation operations and the replacement of small standard parts not involving complex assembly operations. Appendix A to Part 43 provides a list of work that is considered preventive maintenance. Part 43 allows a person who holds a pilot certificate to perform preventive maintenance on any aircraft owned or operated by that pilot which is not used in air carrier service.
   Answer (A) is incorrect. FAR 91.403 provides the general operating rules relating to maintenance, not those relating to who can perform preventive maintenance. Answer (C) is incorrect. FAR 61.113 provides the limitations and privileges of a private pilot as pilot in command.

43.7 Persons Authorized to Approve Aircraft Airframes, Aircraft Engines, Propellers, Appliances, or Component Parts for Return to Service after Maintenance, Preventive Maintenance, Rebuilding, or Alteration

18. Who may perform preventive maintenance on an aircraft and approve it for return to service?

   A. Student or Recreational pilot.

   B. Private or Commercial pilot.

   C. None of the above.

Answer (B) is correct. *(FAR 43.7)*
   DISCUSSION: A person who holds a pilot certificate issued under Part 61 may perform preventive maintenance on any airplane owned or operated by that pilot which is not used in air carrier service. To approve the airplane for return to service after preventive maintenance is performed by a pilot, the pilot must hold at least a private pilot certificate.
   Answer (A) is incorrect. While a student or recreational pilot may perform preventive maintenance on any airplane owned or operated by that pilot, the pilot must hold at least a private pilot certificate to approve the airplane's return to service. Answer (C) is incorrect. Any pilot may perform preventive maintenance on an airplane owned or operated by that pilot, but the pilot must hold at least a private pilot certificate to approve the airplane's return to service.

43.9 Maintenance Records

19. Preventive maintenance has been performed on an aircraft. What paperwork is required?

   A. A full, detailed description of the work done must be entered in the airframe logbook.

   B. The date the work was completed, and the name of the person who did the work must be entered in the airframe and engine logbook.

   C. The signature, certificate number, and kind of certificate held by the person approving the work and a description of the work must be entered in the aircraft maintenance records.

Answer (C) is correct. *(FAR 43.9)*
   DISCUSSION: After preventive maintenance has been performed, the signature, certificate number, and kind of certificate held by the person approving the work and a description of the work must be entered in the aircraft maintenance records.
   Answer (A) is incorrect. The signature, certificate number, and kind of certificate, in addition to the description of work performed, must be entered into the maintenance records. Answer (B) is incorrect. A description of work completed, signature, certificate number, and kind of certificate held by the person approving the work (if different than the person who did the work), in addition to the date the work was completed, must be entered into the maintenance records.

Part 43 Appendix A. Major Alterations and Repairs and Preventive Maintenance

20. Which operation would be described as preventive maintenance?

   A. Servicing landing gear wheel bearings.

   B. Alteration of main seat support brackets.

   C. Engine adjustments to allow automotive gas to be used.

Answer (A) is correct. *(Appendix A to Part 43)*
   DISCUSSION: Appendix A to Part 43 provides a list of work that is considered preventive maintenance. Preventive maintenance means simple or minor preservation operations and the replacement of small standard parts not involving complex assembly operations. Servicing landing gear wheel bearings, such as cleaning and greasing, is considered preventive maintenance.
   Answer (B) is incorrect. The alteration of main seat support brackets is considered an airframe major repair, not preventive maintenance. Answer (C) is incorrect. Engine adjustments to allow automotive gas to be used is considered a powerplant major alteration, not preventive maintenance.

21. Which operation would be described as preventive maintenance?

    A. Repair of landing gear brace struts.

    B. Replenishing hydraulic fluid.

    C. Repair of portions of skin sheets by making additional seams.

Answer (B) is correct.   *(Appendix A to Part 43)*
    DISCUSSION: Appendix A to Part 43 provides a list of work that is considered preventive maintenance. Preventive maintenance means simple or minor preservation operations and the replacement of small standard parts not involving complex assembly operations. An example of preventive maintenance is replenishing hydraulic fluid.
    Answer (A) is incorrect. The repair of landing gear brace struts is considered an airframe major repair, not preventive maintenance. Answer (C) is incorrect. The repair of portions of skin sheets by making additional seams is considered an airframe major repair, not preventive maintenance.

4.5 FAR Part 61

61.3 Requirements for Certificates, Ratings, and Authorizations

22. When must a current pilot certificate be in the pilot's personal possession or readily accessible in the aircraft?

    A. When acting as a crew chief during launch and recovery.

    B. Only when passengers are carried.

    C. Anytime when acting as pilot in command or as a required crewmember.

Answer (C) is correct.   *(FAR 61.3)*
    DISCUSSION: Current and appropriate pilot and medical certificates must be in your personal possession or readily accessible in the aircraft when you act as pilot in command (PIC) or as a required pilot flight crewmember.
    Answer (A) is incorrect. A current pilot certificate must be in your personal possession when acting as a PIC or as a required crewmember of an aircraft, not when acting as a crew chief during launch and recovery of an airship. Answer (B) is incorrect. Anytime you fly as PIC or as a required crewmember, you must have a current pilot certificate in your personal possession regardless of whether passengers are carried or not.

23. A recreational or private pilot acting as pilot in command, or in any other capacity as a required pilot flight crewmember, must have in his or her personal possession or readily accessible in the aircraft a current

    A. logbook endorsement to show that a flight review has been satisfactorily accomplished.

    B. medical certificate if required and an appropriate pilot certificate.

    C. endorsement on the pilot certificate to show that a flight review has been satisfactorily accomplished.

Answer (B) is correct.   *(FAR 61.3)*
    DISCUSSION: Current and appropriate pilot and medical certificates must be in your personal possession or readily accessible in the aircraft when you act as pilot in command (PIC) or as a required pilot flight crewmember.
    Answer (A) is incorrect. As a private pilot, you need not have your logbook in your possession or readily accessible aboard the airplane. A recreational pilot must carry a logbook to show evidence of an endorsement that permits certain flights, not to show completion of a flight review. Answer (C) is incorrect. The endorsement after satisfactorily completing a flight review is made in your pilot logbook, not on your pilot certificate.

24. What document(s) must be in your personal possession or readily accessible in the aircraft while operating as pilot in command of an aircraft?

    A. Certificates showing accomplishment of a checkout in the aircraft and a current biennial flight review.

    B. A pilot certificate with an endorsement showing accomplishment of an annual flight review and a pilot logbook showing recency of experience.

    C. An appropriate pilot certificate and an appropriate current medical certificate if required.

Answer (C) is correct.   *(FAR 61.3)*
    DISCUSSION: Current and appropriate pilot and medical certificates must be in your personal possession or readily accessible in the aircraft when you act as pilot in command (PIC) or as a required pilot flight crewmember.
    Answer (A) is incorrect. Flight reviews and checkouts in aircraft are documented in your logbook rather than on separate certificates and need not be in your personal possession. Answer (B) is incorrect. The endorsement after satisfactorily completing a flight review is made in your logbook, not on your pilot certificate. You are not required to have your pilot logbook in your personal possession while acting as pilot in command.

25. Each person who holds a pilot certificate or a medical certificate shall present it for inspection upon the request of the Administrator, the National Transportation Safety Board, or any

- A. authorized representative of the Department of Transportation.
- B. person in a position of authority.
- C. federal, state, or local law enforcement officer.

Answer (C) is correct. *(FAR 61.3)*
DISCUSSION: Each person who holds a pilot certificate, flight instructor certificate, medical certificate, authorization, or license required by the FARs shall present it for inspection upon the request of the Administrator (of the FAA), an authorized representative of the National Transportation Safety Board, or any federal, state, or local law enforcement officer.
Answer (A) is incorrect. An authorized representative of the Department of Transportation is just one example of those who can ask to inspect pilot and medical certificates. Answer (B) is incorrect. Not just any person with any kind of authority, such as a foreman, can inspect your pilot certificate or medical certificate.

61.15 Offenses Involving Alcohol or Drugs

26. How soon after the conviction for driving while intoxicated by alcohol or drugs shall it be reported to the FAA, Civil Aviation Security Division?

- A. No later than 60 days after the motor vehicle action.
- B. No later than 30 working days after the motor vehicle action.
- C. Required to be reported upon renewal of medical certificate.

Answer (A) is correct. *(FAR 61.15)*
DISCUSSION: Each person holding a certificate under Part 61 shall provide a written report of each motor vehicle action involving alcohol or drugs to the FAA, Civil Aviation Security Division, no later than 60 days after the motor vehicle action.
Answer (B) is incorrect. FAR 61.15 allows a person 60 days after the motor vehicle action, not 30 days. Answer (C) is incorrect. A person must notify the Civil Aviation Security Division no later than 60 days after a motor vehicle action involving drugs or alcohol.

27. How soon after the conviction for driving while intoxicated by alcohol or drugs shall it be reported to the FAA, and which division should this be reported to?

- A. Within 60 days to the Airmen Records Division.
- B. Within 60 days to the Civil Aviation Security Division.
- C. Within 60 days to the Regulatory Support Division.

Answer (B) is correct. *(FAR 61.15)*
DISCUSSION: Each person holding a certificate under Part 61 shall provide a written report of each motor vehicle action involving alcohol or drugs to the FAA, Civil Aviation Security Division, no later than 60 days after the motor vehicle action.
Answer (A) is incorrect. The Airmen Records Division handles the certification of airmen, not alcohol and/or drug convictions. Answer (C) is incorrect. The Regulatory Support Division is involved in airmen testing, designee standardization, and the management of aviation data systems.

61.23 Medical Certificates: Requirement and Duration

28. A Third-Class Medical Certificate is issued to a 36-year-old pilot on August 10, this year. To exercise the privileges of a Private Pilot Certificate, the medical certificate will be valid until midnight on

- A. August 10, 3 years later.
- B. August 31, 5 years later.
- C. August 31, 3 years later.

Answer (B) is correct. *(FAR 61.23)*
DISCUSSION: A pilot may exercise the privileges of a private pilot certificate under a third-class medical certificate until it expires at the end of the last day of the month 5 years after it was issued, for pilots less than 40 years old on the date of the medical examination. A third-class medical certificate issued to a 36-year-old pilot on Aug. 10 will be valid until midnight on Aug. 31, 5 years later.
Answer (A) is incorrect. Medical certificates expire at the last day of the month. Thus, a medical certificate issued on Aug. 10 will expire on Aug. 31, not Aug. 10. Additionally, since the pilot is less than 40 years old, the third-class medical certificate is valid for 5 years, not 3 years. Answer (C) is incorrect. A pilot may exercise the privileges of a private pilot certificate under a third-class medical certificate until it expires at the end of the last day of the month 5 years later if the pilot was less than 40 years old on the date of the medical examination.

29. A Third-Class Medical Certificate is issued to a 51-year-old pilot on May 3, this year. To exercise the privileges of a Private Pilot Certificate, the medical certificate will be valid until midnight on

    A. May 3, 1 year later.

    B. May 31, 1 year later.

    C. May 31, 2 years later.

Answer (C) is correct. *(FAR 61.23)*
    DISCUSSION: A pilot may exercise the privileges of a private pilot certificate under a third-class medical certificate until it expires at the end of the last day of the month 2 years after it was issued, for pilots 40 years old or older on the date of the medical examination. A third-class medical certificate issued to a 51-year-old pilot on May 3 will be valid until midnight on May 31, 2 years later.
    Answer (A) is incorrect. Medical certificates expire on the last day of the month. Thus, a medical certificate issued on May 3 will expire on May 31, not May 3. Additionally, a third-class medical certificate is valid for 2 years, not 1 year, if the pilot is over 40 years old. Answer (B) is incorrect. A pilot may exercise the privileges of a private pilot certificate under a third-class medical certificate until it expires at the end of the last day of the month, 2 years, not 1 year, later if the pilot was 40 years old or older on the date of the examination.

30. For private pilot operations, a Second-Class Medical Certificate issued to a 42-year-old pilot on July 15, this year, will expire at midnight on

    A. July 15, 2 years later.

    B. July 31, 1 year later.

    C. July 31, 2 years later.

Answer (C) is correct. *(FAR 61.23)*
    DISCUSSION: For private pilot operations, a second-class medical certificate will expire at the end of the last day of the month, 2 years after it was issued, for pilots 40 years old or older on the date of the medical examination. For private pilot operations, a second-class medical certificate issued to a 42-year-old pilot on July 15 will be valid until midnight on July 31, 2 years later.
    Answer (A) is incorrect. A medical certificate expires on the last day of the month. Thus, a medical certificate issued on July 15 will expire on July 31, not July 15. Answer (B) is incorrect. A second-class medical certificate is valid for 1 year for operations requiring a commercial pilot certificate.

31. For private pilot operations, a First-Class Medical Certificate issued to a 23-year-old pilot on October 21, this year, will expire at midnight on

    A. October 21, 2 years later.

    B. October 31, next year.

    C. October 31, 5 years later.

Answer (C) is correct. *(FAR 61.23)*
    DISCUSSION: For private pilot operations, a first-class medical certificate will expire at the end of the last day of the month, 5 years after it was issued, for pilots less than 40 years old on the date of the medical examination. For private pilot operations, a first-class medical certificate issued to a 23-year-old pilot on Oct. 21 will be valid until midnight on Oct. 31, 5 years later.
    Answer (A) is incorrect. A medical certificate expires on the last day of the month. Thus, a medical certificate issued on Oct. 21 will expire on Oct. 31, not Oct. 21. Additionally, for private pilot operations, the medical certificate is valid for 5 years, not 2 years, for a pilot less than 40 years old on the date of the medical examination. Answer (B) is incorrect. A first-class medical certificate is valid for 1 year for operations requiring a commercial pilot certificate.

32. A Third-Class Medical Certificate was issued to a 19-year-old pilot on August 10, this year. To exercise the privileges of a recreational or private pilot certificate, the medical certificate will expire at midnight on

    A. August 10, 2 years later.

    B. August 31, 5 years later.

    C. August 31, 2 years later.

Answer (B) is correct. *(FAR 61.23)*
    DISCUSSION: A pilot may exercise the privileges of a recreational or private pilot certificate under a third-class medical certificate until it expires at the end of the last day of the month 5 years after it was issued, for pilots less than 40 years old at the time of the medical examination. A third-class medical certificate issued to a 19-year-old pilot on Aug. 10 will be valid until midnight on Aug. 31, 5 years later.
    Answer (A) is incorrect. Medical certificates expire at the end of the month. A medical certificate issued on Aug. 10 will expire at midnight on Aug. 31, not Aug. 10. Additionally, since the pilot is less than 40 years old, the third-class medical certificate is valid for 5 years, not 3 years. Answer (C) is incorrect. A pilot may exercise the privileges of a recreational or private pilot certificate under a third-class medical certificate until it expires at the end of the last day of the month 2 years later if the pilot was 40 years old or older, not less than 40 years old, on the date of the medical examination.

61.31 Type Rating Requirements, Additional Training, and Authorization Requirements

33. Before a person holding a private pilot certificate may act as pilot in command of a high-performance airplane, that person must have

A. passed a flight test in that airplane from an FAA inspector.

B. an endorsement in that person's logbook that he or she is competent to act as pilot in command.

C. received ground and flight instruction from an authorized flight instructor who then endorses that person's logbook.

Answer (C) is correct. *(FAR 61.31)*
DISCUSSION: A private pilot may not act as pilot in command of a high-performance airplane (an airplane with an engine of more than 200 horsepower) unless (s)he has received and logged ground and flight training from an authorized instructor who has certified in his/her logbook that (s)he is proficient to operate a high-performance airplane.
Answer (A) is incorrect. No FAA flight test is required, only ground and flight training and an endorsement from an authorized flight instructor. Answer (B) is incorrect. The ground and flight training and endorsement must be by an authorized flight instructor.

34. What is the definition of a high-performance airplane?

A. An airplane with 180 horsepower, or retractable landing gear, flaps, and a fixed-pitch propeller.

B. An airplane with a normal cruise speed in excess of 200 knots.

C. An airplane with an engine of more than 200 horsepower.

Answer (C) is correct. *(FAR 61.31)*
DISCUSSION: A high-performance airplane is defined as an airplane with an engine of more than 200 horsepower.
Answer (A) is incorrect. A high-performance airplane is an airplane with an engine of more than 200 horsepower, not an airplane with 180 horsepower, or retractable landing gear, flaps, and a fixed-pitch propeller. Answer (B) is incorrect. A high-performance airplane is an airplane with an engine of more than 200 horsepower, not an airplane with a normal cruise speed in excess of 200 knots.

35. The pilot in command is required to hold a type rating in which aircraft?

A. Aircraft operated under an authorization issued by the Administrator.

B. Aircraft having a gross weight of more than 12,500 pounds.

C. Aircraft involved in ferry flights, training flights, or test flights.

Answer (B) is correct. *(FAR 61.31)*
DISCUSSION: A person may not act as pilot in command of any of the following aircraft unless (s)he holds a type rating for that aircraft:

(1) A large aircraft (except lighter-than-air), i.e., over 12,500 lb. gross weight
(2) A turbojet-powered airplane
(3) Other aircraft specified by the FAA through aircraft type certificate procedures

Answer (A) is incorrect. All aircraft, not only those requiring type ratings, are operated under an authorization issued by the Administrator of the FAA. Answer (C) is incorrect. Any type of aircraft can be involved in ferry flights, training flights, or test flights and may not require a type rating (e.g., your single-engine trainer airplane on a training flight).

36. In order to act as pilot in command of a high-performance airplane, a pilot must have

A. made and logged three solo takeoffs and landings in a high-performance airplane.

B. received and logged ground and flight instruction in an airplane that has more than 200 horsepower.

C. passed a flight test in a high-performance airplane.

Answer (B) is correct. *(FAR 61.31)*
DISCUSSION: Prior to acting as pilot in command of an airplane with an engine of more than 200 horsepower, a person is required to receive and log ground and flight training in such an airplane from an authorized flight instructor who has certified in the pilot's logbook that the individual is proficient to operate a high-performance airplane.
Answer (A) is incorrect. In order to act as pilot in command of a high-performance airplane, you must have received and logged ground and flight training from an authorized flight instructor, not have made three solo takeoffs and landings. Answer (C) is incorrect. You must have received and logged ground and flight training, and a logbook endorsement from an authorized flight instructor, not a flight test, is required prior to acting as a pilot in command of a high-performance airplane.

61.56 Flight Review

37. To act as pilot in command of an aircraft carrying passengers, a pilot must show by logbook endorsement the satisfactory completion of a flight review or completion of a pilot proficiency check within the preceding

A. 6 calendar months.

B. 12 calendar months.

C. 24 calendar months.

Answer (C) is correct. *(FAR 61.56)*
DISCUSSION: To act as pilot in command of an aircraft (whether carrying passengers or not), a pilot must show by logbook endorsement the satisfactory completion of a flight review or completion of a pilot proficiency check within the preceding 24 calendar months.
Answer (A) is incorrect. A pilot must have satisfactorily completed a flight review or completion of a pilot proficiency check within the preceding 24 (not 6) calendar months. Answer (B) is incorrect. A pilot must have satisfactorily completed a flight review or completion of a pilot proficiency check within the preceding 24 (not 12) calendar months.

38. If a recreational or private pilot had a flight review on August 8, this year, when is the next flight review required?

A. August 8, 2 years later.

B. August 31, next year.

C. August 31, 2 years later.

Answer (C) is correct. *(FAR 61.56)*
DISCUSSION: A pilot is required to have a flight review within the preceding 24 calendar months before the month in which the pilot acts as pilot in command. Thus, a pilot who had a flight review on Aug. 8 of this year must have a flight review completed by Aug. 31, 2 years later.
Answer (A) is incorrect. Flight reviews expire at the end of the month. Thus, a flight review on Aug. 8 will expire on Aug. 31, not Aug. 8. Answer (B) is incorrect. A flight review is valid for 2 years, not 1 year.

39. Each recreational or private pilot is required to have

A. a biennial flight review.

B. an annual flight review.

C. a semiannual flight review.

Answer (A) is correct. *(FAR 61.56)*
DISCUSSION: Each recreational or private pilot is required to have a biennial (every 2 years) flight review.
Answer (B) is incorrect. Each pilot is required to have a biennial, not annual, flight review. Answer (C) is incorrect. Each pilot is required to have a biennial, not semiannual, flight review.

40. If a recreational or private pilot had a flight review on August 8, this year, when is the next flight review required?

A. August 8, next year.

B. August 31, 1 year later.

C. August 31, 2 years later.

Answer (C) is correct. *(FAR 61.56)*
DISCUSSION: A pilot is required to have a flight review within the preceding 24 calendar months before the month in which the pilot acts as pilot in command. Thus, a recreational or private pilot who had a flight review on Aug. 8 of this year must have a flight review completed by Aug. 31, 2 years later.
Answer (A) is incorrect. Flight reviews expire at the end of the month. Thus, a flight review on Aug. 8 will expire on Aug. 31, not Aug. 8. Answer (B) is incorrect. A flight review is valid for 2 years, not 1 year.

61.57 Recent Flight Experience: Pilot in Command

41. To act as pilot in command of an aircraft carrying passengers, the pilot must have made at least three takeoffs and three landings in an aircraft of the same category, class, and if a type rating is required, of the same type, within the preceding

A. 90 days.

B. 12 calendar months.

C. 24 calendar months.

Answer (A) is correct. *(FAR 61.57)*
DISCUSSION: To act as pilot in command of an airplane with passengers aboard, you must have made at least three takeoffs and three landings (to a full stop if in a tailwheel airplane) in an airplane of the same category, class, and, if a type rating is required, of the same type within the preceding 90 days. Category refers to airplane, rotorcraft, etc.; class refers to single- or multi-engine, land or sea.
Answer (B) is incorrect. A flight review, not recency experience, is required every 24, not 12, calendar months. Answer (C) is incorrect. A flight review, not recency experience, is normally required of all pilots every 24 months.

129

42. If recency of experience requirements for night flight are not met and official sunset is 1830, the latest time passengers may be carried is

A. 1829.

B. 1859.

C. 1929.

Answer (C) is correct. *(FAR 61.57)*
DISCUSSION: For the purpose of night recency experience flight time, night is defined as the period beginning 1 hr. after sunset and ending 1 hr. before sunrise. If you have not met the night experience requirements and official sunset is 1830, a landing must be accomplished at or before 1929 if passengers are carried.
Answer (A) is incorrect. This is the time that night begins for the purpose of turning on aircraft position (navigation) lights. Answer (B) is incorrect. There is no regulation concerning the time 30 min. after official sunset.

43. To act as pilot in command of an aircraft carrying passengers, the pilot must have made three takeoffs and three landings within the preceding 90 days in an aircraft of the same

A. make and model.

B. category and class, but not type.

C. category, class, and type, if a type rating is required.

Answer (C) is correct. *(FAR 61.57)*
DISCUSSION: No one may act as pilot in command of an airplane carrying passengers unless within the preceding 90 days (s)he has made three takeoffs and three landings as sole manipulator of the controls in an aircraft of the same category and class and, if a type rating is required, the same type. If the aircraft is a tailwheel airplane, the landings must have been to a full stop.
Answer (A) is incorrect. It must be the same category and class (not make and model) and, if a type rating is required, the same type. Answer (B) is incorrect. It must be the same type aircraft if a type rating is required for that aircraft.

44. The three takeoffs and landings that are required to act as pilot in command at night must be done during the time period from

A. sunset to sunrise.

B. 1 hour after sunset to 1 hour before sunrise.

C. the end of evening civil twilight to the beginning of morning civil twilight.

Answer (B) is correct. *(FAR 61.57)*
DISCUSSION: No one may act as pilot in command of an aircraft carrying passengers at night (i.e., the period from 1 hr. after sunset to 1 hr. before sunrise as published in the American Air Almanac) unless (s)he has made three takeoffs and three landings to a full stop within the preceding 90 days, at night, in the category and class of aircraft to be used.
Answer (A) is incorrect. The period from sunset to sunrise is the time that aircraft lights are required to be on, not the time period for night recency experience. Answer (C) is incorrect. The end of evening civil twilight to the beginning of morning civil twilight is the definition of night, not the time period for night recency experience.

45. To meet the recency of experience requirements to act as pilot in command carrying passengers at night, a pilot must have made at least three takeoffs and three landings to a full stop within the preceding 90 days in

A. the same category and class of aircraft to be used.

B. the same type of aircraft to be used.

C. any aircraft.

Answer (A) is correct. *(FAR 61.57)*
DISCUSSION: No one may act as pilot in command of an aircraft carrying passengers at night (i.e., the period from 1 hr. after sunset to 1 hr. before sunrise) unless (s)he has made three takeoffs and three landings to a full stop within the preceding 90 days, at night, in the category and class of aircraft to be used.
Answer (B) is incorrect. Unless a type-rating is required, it does not have to be the same type of aircraft. ("Type" refers to a specific make and general model, e.g., Cessna 152/172.) Answer (C) is incorrect. It must be the same category and class (not any) of aircraft to be used.

46. The takeoffs and landings required to meet the recency of experience requirements for carrying passengers in a tailwheel airplane

A. may be touch and go or full stop.

B. must be touch and go.

C. must be to a full stop.

Answer (C) is correct. *(FAR 61.57)*
DISCUSSION: To comply with recency requirements for carrying passengers in a tailwheel airplane, one must have made three takeoffs and landings to a full stop within the past 90 days.
Answer (A) is incorrect. In a tailwheel airplane, the takeoffs and landings must be to a full stop only, not touch and go. Answer (B) is incorrect. In a tailwheel airplane, the takeoffs and landings must be to a full stop, not touch and go.

61.60 Change of Address

47. If a certificated pilot changes permanent mailing address and fails to notify the FAA Airmen Certification Branch of the new address, the pilot is entitled to exercise the privileges of the pilot certificate for a period of only

A. 30 days after the date of the move.

B. 60 days after the date of the move.

C. 90 days after the date of the move.

Answer (A) is correct. *(FAR 61.60)*
  DISCUSSION: If you have changed your permanent mailing address, you may not exercise the privileges of your pilot certificate after 30 days from the date of the address change unless you have notified the FAA in writing of the change. You are required to notify the Airmen Certification Branch at P.O. Box 25082, Oklahoma City, OK 73125.
  Note: While you must notify the FAA if your address changes, you are not required to carry a certificate that shows your current address. The FAA will not issue a new certificate upon receipt of your new address unless you send a written request and $2 to the address shown above.
  Answer (B) is incorrect. If you change your permanent mailing address, you may exercise the privileges of your pilot certificate for a period of only 30 (not 60) days after the date you move unless you notify the FAA in writing of the change. Answer (C) is incorrect. If you change your permanent mailing address, you may exercise the privileges of your pilot certificate for a period of only 30 (not 90) days after the date you move unless you notify the FAA in writing of the change.

61.69 Glider Towing: Experience and Training Requirements

48. A certificated private pilot may not act as pilot in command of an aircraft towing a glider unless there is entered in the pilot's logbook a minimum of

A. 100 hours of pilot flight time in any aircraft the pilot is using to tow a glider.

B. 100 hours of pilot-in-command time in the aircraft category, class, and type, if required, that the pilot is using to tow a glider.

C. 200 hours of pilot-in-command time in the aircraft category, class, and type, if required, that the pilot is using to tow a glider.

Answer (B) is correct. *(FAR 61.69[a][2])*
  DISCUSSION: As a private pilot, you may not act as pilot in command of an aircraft towing a glider unless you have had, and entered in your logbook, at least 100 hr. of pilot-in-command time in the aircraft category, class, and type, if required, that you are using to tow a glider.
  Answer (A) is incorrect. You must have logged at least 100 hr. as pilot in command in the aircraft category, class, and type, if required, that you are using to tow a glider, not just any aircraft. Answer (C) is incorrect. You must have logged at least 100 hr., not 200 hr., as pilot in command.

49. To act as pilot in command of an aircraft towing a glider, a pilot is required to have made within the preceding 12 months

A. at least three flights as observer in a glider being towed by an aircraft.

B. at least three flights in a powered glider.

C. at least three actual or simulated glider tows while accompanied by a qualified pilot.

Answer (C) is correct. *(FAR 61.69)*
  DISCUSSION: To act as pilot in command of an aircraft towing a glider, you are required to have made, in the preceding 12 months,
  (1) At least three actual or simulated glider tows while accompanied by a qualified pilot or
  (2) At least three flights as pilot in command of a glider towed by an aircraft.
  Answer (A) is incorrect. You are required to have made within the preceding 12 months at least three flights as pilot in command, not as an observer, of a glider being towed by an aircraft. Answer (B) is incorrect. You are required to have made within the preceding 12 months at least three flights as pilot in command of a glider towed by an aircraft, not three flights in a powered glider.

61.113 Private Pilot Privileges and Limitations: Pilot in Command

50. In regard to privileges and limitations, a private pilot may

A. act as pilot in command of an aircraft carrying a passenger for compensation if the flight is in connection with a business or employment.

B. not pay less than the pro rata share of the operating expenses of a flight with passengers provided the expenses involve only fuel, oil, airport expenditures, or rental fees.

C. not be paid in any manner for the operating expenses of a flight.

Answer (B) is correct. *(FAR 61.113)*
DISCUSSION: A private pilot may not pay less than an equal (pro rata) share of the operating expenses of a flight with passengers. These expenses may involve only fuel, oil, airport expenditures (e.g., landing fees, tie-down fees, etc.), or rental fees.
Answer (A) is incorrect. A private pilot cannot act as pilot in command of an aircraft carrying a passenger for compensation. Answer (C) is incorrect. A private pilot may equally share the operating expenses of a flight with his/her passengers.

51. According to regulations pertaining to privileges and limitations, a private pilot may

A. be paid for the operating expenses of a flight if at least three takeoffs and three landings were made by the pilot within the preceding 90 days.

B. not pay less than the pro rata share of the operating expenses of a flight with passengers provided the expenses involve only fuel, oil, airport expenditures, or rental fees.

C. not be paid in any manner for the operating expenses of a flight.

Answer (B) is correct. *(FAR 61.113)*
DISCUSSION: A private pilot may not pay less than an equal (pro rata) share of the operating expenses of a flight with passengers. These expenses may involve only fuel, oil, airport expenditures (e.g., landing fees, tie-down fees, etc.), or rental fees.
Answer (A) is incorrect. A private pilot may be paid for the operating expenses of a flight in connection with any business or employment if the flight is only incidental to that business or employment and no passengers or property are carried for compensation or hire, not if the pilot has made three takeoffs and landings in the preceding 90 days. Answer (C) is incorrect. A private pilot may equally share the operating expenses of a flight with his/her passengers.

52. What exception, if any, permits a private pilot to act as pilot in command of an aircraft carrying passengers who pay for the flight?

A. If the passengers pay all the operating expenses.

B. If a donation is made to a charitable organization for the flight.

C. There is no exception.

Answer (B) is correct. *(FAR 61.113)*
DISCUSSION: A private pilot may act as pilot in command of an airplane used in a passenger-carrying airlift sponsored by a charitable organization for which passengers make donations to the organization, provided the following requirements are met: the local FSDO is notified at least 7 days before the flight, the flight is conducted from an adequate public airport, the pilot has logged at least 500 hr., no acrobatic or formation flights are performed, the 100-hr. inspection of the airplane requirement is complied with, and the flight is day-VFR.
Answer (A) is incorrect. A private pilot may only share the operating costs, not have the passengers pay for all the operating costs. Answer (C) is incorrect. The exception is a passenger-carrying airlift sponsored by a charitable organization.

4.6 Recreational Pilot Related FARs

61.101 Recreational Pilot Privileges and Limitations

NOTE: Questions 53 through 79 are not included on the private pilot knowledge test.

53. A recreational pilot acting as pilot in command must have in his or her personal possession while aboard the aircraft

A. a current logbook endorsement to show that a flight review has been satisfactorily accomplished.

B. the pilot logbook to show recent experience requirements to serve as pilot in command have been met.

C. a current logbook endorsement that permits flight within 50 nautical miles from the departure airport.

Answer (C) is correct. *(FAR 61.101)*
DISCUSSION: A recreational pilot acting as pilot in command must have in his/her personal possession while aboard the airplane a current logbook endorsement that permits flight within 50 NM from the departure airport.
Answer (A) is incorrect. To act as pilot in command, a recreational pilot must have in his/her personal possession while aboard the aircraft a current logbook endorsement that permits flight within 50 NM from the departure airport, not an endorsement to show that a flight review has been satisfactorily completed. Answer (B) is incorrect. To act as pilot in command, a recreational pilot must have in his/her personal possession while aboard the aircraft a current logbook endorsement that permits flight within 50 NM from the departure airport, not a logbook showing that the recent experience requirements have been met.

54. How many passengers is a recreational pilot allowed to carry on board?

A. One.

B. Two.

C. Three.

Answer (A) is correct. *(FAR 61.101)*
 DISCUSSION: Recreational pilots may carry not more than one passenger.
 Answer (B) is incorrect. A recreational pilot is allowed to carry only one passenger (not two). Answer (C) is incorrect. A recreational pilot is allowed to carry only one passenger (not three).

55. When may a recreational pilot act as pilot in command of an aircraft at night?

A. When obtaining an additional certificate or rating under the supervision of an authorized instructor, provided the surface or flight visibility is at least 1 statute mile.

B. When obtaining an additional certificate or rating under the supervision of an authorized instructor, provided the surface or flight visibility is at least 3 statute miles.

C. When obtaining an additional certificate or rating under the supervision of an authorized instructor, provided the surface or flight visibility is at least 5 statute miles.

Answer (C) is correct. *(FAR 61.101)*
 DISCUSSION: A recreational pilot obtaining an additional certificate or rating under the supervision of an authorized instructor is treated as a student pilot with respect to flight at night, since recreational pilots do not have night-flying privilege. Therefore, a recreational pilot may act as pilot in command of an aircraft at night when obtaining an additional certificate or rating under the supervision of an authorized instructor, provided the surface or flight visibility is at least 5 statute miles.
 Answer (A) is incorrect. A recreational pilot may act as pilot in command of an aircraft at night when obtaining an additional certificate or rating under the supervision of an authorized instructor, provided the surface or flight visibility is at least 5 statute miles, not 1 statute mile. Answer (B) is incorrect. A recreational pilot may act as pilot in command of an aircraft at night when obtaining an additional certificate or rating under the supervision of an authorized instructor, provided the surface or flight visibility is at least 5 statute miles, not 3 statute miles.

56. According to regulations pertaining to privileges and limitations, a recreational pilot may

A. be paid for the operating expenses of a flight.

B. not pay less than the pro rata share of the operating expenses of a flight with a passenger.

C. not be paid in any manner for the operating expenses of a flight.

Answer (B) is correct. *(FAR 61.101)*
 DISCUSSION: A recreational pilot may not pay less than an equal (pro rata) share of the operating expenses of the flight with a passenger. These expenses may involve only fuel, oil, airport expenditures (e.g., landing fees, tie-down fees, etc.), or rental fees.
 Answer (A) is incorrect. Operating expenses can only be shared by a passenger, not paid for by a passenger; i.e., a recreational pilot cannot carry a passenger or property for compensation or hire. Answer (C) is incorrect. A recreational pilot may equally share operating expenses of the flight with the passenger.

57. When may a recreational pilot act as pilot in command on a cross-country flight that exceeds 50 nautical miles from the departure airport?

A. After receiving ground and flight instructions on cross-country training and a logbook endorsement.

B. After attaining 100 hours of pilot-in-command time and a logbook endorsement.

C. 12 calendar months after receiving his or her recreational pilot certificate and a logbook endorsement.

Answer (A) is correct. *(FAR 61.101)*
 DISCUSSION: A recreational pilot may act as pilot in command on a cross-country flight that exceeds 50 NM from the departure airport, provided that person has received ground and flight training from an authorized instructor on the cross-country training requirements for a private pilot certificate and has received a logbook endorsement, which is in the person's possession in the aircraft, certifying the person is proficient in cross-country flying.
 Answer (B) is incorrect. A recreational pilot may act as pilot in command on a cross-country flight that exceeds 50 NM from the departure airport after receiving ground and flight training from an authorized instructor on the cross-country training requirements for a private pilot certificate and after receiving a logbook endorsement, not after attaining 100 hr. as pilot-in-command time and a logbook endorsement. Answer (C) is incorrect. A recreational pilot may act as pilot in command on a cross-country flight after receiving ground and flight training from an authorized instructor on the cross-country training requirements for a private pilot certificate and after receiving a logbook endorsement, not 12 months after receiving his/her recreational pilot certificate and a logbook endorsement.

58. In regard to privileges and limitations, a recreational pilot may

    A. fly for compensation or hire within 50 nautical miles from the departure airport with a logbook endorsement.

    B. not pay less than the pro rata share of the operating expenses of a flight with a passenger.

    C. not be paid in any manner for the operating expenses of a flight from a passenger.

Answer (B) is correct. *(FAR 61.101)*
    DISCUSSION: A recreational pilot may not pay less than an equal (pro rata) share of the operating expenses of the flight with a passenger. These expenses may involve only fuel, oil, airport expenditures (e.g., landing fees, tie-down fees, etc.), or rental fees.
    Answer (A) is incorrect. Recreational pilots may not fly for compensation or hire. Answer (C) is incorrect. A recreational pilot may equally share operating expenses of the flight with a passenger.

59. A recreational pilot may act as pilot in command of an aircraft that is certificated for a maximum of how many occupants?

    A. Two.

    B. Three.

    C. Four.

Answer (C) is correct. *(FAR 61.101)*
    DISCUSSION: Recreational pilots may not act as pilot in command of an aircraft that is certificated for more than four occupants. Note, however, that only two occupants are permitted, the recreational pilot and a passenger.
    Answer (A) is incorrect. A recreational pilot can act as pilot in command of an aircraft that is certificated for up to four (not two) occupants. Answer (B) is incorrect. Recreational pilots can act as pilot in command of an aircraft certificated for up to four (not three) occupants.

60. A recreational pilot may act as pilot in command of an aircraft with a maximum engine horsepower of

    A. 160.

    B. 180.

    C. 200.

Answer (B) is correct. *(FAR 61.101)*
    DISCUSSION: A recreational pilot may act as pilot in command of an aircraft with a maximum engine horsepower of 180.
    Answer (A) is incorrect. A recreational pilot may act as pilot in command of an aircraft with a maximum horsepower of 180 (not 160). Answer (C) is incorrect. A recreational pilot may act as pilot in command of an aircraft with a maximum horsepower of 180 (not 200).

61. With respect to daylight hours, what is the earliest time a recreational pilot may take off?

    A. One hour before sunrise.

    B. At sunrise.

    C. At the beginning of morning civil twilight.

Answer (B) is correct. *(FAR 61.101)*
    DISCUSSION: A recreational pilot may not act as pilot in command of an airplane between sunset and sunrise. Thus, the earliest time a recreational pilot may take off is at sunrise.
    Answer (A) is incorrect. The earliest time a recreational pilot may take off is at sunrise, not 1 hr. before sunrise. Answer (C) is incorrect. The earliest time a recreational pilot may take off is at sunrise, not at the beginning of morning civil twilight.

62. What exception, if any, permits a recreational pilot to act as pilot in command of an aircraft carrying a passenger for hire?

    A. If the passenger pays no more than the operating expenses.

    B. If a donation is made to a charitable organization for the flight.

    C. There is no exception.

Answer (C) is correct. *(FAR 61.101)*
    DISCUSSION: Recreational pilots may not act as pilot in command of an aircraft for compensation or hire. There is no exception.
    Answer (A) is incorrect. A passenger may share expenses, but not pay for the flight. Answer (B) is incorrect. Acting as pilot in command of an aircraft used in a passenger carrying airlift and sponsored by a charitable organization is specifically prohibited by FAR 61.101.

63. May a recreational pilot act as pilot in command of an aircraft in furtherance of a business?

    A. Yes, if the flight is only incidental to that business.

    B. Yes, providing the aircraft does not carry a person or property for compensation or hire.

    C. No, it is not allowed.

Answer (C) is correct. *(FAR 61.101)*
    DISCUSSION: Recreational pilots may not act as pilot in command of an aircraft that is used in furtherance of a business. There is no exception.
    Answer (A) is incorrect. There is no exception to permit recreational pilots to use aircraft in furtherance of a business. Answer (B) is incorrect. There is no exception to permit recreational pilots to use aircraft in furtherance of a business.

64. If sunset is 2021 and the end of evening civil twilight is 2043, when must a recreational pilot terminate the flight?

A. 2021.

B. 2043.

C. 2121.

Answer (A) is correct. *(FAR 61.101)*
DISCUSSION: A recreational pilot may not act as pilot in command of an airplane between sunset and sunrise. Thus, if sunset is 2021, the recreational pilot must terminate the flight at 2021.
Answer (B) is incorrect. The requirements regarding recreational pilots are in terms of sunset and sunrise, not evening civil twilight. Answer (C) is incorrect. A recreational pilot must stop flying at sunset, not 1 hr. after sunset.

65. When may a recreational pilot operate to or from an airport that lies within Class C airspace?

A. Anytime the control tower is in operation.

B. When the ceiling is at least 1,000 feet and the surface visibility is at least 2 miles.

C. After receiving training and a logbook endorsement from an authorized instructor.

Answer (C) is correct. *(FAR 61.101)*
DISCUSSION: For the purpose of obtaining an additional certificate or rating while under the supervision of an authorized flight instructor, a recreational pilot may fly as sole occupant of an airplane within airspace that requires communication with ATC, such as Class C airspace. [Note that in this situation, (s)he is active as a student pilot, not a recreational pilot.]
Answer (A) is incorrect. A recreational pilot may only fly within airspace that requires communication with ATC (e.g., Class C airspace) when under the supervision of an authorized flight instructor for the purpose of obtaining an additional certificate or rating, regardless of whether the control tower is operating. Answer (B) is incorrect. A recreational pilot may only fly within airspace that requires communication with ATC (e.g., Class C airspace) when under the supervision of an authorized flight instructor for the purpose of obtaining an additional certificate or rating, not simply when the ceiling is at least 1,000 ft. and the surface visibility is at least 3 statute miles.

66. Under what conditions may a recreational pilot operate at an airport that lies within Class D airspace and that has a part-time control tower in operation?

A. Between sunrise and sunset when the tower is in operation, the ceiling is at least 2,500 feet, and the visibility is at least 3 miles.

B. Any time when the tower is in operation, the ceiling is at least 3,000 feet, and the visibility is more than 1 mile.

C. Between sunrise and sunset when the tower is closed, the ceiling is at least 1,000 feet, and the visibility is at least 3 miles.

Answer (C) is correct. *(FAR 61.101)*
DISCUSSION: A recreational pilot may not operate in airspace in which communication with ATC is required, e.g., Class D airspace. When a part-time control tower at an airport in Class D airspace is closed, the Class D airspace is classified as either Class E or Class G airspace, which does not require communication with ATC. A recreational pilot must maintain flight or surface visibility of 3 SM or greater, and the flight must be during the day. To operate at an airport in Class E airspace, the ceiling must be at least 1,000 ft. and the visibility at least 3 SM (FAR 91.155).
Answer (A) is incorrect. The condition to operate at an airport with a part-time control tower is that the control tower not be in operation. The ceiling must be at least 1,000 ft., not 2,500 ft., at an airport in Class E airspace. Answer (B) is incorrect. The condition to operate at an airport with a part-time control tower is that the control tower not be in operation. The ceiling must be at least 1,000 ft., not 3,000 ft., at an airport in Class E airspace, and visibility for a recreational pilot must be at least 3 SM, not 1 statute mile.

67. (Refer to Figure 23 on page 286.) (Refer to area 1.) The visibility and cloud clearance requirements to operate over Sandpoint Airport at less than 700 feet AGL are

A. 3 miles and clear of clouds.

B. 3 miles and 1,000 feet above, 500 feet below, and 2,000 feet horizontally from each cloud.

C. 1 mile and 1,000 feet above, 500 feet below, and 2,000 feet horizontally from each cloud.

Answer (A) is correct. *(FAR 61.101 and 91.155)*
DISCUSSION: Sandpoint Airport is about 1 in. above the number 1 in Fig. 23. The airspace around Sandpoint Airport is Class G from the surface to 2,827 ft. MSL (700 ft. AGL). For a recreational pilot to operate over Sandpoint Airport at less than 700 ft. AGL, the visibility and cloud clearance requirements are 3 SM and clear of clouds.
Answer (B) is incorrect. The cloud clearance of 1,000 ft. above, 500 ft. below, and 2,000 ft. horizontally is the requirement in Class G airspace above 1,200 ft. AGL, not less than 700 ft. AGL. Answer (C) is incorrect. A recreational pilot must have a visibility of, at minimum, 3 SM, not 1 SM, and the cloud clearance of 1,000 ft. above, 500 ft. below, and 2,000 ft. horizontally is the requirement in Class G airspace above 1,200 ft. AGL, not less than 700 ft. AGL.

68. When may a recreational pilot fly above 10,000 feet MSL?

A. When 2,000 feet AGL or below.

B. When 2,500 feet AGL or below.

C. When outside of controlled airspace.

Answer (A) is correct. *(FAR 61.101)*
   DISCUSSION: Recreational pilots may not act as pilot in command of an aircraft at an altitude of more than 10,000 ft. MSL or 2,000 ft. AGL, whichever is higher. Thus, an airplane may fly above 10,000 ft. MSL only if below 2,000 ft. AGL.
   Answer (B) is incorrect. A recreational pilot may fly above 10,000 ft. MSL when 2,000 (not 2,500) ft. AGL or below. Answer (C) is incorrect. The higher of 10,000 ft. MSL or 2,000 ft. AGL limitation on recreational pilots is both in and out of controlled airspace.

69. During daytime, what is the minimum flight or surface visibility required for recreational pilots in Class G airspace below 10,000 feet MSL?

A. 1 mile.

B. 3 miles.

C. 5 miles.

Answer (B) is correct. *(FAR 61.101)*
   DISCUSSION: The minimum flight or surface visibility required for recreational pilots in Class G airspace below 10,000 ft. MSL during the day is 3 statute miles.
   Answer (A) is incorrect. The minimum daytime flight or surface visibility for recreational pilots in Class G airspace below 10,000 ft. MSL is 3, not 1, statute miles. Answer (C) is incorrect. The minimum daytime flight or surface visibility for recreational pilots in Class G airspace below 10,000 ft. MSL is 3, not 5, statute miles.

70. (Refer to Figure 27 on page 290.) (Refer to area 2.) The visibility and cloud clearance requirements to operate over the town of Cooperstown below 700 feet AGL are

A. 1 mile and 1,000 feet above, 500 feet below, and 2,000 feet horizontally from clouds.

B. 3 miles and clear of clouds.

C. 1 mile and clear of clouds.

Answer (B) is correct. *(FAR 61.101 and 91.155)*
   DISCUSSION: The town of Cooperstown is about 3/4 in. above and to the right of the number 2 in Fig. 27. The airspace over the town of Cooperstown (yellow color) is Class G from the surface to 2,124 ft. MSL (700 ft. AGL) since the town lies inside the magenta shaded area. For a recreational pilot to operate over the town of Cooperstown below 700 ft. AGL, the minimum visibility is 3 SM and the cloud clearance requirement is to remain clear of clouds.
   Answer (A) is incorrect. The minimum visibility requirement for a recreational pilot is 3 SM, not 1 statute mile. Additionally, the cloud clearance requirement in Class G airspace below 1,200 ft. AGL is to remain clear of clouds, not to remain 1,000 ft. above, 500 ft. below, and 2,000 ft. horizontally. Answer (C) is incorrect. The minimum visibility requirement for a recreational pilot is 3 SM, not 1 statute mile.

71. (Refer to Figure 27 on page 290.) (Refer to area 2.) The day VFR visibility and cloud clearance requirements to operate over the town of Cooperstown, after departing and climbing out of the Cooperstown Airport at or below 700 feet AGL are

A. 1 mile and clear of clouds.

B. 1 mile and 1,000 feet above, 500 feet below, and 2,000 feet horizontally from clouds.

C. 3 miles and clear of clouds.

Answer (C) is correct. *(FAR 61.101 and 91.155)*
   DISCUSSION: The magenta ring around the Cooperstown Airport indicates that Class E airspace in the area begins 700 ft. above the surface. The airspace underlying this ring is Class G airspace; normally, the day VFR visibility and cloud clearance requirements for operating in Class G airspace are 1 mile and clear of clouds. However, a recreational pilot must maintain 3 SM visibility and remain clear of clouds.
   Answer (A) is incorrect. The minimum visibility requirement for a recreational pilot is 3 SM, not 1 statute mile. Answer (B) is incorrect. These are the minimums for Class C airspace, Class D airspace, and Class E airspace below 10,000 ft. MSL, while the question asks about the Class G airspace minimums. Additionally, a recreational pilot must maintain 3 SM visibility, not 1 SM.

72. During daytime, what is the minimum flight visibility required for recreational pilots in controlled airspace below 10,000 feet MSL?

A. 1 mile.

B. 3 miles.

C. 5 miles.

Answer (B) is correct. *(FAR 61.101)*
   DISCUSSION: The minimum flight visibility for recreational pilots in Class E airspace below 10,000 ft. MSL during the day is 3 statute miles.
   Answer (A) is incorrect. The minimum daytime flight or surface visibility for recreational pilots in Class E airspace below 10,000 ft. MSL is 3 SM, not 1 statute mile. Answer (C) is incorrect. The minimum daytime flight or surface visibility for recreational pilots in Class E airspace below 10,000 ft. MSL is 3 SM, not 5 statute miles.

73. When must a recreational pilot have a pilot-in-command flight check?

A. Every 400 hours.

B. Every 180 days.

C. If the pilot has less than 400 total flight hours and has not flown as pilot in command in an aircraft within the preceding 180 days.

Answer (C) is correct. *(FAR 61.101)*
DISCUSSION: The recreational pilot who has logged fewer than 400 flight hr. and has not logged pilot in command time in an aircraft within the preceding 180 days may not act as pilot in command of an aircraft until the pilot has received flight instruction from an authorized flight instructor who certifies in the pilot's logbook that the pilot is competent to act as pilot in command of the aircraft.
Answer (A) is incorrect. A recreational pilot must have a pilot-in-command check if the pilot has less than 400 total flight hr. and has not flown as pilot in command in an aircraft within the preceding 180 days, not every 400 hours. Answer (B) is incorrect. The 180 days refers to the time interval from the most recent time the recreational pilot (with less than 400 flight hr.) acted as pilot in command.

74. Under what conditions, if any, may a recreational pilot demonstrate an aircraft in flight to a prospective buyer?

A. The buyer pays all the operating expenses.

B. The flight is not outside the United States.

C. None.

Answer (C) is correct. *(FAR 61.101)*
DISCUSSION: Recreational pilots may not act as pilot in command of an aircraft to demonstrate that aircraft in flight to a prospective buyer.
Answer (A) is incorrect. It is prohibited. A passenger may only share, not pay all of, the operating expenses. Answer (B) is incorrect. Recreational pilots may not act as pilot in command to demonstrate an aircraft to a prospective buyer.

75. When, if ever, may a recreational pilot act as pilot in command in an aircraft towing a banner?

A. If the pilot has logged 100 hours of flight time in powered aircraft.

B. If the pilot has an endorsement in his/her pilot logbook from an authorized flight instructor.

C. It is not allowed.

Answer (C) is correct. *(FAR 61.101)*
DISCUSSION: Recreational pilots may not act as pilot in command of an aircraft that is towing any object.
Answer (A) is incorrect. A recreational pilot may not act as pilot in command of an airplane that is towing a banner regardless of the amount of flight time. Answer (B) is incorrect. A recreational pilot may not act as pilot in command of an airplane that is towing a banner.

76. A recreational pilot may fly as sole occupant of an aircraft at night while under the supervision of a flight instructor provided the flight or surface visibility is at least

A. 3 miles.

B. 4 miles.

C. 5 miles.

Answer (C) is correct. *(FAR 61.101)*
DISCUSSION: For the purposes of obtaining additional certificates or ratings, a recreational pilot may fly as sole occupant in the aircraft between sunset and sunrise while under the supervision of an authorized flight instructor, providing the flight or surface visibility is at least 5 statute miles.
Answer (A) is incorrect. A recreational pilot may fly as sole occupant of an airplane at night while under the supervision of a flight instructor provided the flight or surface visibility is at least 5 SM, not 3 statute miles. Answer (B) is incorrect. A recreational pilot may fly as sole occupant of an airplane at night while under the supervision of a flight instructor provided the flight or surface visibility is at least 5 SM, not 4 statute miles.

77. What minimum visibility and clearance from clouds are required for a recreational pilot in Class G airspace at 1,200 feet AGL or below during daylight hours?

A. 1 mile visibility and clear of clouds.

B. 3 miles visibility and clear of clouds.

C. 3 miles visibility, 500 feet below the clouds.

Answer (B) is correct. *(FAR 61.101 and 91.155)*
DISCUSSION: Recreational pilots may not act as pilot in command of an aircraft when the visibility is less than 3 statute miles. Additionally, FAR 91.155 specifies basic VFR weather minimums which permit pilots to fly in Class G airspace 1,200 ft. AGL or below at 1 SM clear of clouds. Thus, the 3-SM recreational pilot limitation and the clear of clouds situation apply.
Answer (A) is incorrect. Recreational pilots may never fly when visibility is less than 3 statute miles. Answer (C) is incorrect. At 1,200 ft. AGL or below in Class G airspace, there is no separation from cloud requirement. One must only remain clear of clouds.

78. Outside controlled airspace, the minimum flight visibility requirement for a recreational pilot flying VFR above 1,200 feet AGL and below 10,000 feet MSL during daylight hours is

A. 1 mile.

B. 3 miles.

C. 5 miles.

Answer (B) is correct. *(FAR 61.101 and 91.155)*
DISCUSSION: Recreational pilots may not act as pilot in command of an aircraft when the visibility is less than 3 statute miles.
Answer (A) is incorrect. The minimum flight visibility requirement for a recreational pilot is 3 SM, not 1 statute mile. Answer (C) is incorrect. The minimum flight visibility requirement for a recreational pilot is 3 SM, not 5 statute miles.

79. Unless otherwise authorized, if flying a transponder equipped aircraft, a recreational pilot should squawk which VFR code?

A. 1200.

B. 7600.

C. 7700.

Answer (A) is correct. *(AIM Para 4-1-19)*
DISCUSSION: A recreational pilot flying a transponder-equipped aircraft should set that transponder on code (squawk) 1200, which is the VFR code.
Answer (B) is incorrect. This is the lost communication code. Answer (C) is incorrect. This is the general emergency code.

4.7 FAR Part 71

71.71 Extent of Federal Airways

80. The width of a Federal Airway from either side of the centerline is

A. 4 nautical miles.

B. 6 nautical miles.

C. 8 nautical miles.

Answer (A) is correct. *(FAR 71.71)*
DISCUSSION: The width of a Federal Airway from either side of the centerline is 4 nautical miles.
Answer (B) is incorrect. The width of a Federal Airway from either side of the centerline is 4 NM, not 6 nautical miles. Answer (C) is incorrect. The width of a Federal Airway from either side of the centerline is 4 NM, not 8 nautical miles.

81. Unless otherwise specified, Federal Airways include that Class E airspace extending upward from

A. 700 feet above the surface, up to and including 17,999 feet MSL.

B. 1,200 feet above the surface, up to and including 17,999 feet MSL.

C. the surface, up to and including 18,000 feet MSL.

Answer (B) is correct. *(FAR 71.71)*
DISCUSSION: Unless otherwise specified, Federal Airways include that Class E airspace extending from 1,200 ft. above the surface, up to and including 17,999 feet.
Answer (A) is incorrect. Federal Airways extend from 1,200 (not 700) ft. above the surface, up to and including 17,999 ft. MSL. Answer (C) is incorrect. Federal Airways extend from 1,200 ft. above the surface, up to and including 17,999 ft. MSL, not 18,000 ft. MSL. The airspace that extends upward from 18,000 ft. MSL is Class A airspace.

4.8 FAR Part 91: 91.3 – 91.151

91.3 Responsibility and Authority of the Pilot in Command

82. The final authority as to the operation of an aircraft is the

A. Federal Aviation Administration.

B. pilot in command.

C. aircraft manufacturer.

Answer (B) is correct. *(FAR 91.3)*
DISCUSSION: The final authority as to the operation of an aircraft is the pilot in command.
Answer (A) is incorrect. The final authority as to the operation of an aircraft is the pilot in command, not the FAA. Answer (C) is incorrect. The final authority as to the operation of an aircraft is the pilot in command, not the aircraft manufacturer.

91.7 Civil Aircraft Airworthiness

83. Who is responsible for determining if an aircraft is in condition for safe flight?

    A. A certificated aircraft mechanic.

    B. The pilot in command.

    C. The owner or operator.

Answer (B) is correct. *(FAR 91.7)*
DISCUSSION: The pilot in command of an aircraft is directly responsible for, and is the final authority for, determining whether the airplane is in condition for safe flight.
Answer (A) is incorrect. The pilot in command (not a certificated aircraft mechanic) is responsible for determining if an aircraft is in condition for safe flight. Answer (C) is incorrect. The pilot in command (not the owner or operator) is responsible for determining if an aircraft is in condition for safe flight.

91.9 Civil Aircraft Flight Manual, Marking, and Placard Requirements

84. Where may an aircraft's operating limitations be found?

    A. On the Airworthiness Certificate.

    B. In the current, FAA-approved flight manual, approved manual material, markings, and placards, or any combination thereof.

    C. In the aircraft airframe and engine logbooks.

Answer (B) is correct. *(FAR 91.9)*
DISCUSSION: An aircraft's operating limitations may be found in the current, FAA-approved flight manual, approved manual material, markings, and placards, or any combination thereof.
Answer (A) is incorrect. The airworthiness certificate only indicates the airplane was in an airworthy condition when delivered from the factory, not its operating limitations. Answer (C) is incorrect. The airframe and engine logbooks contain the airplane's maintenance record, not its operating limitations.

85. Where may an aircraft's operating limitations be found if the aircraft has an Experimental or Special light-sport airworthiness certificate?

    A. Attached to the Airworthiness Certificate.

    B. In the current, FAA-approved flight manual.

    C. In the aircraft airframe and engine logbooks.

Answer (B) is correct. *(FAR 91.9)*
DISCUSSION: An aircraft's operating limitations may be found in the current, FAA-approved flight manual, approved manual material, markings, and placards, or any combination thereof.
Answer (A) is incorrect. The airworthiness certificate only indicates the airplane was in an airworthy condition when delivered from the factory, not its operating limitations. Answer (C) is incorrect. The airframe and engine logbooks contain the airplane's maintenance record, not its operating limitations.

91.15 Dropping Objects

86. Under what conditions may objects be dropped from an aircraft?

    A. Only in an emergency.

    B. If precautions are taken to avoid injury or damage to persons or property on the surface.

    C. If prior permission is received from the Federal Aviation Administration.

Answer (B) is correct. *(FAR 91.15)*
DISCUSSION: No pilot in command of a civil aircraft may allow any object to be dropped from that aircraft in flight that creates a hazard to persons or property. However, this section does not prohibit the dropping of any object if reasonable precautions are taken to avoid injury or damage to persons or property.
Answer (A) is incorrect. Objects may be dropped from an aircraft if precautions are taken to avoid injury or damage to persons or property on the surface, not only in an emergency. Answer (C) is incorrect. Objects may be dropped from an aircraft if precautions are taken to avoid injury or damage to persons or property on the surface. Prior permission from the FAA is not required.

91.17 Alcohol or Drugs

87. No person may attempt to act as a crewmember of a civil aircraft with

    A. .008 percent by weight or more alcohol in the blood.

    B. .004 percent by weight or more alcohol in the blood.

    C. .04 percent by weight or more alcohol in the blood.

Answer (C) is correct. *(FAR 91.17)*
DISCUSSION: No person may act or attempt to act as a crewmember of a civil aircraft while having a .04% by weight or more alcohol in the blood.
Answer (A) is incorrect. No person may attempt to act as a crewmember of a civil aircraft with .04% (not .008%) by weight or more alcohol in the blood. Answer (B) is incorrect. No person may attempt to act as a crewmember of a civil aircraft with .04% (not .004%) by weight or more alcohol in the blood.

88. Under what condition, if any, may a pilot allow a person who is obviously under the influence of drugs to be carried aboard an aircraft?

A. In an emergency or if the person is a medical patient under proper care.

B. Only if the person does not have access to the cockpit or pilot's compartment.

C. Under no condition.

Answer (A) is correct. *(FAR 91.17)*
   DISCUSSION: No pilot of a civil aircraft may allow a person who demonstrates by manner or physical indications that the individual is under the influence of drugs to be carried in that aircraft, except in an emergency or if the person is a medical patient under proper care.
   Answer (B) is incorrect. No pilot may allow a person who is obviously under the influence of drugs to be carried aboard an aircraft except in an emergency or if the person is a medical patient under proper care, not if that person does not have access to the cockpit or pilot's compartment. Answer (C) is incorrect. A pilot may allow a person who is obviously under the influence of drugs to be carried aboard an aircraft in an emergency or if the person is a medical patient under proper care.

89. A person may not act as a crewmember of a civil aircraft if alcoholic beverages have been consumed by that person within the preceding

A. 8 hours.

B. 12 hours.

C. 24 hours.

Answer (A) is correct. *(FAR 91.17)*
   DISCUSSION: No person may act as a crewmember of a civil aircraft if alcoholic beverages have been consumed by that person within the preceding 8 hours.
   Answer (B) is incorrect. No person may act as a crewmember of a civil aircraft within 8 (not 12) hr. after the consumption of any alcoholic beverage. Answer (C) is incorrect. No person may act as a crewmember of a civil aircraft within 8 (not 24) hr. after the consumption of any alcoholic beverage.

91.103 Preflight Action

90. Preflight action, as required for all flights away from the vicinity of an airport, shall include

A. the designation of an alternate airport.

B. a study of arrival procedures at airports/heliports of intended use.

C. an alternate course of action if the flight cannot be completed as planned.

Answer (C) is correct. *(FAR 91.103)*
   DISCUSSION: Preflight actions for flights not in the vicinity of an airport include checking weather reports and forecasts, fuel requirements, alternatives available if the planned flight cannot be completed, and any known traffic delays.
   Answer (A) is incorrect. Preflight action, as required for all flights away from the vicinity of an airport, shall include an alternate course of action if the flight cannot be completed as planned, not just the designation of an alternate airport. Answer (B) is incorrect. Preflight action, as required for all flights away from the vicinity of an airport, shall include an alternate course of action if the flight cannot be completed as planned, not simply a study of arrival procedures at airports of intended use.

91. In addition to other preflight actions for a VFR flight away from the vicinity of the departure airport, regulations specifically require the pilot in command to

A. review traffic control light signal procedures.

B. check the accuracy of the navigation equipment and the emergency locator transmitter (ELT).

C. determine runway lengths at airports of intended use and the aircraft's takeoff and landing distance data.

Answer (C) is correct. *(FAR 91.103)*
   DISCUSSION: Preflight actions for a VFR flight away from the vicinity of the departure airport specifically require the pilot in command to determine runway lengths at airports of intended use and the aircraft's takeoff and landing distance data.
   Answer (A) is incorrect. Preflight actions for a VFR flight away from the vicinity of an airport require the pilot in command to determine runway lengths at airports of intended use and takeoff and landing distance data, not to review traffic control light signal procedures. Answer (B) is incorrect. Preflight actions for a VFR flight away from the vicinity of an airport require the pilot in command to determine runway lengths at airports of intended use and takeoff and landing distance data, not to check navigation equipment accuracy and the ELT.

92. Which preflight action is specifically required of the pilot prior to each flight?

A. Check the aircraft logbooks for appropriate entries.

B. Become familiar with all available information concerning the flight.

C. Review wake turbulence avoidance procedures.

Answer (B) is correct. *(FAR 91.103)*
   DISCUSSION: Each pilot in command will, before beginning a flight, become familiar with all available information concerning that flight.
   Answer (A) is incorrect. During preflight action, the pilot is required to become familiar with all available information concerning the flight, not just to check the aircraft logbook for appropriate entries. Answer (C) is incorrect. During preflight action, the pilot is required to become familiar with all available information concerning the flight, not simply review wake turbulence avoidance procedures.

91.105 Flight Crewmembers at Stations

93. Flight crewmembers are required to keep their safety belts and shoulder harnesses fastened during

    A. takeoffs and landings.

    B. all flight conditions.

    C. flight in turbulent air.

Answer (A) is correct. *(FAR 91.105)*
    DISCUSSION: During takeoff and landing and while en route, each required flight crewmember shall keep his/her safety belt fastened while at the crewmember station. If shoulder harnesses are available, they must be used by crewmembers during takeoff and landing.
    Answer (B) is incorrect. Flight crewmembers are required to keep their shoulder harnesses fastened only during takeoffs and landings, not during all flight conditions. Answer (C) is incorrect. Flight crewmembers are required to keep their shoulder harnesses fastened only during takeoffs and landings, not during flight in turbulent air.

94. Which best describes the flight conditions under which flight crewmembers are specifically required to keep their safety belts and shoulder harnesses fastened?

    A. Safety belts during takeoff and landing; shoulder harnesses during takeoff and landing.

    B. Safety belts during takeoff and landing; shoulder harnesses during takeoff and landing and while en route.

    C. Safety belts during takeoff and landing and while en route; shoulder harnesses during takeoff and landing.

Answer (C) is correct. *(FAR 91.105)*
    DISCUSSION: During takeoff and landing and while en route, each required flight crewmember shall keep his/her safety belt fastened while at the crewmember station. If shoulder harnesses are available, they must be used by crewmembers during takeoff and landing.
    Answer (A) is incorrect. Safety belts must be worn while en route. Answer (B) is incorrect. Safety belts (not shoulder harnesses) are required to be fastened while en route.

91.107 Use of Safety Belts, Shoulder Harnesses, and Child Restraint Systems

95. With respect to passengers, what obligation, if any, does a pilot in command have concerning the use of safety belts?

    A. The pilot in command must instruct the passengers to keep their safety belts fastened for the entire flight.

    B. The pilot in command must brief the passengers on the use of safety belts and notify them to fasten their safety belts during taxi, takeoff, and landing.

    C. The pilot in command has no obligation in regard to passengers' use of safety belts.

Answer (B) is correct. *(FAR 91.107)*
    DISCUSSION: The pilot in command is required to brief the passengers on the use of safety belts and notify them to fasten their safety belts during taxi, takeoff, and landing.
    Answer (A) is incorrect. The pilot in command is only required to notify the passengers to fasten their safety belts during taxi, takeoff, and landing, not during the entire flight. Answer (C) is incorrect. The pilot in command has the obligation both to instruct passengers on the use of safety belts and to require their use during taxi, takeoffs, and landings.

96. Safety belts are required to be properly secured about which persons in an aircraft and when?

    A. Pilots only, during takeoffs and landings.

    B. Passengers, during taxi, takeoffs, and landings only.

    C. Each person on board the aircraft during the entire flight.

Answer (B) is correct. *(FAR 91.107)*
    DISCUSSION: Regulations require that safety belts in an airplane be properly secured about all passengers during taxi, takeoffs, and landings.
    Answer (A) is incorrect. Regulations require passengers as well as crewmembers to wear safety belts during takeoffs and landings. Answer (C) is incorrect. Although it is a good procedure, safety belts are required only for passengers during taxi, takeoffs, and landings.

97. With certain exceptions, safety belts are required to be secured about passengers during

A. taxi, takeoffs, and landings.

B. all flight conditions.

C. flight in turbulent air.

Answer (A) is correct. *(FAR 91.107)*
DISCUSSION: During the taxi, takeoff, and landing of U.S.-registered civil aircraft, each person on board that aircraft must occupy a seat or berth with a safety belt and shoulder harness, if installed, properly secured about him/her. However, a person who has not reached his/her second birthday may be held by an adult who is occupying a seat or berth, and a person on board for the purpose of engaging in sport parachuting may use the floor of the aircraft as a seat (but is still required to use approved safety belts for takeoff).
Answer (B) is incorrect. Safety belts are required to be secured about passengers only during taxi, takeoffs, and landings, not during all flight conditions. Answer (C) is incorrect. Safety belts are required to be secured about passengers during taxi, takeoffs, and landings, not during flight in turbulent air.

91.111 Operating near Other Aircraft

98. No person may operate an aircraft in formation flight

A. over a densely populated area.

B. in Class D airspace under special VFR.

C. except by prior arrangement with the pilot in command of each aircraft.

Answer (C) is correct. *(FAR 91.111)*
DISCUSSION: No person may operate in formation flight except by arrangement with the pilot in command of each aircraft in formation.
Answer (A) is incorrect. No person may operate an aircraft in formation flight except by prior arrangement with the pilot in command of each aircraft. There are no restrictions about formation flights over a densely populated area. Answer (B) is incorrect. No person may operate an aircraft in formation flight except by prior arrangement with the pilot in command of each aircraft. There are no restrictions about formation flight in Class D airspace under special VFR.

91.113 Right-of-Way Rules: Except Water Operations

99. An airplane and an airship are converging. If the airship is left of the airplane's position, which aircraft has the right-of-way?

A. The airship.

B. The airplane.

C. Each pilot should alter course to the right.

Answer (A) is correct. *(FAR 91.113)*
DISCUSSION: When aircraft of different categories are converging, the less maneuverable aircraft has the right-of-way. Thus, the airship has the right-of-way in this question.
Answer (B) is incorrect. When converging, the airship has the right-of-way over an airplane or rotorcraft. Answer (C) is incorrect. Each pilot would alter course to the right if the airship and airplane were approaching head-on, or nearly so, not converging.

100. When two or more aircraft are approaching an airport for the purpose of landing, the right-of-way belongs to the aircraft

A. that has the other to its right.

B. that is the least maneuverable.

C. at the lower altitude, but it shall not take advantage of this rule to cut in front of or to overtake another.

Answer (C) is correct. *(FAR 91.113)*
DISCUSSION: When two or more aircraft are approaching an airport for the purpose of landing, the aircraft at the lower altitude has the right-of-way, but it shall not take advantage of this rule to cut in front of or overtake another aircraft.
Answer (A) is incorrect. When two or more aircraft are approaching an airport for the purpose of landing, the right-of-way belongs to the aircraft at the lower altitude, not the aircraft that has the other to the right. Answer (B) is incorrect. When two or more aircraft are approaching an airport for the purpose of landing, the right-of-way belongs to the aircraft at the lower altitude, not the aircraft that is the least maneuverable.

101. Which aircraft has the right-of-way over the other aircraft listed?

A. Glider.

B. Airship.

C. Aircraft refueling other aircraft.

Answer (A) is correct. *(FAR 91.113)*
DISCUSSION: If aircraft of different categories are converging, the right-of-way depends upon who has the least maneuverability. A glider has right-of-way over an airship, airplane or rotorcraft.
Answer (B) is incorrect. An airship has the right-of-way over an airplane or rotorcraft, not a glider. Answer (C) is incorrect. Aircraft refueling have right-of-way over all engine-driven aircraft. A glider has no engine.

102. What action should the pilots of a glider and an airplane take if on a head-on collision course?

   A. The airplane pilot should give way to the left.

   B. The glider pilot should give way to the right.

   C. Both pilots should give way to the right.

Answer (C) is correct. *(FAR 91.113)*
   DISCUSSION: When aircraft are approaching head-on, or nearly so (regardless of category), each aircraft shall alter course to the right.
   Answer (A) is incorrect. The glider has the right-of-way unless the two aircraft are approaching head-on, in which case both pilots should give way by turning to the right. Answer (B) is incorrect. Both pilots of a glider and an airplane should give way to the right, not only the glider pilot.

103. What action is required when two aircraft of the same category converge, but not head-on?

   A. The faster aircraft shall give way.

   B. The aircraft on the left shall give way.

   C. Each aircraft shall give way to the right.

Answer (B) is correct. *(FAR 91.113)*
   DISCUSSION: When two aircraft of the same category converge (but not head-on), the aircraft to the other's right has the right-of-way. Thus, an airplane on the left gives way to the airplane on the right.
   Answer (A) is incorrect. When two aircraft of the same category converge (but not head-on), the aircraft on the left (not the faster aircraft) shall give way. Answer (C) is incorrect. The required action when two aircraft are approaching head-on or nearly so is for each aircraft to give way to the right.

104. Which aircraft has the right-of-way over the other aircraft listed?

   A. Airship.

   B. Aircraft towing other aircraft.

   C. Gyroplane.

Answer (B) is correct. *(FAR 91.113)*
   DISCUSSION: An aircraft towing or refueling another aircraft has the right-of-way over all engine-driven aircraft. An airship is an engine-driven, lighter-than-air aircraft that can be steered.
   Answer (A) is incorrect. An airship has the right-of-way over an airplane or rotorcraft, but not an aircraft towing other aircraft. Answer (C) is incorrect. A gyroplane (which is a rotorcraft) must give way to both an airship and aircraft towing other aircraft.

105. Which aircraft has the right-of-way over all other air traffic?

   A. A balloon.

   B. An aircraft in distress.

   C. An aircraft on final approach to land.

Answer (B) is correct. *(FAR 91.113)*
   DISCUSSION: An aircraft in distress has the right-of-way over all other aircraft.
   Answer (A) is incorrect. An aircraft in distress (not a balloon) has the right-of-way over all other air traffic. Answer (C) is incorrect. An aircraft in distress (not an aircraft on final approach to land) has the right-of-way over all other air traffic.

91.115 Right-of-Way Rules: Water Operations

106. A seaplane and a motorboat are on crossing courses. If the motorboat is to the left of the seaplane, which has the right-of-way?

   A. The motorboat.

   B. The seaplane.

   C. Both should alter course to the right.

Answer (B) is correct. *(FAR 91.115)*
   DISCUSSION: When aircraft, or an aircraft and a vessel (e.g., a motorboat), are on crossing courses, the aircraft or vessel to the other's right has the right-of-way. Since the seaplane is to the motorboat's right, the seaplane has the right-of-way.
   Answer (A) is incorrect. On crossing courses, the aircraft or vessel to the other's right has the right-of-way. Since the seaplane is to the right of the motorboat, the seaplane (not the motorboat) has the right-of-way. Answer (C) is incorrect. Both would alter course to the right only if they were approaching head-on, or nearly so.

91.117 Aircraft Speed

107. When flying in a VFR corridor designated through Class B airspace, the maximum speed authorized is

   A. 180 knots.

   B. 200 knots.

   C. 250 knots.

Answer (B) is correct. *(FAR 91.117)*
   DISCUSSION: No person may operate an airplane in a VFR corridor designated through Class B airspace at an indicated airspeed of more than 200 kt. (230 MPH).
   Answer (A) is incorrect. When flying in a VFR corridor designated through Class B airspace, the maximum speed authorized is 200 (not 180) knots. Answer (C) is incorrect. This is the maximum speed authorized below 10,000 ft. MSL, not when flying in a VFR corridor through Class B airspace.

108. Unless otherwise authorized, what is the maximum indicated airspeed at which a person may operate an aircraft below 10,000 feet MSL?

A. 200 knots.

B. 250 knots.

C. 288 knots.

Answer (B) is correct. *(FAR 91.117)*
  DISCUSSION: Unless otherwise authorized by ATC, no person may operate an aircraft below 10,000 ft. MSL at an indicated airspeed of more than 250 kt. (288 MPH).
  Answer (A) is incorrect. This is the maximum indicated airspeed when at or below 2,500 ft. above the surface and within 4 NM of the primary airport of a Class C or Class D airspace area, not the maximum indicated airspeed for operations below 10,000 ft. MSL. Answer (C) is incorrect. The maximum indicated airspeed below 10,000 ft. MSL is 288 MPH, not 288 knots.

109. When flying in the airspace underlying Class B airspace, the maximum speed authorized is

A. 200 knots.

B. 230 knots.

C. 250 knots.

Answer (A) is correct. *(FAR 91.117)*
  DISCUSSION: No person may operate an airplane in the airspace underlying Class B airspace at an indicated airspeed of more than 200 kt. (230 MPH).
  Answer (B) is incorrect. The maximum indicated airspeed authorized in the airspace underlying Class B airspace is 230 MPH, not 230 knots. Answer (C) is incorrect. This is the maximum indicated airspeed when operating an airplane below 10,000 ft. MSL, not in the airspace underlying Class B airspace.

110. Unless otherwise authorized, the maximum indicated airspeed at which aircraft may be flown when at or below 2,500 feet AGL and within 4 nautical miles of the primary airport of Class C airspace is

A. 200 knots.

B. 230 knots.

C. 250 knots.

Answer (A) is correct. *(FAR 91.117)*
  DISCUSSION: Unless otherwise authorized, the maximum indicated airspeed at which an airplane may be flown when at or below 2,500 ft. AGL and within 4 NM of the primary airport of Class C airspace is 200 kt. (230 MPH).
  Answer (B) is incorrect. The figure of 230 MPH, not 230 kt., is the maximum indicated airspeed at which an airplane may be flown when at or below 2,500 ft. AGL and within 4 NM of the primary airport of a Class C airspace area. Answer (C) is incorrect. The figure of 250 kt. is the maximum indicated airspeed at which an airplane may be flown below 10,000 ft. MSL or in Class B airspace, not when at or below 2,500 ft. AGL and within 4 NM of the primary airport of a Class C airspace area.

91.119 Minimum Safe Altitudes: General

111. Except when necessary for takeoff or landing, what is the minimum safe altitude for a pilot to operate an aircraft anywhere?

A. An altitude allowing, if a power unit fails, an emergency landing without undue hazard to persons or property on the surface.

B. An altitude of 500 feet above the surface and no closer than 500 feet to any person, vessel, vehicle, or structure.

C. An altitude of 500 feet above the highest obstacle within a horizontal radius of 1,000 feet.

Answer (A) is correct. *(FAR 91.119)*
  DISCUSSION: Except when necessary for takeoff or landing, no person may operate an aircraft anywhere below an altitude allowing, if a power unit fails, an emergency landing without undue hazard to persons or property on the surface.
  Answer (B) is incorrect. An altitude of 500 ft. above the surface is the minimum safe altitude over other than congested areas and no closer than 500 ft. to any person, vessel, vehicle, or structure is the minimum safe altitude over open water or sparsely populated areas. Answer (C) is incorrect. The minimum safe altitude anywhere is an altitude that allows an emergency landing to be made without undue hazards to persons or property on the surface, not 500 ft. above the highest obstacle within a horizontal radius of 1,000 feet.

112. Except when necessary for takeoff or landing, what is the minimum safe altitude required for a pilot to operate an aircraft over congested areas?

A. An altitude of 1,000 feet above any person, vessel, vehicle, or structure.

B. An altitude of 500 feet above the highest obstacle within a horizontal radius of 1,000 feet of the aircraft.

C. An altitude of 1,000 feet above the highest obstacle within a horizontal radius of 2,000 feet of the aircraft.

Answer (C) is correct. *(FAR 91.119)*
  DISCUSSION: When operating an aircraft over any congested area of a city, town, or settlement, or over an open-air assembly of persons, a pilot must remain at an altitude of 1,000 ft. above the highest obstacle within a horizontal radius of 2,000 ft. of the aircraft.
  Answer (A) is incorrect. The minimum safe altitude to operate an aircraft over a congested area is an altitude of 1,000 ft. above the highest obstacle (not above any person, vessel, vehicle, or structure) within a horizontal distance of 2,000 ft. Answer (B) is incorrect. The minimum safe altitude to operate an aircraft over a congested area is an altitude of 1,000 (not 500) ft. above the highest obstacle within a horizontal radius of 2,000 (not 1,000) ft. of the aircraft.

113. Except when necessary for takeoff or landing, what is the minimum safe altitude required for a pilot to operate an aircraft over other than a congested area?

A. An altitude allowing, if a power unit fails, an emergency landing without undue hazard to persons or property on the surface.

B. An altitude of 500 feet AGL, except over open water or a sparsely populated area, which requires 500 feet from any person, vessel, vehicle, or structure.

C. An altitude of 500 feet above the highest obstacle within a horizontal radius of 1,000 feet.

Answer (B) is correct. *(FAR 91.119)*
    DISCUSSION: Over other than congested areas, an altitude of 500 ft. above the surface is required. Over open water and sparsely populated areas, a distance of 500 ft. from any person, vessel, vehicle, or structure must be maintained.
    Answer (A) is incorrect. An altitude allowing, if a power unit fails, an emergency landing without undue hazard to persons or property on the surface is the general minimum safe altitude for anywhere, not specifically for operation over other than a congested area. Answer (C) is incorrect. The minimum safe altitude over other than a congested area is an altitude of 500 ft. AGL (not above the highest obstacle within a horizontal radius of 1,000 ft.), except over open water or a sparsely populated area, which requires 500 ft. from any person, vessel, vehicle, or structure.

114. Except when necessary for takeoff or landing, an aircraft may not be operated closer than what distance from any person, vessel, vehicle, or structure?

A. 500 feet.

B. 700 feet.

C. 1,000 feet.

Answer (A) is correct. *(FAR 91.119)*
    DISCUSSION: Over other than congested areas, an altitude of 500 ft. above the surface is required. Over open water and sparsely populated areas, a distance of 500 ft. from any person, vessel, vehicle, or structure must be maintained.
    Answer (B) is incorrect. An aircraft may not be operated closer than 500 (not 700) ft. from any person, vessel, vehicle, or structure. Answer (C) is incorrect. An aircraft may not be operated closer than 500 (not 1,000) ft. from any person, vessel, vehicle, or structure.

91.121 Altimeter Settings

115. Prior to takeoff, the altimeter should be set to which altitude or altimeter setting?

A. The current local altimeter setting, if available, or the departure airport elevation.

B. The corrected density altitude of the departure airport.

C. The corrected pressure altitude for the departure airport.

Answer (A) is correct. *(FAR 91.121)*
    DISCUSSION: Prior to takeoff, the altimeter should be set to the local altimeter setting, or to the departure airport elevation.
    Answer (B) is incorrect. Density altitude is pressure altitude corrected for nonstandard temperature variations and is determined from flight computers or graphs, not an altimeter. Answer (C) is incorrect. Pressure altitude is only used at or above 18,000 ft. MSL.

116. If an altimeter setting is not available before flight, to which altitude should the pilot adjust the altimeter?

A. The elevation of the nearest airport corrected to mean sea level.

B. The elevation of the departure area.

C. Pressure altitude corrected for nonstandard temperature.

Answer (B) is correct. *(FAR 91.121)*
    DISCUSSION: When the local altimeter setting is not available at takeoff, the pilot should adjust the altimeter to the elevation of the departure area.
    Answer (A) is incorrect. Airport elevation is always expressed in true altitude, or feet above MSL. Answer (C) is incorrect. Pressure altitude adjusted for nonstandard temperature is density altitude, not true altitude.

117. At what altitude shall the altimeter be set to 29.92 when climbing to cruising flight level?

A. 14,500 feet MSL.

B. 18,000 feet MSL.

C. 24,000 feet MSL.

Answer (B) is correct. *(FAR 91.121)*
    DISCUSSION: Pressure altitude is the altitude used for all flights at and above 18,000 ft. MSL, i.e., in Class A airspace. When climbing to or above 18,000 ft. MSL, one does not use local altimeter settings, but rather 29.92" Hg after reaching 18,000 ft. MSL.
    Answer (A) is incorrect. This is the base of Class E airspace unless otherwise indicated. Answer (C) is incorrect. This is the altitude above which DME is required aboard the airplane.

91.123 Compliance with ATC Clearances and Instructions

118. When must a pilot who deviates from a regulation during an emergency send a written report of that deviation to the Administrator?

A. Within 7 days.

B. Within 10 days.

C. Upon request.

Answer (C) is correct. *(FAR 91.123)*
DISCUSSION: A pilot who deviates from a regulation during an emergency must send a written report of that deviation to the Administrator of the FAA only upon request.
Answer (A) is incorrect. A written report of a deviation from a regulation during an emergency must be sent to the Administrator upon request, not within 7 days. Answer (B) is incorrect. A written report of a deviation from a regulation during an emergency must be sent to the Administrator upon request, not within 10 days.

119. When would a pilot be required to submit a detailed report of an emergency which caused the pilot to deviate from an ATC clearance?

A. Within 48 hours if requested by ATC.

B. Immediately.

C. Within 7 days.

Answer (A) is correct. *(FAR 91.123)*
DISCUSSION: Each pilot in command who is given priority by ATC in an emergency shall, if requested by ATC, submit a detailed report within 48 hrs. to the manager of that ATC facility.
Answer (B) is incorrect. A pilot would be required to submit a detailed report of an emergency when requested by ATC (not immediately). Answer (C) is incorrect. A pilot would be required to submit a detailed report of an emergency when requested by ATC (not within 7 days).

120. If an in-flight emergency requires immediate action, the pilot in command may

A. deviate from any rule of 14 CFR part 91 to the extent required to meet the emergency, but must submit a written report to the Administrator within 24 hours.

B. deviate from any rule of 14 CFR part 91 to the extent required to meet that emergency.

C. not deviate from any rule of 14 CFR part 91 unless prior to the deviation approval is granted by the Administrator.

Answer (B) is correct. *(FAR 91.123)*
DISCUSSION: In an in-flight emergency requiring immediate action, the pilot in command may deviate from any rule of 14 CFR part 91 to the extent required to meet that emergency. A written report of the deviation must be sent to the Administrator of the FAA only if requested.
Answer (A) is incorrect. A written report must be sent to the Administrator of the FAA only upon request. Answer (C) is incorrect. The pilot in command may deviate from any rule of 14 CFR part 91 to the extent required to meet that emergency without the approval of the Administrator of the FAA.

121. As Pilot in Command of an aircraft, under which situation can you deviate from an ATC clearance?

A. When operating in Class A airspace at night.

B. If an ATC clearance is not understood and in VFR conditions.

C. In response to a traffic alert and collision avoidance system resolution advisory.

Answer (C) is correct. *(FAR 91.123)*
DISCUSSION: No pilot may deviate from an ATC clearance unless an amended clearance is obtained, an emergency exists, or the deviation is in response to a traffic alert and collision avoidance system resolution advisory. A written report of the deviation must be sent to the Administrator of the FAA only if requested.
Answer (A) is incorrect. A pilot cannot deviate from an ATC clearance in any airspace unless it is an emergency or in response to a traffic alert and collision avoidance resolution advisory. Answer (B) is incorrect. When a pilot is uncertain of an ATC clearance, that pilot shall immediately request clarification from ATC.

122. When an ATC clearance has been obtained, no pilot in command may deviate from that clearance unless that pilot obtains an amended clearance. The one exception to this regulation is

A. when the clearance states, "at pilot's discretion."

B. an emergency.

C. if the clearance contains a restriction.

Answer (B) is correct. *(FAR 91.123)*
DISCUSSION: When an ATC clearance has been obtained, no pilot in command may deviate from that clearance, except in an emergency, unless an amended clearance is obtained.
Answer (A) is incorrect. The words "at the pilot's discretion" are part of an ATC clearance, so this is not an exception. Answer (C) is incorrect. Any restriction is still part of the clearance, so this is not an exception.

123. What action, if any, is appropriate if the pilot deviates from an ATC instruction during an emergency and is given priority?

    A. Take no special action since you are pilot in command.

    B. File a detailed report within 48 hours to the chief of the appropriate ATC facility, if requested.

    C. File a report to the FAA Administrator, as soon as possible.

Answer (B) is correct. *(FAR 91.123)*
    DISCUSSION: Each pilot in command who is given priority by ATC in an emergency shall, if requested by ATC, submit a detailed report within 48 hrs. to the manager of that ATC facility.
    Answer (A) is incorrect. As pilot in command, you must file a detailed report within 48 hr. to the chief of the appropriate ATC facility, if requested. Answer (C) is incorrect. A detailed report must be filed to the chief of the appropriate ATC facility (not the FAA Administrator) if requested (not as soon as possible).

91.130 Operations in Class C Airspace

124. Two-way radio communication must be established with the Air Traffic Control facility having jurisdiction over the area prior to entering which class airspace?

    A. Class C.

    B. Class E.

    C. Class G.

Answer (A) is correct. *(FAR 91.130)*
    DISCUSSION: No person may operate an aircraft in Class C airspace unless two-way radio communication is established with the ATC facility having jurisdiction over the airspace prior to entering that area.
    Answer (B) is incorrect. While a Class E airspace area is controlled airspace, two-way radio communication is not required to be established with ATC prior to entering VFR weather conditions. Answer (C) is incorrect. Two-way radio communication with ATC is not required for any operations in Class G airspace.

91.131 Operations in Class B Airspace

125. With certain exceptions, all aircraft within 30 miles of a Class B primary airport from the surface upward to 10,000 feet MSL must be equipped with

    A. an operable VOR or TACAN receiver and an ADF receiver.

    B. instruments and equipment required for IFR operations.

    C. an operable transponder having either Mode S or 4096-code capability with Mode C automatic altitude reporting capability.

Answer (C) is correct. *(FAR 91.131)*
    DISCUSSION: All aircraft within 30 NM of a Class B primary airport must be equipped with an operable transponder having either Mode S or 4096-code capability with Mode C automatic altitude reporting capability. The exception is any aircraft that was not originally certificated with an engine-driven electrical system or that has not subsequently been certified with such a system installed, balloon, or glider may conduct operations in the airspace within 30 NM of a Class B airspace primary airport provided such operations are conducted (1) outside any Class A, Class B, or Class C airspace area; and (2) below the altitude of the ceiling of a Class B or Class C airspace area or 10,000 ft. MSL, whichever is lower.
    Answer (A) is incorrect. An operable VOR or TACAN receiver and an ADF receiver are not required within 30 NM of a Class B primary airport, only an operable transponder having either Mode S or 4096-code capability with Mode C automatic altitude reporting capability. Answer (B) is incorrect. An operable transponder having either Mode S or 4096-code capability with Mode C automatic altitude reporting capability is required within 30 NM of a Class B primary airport, not instruments and equipment required for IFR operations.

126. What minimum pilot certification is required for operation within Class B airspace?

    A. Recreational Pilot Certificate.

    B. Private Pilot Certificate or Student Pilot Certificate with appropriate logbook endorsements.

    C. Private Pilot Certificate with an instrument rating.

Answer (B) is correct. *(FAR 91.131)*
    DISCUSSION: No person may take off or land aircraft at an airport within Class B airspace or operate an aircraft within Class B airspace unless (s)he is at least a private pilot or, if a student pilot, (s)he has the appropriate logbook endorsement required by FAR 61.95.
    Answer (A) is incorrect. A recreational pilot is restricted from operating in airspace (e.g., Class B airspace) that requires communication with ATC. Answer (C) is incorrect. An instrument rating is not required to operate in Class B airspace.

127. What minimum pilot certification is required for operation within Class B airspace?

A. Private Pilot Certificate or Student Pilot Certificate with appropriate logbook endorsements.

B. Commercial Pilot Certificate.

C. Private Pilot Certificate with an instrument rating.

Answer (A) is correct. *(FAR 91.131)*
    DISCUSSION: No person may take off or land aircraft at an airport within Class B airspace or operate an aircraft within Class B airspace unless (s)he is at least a private pilot or, if a student pilot, (s)he has the appropriate logbook endorsement required by FAR 61.95.
    Answer (B) is incorrect. The minimum pilot certification to operate in Class B airspace is a private pilot certificate or student pilot certificate with appropriate logbook endorsements, not a commercial pilot certificate. Answer (C) is incorrect. An instrument rating is not required to operate in Class B airspace.

91.133 Restricted and Prohibited Areas

128. Under what condition, if any, may pilots fly through a restricted area?

A. When flying on airways with an ATC clearance.

B. With the controlling agency's authorization.

C. Regulations do not allow this.

Answer (B) is correct. *(FAR 91.133)*
    DISCUSSION: An aircraft may not be operated within a restricted area unless permission has been obtained from the controlling agency. Frequently, the ATC within the area acts as the controlling agent's authorization; e.g., an approach control in a military restricted area can permit aircraft to enter it when the restricted area is not active.
    Answer (A) is incorrect. Airways do not penetrate restricted areas. Answer (C) is incorrect. Restricted areas may be entered with proper authorization.

91.135 Operations in Class A Airspace

129. In which type of airspace are VFR flights prohibited?

A. Class A.

B. Class B.

C. Class C.

Answer (A) is correct. *(FAR 91.135)*
    DISCUSSION: Class A airspace (from 18,000 ft. MSL up to and including FL 600) requires operation under IFR at specific flight levels assigned by ATC. Accordingly, VFR flights are prohibited.
    Answer (B) is incorrect. VFR flights are prohibited in Class A, not Class B, airspace. Answer (C) is incorrect. VFR flights are prohibited in Class A, not Class C, airspace.

91.151 Fuel Requirements for Flight in VFR Conditions

130. What is the specific fuel requirement for flight under VFR at night in an airplane?

A. Enough to complete the flight at normal cruising speed with adverse wind conditions.

B. Enough to fly to the first point of intended landing and to fly after that for 30 minutes at normal cruising speed.

C. Enough to fly to the first point of intended landing and to fly after that for 45 minutes at normal cruising speed.

Answer (C) is correct. *(FAR 91.151)*
    DISCUSSION: The night VFR requirement is enough fuel to fly to the first point of intended landing and to fly thereafter for 45 min. at normal cruising speed given forecast conditions.
    Answer (A) is incorrect. The fuel requirements are based upon the wind conditions existing that day plus the 45-min. reserve. Answer (B) is incorrect. A 30-min. reserve is the requirement for day flight.

131. What is the specific fuel requirement for flight under VFR during daylight hours in an airplane?

A. Enough to complete the flight at normal cruising speed with adverse wind conditions.

B. Enough to fly to the first point of intended landing and to fly after that for 30 minutes at normal cruising speed.

C. Enough to fly to the first point of intended landing and to fly after that for 45 minutes at normal cruising speed.

Answer (B) is correct. *(FAR 91.151)*
    DISCUSSION: The day-VFR requirement is enough fuel to fly to the first point of intended landing and thereafter for 30 min. at normal cruising speed.
    Answer (A) is incorrect. The fuel requirements are based upon the wind conditions existing that day plus the 30-min. reserve. Answer (C) is incorrect. A 45-min. reserve is the requirement for night flight.

4.9 FAR Part 91: 91.155 – 91.519

91.155 Basic VFR Weather Minimums

132. The basic VFR weather minimums for operating an aircraft within Class D airspace are

    A. 500-foot ceiling and 1 mile visibility.

    B. 1,000-foot ceiling and 3 miles visibility.

    C. clear of clouds and 2 miles visibility.

Answer (B) is correct. *(FAR 91.155)*
    DISCUSSION: The basic VFR weather minimums for operating an aircraft within Class D airspace are 1,000-ft. ceiling and 3 SM visibility.
    Answer (A) is incorrect. The basic VFR weather minimums for operating an aircraft in Class D airspace are 1,000-ft., not 500-ft., ceiling and 3, not 1, SM visibility. Answer (C) is incorrect. The basic VFR weather minimums for operating an aircraft in Class D airspace are 1,000-ft. ceiling, not clear of clouds, and 3, not 2, SM visibility.

133. The minimum flight visibility required for VFR flights above 10,000 feet MSL and more than 1,200 feet AGL in controlled airspace is

    A. 1 mile.

    B. 3 miles.

    C. 5 miles.

Answer (C) is correct. *(FAR 91.155)*
    DISCUSSION: Controlled airspace is the generic term for Class A, B, C, D, or E airspace. Of these, only in Class E airspace is the minimum flight visibility 5 SM for VFR flights at or above 10,000 ft. MSL.
    Note: AGL altitudes are not used in controlled airspace. In Class E airspace, the visibility and distance from clouds are given for (1) below 10,000 ft. MSL and (2) at or above 10,000 ft. MSL.
    Answer (A) is incorrect. This is the visibility in Class G, not Class E, airspace when more than 1,200 ft. AGL but less, not more, than 10,000 ft. MSL. Answer (B) is incorrect. This is the minimum visibility in Class E airspace when below, not at or above, 10,000 ft. MSL.

134. VFR flight in controlled airspace above 1,200 feet AGL and below 10,000 feet MSL requires a minimum visibility and vertical cloud clearance of

    A. 3 miles, and 500 feet below or 1,000 feet above the clouds in controlled airspace.

    B. 5 miles, and 1,000 feet below or 1,000 feet above the clouds at all altitudes.

    C. 5 miles, and 1,000 feet below or 1,000 feet above the clouds only in Class A airspace.

Answer (A) is correct. *(FAR 91.155)*
    DISCUSSION: Controlled airspace is the generic term for Class A, B, C, D, or E airspace. Only in Class C, D, or below 10,000 ft. MSL in Class E airspace are the minimum flight visibility and vertical distance from cloud for VFR flight required to be 3 SM, and 500 ft. below or 1,000 ft. above the clouds.
    Note: AGL altitudes are not used in controlled airspace. In Class E airspace, the visibility and distance from clouds are given for (1) below 10,000 ft. MSL and (2) at or above 10,000 ft. MSL.
    Answer (B) is incorrect. Five SM and 1,000 ft. above and below the clouds is the minimum visibility and vertical cloud clearance in Class E airspace at altitudes at or above, not below, 10,000 ft. MSL. Answer (C) is incorrect. VFR flight in Class A airspace is prohibited.

135. For VFR flight operations above 10,000 feet MSL and more than 1,200 feet AGL, the minimum horizontal distance from clouds required is

    A. 1,000 feet.

    B. 2,000 feet.

    C. 1 mile.

Answer (C) is correct. *(FAR 91.155)*
    DISCUSSION: For VFR flight operations in Class G airspace at altitudes more than 1,200 ft. AGL and at or above 10,000 ft. MSL, the minimum horizontal distance from clouds required is 1 statute mile.
    Note: The FAA question fails to specify what type of airspace. Since AGL altitudes are not used in controlled airspace (Class A, B, C, D, or E), that implies Class G airspace.
    Answer (A) is incorrect. This is the minimum vertical, not horizontal, distance from the clouds. Answer (B) is incorrect. This is the minimum horizontal distance from clouds in Class G airspace at night below, not above, 10,000 ft. MSL and when at altitudes more than 1,200 ft. AGL but less, not more, than 10,000 ft. MSL.

136. The minimum distance from clouds required for VFR operations on an airway below 10,000 feet MSL is

    A. remain clear of clouds.

    B. 500 feet below, 1,000 feet above, and 2,000 feet horizontally.

    C. 500 feet above, 1,000 feet below, and 2,000 feet horizontally.

Answer (B) is correct. *(FAR 91.155)*
    DISCUSSION: An airway includes that Class E airspace extending upward from 1,200 ft. AGL to, but not including, 18,000 ft. MSL. The minimum distance from clouds below 10,000 ft. MSL in Class E airspace is 500 ft. below, 1,000 ft. above, and 2,000 ft. horizontally.
    Answer (A) is incorrect. Clear of clouds is the minimum distance from clouds required in Class B, not Class E, airspace. Answer (C) is incorrect. The minimum distance from clouds required for VFR operations in Class E airspace below 10,000 ft. MSL is 500 ft. below, not above; 1,000 ft. above, not below; and 2,000 ft. horizontally.

137. What minimum visibility and clearance from clouds are required for VFR operations in Class G airspace at 700 feet AGL or below during daylight hours?

A. 1 mile visibility and clear of clouds.

B. 1 mile visibility, 500 feet below, 1,000 feet above, and 2,000 feet horizontal clearance from clouds.

C. 3 miles visibility and clear of clouds.

Answer (A) is correct. *(FAR 91.155)*
DISCUSSION: Below 1,200 ft. AGL in Class G airspace during daylight hours, the VFR weather minimum is 1 SM visibility and clear of clouds.
    Answer (B) is incorrect. One SM visibility, 500 ft. below, 1,000 ft. above, and 2,000 ft. horizontal clearance from clouds is the minimum visibility and clearance from clouds in Class G airspace at more than 1,200 ft. AGL but less than 10,000 ft. MSL, not at 700 ft. AGL. At night the requirement is 3 SM and 500 ft. below, 1,000 ft. above, and 2,000 ft. horizontal from clouds. Answer (C) is incorrect. Three SM visibility and clear of clouds are the visibility and clearance from clouds requirements in Class B, not Class G, airspace.

138. What minimum flight visibility is required for VFR flight operations on an airway below 10,000 feet MSL?

A. 1 mile.

B. 3 miles.

C. 4 miles.

Answer (B) is correct. *(FAR 91.155)*
DISCUSSION: An airway includes that Class E airspace extending upward from 1,200 ft. AGL to, but not including, 18,000 ft. MSL. The minimum flight visibility for VFR flight operations in Class E airspace less than 10,000 ft. MSL is 3 statute miles.
    Answer (A) is incorrect. One SM is the minimum daytime visibility for a VFR flight below 10,000 ft. MSL in Class G, not Class E, airspace. Answer (C) is incorrect. The minimum flight visibility for VFR flight operations in Class E airspace below 10,000 ft. MSL is 3 SM, not 4 statute miles.

139. During operations outside controlled airspace at altitudes of more than 1,200 feet AGL, but less than 10,000 feet MSL, the minimum flight visibility for VFR flight at night is

A. 1 mile.

B. 3 miles.

C. 5 miles.

Answer (B) is correct. *(FAR 91.155)*
DISCUSSION: When operating outside controlled airspace (i.e., Class G airspace) at night at altitudes of more than 1,200 ft. AGL, but less than 10,000 ft. MSL, the minimum flight visibility is 3 statute miles.
    Answer (A) is incorrect. One SM is the minimum day, not night, flight visibility in Class G airspace at altitudes of more than 1,200 ft. AGL, but less than 10,000 ft. MSL. Answer (C) is incorrect. Five SM is for operations more than 1,200 ft. AGL and at or above, not below, 10,000 ft. MSL in Class G airspace.

140. During operations within controlled airspace at altitudes of more than 1,200 feet AGL, but less than 10,000 feet MSL, the minimum distance above clouds requirement for VFR flight is

A. 500 feet.

B. 1,000 feet.

C. 1,500 feet.

Answer (B) is correct. *(FAR 91.155)*
DISCUSSION: Controlled airspace is the generic term for Class A, B, C, D, or E airspace. Only in Class C, D, or below 10,000 ft. MSL in Class E airspace are the minimum flight visibility and vertical distance from clouds for VFR flight required to be 3 SM, and 500 ft. below or 1,000 ft. above the clouds.
    Note: AGL altitudes are not used in controlled airspace. In Class E airspace, the visibility and distance from clouds are given for (1) below 10,000 ft. MSL and (2) at or above 10,000 ft. MSL.
    Answer (A) is incorrect. This is the minimum distance below, not above, clouds requirement for VFR flight in Class E airspace at altitudes of less than 10,000 ft. MSL. Answer (C) is incorrect. The minimum distance above clouds requirement for VFR flight in Class E airspace at altitudes of less than 10,000 ft. MSL is 1,000 ft., not 1,500 feet.

141. No person may take off or land an aircraft under basic VFR at an airport that lies within Class D airspace unless the

A. flight visibility at that airport is at least 1 mile.

B. ground visibility at that airport is at least 1 mile.

C. ground visibility at that airport is at least 3 miles.

Answer (C) is correct. *(FAR 91.155)*
DISCUSSION: No person may take off or land an aircraft at any airport that lies within Class D airspace under basic VFR unless the ground visibility is 3 statute miles. If ground visibility is not reported, flight visibility during landing or takeoff, or while operating in the traffic pattern, must be at least 3 statute miles.
    Answer (A) is incorrect. Flight visibility during landing or takeoff under basic VFR must be at least 3, not 1, statute miles. Answer (B) is incorrect. Ground visibility during landing or takeoff under basic VFR must be at least 3, not 1, statute miles.

142. During operations at altitudes of more than 1,200 feet AGL and at or above 10,000 feet MSL, the minimum distance above clouds requirement for VFR flight is

A. 500 feet.

B. 1,000 feet.

C. 1,500 feet.

Answer (B) is correct. *(FAR 91.155)*
  DISCUSSION: During operations in Class G airspace at altitudes of more than 1,200 ft. AGL and at or above 10,000 ft. MSL, the minimum distance above clouds requirement for VFR flight is 1,000 feet.
  Note: The FAA question fails to specify what type of airspace. Since AGL altitudes are not used in controlled airspace (Class A, B, C, D, and E), that implies Class G airspace.
  Answer (A) is incorrect. This is the vertical distance below, not above, the clouds for VFR operations below, not at or above, 10,000 ft. MSL and above 1,200 ft. AGL in Class G airspace. Answer (C) is incorrect. The figure of 1,000 ft., not 1,500 ft., is the vertical distance required above the clouds for VFR operations above 1,200 ft. AGL and at or above 10,000 ft. MSL in Class G airspace.

143. Outside controlled airspace, the minimum flight visibility requirement for VFR flight above 1,200 feet AGL and below 10,000 feet MSL during daylight hours is

A. 1 mile.

B. 3 miles.

C. 5 miles.

Answer (A) is correct. *(FAR 91.155)*
  DISCUSSION: Outside controlled airspace (i.e., Class G airspace) at altitudes above 1,200 ft. AGL and below 10,000 ft. MSL, the minimum flight visibility requirement for VFR flight during the day is 1 statute mile.
  Answer (B) is incorrect. Three SM is the minimum VFR flight visibility required at night, not day, for flights in Class G airspace at altitudes below 10,000 ft. MSL. Answer (C) is incorrect. Five SM is the minimum VFR flight visibility required for flights in Class G airspace above 1,200 ft. AGL and at or above, not below, 10,000 ft. MSL.

144. During operations outside controlled airspace at altitudes of more than 1,200 feet AGL, but less than 10,000 feet MSL, the minimum distance below clouds requirement for VFR flight at night is

A. 500 feet.

B. 1,000 feet.

C. 1,500 feet.

Answer (A) is correct. *(FAR 91.155)*
  DISCUSSION: Outside controlled airspace (i.e., Class G airspace) at altitudes above 1,200 ft. AGL and less than 10,000 ft. MSL, the minimum distance below clouds requirement for VFR flight at night is 500 feet.
  Answer (B) is incorrect. This is the minimum distance above, not below, the clouds. Answer (C) is incorrect. The minimum distance below the clouds is 500 ft., not 1,500 feet.

145. During operations within controlled airspace at altitudes of less than 1,200 feet AGL, the minimum horizontal distance from clouds requirement for VFR flight is

A. 1,000 feet.

B. 1,500 feet.

C. 2,000 feet.

Answer (C) is correct. *(FAR 91.155)*
  DISCUSSION: Controlled airspace is the generic term for Class A, B, C, D, or E airspace. Only in Class C, D, or below 10,000 ft. MSL in Class E airspace is the minimum horizontal distance from clouds for VFR flight required to be 2,000 feet.
  Note: AGL altitudes are not used in controlled airspace. In Class E airspace, the visibility and distance from clouds are given for (1) below 10,000 ft. MSL and (2) at or above 10,000 ft. MSL.
  Answer (A) is incorrect. This is the minimum vertical, not horizontal, distance above the clouds in Class E airspace below 10,000 ft. MSL. Answer (B) is incorrect. The minimum horizontal distance is 2,000 ft., not 1,500 feet.

146. Normal VFR operations in Class D airspace with an operating control tower require the visibility and ceiling to be at least

A. 1,000 feet and 1 mile.

B. 1,000 feet and 3 miles.

C. 2,500 feet and 3 miles.

Answer (B) is correct. *(FAR 91.155)*
  DISCUSSION: The basic VFR weather minimums for operating an aircraft within Class D airspace are a 1,000-ft. ceiling and 3 SM visibility.
  Answer (A) is incorrect. The basic VFR weather minimums for operating an aircraft in Class D airspace are a 1,000-ft. ceiling and 3 SM, not 1 SM, visibility. Answer (C) is incorrect. The basic VFR weather minimums for operating an aircraft in Class D airspace are a 1,000-ft., not 2,500-ft., ceiling and 3 SM visibility.

91.157 Special VFR Weather Minimums

147. What ATC facility should the pilot contact to receive a special VFR departure clearance in Class D airspace?

  A. Automated Flight Service Station.
  B. Air Traffic Control Tower.
  C. Air Route Traffic Control Center.

Answer (B) is correct. *(FAR 91.157)*
  DISCUSSION: When special VFR is needed, the pilot should contact the Air Traffic Control Tower to receive a departure clearance in Class D airspace.
  Answer (A) is incorrect. A pilot may request a clearance through an FSS for a special VFR clearance in Class E, not Class D, airspace. The FSS would only act as a relay point between the pilot and the ATC facility responsible for the Class E airspace (i.e., FSS personnel cannot issue a clearance). Answer (C) is incorrect. An Air Route Traffic Control Center can issue a clearance for a special VFR for an airport in Class E, not Class D, airspace.

148. A special VFR clearance authorizes the pilot of an aircraft to operate VFR while within Class D airspace when the visibility is

  A. less than 1 mile and the ceiling is less than 1,000 feet.
  B. at least 1 mile and the aircraft can remain clear of clouds.
  C. at least 3 miles and the aircraft can remain clear of clouds.

Answer (B) is correct. *(FAR 91.157)*
  DISCUSSION: To operate within Class D airspace under special VFR clearance, visibility must be at least 1 statute mile. There is no ceiling requirement, but the aircraft must remain clear of clouds.
  Answer (A) is incorrect. A special VFR clearance authorizes the pilot to operate VFR within Class D airspace if the visibility is at least, not below, 1 statute mile. Answer (C) is incorrect. A special VFR clearance requires the pilot to maintain at least 1 SM, not 3 SM, visibility and remain clear of clouds.

149. No person may operate an airplane within Class D airspace at night under special VFR unless the

  A. flight can be conducted 500 feet below the clouds.
  B. airplane is equipped for instrument flight.
  C. flight visibility is at least 3 miles.

Answer (B) is correct. *(FAR 91.157)*
  DISCUSSION: To operate under special VFR within Class D airspace at night, the pilot must be instrument rated and the airplane equipped for instrument flight.
  Answer (A) is incorrect. The only additional requirement at night for special VFR in Class D airspace is that the airplane be IFR equipped and the pilot be instrument rated, not that the flight be conducted 500 ft. below the clouds. Answer (C) is incorrect. For special VFR at night in Class D airspace, the flight visibility must be at least 1 SM, not 3 statute miles.

150. What are the minimum requirements for airplane operations under special VFR in Class D airspace at night?

  A. The airplane must be under radar surveillance at all times while in Class D airspace.
  B. The airplane must be equipped for IFR with an altitude reporting transponder.
  C. The pilot must be instrument rated, and the airplane must be IFR equipped.

Answer (C) is correct. *(FAR 91.157)*
  DISCUSSION: To operate under special VFR within Class D airspace at night, the pilot must be instrument rated and the airplane must be IFR equipped.
  Answer (A) is incorrect. To operate an airplane under special VFR at night in Class D airspace, there is no requirement that the airplane be under radar surveillance. Answer (B) is incorrect. There is no requirement for an altitude reporting transponder for special VFR at night in Class D airspace, but the pilot must be instrument rated.

151. What is the minimum weather condition required for airplanes operating under special VFR in Class D airspace?

  A. 1 mile flight visibility.
  B. 1 mile flight visibility and 1,000-foot ceiling.
  C. 3 miles flight visibility and 1,000-foot ceiling.

Answer (A) is correct. *(FAR 91.157)*
  DISCUSSION: To operate within Class D airspace under special VFR clearance, visibility must be at least 1 statute mile. There is no ceiling requirement, but the aircraft must remain clear of clouds.
  Answer (B) is incorrect. To operate within Class D airspace under a special VFR clearance, there is no (not 1,000-ft.) ceiling requirement other than to remain clear of clouds. Answer (C) is incorrect. Three SM flight visibility and 1,000-ft. ceiling are basic (not special) VFR weather minimums to operate an airplane in Class D airspace.

91.159 VFR Cruising Altitude or Flight Level

152. Which VFR cruising altitude is acceptable for a flight on a Victor Airway with a magnetic course of 175°? The terrain is less than 1,000 feet.

A. 4,500 feet.

B. 5,000 feet.

C. 5,500 feet.

Answer (C) is correct. *(FAR 91.159)*
DISCUSSION: When operating a VFR flight above 3,000 ft. AGL on a magnetic course of 0° through 179°, fly any odd thousand-ft. MSL altitude plus 500 feet. Thus, on a magnetic course of 175°, an appropriate VFR cruising altitude is 5,500 feet.
Answer (A) is incorrect. This would be an acceptable VFR cruising altitude if you were on a magnetic course of 180° to 359°, not 175°. Answer (B) is incorrect. On a magnetic course of 175°, the acceptable VFR cruising altitude is an odd thousand plus 500 ft. (5,500 ft., not 5,000 feet).

153. Which cruising altitude is appropriate for a VFR flight on a magnetic course of 135°?

A. Even thousand.

B. Even thousand plus 500 feet.

C. Odd thousand plus 500 feet.

Answer (C) is correct. *(FAR 91.159)*
DISCUSSION: When operating a VFR flight above 3,000 ft. AGL on a magnetic course of 0° through 179°, fly any odd thousand-ft. MSL altitude plus 500 feet. Thus, on a magnetic course of 135°, an appropriate VFR cruising altitude is an odd thousand plus 500 feet.
Answer (A) is incorrect. A VFR flight on a magnetic course of 135° will use an odd (not even) thousand, plus 500 ft. altitude. Answer (B) is incorrect. A VFR flight on a magnetic course of 135° will use an odd (not even) thousand plus 500 ft. altitude.

154. Which VFR cruising altitude is appropriate when flying above 3,000 feet AGL on a magnetic course of 185°?

A. 4,000 feet.

B. 4,500 feet.

C. 5,000 feet.

Answer (B) is correct. *(FAR 91.159)*
DISCUSSION: When operating a VFR flight above 3,000 ft. AGL on a magnetic course of 180° through 359°, fly any even thousand-ft. MSL altitude, plus 500 ft. Thus, on a magnetic course of 185°, an appropriate VFR cruising altitude is 4,500 feet.
Answer (A) is incorrect. On a magnetic course of 185° the appropriate VFR cruising altitude is an even thousand-ft. plus 500 ft. altitude (4,500 ft., not 4,000 feet). Answer (C) is incorrect. On a magnetic course of 185° the appropriate VFR cruising altitude is an even (not odd) thousand-ft., plus 500 ft. (4,500 ft., not 5,000 feet).

155. Each person operating an aircraft at a VFR cruising altitude shall maintain an odd-thousand plus 500-foot altitude while on a

A. magnetic heading of 0° through 179°.

B. magnetic course of 0° through 179°.

C. true course of 0° through 179°.

Answer (B) is correct. *(FAR 91.159)*
DISCUSSION: When operating above 3,000 ft. AGL but less than 18,000 ft. MSL on a magnetic course of 0° to 179°, fly at an odd thousand-ft. MSL altitude plus 500 feet.
Answer (A) is incorrect. A magnetic heading includes wind correction, and VFR cruising altitudes are based on magnetic course, i.e., without wind correction. Answer (C) is incorrect. True course does not include an adjustment for magnetic variation.

91.203 Civil Aircraft: Certifications Required

156. In addition to a valid Airworthiness Certificate, what documents or records must be aboard an aircraft during flight?

A. Aircraft engine and airframe logbooks, and owner's manual.

B. Radio operator's permit, and repair and alteration forms.

C. Operating limitations and Registration Certificate.

Answer (C) is correct. *(FAR 91.203 and 91.9)*
DISCUSSION: FAR 91.203 requires both an Airworthiness Certificate and a Registration Certificate to be aboard aircraft during flight. FAR 91.9 requires that operating limitations be available in the aircraft in an approved Airplane Flight Manual, approved manual material, markings, and placards, or any combination thereof.
Answer (A) is incorrect. The airframe and engine logbooks are usually maintained and stored on the ground. Answer (B) is incorrect. Repair and alteration forms are handled in the maintenance shop. Also, the Radio Operator's permit, although carried by the pilot, is an FCC requirement. A pilot may still fly without it as long as (s)he does not use any radio equipment that transmits a signal (e.g., communication, DME, or transponder).

91.207 Emergency Locator Transmitters

157. When must batteries in an emergency locator transmitter (ELT) be replaced or recharged, if rechargeable?

A. After any inadvertent activation of the ELT.

B. When the ELT has been in use for more than 1 cumulative hour.

C. When the ELT can no longer be heard over the airplane's communication radio receiver.

Answer (B) is correct. *(FAR 91.207)*
DISCUSSION: ELT batteries must be replaced or recharged (if rechargeable) when the transmitter has been in use for more than 1 cumulative hr. or when 50% of their useful life (or useful life of charge) has expired.
Answer (A) is incorrect. The batteries in an ELT must be replaced (or recharged, if rechargeable) only after the transmitter has been used for more than 1 cumulative hr., not after any inadvertent activation of the transmitter. Answer (C) is incorrect. ELT batteries are replaced (or recharged, if rechargeable) based on use or useful life, not when an ELT can no longer be heard over the airplane's communication radio receiver.

158. When may an emergency locator transmitter (ELT) be tested?

A. Anytime.

B. At 15 and 45 minutes past the hour.

C. During the first 5 minutes after the hour.

Answer (C) is correct. *(AIM Para 6-2-5)*
DISCUSSION: Emergency locator transmitters (ELT) may only be tested on the ground during the first 5 min. after the hour. Other times it is only allowed with prior arrangement with the nearest FAA Control Tower or FSS. No airborne checks are allowed.
Answer (A) is incorrect. An ELT should only be tested during the first 5 min. after the hour, not anytime. Answer (B) is incorrect. An ELT should only be tested during the first 5 min. after the hour, not at 15 and 45 min. past the hour.

159. When are non-rechargeable batteries of an emergency locator transmitter (ELT) required to be replaced?

A. Every 24 months.

B. When 50 percent of their useful life expires.

C. At the time of each 100-hour or annual inspection.

Answer (B) is correct. *(FAR 91.207)*
DISCUSSION: Non-rechargeable batteries of an ELT must be replaced when 50% of their useful life expires or after the transmitter has been in use for more than 1 cumulative hour.
Answer (A) is incorrect. Every 24 months is the requirement for the transponder to be tested and inspected, not when non-rechargeable ELT batteries are to be replaced. Answer (C) is incorrect. Non-rechargeable ELT batteries are replaced when 50% of their useful life expires or after 1 cumulative hr. of use, not necessarily at the time of each 100-hr. or annual inspection.

160. When must the battery in an emergency locator transmitter (ELT) be replaced (or recharged if the battery is rechargeable)?

A. After one-half the battery's useful life.

B. During each annual and 100-hour inspection.

C. Every 24 calendar months.

Answer (A) is correct. *(FAR 91.207)*
DISCUSSION: Emergency locator transmitter (ELT) batteries must be replaced or recharged after 50% of their useful life has expired or when the transmitter has been in use for more than 1 cumulative hour.
Answer (B) is incorrect. ELT batteries must be replaced (or recharged) after one-half the battery's useful life has expired, not during each annual and 100-hr. inspection. Answer (C) is incorrect. A transponder (not an ELT battery) must be tested and inspected every 24 calendar months.

91.209 Aircraft Lights

161. Except in Alaska, during what time period should lighted position lights be displayed on an aircraft?

A. End of evening civil twilight to the beginning of morning civil twilight.

B. 1 hour after sunset to 1 hour before sunrise.

C. Sunset to sunrise.

Answer (C) is correct. *(FAR 91.209)*
DISCUSSION: Except in Alaska, no person may operate an aircraft during the period from sunset to sunrise unless the aircraft's lighted position lights are on.
Answer (A) is incorrect. End of evening civil twilight to the beginning of morning civil twilight is the definition of night, not the time period in which lighted position lights be displayed on an aircraft. Answer (B) is incorrect. The period from 1 hr. after sunset to 1 hr. before sunrise is the time used to meet night recency requirements, not when the aircraft position lights should be on.

91.211 Supplemental Oxygen

162. Unless each occupant is provided with supplemental oxygen, no person may operate a civil aircraft of U.S. registry above a maximum cabin pressure altitude of

A. 12,500 feet MSL.

B. 14,000 feet MSL.

C. 15,000 feet MSL.

Answer (C) is correct. *(FAR 91.211)*
DISCUSSION: No person may operate a civil aircraft of U.S. registry at cabin pressure altitudes above 15,000 ft. MSL unless each occupant is provided with supplemental oxygen.
Answer (A) is incorrect. At cabin pressure altitudes above 12,500 ft. MSL, up to and including 14,000 ft. MSL, only the minimum required flight crew, not each occupant, must be provided with and use supplemental oxygen after 30 min. at those altitudes. Answer (B) is incorrect. At cabin pressure altitudes above 14,000 ft. MSL, only the minimum required flight crew, not each occupant, must be provided with and continuously use supplemental oxygen at those altitudes.

163. When operating an aircraft at cabin pressure altitudes above 12,500 feet MSL up to and including 14,000 feet MSL, supplemental oxygen shall be used during

A. the entire flight time at those altitudes.

B. that flight time in excess of 10 minutes at those altitudes.

C. that flight time in excess of 30 minutes at those altitudes.

Answer (C) is correct. *(FAR 91.211)*
DISCUSSION: At cabin pressure altitudes above 12,500 ft. MSL, up to and including 14,000 ft. MSL, the required minimum flight crew must use supplemental oxygen only after 30 min. at those altitudes.
Answer (A) is incorrect. At cabin pressure altitudes above 12,500 ft. MSL up to and including 14,000 ft. MSL, supplemental oxygen shall be used during that time in excess of 30 min. (not the entire flight time) at those altitudes. Answer (B) is incorrect. At cabin pressure altitudes above 12,500 ft. MSL up to and including 14,000 ft. MSL, supplemental oxygen shall be used during that time in excess of 30 (not 10) min. at those altitudes.

91.215 ATC Transponder and Altitude Reporting Equipment and Use

164. An operable 4096-code transponder with an encoding altimeter is required in which airspace?

A. Class A, Class B (and within 30 miles of the Class B primary airport), and Class C.

B. Class D and Class E (below 10,000 feet MSL).

C. Class D and Class G (below 10,000 feet MSL).

Answer (A) is correct. *(FAR 91.215)*
DISCUSSION: An operable transponder with an encoding altimeter (Mode C) is required in Class A, Class B (and within 30 NM of the Class B primary airport), and Class C airspace, and at or above 10,000 ft. MSL excluding that airspace below 2,500 ft. AGL.
Answer (B) is incorrect. An operable 4096-code transponder with an encoding altimeter is not required to operate in Class D or Class E (below 10,000 ft. MSL) airspace. Answer (C) is incorrect. An operable 4096-code transponder with an encoding altimeter is not required to operate in Class D or Class G (below 10,000 ft. MSL) airspace.

165. An operable 4096-code transponder and Mode C encoding altimeter are required in

A. Class B airspace and within 30 miles of the Class B primary airport.

B. Class D airspace.

C. Class E airspace below 10,000 feet MSL.

Answer (A) is correct. *(FAR 91.215)*
DISCUSSION: An operable 4096-code transponder and Mode C encoding altimeter are required in Class B airspace and within 30 NM of the Class B primary airport.
Answer (B) is incorrect. An operable 4096-code transponder and Mode C encoding altimeter are required in Class B airspace and within 30 NM of the Class B primary airport, not Class D airspace. Answer (C) is incorrect. An operable 4096-code transponder and Mode C encoding altimeter are required in Class B airspace and within 30 NM of the Class B primary airport, not Class E airspace below 10,000 ft. MSL.

91.303 Aerobatic Flight

166. In which class of airspace is acrobatic flight prohibited?

A. Class E airspace not designated for Federal Airways above 1,500 feet AGL.

B. Class E airspace below 1,500 feet AGL.

C. Class G airspace above 1,500 feet AGL.

Answer (B) is correct. *(FAR 91.303)*
DISCUSSION: No person may operate an aircraft in acrobatic flight below an altitude of 1,500 ft. AGL.
Answer (A) is incorrect. Acrobatic flight is prohibited in Class E airspace within 4 NM of the centerline of a Federal Airway, not in all Class E airspace. Answer (C) is incorrect. Acrobatic flight is only prohibited below 1,500 ft. AGL.

167. No person may operate an aircraft in acrobatic flight when the flight visibility is less than

    A. 3 miles.

    B. 5 miles.

    C. 7 miles.

Answer (A) is correct. *(FAR 91.303)*
    DISCUSSION: No person may operate an aircraft in acrobatic flight when the flight visibility is less than 3 statute miles.
    Answer (B) is incorrect. The minimum flight visibility for acrobatic flight is 3 SM, not 5 statute miles. Answer (C) is incorrect. The minimum flight visibility for acrobatic flight is 3 SM, not 7 statute miles.

168. What is the lowest altitude permitted for acrobatic flight?

    A. 1,000 feet AGL.

    B. 1,500 feet AGL.

    C. 2,000 feet AGL.

Answer (B) is correct. *(FAR 91.303)*
    DISCUSSION: No person may operate an aircraft in acrobatic flight below 1,500 ft. AGL.
    Answer (A) is incorrect. The figure of 1,500 ft. AGL, not 1,000 ft. AGL, is the lowest altitude permitted for acrobatic flight. Answer (C) is incorrect. The figure of 1,500 ft. AGL, not 2,000 ft. AGL, is the lowest altitude permitted for acrobatic flight.

169. No person may operate an aircraft in acrobatic flight when

    A. flight visibility is less than 5 miles.

    B. over any congested area of a city, town, or settlement.

    C. less than 2,500 feet AGL.

Answer (B) is correct. *(FAR 91.303)*
    DISCUSSION: No person may operate an aircraft in acrobatic flight over any congested area of a city, town, or settlement.
    Answer (A) is incorrect. The flight visibility limitation for acrobatic flight is 3 SM, not 5 statute miles. Answer (C) is incorrect. The minimum altitude for acrobatic flight is 1,500 ft. AGL, not 2,500 ft. AGL.

91.307 Parachutes and Parachuting

170. With certain exceptions, when must each occupant of an aircraft wear an approved parachute?

    A. When a door is removed from the aircraft to facilitate parachute jumpers.

    B. When intentionally pitching the nose of the aircraft up or down 30° or more.

    C. When intentionally banking in excess of 30°.

Answer (B) is correct. *(FAR 91.307)*
    DISCUSSION: Unless each occupant of an airplane is wearing an approved parachute, no pilot carrying any other person (other than a crewmember) may execute any intentional maneuver that exceeds a bank of 60° or a nose-up or nose-down attitude of 30° relative to the horizon.
    Answer (A) is incorrect. Pilots of airplanes that are carrying parachute jumpers are not required to use a parachute. Answer (C) is incorrect. A parachute is required when an intentional bank that exceeds 60°, not 30°, is to be made.

171. A parachute composed of nylon, rayon, or other synthetic fibers must have been packed by a certificated and appropriately rated parachute rigger within the preceding

    A. 60 days.

    B. 90 days.

    C. 180 days.

Answer (C) is correct. *(FAR 91.307)*
    DISCUSSION: No pilot of a civil aircraft may allow a parachute that is available for emergency use to be carried in that aircraft unless it is an approved type and, if a chair type, it has been packed by a certificated and appropriately rated parachute rigger within the preceding 180 days, if synthetic fibers are used in its design.
    Answer (A) is incorrect. A parachute constructed with natural fibers, not synthetic fibers, must be repacked within the preceding 60 days. Answer (B) is incorrect. A parachute composed of synthetic fibers must have been repacked within the preceding 180 days, not 90 days.

172. An approved parachute constructed of natural fibers may be carried in an aircraft for emergency use if it has been packed by an appropriately rated parachute rigger within the preceding

    A. 60 days.

    B. 120 days.

    C. 180 days.

Answer (A) is correct. *(FAR 91.307)*
    DISCUSSION: No pilot of a civil aircraft may allow a parachute that is available for emergency use to be carried in that aircraft unless it is an approved type and has been packed by a certificated and appropriately rated parachute rigger within the preceding 60 days, if natural fibers are used in its design.
    Answer (B) is incorrect. A parachute constructed of natural fibers must have been repacked within the preceding 60 days, not 120 days. Answer (C) is incorrect. A parachute constructed from synthetic fibers, not natural fibers, must be repacked every 180 days.

91.313 Restricted Category Civil Aircraft: Operating Limitations

173. Which is normally prohibited when operating a restricted category civil aircraft?

A. Flight under instrument flight rules.

B. Flight over a densely populated area.

C. Flight within Class D airspace.

Answer (B) is correct. *(FAR 91.313)*
  DISCUSSION: Normally, no person may operate a restricted category civil aircraft over a densely populated area.
  Answer (A) is incorrect. Flight over a densely populated area, not IFR flight, is normally prohibited when operating a restricted category civil aircraft. Answer (C) is incorrect. Flight over a densely populated area, not within Class D airspace, is normally prohibited when operating a restricted category civil aircraft.

91.319 Aircraft Having Experimental Certificates: Operating Limitations

174. Unless otherwise specifically authorized, no person may operate an aircraft that has an experimental certificate

A. beneath the floor of Class B airspace.

B. over a densely populated area or in a congested airway.

C. from the primary airport within Class D airspace.

Answer (B) is correct. *(FAR 91.319)*
  DISCUSSION: Unless otherwise specifically authorized, no person may operate an aircraft that has an experimental certificate over a densely populated area or along a congested airway.
  Answer (A) is incorrect. Normally no person may operate an aircraft that has an experimental certificate along a congested airway, not beneath the floor of Class B airspace. Answer (C) is incorrect. A person can operate an aircraft that has an experimental certificate from the primary airport within Class D airspace as long as ATC is notified of the experimental nature of the aircraft.

91.403 General

175. The responsibility for ensuring that an aircraft is maintained in an airworthy condition is primarily that of the

A. pilot in command.

B. owner or operator.

C. mechanic who performs the work.

Answer (B) is correct. *(FAR 91.403)*
  DISCUSSION: The owner or operator of an aircraft is primarily responsible for maintaining that aircraft in an airworthy condition. The term "operator" includes the pilot in command.
  Answer (A) is incorrect. The owner or operator, not only the pilot in command, of an aircraft is responsible for ensuring the airworthiness of the aircraft. Answer (C) is incorrect. Although a mechanic will perform inspections and maintenance, the primary responsibility for an aircraft's airworthiness lies with its owner or operator.

176. Who is responsible for ensuring Airworthiness Directives (AD's) are complied with?

A. Owner or operator.

B. Repair station.

C. Mechanic with inspection authorization (IA).

Answer (A) is correct. *(FAR 91.403)*
  DISCUSSION: Airworthiness Directives (ADs) are regulatory and must be complied with, unless a specific exemption is granted. It is the responsibility of the owner or operator to ensure compliance with all pertinent ADs, including those ADs that require recurrent or continuing action.
  Answer (B) is incorrect. The owner or operator, not a repair station, is responsible for ensuring ADs are complied with. Answer (C) is incorrect. The owner or operator, not a mechanic with inspection authorization, is responsible for ensuring ADs are complied with.

91.405 Maintenance Required

177. The responsibility for ensuring that maintenance personnel make the appropriate entries in the aircraft maintenance records indicating the aircraft has been approved for return to service lies with the

A. owner or operator.

B. pilot in command.

C. mechanic who performed the work.

Answer (A) is correct. *(FAR 91.405)*
  DISCUSSION: Each owner or operator of an aircraft shall ensure that maintenance personnel make the appropriate entries in the aircraft maintenance records indicating the aircraft has been approved for return to service.
  Answer (B) is incorrect. The owner or operator, not only the pilot in command, is responsible for ensuring that maintenance personnel make the proper entries in the aircraft's maintenance records. Answer (C) is incorrect. The owner or operator, not the mechanic who performed the work, is responsible for ensuring that proper entries are made in the aircraft's maintenance records.

178. Who is responsible for ensuring appropriate entries are made in maintenance records indicating the aircraft has been approved for return to service?

A. Owner or operator.

B. Certified mechanic.

C. Repair station.

Answer (A) is correct. *(FAR 91.405)*
DISCUSSION: It is the responsibility of the owner or operator of an aircraft to ensure that appropriate entries are made in maintenance records by maintenance personnel indicating the aircraft has been approved for return to service.
Answer (B) is incorrect. The certified mechanic performing the work must make the entries, but it is the responsibility of the owner or operator to ensure that the entries have been made. Answer (C) is incorrect. It is the responsibility of the owner or operator, not a repair station, to ensure appropriate entries have been made.

91.407 Operation after Maintenance, Preventive Maintenance, Rebuilding, or Alteration

179. If an alteration or repair substantially affects an aircraft's operation in flight, that aircraft must be test flown by an appropriately-rated pilot and approved for return to service prior to being operated

A. by any private pilot.

B. with passengers aboard.

C. for compensation or hire.

Answer (B) is correct. *(FAR 91.407)*
DISCUSSION: If an alteration or repair has been made that substantially affects the airplane's flight characteristics, the airplane must be test flown and approved for return to service by an appropriately rated pilot prior to being operated with passengers aboard. The test pilot must be at least a private pilot and appropriately rated for the airplane being tested and must make an operational check of the alteration or repair made, and log the flight in the aircraft records.
Answer (A) is incorrect. If an alteration or repair substantially affects an aircraft's operation in flight, a private pilot may only test fly that airplane if (s)he is appropriately rated to fly that airplane. Answer (C) is incorrect. After any alteration or repair that substantially affects an aircraft's operation in flight, that aircraft must be test flown and approved for return to service prior to being operated with any passengers aboard, not for compensation or hire.

180. Before passengers can be carried in an aircraft that has been altered in a manner that may have appreciably changed its flight characteristics, it must be flight tested by an appropriately-rated pilot who holds at least a

A. Commercial Pilot Certificate with an instrument rating.

B. Private Pilot Certificate.

C. Commercial Pilot Certificate and a mechanic's certificate.

Answer (B) is correct. *(FAR 91.407)*
DISCUSSION: If an alteration or repair has been made that may have changed an airplane's flight characteristics, the airplane must be test flown and approved for return to service by an appropriately rated pilot prior to being operated with passengers aboard. The test pilot must be at least a private pilot and appropriately rated for the airplane being tested.
Answer (A) is incorrect. The test flight must be made by an appropriately rated pilot who holds at least a private (not commercial) pilot certificate. An instrument rating is not required. Answer (C) is incorrect. The test flight must be made by an appropriately rated pilot who holds at least a private (not commercial) pilot certificate. A mechanic's certificate is not required for the test pilot.

91.409 Inspections

181. A 100-hour inspection was due at 3302.5 hours. The 100-hour inspection was actually done at 3309.5 hours. When is the next 100-hour inspection due?

A. 3312.5 hours.

B. 3402.5 hours.

C. 3409.5 hours.

Answer (B) is correct. *(FAR 91.409)*
DISCUSSION: Since the 100-hr. inspection was due at 3302.5 hr., the next 100-hr. inspection is due at 3402.5 (3302.5 + 100). The excess time used before the 100-hr. inspection was done must be included in computing the next 100 hr. of time in service.
Answer (A) is incorrect. This is the latest time on the tachometer the last 100-hr. inspection could have been completed, not when the next 100-hr. inspection is due. Answer (C) is incorrect. This is 100 hr. from the actual completion time of the last inspection, but the excess time is computed into the next 100 hr. of time in service, or 100 hr. from the last due time.

182. An aircraft's annual condition inspection was performed on July 12, this year. The next annual inspection will be due no later than

    A. July 1, next year.

    B. July 13, next year.

    C. July 31, next year.

Answer (C) is correct. *(FAR 91.409)*
    DISCUSSION: Annual condition inspections expire on the last day of the 12th calendar month after the previous annual condition inspection. If an annual condition inspection is performed on July 12 of this year, it will expire at midnight on July 31 next year.
    Answer (A) is incorrect. Annual condition inspections are due on the last day of the month. Thus, if an annual condition inspection is performed July 12, this year the next annual condition inspection is due July 31 (not July 1), next year. Answer (B) is incorrect. Annual condition inspections are due on the last day of the month. Thus, if an annual condition inspection is performed July 12, this year the next annual condition inspection is due July 31 (not July 13), next year.

183. What aircraft inspections are required for rental aircraft that are also used for flight instruction?

    A. Annual condition and 100-hour inspections.

    B. Biannual condition and 100-hour inspections.

    C. Annual condition and 50-hour inspections.

Answer (A) is correct. *(FAR 91.409)*
    DISCUSSION: All aircraft that are used for hire (e.g., rental) and flight instruction must be inspected on a 100-hr. basis. Also, an annual condition inspection must be completed.
    Answer (B) is incorrect. An annual, not biannual, condition inspection is required for all aircraft. A 100-hr. inspection is also required for aircraft rented for flight instruction. Answer (C) is incorrect. Besides an annual condition inspection, aircraft rented for flight instruction purposes are also required to have a 100- (not 50-) hr. inspection.

184. An aircraft had a 100-hour inspection when the tachometer read 1259.6. When is the next 100-hour inspection due?

    A. 1349.6 hours.

    B. 1359.6 hours.

    C. 1369.6 hours.

Answer (B) is correct. *(FAR 91.409)*
    DISCUSSION: The next 100-hr. inspection is due within 100 hr. of time in service. The 100-hr. may be exceeded by 10 hr. in order to get to a place where the work can be done. However, this additional time is included in computing the next 100 hr. period. Therefore, in this question, add 100 hr. to 1259.6 to get the next inspection, due at 1359.6 hours.
    Answer (A) is incorrect. The 100-hr. inspection is due when the tachometer indicates 1359.6 hr. (1259.6 + 100), not 1349.6 hours. Answer (C) is incorrect. This is 10 hr. over the 100-hr. limitation that is allowed if the aircraft is en route to reach a place where the inspection can be done. The 100-hr. inspection is due at 1359.6 hr. (1259.6 + 100), not 1369.6 hours.

91.413 ATC Transponder Tests and Inspections

185. No person may use an ATC transponder unless it has been tested and inspected within at least the preceding

    A. 6 calendar months.

    B. 12 calendar months.

    C. 24 calendar months.

Answer (C) is correct. *(FAR 91.413)*
    DISCUSSION: No person may use an ATC transponder that is specified in the regulations unless within the preceding 24 calendar months it has been tested and found to comply with its operating specifications.
    Answer (A) is incorrect. An ATC transponder must be tested and inspected every 24 (not 6) calendar months. Answer (B) is incorrect. An ATC transponder must be tested and inspected every 24 (not 12) calendar months.

186. Maintenance records show the last transponder inspection was performed on September 1, 2011. The next inspection will be due no later than

    A. September 30, 2012.

    B. September 1, 2013.

    C. September 30, 2013.

Answer (C) is correct. *(FAR 91.413)*
    DISCUSSION: No person may use an ATC transponder that is specified in the regulations unless within the preceding 24 calendar months it has been tested and found to comply with its operating specifications. Thus, if the last inspection was performed on September 1, 2011, the next inspection will be due no later than September 30, 2013.
    Answer (A) is incorrect. The requirement states that the transponder must be inspected every 24 calendar months, not every 12. Answer (B) is incorrect. The "calendar month" requirement means that the inspection may be done on any date from the first to the last day of the month. Therefore, the transponder must be inspected no later than the end of the month of September, not the beginning.

91.417 Maintenance Records

187. Completion of an annual condition inspection and the return of the aircraft to service should always be indicated by

A. the relicensing date on the Registration Certificate.

B. an appropriate notation in the aircraft maintenance records.

C. an inspection sticker placed on the instrument panel that lists the annual inspection completion date.

Answer (B) is correct. *(FAR 91.417)*
DISCUSSION: Completion of an annual condition inspection and the return of the aircraft to service should always be indicated by an appropriate notation in the aircraft's maintenance records.
Answer (A) is incorrect. The registration certificate shows ownership, not completion of an annual inspection. Answer (C) is incorrect. Maintenance information is found in the airplane logbooks, not on inspection stickers.

188. To determine the expiration date of the last annual aircraft inspection, a person should refer to the

A. Airworthiness Certificate.

B. Registration Certificate.

C. aircraft maintenance records.

Answer (C) is correct. *(FAR 91.417)*
DISCUSSION: After maintenance inspections have been completed, maintenance personnel should make the appropriate entries in the aircraft maintenance records or logbooks. This is where the date of the last annual inspection can be found.
Answer (A) is incorrect. To determine the expiration date of the last annual inspection, a person should refer to the aircraft maintenance records, not the Airworthiness Certificate. Answer (B) is incorrect. To determine the expiration date of the last annual inspection, a person should refer to the aircraft maintenance records, not the Registration Certificate.

189. Which records or documents shall the owner or operator of an aircraft keep to show compliance with an applicable Airworthiness Directive?

A. Aircraft maintenance records.

B. Airworthiness Certificate and Pilot's Operating Handbook.

C. Airworthiness and Registration Certificates.

Answer (A) is correct. *(FAR 91.417)*
DISCUSSION: Aircraft maintenance records must show the current status of applicable airworthiness directives (ADs) including, for each, the method of compliance, the AD number, and revision date. If the AD involves recurring action, the time and date when the next action is required.
Answer (B) is incorrect. Compliance with an AD is found in aircraft maintenance records, not the Airworthiness Certificate and Pilot's Operating Handbook. Answer (C) is incorrect. Compliance with an AD is found in aircraft maintenance records, not in the Airworthiness and Registration Certificates.

190. The airworthiness of an aircraft can be determined by a preflight inspection and a

A. statement from the owner or operator that the aircraft is airworthy.

B. log book endorsement from a flight instructor.

C. review of the maintenance records.

Answer (C) is correct. *(FAR 91.417)*
DISCUSSION: As pilot in command, you are responsible for determining whether your aircraft is in condition for safe flight. Only by conducting a preflight inspection and a review of the maintenance records can you determine whether all required maintenance has been performed and, thus, whether the aircraft is airworthy.
Answer (A) is incorrect. A statement from the owner or operator that the aircraft is airworthy does not ensure that all required maintenance has been performed. Answer (B) is incorrect. A log book endorsement from a flight instructor does not give any assurance that the aircraft has received required maintenance, and it is not required for determining airworthiness.

91.519 Passenger Briefings

191. The party directly responsible for the pre-takeoff briefing of passengers is the

A. pilot in command.

B. safety officer.

C. ground crew.

Answer (A) is correct. *(FAR 91.519)*
DISCUSSION: Before each takeoff, the pilot in command of an airplane carrying passengers shall ensure that all passengers have been orally briefed on smoking, the use of safety belts and shoulder harnesses, location and means of opening a passenger door as a means of emergency exit, location of survival equipment, ditching procedures and the use of the flotation equipment, and the normal and emergency use of oxygen equipment if installed in the airplane.
Answer (B) is incorrect. Passenger briefings are the responsibility of the pilot in command, not a safety officer. Answer (C) is incorrect. Passenger briefings are the responsibility of the pilot in command, not the ground crew.

192. Pre-takeoff briefing of passengers for a flight is the responsibility of

A. all passengers.

B. the pilot.

C. a crewmember.

Answer (B) is correct. *(FAR 91.519)*
DISCUSSION: Before each takeoff, the pilot in command of an airplane carrying passengers shall ensure that all passengers have been orally briefed on smoking, the use of safety belts and shoulder harnesses, location and means of opening a passenger door as a means of emergency exit, location of survival equipment, ditching procedures and the use of the flotation equipment, and the normal and emergency use of oxygen equipment if installed in the airplane.
Answer (A) is incorrect. It is the responsibility of the pilot in command to brief the passengers; passengers cannot self-brief. Answer (C) is incorrect. A pilot in command can delegate the role of the pre-takeoff briefing to a crew member, but (s)he is ultimately the one responsible for ensuring the briefing has been completed.

4.10 NTSB Part 830

830.5 Immediate Notification

193. If an aircraft is involved in an accident which results in substantial damage to the aircraft, the nearest NTSB field office should be notified

A. immediately.

B. within 48 hours.

C. within 7 days.

Answer (A) is correct. *(NTSB 830.5)*
DISCUSSION: The NTSB must be notified immediately and by the most expeditious means possible when an aircraft accident or any of various listed incidents occurs or when an aircraft is overdue and is believed to have been in an accident.
Answer (B) is incorrect. An aircraft involved in an accident must be reported immediately (not within 48 hr.) to the NTSB office. Answer (C) is incorrect. An aircraft accident must be reported immediately (not within 7 days) to the nearest NTSB office.

194. Which incident would necessitate an immediate notification to the nearest NTSB field office?

A. An in-flight generator/alternator failure.

B. An in-flight fire.

C. An in-flight loss of VOR receiver capability.

Answer (B) is correct. *(NTSB 830.5)*
DISCUSSION: The NTSB must be notified immediately and by the most expeditious means possible when an aircraft accident or any of various listed incidents occurs or when an aircraft is overdue and believed to have been in an accident. The following are considered incidents:

1. Flight control system malfunction or failure;
2. Inability of any required flight crewmember to perform normal flight duties as a result of injury or illness;
3. Failure of structural components of a turbine engine, excluding compressor and turbine blades and vanes;
4. In-flight fire; or
5. Aircraft collision in flight.

Answer (A) is incorrect. An in-flight generator/alternator failure does not require immediate notification. Answer (C) is incorrect. An in-flight loss of VOR receiver capability does not require any type of notification to the NTSB.

195. Which incident requires an immediate notification be made to the nearest NTSB field office?

A. An overdue aircraft that is believed to be involved in an accident.

B. An in-flight radio communications failure.

C. An in-flight generator or alternator failure.

Answer (A) is correct. *(NTSB 830.5)*
DISCUSSION: The NTSB must be notified immediately and by the most expeditious means possible when an aircraft is overdue and is believed to have been involved in an accident.
Answer (B) is incorrect. An in-flight radio communications failure does not require notification to the NTSB at any time. Answer (C) is incorrect. An in-flight generator or alternator failure does not require notification to the NTSB at any time.

196. Which incident requires an immediate notification to the nearest NTSB field office?

A. A forced landing due to engine failure.

B. Landing gear damage, due to a hard landing.

C. Flight control system malfunction or failure.

Answer (C) is correct. *(NTSB 830.5)*
DISCUSSION: The NTSB must be notified immediately and by the most expeditious means possible when an aircraft accident or any of various listed incidents occurs or when an aircraft is overdue and believed to have been in an accident. The following are considered incidents:

1. Flight control system malfunction or failure;
2. Inability of any required flight crewmember to perform normal flight duties as a result of injury or illness;
3. Failure of structural components of a turbine engine, excluding compressor and turbine blades and vanes;
4. In-flight fire; or
5. Aircraft collision in flight.

Answer (A) is incorrect. Only failure of structural components of a turbine engine (not a forced landing due to engine failure) must be reported immediately to the nearest NTSB office. Answer (B) is incorrect. Landing gear damage due to a hard landing is not considered an incident that requires immediate notification to the NTSB.

830.10 Preservation of Aircraft Wreckage, Mail, Cargo, and Records

197. May aircraft wreckage be moved prior to the time the NTSB takes custody?

A. Yes, but only if moved by a federal, state, or local law enforcement officer.

B. Yes, but only to protect the wreckage from further damage.

C. No, it may not be moved under any circumstances.

Answer (B) is correct. *(NTSB 830.10)*
DISCUSSION: Prior to the time the Board or its authorized representative takes custody of aircraft wreckage, mail, or cargo, such wreckage, mail, or cargo may not be disturbed or moved except to the extent necessary to

1. Remove persons injured or trapped,
2. Protect the wreckage from further damage, or
3. Protect the public from injury.

Answer (A) is incorrect. Aircraft wreckage can only be moved to protect the wreckage from further damage, protect the public from injury, or to remove persons injured or trapped, not by any federal, state, or local law enforcement officer. Answer (C) is incorrect. Aircraft wreckage may be moved in certain circumstances, such as to remove persons injured or trapped, to protect the wreckage from further damage, or to protect the public from injury.

830.15 Reports and Statements to Be Filed

198. The operator of an aircraft that has been involved in an accident is required to file an NTSB accident report within how many days?

A. 5.

B. 7.

C. 10.

Answer (C) is correct. *(NTSB 830.15)*
DISCUSSION: The operator of an aircraft shall file a report on NTSB Form 6120.1/2 within 10 days after an accident, or after 7 days if an overdue aircraft is still missing. A report on an incident for which notification is required shall be filed only as required.
Answer (A) is incorrect. NTSB Form 6120.1/2 is required within 10 (not 5) days after an accident. Answer (B) is incorrect. NTSB Form 6120.1/2 is required within 10 (not 7) days after an accident.

199. The operator of an aircraft that has been involved in an incident is required to submit a report to the nearest field office of the NTSB

A. within 7 days.

B. within 10 days.

C. when requested.

Answer (C) is correct. *(NTSB 830.15)*
DISCUSSION: The operator of an aircraft shall file a report on NTSB Form 6120.1/2 only when requested. A report is required within 10 days of an accident, or after 7 days if an overdue aircraft is still missing.
Answer (A) is incorrect. Seven days is the time allowed to file a written report on an overdue aircraft that is still missing; an incident requires a report only when requested. Answer (B) is incorrect. A report must be filed within 10 days of an accident; an incident requires a report only when requested.

END OF STUDY UNIT

STUDY UNIT FIVE
AIRPLANE PERFORMANCE AND WEIGHT AND BALANCE

(10 pages of outline)

This study unit contains outlines of major concepts tested, sample test questions and answers regarding airplane performance and weight and balance, and an explanation of each answer. The table of contents above lists each subunit within this study unit, the number of questions pertaining to that particular subunit, and the pages on which the outlines and questions begin, respectively.

CAUTION: Recall that the **sole purpose** of this book is to expedite your passing the FAA pilot knowledge test for the private pilot certificate. Accordingly, all extraneous material (i.e., topics or regulations not directly tested on the FAA pilot knowledge test) is omitted, even though much more information and knowledge are necessary to fly safely. This additional material is presented in *Pilot Handbook* and *Private Pilot Flight Maneuvers and Practical Test Prep*, available from Gleim Publications, Inc. See the order form on page 371.

Many of the topics in this study unit require interpretation of graphs and charts. Graphs and charts pictorially describe the relationship between two or more variables. Thus, they are a substitute for solving one or more equations. Each time you must interpret (i.e., get an answer from) a graph or chart, you should

1. Understand clearly what is required, e.g., landing roll distance, weight, etc.

2. Analyze the chart or graph to determine the variables involved, including

    a. Labeled sides (axes) of the graph or chart
    b. Labeled lines within the graph or chart

3. Plug the data given in the question into the graph or chart.

4. Finally, determine the value of the item required in the question.

5.1 DENSITY ALTITUDE

1. Density altitude is a measurement of the density of the air expressed in terms of altitude.

    a. Air density varies inversely with altitude; i.e., air is very dense at low altitudes and less dense at high altitudes.

    b. Temperature, humidity, and barometric pressure also affect air density.

        1) A scale of air density to altitude has been established using a standard temperature and pressure for each altitude. At sea level, standard is 15°C and 29.92" Hg.

        2) When temperature and pressure are not at standard (which is almost always), density altitude will not be the same as true altitude.

2. You are required to know how barometric pressure, temperature, and humidity affect density altitude. Visualize the following:

    a. As barometric pressure increases, the air becomes more compressed and compact. This is an increase in density. Air density is higher if the pressure is high, so the density altitude is said to be lower.

        1) Density altitude is increased by a decrease in pressure.
        2) Density altitude is decreased by an increase in pressure.

    b. As temperature increases, the air expands and therefore becomes less dense. This decrease in density means a higher density altitude. Remember, air is normally less dense at higher altitudes.

        1) Density altitude is increased by an increase in temperature.
        2) Density altitude is decreased by a decrease in temperature.

    c. As relative humidity increases, the air becomes less dense. A given volume of moist air weighs less than the same volume of dry air. This decrease in density means a higher density altitude.

        1) Density altitude is increased by an increase in humidity.
        2) Density altitude is decreased by a decrease in humidity.

3. Said another way, density altitude varies directly with temperature and humidity, and inversely with barometric pressure:

    a. Cold, dry air and higher barometric pressure = low density altitude.
    b. Hot, humid air and lower barometric pressure = high density altitude.

4. Pressure altitude is based on standard temperature. Therefore, density altitude will exceed pressure altitude if the temperature is above standard.

5. The primary reason for computing density altitude is to determine airplane performance.

    a. High density altitude reduces an airplane's overall performance.

        1) For example, climb performance is less and takeoff distance is longer.
        2) Propellers have less efficiency because there is less air for the propeller to get a grip on.

    b. However, the same indicated airspeed is used for takeoffs and landings regardless of altitude or air density because the airspeed indicator is also directly affected by air density.

5.2 DENSITY ALTITUDE COMPUTATIONS

1. Density altitude is determined most easily by finding the pressure altitude (indicated altitude when your altimeter is set to 29.92) and adjusting for the temperature.

   a. The adjustment may be made using your flight computer or a density altitude chart. This part of the FAA test requires you to use a density altitude chart.

2. When using a density altitude chart (see Figure 8 on page 176),

   a. Adjust the airport elevation to pressure altitude by adding or subtracting the conversion factor for the current altimeter setting.

   b. To adjust the pressure altitude for nonstandard temperature, plot the intersection of the actual air temperature (listed on the horizontal axis of the chart) with the pressure altitude lines that slope diagonally upward. Move left horizontally from the intersection to read density altitude on the vertical axis of the chart.

   c. EXAMPLE: Outside air temperature   90°F
   Altimeter setting   30.20" Hg
   Airport elevation   4,725 ft.

   Referring to Figure 8 on page 176, you determine the density altitude to be approximately 7,400 feet. This is found as follows:

   1) The altimeter setting of 30.20 requires a –257 altitude correction factor.
   2) Subtract 257 from field elevation of 4,725 ft. to obtain pressure altitude of 4,468 feet.
   3) Locate 90°F on the bottom axis of the chart and move up to intersect the diagonal pressure altitude line of 4,468 feet.
   4) Move horizontally to the left axis of the chart to obtain the density altitude of about 7,400 feet.
   5) Note that while true altitude (i.e., airport elevation) is 4,725 ft., density altitude is about 7,400 feet.
   6) Finally, note that you may determine the effects of temperature changes on density altitude simply by following the above chart procedure and substituting different temperatures.

5.3 TAKEOFF DISTANCE

1. Conditions that reduce airplane takeoff and climb performance are

   a. High altitude
   b. High temperature
   c. High humidity

2. Takeoff distance performance is displayed in the airplane operating manual either

   a. In chart form or
   b. On a graph

3. If a graph, it is usually presented in terms of density altitude. Thus, one must first adjust the airport elevation for nonstandard pressure and temperature.

   a. In the graph used on this exam (see Figure 41 on page 178), the first section on the left uses outside air temperature and pressure altitude to obtain density altitude.

      1) The curved line on the left portion is standard atmosphere, which you use when the question calls for standard temperature.

   b. The second section of the graph, to the right of the first reference line, takes the weight in pounds into account.

c.   The third section of the graph, to the right of the second reference line, takes the headwind or tailwind into account.

d.   The fourth section of the graph, at the right margin, takes obstacles into account.

e.   EXAMPLE: Given an outside air temperature of 15°C, a pressure altitude of 5,650 ft., a takeoff weight of 2,950 lb., and a headwind component of 9 kt., find the ground roll and the total takeoff distance over a 50-ft. obstacle. Use Figure 41 on page 178.

f.   The solution to the example problem is marked with the dotted arrows on the graph. Move straight up from 15°C (which is also where the standard temperature line begins) to the pressure altitude of 5,650 ft. and then horizontally to the right to the first reference line. It is not necessary to adjust for weight because the airplane is at maximum weight of 2,950 pounds. Continuing to the next reference line, the headwind component of 9 kt. means an adjustment downward in the wind component section (parallel to the guidelines). Finally, moving straight to the right gives the ground roll of 1,375 feet. The total takeoff distance over a 50-ft. obstacle, following parallel to the guideline up and to the right, is 2,300 feet.

5.4 CRUISE POWER SETTINGS

1.   Cruise power settings are found by use of a table (see Figure 36 on page 180).

a.   It is based on 65% power.

b.   It consists of three sections to adjust for varying temperatures:

1)   Standard temperature (in middle)
2)   ISA –20°C (on left)
3)   ISA +20°C (on right)

c.   Values found on the table based on various pressure altitudes and temperatures include

1)   Engine RPM
2)   Manifold pressure (in. Hg)
3)   Fuel flow in gal. per hr. (with the expected fuel pressure gauge indication in pounds per square inch)
4)   True airspeed (kt. and MPH)

2.   The FAA test questions gauge your ability to find values on the chart and interpolate between lines (see Figure 36 on page 180).

a.   EXAMPLE: A value for 9,500 ft. would be 75% of the distance between the number for 8,000 ft. and the number for 10,000 feet.

b.   EXAMPLE: At a pressure altitude of 6,000 ft. and a temperature of 26°C and with no wind, a 1,000-NM trip would take 71.42 gal. of fuel (1,000 ÷ 161 kt. = 6.21 hr.) (6.21 hr. × 11.5 gph = 71.42 gallons).

5.5 CROSSWIND COMPONENTS

1.   Airplanes have a limit to the amount of direct crosswind in which they can land. When the wind is not directly across the runway (i.e., quartering), a crosswind component chart may be used to determine the amount of direct crosswind. Variables on the crosswind component charts are

a.   Angle between wind and runway
b.   Wind velocity

NOTE: The coordinates on the vertical and horizontal axes of the graph will indicate the headwind and crosswind components of a quartering wind.

2. Refer to the crosswind component graph, which is Figure 37 on page 182.

    a. Note the example on the chart of a 40-kt. wind at a 30° angle.

    b. Find the 30° wind angle line. This is the angle between the wind direction and runway direction, e.g., runway 18 and wind from 210°.

    c. Find the 40-kt. wind velocity arc. Note the intersection of the wind arc and the 30° angle line.

        1) Drop straight down to determine the crosswind component of 20 kt.; i.e., landing in this situation would be like having a direct crosswind of 20 knots.

        2) Move horizontally to the left to determine the headwind component of 35 kt.; i.e., landing in this situation would be like having a headwind of 35 knots.

        3) EXAMPLE: You have been given 20 kt. as the maximum crosswind component for the airplane, and the angle between the runway and the wind is 30°. What is the maximum wind velocity without exceeding the 20-kt. crosswind component? Find where the 20-kt. crosswind line from the bottom of the chart crosses the 30° angle line, and note that it intersects the 40-kt. wind velocity line. This means you can land an airplane with a 20-kt. maximum crosswind component in a 40-kt. wind from a 30° angle to the runway.

5.6 LANDING DISTANCE

1. Required landing distances differ at various altitudes and temperatures due to changes in air density.

    a. However, indicated airspeed for landing is the same at all altitudes.

2. Landing distance information is given in airplane operating manuals in chart or graph form to adjust for headwind, temperature, and dry grass runways.

3. If an emergency situation requires a downwind landing, pilots should expect a higher groundspeed at touchdown, a longer ground roll, and the likelihood of overshooting the desired touchdown point.

4. It is imperative that you distinguish between distances for clearing a 50-ft. obstacle and distances without a 50-ft. obstacle at the beginning of the runway (the latter is described as the ground roll).

5. See Figure 38 on page 184 for an example landing distance graph. It is used in the same manner as the takeoff distance graph (Figure 41) discussed on page 165 and printed on page 178.

6. Refer to Figure 39 on page 187, which is a landing distance table.

    a. It has been computed for landing with no wind, at standard temperature, and at pressure altitude.

    b. The bottom "notes" tell you how to adjust for wind, nonstandard temperature, and a grass runway.

        1) Note 1 says to decrease the distance for a headwind. Note that tailwind hurts much more than headwind helps, so you cannot use the headwind formula in reverse.

    c. EXAMPLE: Given standard air temperature, 8-kt. headwind, and pressure altitude of 2,500 ft., find both the ground roll and the landing distance to clear a 50-ft. obstacle.

        1) On the table (Figure 39) for 2,500 ft., at standard temperature with no wind, the ground roll is 470 ft., and the distance to clear a 50-ft. obstacle is 1,135 feet. These amounts must be decreased by 20% because of the headwind (8 kt. ÷ 4 × 10% = 20%). Therefore, the ground roll is 376 ft. (470 × 80%) and the distance to clear a 50-ft. obstacle is 908 ft. (1,135 × 80%).

5.7 WEIGHT AND BALANCE DEFINITIONS

1. **Empty weight** consists of the airframe, engine, and all items of operating equipment permanently installed in the airplane, including optional special equipment, fixed ballast, hydraulic fluid, unusable fuel, and undrainable (or, in some aircraft, all) oil.

2. Standard weights have been established for numerous items involved in weight and balance computations.

    a. The standard weight for aviation gasoline (AVGAS) is 6 lb./gallon.

      1) EXAMPLE: 90 lb. of gasoline is equal to 15 gal. (90 ÷ 6).

3. The **center of gravity** (CG) is the point of balance along the airplane's longitudinal axis. By multiplying the weight of each component of the airplane by its arm (distance from an arbitrary reference point, called the reference datum), that component's moment is determined. The CG of the airplane is the sum of all the moments divided by the total weight.

5.8 CENTER OF GRAVITY CALCULATIONS

1. The basic formula for weight and balance is

    Weight × Arm = Moment

    a. Arm is the distance of the weight from the datum (a fixed position on the longitudinal axis of the airplane).

    b. The weight/arm/moment calculation computes where the CG is.

      1) Multiply the weight of each item loaded into the airplane by its arm (distance from datum) to determine moment.

      2) Add moments.

      3) Divide total moments by total weight to obtain CG (expressed in distance from the datum).

    c. EXAMPLE: You have items A, B, and C in the airplane. Note the airplane's empty weight is given as 1,500 lb. with a 20-in. arm.

	Weight		Arm		Moment
Empty airplane	1,500	×	20	=	30,000
A (pilot and passenger)	300	×	25	=	7,500
B (25 gal. of fuel × 6 lb./gal.)	150	×	30	=	4,500
C (baggage)	100	×	40	=	4,000
	2,050				46,000

The total loaded weight of the airplane is 2,050 pounds. Take the total moments of 46,000 lb.-in. divided by the total weight of 2,050 lb. to obtain the CG of 22.44 inches.

The weight and the CG are then checked to see whether they are within allowable limits.

2. The moment of an object is a measure of the force that causes a tendency of the object to rotate about a point or axis. It is usually expressed in pound-inches. In the figures on the next page, assume that a weight of 50 lb. is placed on the board at a point (station) 100 in. from the datum (fulcrum). The downward force of the weight at that spot can be determined by multiplying 50 lb. by 100 in., which produces a moment of 5,000 pound-inches.

    a. To establish a balance, a total moment of 5,000 lb.-in. must be applied to the other end of the board. Any combination of weight and distance that, when multiplied, produces 5,000 lb.-in. moment to the left of the datum will balance the board.

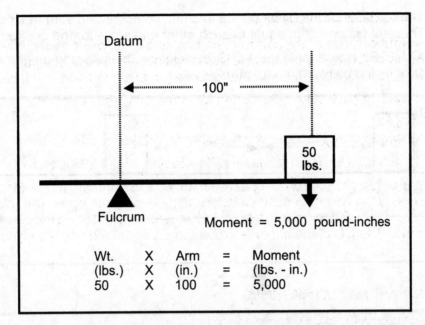

b. If a 100-lb. weight is placed at a point (station) 25 in. on the other side of the datum, and a second 50-lb. weight is placed at a point (station) 50 in. on the other side of the datum, the sum of the products of these two weights and their distances will total a moment of 5,000 lb.-in., which will balance the board. (See the figure below.)

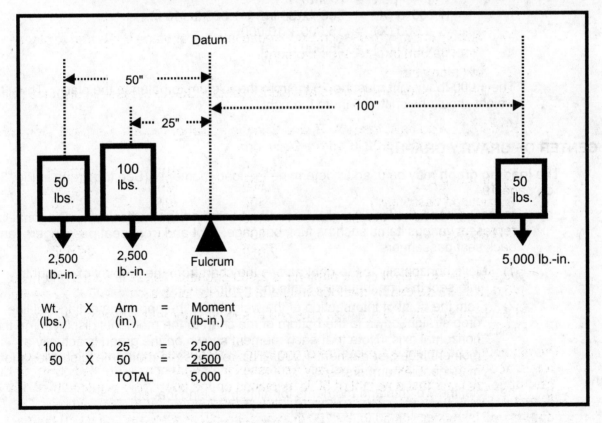

3.   When asked to balance the plank on the fulcrum, compute and sum the moments left and right. Then set left and right equal to each other and solve for the desired variable.

a.   EXAMPLE: How should the 1,000-lb. weight in the following diagram be shifted to balance the plank on the fulcrum?

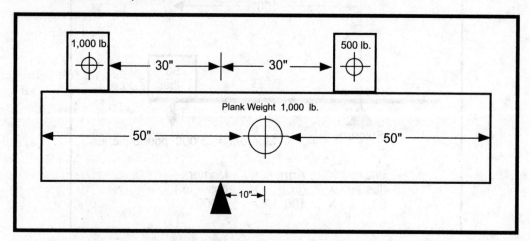

Compute and sum the moments left and right. Note that the plank itself weighs 1,000 lb. and that its CG is 10 in. right of the fulcrum. Set them equal to one another and solve for the desired variable:

$$
\begin{aligned}
\text{Left} &= \text{Right} \\
1{,}000 \text{ lb.}(X) &= 500 \text{ lb.}(30 \text{ in.}) + 1{,}000 \text{ lb.}(10 \text{ in.}) \\
1{,}000 \ (X) &= 15{,}000 + 10{,}000 \\
1{,}000 \ (X) &= 25{,}000 \\
X &= 25 \text{ in.}
\end{aligned}
$$

The 1,000-lb. weight must be 25 in. from the fulcrum to balance the plank. Thus, the weight should be shifted 5 in. to the right.

5.9 CENTER OF GRAVITY GRAPHS

1.   The **loading graph** may be used to determine the load moment. (See top graph in Fig. 35 on page 191.)

a.   On most graphs, the load weight in pounds is listed on the vertical axis. Diagonal lines represent various items such as fuel, baggage, pilot and front seat passengers, and back seat passengers.

1)   Move horizontally to the right across the chart from the amount of weight to intersect the line that represents the particular item.

2)   From the point of intersection of the weight with the appropriate diagonal line, drop straight down to the bottom of the chart to the moments displayed on the horizontal axis. Note that each moment shown on the graph is actually a moment index, or moment/1,000. This reduces the moments to smaller, more manageable numbers.

b.   Then total the weights and moments for all items being loaded.

c.   EXAMPLE: Determine the load (total) moment/1,000 in the following situation:

	Weight (lb.)	Moment/1,000 (lb.-in.)
Empty weight	1,350	51.5
Pilot & front seat passenger	400	?
Baggage	120	?
Usable fuel (38 gal. × 6 lb./gal.)	228	?
Oil (8 qt.)	15	−0.2

1)   Compute the moment of the pilot and front seat passenger by referring to the loading graph, and locate 400 on the weight scale. Move horizontally across the graph to intersect the diagonal line representing the pilot and front passenger, and then to the bottom scale, which indicates a moment of approximately 15.0.

2)   Locate 120 on the weight scale for the baggage. Move horizontally across the graph to intersect the diagonal line that represents baggage, then down vertically to the bottom scale, which indicates a moment of approximately 11.5.

3)   Locate 228 on the weight scale for the usable fuel. Move horizontally across the graph to intersect the diagonal line representing fuel, then down vertically to the bottom scale, which indicates a moment of 11.0.

4)   Notice a −0.2 moment for the engine oil (see note 2 on the top graph in Fig. 35 on page 191). Add all moments except this negative moment, and obtain a total of 89.0. Then subtract the negative moment to obtain a total aircraft moment of 88.8.

d.   Now add all the weights to determine that the airplane's maximum gross weight is not exceeded.

	Weight (lb.)	Moment/1,000 (lb.-in.)
Empty weight	1,350	51.5
Pilot & passengers	400	15.0
Baggage	120	11.5
Fuel	228	11.0
Oil	15	−0.2
	2,113	88.8

2.   The **center of gravity moment envelope chart** (see bottom graph in Fig. 35 on page 191) is a graph showing CG moment limits for various gross weights. Acceptable limits are established as an area on the graph. This area is called the envelope. Weight is on the vertical axis, and moments are on the horizontal axis.

a.   Identify the center of gravity point on the center of gravity moment envelope graph by plotting the total loaded aircraft weight across to the right.

b.   Plot the total moment upward from the bottom.

c.   The intersection will be within the CG moment envelope if the airplane has been loaded within limits.

d.   EXAMPLE: Using the data above, locate the weight of 2,113 lb. on the vertical axis, and then move across the chart to the moment line of 88.8. The point of intersection will indicate that the aircraft is within both CG (i.e., normal category) and gross weight (i.e., less than 2,300 lb.) limits.

5.10 CENTER OF GRAVITY TABLES

1.   Another approach to determining weight and CG limits is to use tables.

2.   First, determine the total moment from the Useful Load Weights and Moments table (Fig. 33 on page 194).

    a.   Moments can be read directly from the table for a specific weight.

    b.   If weight is between values, you can use the basic formula to determine the moment:

$$Weight \times Arm = Moment$$

       1)   Then divide by 100 to determine moment/100.

3.   Then use the Moment Limits vs. Weight table (Fig. 34 on page 195) to see if the total moment is within maximum and minimum limits for the gross weight.

QUESTIONS AND ANSWER EXPLANATIONS

    All of the private pilot knowledge test questions chosen by the FAA for release as well as additional questions selected by Gleim relating to the material in the previous outlines are reproduced on the following pages. These questions have been organized into the same subunits as the outlines. To the immediate right of each question are the correct answer and answer explanation. You should cover these answers and answer explanations while responding to the questions. Refer to the general discussion in the Introduction on how to take the FAA pilot knowledge test.

    Remember that the questions from the FAA pilot knowledge test bank have been reordered by topic and organized into a meaningful sequence. Also, the first line of the answer explanation gives the citation of the authoritative source for the answer.

QUESTIONS

5.1 **Density Altitude**

1. What are the standard temperature and pressure values for sea level?

    A. 15°C and 29.92" Hg.

    B. 59°C and 1013.2 millibars.

    C. 59°F and 29.92 millibars.

Answer (A) is correct. *(AvW Chap 3)*
    DISCUSSION: The standard temperature and pressure values for sea level are 15°C and 29.92" Hg. This is equivalent to 59°F and 1013.2 millibars of mercury.
    Answer (B) is incorrect. Standard temperature is 59°F (not 59°C). Answer (C) is incorrect. Standard pressure is 29.92" Hg (not 29.92 millibars).

2. What effect, if any, does high humidity have on aircraft performance?

    A. It increases performance.

    B. It decreases performance.

    C. It has no effect on performance.

Answer (B) is correct. *(PHAK Chap 10)*
    DISCUSSION: As the air becomes more humid, it becomes less dense. This is because a given volume of moist air weighs less than the same volume of dry air. Less dense air reduces aircraft performance.
    Answer (A) is incorrect. High humidity reduces (not increases) performance. Answer (C) is incorrect. The three factors that affect aircraft performance are pressure, temperature, and humidity.

3. Which factor would tend to increase the density altitude at a given airport?

    A. An increase in barometric pressure.

    B. An increase in ambient temperature.

    C. A decrease in relative humidity.

Answer (B) is correct. *(AvW Chap 3)*
    DISCUSSION: When air temperature increases, density altitude increases because, at a higher temperature, the air is less dense.
    Answer (A) is incorrect. Density altitude decreases as barometric pressure increases. Answer (C) is incorrect. Density altitude decreases as relative humidity decreases.

4. What effect does high density altitude, as compared to low density altitude, have on propeller efficiency and why?

    A. Efficiency is increased due to less friction on the propeller blades.

    B. Efficiency is reduced because the propeller exerts less force at high density altitudes than at low density altitudes.

    C. Efficiency is reduced due to the increased force of the propeller in the thinner air.

Answer (B) is correct. *(AvW Chap 3)*
    DISCUSSION: The propeller produces thrust in proportion to the mass of air being accelerated through the rotating propeller. If the air is less dense, the propeller efficiency is decreased. Remember, higher density altitude refers to less dense air.
    Answer (A) is incorrect. There is decreased, not increased, efficiency. Answer (C) is incorrect. The propeller exerts less (not more) force on the air when the air is thinner, i.e., at higher density altitudes.

5. What effect does high density altitude have on aircraft performance?

    A. It increases engine performance.

    B. It reduces climb performance.

    C. It increases takeoff performance.

Answer (B) is correct. *(PHAK Chap 10)*
    DISCUSSION: High density altitude reduces all aspects of an airplane's performance, including takeoff and climb performance.
    Answer (A) is incorrect. Engine performance is decreased (not increased). Answer (C) is incorrect. Takeoff runway length is increased, i.e., reduces takeoff performance.

6. Which combination of atmospheric conditions will reduce aircraft takeoff and climb performance?

    A. Low temperature, low relative humidity, and low density altitude.

    B. High temperature, low relative humidity, and low density altitude.

    C. High temperature, high relative humidity, and high density altitude.

Answer (C) is correct. *(PHAK Chap 10)*
    DISCUSSION: Takeoff and climb performance are reduced by high density altitude. High density altitude is a result of high temperatures and high relative humidity.
    Answer (A) is incorrect. Low temperature, low relative humidity, and low density altitude all improve airplane performance. Answer (B) is incorrect. Low relative humidity and low density altitude both improve airplane performance.

7. If the outside air temperature (OAT) at a given altitude is warmer than standard, the density altitude is

    A. equal to pressure altitude.

    B. lower than pressure altitude.

    C. higher than pressure altitude.

Answer (C) is correct. *(PHAK Chap 10)*
    DISCUSSION: When temperature increases, the air expands and therefore becomes less dense. This decrease in density means a higher density altitude. Pressure altitude is based on standard temperature. Thus, density altitude exceeds pressure altitude when the temperature is warmer than standard.
    Answer (A) is incorrect. Density altitude equals pressure altitude only when temperature is standard. Answer (B) is incorrect. Density altitude is lower than pressure altitude when the temperature is below standard.

8. You have planned a cross-country flight on a warm spring morning. Your course includes a mountain pass, which is at 11,500 feet MSL. The service ceiling of your airplane is 14,000 feet MSL. After checking the local weather report, you are able to calculate the density altitude of the mountain pass as 14,800 feet MSL. Which of the following is the correct action to take?

    A. Replan your journey to avoid the mountain pass.

    B. Continue as planned since density altitude is only a factor for takeoff.

    C. Continue as planned because mountain thermals will assist your climb.

Answer (A) is correct. *(PHAK Chap 10)*
    DISCUSSION: Because the density altitude through the mountain pass is higher than the service ceiling of the aircraft, it will be impossible to fly through the pass given the current conditions. You must replan your journey to avoid the mountain pass.
    Answer (B) is incorrect. Density altitude affects all aspects of aircraft performance, not just takeoff performance. Answer (C) is incorrect. Mountain thermals cannot be relied upon to safely carry you through the mountain pass.

9. A pilot and two passengers landed on a 2,100 foot east-west gravel strip with an elevation of 1,800 feet. The temperature is warmer than expected and after computing the density altitude it is determined the takeoff distance over a 50 foot obstacle is 1,980 feet. The airplane is 75 pounds under gross weight. What would be the best choice?

    A. Takeoff to the west because the headwind will give the extra climb-out time needed.

    B. Try a takeoff without the passengers to make sure the climb is adequate.

    C. Wait until the temperature decreases, and recalculate the takeoff performance.

Answer (C) is correct. *(PHAK Chap 10)*
    DISCUSSION: The majority of pilot-induced accidents occur during the takeoff and landing phases of flight. In this instance, the pilot in command of this aircraft has an important decision to make. The takeoff distance over a 50-foot obstacle appears on initial inspection to be possible (1,980 feet on a 2,100-foot runway). It is important to remember, however, the performance charts are based on ideal conditions and created by testing brand new aircraft with optimal performance and highly experienced test pilots at the controls. It would be ill-advised for this pilot to attempt to take off. The pilot should wait for the temperature to decrease and recalculate the takeoff performance.
    Answer (A) is incorrect. There are no winds provided in this question and no guarantee the takeoff performance to the west would be improved in any way. Answer (B) is incorrect. The decision to attempt a takeoff without the passengers and ensure climb performance is flawed in a few ways. Charts provided in the aircraft information manual should be used to determine climb performance prior to a flight. Coming to the realization the climb performance is not sufficient to clear terrain features and obstacles once airborne is a position no pilot wants to find him/herself in. If the pilot did attempt to take off without the passengers and the climb performance was adequate, there is absolutely no reason to believe the performance would be sufficient when the passengers are added and the weight of the aircraft is increased.

5.2 Density Altitude Computations

10. (Refer to Figure 8 on page 176.) Determine the density altitude for these conditions:

Altimeter setting 30.35
Runway temperature . . . . +25°F
Airport elevation . . . . 3,894 ft. MSL

    A. 2,000 feet MSL.

    B. 2,900 feet MSL.

    C. 3,500 feet MSL.

Answer (A) is correct. *(PHAK Chap 10)*
    DISCUSSION: With an altimeter setting of 30.35" Hg, 394 ft. must be subtracted from a field elevation of 3,894 to obtain a pressure altitude of 3,500 feet. Note that the higher-than-normal pressure of 30.35 means the pressure altitude will be less than true altitude. The 394 ft. was found by interpolation: 30.3 on the graph is –348, and 30.4 was –440 feet. Adding one-half the –92 ft. difference (–46 ft.) to –348 ft. results in –394 feet. Once you have found the pressure altitude, use the chart to plot 3,500 ft. pressure altitude at 25°F, to reach 2,000 ft. density altitude. Note that since the temperature is lower than standard, the density altitude is lower than the pressure altitude.
    Answer (B) is incorrect. This would be the density altitude if you added (not subtracted) 394 ft. to 3,894 feet. Answer (C) is incorrect. This is pressure (not density) altitude.

11. (Refer to Figure 8 on page 176.) What is the effect of a temperature increase from 30 to 50 °F on the density altitude if the pressure altitude remains at 3,000 feet MSL?

    A. 900-foot increase.

    B. 1,100-foot decrease.

    C. 1,300-foot increase.

Answer (C) is correct. *(PHAK Chap 10)*
    DISCUSSION: Increasing the temperature from 30°F to 50°F, given a constant pressure altitude of 3,000 ft., requires you to find the 3,000-ft. line on the density altitude chart at the 30°F level. At this point, the density altitude is approximately 1,650 feet. Then move up the 3,000-ft. line to 50°F, where the density altitude is approximately 2,950 feet. There is an approximate 1,300-ft. increase (2,950 – 1,650 feet). Note that 50°F is just about standard and pressure altitude is very close to density altitude.
    Answer (A) is incorrect. A 900-ft. increase would be caused by a temperature increase of 14°F (not 20°F). Answer (B) is incorrect. A decrease in density altitude would be caused by a decrease, not an increase, in temperature.

12. (Refer to Figure 8 on page 176.) What is the effect of a temperature increase from 35 to 50 °F on the density altitude if the pressure altitude remains at 3,000 feet MSL?

    A. 1,000-foot increase.

    B. 1,100-foot decrease.

    C. 1,300-foot increase.

Answer (A) is correct. *(PHAK Chap 10)*
    DISCUSSION: Increasing the temperature from 35°F to 50°F, given a constant pressure altitude of 3,000 ft., requires you to find the 3,000-ft. line on the density altitude chart at the 35°F level. At this point, the density altitude is approximately 1,950 feet. Then move up the 3,000-ft. line to 50°F, where the density altitude is approximately 2,950 feet. There is an approximate 1,000-ft. increase (2,950 – 1,950 feet). Note that 50°F is just about standard, and pressure altitude is very close to density altitude.
    Answer (B) is incorrect. An 1,100-foot decrease would require a temperature decrease of 18°F to 17°F, not a 15°F increase to 50°F. Answer (C) is incorrect. An 1,300-ft. increase would be caused by a temperature increase of 20°F (not 15°F).

13. (Refer to Figure 8 on page 176.) Determine the pressure altitude at an airport that is 3,563 feet MSL with an altimeter setting of 29.96.

    A. 3,527 feet MSL.

    B. 3,556 feet MSL.

    C. 3,639 feet MSL.

Answer (A) is correct. *(PHAK Chap 10)*
    DISCUSSION: Note that the question asks only for pressure altitude, not density altitude. Pressure altitude is determined by adjusting the altimeter setting to 29.92" Hg, i.e., adjusting for nonstandard pressure. This is the true altitude plus or minus the pressure altitude conversion factor (based on current altimeter setting). On the chart, an altimeter setting of 30.0 requires you to subtract 73 ft. to determine pressure altitude (note that at 29.92, nothing is subtracted because that is pressure altitude). Since 29.96 is halfway between 29.92 and 30.0, you need only subtract 36 (–73/2) from 3,563 ft. to obtain a pressure altitude of 3,527 ft. (3,563 – 36). Note that a higher-than-standard barometric pressure means pressure altitude is lower than true altitude.
    Answer (B) is incorrect. You must subtract 36 (not 7) from 3,563 ft. to obtain the correct pressure altitude. Answer (C) is incorrect. You must subtract 36 (not add 76) from 3,563 ft. to obtain the correct pressure altitude.

14. (Refer to Figure 8 below.) What is the effect of a temperature decrease and a pressure altitude increase on the density altitude from 90°F and 1,250 feet pressure altitude to 55°F and 1,750 feet pressure altitude?

A. 1,700-foot increase.

B. 1,300-foot decrease.

C. 1,700-foot decrease.

Answer (C) is correct.   *(PHAK Chap 10)*
DISCUSSION: The requirement is the effect of a temperature decrease and a pressure altitude increase on density altitude. First, find the density altitude at 90°F and 1,250 ft. (approximately 3,600 feet). Then find the density altitude at 55°F and 1,750 ft. pressure altitude (approximately 1,900 feet). Next, subtract the two numbers. Subtracting 1,900 ft. from 3,600 ft. equals a 1,700-ft. decrease in density altitude.
Answer (A) is incorrect. Such a large decrease in temperature would decrease, not increase, density altitude. Answer (B) is incorrect. Density altitude would decrease 1,300 ft. if the temperature decreased to 60°F, not 55°F.

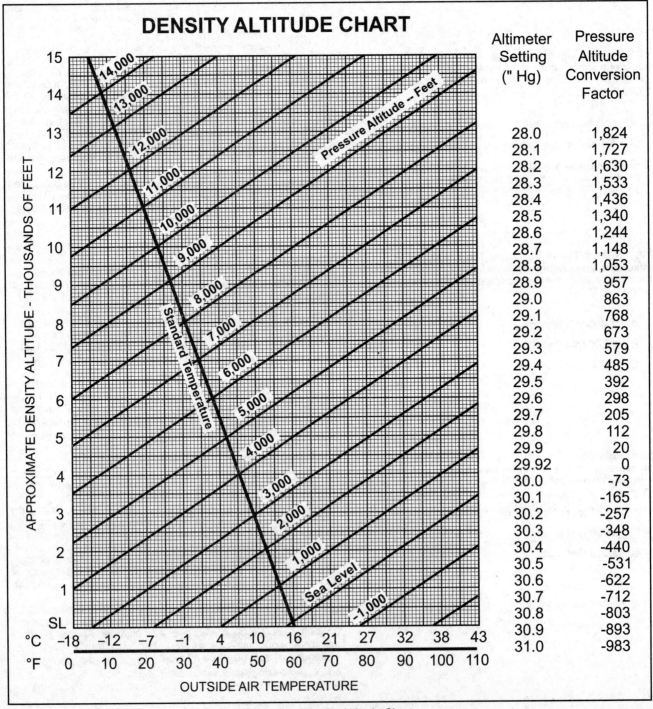

Figure 8. – Density Altitude Chart.

15. (Refer to Figure 8 on page 176.) Determine the pressure altitude at an airport that is 1,386 feet MSL with an altimeter setting of 29.97.

    A. 1,341 feet MSL.

    B. 1,451 feet MSL.

    C. 1,562 feet MSL.

Answer (A) is correct. *(PHAK Chap 10)*
    DISCUSSION: Pressure altitude is determined by adjusting the altimeter setting to 29.92" Hg. This is the true altitude plus or minus the pressure altitude conversion factor (based on current altimeter setting). Since 29.97 is not a number given on the conversion chart, you must interpolate. Compute 5/8 of –73 (since 29.97 is 5/8 of the way between 29.92 and 30.0), which is 45. Subtract 45 ft. from 1,386 ft. to obtain a pressure altitude of 1,341 feet. Note that if the altimeter setting is greater than standard (e.g., 29.97), the pressure altitude (i.e., altimeter set to 29.92) will be less than true altitude.
    Answer (B) is incorrect. You must subtract 45 ft. (not add 65) from 1,386 ft. to obtain the correct pressure altitude. Answer (C) is incorrect. You must subtract 45 ft. (not add 176) from 1,386 ft. to obtain the correct pressure altitude.

16. (Refer to Figure 8 on page 176.) What is the effect of a temperature increase from 25 to 50° F on the density altitude if the pressure altitude remains at 5,000 feet?

    A. 1,200-foot increase.

    B. 1,400-foot increase.

    C. 1,650-foot increase.

Answer (C) is correct. *(PHAK Chap 10)*
    DISCUSSION: Increasing the temperature from 25°F to 50°F, given a pressure altitude of 5,000 ft., requires you to find the 5,000-ft. line on the density altitude chart at the 25°F level. At this point, the density altitude is approximately 3,850 feet. Then move up the 5,000-ft. line to 50°F, where the density altitude is approximately 5,500 feet. There is about a 1,650-ft. increase (5,500 – 3,850 feet). As temperature increases, so does density altitude; i.e., the atmosphere becomes thinner (less dense).
    Answer (A) is incorrect. An 1,200-ft. increase would result from a temperature increase of 18°F (not 25°F). Answer (B) is incorrect. An 1,400-ft. increase would result from a temperature increase of 20°F (not 25°F).

17. (Refer to Figure 8 on page 176.) Determine the pressure altitude with an indicated altitude of 1,380 feet MSL with an altimeter setting of 28.22 at standard temperature.

    A. 3,010 feet MSL.

    B. 2,991 feet MSL.

    C. 2,913 feet MSL.

Answer (B) is correct. *(PHAK Chap 10)*
    DISCUSSION: Pressure altitude is determined by adjusting the altimeter setting to 29.92" Hg, i.e., adjusting for nonstandard pressure. This is the indicated altitude of 1,380 ft. plus or minus the pressure altitude conversion factor (based on the current altimeter setting).
    On the right side of Fig. 8 is a pressure altitude conversion factor schedule. Add 1,533 ft. for an altimeter setting of 28.30 and 1,630 ft. for an altimeter setting of 28.20. Using interpolation, you must subtract 20% of the difference between 28.3 and 28.2 from 1,630 ft. (1,630 – 1,533 = 97 × .2 = 19). Since 1,630 – 19 = 1,611, add 1,611 ft. to 1,380 ft. to get the pressure altitude of 2,991 feet.
    Answer (A) is incorrect. This figure is obtained by adding the conversion factor for an altimeter setting of 28.20, not an altimeter setting of 28.22, to the indicated altitude. Answer (C) is incorrect. This figure is obtained by adding the conversion factor for an altimeter setting of 28.30, not an altimeter setting of 28.22, to the indicated altitude.

18. (Refer to Figure 8 on page 176.) Determine the density altitude for these conditions:

Altimeter setting 29.25
Runway temperature . . . . +81°F
Airport elevation . . . . 5,250 ft MSL

    A. 4,600 feet MSL.

    B. 5,877 feet MSL.

    C. 8,500 feet MSL.

Answer (C) is correct. *(PHAK Chap 10)*
    DISCUSSION: With an altimeter setting of 29.25" Hg, about 626 ft. (579 plus 1/2 the 94-ft. pressure altitude conversion factor difference between 29.2 and 29.3) must be added to the field elevation of 5,250 ft. to obtain the pressure altitude, or 5,876 feet. Note that barometric pressure is less than standard and pressure altitude is greater than true altitude. Next, convert pressure altitude to density altitude. On the chart, find the point at which the pressure altitude line for 5,876 ft. crosses the 81°F line. The density altitude at that spot shows somewhere in the mid-8,000s of feet. The closest answer choice is 8,500 feet. Note that, when temperature is higher than standard, density altitude exceeds pressure altitude.
    Answer (A) is incorrect. This would be pressure altitude if 650 ft. were subtracted from, not added to, 5,250 ft. MSL. Answer (B) is incorrect. This is pressure altitude, not density altitude.

5.3 Takeoff Distance

19. (Refer to Figure 41 below.) Determine the approximate ground roll distance required for takeoff.

OAT . . . . 100°F
Pressure altitude . . . . 2,000 ft
Takeoff weight . . . . 2,750 lb
Headwind component . . . . Calm

    A. 1,150 feet.

    B. 1,300 feet.

    C. 1,800 feet.

Answer (A) is correct.   *(PHAK Chap 10)*
DISCUSSION: Begin on the left section of Fig. 41 at 100°F (see outside air temperature at the bottom). Move up vertically to the pressure altitude of 2,000 feet. Then proceed horizontally to the first reference line. Since takeoff weight is 2,750, move parallel to the closest guideline to 2,750 pounds. Then proceed horizontally to the second reference line. Since the wind is calm, proceed again horizontally to the right-hand margin of the diagram (ignore the third reference line because there is no obstacle, i.e., ground roll is desired), which will be at 1,150 feet.
Answer (B) is incorrect. This would be the ground roll distance required at maximum takeoff weight. Answer (C) is incorrect. This would be the total distance required to clear a 50-ft. obstacle.

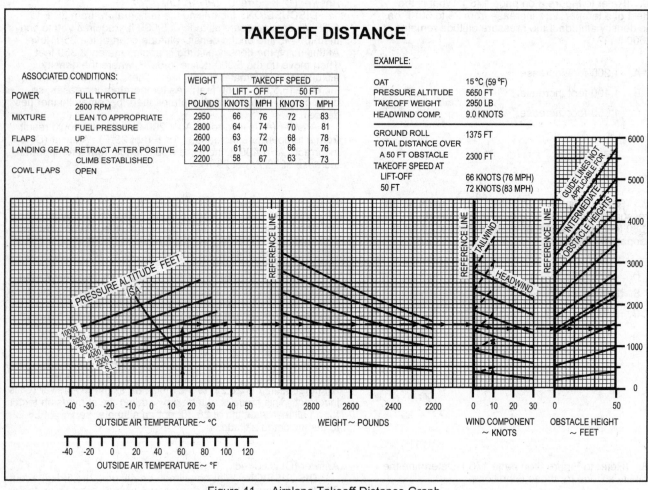

Figure 41. – Airplane Takeoff Distance Graph.

20. (Refer to Figure 41 on page 178.) Determine the total distance required for takeoff to clear a 50-foot obstacle.

OAT . . . . Std
Pressure altitude . . . . Sea level
Takeoff weight . . . . 2,700 lb
Headwind component . . . . Calm

    A. 1,000 feet.

    B. 1,400 feet.

    C. 1,700 feet.

Answer (B) is correct. *(PHAK Chap 10)*
    DISCUSSION: Begin in the left section of Fig. 41 by finding the intersection of the sea level pressure altitude and standard temperature (59°F) and proceed horizontally to the right to the first reference line. Then proceed parallel to the closest guideline, to 2,700 pounds. From there, proceed horizontally to the right to the third reference line. You skip the second reference line because the wind is calm. Then proceed upward parallel to the closest guideline to the far right side. To clear the 50-ft. obstacle, you need a takeoff distance of about 1,400 feet.
    Answer (A) is incorrect. This would be the total distance required at 2,200 lb. takeoff weight. Answer (C) is incorrect. This would be the total distance required at maximum takeoff weight.

21. (Refer to Figure 41 on page 178.) Determine the total distance required for takeoff to clear a 50-foot obstacle.

OAT . . . . Std
Pressure altitude . . . . 4,000 ft
Takeoff weight . . . . 2,800 lb
Headwind component . . . . Calm

    A. 1,500 feet.

    B. 1,750 feet.

    C. 2,000 feet.

Answer (B) is correct. *(PHAK Chap 10)*
    DISCUSSION: The takeoff distance to clear a 50-ft. obstacle is required. Begin on the left side of the graph at standard temperature (as represented by the curved line labeled "ISA"). From the intersection of the standard temperature line and the 4,000-ft. pressure altitude, proceed horizontally to the right to the first reference line, and then move parallel to the closest guideline to 2,800 pounds. From there, proceed horizontally to the right to the third reference line (skip the second reference line because there is no wind), and move upward parallel to the closest guideline all the way to the far right. You are at 1,750 ft., which is the takeoff distance to clear a 50-ft. obstacle.
    Answer (A) is incorrect. This would be the total distance required with a 10-kt. headwind. Answer (C) is incorrect. This would be the total distance required at maximum takeoff weight.

22. (Refer to Figure 41 on page 178.) Determine the approximate ground roll distance required for takeoff.

OAT . . . . 90°F
Pressure altitude . . . . 2,000 ft
Takeoff weight . . . . 2,500 lb
Headwind component . . . . 20 kts

    A. 650 feet.

    B. 850 feet.

    C. 1,000 feet.

Answer (A) is correct. *(PHAK Chap 10)*
    DISCUSSION: Begin with the intersection of the 2,000-ft. pressure altitude curve and 90°F in the left section of Fig. 41. Move horizontally to the right to the first reference line, and then parallel to the closest guideline to 2,500 pounds. Then move horizontally to the right to the second reference line, and then parallel to the closest guideline to the right to 20 knots. Then move horizontally to the right, directly to the right margin because there is no obstacle clearance. You should end up at about 650 ft., which is the required ground roll when there is no obstacle to clear.
    Answer (B) is incorrect. This would be the ground roll distance required if the wind were calm. Answer (C) is incorrect. This would be the ground roll distance required at maximum takeoff weight.

5.4 Cruise Power Settings

23. (Refer to Figure 36 below.) What fuel flow should a pilot expect at 11,000 feet on a standard day with 65 percent maximum continuous power?

    A. 10.6 gallons per hour.

    B. 11.2 gallons per hour.

    C. 11.8 gallons per hour.

Answer (B) is correct. *(PHAK Chap 10)*
    DISCUSSION: Note that the entire chart applies to 65% maximum continuous power (regardless of the throttle), so use the middle section of the chart, which is labeled a standard day.
    The fuel flow at 11,000 ft. on a standard day would be 1/2 of the way between the fuel flow at 10,000 ft. (11.5 GPH) and the fuel flow at 12,000 ft. (10.9 GPH). Thus, the fuel flow at 11,000 ft. would be 11.5 – 0.3, or 11.2 GPH.
    Answer (A) is incorrect. You must add (not subtract) 0.3 to 10.9 to obtain the correct fuel flow. Answer (C) is incorrect. You must subtract (not add) 0.3 from 11.5 to obtain the correct fuel flow.

24. (Refer to Figure 36 below.) What is the expected fuel consumption for a 1,000-nautical mile flight under the following conditions?

Pressure altitude . . . . 8,000 ft
Temperature . . . . 22°C
Manifold pressure . . . . 20.8" Hg
Wind . . . . Calm

    A. 60.2 gallons.

    B. 70.1 gallons.

    C. 73.2 gallons.

Answer (B) is correct. *(PHAK Chap 10)*
    DISCUSSION: To determine the fuel consumption, you need to know the number of hours the flight will last and the gallons per hour the airplane will use. The chart is divided into three sections. They differ based on air temperature. Use the right section of the chart, as the temperature at 8,000 ft. is 22°C.
    At a pressure altitude of 8,000 ft., 20.8" Hg manifold pressure, and 22°C, the fuel flow is 11.5 GPH and the true airspeed is 164 knots. Given a calm wind, the 1,000-NM trip will take 6.09 hr. (1,000 NM ÷ 164 knots).

$$6.09 \text{ hr.} \times 11.5 \text{ GPH} = 70.1 \text{ gal.}$$

    Answer (A) is incorrect. This is the expected fuel consumption for a 1,000-NM flight with a true airspeed of 189 (not 164) knots. Answer (C) is incorrect. This is the expected fuel consumption for a 1,000-NM flight with a true airspeed of 157 (not 164) knots.

CRUISE POWER SETTINGS

65% MAXIMUM CONTINUOUS POWER (OR FULL THROTTLE)
2800 POUNDS

PRESS ALT.	ISA –20 °C (–36 °F)							STANDARD DAY (ISA)							ISA +20 °C (+36 °F)						
	IOAT		ENGINE SPEED	MAN. PRESS	FUEL FLOW PER ENGINE		TAS		IOAT		ENGINE SPEED	MAN. PRESS	FUEL FLOW PER ENGINE		TAS		IOAT		ENGINE SPEED	MAN. PRESS	FUEL FLOW PER ENGINE		TAS	
FEET	°F	°C	RPM	IN HG	PSI	GPH	KTS	MPH	°F	°C	RPM	IN HG	PSI	GPH	KTS	MPH	°F	°C	RPM	IN HG	PSI	GPH	KTS	MPH
SL	27	-3	2450	20.7	6.6	11.5	147	169	63	17	2450	21.2	6.6	11.5	150	173	99	37	2450	21.8	6.6	11.5	153	176
2000	19	-7	2450	20.4	6.6	11.5	149	171	55	13	2450	21.0	6.6	11.5	153	176	91	33	2450	21.5	6.6	11.5	156	180
4000	12	-11	2450	20.1	6.6	11.5	152	175	48	9	2450	20.7	6.6	11.5	156	180	84	29	2450	21.3	6.6	11.5	159	183
6000	5	-15	2450	19.8	6.6	11.5	155	178	41	5	2450	20.4	6.6	11.5	158	182	79	26	2450	21.0	6.6	11.5	161	185
8000	-2	-19	2450	19.5	6.6	11.5	157	181	36	2	2450	20.2	6.6	11.5	161	185	72	22	2450	20.8	6.6	11.5	164	189
10000	-8	-22	2450	19.2	6.6	11.5	160	184	28	-2	2450	19.9	6.6	11.5	163	188	64	18	2450	20.3	6.5	11.4	166	191
12000	-15	-26	2450	18.8	6.4	11.3	162	186	21	-6	2450	18.8	6.1	10.9	163	188	57	14	2450	18.8	5.9	10.6	163	188
14000	-22	-30	2450	17.4	5.8	10.5	159	183	14	-10	2450	17.4	5.6	10.1	160	184	50	10	2450	17.4	5.4	9.8	160	184
16000	-29	-34	2450	16.1	5.3	9.7	156	180	7	-14	2450	16.1	5.1	9.4	156	180	43	6	2450	16.1	4.9	9.1	155	178

NOTES: 1. Full throttle manifold pressure settings are approximate.
           2. Shaded area represents operation with full throttle.

Figure 36. – Airplane Power Setting Table.

25. (Refer to Figure 36 on page 180.) What is the expected fuel consumption for a 500-nautical mile flight under the following conditions?

Pressure altitude . . . . 4,000 ft
Temperature . . . . +29°C
Manifold pressure . . . . 21.3" Hg
Wind . . . . Calm

    A. 31.4 gallons.

    B. 36.1 gallons.

    C. 40.1 gallons.

Answer (B) is correct. *(PHAK Chap 10)*
  DISCUSSION: At 4,000 ft., 21.3" Hg manifold pressure, and 29°C (use the section on the right), the fuel flow will be 11.5 GPH, and the true airspeed will be 159 knots. The 500-NM trip will take 3.14 hr. (500 NM ÷ 159 knots).

3.14 hr. × 11.5 GPH = 36.1 gal.

  Answer (A) is incorrect. This is the expected fuel consumption for a 500-NM flight with a true airspeed of 183 (not 159) knots. Answer (C) is incorrect. This is the expected fuel consumption for a 500-NM flight with a true airspeed of 143 (not 159) knots.

26. (Refer to Figure 36 on page 180.) Determine the approximate manifold pressure setting with 2,450 RPM to achieve 65 percent maximum continuous power at 6,500 feet with a temperature of 36°F higher than standard.

    A. 19.8" Hg.

    B. 20.8" Hg.

    C. 21.0" Hg.

Answer (C) is correct. *(PHAK Chap 10)*
  DISCUSSION: The part of the chart on the right is for temperatures 36°F greater than standard. At 6,500 ft. with a temperature of 36°F higher than standard, the required manifold pressure change is 1/4 of the difference between the 21.0" Hg at 6,000 ft. and the 20.8" Hg at 8,000 ft., or slightly less than 21.0. Thus, 21.0 is the best answer given. The manifold pressure is closer to 21.0 than 20.8.
  Answer (A) is incorrect. This setting would achieve 65% power at 36°F below (not above) standard temperature. Answer (B) is incorrect. The manifold pressure at 6,500 ft. is closer to 21.0 than 20.8.

27. (Refer to Figure 36 on page 180.) Approximately what true airspeed should a pilot expect with 65 percent maximum continuous power at 9,500 feet with a temperature of 36°F below standard?

    A. 178 MPH.

    B. 181 MPH.

    C. 183 MPH.

Answer (C) is correct. *(PHAK Chap 10)*
  DISCUSSION: The left part of the chart applies to 36°F below standard. At 8,000 ft., TAS is 181 MPH. At 10,000 ft., TAS is 184 MPH. At 9,500 ft., with a temperature 36°F below standard, the expected true airspeed is 75% above the 181 MPH at 8,000 ft. toward the 184 MPH at 10,000 ft., i.e., approximately 183 MPH.
  Answer (A) is incorrect. This is the expected TAS at 6,000 ft. Answer (B) is incorrect. This is the expected TAS at 8,000 ft.

5.5 Crosswind Components

28. (Refer to Figure 37 below.) What is the crosswind component for a landing on Runway 18 if the tower reports the wind as 220° at 30 knots?

A. 19 knots.

B. 23 knots.

C. 30 knots.

Answer (A) is correct.   *(PHAK Chap 10)*
   DISCUSSION: The requirement is the crosswind component, which is found on the horizontal axis of the graph. You are given a 30-kt. wind speed (the wind speed is shown on the circular lines or arcs). First, calculate the angle between the wind and the runway (220° − 180° = 40°). Next, find the intersection of the 40° line and the 30-kt. headwind arc. Then, proceed downward to determine a crosswind component of 19 knots.
   Note the crosswind component is on the horizontal axis and the headwind component is on the vertical axis.
   Answer (B) is incorrect. This is the headwind (not crosswind) component. Answer (C) is incorrect. This is the total wind (not crosswind component).

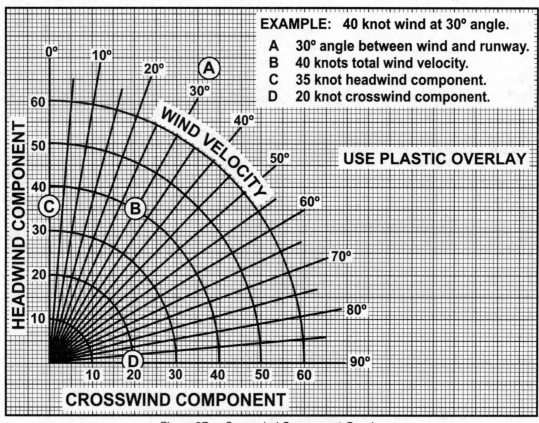

Figure 37. – Crosswind Component Graph.

29. (Refer to Figure 37 above.) What is the headwind component for a landing on Runway 18 if the tower reports the wind as 220° at 30 knots?

A. 19 knots.

B. 23 knots.

C. 26 knots.

Answer (B) is correct.   *(PHAK Chap 10)*
   DISCUSSION: The headwind component is on the vertical axis (left-hand side of the graph). Find the same intersection as in the preceding question, i.e., the 30-kt. wind speed arc, and the 40° angle between wind direction and flight path (220° − 180°). Then move horizontally to the left and read approximately 23 knots.
   Answer (A) is incorrect. This is the crosswind (not headwind) component. Answer (C) is incorrect. This would be the headwind component if the wind were 30° (not 40°) off the runway.

30. (Refer to Figure 37 on page 182.) Determine the maximum wind velocity for a 45° crosswind if the maximum crosswind component for the airplane is 25 knots.

A. 25 knots.

B. 29 knots.

C. 35 knots.

Answer (C) is correct. *(PHAK Chap 10)*
   DISCUSSION: Start on the bottom of the graph's horizontal axis at 25 kt. and move straight upward to the 45° angle between wind direction and flight path line (halfway between the 40° and 50° lines). Note that you are halfway between the 30 and 40 arc-shaped wind speed lines, which means that the maximum wind velocity for a 45° crosswind is 35 kt. if the airplane is limited to a 25-kt. crosswind component.
   Answer (A) is incorrect. This would be the maximum wind velocity for a 90° (not 45°) crosswind. Answer (B) is incorrect. This would be the maximum wind velocity for a 60° (not 45°) crosswind.

31. (Refer to Figure 37 on page 182.) With a reported wind of north at 20 knots, which runway is acceptable for use for an airplane with a 13-knot maximum crosswind component?

A. Runway 6.

B. Runway 29.

C. Runway 32.

Answer (C) is correct. *(PHAK Chap 10)*
   DISCUSSION: If the wind is from the north (i.e., either 360° or 0°) at 20 kt., runway 32, i.e., 320°, would provide a 40° crosswind component (360° – 320°). Given a 20-kt. wind, find the intersection between the 20-kt. arc and the angle between wind direction and the flight path of 40°. Dropping straight downward to the horizontal axis gives 13 kt., which is the maximum crosswind component of the example airplane.
   Answer (A) is incorrect. Runway 6 would have a crosswind component of approximately 17 knots. Answer (B) is incorrect. Runway 29 would have a crosswind component of 19 knots.

32. (Refer to Figure 37 on page 182.) What is the maximum wind velocity for a 30° crosswind if the maximum crosswind component for the airplane is 12 knots?

A. 16 knots.

B. 20 knots.

C. 24 knots.

Answer (C) is correct. *(PHAK Chap 10)*
   DISCUSSION: Start on the graph's horizontal axis at 12 kt. and move upward to the 30° angle between wind direction and flight path line. Note that you are almost halfway between the 20 and 30 arc-shaped wind speed lines, which means that the maximum wind velocity for a 30° crosswind is approximately 24 kt. if the airplane is limited to a 12-kt. crosswind component.
   Answer (A) is incorrect. This would be the maximum wind velocity for a 50° (not 30°) crosswind. Answer (B) is incorrect. This would be the maximum wind velocity for a 40° (not 30°) crosswind.

33. (Refer to Figure 37 on page 182.) With a reported wind of south at 20 knots, which runway is appropriate for an airplane with a 13-knot maximum crosswind component?

A. Runway 10.

B. Runway 14.

C. Runway 24.

Answer (B) is correct. *(PHAK Chap 10)*
   DISCUSSION: If the wind is from the south at 20 kt., runway 14, i.e., 140°, would provide a 40° crosswind component (180° – 140°). Given a 20-kt. wind, find the intersection between the 20-kt. arc and the angle between wind direction and the flight path of 40°. Dropping straight downward to the horizontal axis gives 13 kt., which is the maximum crosswind component of the example airplane.
   Answer (A) is incorrect. Runway 10 would have a crosswind component of 20 knots. Answer (C) is incorrect. Runway 24 would have a crosswind component of approximately 17 knots.

5.6 Landing Distance

34. (Refer to Figure 38 below.) Determine the total distance required to land.

OAT	Std
Pressure altitude	10,000 ft
Weight	2,400 lb
Wind component	Calm
Obstacle	50 ft

    A. 750 feet.

    B. 1,925 feet.

    C. 1,450 feet.

Answer (B) is correct. *(PHAK Chap 10)*
    DISCUSSION: The landing distance graphs are very similar to the takeoff distance graphs. Begin with the pressure altitude line of 10,000 ft. and the intersection with the standard temperature line, which begins at 20°C and slopes up and to the left; i.e., standard temperature decreases as pressure altitude increases. Then move horizontally to the right to the first reference line. Proceed parallel to the closest guideline to 2,400 pounds. Proceed horizontally to the right to the second reference line. Since the wind is calm, proceed horizontally to the third reference line. Given a 50-ft. obstacle, proceed parallel to the closest guideline to the right margin to determine a distance of approximately 1,900 feet.
    Answer (A) is incorrect. This is the total distance required to land with a 30-kt. headwind, not a calm wind, and without an obstacle, not with a 50-ft. obstacle. Answer (C) is incorrect. This is the approximate total distance required to land at a pressure altitude of 2,000 ft., not 10,000 ft., and a weight of 2,300 lb., not 2,400 pounds.

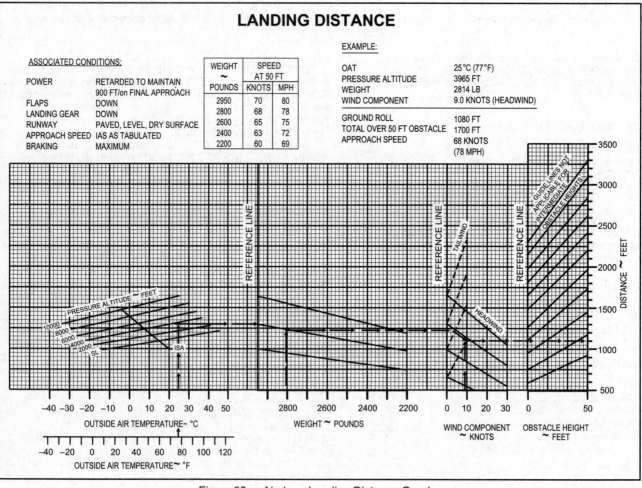

Figure 38. – Airplane Landing Distance Graph.

35. (Refer to Figure 38 on page 184.) Determine the approximate total distance required to land over a 50-ft. obstacle.

OAT . . . . 90°F
Pressure altitude . . . . 4,000 ft
Weight . . . . 2,800 lb
Headwind component . . . . 10 kts

   A. 1,525 feet.

   B. 1,775 feet.

   C. 1,950 feet.

Answer (B) is correct. *(PHAK Chap 10)*
   DISCUSSION: To determine the total landing distance, begin at the left side of Fig. 38 on the 4,000-ft. pressure altitude line at the intersection of 90°F. Proceed horizontally to the right to the first reference line. Proceed parallel to the closest guideline to 2,800 lb., and then straight across to the second reference line. Since the headwind component is 10 kt., proceed parallel to the closest headwind guideline to the 10-kt. line. Then move directly to the right, to the third reference line. Given a 50-ft. obstacle, proceed parallel to the closest guideline for obstacles to find the total distance of approximately 1,775 feet.
   Answer (A) is incorrect. A distance of 1,525 ft. would be the total distance required with an 18-kt. headwind, not a 10-kt. headwind. Answer (C) is incorrect. A distance of 1,950 ft. would be the total distance required with calm wind conditions, not with a 10-kt. headwind.

36. If an emergency situation requires a downwind landing, pilots should expect a faster

   A. airspeed at touchdown, a longer ground roll, and better control throughout the landing roll.

   B. groundspeed at touchdown, a longer ground roll, and the likelihood of overshooting the desired touchdown point.

   C. groundspeed at touchdown, a shorter ground roll, and the likelihood of undershooting the desired touchdown point.

Answer (B) is correct. *(AFH Chap 8)*
   DISCUSSION: A downwind landing, in an emergency or other situation, will result in a faster groundspeed at touchdown, which means a longer ground roll, which in turn increases the likelihood of overshooting the desired touchdown point.
   Answer (A) is incorrect. The airspeed will probably be the same even though the groundspeed is greater, and the control during the landing roll will be less due to the high groundspeed. Answer (C) is incorrect. The ground roll is longer, not shorter, and there is a greater likelihood of overshooting, not undershooting, the touchdown point due to the faster groundspeed.

37. (Refer to Figure 38 on page 184.) Determine the total distance required to land.

OAT . . . . 90°F
Pressure altitude . . . . 3,000 ft
Weight . . . . 2,900 lb
Headwind component . . . . 10 kts
Obstacle . . . . 50 ft

   A. 1,450 feet.

   B. 1,550 feet.

   C. 1,725 feet.

Answer (C) is correct. *(PHAK Chap 10)*
   DISCUSSION: To determine the total landing distance, begin with pressure altitude of 3,000 ft. (between the 2,000- and 4,000-ft. lines) at its intersection with 90°F. Proceed horizontally to the right to the first reference line, and then parallel to the closest guideline to 2,900 pounds. From that point, proceed horizontally to the second reference line. Since there is a headwind component of 10 kt., proceed parallel to the closest headwind guideline down to 10 kt. and then horizontally to the right to the third reference line. Given a 50-ft. obstacle, proceed parallel to the closest guideline for obstacles to find the landing distance of approximately 1,725 feet.
   Answer (A) is incorrect. This would be the total distance required with a 20-kt., not 10-kt., headwind. Answer (B) is incorrect. This would be the total distance required at a pressure altitude of 2,000 ft., not 3,000 feet.

38. (Refer to Figure 38 on page 184.) Determine the total distance required to land.

OAT . . . . 32°F
Pressure altitude . . . . 8,000 ft
Weight . . . . 2,600 lb
Headwind component . . . . 20 kts
Obstacle . . . . 50 ft

   A. 850 feet.

   B. 1,400 feet.

   C. 1,750 feet.

Answer (B) is correct. *(PHAK Chap 10)*
   DISCUSSION: To determine the total landing distance, begin with the pressure altitude of 8,000 ft. at its intersection with 32°F (0°C). Proceed horizontally to the first reference line, and then parallel to the closest guideline to 2,600 pounds. From that point, proceed horizontally to the second reference line. Since there is a headwind component of 20 kt., follow parallel to the closest headwind guideline down to 20 kt., and then horizontally to the right to the third reference line. Given a 50-ft. obstacle, proceed parallel to the closest guideline for obstacles to find the landing distance of approximately 1,400 feet.
   Answer (A) is incorrect. This would be the ground roll with no obstacle. Answer (C) is incorrect. This would be the total distance required at maximum landing weight.

39. (Refer to Figure 39 on page 187.) Determine the approximate landing ground roll distance.

Pressure altitude .... Sea level
Headwind .... 4 kts
Temperature .... Std

    A. 356 feet.

    B. 401 feet.

    C. 490 feet.

Answer (B) is correct. *(PHAK Chap 10)*
    DISCUSSION: At sea level, the ground roll is 445 feet. The standard temperature needs no adjustment. According to Note 1 in Fig. 39, the distance should be decreased 10% for each 4 kt. of headwind, so the headwind of 4 kt. means that the landing distance is reduced by 10%. The result is 401 ft. (445 ft. × 90%).
    Answer (A) is incorrect. This would be the ground roll with an 8-kt., not a 4-kt., headwind. Answer (C) is incorrect. Ground roll is reduced, not increased, to account for headwind.

40. (Refer to Figure 39 on page 187.) Determine the total distance required to land over a 50-ft. obstacle.

Pressure altitude .... 3,750 ft
Headwind .... 12 kts
Temperature .... Std

    A. 794 feet.

    B. 836 feet.

    C. 816 feet.

Answer (C) is correct. *(PHAK Chap 10)*
    DISCUSSION: The total distance to clear a 50-ft. obstacle for a 3,750-ft. pressure altitude is required. Note that this altitude lies halfway between 2,500 ft. and 5,000 feet. Halfway between the total distance at 2,500 ft. of 1,135 ft. and the total distance at 5,000 ft. of 1,195 ft. is 1,165 feet. Since the headwind is 12 kt., the total distance must be reduced by 30% (10% for each 4 kt.).

            70% × 1,165 = 816 ft.

    Answer (A) is incorrect. This would be the total distance to land at a pressure altitude of 2,500 ft., not 3,750 ft., with a 12-kt. headwind and standard temperature. Answer (B) is incorrect. This would be the total distance to land at a pressure altitude of 5,000 ft., not 3,750 ft., with a 12-kt. headwind and standard temperature.

41. (Refer to Figure 39 on page 187.) Determine the approximate landing ground roll distance.

Pressure altitude .... 5,000 ft
Headwind .... Calm
Temperature .... 101°F

    A. 495 feet.

    B. 545 feet.

    C. 445 feet.

Answer (B) is correct. *(PHAK Chap 10)*
    DISCUSSION: The ground roll distance at 5,000 ft. is 495 feet. According to Note 2 in Fig. 39, since the temperature is 60°F above standard, the distance should be increased by 10%.

          495 ft. × 110% = 545 ft.

    Answer (A) is incorrect. This would be ground roll if the temperature were 41°F, not 101°F. Answer (C) is incorrect. This is obtained by decreasing, not increasing, the distance for a temperature 60°F above standard.

42. (Refer to Figure 39 on page 187.) Determine the approximate landing ground roll distance.

Pressure altitude .... 1,250 ft
Headwind .... 8 kts
Temperature .... Std

    A. 275 feet.

    B. 366 feet.

    C. 470 feet.

Answer (B) is correct. *(PHAK Chap 10)*
    DISCUSSION: The landing ground roll at a pressure altitude of 1,250 ft. is required. The difference between landing distance at sea level and 2,500 ft. is 25 ft. (470 – 445). One-half of this distance (12) plus the 445 ft. at sea level is 457 feet. The temperature is standard, requiring no adjustment. The headwind of 8 kt. requires the distance to be decreased by 20%. Thus, the distance required will be 366 ft. (457 × 80%).
    Answer (A) is incorrect. The distance should be decreased by 20% (not 40%). Answer (C) is incorrect. This is the distance required at 2,500 ft. in a calm wind.

43. (Refer to Figure 39 on page 187.) Determine the total distance required to land over a 50-foot obstacle.

Pressure altitude .... 5,000 ft
Headwind .... 8 kts
Temperature .... 41°F
Runway .... Hard surface

    A. 837 feet.

    B. 956 feet.

    C. 1,076 feet.

Answer (B) is correct. *(PHAK Chap 10)*
    DISCUSSION: Under standard conditions, the distance to land over a 50-ft. obstacle at 5,000 ft. is 1,195 feet. The temperature is standard, requiring no adjustment. The headwind of 8 kt., however, requires that the distance be decreased by 20% (10% for each 4 kt. headwind). Thus, the landing ground roll will be 956 ft. (80% of 1,195).
    Answer (A) is incorrect. The distance should be decreased by 20% (not 30%). Answer (C) is incorrect. The distance should be decreased by 20% (not 10%).

44. (Refer to Figure 39 below.) Determine the total distance required to land over a 50-foot obstacle.

Pressure altitude . . . . 7,500 ft
Headwind . . . . 8 kts
Temperature . . . . 32°F
Runway . . . . Hard surface

    A. 1,004 feet.

    B. 1,205 feet.

    C. 1,506 feet.

Answer (A) is correct. *(PHAK Chap 10)*
    DISCUSSION: Under normal conditions, the total landing distance required to clear a 50-ft. obstacle is 1,255 feet. The temperature is standard (32°F), requiring no adjustment. The headwind of 8 kt. reduces the 1,255 by 20% (10% for each 4 knots). Thus, the total distance required will be 1,004 ft. (1,255 × 80%).
    Answer (B) is incorrect. This results from incorrectly assuming that an adjustment for a dry grass runway is necessary and then applying that adjustment (an increase of 20%) to 1,004 ft. than to the total landing distance required to clear a 50-ft. obstacle as stated in Note 3, which is 1,255 feet. Answer (C) is incorrect. This is obtained by increasing, not decreasing, the distance for the headwind.

LANDING DISTANCE

FLAPS LOWERED TO 40° - POWER OFF
HARD SURFACE RUNWAY - ZERO WIND

GROSS WEIGHT LB	APPROACH SPEED, IAS, MPH	AT SEA LEVEL & 59 °F		AT 2500 FT & 50 °F		AT 5000 FT & 41 °F		AT 7500 FT & 32 °F	
		GROUND ROLL	TOTAL TO CLEAR 50 FT OBS	GROUND ROLL	TOTAL TO CLEAR 50 FT OBS	GROUND ROLL	TOTAL TO CLEAR 50 FT OBS	GROUND ROLL	TOTAL TO CLEAR 50 FT OBS
1600	60	445	1075	470	1135	495	1195	520	1255

NOTES: 1. Decrease the distances shown by 10% for each 4 knots of headwind.
2. Increase the distance by 10% for each 60° F temperature increase above standard.
3. For operation on a dry, grass runway, increase distances (both "ground roll" and "total to clear 50 ft obstacle") by 20% of the "total to clear 50 ft obstacle" figure.

Figure 39. – Airplane Landing Distance Graph.

5.7 Weight and Balance Definitions

45. Which items are included in the empty weight of an aircraft?

    A. Unusable fuel and undrainable oil.

    B. Only the airframe, powerplant, and optional equipment.

    C. Full fuel tanks and engine oil to capacity.

Answer (A) is correct. *(PHAK Chap 9)*
    DISCUSSION: The empty weight of an airplane includes airframe, engines, and all items of operating equipment that have fixed locations and are permanently installed. It includes optional and special equipment, fixed ballast, hydraulic fluid, unusable fuel, and undrainable oil.
    Answer (B) is incorrect. Unusable and undrainable fuel and oil and permanently installed optional equipment are also included in empty weight. Answer (C) is incorrect. Usable fuel (included in full fuel) and full engine oil are not components of basic empty weight.

46. An aircraft is loaded 110 pounds over maximum certificated gross weight. If fuel (gasoline) is drained to bring the aircraft weight within limits, how much fuel should be drained?

    A. 15.7 gallons.

    B. 16.2 gallons.

    C. 18.4 gallons.

Answer (C) is correct. *(PHAK Chap 9)*
    DISCUSSION: Fuel weighs 6 lb./gallon. If an airplane is 110 lb. over maximum gross weight, 18.4 gal. (110 lb. ÷ 6) must be drained to bring the airplane weight within limits.
    Answer (A) is incorrect. Fuel weighs 6 (not 7) lb./gallon. Answer (B) is incorrect. Fuel weighs 6 (not 6.8) lb./gallon.

47. If an aircraft is loaded 90 pounds over maximum certificated gross weight and fuel (gasoline) is drained to bring the aircraft weight within limits, how much fuel should be drained?

    A. 10 gallons.

    B. 12 gallons.

    C. 15 gallons.

Answer (C) is correct. *(PHAK Chap 9)*
    DISCUSSION: Since fuel weighs 6 lb./gal., draining 15 gal. (90 lb. ÷ 6) will reduce the weight of an airplane that is 90 lb. over maximum gross weight to the acceptable amount.
    Answer (A) is incorrect. Fuel weighs 6 (not 9) lb./gallon. Answer (B) is incorrect. Fuel weighs 6 (not 7.5) lb./gallon.

48. GIVEN:

	WEIGHT (LB)	ARM (IN)	MOMENT (LB-IN)
Empty weight	1,495.0	101.4	151,593.0
Pilot and passengers	380.0	64.0	---
Fuel (30 gal usable no reserve)	---	96.0	---

The CG is located how far aft of datum?

A. CG 92.44.

B. CG 94.01.

C. CG 119.8.

Answer (B) is correct. *(PHAK Chap 9)*
DISCUSSION: To compute the CG, you must first multiply each weight by the arm to get the moment. Note that the fuel is given as 30 gallons. To get the weight, multiply the 30 by 6 lb. per gal. (30 × 6) = 180 pounds.

	Weight (lb.)	Arm (in.)	Moment (lb.-in.)
Empty weight	1,495.0	101.4	151,593.0
Pilot and passengers	380.0	64.0	24,320.0
Fuel (30 × 6)	180.0	96.0	17,280.0
	2,055.0		193,193.0

Now add the weights and moments. To get CG, you divide total moment by total weight (193,193 ÷ 2,055.0) = a CG of 94.01 inches.
     Answer (A) is incorrect. The total moment must be divided by the total weight to obtain the correct CG. Answer (C) is incorrect. The total moment must be divided by the total weight to obtain the correct CG.

49. (Refer to Figure 35 on page 191.) What is the maximum amount of fuel that may be aboard the airplane on takeoff if loaded as follows?

	WEIGHT (LB)	MOM/1000
Empty weight	1,350	51.5
Pilot and front passenger	340	---
Rear passengers	310	---
Baggage	45	---
Oil, 8 qt.	---	---

A. 24 gallons.

B. 32 gallons.

C. 40 gallons.

Answer (C) is correct. *(PHAK Chap 9)*
DISCUSSION: To find the maximum amount of fuel this airplane can carry, add the empty weight (1,350), pilot and front passenger weight (340), rear passengers (310), baggage (45), and oil (15), for a total of 2,060 pounds. (Find the oil weight and moment by consulting Note (2) on Fig. 35. It is 15 lb. and –0.2 moments.) Gross weight maximum on the center of gravity moment envelope chart is 2,300. Thus, 240 lb. of weight (2,300 – 2,060) is available for fuel. Since each gallon of fuel weighs 6 lb., this airplane can carry 40 gal. of fuel (240 ÷ 6 lb. per gal.) if its center of gravity moments do not exceed the limit. Note that long-range tanks were not mentioned; assume they exist.
     Compute the moments for each item. The empty weight moment is given as 51.5. Calculate the moment for the pilot and front passenger as 12.5, the rear passengers as 22.5, the fuel as 11.5, the baggage as 4.0, and the oil as –0.2. These total to 101.8, which is within the envelope, so 40 gal. of fuel may be carried.

	Weight	Moment/1000 lb.-in.
Empty weight	1,350	51.5
Pilot and front seat passenger	340	12.5
Rear passengers	310	22.5
Baggage	45	4.0
Fuel (40 gal. × 6 lb./gal.)	240	11.5
Oil	15	–0.2
	2,300	101.8

Answer (A) is incorrect. More than 24 gal. of fuel may be carried. Answer (B) is incorrect. More than 32 gal. of fuel may be carried.

5.8 Center of Gravity Calculations

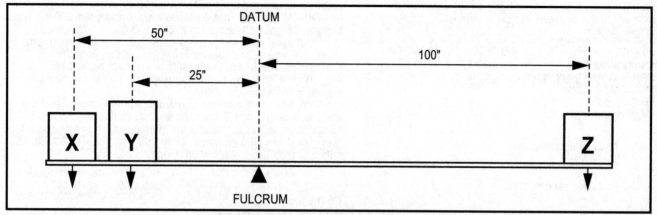

Figure 62. – Weight and Balance Diagram.

50. (Refer to Figure 62 above.) If 50 pounds of weight is located at point X and 100 pounds at point Z, how much weight must be located at point Y to balance the plank?

A. 30 pounds.

B. 50 pounds.

C. 300 pounds.

Answer (C) is correct. *(PHAK Chap 9)*
DISCUSSION: Compute and sum the moments left and right of the fulcrum. Set them equal to one another and solve for the desired variable:

$$
\begin{aligned}
\text{left} &= \text{right} \\
50\ \text{lb.}(50\ \text{in.}) + Y(25\ \text{in.}) &= 100\ \text{lb.}(100\ \text{in.}) \\
2{,}500 + 25Y &= 10{,}000 \\
25Y &= 7{,}500 \\
Y &= 300\ \text{lb.}
\end{aligned}
$$

Answer (A) is incorrect. This weight in the place of Y would cause the plank to be heavier on the right side. Answer (B) is incorrect. This weight in the place of Y would cause the plank to be heavier on the right side.

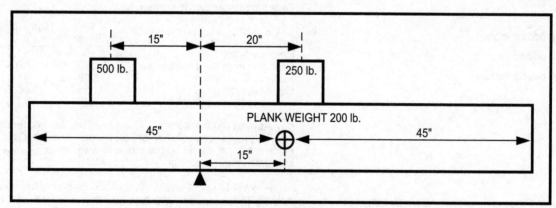

Figure 61. – Weight and Balance Diagram.

51. (Refer to Figure 61 above.) How should the 500-pound weight be shifted to balance the plank on the fulcrum?

A. 1 inch to the left.

B. 1 inch to the right.

C. 4.5 inches to the right.

Answer (A) is correct. *(PHAK Chap 9)*
DISCUSSION: To find the desired location of the 500-lb. weight, compute and sum the moments left and right of the fulcrum. Set them equal to one another and solve for the desired variable:

$$
\begin{aligned}
\text{left} &= \text{right} \\
500\ \text{lb.}(X) &= 250\ \text{lb.}(20\ \text{in.}) + 200\ \text{lb.}(15\ \text{in.}) \\
500X &= 8{,}000 \\
X &= 16\ \text{in.}
\end{aligned}
$$

The 500-lb. weight must be 16 in. from the fulcrum to balance the plank. The weight should be shifted 1 in. to the left.
Answer (B) is incorrect. The 500-lb. weight should be 16 in. from the fulcrum; thus, it must be moved 1 in. to the left, not right, to balance the plank. Answer (C) is incorrect. Shifting the 500-lb. weight 4.5 in. to the right would cause the plank to be heavier on the right side.

5.9   Center of Gravity Graphs

52. (Refer to Figure 35 on page 191.) Calculate the moment of the airplane and determine which category is applicable.

	WEIGHT (LB)	MOM/1000
Empty weight	1,350	51.5
Pilot and front passenger	310	---
Rear passengers	96	---
Fuel, 38 gal.	---	---
Oil, 8 qt.	---	–0.2

    A. 79.2, utility category.

    B. 80.8, utility category.

    C. 81.2, normal category.

Answer (B) is correct. *(PHAK Chap 9)*

    DISCUSSION: First, total the weight and get 1,999 lb. Note that the 38 gal. of fuel weighs 228 lb. (38 gal. × 6 lb./gallon).

    Find the moments for the pilot and front seat passengers, rear passengers, and fuel by using the loading graph in Fig. 35. Find the oil weight and moment by consulting Note (2) on Fig. 35. It is 15 lb. and –0.2 moments. Total the moments as shown in the schedule below.

    Now refer to the center of gravity moment envelope. Find the gross weight of 1,999 on the vertical scale, and move horizontally across the chart until intersecting the vertical line that represents the 80.8 moment. Note that a moment of 80.8 lb.-in. falls into the utility category envelope.

	Weight	Moment/1000 lb.-in.
Empty weight	1,350	51.5
Pilot and front seat passenger	310	11.5
Rear passengers	96	7.0
Fuel (38 gal. × 6 lb./gal.)	228	11.0
Oil	15	–0.2
	1,999	80.8

    Answer (A) is incorrect. A moment of 79.2 is 1.6 less than the correct moment of 80.8 pound-inches. Answer (C) is incorrect. The moment of the oil must be subtracted, not added.

53. (Refer to Figure 35 on page 191.) Determine the moment with the following data:

	WEIGHT (LB)	MOM/1000
Empty weight	1,350	51.5
Pilot and front passenger	340	---
Fuel (std tanks)	Capacity	---
Oil, 8 qt.	---	---

    A. 69.9 pound-inches.

    B. 74.9 pound-inches.

    C. 77.6 pound-inches.

Answer (B) is correct. *(PHAK Chap 9)*

    DISCUSSION: To find the CG moment/1000, find the moments for each item and total the moments as shown in the schedule below. For the fuel, the loading graph shows the maximum as 38 gal. for standard tanks (38 gal. × 6 lb. = 228 pounds). (Find the oil weight and moment by consulting Note (2) on Fig. 35; it is 15 lb. and –0.2. moments.) These total 74.9, so this answer is correct.

	Weight	Moment/1000 lb.-in.
Empty weight	1,350	51.5
Pilot and front seat passenger	340	12.6
Fuel	228	11.0
Oil	15	–0.2
	1,933	74.9

    Answer (A) is incorrect. This would be the moment with only 20 gal. (not full capacity) of fuel on board. Answer (C) is incorrect. This would be the moment with full long-range (not standard) tanks on board.

54. (Refer to Figure 35 on page 191.) Determine the aircraft loaded moment and the aircraft category.

	WEIGHT (LB)	MOM/1000
Empty weight	1,350	51.5
Pilot and front passenger	380	---
Fuel, 48 gal	288	---
Oil, 8 qt.	---	---

    A. 78.2, normal category.

    B. 79.2, normal category.

    C. 80.4, utility category.

Answer (B) is correct. *(PHAK Chap 9)*

    DISCUSSION: The moments for the pilot, front passenger, fuel, and oil must be found on the loading graph in Fig. 35. Total all the moments and the weight as shown in the schedule below.

    Now refer to the center of gravity moment envelope graph. Find the gross weight of 2,033 on the vertical scale, and move horizontally across the graph until intersecting the vertical line that represents the 79.2 moment. A moment of 79.2 lb.-in. falls into the normal category envelope.

	Weight	Moment/1000 lb.-in.
Empty weight	1,350	51.5
Pilot and front seat passenger	380	14.2
Fuel (capacity)	288	13.7
Oil	15	–0.2
	2,033	79.2

    Answer (A) is incorrect. A moment of 78.2 lb.-in. is 1.0 less than the correct moment of 79.2 pound-inches. Answer (C) is incorrect. A moment of 80.4 lb.-in. is 1.2 more than the correct moment of 79.2 pound-inches.

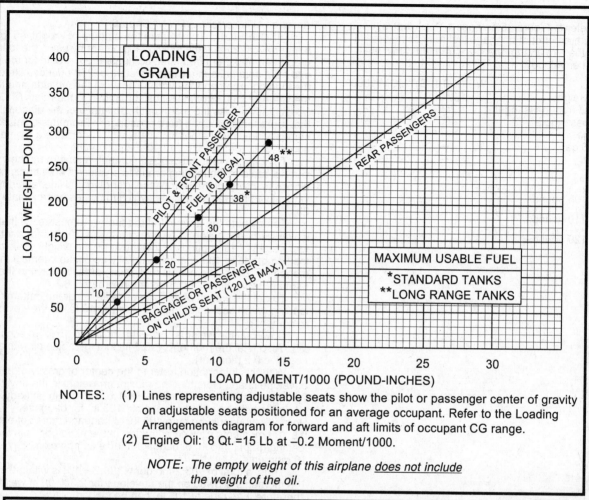

NOTES: (1) Lines representing adjustable seats show the pilot or passenger center of gravity on adjustable seats positioned for an average occupant. Refer to the Loading Arrangements diagram for forward and aft limits of occupant CG range.

(2) Engine Oil: 8 Qt.=15 Lb at –0.2 Moment/1000.

NOTE: *The empty weight of this airplane does not include the weight of the oil.*

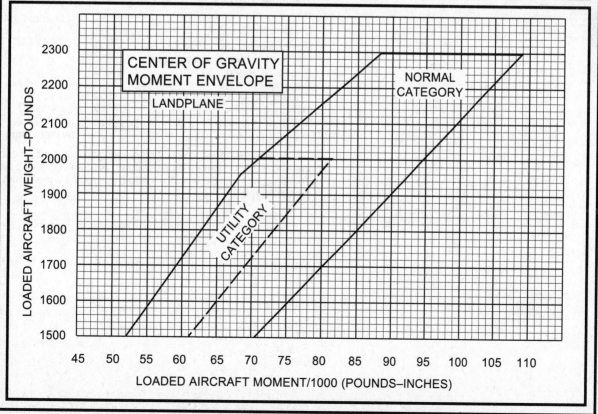

Figure 35. – Airplane Weight and Balance Graphs.

55. (Refer to Figure 35 on page 191.) What is the maximum amount of baggage that may be loaded aboard the airplane for the CG to remain within the moment envelope?

	WEIGHT (LB)	MOM/1000
Empty weight	1,350	51.5
Pilot and front passenger	250	---
Rear passengers	400	---
Baggage	---	---
Fuel, 30 gal.	---	---
Oil, 8 qt.	---	−0.2

   A. 105 pounds.

   B. 110 pounds.

   C. 120 pounds.

Answer (A) is correct. *(PHAK Chap 9)*

DISCUSSION: To compute the amount of weight left for baggage, compute each individual moment by using the loading graph and add them up. First, compute the moment for the pilot and front seat passenger with a weight of 250 pounds. Refer to the loading graph and the vertical scale at the left side and find the value of 250. From this position, move to the right horizontally across the graph until you intersect the diagonal line that represents pilot and front passenger. From this point, move vertically down to the bottom scale, which indicates a moment of about 9.2.

To compute rear passenger moment, measure up the vertical scale of the loading graph to a value of 400, horizontally across to intersect the rear passenger diagonal line, and down vertically to the moment scale, which indicates approximately 29.0.

To compute the moment of the fuel, you must recall that fuel weighs 6 lb. per gallon. The question gives 30 gal., for a total fuel weight of 180 pounds. Now move up the weight scale on the loading graph to 180, then horizontally across to intersect the diagonal line that represents fuel, then vertically down to the moment scale, which indicates approximately 8.7.

To get the weight of the oil, see Note (2) at the bottom of the loading graph section of Fig. 35. It gives 15 lb. as the weight with a moment of −0.2.

Now total the weights (2,195 lb. including 15 lb. of engine oil). Also total the moments (98.2 including engine oil with a negative 0.2 moment).

With this information, refer to the center of gravity moment envelope chart. Note that the maximum weight in the envelope is 2,300 pounds. The amount of 2,300 lb. − 2,195 lb. already totaled leaves a maximum possible 105 lb. for baggage. However, you must be sure 105 lb. of baggage does not exceed the 109 moments allowed at the top of the envelope. On the loading graph, 105 lb. of baggage indicates approximately 10 moments.

Thus, a total of 108.2 moments (98.2 + 10) is within the 109 moments allowed on the envelope for 2,300 lb. of weight. Therefore, baggage of 105 lb. can be loaded.

	Weight	Moment/1000 lb.-in.
Empty weight	1,350	51.5
Pilot and front seat passenger	250	9.2
Rear passengers	400	29.0
Baggage	?	?
Fuel (30 gal. × 6 lb./gal.)	180	8.7
Oil	15	−0.2
	2,195	98.2
		(without baggage)

Answer (B) is incorrect. This baggage weight would exceed the airplane's maximum gross weight. Answer (C) is incorrect. This baggage weight would exceed the airplane's maximum gross weight.

5.10 Center of Gravity Tables

56. (Refer to Figure 33 on page 194 and Figure 34 on page 195.) Determine if the airplane weight and balance is within limits.

Front seat occupants . . . . 340 lb
Rear seat occupants . . . . 295 lb
Fuel (main wing tanks) . . . . 44 gal
Baggage . . . . 56 lb

    A. 20 pounds overweight, CG aft of aft limits.

    B. 20 pounds overweight, CG within limits.

    C. 20 pounds overweight, CG forward of forward limits.

Answer (B) is correct. *(PHAK Chap 9)*
    DISCUSSION: Both the total weight and the total moment must be calculated. As in most weight and balance problems, you should begin by setting up a schedule as below. Note that the empty weight in Fig. 33 is given as 2,015 with a moment/100 in. of 1,554 (note the change to moment/100 on this chart) and that empty weight includes the oil.
    The next step is to compute the moment/100 for each item. The front seat occupants' moment/100 is 289 (340 × 85 ÷ 100). The rear seat occupants' moment/100 is 357 (295 × 121 ÷ 100). The fuel (main tanks) weight of 264 lb. and moment/100 of 198 is read directly from the table. The baggage moment/100 is 78 (56 × 140 ÷ 100).
    The last step is to go to the Moment Limits vs. Weight chart (Fig. 34) and note that the maximum weight allowed is 2,950, which means that the plane is 20 lb. over. At a moment/100 of 2,476, the plane is within the CG limits because the moments/100 may be from 2,422 to 2,499 at 2,950 pounds.

	Weight	Moment/100 lb.-in.
Empty weight w/oil	2,015	1,554
Front seat	340	289
Rear seat	295	357
Fuel (44 gal. × 6 lb/gal)	264	198
Baggage	56	78
	2,970	2,476

    Answer (A) is incorrect. The total moment of 2,476 lb.-in. is less (not more) than the aft limit of 2,499 lb.-in. at 2,950 pounds. Answer (C) is incorrect. The total moment of 2,476 lb.-in. is more (not less) than the forward limit of 2,422 lb.-in. at 2,900 pounds.

57. (Refer to Figure 33 on page 194 and Figure 34 on page 195.) Calculate the weight and balance and determine if the CG and the weight of the airplane are within limits.

Front seat occupants . . . . 350 lb
Rear seat occupants . . . . 325 lb
Baggage . . . . 27 lb
Fuel . . . . 35 gal

    A. CG 81.7, out of limits forward.

    B. CG 83.4, within limits.

    C. CG 84.1, within limits.

Answer (B) is correct. *(PHAK Chap 9)*
    DISCUSSION: Total weight, total moment, and CG must all be calculated. As in most weight and balance problems, you should begin by setting up the schedule as shown below.
    Next, go to the Moment Limits vs. Weight chart (Fig. 34), and note that the maximum weight allowed is 2,950, which means that this airplane is 23 lb. under maximum weight. At a total moment of 2,441, it is also within the CG limits (2,399 to 2,483) at that weight.
    Finally, compute the CG. Recall that Fig. 33 gives moment per 100 inches. The total moment is therefore 244,100 (2,441 × 100). The CG is 244,100/2,927 = 83.4.

	Weight	Moment/100 lb.-in.
Empty weight w/oil	2,015	1,554
Front seat	350	298
Rear seat	325	393
Fuel, main (35 gal.)	210	158
Baggage	27	38
	2,927	2,441

    Answer (A) is incorrect. The correct moment of 2,441 lb.-in./100 is within CG limits. Answer (C) is incorrect. You must divide the total moment by the total weight to arrive at the correct CG of 83.4 inches.

USEFUL LOAD WEIGHTS AND MOMENTS

OCCUPANTS

FRONT SEATS ARM 85		REAR SEATS ARM 121	
Weight	Moment 100	Weight	Moment 100
120	102	120	145
130	110	130	157
140	119	140	169
150	128	150	182
160	136	160	194
170	144	170	206
180	153	180	218
190	162	190	230
200	170	200	242

BAGGAGE OR 5TH SEAT OCCUPANT ARM 140

Weight	Moment 100
10	14
20	28
30	42
40	56
50	70
60	84
70	98
80	112
90	126
100	140
110	154
120	168
130	182
140	196
150	210
160	224
170	238
180	252
190	266
200	280
210	294
220	308
230	322
240	336
250	350
260	364
270	378

USABLE FUEL

MAIN WING TANKS ARM 75

Gallons	Weight	Moment 100
5	30	22
10	60	45
15	90	68
20	120	90
25	150	112
30	180	135
35	210	158
40	240	180
44	264	198

AUXILIARY WING TANKS ARM 94

Gallons	Weight	Moment 100
5	30	28
10	60	56
15	90	85
19	114	107

*OIL

Quarts	Weight	Moment 100
10	19	5

*Included in basic Empty Weight

Empty Weight ~ 2015

MOM / 100 ~ 1554

MOMENT LIMITS vs WEIGHT

Moment limits are based on the following weight and center of gravity limit data (landing gear down).

WEIGHT CONDITION	FORWARD CG LIMIT	AFT CG LIMIT
2950 lb (takeoff or landing)	82.1	84.7
2525 lb	77.5	85.7
2475 lb or less	77.0	85.7

Figure 33. – Airplane Weight and Balance Tables.

MOMENT LIMITS vs WEIGHT (Continued)

Weight	Minimum Moment 100	Maximum Moment 100	Weight	Minimum Moment 100	Maximum Moment 100
2100	1617	1800	2600	2037	2224
2110	1625	1808	2610	2048	2232
2120	1632	1817	2620	2058	2239
2130	1640	1825	2630	2069	2247
2140	1648	1834	2640	2080	2255
2150	1656	1843	2650	2090	2263
2160	1663	1851	2660	2101	2271
2170	1671	1860	2670	2112	2279
2180	1679	1868	2680	2123	2287
2190	1686	1877	2690	2133	2295
2200	1694	1885	2700	2144	2303
2210	1702	1894	2710	2155	2311
2220	1709	1903	2720	2166	2319
2230	1717	1911	2730	2177	2326
2240	1725	1920	2740	2188	2334
2250	1733	1928	2750	2199	2342
2260	1740	1937	2760	2210	2350
2270	1748	1945	2770	2221	2358
2280	1756	1954	2780	2232	2366
2290	1763	1963	2790	2243	2374
2300	1771	1971			
2310	1779	1980	2800	2254	2381
2320	1786	1988	2810	2265	2389
2330	1794	1997	2820	2276	2397
2340	1802	2005	2830	2287	2405
2350	1810	2014	2840	2298	2413
2360	1817	2023	2850	2309	2421
2370	1825	2031	2860	2320	2428
2380	1833	2040	2870	2332	2436
2390	1840	2048	2880	2343	2444
			2890	2354	2452
2400	1848	2057	2900	2365	2460
2410	1856	2065	2910	2377	2468
2420	1863	2074	2920	2388	2475
2430	1871	2083	2930	2399	2483
2440	1879	2091	2940	2411	2491
2450	1887	2100	2950	2422	2499
2460	1894	2108			
2470	1902	2117			
2480	1911	2125			
2490	1921	2134			
2500	1932	2143			
2510	1942	2151			
2520	1953	2160			
2530	1963	2168			
2540	1974	2176			
2550	1984	2184			
2560	1995	2192			
2570	2005	2200			
2580	2016	2208			
2590	2026	2216			

Figure 34. – Airplane Weight and Balance Tables.

58. (Refer to Figure 33 on page 194 and Figure 34 on page 195.) What is the maximum amount of baggage that can be carried when the airplane is loaded as follows?

Front seat occupants . . . . 387 lb
Rear seat occupants . . . . 293 lb
Fuel . . . . 35 gal

   A. 45 pounds.

   B. 63 pounds.

   C. 220 pounds.

Answer (A) is correct. *(PHAK Chap 9)*
DISCUSSION: The maximum allowable weight on the Moment Limits vs. Weight chart (Fig. 34) is 2,950 pounds. The total of the given weights is 2,905 lb. (including the empty weight of the airplane at 2,015 lb. and the fuel at 6 lb./gal.), so baggage cannot weigh more than 45 pounds.

It is still necessary to compute total moments to verify that the position of these weights does not move the CG out of CG limits.

The total moment of 2,460 lies safely between the moment limits of 2,422 and 2,499 on Fig. 34, at the maximum weight, so this airplane can carry as much as 45 lb. of baggage when loaded in this manner.

	Weight	Moment/100 lb.-in.
Empty weight w/oil	2,015	1,554
Front seat	387	330
Rear seat	293	355
Fuel, main (35 gal.)	210	158
Baggage	45	63
	2,950	2,460

Answer (B) is incorrect. This baggage weight would load the airplane above its maximum gross weight. Answer (C) is incorrect. This baggage weight would load the airplane above its maximum gross weight.

59. (Refer to Figure 33 on page 194 and Figure 34 on page 195.) Determine if the airplane weight and balance is within limits.

Front seat occupants . . . . 415 lb
Rear seat occupants . . . . 110 lb
Fuel, main tanks . . . . 44 gal
Fuel, aux. tanks . . . . 19 gal
Baggage . . . . 32 lb

   A. 19 pounds overweight, CG within limits.

   B. 19 pounds overweight, CG out of limits forward.

   C. Weight within limits, CG out of limits.

Answer (C) is correct. *(PHAK Chap 9)*
DISCUSSION: Both the weight and the total moment must be calculated. Begin by setting up the schedule shown below. The fuel must be separated into main and auxiliary tanks, but weights and moments for both tanks are provided in Fig. 33.

Since 415 lb. is not shown on the front seat table, simply multiply the weight by the arm shown at the top of the table (415 lb. × 85 in. = 35,275 lb.-in.) and divide by 100 for moment/100 of 353 (35,275 ÷ 100 = 352.75). The rear seat moment must also be multiplied (110 lb. × 121 in. = 13,310 pound-inches). Divide by 100 to get 133.1, or 133 lb.-in. ÷ 100. The last step is to go to the Moment Limits vs. Weight chart (Fig. 34). The maximum weight allowed is 2,950, which means that the airplane weight is within the limits. However, the CG is out of limits because the minimum moment/100 for a weight of 2,950 lb. is 2,422.

	Weight	Moment/100 lb.-in.
Empty weight w/oil	2,015	1,554
Front seat	415	353
Rear seat	110	133
Fuel, main	264	198
Fuel, aux.	114	107
Baggage	32	45
	2,950	2,390

Answer (A) is incorrect. The airplane's weight of 2,950 lb. is within limits. Answer (B) is incorrect. The airplane's weight of 2,950 lb. is within limits.

60. (Refer to Figure 33 on page 194 and Figure 34 on page 195.) Which action can adjust the airplane's weight to maximum gross weight and the CG within limits for takeoff?

Front seat occupants . . . . 425 lb
Rear seat occupants . . . . 300 lb
Fuel, main tanks . . . . 44 gal

A. Drain 12 gallons of fuel.

B. Drain 9 gallons of fuel.

C. Transfer 12 gallons of fuel from the main tanks to the auxiliary tanks.

Answer (B) is correct. *(PHAK Chap 9)*
DISCUSSION: First, determine the total weight to see how much must be reduced. As shown below, this original weight is 3,004 pounds. Fig. 34 shows the maximum weight as 2,950 pounds. Thus, you must adjust the total weight by removing 54 lb. (3,004 – 2,950). Since fuel weighs 6 lb./gal., you must drain at least 9 gallons.

To check for CG, recompute the total moment using a new fuel moment of 158 (from the chart) for 210 pounds. The plane now weighs 2,950 lb. with a total moment of 2,437, which falls within the moment limits on Fig. 34.

	Original Weight	Adjusted Weight	Moment/100 lb.-in.
Empty weight with oil	2,015	2,015	1,554
Front seat	425	425	362
Rear seat	300	300	363
Fuel	264	210	158
	3,004	2,950	2,437

Answer (A) is incorrect. It is not necessary to drain 12 gal., only 9 gallons. Answer (C) is incorrect. Transferring fuel to auxiliary tanks will only affect the moment, not the total weight.

61. (Refer to Figure 33 on page 194 and Figure 34 on page 195.) With the airplane loaded as follows, what action can be taken to balance the airplane?

Front seat occupants . . . . 411 lb
Rear seat occupants . . . . 100 lb
Main wing tanks . . . . 44 gal

A. Fill the auxiliary wing tanks.

B. Add a 100-pound weight to the baggage compartment.

C. Transfer 10 gallons of fuel from the main tanks to the auxiliary tanks.

Answer (B) is correct. *(PHAK Chap 9)*
DISCUSSION: You need to calculate the weight and moment. The weight of the empty plane including oil is 2,015, with a moment of 1,554. The 411 lb. in the front seats has a total moment of 349.35 [411 × 85 (ARM) = 34,935 ÷ 100 = 349.35]. The rear seat occupants have a weight of 100 lb. and a moment of 121.0 [100 × 121 (ARM) = 12,100 ÷ 100 = 121.0]. The fuel weight is given on the chart as 264 lb. with a moment of 198.

	Weight	Moment/100 lb.-in.
Empty weight	2,015	1,554
Front seat	411	349.35
Rear seat	100	121.0
Fuel	264	198.0
	2,790	2,222.35

On the Fig. 34 chart, the minimum moment for 2,790 lb. is 2,243. Thus, the CG of 2,222.35 is forward. Evaluate A, B, and C to see which puts the CG within limits.

	Weight	Moment/100
A	+114	+107
B	+100	+140
C	+60	+56
	−60	−45
	0	+11

This answer is correct because, at 2,890 lb. (2,790 + 100), moment/100 of 2,362.35 (2,222.35 + 140) is over the minimum moment/100 of 2,354.

Answer (A) is incorrect. At 2,904 lb. (2,790 + 114), the calculated moment/100 of 2,329.35 (2,222.35 + 107) does not reach the minimum required moment/100 of 2,370 for that weight. Answer (C) is incorrect. At 2,790 lb., an increase of 11 moment/100 does not reach the minimum of 2,243.

62. (Refer to Figure 33 on page 194 and Figure 34 on page 195.) Upon landing, the front passenger (180 pounds) departs the airplane. A rear passenger (204 pounds) moves to the front passenger position. What effect does this have on the CG if the airplane weighed 2,690 pounds and the MOM/100 was 2,260 just prior to the passenger transfer?

   A. The CG moves forward approximately 3 inches.

   B. The weight changes, but the CG is not affected.

   C. The CG moves forward approximately 0.1 inch.

Answer (A) is correct. *(AWBH Chap 2)*
  DISCUSSION: The requirement is the effect of a change in loading. Look at Fig. 33 for occupants. Losing the 180-lb. passenger from the front seat reduces the MOM/100 by 153. Moving the 204-lb. passenger from the rear seat to the front reduces the MOM/100 by about 74 (247 – 173). The total moment reduction is thus about 227 (153 + 74). As calculated below, the CG moves forward from 84.01 to 81.00 inches.

$$\text{Old CG} = \frac{226,000 \text{ lb.-in.}}{2,690 \text{ lb.}} = 84.01 \text{ in.}$$

$$\text{New CG} = \frac{203,300 \text{ lb.-in.}}{2,510 \text{ lb.}} = 81.00 \text{ in.}$$

  Answer (B) is incorrect. Intuitively, one can see that the CG will be affected. Answer (C) is incorrect. Intuitively, one can see that the CG will move forward more than only 0.1 inch.

63. (Refer to Figure 33 on page 194 and Figure 34 on page 195.) What effect does a 35-gallon fuel burn (main tanks) have on the weight and balance if the airplane weighed 2,890 pounds and the MOM/100 was 2,452 at takeoff?

   A. Weight is reduced by 210 pounds and the CG is aft of limits.

   B. Weight is reduced by 210 pounds and the CG is unaffected.

   C. Weight is reduced to 2,680 pounds and the CG moves forward.

Answer (A) is correct. *(AWBH Chap 2)*
  DISCUSSION: The effect of a 35-gal. fuel burn on weight balance is required. Burning 35 gal. of fuel will reduce weight by 210 lb. and moment by 158. At 2,680 lb. (2,890 – 210), the 2,294 MOM/100 (2,452 – 158) is above the maximum moment of 2,287; i.e., CG is aft of limits. This is why weight and balance should always be computed for the beginning and end of each flight.
  Answer (B) is incorrect. Intuitively, one can see that the CG would be affected. Answer (C) is incorrect. Although the moment has decreased, the CG (moment divided by weight) has moved aft.

END OF STUDY UNIT

STUDY UNIT SIX
AEROMEDICAL FACTORS AND
AERONAUTICAL DECISION MAKING (ADM)

(3 pages of outline)

This study unit contains outlines of major concepts tested, sample test questions and answers regarding aeromedical factors and aeronautical decision making (ADM), and an explanation of each answer. The table of contents above lists each subunit within this study unit, the number of questions pertaining to that particular subunit, and the pages on which the outlines and questions begin, respectively.

CAUTION: Recall that the **sole purpose** of this book is to expedite your passing the FAA pilot knowledge test for the private pilot certificate. Accordingly, all extraneous material (i.e., topics or regulations not directly tested on the FAA pilot knowledge test) is omitted, even though much more information and knowledge are necessary to fly safely. This additional material is presented in *Pilot Handbook* and *Private Pilot Flight Maneuvers and Practical Test Prep*, available from Gleim Publications, Inc. See the order form on page 371.

6.1 HYPOXIA

1. The following are four types of hypoxia based on their causes:

    a. **Hypoxic hypoxia** is a result of insufficient oxygen available to the body as a whole.

        1) EXAMPLE: Reduction of partial pressure at high altitude, a blocked airway, or drowning

    b. **Hypemic hypoxia** occurs when the blood is not able to take up and transport a sufficient amount of oxygen to the cells in the body. The oxygen deficiency is in the blood, not the result of too little inhaled oxygen.

        1) EXAMPLE: Carbon monoxide poisoning

    c. **Stagnant hypoxia** results when oxygen-rich blood in the lungs is not moving.

        1) EXAMPLE: Shock, reduced circulation due to extreme cold, or pulling excessive Gs in flight

    d. **Histotoxic hypoxia** is the inability of cells to effectively use oxygen.

        1) EXAMPLE: Impairment due to alcohol and drugs

2. Symptoms of hypoxia include an initial feeling of euphoria but lead to more serious concerns such as headache, decreased reaction time, visual impairment, and eventual unconsciousness.

3. The correct response to counteract feelings of hypoxia is to lower altitude or use supplemental oxygen.

6.2 HYPERVENTILATION

1. Hyperventilation occurs when an excessive amount of air is breathed out of the lungs, e.g., when one becomes excited or undergoes stress, tension, fear, or anxiety.

    a. This results in an excessive amount of carbon dioxide passed out of the body and too much oxygen retained.

    b. The symptoms are dizziness, hot and cold sensations, nausea, etc.

2. Overcome hyperventilation symptoms by slowing the breathing rate, breathing into a bag, or talking aloud.

6.3 SPATIAL DISORIENTATION

1. Spatial disorientation, e.g., not knowing whether you are going up, going down, or turning, is a state of temporary confusion resulting from misleading information being sent to the brain by various sensory organs.

2. If you lose outside visual references and become disoriented, you are experiencing spatial disorientation. This occurs when you rely on the sensations of muscles and the inner ear to tell you what the airplane's attitude is.

    a. This might occur during a night flight, in clouds, or in dust.

3. Ways to overcome the effects of spatial disorientation include relying on the airplane instruments, avoiding sudden head movements, and ensuring that outside visual references are fixed points on the surface.

6.4 VISION

1. Pilots should adapt their eyes for night flying by avoiding bright white lights for 30 min. prior to flight.

2. Due to the eye's physiology, off-center eyesight is better than direct at night. Pilots should scan slowly at night to permit off-center viewing.

3. Scanning for traffic is best accomplished by bringing small portions of the sky into the central field of vision slowly and in succession.

4. Haze can create the illusion of traffic or terrain being farther away than they actually are.

6.5 CARBON MONOXIDE

1. Carbon monoxide (CO) is a colorless, odorless gas produced by all combustion engines.

2. CO can enter a cockpit or cabin through heater and defrost vents.

    a. If a leak is detected or a pilot smells gas fumes or exhaust, immediate corrective action should be taken. This could include turning off the heater, opening air vents or windows, if appropriate, and using supplemental oxygen, if available.

3. Blurred (hazy) thinking and vision, uneasiness, dizziness, and tightness across the forehead are early symptoms of carbon monoxide poisoning. They are followed by a headache and, with large accumulations of carbon monoxide, a loss of muscle power.

    a. Tobacco smoke also causes CO poisoning and physiological debilitation, which are medically disqualifying for pilots.

4. Increases in altitude increase susceptibility to carbon monoxide poisoning because of decreased oxygen availability.

6.6 AERONAUTICAL DECISION MAKING (ADM)

1. **Aeronautical decision making (ADM)** is a systematic approach to the mental process used by pilots to consistently determine the best course of action in response to a given set of circumstances.

2. Risk management is the part of the decision-making process that relies on situational awareness, problem recognition, and good judgment to reduce risks associated with each flight.

    a. The four fundamental risk elements in the ADM process that comprise any given aviation situation are the

        1) Pilot
        2) Aircraft
        3) Environment
        4) Mission (type of operation)

3. Most pilots have fallen prey to dangerous tendencies or behavioral problems at some time. Scud running, continuing visual flight into instrument conditions, and neglecting checklists are three examples of these dangerous tendencies or behavioral problems that must be identified and eliminated.

    a. In scud running, a pilot pushes his/her capabilities and the airplane to the limits by trying to maintain visual contact with the terrain while trying to avoid contact with it during low visibility and ceilings.

    b. Continuing visual flight into instrument conditions often leads to spatial disorientation or collision with the ground or obstacles.

    c. Neglect of checklists is an example of a pilot's unjustified reliance on his/her short- and long-term memory for repetitive tasks.

4. ADM addresses five hazardous attitudes that contribute to poor pilot judgment.

    a. Recognition of hazardous attitudes (thoughts) is the first step in neutralizing them in the ADM process.

    b. When you recognize a hazardous attitude, you should label it as hazardous and then correct it by stating the corresponding antidote, as shown below.

Hazardous Attitude	Antidote
Antiauthority: *Don't tell me!*	Follow the rules. They are usually right.
Impulsivity: *Do something quickly!*	Not so fast. Think first.
Invulnerability: *It won't happen to me.*	It could happen to me.
Macho: *I can do it.*	Taking chances is foolish.
Resignation: *What's the use?*	I'm not helpless. I can make a difference.

5. You are responsible for determining whether or not you are fit to fly for a particular flight.

    a. You should ask, "Could I pass my medical examination right now?" If you cannot answer with an absolute yes, you should not fly.

6. Human error is the one common factor of most preventable accidents.

    a. A pilot who is involved in an accident usually knows what went wrong and was aware of the possible hazards when (s)he was making the decision that led to the wrong course of action.

QUESTIONS AND ANSWER EXPLANATIONS

All of the private pilot knowledge test questions chosen by the FAA for release as well as additional questions selected by Gleim relating to the material in the previous outlines are reproduced on the following pages. These questions have been organized into the same subunits as the outlines. To the immediate right of each question are the correct answer and answer explanation. You should cover these answers and answer explanations while responding to the questions. Refer to the general discussion in the Introduction on how to take the FAA pilot knowledge test.

Remember that the questions from the FAA pilot knowledge test bank have been reordered by topic and organized into a meaningful sequence. Also, the first line of the answer explanation gives the citation of the authoritative source for the answer.

QUESTIONS

6.1 Hypoxia

1. Which statement best defines hypoxia?

A. A state of oxygen deficiency in the body.

B. An abnormal increase in the volume of air breathed.

C. A condition of gas bubble formation around the joints or muscles.

Answer (A) is correct. *(AIM Para 8-1-2)*
DISCUSSION: Hypoxia is oxygen deficiency in the bloodstream and may cause lack of clear thinking, fatigue, euphoria and, shortly thereafter, unconsciousness.
Answer (B) is incorrect. It describes a cause of hyperventilation. Answer (C) is incorrect. It describes decompression sickness after scuba diving.

2. Which is not a type of hypoxia?

A. Histotoxic.

B. Hypoxic.

C. Hypertoxic.

Answer (C) is correct. *(PHAK Chap 16)*
DISCUSSION: There is no such thing as hypertoxic hypoxia. The four types of hypoxia are histotoxic, hypoxic, hypemic, and stagnant hypoxia.
Answer (A) is incorrect. The four types of hypoxia are histotoxic, hypoxic, hypemic, and stagnant hypoxia. Answer (B) is incorrect. The four types of hypoxia are histotoxic, hypoxic, hypemic, and stagnant hypoxia.

3. Which of the following is a correct response to counteract the feelings of hypoxia in flight?

A. Promptly descend altitude.

B. Increase cabin air flow.

C. Avoid sudden inhalations.

Answer (A) is correct. *(PHAK Chap 16)*
DISCUSSION: The correct response to counteract feelings of hypoxia is to lower altitude or use supplemental oxygen, if the aircraft is so equipped.
Answer (B) is incorrect. Increasing the amount of air flowing inside an aircraft will not help counteract hypoxia. Because of the reduction of partial pressure at higher altitudes, there is less oxygen in the air to draw from. Answer (C) is incorrect. Breathing deeply or suddenly will not counteract feelings of hypoxia.

6.2 Hyperventilation

4. A pilot should be able to overcome the symptoms or avoid future occurrences of hyperventilation by

A. closely monitoring the flight instruments to control the airplane.

B. slowing the breathing rate, breathing into a bag, or talking aloud.

C. increasing the breathing rate in order to increase lung ventilation.

Answer (B) is correct. *(AIM Para 8-1-3)*
DISCUSSION: To recover from hyperventilation, the pilot should slow the breathing rate, breathe into a bag, or talk aloud.
Answer (A) is incorrect. Closely monitoring the flight instruments is used to overcome vertigo (spatial disorientation). Answer (C) is incorrect. Increased breathing aggravates hyperventilation.

5. Rapid or extra deep breathing while using oxygen can cause a condition known as

A. hyperventilation.

B. aerosinusitis.

C. aerotitis.

Answer (A) is correct. *(AIM Para 8-1-3)*
DISCUSSION: Hyperventilation occurs when an excessive amount of carbon dioxide is passed out of the body and too much oxygen is retained. This occurs when breathing rapidly and especially when using oxygen.
Answer (B) is incorrect. Aerosinusitis is an inflammation of the sinuses caused by changes in atmospheric pressure. Answer (C) is incorrect. Aerotitis is an inflammation of the inner ear caused by changes in atmospheric pressure.

6. When a stressful situation is encountered in flight, an abnormal increase in the volume of air breathed in and out can cause a condition known as

A. hyperventilation.

B. aerosinusitis.

C. aerotitis.

Answer (A) is correct. *(AIM Para 8-1-3)*
DISCUSSION: Hyperventilation occurs when an excessive amount of carbon dioxide is passed out of the body and too much oxygen is retained. This occurs when breathing rapidly and especially when using oxygen.
Answer (B) is incorrect. Aerosinusitis is an inflammation of the sinuses caused by changes in atmospheric pressure. Answer (C) is incorrect. Aerotitis is an inflammation of the inner ear caused by changes in atmospheric pressure.

7. Which would most likely result in hyperventilation?

A. Emotional tension, anxiety, or fear.

B. The excessive consumption of alcohol.

C. An extremely slow rate of breathing and insufficient oxygen.

Answer (A) is correct. *(AIM Para 8-1-3)*
DISCUSSION: Hyperventilation usually occurs when one becomes excited or undergoes stress, which results in an increase in one's rate of breathing.
Answer (B) is incorrect. Hyperventilation is usually caused by some type of stress, not by alcohol. Answer (C) is incorrect. The opposite is true: Hyperventilation is an extremely fast rate of breathing that produces excessive oxygen.

8. A pilot experiencing the effects of hyperventilation should be able to restore the proper carbon dioxide level in the body by

A. slowing the breathing rate, breathing into a paper bag, or talking aloud.

B. breathing spontaneously and deeply or gaining mental control of the situation.

C. increasing the breathing rate in order to increase lung ventilation.

Answer (A) is correct. *(PHAK Chap 16)*
DISCUSSION: A stressful situation can often lead to hyperventilation, which results from an increased rate and depth of respiration that leads to an abnormally low amount of carbon dioxide in the bloodstream. By slowing the breathing rate, breathing into a paper bag, or talking aloud, a pilot can overcome the effects of hyperventilation and return the carbon dioxide level in the bloodstream to normal.
Answer (B) is incorrect. Breathing deeply further aggravates the effects of hyperventilation. Answer (C) is incorrect. Increasing the rate of breathing will further aggravate the effects of hyperventilation.

6.3 Spatial Disorientation

9. Pilots are more subject to spatial disorientation if

A. they ignore the sensations of muscles and inner ear.

B. visual cues are taken away, as they are in instrument meteorological conditions (IMC).

C. eyes are moved often in the process of cross-checking the flight instruments.

Answer (B) is correct. *(AIM Para 8-1-5)*
DISCUSSION: Spatial disorientation is a state of temporary confusion resulting from misleading information being sent to the brain by various sensory organs. Thus, the pilot should ignore sensations of muscles and the inner ear and kinesthetic senses (those that sense motion), especially during flight in IMC when outside visual cues are taken away.
Answer (A) is incorrect. Ignoring the sensations of muscles and the inner ear will help overcome spatial disorientation. Answer (C) is incorrect. Cross-checking the flight instruments will help prevent spatial disorientation.

10. Pilots are more subject to spatial disorientation if

A. they ignore the sensations of muscles and inner ear.

B. body signals are used to interpret flight attitude.

C. eyes are moved often in the process of cross-checking the flight instruments.

Answer (B) is correct. *(AIM Para 8-1-5)*
DISCUSSION: Spatial disorientation is a state of temporary confusion resulting from misleading information being sent to the brain by various sensory organs. Thus, the pilot should ignore sensations of muscles and the inner ear and kinesthetic senses (those that sense motion).
Answer (A) is incorrect. Ignoring the sensations of muscles and the inner ear will help overcome spatial disorientation. Answer (C) is incorrect. Cross-checking the flight instruments will help prevent spatial disorientation.

11. If a pilot experiences spatial disorientation during flight in a restricted visibility condition, the best way to overcome the effect is to

A. rely upon the aircraft instrument indications.

B. concentrate on yaw, pitch, and roll sensations.

C. consciously slow the breathing rate until symptoms clear and then resume normal breathing rate.

Answer (A) is correct.   *(AIM Para 8-1-5)*
DISCUSSION: The best way to overcome the effects of spatial disorientation is to rely entirely on the aircraft's instrument indications and not upon body sensations. Sight of the horizon also overrides inner ear sensations. Thus, in areas of poor visibility, especially, such bodily signals should be ignored.
Answer (B) is incorrect. Yaw, pitch, and roll sensations should be ignored. Answer (C) is incorrect. A decrease in breathing rate is the proper treatment for hyperventilation, not spatial disorientation.

12. A lack of orientation with regard to the position, attitude, or movement of the aircraft in space is defined as

A. spatial disorientation.

B. hyperventilation.

C. hypoxia.

Answer (A) is correct.   *(PHAK Chap 16)*
DISCUSSION: Spatial disorientation is a state of temporary confusion resulting from misleading information being sent to the brain by various sensory organs. Thus, the pilot should ignore sensations of muscles and the inner ear and kinesthetic senses (those that sense motion), especially during flight in IMC when outside visual cues are taken away.
Answer (B) is incorrect. Hyperventilation occurs when an excessive amount of carbon dioxide is passed out of the body and too much oxygen is retained. This occurs when breathing rapidly and especially when using supplemental oxygen. Answer (C) is incorrect. Hypoxia is the result of an oxygen deficiency in the bloodstream and may cause lack of clear thinking, fatigue, euphoria, and, shortly thereafter, unconsciousness.

13. A state of temporary confusion resulting from misleading information being sent to the brain by various sensory organs is defined as

A. spatial disorientation.

B. hyperventilation.

C. hypoxia.

Answer (A) is correct.   *(AIM Para 8-1-5)*
DISCUSSION: A state of temporary confusion resulting from misleading information being sent to the brain by various sensory organs is defined as vertigo (spatial disorientation). Put simply, the pilot cannot determine his/her relationship to the earth's horizon.
Answer (B) is incorrect. Hyperventilation causes excessive oxygen and/or a decrease in carbon dioxide in the bloodstream. Answer (C) is incorrect. Hypoxia occurs when there is insufficient oxygen in the bloodstream.

14. The danger of spatial disorientation during flight in poor visual conditions may be reduced by

A. shifting the eyes quickly between the exterior visual field and the instrument panel.

B. having faith in the instruments rather than taking a chance on the sensory organs.

C. leaning the body in the opposite direction of the motion of the aircraft.

Answer (B) is correct.   *(AIM Para 8-1-5)*
DISCUSSION: Various complex motions and forces and certain visual scenes encountered in flight can create illusions of motion and position. Spatial disorientation from these illusions can be prevented only by visual reference to reliable fixed points on the ground and horizon or to flight instruments.
Answer (A) is incorrect. In poor visual conditions, reliable exterior references are not available. Answer (C) is incorrect. To avoid spatial disorientation, the pilot should avoid undue head and body movements and rely totally on the flight instruments. By moving the body in response to perceived motion, the conflicting signals reaching the brain will cause spatial disorientation.

6.4 Vision

15. Which technique should a pilot use to scan for traffic to the right and left during straight-and-level flight?

A. Systematically focus on different segments of the sky for short intervals.

B. Concentrate on relative movement detected in the peripheral vision area.

C. Continuous sweeping of the windshield from right to left.

Answer (A) is correct.   *(AIM Para 8-1-6)*
DISCUSSION: Due to the fact that eyes can focus only on a narrow viewing area, effective scanning is accomplished with a series of short, regularly spaced eye movements that bring successive areas of the sky into the central vision field.
Answer (B) is incorrect. It concerns scanning for traffic at night. Answer (C) is incorrect. You must continually scan successive small portions of the sky. The eyes can focus only on a narrow viewing area and require at least 1 sec. to detect a faraway object.

16. What effect does haze have on the ability to see traffic or terrain features during flight?

    A. Haze causes the eyes to focus at infinity.

    B. The eyes tend to overwork in haze and do not detect relative movement easily.

    C. All traffic or terrain features appear to be farther away than their actual distance.

Answer (C) is correct. *(AIM Para 8-1-5)*
    DISCUSSION: Atmospheric haze can create the illusion of being at a greater distance from traffic or terrain than you actually are. This is especially prevalent on landings.
    Answer (A) is incorrect. In haze, the eyes focus at a comfortable distance, which may be only 10 to 30 ft. outside of the cockpit. Answer (B) is incorrect. In haze, the eyes relax and tend to stare outside without focusing or looking for common visual cues.

17. What preparation should a pilot make to adapt the eyes for night flying?

    A. Wear sunglasses after sunset until ready for flight.

    B. Avoid red lights at least 30 minutes before the flight.

    C. Avoid bright white lights at least 30 minutes before the flight.

Answer (C) is correct. *(AFH Chap 10)*
    DISCUSSION: Prepare for night flying by letting your eyes adapt to darkness, including avoiding bright white light for at least 30 min. prior to night flight.
    Answer (A) is incorrect. Preparation does not involve wearing sunglasses but rather avoiding bright white lights. Answer (B) is incorrect. White, not red, lights impair night vision.

18. What is the most effective way to use the eyes during night flight?

    A. Look only at far away, dim lights.

    B. Scan slowly to permit off-center viewing.

    C. Concentrate directly on each object for a few seconds.

Answer (B) is correct. *(AFH Chap 10)*
    DISCUSSION: Physiologically, the eyes are most effective at seeing objects off-center at night. Accordingly, pilots should scan slowly to permit off-center viewing.
    Answer (A) is incorrect. Pilots must look at their gauges and instruments, which are 2 ft. in front of them. Answer (C) is incorrect. Peripheral (off-center) vision is more effective at night.

19. The best method to use when looking for other traffic at night is to

    A. look to the side of the object and scan slowly.

    B. scan the visual field very rapidly.

    C. look to the side of the object and scan rapidly.

Answer (A) is correct. *(AFH Chap 10)*
    DISCUSSION: Physiologically, the eyes are most effective at seeing objects off-center at night. Accordingly, pilots should scan slowly to permit off-center viewing.
    Answer (B) is incorrect. Scanning should always be done slowly and methodically. Answer (C) is incorrect. Scanning should always be done slowly and methodically.

6.5 Carbon Monoxide

20. Large accumulations of carbon monoxide in the human body result in

    A. tightness across the forehead.

    B. loss of muscular power.

    C. an increased sense of well-being.

Answer (B) is correct. *(AC 20-32B)*
    DISCUSSION: Carbon monoxide reduces the ability of the blood to carry oxygen. Large accumulations result in loss of muscular power.
    Answer (A) is incorrect. It describes an early symptom, not the effect of large accumulations. Answer (C) is incorrect. Euphoria is a result of the lack of sufficient oxygen, not specifically an accumulation of carbon monoxide.

21. Susceptibility to carbon monoxide poisoning increases as

    A. altitude increases.

    B. altitude decreases.

    C. air pressure increases.

Answer (A) is correct. *(AIM Para 8-1-4)*
    DISCUSSION: Carbon monoxide poisoning results in an oxygen deficiency. Since there is less oxygen available at higher altitudes, carbon monoxide poisoning can occur with lesser amounts of carbon monoxide as altitude increases.
    Answer (B) is incorrect. There is more available oxygen at lower altitudes. Answer (C) is incorrect. There is more available oxygen at higher air pressures.

22. What is a correct response if an exhaust leak were to be detected while in flight?

A. Increase altitude so the effects of CO would be decreased.

B. Take deep breaths so as to inhale more oxygen.

C. Open air vents or windows.

Answer (C) is correct. *(PHAK Chap 16)*
  DISCUSSION: Taking corrective steps such as turning off the heater, opening air vents or windows, and using supplemental oxygen are the correct responses if a pilot smells gas fumes or otherwise detects increased amounts of CO.
  Answer (A) is incorrect. An increase in altitude increases the susceptibility of CO poisoning because of the decreased oxygen available. Answer (B) is incorrect. Inhaling more CO-tainted air would be detrimental to a pilot's health and is not a positive corrective action.

23. Effects of carbon monoxide poisoning include

A. dizziness, blurred vision, and loss of muscle power.

B. sweating, increased breathing, and paleness.

C. motion sickness, tightness across the forehead, and drowsiness.

Answer (A) is correct. *(PHAK Chap 16)*
  DISCUSSION: Effects of CO poisoning include headache, blurred vision, dizziness, drowsiness, and loss of muscle control.
  Answer (B) is incorrect. Sweating, increased breathing, and paleness are symptoms of motion sickness, not CO poisoning. Answer (C) is incorrect. Motion sickness is not an effect or characteristic of carbon monoxide poisoning.

6.6 Aeronautical Decision Making (ADM)

24. Risk management, as part of the aeronautical decision making (ADM) process, relies on which features to reduce the risks associated with each flight?

A. Application of stress management and risk element procedures.

B. The mental process of analyzing all information in a particular situation and making a timely decision on what action to take.

C. Situational awareness, problem recognition, and good judgment.

Answer (C) is correct. *(AC 60-22)*
  DISCUSSION: Risk management is that part of the ADM process that relies on situational awareness, problem recognition, and good judgment to reduce risks associated with each flight.
  Answer (A) is incorrect. Risk management relies on situational awareness, problem recognition, and good judgment, not the application of stress management and risk-element procedures, to reduce the risks associated with each flight. Answer (B) is incorrect. Judgment, not risk management, is the mental process of analyzing all information in a particular situation and making a timely decision on what action to take.

25. What is it often called when a pilot pushes his or her capabilities and the aircraft's limits by trying to maintain visual contact with the terrain in low visibility and ceiling?

A. Scud running.

B. Mind set.

C. Peer pressure.

Answer (A) is correct. *(AC 60-22)*
  DISCUSSION: Scud running refers to a pilot pushing his/her capabilities and the aircraft's limits by trying to maintain visual contact with the terrain while flying with a low visibility or ceiling. Scud running is a dangerous (and often illegal) practice that may lead to a mishap. This dangerous tendency must be identified and eliminated.
  Answer (B) is incorrect. Mind-set may produce an inability to recognize and cope with changes in the situation requiring actions different from those anticipated or planned. Answer (C) is incorrect. Peer pressure may produce poor decision making based upon an emotional response to peers rather than an objective evaluation of a situation.

26. What often leads to spatial disorientation or collision with ground/obstacles when flying under Visual Flight Rules (VFR)?

A. Continual flight into instrument conditions.

B. Getting behind the aircraft.

C. Duck-under syndrome.

Answer (A) is correct. *(AC 60-22)*
  DISCUSSION: Continuing VFR flight into instrument conditions often leads to spatial disorientation or collision with ground/obstacles due to the loss of outside visual references. It is even more dangerous if the pilot is not instrument qualified or current.
  Answer (B) is incorrect. Getting behind the aircraft results in allowing events or the situation to control your actions, rather than the other way around. Answer (C) is incorrect. Duck-under syndrome is the tendency to descend below minimums during an approach based on the belief that there is always a fudge factor built in; it occurs during IFR, not VFR, flight.

27. What is one of the neglected items when a pilot relies on short and long term memory for repetitive tasks?

A. Checklists.

B. Situation awareness.

C. Flying outside the envelope.

Answer (A) is correct. *(AC 60-22)*
   DISCUSSION: Neglect of checklists, flight planning, preflight inspections, etc., is an indication of a pilot's unjustified reliance on his/her short- and long-term memory for repetitive flying tasks.
   Answer (B) is incorrect. Situation awareness suffers when a pilot gets behind the airplane, which results in an inability to recognize deteriorating circumstances and/or misjudgment on the rate of deterioration. Answer (C) is incorrect. Flying outside the envelope occurs when the pilot believes (often in error) that the aircraft's high-performance capability meets the demands imposed by the pilot's (often overestimated) flying skills.

28. Hazardous attitudes occur to every pilot to some degree at some time. What are some of these hazardous attitudes?

A. Antiauthority, impulsivity, macho, resignation, and invulnerability.

B. Poor situational awareness, snap judgments, and lack of a decision making process.

C. Poor risk management and lack of stress management.

Answer (A) is correct. *(AC 60-22)*
   DISCUSSION: The five hazardous attitudes addressed in the ADM process are antiauthority, impulsivity, invulnerability, macho, and resignation.
   Answer (B) is incorrect. Poor situational awareness and snap judgments are indications of the lack of a decision-making process, not hazardous attitudes. Answer (C) is incorrect. Poor risk management and lack of stress management lead to poor ADM and are not considered hazardous attitudes.

29. In the aeronautical decision making (ADM) process, what is the first step in neutralizing a hazardous attitude?

A. Recognizing hazardous thoughts.

B. Recognizing the invulnerability of the situation.

C. Making a rational judgment.

Answer (A) is correct. *(AC 60-22)*
   DISCUSSION: Hazardous attitudes, which contribute to poor pilot judgment, can be effectively counteracted by redirecting that hazardous attitude so that appropriate action can be taken. Recognition of hazardous thoughts is the first step in neutralizing them in the ADM process.
   Answer (B) is incorrect. Invulnerability is a hazardous attitude. The first step in neutralizing a hazardous attitude is to recognize it. Answer (C) is incorrect. Before a rational judgment can be made, the hazardous attitude must be recognized then redirected so that appropriate action can be taken.

30. What is the antidote when a pilot has a hazardous attitude, such as "Antiauthority"?

A. Rules do not apply in this situation.

B. I know what I am doing.

C. Follow the rules.

Answer (C) is correct. *(AC 60-22)*
   DISCUSSION: When you recognize a hazardous thought, you should correct it by stating the corresponding antidote. The antidote for the antiauthority ("Do not tell me!") hazardous attitude is "Follow the rules; they are usually right."
   Answer (A) is incorrect. "Rules do not apply in this situation" is an example of the antiauthority hazardous attitude, not its antidote. Answer (B) is incorrect. "I know what I'm doing" is an example of the macho hazardous attitude, not an antidote to the antiauthority attitude.

31. What is the antidote when a pilot has a hazardous attitude, such as "Impulsivity"?

A. It could happen to me.

B. Do it quickly to get it over with.

C. Not so fast, think first.

Answer (C) is correct. *(AC 60-22)*
   DISCUSSION: When you recognize a hazardous thought, you should correct it by stating the corresponding antidote. The antidote for the impulsivity ("Do something quickly!") hazardous attitude is "Not so fast, think first."
   Answer (A) is incorrect. "It could happen to me" is the antidote for the invulnerability, not impulsivity, hazardous attitude. Answer (B) is incorrect. "Do it quickly and get it over with" is an example of the impulsivity hazardous attitude, not its antidote.

32. What is the antidote when a pilot has a hazardous attitude, such as "Invulnerability"?

   A. It will not happen to me.

   B. It cannot be that bad.

   C. It could happen to me.

Answer (C) is correct. *(AC 60-22)*
   DISCUSSION: When you recognize a hazardous thought, you should correct it by stating the corresponding antidote. The antidote for the invulnerability ("It will not happen to me") hazardous attitude is "It could happen to me."
   Answer (A) is incorrect. "It will not happen to me" is an example of the invulnerability hazardous attitude, not its antidote. Answer (B) is incorrect. "It cannot be that bad" is an example of the invulnerability hazardous attitude, not its antidote.

33. What is the antidote when a pilot has a hazardous attitude, such as "Macho"?

   A. I can do it.

   B. Taking chances is foolish.

   C. Nothing will happen.

Answer (B) is correct. *(AC 60-22)*
   DISCUSSION: When you recognize a hazardous thought, you should correct it by stating the corresponding antidote. The antidote for the macho ("I can do it") hazardous attitude is "Taking chances is foolish."
   Answer (A) is incorrect. "I can do it" is an example of the macho hazardous attitude, not its antidote. Answer (C) is incorrect. "Nothing will happen" is an example of the invulnerability hazardous attitude, not an antidote to the macho attitude.

34. What is the antidote when a pilot has a hazardous attitude, such as "Resignation"?

   A. What is the use?

   B. Someone else is responsible.

   C. I am not helpless.

Answer (C) is correct. *(AC 60-22)*
   DISCUSSION: When you recognize a hazardous thought, you should correct it by stating the corresponding antidote. The antidote for the resignation ("What is the use?") hazardous attitude is "I am not helpless. I can make a difference."
   Answer (A) is incorrect. "What is the use?" is an example of the resignation hazardous attitude, not its antidote. Answer (B) is incorrect. "Someone else is responsible" is an example of the resignation hazardous attitude, not its antidote.

35. Who is responsible for determining whether a pilot is fit to fly for a particular flight, even though he or she holds a current medical certificate?

   A. The FAA.

   B. The medical examiner.

   C. The pilot.

Answer (C) is correct. *(AC 60-22)*
   DISCUSSION: A number of factors, from lack of sleep to illness, can reduce a pilot's fitness to make a particular flight. It is the responsibility of the pilot to determine whether (s)he is fit to make a particular flight, even though (s)he holds a current medical certificate. Additionally, FAR 61.53 prohibits a pilot who possesses a current medical certificate from acting as pilot in command, or in any other capacity as a required pilot flight crewmember, while the pilot has a known medical condition or an aggravation of a known medical condition that would make the pilot unable to meet the standards for a medical certificate.
   Answer (A) is incorrect. The pilot, not the FAA, is responsible for determining whether (s)he is fit for a particular flight. Answer (B) is incorrect. The pilot, not the medical examiner, is responsible for determining whether (s)he is fit for a particular flight.

36. What is the one common factor which affects most preventable accidents?

   A. Structural failure.

   B. Mechanical malfunction.

   C. Human error.

Answer (C) is correct. *(AC 60-22)*
   DISCUSSION: Most preventable accidents, such as fuel starvation or exhaustion, VFR flight into IFR conditions leading to disorientation, and flight into known icing, have one common factor: human error. Pilots who are involved in accidents usually know what went wrong. In the interest of expediency, cost savings, or other often irrelevant factors, the wrong course of action (decision) was chosen.
   Answer (A) is incorrect. Most preventable accidents have human error, not structural failure, as a common factor. Answer (B) is incorrect. Most preventable accidents have human error, not mechanical malfunction, as a common factor.

END OF STUDY UNIT

STUDY UNIT SEVEN
AVIATION WEATHER

(4 pages of outline)

This study unit contains outlines of major concepts tested, sample test questions and answers regarding aviation weather, and an explanation of each answer. The table of contents above lists each subunit within this study unit, the number of questions pertaining to that particular subunit, and the pages on which the outlines and questions begin, respectively.

CAUTION: Recall that the **sole purpose** of this book is to expedite your passing the FAA pilot knowledge test for the private pilot certificate. Accordingly, all extraneous material (i.e., topics or regulations not directly tested on the FAA pilot knowledge test) is omitted, even though much more information and knowledge are necessary to fly safely. This additional material is presented in *Pilot Handbook* and *Private Pilot Flight Maneuvers and Practical Test Prep*, available from Gleim Publications, Inc. See the order form on page 371.

7.1 CAUSES OF WEATHER

1. Every physical process of weather is accompanied by, or is the result of, heat exchanges.

2. Unequal heating of the Earth's surface causes differences in pressure and altimeter settings.

3. The Coriolis force deflects winds to the right in the Northern Hemisphere. It is caused by the Earth's rotation.

    a. The deflections caused by Coriolis force are less at the surface due to the slower wind speed.

    b. The wind speed is slower at the surface due to friction between wind and the Earth's surface.

7.2 CONVECTIVE CURRENTS

1. Sea breezes are caused by cool and more dense air moving inland off the water. Once inland over the warmer land, the air heats up and rises. Currents push the air over the water where it cools and descends, starting the process over again.

2. The development of thermals depends upon solar heating.

7.3 FRONTS

1. A front is the zone of transition (boundary) between two air masses of different density, e.g., the area separating a high pressure system and a low pressure system.

2. There is always a change in wind when flying across a front.

3. The most easily recognizable change when crossing a front is the change in temperature.

7.4 THUNDERSTORMS

1. Thunderstorms have three phases in their life cycle:

    a. Cumulus: The building stage of a thunderstorm when there are continuous updrafts.

    b. Mature: The time of greatest intensity when there are both updrafts and downdrafts (causing severe wind shear and turbulence).

        1) The commencing of rain on the Earth's surface indicates the beginning of the mature stage of a thunderstorm.

    c. Dissipating: When there are only downdrafts, i.e., the storm is raining itself out.

2. Thunderstorms are produced by cumulonimbus clouds. They form when there is

    a. Sufficient water vapor,
    b. An unstable lapse rate, and
    c. An initial upward boost to start the process.

3. Thunderstorms produce wind shear turbulence, a hazardous and invisible phenomenon particularly for airplanes landing and taking off.

    a. Hazardous wind shear near the ground can also be present during periods of strong temperature inversion.

4. The most severe thunderstorm conditions (heavy hail, destructive winds, tornadoes, etc.) are generally associated with squall line thunderstorms.

    a. A squall line is a nonfrontal narrow band of thunderstorms, usually ahead of a cold front.

5. A thunderstorm, by definition, has lightning because that is what causes thunder.

6. Embedded thunderstorms are obscured (i.e., pilots cannot see them) because they occur in very cloudy conditions.

7.5 ICING

1. Structural icing requires two conditions:

    a. Flight through visible moisture and
    b. The temperature at freezing or below.

2. Freezing rain usually causes the greatest accumulation of structural ice.

3. Ice pellets are caused when rain droplets freeze at a higher altitude, i.e., freezing rain exists above.

7.6 MOUNTAIN WAVE

1. Lenticular clouds are almond- or lens-shaped clouds, usually found on the leeward side of a mountain range.

    a. They may contain winds of 50 kt. or more.
    b. They appear stationary as the wind blows through them.

2. Expect mountain wave turbulence when the air is stable and winds of 40 kt. or greater blow across a mountain or ridge.

7.7 WIND SHEAR

1. Wind shear can occur at any altitude and be horizontal and/or vertical, i.e., whenever adjacent air is flowing in different directions or speeds.

2. Expect wind shear in a temperature inversion whenever wind speed at 2,000 to 4,000 ft. AGL is 25 kt. or more.

3. Hazardous wind shear may be expected in areas of low-level temperature inversions, frontal zones, and clear air turbulence.

7.8 TEMPERATURE/DEW POINT AND FOG

1. When the air temperature is within 5°F of the dew point and the spread is decreasing, you should expect fog and/or low clouds.

   a. Dew point is the temperature at which the air will have 100% humidity, i.e., be saturated.
   b. Thus, air temperature determines how much water vapor can be held by the air.
   c. Frost forms when both the collecting surface is below the dew point of the adjacent air and the dew point is below freezing. Frost is the direct sublimation of water vapor to ice crystals.

2. Water vapor becomes visible as it condenses into clouds, fog, or dew.

3. Evaporation is the conversion of liquid to water vapor.

4. Sublimation is the conversion of solids (e.g., ice) to water vapor or water vapor to solids (e.g., frost).

5. Radiation fog (shallow fog) is most likely to occur when there is a clear sky, little or no wind, and a small temperature/dew point spread.

6. Advection fog forms as a result of moist air condensing as it moves over a cooler surface.

7. Upslope fog results from warm, moist air being cooled as it is forced up sloping terrain.

8. Precipitation-induced fog occurs when warm rain or drizzle falls through cool air and evaporation from the precipitation saturates the cool air and forms fog.

   a. Precipitation-induced fog is usually associated with fronts.
   b. Because of this, it is in the proximity of icing, turbulence, and thunderstorms.

9. Steam fog forms in winter when cold, dry air passes from land areas over comparatively warm ocean waters. It is composed entirely of water droplets that often freeze quickly.

   a. Low-level turbulence can occur, and icing can become hazardous in steam fog.

7.9 CLOUDS

1. Clouds are divided into four families based on their height:

   a. High clouds
   b. Middle clouds
   c. Low clouds
   d. Clouds with extensive vertical development

2. The greatest turbulence is in cumulonimbus clouds.

3. Towering cumulus are early stages of cumulonimbus; they usually indicate convective turbulence.

4. Lifting action, unstable air, and moisture are the ingredients for the formation of cumulonimbus clouds.

5. Nimbus means rain cloud.

6. When air rises in a convective current, it cools at the rate of 5.4°F/1,000 ft., and its dew point decreases 1°F/1,000 feet. The temperature and dew point then are converging at 4.4°F/1,000 feet.

    a. Since clouds form when the temperature/dew point spread is 0°, we can use this to estimate the bases of cumulus clouds.

    b. The surface temperature/dew point spread divided by 4.4°F equals the bases of cumulus clouds in thousands of feet above ground level (AGL).

    c. EXAMPLE: A surface dew point of 56°F and a surface temperature of 69°F results in an estimate of cumulus cloud bases at 3,000 ft. AGL: 69°F − 56°F = 13°F temperature/dew point spread; 13°F/4.4°F = approximately 3,000 ft. AGL.

7.10 STABILITY OF AIR MASSES

1. Stable air characteristics

    a. Stratiform clouds
    b. Smooth air
    c. Fair-to-poor visibility in haze and smoke
    d. Continuous precipitation

2. Unstable air characteristics

    a. Cumuliform clouds
    b. Turbulent air
    c. Good visibility
    d. Showery precipitation

3. When air is warmed from below, it rises and causes instability.

4. The lapse rate is the decrease in temperature with increase in altitude. As the lapse rate increases (i.e., air cools more with increases in altitude), air is more unstable.

    a. The lapse rate can be used to determine the stability of air masses.

5. Moist, stable air moving up a mountain slope produces stratus type clouds as it cools.

6. Turbulence and clouds with extensive vertical development result when unstable air rises.

7. Steady precipitation preceding a front is usually an indication of a warm front, which results from warm air being cooled from the bottom by colder air.

    a. This results in stable air with stratiform clouds and little or no turbulence.

7.11 TEMPERATURE INVERSIONS

1. Normally, temperature decreases as altitude increases. A temperature inversion occurs when temperature increases as altitude increases.

2. Temperature inversions usually result in a stable layer of air.

3. A temperature inversion often develops near the ground on clear, cool nights when the wind is light.

    a. It is caused by terrestrial radiation.

4. Smooth air with restricted visibility is usually found beneath a low-level temperature inversion.

QUESTIONS AND ANSWER EXPLANATIONS

All of the private pilot knowledge test questions chosen by the FAA for release as well as additional questions selected by Gleim relating to the material in the previous outlines are reproduced on the following pages. These questions have been organized into the same subunits as the outlines. To the immediate right of each question are the correct answer and answer explanation. You should cover these answers and answer explanations while responding to the questions. Refer to the general discussion in the Introduction on how to take the FAA pilot knowledge test.

Remember that the questions from the FAA pilot knowledge test bank have been reordered by topic and organized into a meaningful sequence. Also, the first line of the answer explanation gives the citation of the authoritative source for the answer.

QUESTIONS

7.1 Causes of Weather

1. Every physical process of weather is accompanied by, or is the result of, a

A. movement of air.

B. pressure differential.

C. heat exchange.

Answer (C) is correct. *(AvW Chap 2)*
DISCUSSION: Every physical process of weather is accompanied by, or is the result of, a heat exchange. A heat differential (difference between the temperatures of two air masses) causes a differential in pressure, which in turn causes movement of air. Heat exchanges occur constantly, e.g., melting, cooling, updrafts, downdrafts, wind, etc.
Answer (A) is incorrect. Movement of air is a result of heat exchange. Answer (B) is incorrect. Pressure differential is a result of heat exchange.

2. What causes variations in altimeter settings between weather reporting points?

A. Unequal heating of the Earth's surface.

B. Variation of terrain elevation.

C. Coriolis force.

Answer (A) is correct. *(AvW Chap 3)*
DISCUSSION: Unequal heating of the Earth's surface causes differences in air pressure, which is reflected in differences in altimeter settings between weather reporting points.
Answer (B) is incorrect. Variations in altimeter settings between stations is a result of unequal heating of the Earth's surface, not variations of terrain elevations. Answer (C) is incorrect. Variations in altimeter settings between stations is a result of unequal heating of the Earth's surface, not the Coriolis force.

3. The wind at 5,000 feet AGL is southwesterly while the surface wind is southerly. This difference in direction is primarily due to

A. stronger pressure gradient at higher altitudes.

B. friction between the wind and the surface.

C. stronger Coriolis force at the surface.

Answer (B) is correct. *(AvW Chap 4)*
DISCUSSION: Winds aloft at 5,000 ft. are largely affected by Coriolis force, which deflects wind to the right, in the Northern Hemisphere. But at the surface, the winds will be more southerly (they were southwesterly aloft) because Coriolis force has less effect at the surface where the wind speed is slower. The wind speed is slower at the surface due to the friction between the wind and the surface.
Answer (A) is incorrect. Pressure gradient is a force that causes wind, not the reason for wind direction differences. Answer (C) is incorrect. The Coriolis force at the surface is weaker (not stronger) with slower wind speed.

7.2 Convective Currents

4. Convective circulation patterns associated with sea breezes are caused by

    A. warm, dense air moving inland from over the water.

    B. water absorbing and radiating heat faster than the land.

    C. cool, dense air moving inland from over the water.

Answer (C) is correct. *(AvW Chap 4)*
    DISCUSSION: Sea breezes are caused by cool and more dense air moving inland off the water. Once over the warmer land, the air heats up and rises. Thus the cooler, more dense air from the sea forces the warmer air up. Currents push the hot air over the water where it cools and descends, starting the cycle over again. This process is caused by land heating faster than water.
    Answer (A) is incorrect. The air over the water is cooler (not warmer). Answer (B) is incorrect. Water absorbs and radiates heat slower (not faster) than land.

5. The development of thermals depends upon

    A. a counterclockwise circulation of air.

    B. temperature inversions.

    C. solar heating.

Answer (C) is correct. *(AvW Chap 4)*
    DISCUSSION: Thermals are updrafts in small-scale convective currents. Convective currents are caused by uneven heating of the Earth's surface. Solar heating is the means of heating the Earth's surface.
    Answer (A) is incorrect. A counterclockwise circulation describes an area of low pressure in the Northern Hemisphere. Answer (B) is incorrect. A temperature inversion is an increase in temperature with height, which hinders the development of thermals.

7.3 Fronts

6. The boundary between two different air masses is referred to as a

    A. frontolysis.

    B. frontogenesis.

    C. front.

Answer (C) is correct. *(AvW Chap 8)*
    DISCUSSION: A front is a surface, interface, or transition zone of discontinuity between two adjacent air masses of different densities. It is the boundary between two different air masses.
    Answer (A) is incorrect. Frontolysis is the dissipation of a front. Answer (B) is incorrect. Frontogenesis is the initial formation of a front or frontal zone.

7. One weather phenomenon which will always occur when flying across a front is a change in the

    A. wind direction.

    B. type of precipitation.

    C. stability of the air mass.

Answer (A) is correct. *(AvW Chap 8)*
    DISCUSSION: The definition of a front is the zone of transition between two air masses of different air pressure or density, e.g., the area separating high and low pressure systems. Due to the difference in changes in pressure systems, there will be a change in wind.
    Answer (B) is incorrect. Frequently, precipitation will exist or not exist for both sides of the front: rain showers before and after or no precipitation before and after a dry front. Answer (C) is incorrect. Fronts separate air masses with different pressures, not stabilities; e.g., both air masses could be either stable or unstable.

8. One of the most easily recognized discontinuities across a front is

    A. a change in temperature.

    B. an increase in cloud coverage.

    C. an increase in relative humidity.

Answer (A) is correct. *(AvW Chap 8)*
    DISCUSSION: Of the many changes that take place across a front, the most easily recognized is the change in temperature. When flying through a front, you will notice a significant change in temperature, especially at low altitudes.
    Answer (B) is incorrect. Although cloud formations may indicate a frontal system, they may not be present or easily recognized across the front. Answer (C) is incorrect. Precipitation is not always associated with a front.

7.4 Thunderstorms

9. If there is thunderstorm activity in the vicinity of an airport at which you plan to land, which hazardous atmospheric phenomenon might be expected on the landing approach?

A. Precipitation static.

B. Wind-shear turbulence.

C. Steady rain.

Answer (B) is correct. *(AvW Chap 11)*
DISCUSSION: The most hazardous atmospheric phenomenon near thunderstorms is wind shear turbulence.
Answer (A) is incorrect. Precipitation static is a steady, high level of noise in radio receivers, which is caused by intense corona discharges from sharp metallic points and edges of flying aircraft. This discharge may be seen at night and is also called St. Elmo's fire. Answer (C) is incorrect. Thunderstorms are usually associated with unstable air, which would produce rain showers (not steady rain).

10. A nonfrontal, narrow band of active thunderstorms that often develop ahead of a cold front is known as a

A. prefrontal system.

B. squall line.

C. dry line.

Answer (B) is correct. *(AvW Chap 11)*
DISCUSSION: A nonfrontal, narrow band of active thunderstorms that often develops ahead of a cold front is known as a squall line.
Answer (A) is incorrect. A prefrontal system is a term that has no meaning. Answer (C) is incorrect. A dry line is a front that seldom has any significant air mass contrast except for moisture.

11. What conditions are necessary for the formation of thunderstorms?

A. High humidity, lifting force, and unstable conditions.

B. High humidity, high temperature, and cumulus clouds.

C. Lifting force, moist air, and extensive cloud cover.

Answer (A) is correct. *(AvW Chap 11)*
DISCUSSION: Thunderstorms form when there is sufficient water vapor, an unstable lapse rate, and an initial upward boost (lifting) to start the storm process.
Answer (B) is incorrect. A high temperature is not required for the formation of thunderstorms. Answer (C) is incorrect. Extensive cloud cover is not necessary for the formation of thunderstorms.

12. During the life cycle of a thunderstorm, which stage is characterized predominately by downdrafts?

A. Cumulus.

B. Dissipating.

C. Mature.

Answer (B) is correct. *(AvW Chap 11)*
DISCUSSION: Thunderstorms have three life cycles: cumulus, mature, and dissipating. It is in the dissipating stage that the storm is characterized by downdrafts as the storm rains itself out.
Answer (A) is incorrect. Cumulus is the building stage when there are updrafts. Answer (C) is incorrect. The mature stage is when there are both updrafts and downdrafts, which create dangerous wind shears.

13. Thunderstorms reach their greatest intensity during the

A. mature stage.

B. downdraft stage.

C. cumulus stage.

Answer (A) is correct. *(AvW Chap 11)*
DISCUSSION: Thunderstorms reach their greatest intensity during the mature stage, where updrafts and downdrafts cause a high level of wind shear.
Answer (B) is incorrect. The downdraft stage is known as the dissipating stage, which is when the thunderstorm rains itself out. Answer (C) is incorrect. The cumulus stage is characterized by continuous updrafts and is not the most intense stage of a thunderstorm.

14. What feature is normally associated with the cumulus stage of a thunderstorm?

A. Roll cloud.

B. Continuous updraft.

C. Frequent lightning.

Answer (B) is correct. *(AvW Chap 11)*
DISCUSSION: The cumulus stage of a thunderstorm has continuous updrafts that build the storm. The water droplets are carried up until they become too heavy. Once they begin falling and creating downdrafts, the storm changes from the cumulus to the mature stage.
Answer (A) is incorrect. The roll cloud is the cloud on the ground, which is formed by the downrushing cold air pushing out from underneath the bottom of the thunderstorm. Answer (C) is incorrect. Frequent lightning is associated with the mature stage, where there is a considerable amount of wind shear and static electricity.

15. Which weather phenomenon signals the beginning of the mature stage of a thunderstorm?

A. The appearance of an anvil top.

B. Precipitation beginning to fall.

C. Maximum growth rate of the clouds.

Answer (B) is correct. *(AvW Chap 11)*
DISCUSSION: The mature stage of a thunderstorm begins when rain begins falling. This means that the downdrafts are occurring sufficiently to carry water all the way through the thunderstorm.
Answer (A) is incorrect. The appearance of an anvil top normally occurs during the dissipating stage when the upper winds blow the top of the cloud downwind. Answer (C) is incorrect. The maximum growth rate of clouds is later in the mature stage and does not necessarily mark the start of the mature stage.

16. Thunderstorms which generally produce the most intense hazard to aircraft are

A. squall line thunderstorms.

B. steady-state thunderstorms.

C. warm front thunderstorms.

Answer (A) is correct. *(AvW Chap 11)*
DISCUSSION: A squall line is a nonfrontal narrow band of active thunderstorms. It often contains severe, steady-state thunderstorms and presents the single most intense weather hazard to airplanes.
Answer (B) is incorrect. Steady-state thunderstorms are normally associated with weather systems and often form into squall lines. Answer (C) is incorrect. Squall line (not warm front) thunderstorms generally produce the most intense hazard to aircraft.

17. Which weather phenomenon is always associated with a thunderstorm?

A. Lightning.

B. Heavy rain.

C. Hail.

Answer (A) is correct. *(AvW Chap 11)*
DISCUSSION: A thunderstorm, by definition, has lightning, because lightning causes the thunder.
Answer (B) is incorrect. While heavy rain showers usually occur, hail may occur instead. Answer (C) is incorrect. Hail is produced only when the lifting action extends above the freezing level and the supercooled water begins to freeze.

18. The destination airport has one runway, 08-26, and the wind is calm. The normal approach in calm wind is a left hand pattern to runway 08. There is no other traffic at the airport. A thunderstorm about 6 miles west is beginning its mature stage, and rain is starting to reach the ground. The pilot decides to

A. fly the pattern to runway 08 since the storm is too far away to affect the wind at the airport.

B. fly the normal pattern to runway 08 since the storm is west and moving north and any unexpected wind will be from the east or southeast toward the storm.

C. fly an approach to runway 26 since any unexpected wind due to the storm will be westerly.

Answer (C) is correct. *(PHAK Chap 11)*
DISCUSSION: Flying in, near, or under thunderstorms can subject an aircraft to rain, hail, lightning, and violent turbulence. It is recommended that pilots avoid severe thunderstorms by 20 NM. In this case, the storm is to the west of the airport and reaching the mature stage. The mature stage is characterized by both strong updrafts and downdrafts. The pilot of this aircraft could expect the potential of a strong wind from the west, making Runway 26 the best option. Runway 26 also keeps the aircraft on the eastern side of the airport away from the storm to the west.
Answer (A) is incorrect. Flying a left pattern for Runway 08 would put the aircraft closer to or even under the storm. Answer (B) is incorrect. Flying a left pattern for Runway 08 would put the aircraft closer to or even under the storm. The pilot can expect the winds to be from the west, not the east. Westerly winds would favor Runway 26.

7.5 Icing

19. One in-flight condition necessary for structural icing to form is

   A. small temperature/dewpoint spread.

   B. stratiform clouds.

   C. visible moisture.

Answer (C) is correct. *(AvW Chap 10)*
   DISCUSSION: Two conditions are necessary for structural icing while in flight. First, the airplane must be flying through visible moisture, such as rain or cloud droplets. Second, the temperature at the point where the moisture strikes the airplane must be freezing or below.
   Answer (A) is incorrect. The temperature dew point spread is not a factor in icing as it is in the formation of fog or clouds. Answer (B) is incorrect. No special cloud formation is necessary for icing as long as visible moisture is present.

20. In which environment is aircraft structural ice most likely to have the highest accumulation rate?

   A. Cumulus clouds with below freezing temperatures.

   B. Freezing drizzle.

   C. Freezing rain.

Answer (C) is correct. *(AvW Chap 10)*
   DISCUSSION: Freezing rain usually causes the highest accumulation rate of structural icing because of the nature of the supercooled water striking the airplane.
   Answer (A) is incorrect. While icing potential is great in cumulus clouds with below freezing temperatures, the highest accumulation rate is in an area with large, supercooled water drops (i.e., freezing rain). Answer (B) is incorrect. Freezing drizzle will not build up ice as quickly as freezing rain.

21. The presence of ice pellets at the surface is evidence that there

   A. are thunderstorms in the area.

   B. has been cold frontal passage.

   C. is a temperature inversion with freezing rain at a higher altitude.

Answer (C) is correct. *(AvW Chap 5)*
   DISCUSSION: Rain falling through colder air may freeze during its descent, falling as ice pellets. Ice pellets always indicate freezing rain at a higher altitude.
   Answer (A) is incorrect. Ice pellets form when rain freezes during its descent, which may or may not be as a result of a thunderstorm. Answer (B) is incorrect. Ice pellets only indicate that rain is freezing at a higher altitude, not that a cold front has passed through an area.

7.6 Mountain Wave

22. An almond or lens-shaped cloud which appears stationary, but which may contain winds of 50 knots or more, is referred to as

   A. an inactive frontal cloud.

   B. a funnel cloud.

   C. a lenticular cloud.

Answer (C) is correct. *(AvW Chap 9)*
   DISCUSSION: Lenticular clouds are lens-shaped clouds, which indicate the crests of standing mountain waves. They form in the updraft and dissipate in the downdraft, so they do not move as the wind blows through them. Lenticular clouds may contain winds of 50 kt. or more and are extremely dangerous.
   Answer (A) is incorrect. Frontal clouds usually do not contain winds of 50 kt. or more, and if they do, they do not appear stationary. Answer (B) is incorrect. A funnel cloud is not stationary.

23. Crests of standing mountain waves may be marked by stationary, lens-shaped clouds known as

   A. mammatocumulus clouds.

   B. standing lenticular clouds.

   C. roll clouds.

Answer (B) is correct. *(AvW Chap 9)*
   DISCUSSION: Lens-shaped clouds, which indicate crests of standing mountain waves, are called standing lenticular clouds. They form in the updraft and dissipate in the downdraft so that they do not move as the wind blows through them.
   Answer (A) is incorrect. Cumulonimbus mamma clouds (also mammatocumulus) are cumulonimbus clouds with pods or circular domes on the bottom that indicate severe turbulence. Answer (C) is incorrect. Roll clouds are low-level, turbulent areas in the shear zone between the plow wind surrounding the outrushing air from a thunderstorm and the surrounding air.

24. Possible mountain wave turbulence could be anticipated when winds of 40 knots or greater blow

    A. across a mountain ridge, and the air is stable.

    B. down a mountain valley, and the air is unstable.

    C. parallel to a mountain peak, and the air is stable.

Answer (A) is correct. *(AvW Chap 9)*
    DISCUSSION: Always anticipate possible mountain wave turbulence when the air is stable and winds of 40 kt. or greater blow across a mountain or ridge.
    Answer (B) is incorrect. The wind must blow across the mountain or ridge before it flows down a valley. The air must also be stable. If the air is unstable and produces convective turbulence, it will rise and disrupt the "wave." Answer (C) is incorrect. Any time the winds are 40 kt. or more and blowing across (not parallel) to the mountains, you should anticipate mountain wave turbulence.

7.7 Wind Shear

25. Where does wind shear occur?

    A. Only at higher altitudes.

    B. Only at lower altitudes.

    C. At all altitudes, in all directions.

Answer (C) is correct. *(AvW Chap 9)*
    DISCUSSION: Wind shear is the eddies in between two wind currents of differing velocities, direction, or both. Wind shear may be associated with either a wind shift or a wind speed gradient at any level in the atmosphere.
    Answer (A) is incorrect. A wind shear may occur at any (not only higher) altitudes. Answer (B) is incorrect. A wind shear may occur at any (not only lower) altitudes.

26. A pilot can expect a wind-shear zone in a temperature inversion whenever the windspeed at 2,000 to 4,000 feet above the surface is at least

    A. 10 knots.

    B. 15 knots.

    C. 25 knots.

Answer (C) is correct. *(AvW Chap 9)*
    DISCUSSION: When taking off or landing in calm wind under clear skies within a few hours before or after sunset, prepare for a temperature inversion near the ground. You can be relatively certain of a shear zone in the inversion if you know the wind is 25 kt. or more at 2,000 to 4,000 feet. Allow a margin of airspeed above normal climb or approach speed to alleviate the danger of stall in the event of turbulence or sudden change in wind velocity.
    Answer (A) is incorrect. A wind shear zone can be expected in a temperature inversion at 2,000 to 4,000 ft. AGL if the wind speed is at least 25 kt., not 10 kt. Answer (B) is incorrect. A wind shear zone can be expected in a temperature inversion at 2,000 to 4,000 ft. AGL if the wind speed is at least 25 kt., not 15 kt.

27. When may hazardous wind shear be expected?

    A. When stable air crosses a mountain barrier where it tends to flow in layers forming lenticular clouds.

    B. In areas of low-level temperature inversion, frontal zones, and clear air turbulence.

    C. Following frontal passage when stratocumulus clouds form indicating mechanical mixing.

Answer (B) is correct. *(AvW Chap 9)*
    DISCUSSION: Wind shear is the abrupt rate of change of wind velocity (direction and/or speed) per unit of distance and is normally expressed as vertical or horizontal wind shear. Hazardous wind shear may be expected in areas of low-level temperature inversion, frontal zones, and clear air turbulence.
    Answer (A) is incorrect. A mountain wave forms when stable air crosses a mountain barrier where it tends to flow in layers forming lenticular clouds. Turbulence, not wind shear, is expected in this area. Answer (C) is incorrect. Mechanical turbulence (not wind shear) may be expected following frontal passage when clouds form, indicating mechanical mixing.

7.8 Temperature/Dew Point and Fog

28. If the temperature/dewpoint spread is small and decreasing, and the temperature is 62°F, what type weather is most likely to develop?

    A. Freezing precipitation.

    B. Thunderstorms.

    C. Fog or low clouds.

Answer (C) is correct. *(AvW Chap 5)*
    DISCUSSION: The difference between the air temperature and dew point is the temperature/dew point spread. As the temperature/dew point spread decreases, fog or low clouds tend to develop.
    Answer (A) is incorrect. There cannot be freezing precipitation if the temperature is 62°F. Answer (B) is incorrect. Thunderstorms have to do with unstable lapse rates, not temperature/dew point spreads.

29. What is meant by the term "dewpoint"?

   A. The temperature at which condensation and evaporation are equal.
   B. The temperature at which dew will always form.
   C. The temperature to which air must be cooled to become saturated.

Answer (C) is correct. *(AvW Chap 5)*
   DISCUSSION: Dew point is the temperature to which air must be cooled to become saturated or have 100% humidity.
   Answer (A) is incorrect. Evaporation is the change from water to water vapor and is not directly related to the dew point. Answer (B) is incorrect. Dew forms only when heat radiates from an object whose temperature lowers below the dew point of the adjacent air.

30. The amount of water vapor which air can hold depends on the

   A. dewpoint.
   B. air temperature.
   C. stability of the air.

Answer (B) is correct. *(AvW Chap 5)*
   DISCUSSION: Air temperature largely determines how much water vapor can be held by the air. Warm air can hold more water vapor than cool air.
   Answer (A) is incorrect. Dew point is the temperature at which air must be cooled to become saturated by the water vapor already present in the air. Answer (C) is incorrect. Air stability is the state of the atmosphere at which vertical distribution of temperature is such that air particles will resist displacement from their initial level.

31. What are the processes by which moisture is added to unsaturated air?

   A. Evaporation and sublimation.
   B. Heating and condensation.
   C. Supersaturation and evaporation.

Answer (A) is correct. *(AvW Chap 5)*
   DISCUSSION: Evaporation is the process of converting a liquid to water vapor, and sublimation is the process of converting ice to water vapor.
   Answer (B) is incorrect. Heating alone does not add moisture. Condensation is the change of water vapor to liquid water. Answer (C) is incorrect. Supersaturation is a nonsense term in this context.

32. Which conditions result in the formation of frost?

   A. The temperature of the collecting surface is at or below freezing when small droplets of moisture fall on the surface.
   B. The temperature of the collecting surface is at or below the dewpoint of the adjacent air and the dewpoint is below freezing.
   C. The temperature of the surrounding air is at or below freezing when small drops of moisture fall on the collecting surface.

Answer (B) is correct. *(AvW Chap 5)*
   DISCUSSION:. Frost forms when both the collecting surface is below the dew point of the adjacent air and the dew point is below freezing. Frost is the direct sublimation of water vapor to ice crystals.
   Answer (A) is incorrect. If small droplets of water fall on the collecting surface, which is at or below freezing, ice (not frost) will form. Answer (C) is incorrect. If small droplets of water fall while the surrounding air is at or below freezing, ice (not frost) will form.

33. Clouds, fog, or dew will always form when

   A. water vapor condenses.
   B. water vapor is present.
   C. relative humidity reaches 100 percent.

Answer (A) is correct. *(AvW Chap 5)*
   DISCUSSION: As water vapor condenses, it becomes visible as clouds, fog, or dew.
   Answer (B) is incorrect. Water vapor is usually always present but does not form clouds, fog, or dew without condensation. Answer (C) is incorrect. Even at 100% humidity, water vapor may not condense, e.g., sufficient condensation nuclei may not be present.

34. Low-level turbulence can occur and icing can become hazardous in which type of fog?

   A. Rain-induced fog.
   B. Upslope fog.
   C. Steam fog.

Answer (C) is correct. *(AvW Chap 14)*
   DISCUSSION: Steam fog forms in winter when cold, dry air passes from land areas over comparatively warm ocean waters and is composed entirely of water droplets that often freeze quickly. Low-level turbulence can occur, and icing can become hazardous.
   Answer (A) is incorrect. Precipitation- (rain-) induced fog is formed when relatively warm rain or drizzle falls through cool air and evaporation from the precipitation saturates the cool air and forms fog. While the hazards of turbulence and icing may occur in the proximity of rain-induced fog, these hazards occur as a result of the steam fog formation process. Answer (B) is incorrect. Upslope fog forms when moist, stable air is cooled as it moves up sloping terrain.

35. In which situation is advection fog most likely to form?

    A. A warm, moist air mass on the windward side of mountains.

    B. An air mass moving inland from the coast in winter.

    C. A light breeze blowing colder air out to sea.

Answer (B) is correct. *(AvW Chap 12)*
    DISCUSSION: Advection fog forms when moist air moves over colder ground or water. It is most common in coastal areas.
    Answer (A) is incorrect. A warm, moist air mass on the windward side of mountains produces rain or upslope fog as it blows upward and cools. Answer (C) is incorrect. A light breeze blowing colder air out to sea causes steam fog.

36. What situation is most conducive to the formation of radiation fog?

    A. Warm, moist air over low, flatland areas on clear, calm nights.

    B. Moist, tropical air moving over cold, offshore water.

    C. The movement of cold air over much warmer water.

Answer (A) is correct. *(AvW Chap 12)*
    DISCUSSION: Radiation fog is shallow fog of which ground fog is one form. It occurs under conditions of clear skies, little or no wind, and a small temperature/dew point spread. The fog forms almost exclusively at night or near dawn as a result of terrestrial radiation cooling the ground and the ground cooling the air on contact with it.
    Answer (B) is incorrect. Moist, tropical air moving over cold, offshore water causes advection fog, not radiation fog. Answer (C) is incorrect. Movement of cold, dry air over much warmer water results in steam fog.

37. What types of fog depend upon wind in order to exist?

    A. Radiation fog and ice fog.

    B. Steam fog and ground fog.

    C. Advection fog and upslope fog.

Answer (C) is correct. *(AvW Chap 14)*
    DISCUSSION: Advection fog forms when moist air moves over colder ground or water. It is most common in coastal areas. Upslope fog forms when wind blows moist air upward over rising terrain and the air cools below its dew point. Both advection fog and upslope fog require wind to move air masses.
    Answer (A) is incorrect. No wind is required for the formation of either radiation (ground) or ice fog. Answer (B) is incorrect. No wind is required for the formation of ground (radiation) fog.

7.9 Clouds

38. Clouds are divided into four families according to their

    A. outward shape.

    B. height range.

    C. composition.

Answer (B) is correct. *(AvW Chap 7)*
    DISCUSSION: The four families of clouds are high clouds, middle clouds, low clouds, and clouds with extensive vertical development. Thus, they are based upon their height range.
    Answer (A) is incorrect. Clouds are divided by their height range, not outward shape. Answer (C) is incorrect. Clouds are divided by their height range, not their composition.

39. The suffix "nimbus," used in naming clouds, means

    A. a cloud with extensive vertical development.

    B. a rain cloud.

    C. a middle cloud containing ice pellets.

Answer (B) is correct. *(AvW Chap 7)*
    DISCUSSION: The suffix "nimbus" or the prefix "nimbo" means a rain cloud.
    Answer (A) is incorrect. Clouds with extensive vertical development are called either towering cumulus or cumulonimbus. Answer (C) is incorrect. A middle cloud has the prefix "alto."

40. The conditions necessary for the formation of cumulonimbus clouds are a lifting action and

- A. unstable air containing an excess of condensation nuclei.
- B. unstable, moist air.
- C. either stable or unstable air.

Answer (B) is correct. *(AvW Chap 11)*
   DISCUSSION: Unstable, moist air, in addition to a lifting action, i.e., convective activity, is needed to form cumulonimbus clouds.
   Answer (A) is incorrect. There must be moisture available to produce the clouds and rain; e.g., in a hot, dry dust storm, there would be no thunderstorm. Answer (C) is incorrect. The air must be unstable or there will be no lifting action.

41. What clouds have the greatest turbulence?

- A. Towering cumulus.
- B. Cumulonimbus.
- C. Nimbostratus.

Answer (B) is correct. *(AvW Chap 7)*
   DISCUSSION: The greatest turbulence occurs in cumulonimbus clouds, which are thunderstorm clouds.
   Answer (A) is incorrect. Towering cumulus clouds are an earlier stage of cumulonimbus clouds. Answer (C) is incorrect. Nimbostratus is a gray or dark, massive cloud layer diffused by continuous rain or ice pellets. It is a middle cloud with very little turbulence but may pose serious icing problems.

42. What cloud types would indicate convective turbulence?

- A. Cirrus clouds.
- B. Nimbostratus clouds.
- C. Towering cumulus clouds.

Answer (C) is correct. *(AvW Chap 7)*
   DISCUSSION: Towering cumulus clouds are an early stage of cumulonimbus clouds, or thunderstorms, that are based on convective turbulence, i.e., an unstable lapse rate.
   Answer (A) is incorrect. Cirrus clouds are high, thin, featherlike ice crystal clouds in patches and narrow bands that are not based on any convective activity. Answer (B) is incorrect. Nimbostratus are gray or dark, massive clouds diffused by continuous rain or ice pellets with very little turbulence.

43. At approximately what altitude above the surface would the pilot expect the base of cumuliform clouds if the surface air temperature is 82°F and the dewpoint is 38°F?

- A. 9,000 feet AGL.
- B. 10,000 feet AGL.
- C. 11,000 feet AGL.

Answer (B) is correct. *(AvW Chap 6)*
   DISCUSSION: The height of cumuliform cloud bases can be estimated using surface temperature/dew point spread. Unsaturated air in a convective current cools at about 5.4°F/1,000 ft., and dew point decreases about 1°F/1,000 feet. In a convective current, temperature and dew point converge at about 4.4°F/1,000 feet. Thus, if the temperature/dew point spread is 44°F (82°F – 38°F), divide 44 by 4.4 to obtain 10,000 ft. AGL.
   Answer (A) is incorrect. This is the approximate height of the base of cumuliform clouds if the temperature/dew point spread is 40°F. Answer (C) is incorrect. This is the approximate height of the base of cumuliform clouds if the temperature/dew point spread is 48°F.

44. What is the approximate base of the cumulus clouds if the surface air temperature at 1,000 feet MSL is 70°F and the dewpoint is 48°F?

- A. 4,000 feet MSL.
- B. 5,000 feet MSL.
- C. 6,000 feet MSL.

Answer (C) is correct. *(AvW Chap 6)*
   DISCUSSION: The height of cumuliform cloud bases can be estimated using surface temperature/dew point spread. Unsaturated air in a convective current cools at about 5.4°F/1,000 ft., and dew point decreases about 1°F/1,000 feet. In a convective current, temperature and dew point converge at about 4.4°F/1,000 feet. Thus, if the temperature and dew point are 70°F and 48°F, respectively, at 1,000 ft. MSL, there would be a 22°F spread that, divided by the lapse rate of 4.4, is approximately 5,000 ft. AGL, or 6,000 ft. MSL (5,000 + 1,000).
   Answer (A) is incorrect. This is the approximate base of the cumulus clouds if the temperature at 1,000 ft. MSL is 61°F, not 70°F. Answer (B) is incorrect. The figure of 5,000 ft. AGL, not MSL, is the approximate base of the cumulus clouds.

7.10 Stability of Air Masses

45. What is a characteristic of stable air?

A. Stratiform clouds.

B. Unlimited visibility.

C. Cumulus clouds.

Answer (A) is correct.   *(AvW Chap 8)*
    DISCUSSION: Characteristics of a stable air mass include stratiform clouds, continuous precipitation, smooth air, and fair to poor visibility in haze and smoke.
    Answer (B) is incorrect. Restricted, not unlimited, visibility is an indication of stable air. Answer (C) is incorrect. Fair weather cumulus clouds indicate unstable conditions, not stable conditions.

46. Moist, stable air flowing upslope can be expected to

A. produce stratus type clouds.

B. cause showers and thunderstorms.

C. develop convective turbulence.

Answer (A) is correct.   *(AvW Chap 6)*
    DISCUSSION: Moist, stable air flowing upslope can be expected to produce stratus type clouds as the air cools adiabatically as it moves up sloping terrain.
    Answer (B) is incorrect. Showers and thunderstorms are characteristics of unstable (not stable) air. Answer (C) is incorrect. Convective turbulence is a characteristic of unstable (not stable) air.

47. If an unstable air mass is forced upward, what type clouds can be expected?

A. Stratus clouds with little vertical development.

B. Stratus clouds with considerable associated turbulence.

C. Clouds with considerable vertical development and associated turbulence.

Answer (C) is correct.   *(AvW Chap 6)*
    DISCUSSION: When unstable air is lifted, it usually results in considerable vertical development and associated turbulence, i.e., convective activity.
    Answer (A) is incorrect. Stable rather than unstable air creates stratus clouds with little vertical development.
Answer (B) is incorrect. Stratus (layer-type) clouds usually have little turbulence unless they are either lenticular clouds created by mountain waves or other high-altitude clouds associated with high winds near or in the jet stream.

48. What are characteristics of unstable air?

A. Turbulence and good surface visibility.

B. Turbulence and poor surface visibility.

C. Nimbostratus clouds and good surface visibility.

Answer (A) is correct.   *(AvW Chap 8)*
    DISCUSSION: Characteristics of an unstable air mass include cumuliform clouds, showery precipitation, turbulence, and good visibility, except in blowing obstructions.
    Answer (B) is incorrect. Poor surface visibility is a characteristic of stable (not unstable) air. Answer (C) is incorrect. Stratus clouds are characteristic of stable (not unstable) air.

49. A stable air mass is most likely to have which characteristic?

A. Showery precipitation.

B. Turbulent air.

C. Poor surface visibility.

Answer (C) is correct.   *(AvW Chap 8)*
    DISCUSSION: Characteristics of a stable air mass include stratiform clouds and fog, continuous precipitation, smooth air, and fair to poor visibility in haze and smoke.
    Answer (A) is incorrect. Showery precipitation is a characteristic of an unstable (not stable) air mass. Answer (B) is incorrect. Turbulent air is a characteristic of an unstable (not stable) air mass.

50. Steady precipitation preceding a front is an indication of

A. stratiform clouds with moderate turbulence.

B. cumuliform clouds with little or no turbulence.

C. stratiform clouds with little or no turbulence.

Answer (C) is correct.   *(AvW Chap 8)*
    DISCUSSION: Steady precipitation preceding a front is usually an indication of a warm front, which results from warm air being cooled from the bottom by colder air. This results in stratiform clouds with little or no turbulence.
    Answer (A) is incorrect. Stratiform clouds usually are not turbulent. Answer (B) is incorrect. Cumuliform clouds have showery rather than steady precipitation.

51. What are characteristics of a moist, unstable air mass?

    A. Cumuliform clouds and showery precipitation.

    B. Poor visibility and smooth air.

    C. Stratiform clouds and showery precipitation.

Answer (A) is correct. *(AvW Chap 8)*
    DISCUSSION: Characteristics of an unstable air mass include cumuliform clouds, showery precipitation, turbulence, and good visibility, except in blowing obstructions.
    Answer (B) is incorrect. Poor visibility and smooth air are characteristics of stable (not unstable) air. Answer (C) is incorrect. Stratiform clouds and continuous precipitation are characteristics of stable (not unstable) air.

52. What measurement can be used to determine the stability of the atmosphere?

    A. Atmospheric pressure.

    B. Actual lapse rate.

    C. Surface temperature.

Answer (B) is correct. *(AvW Chap 6)*
    DISCUSSION: The stability of the atmosphere is determined by vertical movements of air. Warm air rises when the air above is cooler. The actual lapse rate, which is the decrease of temperature with altitude, is therefore a measure of stability.
    Answer (A) is incorrect. Atmospheric pressure is the pressure exerted by the atmosphere as a consequence of gravitational attraction exerted upon the "column" of air lying directly above the point in question. It cannot be used to determine stability. Answer (C) is incorrect. While the surface temperature may have some effect on temperature changes and air movements, it is the actual lapse rate that determines the stability of the atmosphere.

53. What would decrease the stability of an air mass?

    A. Warming from below.

    B. Cooling from below.

    C. Decrease in water vapor.

Answer (A) is correct. *(AvW Chap 6)*
    DISCUSSION: When air is warmed from below, even though cooling adiabatically, it remains warmer than the surrounding air. The colder, more dense surrounding air forces the warmer air upward, and an unstable condition develops.
    Answer (B) is incorrect. Cooling from below means the surrounding air is warmer, which would increase (not decrease) the stability of an air mass. Answer (C) is incorrect. As water vapor in air decreases, the air mass tends to increase (not decrease) stability.

7.11 Temperature Inversions

54. What feature is associated with a temperature inversion?

    A. A stable layer of air.

    B. An unstable layer of air.

    C. Chinook winds on mountain slopes.

Answer (A) is correct. *(AvW Chap 6)*
    DISCUSSION: A temperature inversion is associated with an increase in temperature with height, a reversal of normal decrease in temperature with height. Thus, any warm air rises to where it is the same temperature and forms a stable layer of air.
    Answer (B) is incorrect. Instability is a result of rising air remaining warmer than the surrounding air aloft, which would not occur with a temperature inversion. Answer (C) is incorrect. A Chinook wind is a warm, dry downslope wind blowing down the eastern slopes of the Rocky Mountains over the adjacent plains in the U.S. and Canada.

55. The most frequent type of ground or surface-based temperature inversion is that which is produced by

    A. terrestrial radiation on a clear, relatively still night.

    B. warm air being lifted rapidly aloft in the vicinity of mountainous terrain.

    C. the movement of colder air under warm air, or the movement of warm air over cold air.

Answer (A) is correct. *(AvW Chap 2)*
    DISCUSSION: An inversion often develops near the ground on clear, cool nights when wind is light. The ground loses heat and cools the air near the ground while the temperature a few hundred feet above changes very little. Thus, temperature increases in height, which is an inversion.
    Answer (B) is incorrect. Warm air being lifted rapidly aloft in the vicinity of mountainous terrain describes convective activity. Answer (C) is incorrect. The movement of colder air under warm air, which causes an inversion, is caused by a cold front, not terrestrial radiation (warm air moving over cold air is a warm front).

56. A temperature inversion would most likely result in which weather condition?

    A. Clouds with extensive vertical development above an inversion aloft.

    B. Good visibility in the lower levels of the atmosphere and poor visibility above an inversion aloft.

    C. An increase in temperature as altitude is increased.

Answer (C) is correct.   *(AvW Chap 2)*
    DISCUSSION: By definition, a temperature inversion is a situation in which the temperature increases as altitude increases. The normal situation is that the temperature decreases as altitude increases.
    Answer (A) is incorrect. Vertical development does not occur in an inversion situation because the warm air cannot rise when the air above is warmer. Answer (B) is incorrect. The inversion traps dust, smoke, and other nuclei beneath the inversion, which reduces visibility.

57. Which weather conditions should be expected beneath a low-level temperature inversion layer when the relative humidity is high?

    A. Smooth air, poor visibility, fog, haze, or low clouds.

    B. Light wind shear, poor visibility, haze, and light rain.

    C. Turbulent air, poor visibility, fog, low stratus type clouds, and showery precipitation.

Answer (A) is correct.   *(AvW Chap 12)*
    DISCUSSION: Beneath temperature inversions, there is usually smooth air because there is little vertical movement due to the inversion. There is also poor visibility due to fog, haze, and low clouds (when there is high relative humidity).
    Answer (B) is incorrect. Wind shears usually do not occur below a low-level temperature inversion. They occur at or just above the inversion. Answer (C) is incorrect. Turbulent air and showery precipitation are not present with low-level temperature inversions.

END OF STUDY UNIT

STUDY UNIT EIGHT
AVIATION WEATHER SERVICES

(9 pages of outline)

This study unit contains outlines of major concepts tested, sample test questions and answers regarding aviation weather services, and an explanation of each answer. The table of contents above lists each subunit within this study unit, the number of questions pertaining to that particular subunit, and the pages on which the outlines and questions begin, respectively.

CAUTION: Recall that the **sole purpose** of this book is to expedite your passing the FAA pilot knowledge test for the private pilot certificate. Accordingly, all extraneous material (i.e., topics or regulations not directly tested on the FAA pilot knowledge test) is omitted, even though much more information and knowledge are necessary to fly safely. This additional material is presented in *Pilot Handbook* and *Private Pilot Flight Maneuvers and Practical Test Prep*, available from Gleim Publications, Inc. See the order form on page 371.

8.1 WEATHER BRIEFINGS

1. When requesting a telephone weather briefing, you should identify

    a. Yourself as a pilot
    b. Your intended route
    c. Your intended destination
    d. Whether you are flying VFR or IFR
    e. Type of aircraft
    f. Proposed departure time and time en route

2. A standard briefing should be obtained before every flight. This briefing will provide all the necessary information for a safe flight.

3. An outlook briefing is provided when it is 6 or more hours before proposed departure time.

4. An abbreviated briefing will be provided when the user requests information to

    a. Supplement mass disseminated data,
    b. Update a previous briefing, or
    c. Be limited to specific information.

8.2 AVIATION ROUTINE WEATHER REPORT (METAR)

1. Aviation routine weather reports (METARs) are actual weather observations at the time indicated on the report. There are two types of reports:

    a. **METAR** is a routine weather report.
    b. **SPECI** is a nonroutine weather report.

2. Following the type of report are the elements listed below:

    a. The four-letter ICAO station identifier.

        1) In the contiguous 48 states, the three-letter domestic identifier is prefixed with a "K."

    b. Date and time of report. It is appended with a "Z" to denote Coordinated Universal Time (UTC).

    c. Modifier (if required).

    d. Wind. Wind is reported as a five-digit group (six digits if the wind speed is greater than 99 knots). It is appended with the abbreviation KT to denote the use of knots for wind speed.

        1) If the wind is gusty, it is reported as a "G" after the speed, followed by the highest gust reported.

        2) EXAMPLE: **11012G18KT** means wind from 110° true at 12 kt. with gusts to 18 knots.

    e. Visibility. Prevailing visibility is reported in statute miles with "SM" appended to it.

        1) EXAMPLE: **1 1/2SM** means visibility 1 1/2 statute miles.

    f. Runway visual range.

    g. Weather phenomena.

        1) **RA** is used to indicate rain.

    h. Sky conditions.

        1) The ceiling is the lowest broken or overcast layer, or vertical visibility into an obscuration.

        2) Cloud bases are reported with three digits in hundreds of feet AGL.

            a) EXAMPLE: **OVC007** means overcast cloud layer at 700 ft. AGL.

    i. Temperature/dew point. They are reported in a two-digit form in whole degrees Celsius separated by a solidus, "/."

    j. Altimeter.

    k. Remarks (RMK).

        1) **RAB35** means rain began at 35 min. past the hour.

3. EXAMPLE: METAR KAUS 301651Z 12008KT 4SM -RA HZ BKN010 OVC023 21/17 A3005 RMK RAB25

    a. **METAR** is a routine weather observation.
    b. **KAUS** is Austin, TX.
    c. **301651Z** means the observation was taken on the 30th day at 1651 UTC (or Z).
    d. **12008KT** means the wind is from 120° true at 8 knots.
    e. **4SM** means the visibility is 4 statute miles.
    f. **-RA HZ** means light rain and haze.
    g. **BKN010 OVC023** means ceiling 1,000 ft. broken, 2,300 ft. overcast.
    h. **21/17** means the temperature is 21°C and the dew point is 17°C.
    i. **A3005** means the altimeter setting is 30.05 in. of Hg.
    j. **RMK RAB25** means remarks, rain began at 25 min. past the hour., i.e., 1625 UTC.

8.3 PILOT WEATHER REPORT (PIREP)

1. No observation is more timely or needed than the one you make from the cockpit.

2. PIREPs are transmitted in the format illustrated below.

UUA/UA	Type of report:
	URGENT (UUA) - Any PIREP that contains any of the following weather phenomena: tornadoes, funnel clouds, or waterspouts; severe or extreme turbulence, including clear air turbulence (CAT); severe icing; hail; low-level wind shear (LLWS) (pilot reports air speed fluctuations of 10 knots or more within 2,000 feet of the surface); any other weather phenomena reported that are considered by the controller to be hazardous, or potentially hazardous, to flight operations. ROUTINE (UA) - Any PIREP that contains weather phenomena not listed above, including low-level wind shear reports with air speed fluctuations of less than 10 knots.
/OV	Location: Use VHF NAVAID(s) or an airport using the three- or four-letter location identifier. Position can be over a site, at some location relative to a site, or along a route. Ex: /OV KABC; /OV KABC090025; /OV KABC045020-DEF; /OV KABC-KDEF
/TM	Time: Four digits in UTC. Ex: /TM 0915
/FL	Altitude/Flight level: Three digits for hundreds of feet with no space between FL and altitude. If not known, use UNKN. Ex: /FL095; /FL310; /FLUNKN
/TP	Aircraft type: Four digits maximum; if not known, use UNKN. Ex: /TP L329; /TP B737; /TP UNKN
/SK	Sky cover: Describes cloud amount, height of cloud bases, and height of cloud tops. If unknown, use UNKN. Ex: /SK SCT040-TOP080; /SK BKNUNKN-TOP075; /SK BKN-OVC050-TOPUNKN; /SK OVCUNKN-TOP085
/WX	Flight visibility and weather: Flight visibility (FV) reported first and use standard METAR weather symbols. Intensity (− for light, no qualifier for moderate, and + for heavy) shall be coded for all precipitation types except ice crystals and hail. Ex: /WX FV05SM -RA; /WX FV01 SN BR; /WX RA
/TA	Temperature (Celsius): If below zero, prefix with an "M." Temperature should also be reported if icing is reported. Ex: /TA 15; /TA M06
/WV	Wind: Direction from which the wind is blowing, coded in tens of degrees using three digits. Directions of less than 100 degrees shall be preceded by a zero. The wind speed shall be entered as a two- or three-digit group immediately following the direction, coded in whole knots using the hundreds, tens, and units digits. Ex: /WV 27045KT; /WV 280110KT
/TB	Turbulence: Use standard contractions for intensity and type (CAT or CHOP when appropriate). Include altitude only if different from FL. Ex: /TB EXTRM; /TB OCNL LGT-MDT BLO 090; /TB MOD-SEV CHOP 080-110
/IC	Icing: Describe using standard intensity and type contractions. Include altitude only if different from FL. Ex: /IC LGT-MDT RIME; /IC SEV CLR 028-045
/RM	Remarks: Use free form to clarify the report, putting hazardous elements first Ex: /RM LLWS −15 KT SFC-030 DURGC RY 22 JFK

3. All heights are given as MSL. To determine AGL, subtract the field height from the given height.

4.   Turbulence is reported as

    a.   Light = LGT
    b.   Moderate = MDT
    c.   Severe = SVR

5.   Icing is reported as

    a.   Clear = CLR
    b.   Rime = RIME

6.   Cloud layers are reported with heights for bases, tops, and layer type if available. "No entry" means that information was not given.

    a.   EXAMPLE: SK 024 BKN 032/042 BKN-OVC decoded means a broken layer 2,400 ft. MSL to 3,200 ft. MSL. A second layer is broken to overcast starting at 4,200 ft. MSL.

7.   Wind direction and velocity are given as a five- or six-digit code (e.g., **/WV 27045** means 270° at 45 knots).

8.   Air temperature is expressed in degrees Celsius (°C).

8.4 AVIATION AREA FORECAST

1.   Aviation area forecasts (FAs) are forecasts of visual meteorological conditions (VMC), clouds, and general weather conditions for several states and/or portions of states, generally for an area greater than 3,000 square miles. They can be used to interpolate conditions at airports that have no terminal forecasts. FAs are issued three times a day and consist of

    a.   A 12-hr. forecast, and
    b.   An additional 6-hr. categorical outlook.

2.   FA weather format. An example is presented on page 240 for questions 26 through 29. It is presented in abbreviations. You will see this same FA utilizing abbreviations on your pilot knowledge test.

3.   There are four sections in an FA:

    a.   Communication and product header section
    b.   Precautionary statements section
    c.   Synopsis section (for the purposes of the test, not considered a forecast)
    d.   VFR clouds/weather section (VFR CLDS/WX), referred to as the forecast section

        1)   Included in the VFR CLDS/WX section is a categorical outlook that is valid for an additional 6 hours. For the purposes of the test, the categorical outlook is not considered a forecast.

4.   In order to get a complete weather picture, including icing, turbulence, and IFR conditions, an FA must be supplemented by In-Flight Aviation Weather Advisories (AIRMETs Zulu, Tango, and Sierra).

    a.   A pilot should refer to the In-Flight Aviation Weather Advisories to determine the freezing level and areas of probable icing aloft.

8.5 TERMINAL AERODROME FORECAST (TAF)

1. Terminal aerodrome forecasts (TAFs) are weather forecasts for selected airports throughout the country.

2. The elements of a TAF are listed below:

    a. Type of report

        1) **TAF** is a routine forecast.
        2) **TAF AMD** is an amended forecast.

    b. ICAO station identifier

    c. Date and time the forecast is actually prepared

    d. Valid period of the forecast

    e. Forecast meteorological conditions. This is the body of the forecast and includes the following:

        1) Wind
        2) Visibility
        3) Weather
        4) Sky condition

            a) Cumulonimbus clouds (CB) are the only cloud type forecast in TAFs.

3. EXAMPLE:
    TAF
    KBRO 300545Z 300606 VRB04KT 3SM SCT040 OVC150 TEMPO 2124 SHRA
       FM0200 10010KT P6SM OVC020 BECMG0306 NSW BKN020=

    a. **TAF** is a routine forecast.

    b. **KBRO** is Brownsville, TX.

    c. **300545Z** means the forecast was prepared on the 30th day at 0545 UTC.

    d. **300606** means the forecast is valid from the 30th day at 0600 UTC until 0600 UTC the following day.

    e. **VRB04KT 3SM SCT040 OVC150 TEMPO 2124 SHRA** means the forecast from 0600 until 0200 UTC is wind variable in direction at 4 kt., visibility 3 SM, scattered cloud layer at 4,000 ft., ceiling 15,000 ft. overcast, with occasional rain showers between 2100 and 2400 UTC.

    f. **FM0200 10010KT P6SM OVC020 BECMG0306 NSW BKN020=** means the forecast from 0200 until 0300 is wind 100° true at 10 kt., visibility greater than 6 SM, ceiling 2,000 ft. overcast then becoming no significant weather, ceiling 2,000 ft. broken between 0300 and 0600 UTC.

        1) Note that, since the becoming group (BECMG) did not forecast wind and visibility, they are the same as the previous forecast group, i.e., wind 100° true at 10 kt., visibility greater than 6 statute miles.

8.6 WEATHER DEPICTION CHARTS

1. A weather depiction chart is an outline of the United States depicting sky conditions at the time stated on the chart based on METAR reports.

    a. Reporting stations are marked with a little circle.

        1) If the sky is clear, the circle is open; if overcast, the circle is solid; if scattered, the circle is 1/4 solid; if broken, the circle is 3/4 solid. If the sky is obscured, there is an "X" in the circle.

        2) The height of clouds is expressed in hundreds of feet above ground level; e.g., 120 means 12,000 ft. AGL.

2. Areas with ceilings below 1,000 ft. and/or visibility less than 3 SM, i.e., below VFR, are bracketed with solid black contour lines and are shaded.

    a. Visibility is indicated next to the circle; e.g., 2 stands for 2 SM visibility.

        1) If the visibility is greater than 6 SM, it is not reported.

    b. Areas of marginal VFR with ceilings of 1,000 to 3,000 ft. and/or visibility at 3 to 5 SM are bracketed by solid black contour lines and are unshaded.

    c. Ceilings greater than 3,000 ft. and visibility greater than 5 SM are not indicated by contour lines on weather depiction charts.

3. Station models and significant weather are indicated by the following symbols:

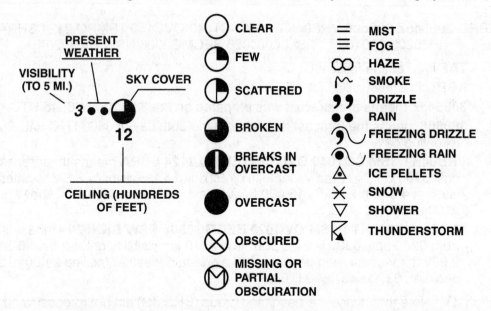

4. The weather depiction chart quickly shows pilots where weather conditions reported are above or below VFR minimums.

5. The weather depiction chart displays recent positions of frontal systems and indicates the type of front by symbols.

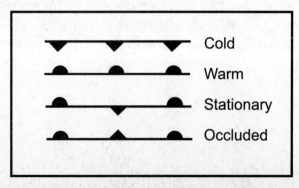

8.7 RADAR SUMMARY CHARTS AND RADAR WEATHER REPORTS

1. Radar summary charts graphically display a collection of radar reports concerning the type, intensity, and movement of precipitation, e.g., squall lines, specific thunderstorm cells, and other areas of hazardous precipitation.

    a. Lines and cells of hazardous thunderstorms can be seen on radar summary charts and are not shown on other weather charts.

2. The symbols below are used on radar summary charts.

Digit	PRECIPITATION INTENSITY	RAINFALL RATE in/hr STRATIFORM	RAINFALL RATE in/hr CONVECTIVE
1	LIGHT	LESS THAN 0.1	LESS THAN 0.2
2	MODERATE	0.1 - 0.5	0.2 - 1.1
3	HEAVY	0.5 - 1.0	1.1 - 2.2
4	VERY HEAVY	1.0 - 2.0	2.2 - 4.5
5	INTENSE	2.0 - 5.0	4.5 - 7.1
6	EXTREME	MORE THAN 5.0	MORE THAN 7.1

450

Highest precipitation top in area in hundreds of feet MSL (45,000 feet MSL).

———————— SYMBOLS USED ON CHART ————————

SYMBOL	SYMBOL MEANING	SYMBOL	SYMBOL MEANING
R	RAIN	35 (arrow)	CELL MOVEMENT TO THE NORTHEAST AT 35 KNOTS
RW	RAIN SHOWER		
S	SNOW	LM	LITTLE MOVEMENT
SW	SNOW SHOWER	WS999	SEVERE THUNDERSTORM WATCH NUMBER 999
T	THUNDERSTORM	WT210	TORNADO WATCH NUMBER 210
NA	NOT AVAILABLE	SLD	8/10 OR GREATER COVERAGE IN A LINE
NE	NO ECHOES	/	LINE OF ECHOES
OM	OUT FOR MAINTENANCE		

3. Severe weather watch areas are enclosed by a heavy dashed line, usually in the form of a rectangular box.

4. Radar weather reports are textual reports of weather radar observations.

    a. They include the type, intensity, location, and cell movement of precipitation.

5. Finally, it is important to remember that the intensity trend (increasing or weakening) is no longer coded on either the radar summary chart or radar weather report (SD/ROB).

8.8   EN ROUTE FLIGHT ADVISORY SERVICE (EFAS)

1.   En Route Flight Advisory Service (EFAS) provides weather advisories on 122.0 MHz below FL 180. It is called Flight Watch.

    a.   Generally, service is available from 6 a.m. to 10 p.m. local time.

    b.   EFAS provides information regarding actual weather and thunderstorm activity along a proposed route.

2.   It is designed to be a continual exchange of information on winds, turbulence, visibility, icing, etc., between pilots and weather briefers.

8.9   WIND AND TEMPERATURE ALOFT FORECASTS (FB)

1.   Forecast winds and temperatures are provided at specified altitudes for specific locations in the United States.

2.   A four-digit group (used when temperatures are not forecast) shows wind direction with reference to **true** north and the wind speed in **knots**.

    a.   The first two digits indicate the wind direction after a zero is added.

    b.   The next two digits indicate the wind speed.

    c.   No temperature is forecast for the 3,000-ft. level or for a level within 2,500 ft. AGL of the station.

3.   A six-digit group includes the forecast temperature aloft.

    a.   The last two digits indicate the temperature in degrees Celsius.

    b.   Plus or minus is indicated before the temperature, except at higher altitudes (above 24,000 ft. MSL) where it is always below freezing.

4.   When the wind speed is less than 5 kt., the forecast is coded 9900, which means that the wind is light and variable.

5.   When the wind speed is over 100 kt., the forecaster adds 50 to the direction and subtracts 100 from the speed. To decode, you must reverse the process. For example, 730649 = 230° (73–50) at 106 kt. (100 + 06) and –49° (above 24,000 ft.).

6.   An example forecast is provided on page 249 for questions 56 through 60.

8.10   SIGNIFICANT WEATHER PROGNOSTIC CHARTS

1.   Significant Weather Prognostic Charts contain four charts (panels).

    a.   The two upper panels forecast significant weather from the surface up to 24,000 ft.: one for 12 hr. and the other for 24 hr. from the time of issuance.

    b.   The two lower panels forecast surface conditions: one for 12 hr. and the other for 24 hr. from the time of issuance.

2.   The top panels show

    a.   Ceilings less than 1,000 ft. and/or visibility less than 3 SM (IFR) by a solid line around the area

    b.   Ceilings 1,000 to 3,000 ft. and/or visibility 3 to 5 SM (MVFR) by a scalloped line around the area

    c.   Moderate or greater turbulence by a broken line around the area

        1)   A peaked hat ⚞⚟ indicates moderate turbulence.

        2)   Altitudes are indicated on the chart; e.g., 180 means from surface to 18,000 feet.

    d.   Freezing levels, given by a dashed line corresponding to the height of the freezing level

3.   The bottom panels show the location of

   a.   Highs, lows, fronts

   b.   Other areas of significant weather

      1)   Unshaded outlined areas indicate precipitation covering half or less of the area.

      2)   Shaded outlined areas indicate precipitation covering more than half of the area.

      3)   Precipitation type and intensity is reported with standard symbols. Examples include

         a)   thunderstorms embedded in a larger area of continuous moderate rain

         b)   thunderstorms embedded in a larger area of intermittent moderate rain

         c)   continuous light to moderate snow

         d)   intermittent light to moderate snow

      4)   Precipitation symbols may be connected to an area of precipitation by an arrow if there is not sufficient room to place them in that area.

4.   These charts are used to determine areas to avoid (freezing levels and turbulence).

8.11 AIRMETs AND SIGMETs

1.   SIGMETs and AIRMETs are issued to notify pilots en route of the possibility of encountering hazardous flying conditions.

2.   SIGMET advisories include weather phenomena that are potentially hazardous to all aircraft.

   a.   Convective SIGMETs include

      1)   Tornadoes

      2)   Lines of thunderstorms

      3)   Embedded thunderstorms

      4)   Thunderstorm areas greater than or equal to thunderstorm intensity level 4 with an area coverage of 40% or more

      5)   Hail greater than or equal to 3/4 in. diameter

   b.   SIGMETs include

      1)   Severe or extreme turbulence or clear air turbulence (CAT) not associated with thunderstorms

      2)   Severe icing not associated with thunderstorms

      3)   Duststorms, sandstorms, or volcanic ash lowering visibility to less than 3 SM

      4)   Volcanic eruption

3.   AIRMETs apply to light (e.g., small single-engine) aircraft to notify of

   a.   Moderate icing

   b.   Moderate turbulence

   c.   Visibility less than 3 SM or ceilings less than 1,000 ft.

   d.   Sustained winds of 30 kt. or more at the surface

   e.   Extensive mountain obscurement

QUESTIONS AND ANSWER EXPLANATIONS

All of the private pilot knowledge test questions chosen by the FAA for release as well as additional questions selected by Gleim relating to the material in the previous outlines are reproduced on the following pages. These questions have been organized into the same subunits as the outlines. To the immediate right of each question are the correct answer and answer explanation. You should cover these answers and answer explanations while responding to the questions. Refer to the general discussion in the Introduction on how to take the FAA pilot knowledge test.

Remember that the questions from the FAA pilot knowledge test bank have been reordered by topic and organized into a meaningful sequence. Also, the first line of the answer explanation gives the citation of the authoritative source for the answer.

QUESTIONS

8.1 Weather Briefings

1. To get a complete weather briefing for the planned flight, the pilot should request

A. a general briefing.

B. an abbreviated briefing.

C. a standard briefing.

Answer (C) is correct. *(AWS Sect 1)*
DISCUSSION: To get a complete briefing before a planned flight, the pilot should request a standard briefing. This will include all pertinent information needed for a safe flight.
Answer (A) is incorrect. A general briefing is not standard terminology for any type of weather briefing. Answer (B) is incorrect. An abbreviated briefing is provided as a supplement to mass disseminated data or a previous briefing. It can also be used to obtain specific information.

2. Which type weather briefing should a pilot request, when departing within the hour, if no preliminary weather information has been received?

A. Outlook briefing.

B. Abbreviated briefing.

C. Standard briefing.

Answer (C) is correct. *(AWS Sect 1)*
DISCUSSION: A pilot should request a standard briefing anytime (s)he is planning a flight and has not received a previous briefing or has not received preliminary information through mass dissemination media (e.g., TWEB, PATWAS, etc.).
Answer (A) is incorrect. Outlook briefings are for flights 6 hr. or more in the future. Answer (B) is incorrect. Abbreviated briefings are to update previous briefings, supplement other data, or answer a specific inquiry.

3. Which type of weather briefing should a pilot request to supplement mass disseminated data?

A. An outlook briefing.

B. A supplemental briefing.

C. An abbreviated briefing.

Answer (C) is correct. *(AWS Sect 1)*
DISCUSSION: An abbreviated briefing will be provided when the user requests information to supplement mass disseminated data, to update a previous briefing, or to obtain specific information.
Answer (A) is incorrect. An outlook briefing should be requested if the proposed departure time is 6 hr. or more in the future. Answer (B) is incorrect. A supplemental briefing is not a standard type of briefing.

4. A weather briefing that is provided when the information requested is 6 or more hours in advance of the proposed departure time is

A. an outlook briefing.

B. a forecast briefing.

C. a prognostic briefing.

Answer (A) is correct. *(AWS Sect 1)*
DISCUSSION: An outlook briefing is given when the briefing is 6 or more hours before the proposed departure time.
Answer (B) is incorrect. A forecast briefing is not a type of weather briefing. Answer (C) is incorrect. A prognostic briefing is not a type of weather briefing.

5. What should pilots state initially when telephoning a weather briefing facility for preflight weather information?

    A. Tell the number of occupants on board.

    B. State their total flight time.

    C. Identify themselves as pilots.

Answer (C) is correct. *(AWS Sect 1)*
    DISCUSSION: When telephoning for a weather briefing, you should identify yourself as a pilot so the person can give you an aviation-oriented briefing. Many nonpilots call weather briefing facilities to get the weather for other activities.
    Answer (A) is incorrect. The number of occupants on board is information needed for a flight plan, not for a weather briefing. Answer (B) is incorrect. Total flight time is a question asked by insurance companies, not information needed for a weather briefing.

6. What should pilots state initially when telephoning a weather briefing facility for preflight weather information?

    A. The intended route of flight radio frequencies.

    B. The intended route of flight and destination.

    C. The address of the pilot in command.

Answer (B) is correct. *(AWS Sect 1)*
    DISCUSSION: By telling the briefer your intended route and destination, the briefer will be able to provide you a more relevant briefing.
    Answer (A) is incorrect. The radio frequencies to be used are the pilot's preflight responsibility, not the weather briefer's. Answer (C) is incorrect. The address of the pilot in command is information needed for a flight plan, not for a weather briefing.

7. When telephoning a weather briefing facility for preflight weather information, pilots should state

    A. the aircraft identification or the pilot's name.

    B. true airspeed.

    C. fuel on board.

Answer (A) is correct. *(AWS Sect 1)*
    DISCUSSION: When requesting a briefing, you should provide the briefer with the following information: VFR or IFR, aircraft identification or the pilot's name, aircraft type, departure point, route of flight, destination, altitude, estimated time of departure, and time en route or estimated time of arrival.
    Answer (B) is incorrect. True airspeed is information provided on a flight plan. Answer (C) is incorrect. Fuel on board is information provided on a flight plan.

8. When telephoning a weather briefing facility for preflight weather information, pilots should state

    A. the full name and address of the formation commander.

    B. that they possess a current pilot certificate.

    C. whether they intend to fly VFR only.

Answer (C) is correct. *(AWS Sect 1)*
    DISCUSSION: When telephoning for a weather briefing, you should identify yourself as a pilot and state the route, destination, type of airplane, and whether you intend to fly VFR or IFR to permit the weather briefer to give you the most complete briefing.
    Answer (A) is incorrect. The full name and address of the formation commander is information provided on a flight plan. Answer (B) is incorrect. You should state that you are a pilot, not that you possess a current pilot certificate.

9. To update a previous weather briefing, a pilot should request

    A. an abbreviated briefing.

    B. a standard briefing.

    C. an outlook briefing.

Answer (A) is correct. *(AWS Sect 1)*
    DISCUSSION: An abbreviated briefing will be provided when the user requests information (1) to supplement mass disseminated data, (2) to update a previous briefing, or (3) to be limited to specific information.
    Answer (B) is incorrect. A standard briefing is a complete preflight briefing to include all (not update) information pertinent to a safe flight. Answer (C) is incorrect. An outlook briefing is for a flight at least 6 hr. in the future.

10. When requesting weather information for the following morning, a pilot should request

    A. an outlook briefing.

    B. a standard briefing.

    C. an abbreviated briefing.

Answer (A) is correct. *(AWS Sect 1)*
    DISCUSSION: An outlook briefing should be requested when the briefing is 6 or more hours in advance of the proposed departure.
    Answer (B) is incorrect. A standard briefing should be requested if the proposed departure time is less than 6 hr. in the future and if you have not received a previous briefing or have received information through mass dissemination media. Answer (C) is incorrect. An abbreviated briefing is provided as a supplement to mass disseminated data, to update a previous briefing, or to obtain specific information.

8.2   Aviation Routine Weather Report (METAR)

11. For aviation purposes, ceiling is defined as the height above the Earth's surface of the

    A.   lowest reported obscuration and the highest layer of clouds reported as overcast.

    B.   lowest broken or overcast layer or vertical visibility into an obscuration.

    C.   lowest layer of clouds reported as scattered, broken, or thin.

Answer (B) is correct. *(AWS Sect 3)*
    DISCUSSION: A ceiling layer is not designated in the METAR code. For aviation purposes, the ceiling is the lowest broken or overcast layer, or vertical visibility into an obscuration.
    Answer (A) is incorrect. A ceiling is the lowest, not highest, broken or overcast layer, or the vertical visibility into an obscuration, not the lowest obscuration. Answer (C) is incorrect. A ceiling is the lowest broken or overcast, not scattered, layer. Also, there is no provision for reporting thin layers in the METAR code.

METAR KINK 121845Z 11012G18KT 15SM SKC 25/17 A3000

METAR KBOI 121854Z 13004KT 30SM SCT150 17/6 A3015

METAR KLAX 121852Z 25004KT 6SM BR SCT007 SCT250 16/15 A2991

SPECI KMDW 121856Z 32005KT 1 1/2SM RA OVC007 17/16 A2980 RMK RAB35

SPECI KJFK 121853Z 18004KT 1/2SM FG R04/2200 OVC005 20/18 A3006

Figure 12. – Aviation Routine Weather Reports (METAR).

---

12. (Refer to Figure 12 above.) Which of the reporting stations have VFR weather?

    A.   All.

    B.   KINK, KBOI, and KJFK.

    C.   KINK, KBOI, and KLAX.

Answer (C) is correct. *(AWS Sect 3)*
    DISCUSSION: KINK is reporting visibility of 15 SM and sky clear (15SM SKC); KBOI is reporting visibility of 30 SM and a scattered cloud layer base at 15,000 ft. (30SM SCT150); and KLAX is reporting visibility of 6SM in mist (fog) with a scattered cloud layer at 700 ft. and another one at 25,000 ft. (6SM BR SCT007 SCT250). All of these conditions are above VFR weather minimums of 1,000-ft. ceiling and/or 3-SM visibility.
    Answer (A) is incorrect. KMDW is reporting a visibility of 1 1/2 SM in rain and a ceiling of 700 ft. overcast (1 1/2SM RA OVC007), and KJFK is reporting a visibility of 1/2SM in fog and a ceiling of 500 ft. overcast (1/2SM FG OVC005). Both of these are below VFR weather minimums of 1,000-ft. ceiling and/or 3-SM visibility. Answer (B) is incorrect. KJFK is reporting a visibility of 1/2 SM in fog and a ceiling of 500 ft. overcast (1/2SM FG OVC005), which is below the VFR weather minimums of 1,000-ft. ceiling and/or 3-SM visibility.

13. (Refer to Figure 12 on page 236.) What are the current conditions depicted for Chicago Midway Airport (KMDW)?

A. Sky 700 feet overcast, visibility 1-1/2 SM, rain.

B. Sky 7000 feet overcast, visibility 1-1/2 SM, heavy rain.

C. Sky 700 feet overcast, visibility 11, occasionally 2 SM, with rain.

Answer (A) is correct. *(AWS Sect 3)*
   DISCUSSION: At KMDW a special METAR (SPECI) taken at 1856Z reported wind 320° at 5 kt., visibility 1 1/2 SM in moderate rain, overcast clouds at 700 ft., temperature 17°C, dew point 16°C, altimeter 29.80 in. Hg, remarks follow, rain began at 35 min. past the hour.
   Answer (B) is incorrect. The intensity of the rain is moderate, not heavy. Heavy rain would be coded +RA. Answer (C) is incorrect. Visibility is 1 1/2 SM, not 11 SM with an occasional 2 statute miles.

14. (Refer to Figure 12 on page 236.) The wind direction and velocity at KJFK is from

A. 180° true at 4 knots.

B. 180° magnetic at 4 knots.

C. 040° true at 18 knots.

Answer (A) is correct. *(AWS Sect 3)*
   DISCUSSION: The wind group at KJFK is coded as 18004KT. The first three digits are the direction the wind is blowing from referenced to true north. The next two digits are the speed in knots. Thus, the wind direction and speed at KJFK are 180° true at 4 knots.
   Answer (B) is incorrect. Wind direction is referenced to true, not magnetic, north. Answer (C) is incorrect. The wind direction is 180° true, not 040° true, at 4 knots, not 18 knots.

15. (Refer to Figure 12 on page 236.) What are the wind conditions at Wink, Texas (KINK)?

A. Calm.

B. 110° at 12 knots, gusts 18 knots.

C. 111° at 2 knots, gusts 18 knots.

Answer (B) is correct. *(AWS Sect 3)*
   DISCUSSION: The wind group at KINK is coded as 11012G18KT. The first three digits are the direction the wind is blowing from referenced to true north. The next two digits are the wind speed in knots. If the wind is gusty, it is reported as a "G" after the speed followed by the highest (or peak) gust reported. Thus, the wind conditions at KINK are 110° true at 12 knots, peak gust at 18 knots.
   Answer (A) is incorrect. A calm wind would be reported as 00000KT, not 11012G18KT. Answer (C) is incorrect. The wind conditions at KINK are 110°, not 111°, at 12 knots, not 2 knots.

16. (Refer to Figure 12 on page 236.) The remarks section for KMDW has RAB35 listed. This entry means

A. blowing mist has reduced the visibility to 1-1/2 SM.

B. rain began at 1835Z.

C. the barometer has risen .35" Hg.

Answer (B) is correct. *(AWS Sect 3)*
   DISCUSSION: In the remarks (RMK) section for KMDW, RAB35 means that rain began at 35 min. past the hour. Since the report was taken at 1856Z, rain began at 35 min. past the hour, or 1835Z.
   Answer (A) is incorrect. RAB35 means that rain began at 35 min. past the hour, not that blowing mist has reduced the visibility to 1 1/2 statute miles. Answer (C) is incorrect. RAB35 means that rain began at 35 min. past the hour, not that the barometer has risen .35 in. Hg.

8.3 Pilot Weather Report (PIREP)

17. (Refer to Figure 14 below.) If the terrain elevation is 1,295 feet MSL, what is the height above ground level of the base of the ceiling?

A. 505 feet AGL.

B. 1,295 feet AGL.

C. 6,586 feet AGL.

Answer (A) is correct.   *(AWS Sect 3)*
DISCUSSION: Refer to the PIREP (identified by the letters UA) in Fig. 14. The base of the ceiling is reported in the sky cover (SK) section. The first layer is considered a ceiling (i.e., broken), and the base is 1,800 ft. MSL. The height above ground of the broken base is 505 ft. AGL (1,800 ft. – 1,295 ft.).
Answer (B) is incorrect. The figure of 1,295 ft. MSL (not AGL) is the terrain elevation. Answer (C) is incorrect. The ceiling base is 505 ft. (not 6,586 ft.) AGL.

18. (Refer to Figure 14 below.) The base and tops of the overcast layer reported by a pilot are

A. 1,800 feet MSL and 5,500 feet MSL.

B. 5,500 feet AGL and 7,200 feet MSL.

C. 7,200 feet MSL and 8,900 feet MSL.

Answer (C) is correct.   *(AWS Sect 3)*
DISCUSSION: Refer to the PIREP (identified by the letters UA) in Fig. 14. The base and tops of the overcast layer are reported in the sky conditions (identified by the letters SK). This pilot has reported the base of the overcast layer at 7,200 ft. and the top of the overcast layer at 8,900 ft. (072 OVC 089). All altitudes are stated in MSL unless otherwise noted. Thus, the base and top of the overcast layer are reported as 7,200 ft. MSL and 8,900 ft. MSL, respectively.
Answer (A) is incorrect. The figures of 1,800 ft. MSL and 5,500 ft. MSL are the base and top of the broken (BKN), not overcast (OVC), layer. Answer (B) is incorrect. The figure of 5,500 ft. MSL (not AGL) is the top of the broken (BKN) layer, not the base of the overcast (OVC) layer.

19. (Refer to Figure 14 below.) The wind and temperature at 12,000 feet MSL as reported by a pilot are

A. 080° at 21 knots and –7°C.

B. 090° at 21 MPH and –9°F.

C. 090° at 21 knots and –9°C.

Answer (A) is correct.   *(AWS Sect 3)*
DISCUSSION: Refer to the PIREP (identified by the letters UA) in Fig. 14. The wind is reported in the section identified by the letters WV and is presented in five or six digits. The temperature is reported in the section identified by the letters TA in degrees Celsius, and if below 0°C, prefixed with an "M." The wind is reported as 080° at 21 kt. with a temperature of –7°C.
Answer (B) is incorrect. Speed is given in kt., not MPH. Temperature is given in degrees Celsius (not Fahrenheit) and is reported as –7, not –9. Answer (C) is incorrect. The wind is reported as being from 080°, not 090°, and the temperature is reported as –7°C, not –9°C.

20. (Refer to Figure 14 below.) The intensity of the turbulence reported at a specific altitude is

A. moderate at 5,500 feet and at 7,200 feet.

B. moderate from 5,500 feet to 7,200 feet.

C. light from 5,500 feet to 7,200 feet.

Answer (C) is correct.   *(AWS Sect 3)*
DISCUSSION: Refer to the PIREP (identified by the letters UA) in Fig. 14. The turbulence is reported in the section identified by the letters TB. In the PIREP the turbulence is reported as light from 5,500 ft. to 7,200 ft. (TB LGT 055-072).
Answer (A) is incorrect. Turbulence is reported from 5,500 to 7,200 ft. MSL, not only at 5,500 ft. and 7,200 feet. Answer (B) is incorrect. Rime ice (not turbulence) is reported as light to moderate from 7,200 to 8,900 ft. MSL (not 5,500 to 7,200 ft. MSL).

21. (Refer to Figure 14 below.) The intensity and type of icing reported by a pilot is

A. light to moderate.

B. light to moderate clear.

C. light to moderate rime.

Answer (C) is correct.   *(AWS Sect 3)*
DISCUSSION: Refer to the PIREP (identified by the letters UA) in Fig. 14. The icing conditions are reported following the letters IC. In this report, icing is reported as light to moderate rime (LGT-MDT RIME) from 7,200 to 8,900 ft. MSL (072-089).
Answer (A) is incorrect. The question asks not only for the intensity of the icing (light to moderate) but also the type, which is rime (RIME) ice. Answer (B) is incorrect. The type is rime (not clear) ice.

UA/OV KOKC-KTUL/TM 1800/FL120/TP BE90//SK BKN018-TOP055/OVC072-
TOP089/CLR ABV/TA M7/WV 08021/TB LGT 055-072/IC LGT-MOD RIME 072-089

Figure 14. – Pilot Weather Report.

8.4 Aviation Area Forecast

22. To best determine general forecast weather conditions over several states, the pilot should refer to

    A. Aviation Area Forecasts.

    B. Weather Depiction Charts.

    C. Satellite Maps.

Answer (A) is correct. *(AWS Sect 7)*
    DISCUSSION: An Aviation Area Forecast is a prediction of general weather conditions over an area consisting of several states or portions of states. It is used to obtain expected en route weather conditions and also to provide an insight to weather conditions that might be expected at airports where weather reports or forecasts are not issued.
    Answer (B) is incorrect. Weather Depiction Charts are compiled from METAR reports of observed, not forecast, areas. Answer (C) is incorrect. Satellite pictures (maps) are observed pictures used to determine the presence and types of clouds, not forecast conditions.

23. To determine the freezing level and areas of probable icing aloft, the pilot should refer to the

    A. Inflight Aviation Weather Advisories.

    B. Weather Depiction Chart.

    C. Area Forecast.

Answer (A) is correct. *(AWS Sect 7)*
    DISCUSSION: To determine the freezing level and areas of probable icing aloft, you should refer to the Inflight Aviation Weather Advisories (AIRMET Zulu for icing and freezing level; AIRMET Tango for turbulence, strong winds/low-level wind shear; and AIRMET Sierra for IFR conditions and mountain obscuration.) Inflight Aviation Weather Advisories supplement the area forecast.
    Answer (B) is incorrect. The Weather Depiction Chart does not include any icing information. Answer (C) is incorrect. The Area Forecast alone contains no icing information; it must be supplemented by Inflight Aviation Weather Advisories.

24. The section of the Area Forecast entitled "VFR CLDS/WX" contains a general description of

    A. cloudiness and weather significant to flight operations broken down by states or other geographical areas.

    B. forecast sky cover, cloud tops, visibility, and obstructions to vision along specific routes.

    C. clouds and weather which cover an area greater than 3,000 square miles and is significant to VFR flight operations.

Answer (C) is the best answer. *(AWS Sect 7)*
    DISCUSSION: The VFR CLDS/WX section contains a 12-hr. specific forecast plus a categorical outlook section. The specific forecast section gives a general description of clouds and weather that covers an area greater than 3,000 square miles and is significant to VFR flight operations.
    Answer (A) is incorrect. While the VFR CLDS/WX section may be broken down by states or well-known geographical areas, the area of coverage will always be greater than 3,000 square miles. Answer (B) is incorrect. A summary of forecast sky cover, cloud tops, visibility, and obstructions to vision along specific routes is contained in a TWEB route forecast, which is no longer an operational product.

25. From which primary source should information be obtained regarding expected weather at the estimated time of arrival if your destination has no Terminal Forecast?

    A. Low-Level Prognostic Chart.

    B. Weather Depiction Chart.

    C. Area Forecast.

Answer (C) is correct. *(AWS Sect 7)*
    DISCUSSION: An area forecast (FA) is a forecast of general weather conditions over an area the size of several states. It is used to determine forecast en route weather and to interpolate conditions at airports that do not have a TAF issued.
    Answer (A) is incorrect. A Low-Level Prognostic Chart forecasts weather conditions expected to exist 12 hr. and 24 hr. in the future for the entire U.S. Answer (B) is incorrect. A Weather Depiction Chart is a national map prepared from METAR reports that give a broad overview of observed weather conditions as of the time on the chart. It is not a forecast.

```
BOSC FA 241845
SYNOPSIS AND VFR CLDS/WX
SYNOPSIS VALID UNTIL 251300
CLDS/WX VALID UNTIL 250700...OTLK VALID 250700-251300
ME NH VT MA RI CT NY LO NJ PA OH LE WV MD DC DE VA AND CSTL WTRS
.
SEE AIRMET SIERRA FOR IFR CONDS AND MTN OBSCN.
TS IMPLY SEV OR GTR TURB SEV ICE LLWS AND IFR CONDS.
NON MSL HGTS DENOTED BY AGL OR CIG.
.
SYNOPSIS...19Z CDFNT ALG A 160NE ACK-ENE LN...CONTG AS A QSTNRY
FNT ALG AN END-50SW MSS LN. BY 13Z...CDFNT ALG A 140ESE ACK-HTO
LN...CONTG AS A QSTNRY FNT ALG A HTO-SYR-YYZ LN. TROF ACRS CNTRL
PA INTO NRN VA.  ...REYNOLDS...
.
OH LE
NRN HLF OH LE...SCT-BKN025 OVC045. CLDS LYRD 150. SCT SHRA. WDLY
     SCT TSRA. CB TOPS FL350. 23-01Z OVC020-030.  VIS 3SM BR. OCNL -
     RA. OTLK...IFR CIG BR FG.
SWRN QTR OH...BKN050-060 TOPS 100. OTLK...MVFR BR.
SERN QTR OH...SCT-BKN040 BKN070 TOPS 120. WDLY SCT -TSRA. 00Z
     SCT-BKN030 OVC050. WDLY SCT -TSRA. CB TOPS FL350. OTLK...VFR
     SHRA.
.
CHIC FA 241945
SYNOPSIS AND VFR CLDS/WX
SYNOPSIS VALID UNTIL 251400
CLDS/WX VALID UNTIL 250800...OTLK VALID 250800-251400
ND SD NE KS MN IA MO WI LM LS MI LH IL IN KY
.
SEE AIRMET SIERRA FOR IFR CONDS AND MTN OBSCN.
TS IMPLY SEV OR GTR TURB SEV ICE LLWS AND IFR CONDS.
NON MSL HGTS DENOTED BY AGL OR CIG.
.
SYNOPSIS...LOW PRES AREA 20Z CNTRD OVR SERN WI FCST MOV NEWD INTO
LH BY 12Z AND WKN. LOW PRES FCST DEEPEN OVR ERN CO DURG PD AND
MOV NR WRN KS BORDER BY 14Z. DVLPG CDFNT WL MOV EWD INTO S CNTRL
NE-CNTRL KS BY 14Z ..SMITH..
.
UPR MI LS
WRN PTNS...AGL SCT030 SCT-BKN050. TOPS 080. 02-05Z BECMG CIG
     OVC010 VIS 3-5SM BR. OTLK...IFR CIG BR.
ERN PTNS...CIG BKN020 OVC040. OCNL VIS 3-5SM -RA BR. TOPS FL200.
     23Z CIG OVC010 VIS 3-5SM -RA BR. OTLK...IFR CIG BR.
.
LWR MI LM LH
CNTRL/NRN PTNS...CIG OVC010 VIS 3-5SM -RA BR. TOPS FL200.
     OTLK...IFR CIG BR.
.
SRN THIRD...CIG OVC015-025. SCT -SHRA. TOPS 150. 00-02Z BECMG CIG
     OVC010 VIS 3-5SM BR. TOPS 060. OTLK...IFR CIG BR.
.
IN
NRN HALF...CIG BKN035 BKN080. TOPS FL200. SCT -SHRA. 00Z CIG
     BKN-SCT040 BKN-SCT080. TOPS 120. 06Z AGL SCT-BKN030. TOPS 080.
     OCNL VIS 3-5SM BR. OTLK...MVFR CIG BR.
SRN HALF...AGL SCT050 SCT-BKN100. TOPS 120. 07Z AGL SCT 030
     SCT 100. OTLK...VFR.
```

Figure 16. – Area Forecast.

26. (Refer to Figure 16 on page 240.) The Chicago FA forecast section is valid until the twenty-fifth at

    A. 1945Z.

    B. 0800Z.

    C. 1400Z.

Answer (B) is correct. *(AWS Sect 7)*
    DISCUSSION: The Chicago area forecast (FA) is the second of two FAs depicted in Fig. 16. There is a note in the communication and product header section that says "CLDS/WX VALID UNTIL 250800," which means that the VFR clouds and weather section of the FA (the forecast section) is valid until 0800Z on the 25th.
    Answer (A) is incorrect. The note "CHIC FA 241945" in the communication and product header section means that the FA was issued at 1945Z on the 24th, not that it is valid until 1945Z on the 25th. Answer (C) is incorrect. The synopsis and categorical outlook (which are not considered to be forecasts), not the forecast section, are valid until 1400Z on the 25th, as indicated by the notes "SYNOPSIS VALID UNTIL 251400" and "OTLK VALID 250800-251400."

27. (Refer to Figure 16 on page 240.) What sky condition and visibility are forecast for upper Michigan in the eastern portions after 2300Z?

    A. Ceiling 100 feet overcast and 3 to 5 statute miles visibility.

    B. Ceiling 1,000 feet overcast and 3 to 5 nautical miles visibility.

    C. Ceiling 1,000 feet overcast and 3 to 5 statute miles visibility.

Answer (C) is correct. *(AWS Sect 7)*
    DISCUSSION: The Chicago area forecast (FA) is the second of two FAs depicted in Fig. 16. It contains an entry labeled "UPR MI LS," meaning "upper Michigan and Lake Superior." Under this heading is a section labeled "ERN PTNS," meaning "eastern portions." The entry "23Z CIG OVC010 VIS 3-5SM -RA BR" means that from 2300Z, the forecast weather is an overcast ceiling at 1,000 ft. AGL, with 3 to 5 statute miles visibility in light rain and mist.
    Answer (A) is incorrect. The ceiling is forecast to be overcast at 1,000 ft., not 100 ft., which would be coded as "OVC001." Answer (B) is incorrect. Visibilities are always given in statute, not nautical, miles.

28. (Refer to Figure 16 on page 240.) What is the outlook for the southern half of Indiana after 0700Z?

    A. VFR.

    B. Scattered clouds at 3,000 feet AGL.

    C. Scattered clouds at 10,000 feet.

Answer (A) is correct. *(AWS Sect 7)*
    DISCUSSION: The question asks for the outlook for the southern half of Indiana after 0700Z. Indiana (IN) is covered by the Chicago area forecast (FA), which is the second of two FAs depicted in Fig. 16. There is a heading under "IN" labeled "SRN HALF," meaning "southern half." Under this heading is an entry, "OTLK...VFR," meaning that the categorical outlook is for VFR conditions. Note in the communication and product header section that there is a note, "OTLK VALID 250800-251400," meaning that the categorical outlook is valid from 0800Z to 1400Z on the 25th. Therefore, the outlook does not become valid until 1 hour after 0700Z. You should still select "VFR" as the answer for this question because it specifically asks for the outlook after 0700Z, not at 0700Z; 0800Z is after 0700Z.
    Answer (B) is incorrect. Scattered clouds at 3,000 ft. AGL is a forecast sky condition from 0700Z to 0800Z (when the VFR CLDS/WX section becomes invalid); it is not an outlook, which would simply indicate whether VFR, MVFR, or IFR conditions are expected. Answer (C) is incorrect. Scattered clouds at 10,000 ft. is a forecast sky condition from 0700Z to 0800Z (when the VFR CLDS/WX section becomes invalid); it is not an outlook, which would simply indicate whether VFR, MVFR, or IFR conditions are expected.

29. (Refer to Figure 16 on page 240.) What sky condition and type obstructions to vision are forecast for upper Michigan in the Western portions from 0200Z until 0500Z?

    A. Ceiling becoming 1,000 feet overcast with visibility 3 to 5 statute miles in mist.

    B. Ceiling becoming 100 feet overcast with visibility 3 to 5 statute miles in mist.

    C. Ceiling becoming 1,000 feet overcast with visibility 3 to 5 nautical miles in mist.

Answer (A) is correct. *(AWS Sect 7)*
    DISCUSSION: The Chicago area forecast (FA) is the second of two FAs depicted in Fig. 16. It contains an entry labeled "UPR MI LS," meaning "upper Michigan and Lake Superior." Under this heading is a section labeled "WRN PTNS," meaning "western portions." The entry "02-05Z BECMG CIG OVC 010 VIS 3-5SM BR" means that between 0200Z and 0500Z, the weather conditions are forecast to become an overcast ceiling at 1,000 ft., with 3-5 statute miles visibility in mist.
    Answer (B) is incorrect. The ceiling is forecast to become overcast at 1,000 ft., not 100 ft., which would be coded as "OVC001." Answer (C) is incorrect. Visibilities are always given in statute, not nautical, miles.

8.5 Terminal Aerodrome Forecast (TAF)

30. (Refer to Figure 15 on page 243.) In the TAF for KMEM, what does "SHRA" stand for?

    A. Rain showers.

    B. A shift in wind direction is expected.

    C. A significant change in precipitation is possible.

Answer (A) is correct. *(AWS Sect 7)*
    DISCUSSION: SHRA is a coded group of forecast weather. SH is a descriptor that means showers. RA is a type of precipitation that means rain. Thus, SHRA means rain showers.
    Answer (B) is incorrect. SHRA means rain showers, not that a shift in wind direction is expected. A change in wind direction would be reflected by a forecast wind. Answer (C) is incorrect. SHRA means rain showers, not that a significant change in precipitation is possible.

31. (Refer to Figure 15 on page 243.) During the time period from 0600Z to 0800Z, what visibility is forecast for KOKC?

    A. Greater than 6 statute miles.

    B. Possibly 6 statute miles.

    C. Not forecasted.

Answer (A) is correct. *(AWS Sect 7)*
    DISCUSSION: At KOKC, between 0600Z and 0800Z, conditions are forecast to become wind 210° at 15 kt., visibility greater than 6 SM (P6SM), scattered clouds at 4,000 ft. with conditions continuing until the end of the forecast (1200Z).
    Answer (B) is incorrect. Between 0600Z and 0800Z, the visibility is forecast to be greater than, not possibly, 6 statute miles. Answer (C) is incorrect. Between 0600Z and 0800Z, the visibility is forecast to be greater than 6 statute miles (P6SM).

32. (Refer to Figure 15 on page 243.) In the TAF from KOKC, the clear sky becomes

    A. overcast at 2,000 feet during the forecast period between 2200Z and 2400Z.

    B. overcast at 200 feet with a 40 percent probability of becoming overcast at 600 feet during the forecast period between 2200Z and 2400Z.

    C. overcast at 200 feet with the probability of becoming overcast at 400 feet during the forecast period between 2200Z and 2400Z.

Answer (A) is correct. *(AWS Sect 7)*
    DISCUSSION: In the TAF for KOKC, from 2200Z to 2400Z, the conditions are forecast to gradually become wind 200° at 13 kt. with gusts to 20 kt., visibility 4 SM in moderate rain showers, overcast clouds at 2,000 feet. Between the hours of 0000Z and 0600Z, a chance (40 percent) exists of visibility 2 SM in thunderstorm with moderate rain, and 800 ft. overcast, cumulus clouds.
    Answer (B) is incorrect. Between 2200Z and 2400Z, the coded sky condition of OVC020 means overcast clouds at 2,000 ft., not 200 feet. Answer (C) is incorrect. Between 2200Z and 2400Z, the coded sky condition of OVC020 means overcast clouds at 2,000 ft., not 200 feet.

33. (Refer to Figure 15 on page 243.) What is the valid period for the TAF for KMEM?

    A. 1200Z to 1200Z.

    B. 1200Z to 1800Z.

    C. 1800Z to 1800Z.

Answer (C) is correct. *(AWS Sect 7)*
    DISCUSSION: The valid period of a TAF follows the four-letter location identifier and the six-digit issuance date/time. The valid period group is a two-digit date followed by the two-digit beginning hour and the two-digit ending hour. The valid period of the TAF for KMEM is 121818, which means the forecast is valid from the 12th day at 1800Z until the 13th at 1800Z.
    Answer (A) is incorrect. The valid period of the TAF for KOKC, not KMEM, is from 1200Z to 1200Z. Answer (B) is incorrect. The valid period of the TAF for KMEM is from the 12th day, not 1200Z, at 1800Z until the 13th at 1800Z.

34. (Refer to Figure 15 on page 243.) Between 1000Z and 1200Z the visibility at KMEM is forecast to be?

    A. 1/2 statute mile.

    B. 3 statute miles.

    C. 6 statute miles.

Answer (B) is correct. *(AWS Sect 7)*
    DISCUSSION: Between 1000Z and 1200Z, the conditions at KMEM are forecast to gradually become wind calm, visibility 3 SM in mist, sky clear with temporary (occasional) visibility 1/2 SM in fog between 1200Z and 1400Z. Conditions are expected to continue until 1600Z.
    Answer (A) is incorrect. Between the hours of 1200Z and 1400Z, not between 1000Z and 1200Z, the forecast is for temporary (occasional) visibility of 1/2 SM in fog. Answer (C) is incorrect. Between 1000Z and 1200Z, the forecast visibility for KMEM is 3 SM, not 6 SM.

35. (Refer to Figure 15 below.) What is the forecast wind for KMEM from 1600Z until the end of the forecast?

   A. No significant wind.

   B. Variable in direction at 6 knots.

   C. Variable in direction at 4 knots.

Answer (B) is correct. *(AWS Sect 7)*
   DISCUSSION: The forecast for KMEM from 1600Z until the end of the forecast (1800Z) is wind direction variable at 6 kt. (VRB06KT), visibility greater than 6 SM, and sky clear.
   Answer (A) is incorrect. The wind is forecast to be variable in direction at 6 knots. Answer (C) is incorrect. The wind is forecast to be variable in direction at 6 kt., not 4 knots. KMEM of 020° at 8 kt. is for 0600Z until 0800Z, not from 1600Z until the end of the forecast.

36. (Refer to Figure 15 below.) In the TAF from KOKC, the "FM (FROM) Group" is forecast for the hours from 1600Z to 2200Z with the wind from

   A. 160° at 10 knots.

   B. 180° at 10 knots.

   C. 180° at 10 knots, becoming 200° at 13 knots.

Answer (B) is correct. *(AWS Sect 7)*
   DISCUSSION: The FM group states that, from 1600Z until 2200Z (time of next change group), the forecast wind is 180° at 10 knots.
   Answer (A) is incorrect. The forecast wind is 180°, not 160°, at 10 knots. Answer (C) is incorrect. The BECMG (becoming) group is a change group and is not part of the FM forecast group. The wind will gradually become 200° at 13 kt. with gusts to 20 kt., between 2200Z and 2400Z.

37. (Refer to Figure 15 below, and Figure 18 on page 244.) The only cloud type forecast in TAF reports is

   A. Nimbostratus.

   B. Cumulonimbus.

   C. Scattered cumulus.

Answer (B) is correct. *(AWS Sect 7)*
   DISCUSSION: Cumulonimbus clouds are the only cloud type forecast in TAFs. If cumulonimbus clouds are expected at the airport, the contraction CB is appended to the cloud layer that represents the base of the cumulonimbus cloud(s).
   Answer (A) is incorrect. The only cloud type forecast in TAFs is cumulonimbus, not nimbostratus, clouds. Answer (C) is incorrect. The only cloud type forecast in TAFs is cumulonimbus, not scattered cumulus, clouds.

```
TAF

KMEM   121720Z 121818 20012KT 5SM HZ BKN030 PROB40 2022 1SM TSRA OVC008CB
       FM2200 33015G20KT P6SM BKN015 OVC025 PROB40 2202 3SM SHRA
       FM0200 35012KT OVC008 PROB40 0205 2SM -RASN BECMG 0608 02008KT BKN012
       BECMG 1012 00000KT 3SM BR SKC TEMPO 1214 1/2SM FG
       FM1600 VRB06KT P6SM SKC=

KOKC   051130Z 051212 14008KT 5SM BR BKN030 TEMPO 1316 1 1/2SM BR
       FM1600 18010KT P6SM SKC BECMG 2224 20013G20KT 4SM SHRA OVC020
       PROB40 0006 2SM TSRA OVC008CB BECMG 0608 21015KT P6SM SCT040=
```

Figure 15. – Terminal Aerodrome Forecasts (TAF).

8.6 Weather Depiction Charts

38. Of what value is the Weather Depiction Chart to the pilot?

   A. For determining general weather conditions on which to base flight planning.

   B. For a forecast of cloud coverage, visibilities, and frontal activity.

   C. For determining frontal trends and air mass characteristics.

Answer (A) is correct. *(AWS Sect 5)*
   DISCUSSION: The Weather Depiction Chart is prepared from surface aviation weather reports giving a quick picture of weather conditions as of the time stated on the chart. Thus, it presents general weather conditions on which to base flight planning.
   Answer (B) is incorrect. A significant weather prognostic chart can provide a forecast of cloud coverage, visibilities, and frontal activity. A weather depiction chart shows actual, not forecast, conditions. Answer (C) is incorrect. A composite moisture stability chart would be used to determine the characteristics of an air mass.

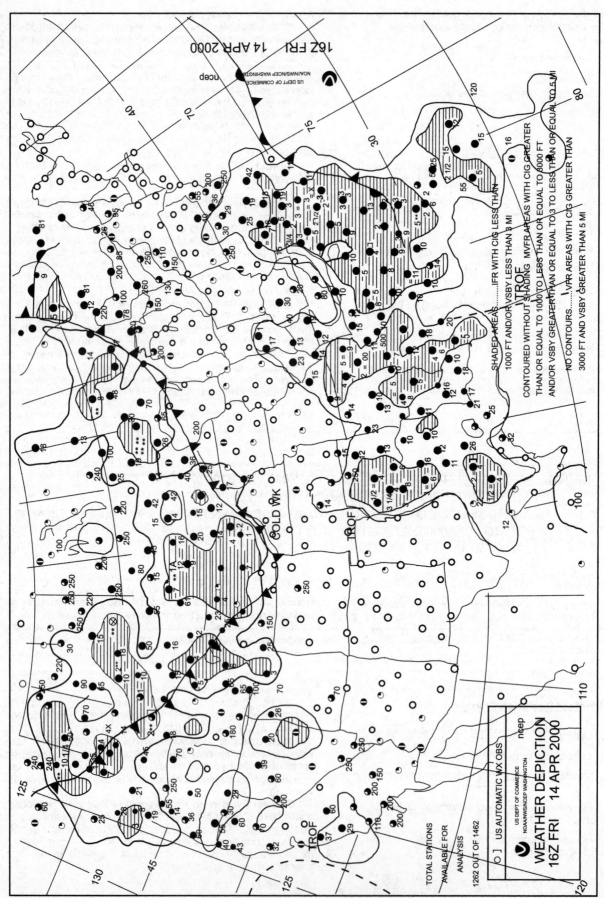

Figure 18. – Weather Depiction Chart.

39. (Refer to Figure 18 on page 244.) The IFR weather in northern Texas is due to

    A. intermittent rain.

    B. low ceilings.

    C. dust devils.

Answer (B) is correct. *(AWS Sect 5)*
    DISCUSSION: Refer to the Weather Depiction Chart in Fig. 18. The shaded area around northern Texas and central Oklahoma indicates that IFR conditions exist. The symbols "3=⚫" and "3=⚫" mean that the visibility is 3 SM in fog (3=), and the sky is overcast at 600 ft. (⚫) to 800 ft. (⚫) AGL. Thus, low ceilings from 600 to 800 ft. are the source of IFR weather conditions
    Answer (A) is incorrect. A solid round dot (●) indicates intermittent rain. Answer (C) is incorrect. The symbol ⚫ indicates dust devils, which are small vigorous whirlwinds, usually of short duration, made visible by dust, sand, or debris picked up from the ground.

40. (Refer to Figure 18 on page 244.) What weather phenomenon is causing IFR conditions in central Oklahoma?

    A. Low visibility only.

    B. Low ceilings and visibility.

    C. Heavy rain showers.

Answer (B) is correct. *(AWS Sect 5)*
    DISCUSSION: Refer to the Weather Depiction Chart in Fig. 18. In central Oklahoma, the IFR conditions are caused by low ceilings and visibility. In the shaded area over central Oklahoma and northern Texas, there are six darkened circles with numbers ranging from one to eight below them, signifying overcast skies with ceilings at 100 to 800 feet. The circles also have numbers ranging from 3/4 to 3 beside them, signifying visibilities between 3/4 and 3 statute miles. The IFR conditions are therefore due to low ceilings and visibility.
    Answer (A) is incorrect. There are also low ceilings in the area from 100 to 800 feet. Answer (C) is incorrect. Heavy rain showers are shown by the symbol ▼.

41. (Refer to Figure 18 on page 244.) What is the status of the front that extends from Nebraska through the upper peninsula of Michigan?

    A. Cold.

    B. Stationary.

    C. Warm.

Answer (A) is correct. *(AWS Sect 5)*
    DISCUSSION: Refer to the Weather Depiction Chart in Fig. 18. The front that extends from Nebraska through the upper peninsula of Michigan is a cold front, as shown by the pointed scallops on the southern side of the frontal line.
    Answer (B) is incorrect. A stationary front has pointed scallops on one side of the frontal line and rounded scallops on the other. Answer (C) is incorrect. A warm front has rounded scallops, not pointed scallops.

42. (Refer to Figure 18 on page 244.) According to the Weather Depiction Chart, the weather for a flight from southern Michigan to north Indiana is ceilings

    A. 1,000 to 3,000 feet and/or visibility 3 to 5 miles.

    B. less than 1,000 feet and/or visibility less than 3 miles.

    C. greater than 3,000 feet and visibility greater than 5 miles.

Answer (C) is correct. *(AWS Sect 5)*
    DISCUSSION: Refer to the Weather Depiction Chart in Fig. 18. The weather from southern Michigan to north Indiana is shown by the lack of shading or contours to have ceilings greater than 3,000 ft. and visibilities greater than 5 miles.
    Answer (A) is incorrect. Ceilings from 1,000 to 3,000 ft. and/or visibilities between 3 and 5 statute miles (MVFR conditions) are indicated on Weather Depiction Charts by an unshaded area surrounded by a contour. Answer (B) is incorrect. Ceilings less than 1,000 ft. and/or visibilities less than 3 statute miles (IFR conditions) are indicated on Weather Depiction Charts by a shaded area surrounded by a contour.

43. (Refer to Figure 18 on page 244.) The marginal weather in central Kentucky is due to low

    A. ceiling.

    B. ceiling and visibility.

    C. visibility.

Answer (A) is correct. *(AWS Sect 5)*
    DISCUSSION: Refer to the Weather Depiction Chart in Fig. 18. The MVFR weather in central Kentucky is indicated by the contour line without shading. The station symbol indicates an overcast ceiling at 3,000 feet. MVFR is ceiling 1,000 ft. to 3,000 ft. and/or visibility 3 to 5 statute miles. Thus, the marginal weather is due to a low ceiling.
    Answer (B) is incorrect. The marginal weather is caused by low ceilings only. Answer (C) is incorrect. The visibility is greater than 6 SM (indicated by the lack of a report). Therefore, the visibility is not a cause of the marginal conditions.

8.7 Radar Summary Charts and Radar Weather Reports

44. (Refer to Figure 19 on page 247.) (Refer to area B.) What is the top for precipitation of the radar return?

A. 24,000 feet AGL.

B. 2,400 feet MSL.

C. 24,000 feet MSL.

Answer (C) is correct. *(AWS Sect 5)*
DISCUSSION: Refer to the Radar Summary Chart in Fig. 19. The radar return at B (northern Nevada) has a "240" with a line under it. This means the maximum top of the precipitation is 24,000 ft. MSL.
Answer (A) is incorrect. The height of precipitation returns is given in MSL, not AGL. Answer (B) is incorrect. The height of precipitation returns is given in hundreds, not tens, of feet MSL. The "240" means 24,000, not 2,400.

45. What does the heavy dashed line that forms a large rectangular box on a Radar Summary Chart refer to?

A. Areas of heavy rain.

B. Severe weather watch area.

C. Areas of hail 1/4 inch in diameter.

Answer (B) is correct. *(AWS Sect 5)*
DISCUSSION: On a Radar Summary Chart, severe weather watch areas are outlined by heavy dashed lines.
Answer (A) is incorrect. Areas of heavy rain would be labeled with "R" for rain and "+" for heavy or increasing in intensity. Answer (C) is incorrect. Hail is denoted by a box with "hail" printed inside (HAIL).

46. (Refer to Figure 19 on page 247.) (Refer to area B.) What type of weather is occurring in the radar return?

A. Continuous rain.

B. Light to moderate rain.

C. Rain showers increasing in intensity.

Answer (B) is correct. *(AWS Sect 5)*
DISCUSSION: The intensity and type of the precipitation are indicated by the contour lines and symbols adjacent to the precipitation areas depicted on the radar summary chart. Next to area B, the intensity of the precipitation is light to moderate (single contour line), and the type is rain (R).
Answer (A) is incorrect. The term "continuous rain" applies to prognostic charts, not radar summary charts. Answer (C) is incorrect. The type of precipitation is rain (R), not showers (RW). Also, the intensity trend (increasing or weakening) is no longer coded on either the radar summary chart or the radar weather report (SD).

47. (Refer to Figure 19 on page 247.) (Refer to area D.) What is the direction and speed of movement of the cell?

A. North at 17 knots.

B. South at 17 knots.

C. North at 17 MPH.

Answer (A) is correct. *(AWS Sect 5)*
DISCUSSION: Refer to the Radar Summary Chart in Fig. 19. The radar return at D (Virginia) has an arrow pointing north with "17" at the point. The movement is thus north at 17 kt.
Answer (B) is incorrect. The arrow above the cell at point D points north, not south. This arrow indicates the direction of the cell's movement. Answer (C) is incorrect. The speed of cell movement is given in knots, not MPH.

48. (Refer to Figure 19 on page 247.) (Refer to area E.) The top of the precipitation of the cell is

A. 16,000 feet AGL.

B. 25,000 feet MSL.

C. 16,000 feet MSL.

Answer (C) is correct. *(AWS Sect 5)*
DISCUSSION: Refer to the Radar Summary Chart in Fig. 19. The cell 1/2 in. below point E (Virginia/North Carolina) has a "160" with a line under it. This means the maximum top of the precipitation is 16,000 ft. MSL.
Answer (A) is incorrect. The height of precipitation returns is given in MSL, not AGL. Answer (B) is incorrect. The "250" with a line under it (indicating 25,000 ft. MSL) extends from the large cell covering Florida, Alabama, and Georgia. This cell is associated with area G, not area E.

49. What information is provided by the Radar Summary Chart that is not shown on other weather charts?

A. Lines and cells of hazardous thunderstorms.

B. Ceilings and precipitation between reporting stations.

C. Types of clouds between reporting stations.

Answer (A) is correct. *(AWS Sect 5)*
DISCUSSION: Radar Summary Charts show lines of thunderstorms and hazardous cells that are not shown on other weather charts.
Answer (B) is incorrect. Weather radar primarily detects particles of precipitation size within a cloud or falling from a cloud, it does not detect clouds and fog. Thus, it cannot determine ceilings. Answer (C) is incorrect. The Radar Summary Chart can provide the type of precipitation (not clouds) between reporting stations.

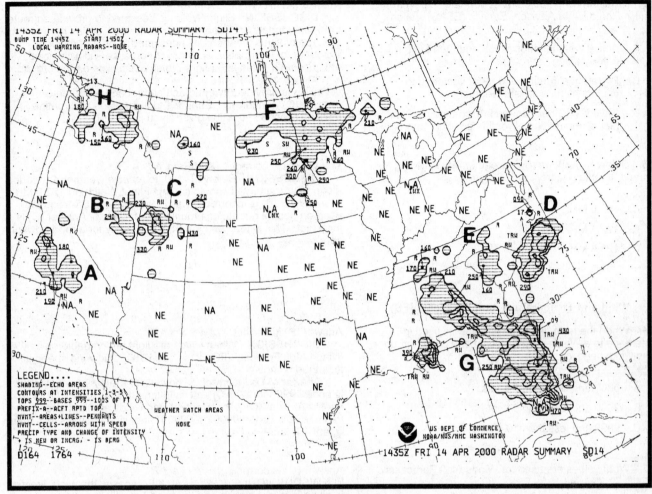

Figure 19. – Radar Summary Chart.

50. Radar weather reports are of special interest to pilots because they indicate

    A. large areas of low ceilings and fog.

    B. location of precipitation along with type, intensity, and cell movement of precipitation.

    C. location of precipitation along with type, intensity, and trend.

Answer (B) is correct. *(AWS Sect 3)*
    DISCUSSION: Radar weather reports are of special interest to pilots because they report the location of precipitation along with type, intensity, and cell movement.
    Answer (A) is incorrect. Weather radar cannot detect clouds or fog, only precipitation size particles. Answer (C) is incorrect. Radar weather reports no longer include trend information.

8.8 En Route Flight Advisory Service (EFAS)

51. How should contact be established with an En Route Flight Advisory Service (EFAS) station, and what service would be expected?

    A. Call EFAS on 122.2 for routine weather, current reports on hazardous weather, and altimeter settings.

    B. Call flight assistance on 122.5 for advisory service pertaining to severe weather.

    C. Call Flight Watch on 122.0 for information regarding actual weather and thunderstorm activity along proposed route.

Answer (C) is correct. *(AIM Para 7-1-5)*
    DISCUSSION: The frequency designed for en route flight advisory stations calling Flight Watch is 122.0 MHz. It is designed to provide en route aircraft with timely and meaningful weather advisories during the route. It is not for complete briefings or random weather reports.
    Answer (A) is incorrect. You would call FSS (not EFAS) on 122.2 for routine weather, current reports on hazardous weather, and altimeter settings. Answer (B) is incorrect. You would possibly call FSS (not Flight Watch) on 122.5 for advisory service pertaining to severe weather.

52. What service should a pilot normally expect from an En Route Flight Advisory Service (EFAS) station?

A. Actual weather information and thunderstorm activity along the route.

B. Preferential routing and radar vectoring to circumnavigate severe weather.

C. Severe weather information, changes to flight plans, and receipt of routine position reports.

Answer (A) is correct.   *(AIM Para 7-1-5)*
   DISCUSSION: Flight Watch is designed to provide en route traffic with timely and meaningful weather advisories pertinent to the type of flight intended. It is designed to be a continuous exchange of information on winds, turbulence, visibility, icing, etc., between pilots and Flight Watch specialists on the ground.
   Answer (B) is incorrect. Preferential routing and radar vectoring is provided by approach control and ATC center. Answer (C) is incorrect. Changes to flight plans and routine position reports should be given to an FSS.

53. Below FL 180, en route weather advisories should be obtained from an FSS on

A. 122.0 MHz.

B. 122.1 MHz.

C. 123.6 MHz.

Answer (A) is correct.   *(AIM Para 7-1-5)*
   DISCUSSION: Below FL 180, to receive weather advisories along your route, you should contact Flight Watch on 122.0 MHz.
   Answer (B) is incorrect. This is the pilot-to-FSS frequency used on duplex remote communication facilities. Answer (C) is incorrect. This is the common FSS frequency for airport advisory service.

8.9 Wind and Temperature Aloft Forecasts (FB)

54. When the term "light and variable" is used in reference to a Winds Aloft Forecast, the coded group and windspeed is

A. 0000 and less than 7 knots.

B. 9900 and less than 5 knots.

C. 9999 and less than 10 knots.

Answer (B) is correct.   *(AWS Sect 7)*
   DISCUSSION: When winds are light and variable on a Winds Aloft Forecast (FB), it is coded 9900 and wind speed is less than 5 knots.
   Answer (A) is incorrect. When winds are light and variable, it is coded 9900 (not 0000) and wind speed is less than 5 (not 7) knots. Answer (C) is incorrect. When winds are light and variable, it is coded 9900 (not 9999) and wind speed is less than 5 (not 10) knots.

55. What values are used for Winds Aloft Forecasts?

A. Magnetic direction and knots.

B. Magnetic direction and miles per hour.

C. True direction and knots.

Answer (C) is correct.   *(AWS Sect 7)*
   DISCUSSION: For Winds Aloft Forecasts, the wind direction is given in true direction and the wind speed is in knots.
   Answer (A) is incorrect. ATC (not Winds Aloft Forecasts) will provide winds in magnetic direction and knots. Answer (B) is incorrect. Winds Aloft Forecast will provide winds based on true (not magnetic) direction and speed in kt. (not MPH).

56. (Refer to Figure 17 on page 249.) What wind is forecast for STL at 12,000 feet?

A. 230° true at 39 knots.

B. 230° true at 56 knots.

C. 230° magnetic at 56 knots.

Answer (A) is correct.   *(AWS Sect 7)*
   DISCUSSION: Refer to the FB forecast in Fig. 17. Locate STL and move right to the 12,000-ft. column. The wind forecast (first four digits) is coded as 2339, which means the wind is 230° true at 39 knots.
   Answer (B) is incorrect. This is the forecast wind direction and speed for 18,000 ft., not 12,000 feet. Answer (C) is incorrect. The first two digits are direction referenced to true (not magnetic) north. Thus, 2356 is 230° true (not magnetic) at 56 kt., which is the forecast wind direction and speed for 18,000 ft., not 12,000 feet.

57. (Refer to Figure 17 on page 249.) Determine the wind and temperature aloft forecast for DEN at 9,000 feet.

A. 230° magnetic at 53 knots, temperature 47°C.

B. 230° true at 53 knots, temperature –47°C.

C. 230° true at 21 knots, temperature –4°C.

Answer (C) is correct.   *(AWS Sect 7)*
   DISCUSSION: Refer to the FB forecast in Fig. 17. Locate DEN on the left side of the chart and move to the right to the 9,000-ft. column. The wind and temperature forecast is coded as 2321-04. The forecast is decoded as 230° true at 21 kt., temperature –4°C.
   Answer (A) is incorrect. The correct measurement is 230° true (not magnetic), and the temperature is –4°C, not 47°C. Answer (B) is incorrect. The temperature is –4°C, not –47°C, which is the temperature for DEN at 30,000 ft., not 9,000 feet.

58. (Refer to Figure 17 below.) What wind is forecast for STL at 12,000 feet?

A. 230° magnetic at 39 knots.

B. 230° true at 39 knots.

C. 230° true at 106 knots.

Answer (B) is correct. *(AWS Sect 7)*
   DISCUSSION: Refer to the FB forecast in Fig. 17. Locate STL on the left side of the chart and move to the right to the 12,000-ft. column. The wind forecast (first four digits) is coded as 2339. The forecast is decoded as 230° true at 39 knots.
   Answer (A) is incorrect. The wind is from 230° true, not magnetic. Answer (C) is incorrect. This is the forecast wind speed and direction for 34,000 ft., not 12,000 ft. (coded as 7306).

59. (Refer to Figure 17 below.) Determine the wind and temperature aloft forecast for MKC at 6,000 ft.

A. 050° true at 7 knots, temperature missing.

B. 200° magnetic at 6 knots, temperature +3°C.

C. 200° true at 6 knots, temperature +3°C.

Answer (C) is correct. *(AWS Sect 7)*
   DISCUSSION: Refer to the FB forecast in Fig. 17. Locate MKC on the left side of the chart and move to the right to the 6,000-ft. column. The wind and temperature forecast is coded as 2006+03, which translates as the forecast wind at 200° true at 6 kt. and a temperature of 3°C.
   Answer (A) is incorrect. This is the forecast for MKC at 3,000 ft., not 6,000 feet. Answer (B) is incorrect. Wind direction is given in true degrees, not magnetic degrees.

60. (Refer to Figure 17 below.) What wind is forecast for STL at 9,000 feet?

A. 230° magnetic at 25 knots.

B. 230° true at 32 knots.

C. 230° true at 25 knots.

Answer (B) is correct. *(AWS Sect 7)*
   DISCUSSION: Refer to the FB forecast in Fig. 17. Locate STL on the left side of the chart and move right to the 9,000-ft. column. The coded wind forecast (first four digits) is 2332. Thus, the forecast wind is 230° true at 32 knots.
   Answer (A) is incorrect. Wind direction is forecast in true (not magnetic) direction. Wind forecast of 230° true at 25 kt. is for STL at 6,000 ft., not 9,000 feet. Answer (C) is incorrect. This is the wind forecast for STL at 6,000 ft., not 9,000 feet.

```
FD WBC 151745
DATA BASED ON 151200Z
VALID 1600Z FOR USE 1800-0300Z.  TEMPS NEG ABV 24000
```

FT	3000	6000	9000	12000	18000	24000	30000	34000	39000
ALS			2420	2635–08	2535–18	2444–30	245945	246755	246862
AMA		2714	2725+00	2625–04	2531–15	2542–27	265842	256352	256762
DEN			2321–04	2532–08	2434–19	2441–31	235347	236056	236262
HLC		1707–01	2113–03	2219–07	2330–17	2435–30	244145	244854	245561
MKC	0507	2006+03	2215–01	2322–06	2338–17	2348–29	236143	237252	238160
STL	2113	2325+07	2332+02	2339–04	2356–16	2373–27	239440	730649	731960

Figure 17. – Winds and Temperatures Aloft Forecast.

8.10 Significant Weather Prognostic Charts

61. How are Significant Weather Prognostic Charts best used by a pilot?

A. For overall planning at all altitudes.

B. For determining areas to avoid (freezing levels and turbulence).

C. For analyzing current frontal activity and cloud coverage.

Answer (B) is correct. *(AWS Sect 8)*
   DISCUSSION: Weather prognostic charts forecast conditions that exist 12 and 24 hr. in the future. They include two types of forecasts: low level significant weather, such as IFR and marginal VFR areas, and moderate or greater turbulence areas and freezing levels.
   Answer (A) is incorrect. A complete set of weather forecasts for overall planning includes terminal forecasts, area forecasts, etc. Answer (C) is incorrect. The weather depiction chart shows analysis of frontal activities, cloud coverage, areas of precipitation, ceilings, etc.

62. (Refer to Figure 20 on page 251.) What weather is forecast for the Florida area just ahead of the stationary front during the first 12 hours?

A. Ceiling 1,000 to 3,000 feet and/or visibility 3 to 5 miles with intermittent precipitation.

B. Ceiling 1,000 to 3,000 feet and/or visibility 3 to 5 miles with continuous precipitation.

C. Ceiling less than 1,000 feet and/or visibility less than 3 miles with continuous precipitation.

Answer (B) is correct. *(AWS Sect 8)*
DISCUSSION: Refer to the Significant Weather Prognostic Chart in Fig. 20. During the first 12 hr. (bottom and top left panels), the weather just ahead of the stationary front that extends from coastal Virginia into the Gulf of Mexico is forecast to have ceilings from 1,000 to 3,000 ft. and/or visibility 3 to 5 SM (as indicated by the scalloped lines) with continuous light to moderate rain covering more than half the area (as indicated by the shading).
Answer (A) is incorrect. Thunderstorms embedded in an area of moderate continuous, not intermittent, precipitation are forecast just ahead of the stationary front, as indicated by the following symbol: ⚡. Thunderstorms embedded in an area of intermittent rain would be indicated with this symbol: ⚡. Answer (C) is incorrect. Marginal VFR conditions (ceilings from 1,000 to 3,000 ft. and/or visibility from 3 to 5 SM), not IFR conditions (ceilings less than 1,000 ft. and/or visibility less than 3 miles) are forecast, as indicated by the scalloped line surrounding the southeastern states on the top left panel.

63. (Refer to Figure 20 on page 251.) Interpret the weather symbol depicted in Utah on the 12-hour Significant Weather Prognostic Chart.

A. Moderate turbulence, surface to 18,000 feet.

B. Thunderstorm tops at 18,000 feet.

C. Base of clear air turbulence, 18,000 feet.

Answer (A) is correct. *(AWS Sect 8)*
DISCUSSION: Refer to the upper left panel of the Significant Weather Prog Chart in Fig. 20. In Utah, the weather symbol indicates moderate turbulence as designated by the symbol of a small peaked hat. Note that the broken line indicates moderate or greater turbulence. The peaked hat is the symbol for moderate turbulence. The 180 means the moderate turbulence extends from the surface upward to 18,000 feet.
Answer (B) is incorrect. The peaked hat symbol denotes moderate turbulence, not thunderstorms. The symbol for thunderstorms is shown by what looks like the letter "R," as shown on the 12-hr. surface prog in the Gulf of Mexico. Answer (C) is incorrect. This is not the base of the clear air turbulence. A line over a number indicates a base.

64. (Refer to Figure 20 on page 251.) At what altitude is the freezing level over the middle of Florida on the 12-hour Significant Weather Prognostic Chart?

A. 4,000 feet.

B. 8,000 feet.

C. 12,000 feet.

Answer (C) is correct. *(AWS Sect 8)*
DISCUSSION: Refer to the upper left panel of the Significant Weather Prog Chart in Fig. 20. On prog charts, the freezing level is indicated by a dashed line, with the height given in hundreds of feet MSL. In Fig. 20, there is a dashed line across the middle of Florida, marked with "120" just off the coast. This signifies that the freezing level is 12,000 ft. MSL.
Answer (A) is incorrect. The freezing level is at 4,000 ft. MSL across the northern U.S. and Canada, not over the middle of Florida. Answer (B) is incorrect. The freezing level is at 8,000 ft. MSL extending from southern California, upward and across the northern U.S., and into New Jersey, not over the middle of Florida.

65. (Refer to Figure 20 on page 251.) The enclosed shaded area associated with the low pressure system over northern Utah is forecast to have

A. continuous snow.

B. intermittent snow.

C. continuous snow showers.

Answer (A) is correct. *(AWS Sect 8)*
DISCUSSION: Refer to the lower left panel of the 24-hr. Significant Weather Prog Chart in Fig. 20. There is a low pressure center over northern Utah, indicated by a bold "L." To the left of the "L" is a shaded area, indicating precipitation covering more than half the area. Just to the right of the "L" is a symbol, ✳, with an arrow pointing to the shaded area. This means that the shaded area is forecast to have continuous light to moderate snow.
Answer (B) is incorrect. Intermittent light to moderate snow would be indicated with the symbol ✳, not ✳. Answer (C) is incorrect. Continuous snow showers are indicated for the unshaded area in southern Utah, not the shaded area in northern Utah, by the symbol ✳.

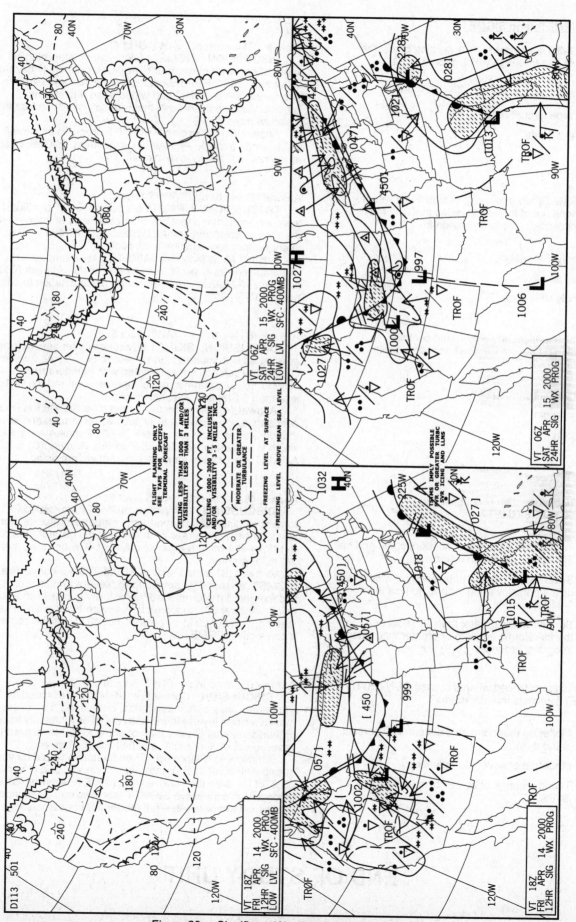

Figure 20. – Significant Weather Prognostic Chart.

8.11 AIRMETs and SIGMETs

66. SIGMETs are issued as a warning of weather conditions hazardous to which aircraft?

    A.   Small aircraft only.

    B.   Large aircraft only.

    C.   All aircraft.

Answer (C) is correct. *(AWS Sect 6)*
    DISCUSSION: SIGMETs (significant meteorological information) warn of weather considered potentially hazardous to all aircraft. SIGMET advisories cover severe and extreme turbulence; severe icing; and widespread duststorms, sandstorms, or volcanic ash that reduce visibility to less than 3 statute miles.
    Answer (A) is incorrect. SIGMETs apply to all aircraft, not just to small aircraft. Answer (B) is incorrect. SIGMETs apply to all aircraft, not just to large aircraft.

67. AIRMETs are advisories of significant weather phenomena but of lower intensities than SIGMETs and are intended for dissemination to

    A.   only IFR pilots.

    B.   all pilots.

    C.   only VFR pilots.

Answer (B) is correct. *(AWS Sect 6)*
    DISCUSSION: AIRMETs are advisories of significant weather phenomena that describe conditions at intensities lower than those which require the issuance of SIGMETs. They are intended for dissemination to all pilots.
    Answer (A) is incorrect. AIRMETs are intended for dissemination to all pilots, not just IFR pilots. Answer (C) is incorrect. AIRMETs are intended for dissemination to all pilots, not just VFR pilots.

68. Which in-flight advisory would contain information on severe icing not associated with thunderstorms?

    A.   Convective SIGMET.

    B.   SIGMET.

    C.   AIRMET.

Answer (B) is correct. *(AWS Sect 6)*
    DISCUSSION: SIGMET advisories cover severe icing not associated with thunderstorms; severe or extreme turbulence or clear air turbulence not associated with thunderstorms; duststorms, sandstorms, or volcanic ash that reduce visibility to less than 3 SM; and volcanic eruption.
    Answer (A) is incorrect. A convective SIGMET is issued concerning convective activity such as tornadoes and severe thunderstorms. Any convective SIGMET implies severe icing, which is associated with thunderstorms. Answer (C) is incorrect. AIRMETs are issued for moderate, not severe, icing.

69. What information is contained in a CONVECTIVE SIGMET?

    A.   Tornadoes, embedded thunderstorms, and hail 3/4 inch or greater in diameter.

    B.   Severe icing, severe turbulence, or widespread dust storms lowering visibility to less than 3 miles.

    C.   Surface winds greater than 40 knots or thunderstorms equal to or greater than video integrator processor (VIP) level 4.

Answer (A) is correct. *(AWS Sect 6)*
    DISCUSSION: Convective SIGMETs are issued for tornadoes, lines of thunderstorms, embedded thunderstorms of any intensity level, areas of thunderstorms greater than or equal to VIP level 4 with an area coverage of 40% or more, and hail 3/4 in. or greater.
    Answer (B) is incorrect. A SIGMET, not a convective SIGMET, is issued for severe icing, severe turbulence, or widespread duststorms lowering visibility to less than 3 statute miles. Answer (C) is incorrect. A severe thunderstorm having surface winds of 50 kt. or greater, not 40 kt., will be contained in a convective SIGMET.

70. What is indicated when a current CONVECTIVE SIGMET forecasts thunderstorms?

    A.   Moderate thunderstorms covering 30 percent of the area.

    B.   Moderate or severe turbulence.

    C.   Thunderstorms obscured by massive cloud layers.

Answer (C) is correct. *(AWS Sect 6)*
    DISCUSSION: Convective SIGMETs are issued for tornadoes, lines of thunderstorms, embedded (i.e., obscured by massive cloud layers) thunderstorms of any intensity level, areas of thunderstorms greater than or equal to VIP level 4 with an area coverage of 40% or more, and hail 3/4 in. or greater.
    Answer (A) is incorrect. Thunderstorms would be very strong (VIP level 4) or greater, not moderate, and cover 40%, not 30%, of the area for a convective SIGMET. Answer (B) is incorrect. A convective SIGMET that is issued for thunderstorms implies severe or greater, not moderate, turbulence.

END OF STUDY UNIT

STUDY UNIT NINE
NAVIGATION: CHARTS AND PUBLICATIONS

(15 pages of outline)

This study unit contains outlines of major concepts tested, sample test questions and answers regarding navigation charts and publications, and an explanation of each answer. The table of contents above lists each subunit within this study unit, the number of questions pertaining to that particular subunit, and the pages on which the outlines and questions begin, respectively.

Many of the questions in this study unit ask about the sectional (aeronautical) charts, which appear as Legend 1 and Figures 21 through 27. The acronym ACL is the question source code used to refer to this aeronautical chart legend. To produce the legend and these charts in color economically, we have put them together on pages 283 through 290. As you will need to turn to these pages frequently, mark them with "dog ears" (fold their corners) or paper clip them. Also, the second subunit, "Airspace and Altitudes," is long (34 questions) and covers a number of diverse topics regarding interpretation of sectional charts. Be prepared.

Note that a number of questions in Study Unit 10, "Navigation Systems," and Study Unit 11, "Cross-Country Flight Planning," will refer you to these same figures. These questions require you to determine your position, compute magnetic heading, true course, time en route, etc.

CAUTION: Recall that the **sole purpose** of this book is to expedite your passing the FAA pilot knowledge test for the private pilot certificate. Accordingly, all extraneous material (i.e., topics or regulations not directly tested on the FAA pilot knowledge test) is omitted, even though much more information and knowledge are necessary to fly safely. This additional material is presented in *Pilot Handbook* and *Private Pilot Flight Maneuvers and Practical Test Prep*, available from Gleim Publications, Inc. See the order form on page 371.

9.1 LONGITUDE AND LATITUDE

1.   The location of an airport can be determined by the intersection of the lines of latitude and longitude.

   a.   Lines of latitude are parallel to the equator, and those north of the equator are numbered from 0° to 90° north latitude.

   b.   Lines of longitude extend from the north pole to the south pole. The prime meridian (which passes through Greenwich, England) is 0° longitude with 180° on both the east and west sides of the prime meridian.

2.   The lines of latitude and longitude are printed on aeronautical charts (e.g., sectional) with each degree subdivided into 60 equal segments called minutes; i.e., 1/2° is 30' (the min. symbol is " ' ").

9.2 AIRSPACE AND ALTITUDES

Cloud Clearance and Visibility Required for VFR

Airspace	Flight Visibility	Distance from Clouds
Class A	Not Applicable	Not applicable
Class B	3 SM	Clear of Clouds
Class C	3 SM	500 ft. below 1,000 ft. above 2,000 ft. horiz.
Class D	3 SM	500 ft. below 1,000 ft. above 2,000 ft. horiz.
Class E		
Less than 10,000 ft. MSL	3 SM	500 ft. below 1,000 ft. above 2,000 ft. horiz.
At or above 10,000 ft. MSL	5 SM	1,000 ft. below 1,000 ft. above 1 SM horiz.

Airspace	Flight Visibility	Distance from Clouds
Class G:		
1,200 ft. or less above the surface (regardless of MSL altitude)		
Day	1 SM	Clear of clouds
Night, except as provided in 1. below	3 SM	500 ft. below 1,000 ft. above 2,000 ft. horiz.
More than 1,200 ft. above the surface but less than 10,000 ft. MSL		
Day	1 SM	500 ft. below 1,000 ft. above 2,000 ft. horiz.
Night	3 SM	500 ft. below 1,000 ft. above 2,000 ft. horiz.
More than 1,200 ft. above the surface and at or above 10,000 ft. MSL	5 SM	1,000 ft. below 1,000 ft. above 1 SM horiz.

1. An airplane may be operated clear of clouds in Class G airspace at night below 1,200 ft. AGL when the visibility is less than 3 SM but more than 1 SM in an airport traffic pattern and within 1/2 NM of the runway.

2. Class G airspace is all navigable airspace that is not classified as Class A, Class B, Class C, Class D, or Class E airspace.

3. On sectional charts, blue airport symbols indicate that the airport has a control tower on the field. Magenta airport symbols indicate non-towered airports.

4. Class E airspace is controlled airspace that is not defined as Class A, Class B, Class C, or Class D.

   a. The lower limits of Class E airspace are specified by markings on terminal and sectional charts.

      1) The surface in areas marked by segmented (dashed) magenta lines.
      2) 700 ft. AGL in areas marked by shaded magenta lines.
      3) 1,200 ft. AGL in areas marked by shaded blue lines.
      4) 1,200 ft. AGL in areas defined as Federal Airways. Blue lines between VOR facilities labeled with the letter "V" followed by numbers, e.g., V-120.
      5) A specific altitude depicted in En Route Domestic Areas denoted by blue "zipper" marks.

   b. If not defined, the floor of Class E airspace begins at 14,500 ft. MSL or 1,200 ft. AGL, whichever is higher.

   c. Class E airspace extends up to, but does not include, 18,000 ft. MSL.

5. Class D airspace is an area of controlled airspace surrounding an airport with an operating control tower, not associated with Class B or Class C airspace areas.

    a. Class D airspace is depicted by a segmented (dashed) blue line on sectional charts.

    b. The height of the Class D airspace is shown in a broken box and is expressed in hundreds of feet MSL.

        1) EXAMPLE: [29] means the height of the Class D airspace is 2,900 ft. MSL.

6. Class C airspace areas are depicted by solid magenta lines on sectional charts.

    a. The surface area (formerly called the inner circle) of Class C airspace, the area within 5 NM from the primary airport, begins at the surface and goes up to 4,000 ft. above the airport. The shelf area (formerly called the outer circle) of a Class C airspace area, the area from 5 NM to 10 NM from the primary airport, begins at about 1,200 ft. AGL and extends to the same altitude as the surface area.

    b. The vertical limits are indicated on the chart within each circle and are expressed in hundreds of feet MSL. The top limit is shown above a straight line and the bottom limit beneath the line.

        1) EXAMPLE: See Fig. 24 on page 287. At the bottom right (area 3) is the Savannah Class C airspace.

            a) $\frac{41}{SFC}$ in the surface area means Class C airspace extends from the surface (SFC) to 4,100 ft. MSL.

            b) $\frac{41}{13}$ in the shelf area means Class C airspace extends from 1,300 ft. MSL to 4,100 ft. MSL.

    c. The minimum equipment needed to operate in Class C airspace

        1) 4096-code transponder
        2) Mode C (altitude encoding) capability
        3) Two-way radio communication capability

    d. You must establish and maintain two-way radio communication with ATC prior to entering Class C airspace.

7. Class B airspace areas are depicted by heavy blue lines on sectional charts.

    a. The vertical limits are shown on the chart in the same manner as the vertical limits for Class C airspace discussed in item 6.b. above.

    b. The minimum equipment needed is

        1) A 4096-code transponder
        2) Mode C capability
        3) Two-way radio communication capability

8. When overlapping airspace designations apply to the same airspace, the more restrictive designation applies. Remember that Class A airspace is the most restrictive, and Class G is the least restrictive.

    a. EXAMPLE: The primary airport of a Class D airspace area underlies Class B airspace. The ceiling of the Class D airspace is 3,100 ft. MSL, and the floor of the Class B airspace is 3,000 ft. MSL. Since Class B is more restrictive than Class D, the overlapping airspace between 3,000 ft. and 3,100 ft. MSL is considered to be Class B airspace.

9.   Special use airspace includes prohibited, restricted, warning, military operations, alert, national security, and controlled firing areas.

    a.   **Restricted areas** denote the existence of unusual, often invisible hazards to aircraft such as military firing, aerial gunnery, or guided missiles.

    b.   **Warning areas** contain activity that may be hazardous to nonparticipating aircraft, e.g., aerial gunnery, guided missiles, etc.

        1)   Warning areas extend from 3 NM outward from the U.S. coast.

        2)   A warning area may be located over domestic air or international waters or both.

    c.   **Military operations areas (MOAs)** denote areas of military training activities.

        1)   Pilots should contact any FSS within 100 NM to determine the MOA hours of operation.

        2)   If it is active, the pilot should contact the controlling agency prior to entering the MOA for traffic advisories because of high-density military training.

        3)   When operating in an MOA, exercise extreme caution when military activity is being conducted.

10.   **Military training routes (MTR)** are established below 10,000 ft. MSL for operations at speeds in excess of 250 kt.

    a.   IR means the routes are made in accordance with IFRs.

        1)   VR means the routes are made in accordance with VFRs.

    b.   MTRs that include one or more segments above 1,500 ft. AGL are identified by a three-digit number.

        1)   MTRs with no segment above 1,500 ft. AGL are identified by a four-digit number.

11.   Information about parachute jumping areas and glider operations is contained in the *Airport/Facility Directory (A/FD)*. Parachute jumping areas are marked on sectional charts with a parachute symbol.

12.   Over national wildlife refuges, pilots are requested to maintain a minimum altitude of 2,000 ft. AGL.

13.   Airport data on sectional charts include the following information:

    a.   The name of the airport.

    b.   The elevation of the airport, followed by the length of the longest hard-surfaced runway. An L between the altitude and length indicates lighting.

        1)   EXAMPLE: 1008 L 70 means 1,008 ft. MSL airport elevation, L is for lighting sunset to sunrise, and the length of the longest hard-surfaced runway is 7,000 ft.

        2)   If the L has an asterisk beside it, airport lighting limitations exist, and you should refer to the *A/FD* for information.

    c.   The UNICOM frequency if one has been assigned (e.g., 122.8) is shown after or underneath the runway length.

    d.   At controlled airports, the tower frequency is usually under the airport name and above the runway information. It is preceded by CT.

        1)   If not a federal control tower, NFCT precedes the CT frequency.

    e.   A small, star-shaped symbol immediately above the airport symbol indicates a rotating beacon from sunset to sunrise.

    f.   The notation "NO SVFR" above the airport name means that fixed-wing special VFR operations are prohibited.

14. Obstructions on sectional charts

    a. Obstructions of a height less than 1,000 ft. AGL have the symbol Λ.

        1) A group of such obstructions has the symbol Ѫ.

    b. Obstructions of a height of 1,000 ft. or more AGL have the symbol ⋀.

        1) A group of such obstructions has the symbol ⋀⋀.

    c. Obstructions with high-intensity lights have arrows, or lightning bolts, projecting from the top of the obstruction symbol.

    d. The actual height of the top of obstructions is listed near the obstruction by two numbers: one in bold print over another in light print with parentheses around it.

        1) The bold number is the elevation of the top of the obstruction in feet above MSL.

        2) The light number in parentheses is the height of the obstruction in feet AGL.

        3) The elevation (MSL) at the base of the obstruction is the bold figures minus the light figures.

            a) Use this computation to compute terrain elevation.

            b) Terrain elevation is also given in the airport identifier for each airport and by the contour lines and color shading on the chart.

    e. You must maintain at least 1,000 ft. above obstructions in congested areas and 500 ft. above obstructions in other areas.

15. Navigational facilities are depicted on sectional charts with various symbols depending on type and services available. These symbols are shown on page 283.

    a. A VORTAC is depicted as a hexagon with a dot in the center and a small solid rectangle attached to three of the six sides.

    b. A VOR/DME is depicted as a hexagon within a square.

    c. A VOR is depicted as a hexagon with a dot in the center.

9.3 IDENTIFYING LANDMARKS

1. On aeronautical charts, magenta (red) flags denote prominent landmarks that may be used as visual reporting checkpoints for VFR traffic when contacting ATC.

2. The word "CAUTION" on aeronautical charts usually has an accompanying explanation of the hazard.

3. Airports with a rotating beacon will have a star at the top of the airport symbol on sectional charts.

4. Airports attended during normal business hours and having fuel service are indicated on airport symbols by the presence of small solid squares at the top and bottom and on both sides (9 o'clock and 3 o'clock) on the airport symbol.

9.4 RADIO FREQUENCIES

1. At airports without operating control towers, you should use the Common Traffic Advisory Frequency (CTAF), marked with a letter C in the airport data on the sectional chart.

    a. The control tower (CT) frequency is usually used for CTAF when the control tower is closed.

    b. At airports without control towers but with FSS at the airport, the FSS airport advisory frequency is usually the CTAF.

    c. At airports without a tower or FSS, the UNICOM frequency is the CTAF.

    d. At airports without a tower, FSS, or UNICOM, the CTAF is MULTICOM, i.e., 122.9.

e.   Inbound and outbound traffic should communicate position and monitor CTAF within a 10-NM radius of the airport and give position reports when in the traffic pattern.

f.   At airports with operating control towers the UNICOM frequency listed on the sectional chart and *A/FD* can be used to request services such as fuel, phone calls, and catering.

2.   Flight Watch is the common term for En Route Flight Advisory Service (EFAS). It specifically provides en route aircraft with current weather along their route of flight.

a.   Flight Watch is available throughout the country on 122.0 between 5,000 ft. MSL and 18,000 ft. MSL.

b.   The name of the nearest Flight Watch facility is sometimes indicated in communications boxes.

3.   Hazardous Inflight Weather Advisory Service (HIWAS) is available from navigation facilities that have a small square inside the upper right corner of the navigation aid identifier box.

9.5 FAA ADVISORY CIRCULARS

1.   The FAA issues advisory circulars to provide a systematic means for the issuance of nonregulatory material of interest to the aviation public.

2.   The circulars are issued in a numbered system of general subject matter areas to correspond with the subject areas in Federal Aviation Regulations (e.g., 60 Airmen, 70 Airspace, 90 Air Traffic Control and General Operation).

3.   FAA Advisory Circulars are available from the FAA and the U.S. Government Printing Office.

a.   An Advisory Circular Checklist (AC 00-2) is available by writing to the U.S. Department of Transportation, Subsequent Distribution Office, SVC-121.23, Ardmore East Business Center, 3341 Q 75th Ave., Landover, MD, 20785. Fax requests can be sent to (301) 386-5394.

9.6 *AIRPORT/FACILITY DIRECTORY*

1.   *Airport/Facility Directories (A/FDs)* are published by the U.S. Department of Commerce every 56 days for each of seven geographical districts of the United States.

a.   *A/FDs* provide information on services available, runways, special conditions at the airport, communications, navigation aids, etc.

2.   The airport name comes first.

3.   The third item on the first line is the number of miles and direction of the airport from the city.

a.   EXAMPLE: **4 NW** means 4 NM northwest of the city.

4.   Right-turn traffic is indicated by "Rgt tfc" following a runway number.

5.   When a control tower is not in operation, the CTAF frequency (found in the section titled **Communications**) should be used for traffic advisories.

6.   Initial communication should be with Approach Control if available where you are landing. The frequency is listed following "APP/DEP CON."

a.   It may be different for approaches from different headings.

b.   It may be operational only for certain hours of the day.

7.   In Class C airspace, VFR aircraft are provided the following radar services:

a.   Sequencing to the primary Class C airport

b.   Approved separation between IFR and VFR aircraft

c.   Basic radar services, including safety alerts, limited vectoring, and traffic advisories

8. A sample *A/FD* legend and explanations are provided below through page 266 from Appendix 1 of the FAA's *Computer Testing Supplement for Recreational Pilot and Private Pilot*. That is, you will have access to them during the examination.

2   **DIRECTORY LEGEND**
                                           SAMPLE

①   ③ ④   ⑤   ⑥   ⑦

CITY NAME
   AIRPORT NAME   (ORL)   4 E   UTC–5(–4DT)   N28° 32.72' W81° 21.17'   JACKSONVILLE
     200 B S4 **FUEL** 100, JET A OX1, 2.3 TPA–1000(800)   AOE   ARFF Index A   Not insp.   **COPTER**
                                                                   H–4G. L–19C
     ⑨ ⑩ ⑪   ⑫   ⑬   ⑭   ⑮ ⑯   ⑰   IAP
                                                                            ⑧

⑱ → **RWY 07-25:** H6000X150 (ASPH-PFC)   S–90, D–160, DT–300–PCN 80 R/B/W/T   HIRL   CL   0.4%up E
     RWY 07: ALSF1. Trees.   **RWY 25:** REIL. Rgt tfc.
   RWY 13-31: H4620X100 (ASPH)   HIRL
     RWY 13: SAVASI(S2L)–GA 3.3° TCH 89'. Pole.   **RWY 31:** PAPI(P2L)–GA 3.1° TCH 36'. Tree. Rgt tfc.
   RUNWAY DECLARED DISTANCE INFORMATION
     RWY 07:   TORA–6000   TODA–6700   ASDA–5700   LDA–5500
     RWY 25:   TORA–6000   TODA–6000   ASDA–6000   LDA–5700
⑲ → **AIRPORT REMARKS:** Special Air Traffic Rules—Part 93, see Regulatory Notices. Attended 1200-0300Z ‡ Parachute Jumping. CAUTION cattle and deer on arpt. Acft 100,000 lbs or over ctc Director of Aviation for approval 305–894-9831. Fee for all airline charters, travel clubs and certain revenue producing acft. Flight Notification Service (ADCUS) available.
⑳ → **WEATHER DATA SOURCES:** AWOS-1 120.3 (202) 426-8000. LLWAS.
㉑ → **COMMUNICATIONS:** ATIS 127.25   UNICOM 122.95
   NAME FSS (ORL) on arpt. 123.65 122.65 122.2. TF 1-800-WX-BRIEF. NOTAM FILE ORL.   ←———②
   Ⓡ **NAME APP/DEP CON** 128.35 (1200-0400Z‡)
   TOWER 118.7   **GND CON** 121.7   **CLNC DEL** 125.55   **PRE TAXI CLNC** 125.5
㉒ → **AIRSPACE: CLASS B** See VFR Terminal Area Chart.
㉓ → **RADIO AIDS TO NAVIGATION:** NOTAM FILE MCO. VHF/DF ctc FSS.
   (H) ABVORTAC 112.2   MCO   Chan 59   N28° 32.55' W81° 20.12'   at fld.   1110/8E.
     TWEB avbl 1300-0100Z‡. VOR unusable 050°–060° beyond 15 NM below 5000'.
   HERNY NDB (LOM) 221   OR   N28° 30.40' W81° 26.05'   067° 5.4 NM to fld.
   ILS 109.9   I-ORL   Rwy 07.   LOM HERNY NDB.
   ASR/PAR (1200–0400Z‡)
㉔ → **COMM/HAVAID REMARKS:** Emerg frequency 121.5 not available at tower.

· ·
   HELIPAD H1: H100X75 (ASPH)
   HELIPAD H2: H60X60 (ASPH)   ①
   HELIPORT REMARKS: Helipad H1 lctd on general aviation side and H2 lctd on air carrier side of arpt.
· ·
   187   TPA 1000(813)
   WATERWAY 13-31: 5000X300 (WATER)
   SEAPLANE REMARKS: Birds roosting and feeding areas along river banks. Seaplanes operating adjacent to NE side of arpt not visible from twr and are required to ctc twr.

D AIRPORT NAME   (MCO)   6 SE   UTC–5(–4DT)   N28° 25.88' W81° 19.48'   JACKSONVILLE
  96   B   FUEL 100, JET A, MOGAS LRA   H–4G. L–19C
  RWY 18R-36L: H12004X300 (CONC-GRVD)   S–100, D–200, DT–400   HIRL   IAP
     RWY 18R: ALSF1. REIL. Rgt tfc.   0.3% up.   **RWY 36L:** ALSF1.   0.4% down.
  RWY 18L-36R: H12004X200 (ASPH)   S–165, D–200, DT–400   HIRL
     RWY 18L: LDIN. ALSF1. TDZL. REIL. VASI(V4L)–GA 3.5° TCH 36'. Thld dsplcd 300'. Trees. Rgt tfc. Arresting device.
  AIRPORT REMARKS: Attended 1200-0300Z‡. ACTIVATE HIRL Rwy 18L–36R–CTAF.
  COMMUNICATIONS: CTAF 124.3   ATIS 127.75   UNICOM 122.8
   NAME FSS (MCO) TF 1-800-WX-BRIEF. LC 894-0869. NOTAM FILE MCO.
   NAME RCO 122.4 112.2T 122.1R (NAME FSS)
  Ⓡ **APP CON** 124.8 (337°–179°)   120.1 (180°–336°)   **DEP CON** 120.15
   TOWER 124.3 NFCT (1200-0400Z‡)   **GND CON** 121.85   **CLNC DEL** 134.7
  AIRSPACE: CLASS D svc 1200–0400Z‡ other times CLASS E.
  RADIO AIDS TO NAVIGATION: NOTAM FILE MCO.
   (H) VORTAC 112.2   MCO   Chan 59   N28° 32.55' W81° 20.12'   173° 5.7 NM to fld. 1110/8E.   **HIWAS.**
   MLS   Chan 514   Rwy 36R.

All Bearings and Radials are Magnetic unless otherwise specified.
All mileages are nautical unless otherwise noted.
All times are UTC except as noted.
The horizontal reference datum of this publication is North American Datum of 1983 (NAD83), which for charting purposes is considered equivalent to World Geodetic System 1984 (WGS 84).

Legend 2. – Airport/Facility Directory.

DIRECTORY LEGEND   3

LEGEND

This Directory is an alphabetical listing of data on record with the FAA on all airports that are open to the public, associated terminal control facilities, air route traffic control centers and radio aids to navigation within the conterminous United States, Puerto Rico and the Virgin Islands. Airports are listed alphabetically by associated city name and cross referenced by airport name. Facilities associated with an airport, but with a different name, are listed individually under their own name, as well as under the airport with which they are associated.

The listing of an airport in this directory merely indicates the airport operator's willingness to accommodate transient aircraft, and does not represent that the facility conforms with any Federal or local standards, or that it has been approved for use on the part of the general public.

The information on obstructions is taken from reports submitted to the FAA. It has not been verified in all cases. Pilots are cautioned that objects not indicated in this tabulation (or on charts) may exist which can create a hazard to flight operation. Detailed specifics concerning services and facilities tabulated within this directory are contained in Aeronautical Information Manual, Basic Flight Information and ATC Procedures.

The legend items that follow explain in detail the contents of this Directory and are keyed to the circled numbers on the sample on the preceding page.

① CITY/AIRPORT NAME

Airports and facilities in this directory are listed alphabetically by associated city and state. Where the city name is different from the airport name the city name will appear on the line above the airport name. Airports with the same associated city name will be listed alphabetically by airport name and will be separated by a dashed rule line. All others will be separated by a solid rule line. (Designated Helipads and Seaplane Landing Areas (Water) associated with a land airport will be separated by a dotted line.)

② NOTAM SERVICE

All public use landing areas are provided NOTAM "D" (distant dissemination) and NOTAM "L" (local dissemination) service. Airport NOTAM file identifier is shown following the associated FSS data for individual airports, e.g. "NOTAM FILE IAD". See AIM, Basic Flight Information and ATC Procedures for detailed description of NOTAM's.

③ LOCATION IDENTIFIER

A three or four character code assigned to airports. These identifiers are used by ATC in lieu of the airport name in flight plans, flight strips and other written records and computer operations.

④ AIRPORT LOCATION

Airport location is expressed as distance and direction from the center of the associated city in nautical miles and cardinal points, i.e., 4 NE.

⑤ TIME CONVERSION

Hours of operation of all facilities are expressed in Coordinated Universal Time (UTC) and shown as "Z" time. The directory indicates the number of hours to be subtracted from UTC to obtain local standard time and local daylight saving time UTC−5(−4DT). The symbol ‡ indicates that during periods of Daylight Saving Time effective hours will be one hour earlier than shown. In those areas where daylight saving time is not observed that (−4DT) and ‡ will not be shown. All states observe daylight savings time except Arizona, Hawaii and that portion of Indiana in the Eastern Time Zone and Puerto Rico and the Virgin Islands.

⑥ GEOGRAPHIC POSITION OF AIRPORT

Positions are shown in degrees, minutes and hundredths of a minute and represent the approximate center of mass of all usable runways.

⑦ CHARTS

The Sectional Chart and Low and High Altitude Enroute Chart and panel on which the airport or facility is located. Helicopter Chart locations will be indicated as, i.e., COPTER.

⑧ INSTRUMENT APPROACH PROCEDURES

IAP indicates an airport for which a prescribed (Public Use) FAA Instrument Approach Procedure has been published.

⑨ ELEVATION

The highest point of an airport's usable runways measured in feet from mean sea level. When elevation is sea level it will be indicated as (00). When elevation is below sea level a minus (−) sign will precede the figure.

⑩ ROTATING LIGHT BEACON

B indicates rotating beacon is available. Rotating beacons operate dusk to dawn unless otherwise indicated in AIRPORT REMARKS.

⑪ SERVICING

S1: Minor airframe repairs.   S3: Major airframe and minor powerplant repairs.
S2: Minor airframe and minor powerplant repairs.   S4: Major airframe and major powerplant repairs.

Legend 3. – Airport/Facility Directory.

```
4                           DIRECTORY LEGEND
```

⑫ FUEL

CODE	FUEL		CODE	FUEL
80	Grade 80 gasoline (Red)		B	Jet B—Wide-cut turbine fuel,
100	Grade 100 gasoline (Green)			freeze point–50° C.
100LL	100LL gasoline (low lead) (Blue)		B+	Jet B—Wide-cut turbine fuel with icing inhibitor,
115	Grade 115 gasoline			freeze point–50° C.
A	Jet A—Kerosene freeze point–40° C.		MOGAS	Automobile gasoline which is to be used
A1	Jet A-1—Kerosene freeze point–50° C.			as aircraft fuel.
A1+	Jet A-1—Kerosene with icing inhibitor, freeze point–50° C.			

NOTE: Automobile Gasoline. Certain automobile gasoline may be used in specific aircraft engines if a FAA supplemental type certificate has been obtained. Automobile gasoline which is to be used in aircraft engines will be identified as "MOGAS", however, the grade/type and other octane rating will not be published.

Data shown on fuel availability represents the most recent information the publisher has been able to acquire. Because of a variety of factors, the fuel listed may not always be obtainable by transient civil pilots. Confirmation of availability of fuel should be made directly with fuel dispensers at locations where refueling is planned.

⑬ OXYGEN

OX 1 High Pressure
OX 2 Low Pressure
OX 3 High Pressure—Replacement Bottles
OX 4 Low Pressure—Replacement Bottles

⑭ TRAFFIC PATTERN ALTITUDE

Traffic Pattern Altitude (TPA)—The first figure shown is TPA above mean sea level. The second figure in parentheses is TPA above airport elevation.

⑮ AIRPORT OF ENTRY, LANDING RIGHTS, AND CUSTOMS USER FEE AIRPORTS

U.S. CUSTOMS USER FEE AIRPORT—Private Aircraft operators are frequently required to pay the costs associated with customs processing.
AOE—Airport of Entry—A customs Airport of Entry where permission from U.S. Customs is not required, however, at least one hour advance notice of arrival must be furnished.
LRA—Landing Rights Airport—Application for permission to land must be submitted in advance to U.S. Customs. At least one hour advance notice of arrival must be furnished.
NOTE: Advance notice of arrival at both an AOE and LRA airport may be included in the flight plan when filed in Canada or Mexico, where Flight Notification Service (ADCUS) is available the airport remark will indicate this service. This notice will also be treated as an application for permission to land in the case of an LRA. Although advance notice of arrival may be relayed to Customs through Mexico, Canadian, and U.S. Communications facilities by flight plan, the aircraft operator is solely responsible for insuring that Customs receives the notification. (See Customs, Immigration and Naturalization, Public Health and Agriculture Department requirements in the International Flight Information Manual for further details.)

⑯ CERTIFICATED AIRPORT (FAR 139)

Airports serving Department of transportation certified carriers and certified under FAR, Part 139, are indicated by the ARFF index; i.e., ARFF Index A, which relates to the availability of crash, fire, rescue equipment.

FAR–PART 139 CERTIFICATED AIRPORTS
INDICES AND AIRCRAFT RESCUE AND FIRE FIGHTING EQUIPMENT REQUIREMENTS

Airport Index	Required No. Vehicles	Aircraft Length	Scheduled Departures	Agent + Water for Foam
A	1	<90'	≥1	500#DC or HALON 1211 or 450#DC + 100 gal H$_2$O
B	1 or 2	≥90', <126'	≥5	Index A + 1500 gal H$_2$O
		≥126', <159'	<5	
C	2 or 3	≥126', <159'	≥5	Index A + 3000 gal H$_2$O
		≥159', <200'	<5	
D	3	≥159', <200'	≥5	Index A + 4000 gal H$_2$O
		>200'	<5	
E	3	≥200'	≥5	Index A + 6000 gal H$_2$O

> Greater Than; < Less Than; ≥ Equal of Greater Than; ≤ Equal of Less Than; H$_2$O–Water; DC–Dry Chemical.

NOTE: The listing of ARFF index does not necessarily assure coverage for non-air carrier operations or at other than prescribed times for air carrier. ARFF Index Ltd.–indicates ARFF coverage may or may not be available, for information contact airport manager prior to flight.

Legend 4. – Airport/Facility Directory.

DIRECTORY LEGEND 5

⑰ FAA INSPECTION

All airports not inspected by FAA will be identified by the note: Not insp. This indicates that the airport information has been provided by the owner or operator of the field.

⑱ RUNWAY DATA

Runway information is shown on two lines. That information common to the entire runway is shown on the first line while information concerning the runway ends are shown on the second or following line. Lengthy information will be placed in the Airport Remarks.

Runway direction, surface, length, width, weight bearing capacity, lighting, slope and appropriate remarks are shown for each runway. Direction, length, width, lighting and remarks are shown for seaplanes. The full dimensions of helipads are shown, i.e., 50x150.

RUNWAY SURFACE AND LENGTH

Runway lengths prefixed by the letter "H" indicate that the runways are hard surfaced (concrete, asphalt). If the runway length is not prefixed, the surface is sod, clay, etc. The runway surface composition is indicated in parentheses after runway length as follows:

(AFSC)—Aggregate friction seal coat
(ASPH)—Asphalt
(CONC)—Concrete
(DIRT)—Dirt

(GRVD)—Grooved
(GRVL)—Gravel, or cinders
(PVC)—Porous friction courses
(PSP)—Pierced steel plank

(RFSC)—Rubberized friction seal coat
(TURF)—Turf
(TRTD)—Treated
(WC)—Wire combed

RUNWAY WEIGHT BEARING CAPACITY

Runway strength data shown in this publication is derived from available information and is a realistic estimate of capability at an average level of activity. It is not intended as a maximum allowable weight or as an operating limitation. Many airport pavements are capable of supporting limited operations with gross weights of 25-50% in excess of the published figures. Permissable operating weights, insofar as runway strengths are concerned, are a matter of agreement between the owner and user. When desiring to operate into any airport at weights in excess of those published in the publication, users should contact the airport management for permission. Add 000 to figure following S, D, DT, DDT, AUW, etc., for gross weight capacity.

S—Single-wheel type landing gear, (DC-3), (C-47), (F-15), etc.
D—Dual-wheel type landing gear, (DC-6), etc.
T—Twin-wheel type landing gear, (DC-6), (C-9A), etc.
ST—Single-tandem type landing gear, (C-130).
SBTT—Single-belly twin tandem landing gear, (KC-10).
DT—Dual-tandem type landing gear, (707), etc.
TT—Twin-tandem type (includes quadricycle) landing gear (707), (B-52), (C-135), etc.
TRT—Triple-tandem landing gear, (C-17).
DDT—Double dual-tandem landing gear, (E4A/747).
TDT—Twin delta-tandem landing gear, (C-5, Concorde).
AUW—All up weight. Maximum weight bearing capacity for any aircraft irrespective of landing gear configuration.
SWL—Single Wheel Loading. (This includes information submitted in terms of Equivalent Single Wheel Loading (ESWL) and Single Isolated Wheel Loading). SWL figures are shown in thousands of pounds with the last three figures being omitted.
PSI—Pounds per square inch. PSI is the actual figure expressing maximum pounds per square inch runway will support, e.g., (SWL 000/PSI 535).

Quadricycle and dual-tandem are considered virtually equal for runway weight bearing consideration, as are single-tandem and dual-wheel. Omission of weight bearing capacity indicates information unknown.

The ACN/PCN System is the ICAO method of reporting pavement stength for pavements with bearing strengths greater than 12,500 pounds. The Pavement Classification Number (PCN) is established by an engineering assessment of the runway. The PCN is for use in conjunction with an Aircraft Classification Number (ACN). Consult the Aircraft Flight Manual or other appropriate source for ACN tables or charts. Currently, ACN data may not be available for all aircraft. If an ACN table or chart is available, the ACN can be calculated by taking into account the aircraft weight, the pavement type, and the subgrade category. For runways that have been evaluated under the ACN/PCN system, the PCN will be shown as a five part code (e.g. PCN 80 R/B/W/T). Details of the coded format are as follows:

(1) The PCN NUMBER—The reported PCN indicates that an aircraft with an ACN equal or less than the reported PCN can operate on the pavement subject to any limitation on the tire pressure.

(2) The type of pavement:
R—Rigid
F—Flexible

(3) The pavement subgrade category:
A—High
B—Medium
C—Low
D—Ultra-low

(4) The maximum tire pressure authorized for the pavement:
W—High, no limit
X—Medium, limited to 217 psi
Y—Low, limited to 145 psi
Z—Very low, limited to 73 psi

(5) Pavement evaluation method:
T—Technical evaluation
U—By experience of aircraft using the pavement

NOTE: Prior permission from the airport controlling authority is required when the ACN of the aircraft exceeds the published PCN or aircraft tire pressure exceeds the published limits.

Legend 5. – Airport/Facility Directory.

6

DIRECTORY LEGEND

RUNWAY LIGHTING

Lights are in operation sunset to sunrise. Lighting available by prior arrangement only or operating part of the night only and/or pilot controlled and with specific operating hours are indicated under airport remarks. Since obstructions are usually lighted, obstruction lighting is not included in this code. Unlighted obstructions on or surrounding an airport will be noted in airport remarks. Runway lights nonstandard (NSTD) are systems for which the light fixtures are not FAA approved L-800 series: color, intensity, or spacing does not meet FAA standards. Nonstandard runway lights, VASI, or any other system not listed below will be shown in airport remarks.

Temporary, emergency or limited runway edge lighting such as flares, smudge pots, lanterns or portable runway lights will also be shown in airport remarks. Types of lighting are shown with the runway or runway end they serve.

NSTD—Light system fails to meet FAA standards.
LIRL—Low Intensity Runway Lights
MIRL—Medium Intensity Runway Lights
HIRL—High Intensity Runway Lights
RAIL—Runway Alignment Indicator Lights
REIL—Runway End Identifier Lights
CL—Centerline Lights
TDZL—Touchdown Zone Lights
ODALS—Omni Directional Approach Lighting System
AF OVRN—Air Force Overrun 1000' Standard
      Approach Lighting System
LDIN—Lead-In Lighting System
MALS—Medium Intensity Approach Lighting System
MALSF—Medium Intensity Approach Lighting System with
      Sequenced Flashing Lights
MALSR—Medium Intensity Approach Lighting System with
      Runway Alignment Indicator Lights

SALS—Short Approach Lighting System
SALSF—Short Approach Lighting System with Sequenced
      Flashing Lights
SSALS—Simplified Short Approach Lighting System
SSALF—Simplified Short Approach Lighting System with
      Sequenced Flashing Lights
SSALR—Simplified Short Approach Lighting System with
      Runway Alignment Indicator Lights
ALSAF—High Intensity Approach Lighting System with
      Sequenced Flashing Lights
ALSF1—High Intensity Approach Lighting System with Se-
      quenced Flashing Lights, Category I, Configuration
ALSF2—High Intensity Approach Lighting System with Se-
      quenced Flashing Lights, Category II, Configuration
VASI—Visual Approach Slope Indicator System

NOTE: Civil ALSF-2 may be operated as SSALR during favorable weather conditions.

VISUAL GLIDESLOPE INDICATORS

APAP—A system of panels, which may or may not be lighted, used for alignment of approach path.
    PNIL   APAP on left side of runway   PNIR   APAP on right side of runway
PAPI—Precision Approach Path Indicator
    P2L   2-identical light units placed on left side of   P4L   4-identical light units placed on left side of
        runway   runway
    P2R   2-identical light units placed on right side of   P4R   4-identical light units placed on right side of
        runway   runway
PVASI—Pulsating/steady burning visual approach slope indicator, normally a single light unit projecting two colors.
    PSIL   PVASI on left side of runway   PSIR   PVASI on right side of runway
SAVASI—Simplified Abbreviated Visual Approach Slope Indicator
    S2L   2-box SAVASI on left side of runway   S2R   2-box SAVASI on right side of runway
TRCV—Tri-color visual approach slope indicator, normally a single light unit projecting three colors.
    TRIL   TRCV on left side of runway   TRIR   TRCV on right side of runway
VASI–   Visual Approach Slope Indicator
    V2L   2-box VASI on left side of runway   V6L   6-box VASI on left side of runway
    V2R   2-box VASI on right side of runway   V6R   6-box VASI on right side of runway
    V4L   4-box VASI on left side of runway   V12   12-box VASI on both sides of runway
    V4R   4-box VASI on right side of runway   V16   16-box VASI on both sides of runway
NOTE: Approach slope angle and threshold crossing height will be shown when available; i.e., –GA 3.5° TCH 37'.

PILOT CONTROL OF AIRPORT LIGHTING

<u>Key Mike</u>	<u>Function</u>
7 times within 5 seconds	Highest intensity available
5 times within 5 seconds	Medium or lower intensity
	(Lower REIL or REIL-Off)
3 times within 5 seconds	Lowest intensity available
	(Lower REIL or REIL-Off)

Available systems will be indicated in the Airport Remarks, as follows:
        ACTIVATE MALSR Rwy 07, HIRL Rwy 07–25–122.8 (or CTAF).
                or
        ACTIVATE MIRL Rwy 18–36–122.8 (or CTAF).
                or
        ACTIVATE VASI and REIL, Rwy 07–122.8 (or CTAF).

Where the airport is not served by an instrument approach procedure and/or has an independent type system of different specification installed by the airport sponsor, descriptions of the type lights, method of control, and operating frequency will be explained in clear text. See AIM, "Basic Flight Information and ATC Procedures," for detailed description of pilot control of airport lighting.

RUNWAY SLOPE

Runway slope will be shown only when it is 0.3 percent or more. On runways less than 8000 feet: When available the direction of the slope upward will be indicated, ie., 0.3% up NW. On runways 8000 feet or greater: When available the slope will be shown on the runway end line, ie., RWY 13: 0.3% up., RWY 21: Pole. Rgt tfc. 0.4% down.

RUNWAY END DATA

Lighting systems such as VASI, MALSR, REIL; obstructions; displaced thresholds will be shown on the specific runway end. "Rgt tfc" —Right traffic indicates right turns should be made on landing and takeoff for specified runway end.

Legend 6. – Airport/Facility Directory.

DIRECTORY LEGEND   7

RUNWAY DECLARED DISTANCE INFORMATION

TORA—Take-off Run Available
TODA—Take-off Distance Available
ASDA—Accelerate-Stop Distance Available
LDA—Landing Distance Available

⑲ AIRPORT REMARKS

Landing Fee indicates landing charges for private or non-revenue producing aircraft, in addition, fees may be charged for planes that remain over a couple of hours and buy no services, or at major airline terminals for all aircraft.
Remarks—Data is confined to operational items affecting the status and usability of the airport.
Parachute Jumping—See "PARACHUTE" tabulation for details.
Unless otherwise stated, remarks including runway ends refer to the runway's approach end.

⑳ WEATHER DATA SOURCES

ASOS—Automated Surface Observing System. Reports the same as an AWOS-3 plus precipitation identification and intensity, and freezing rain occurrence (future enhancement).
AWOS—Automated Weather Observing System

  AWOS-A—reports altimeter setting
  AWOS-1—reports altimeter setting, wind data and usually temperature, dewpoint and density altitude.
  AWOS-2—reports the same as AWOS-1 plus visibility.
  AWOS-3—reports the same as AWOS-1 plus visibility and cloud/ceiling data.
  See AIM, Basic Flight Information and ATC Procedures for detailed description of AWOS.

HIWAS—See RADIO AIDS TO NAVIGATION
LAWRS—Limited Aviation Weather Reporting Station where observers report cloud height, weather, obstructions to vision, temperature and dewpoint (in most cases), surface wind, altimeter and pertinent remarks.
LLWAS—indicates a Low Level Wind Shear Alert System consisting of a center field and several field perimeter anemometers.
SAWRS—identifies airports that have a Supplemental Aviation Weather Reporting Station available to pilots for current weather information.
SWSL—Supplemental Weather Service Location providing current local weather information via radio and telephone.

㉑ COMMUNICATIONS

Communications will be listed in sequence in the order shown below:
Common Traffic Advisory Frequency (CTAF). Automatic Terminal Information Service (ATIS) and Aeronautical Advisory Stations (UNICOM) along with their frequency is shown, where available, on the line following the heading "COMMUNICATIONS." When the CTAF and UNICOM is the same frequency, the frequency will be shown as CTAF/UNICOM freq.
Flight Service Station (FSS) information. The associated FSS will be shown followed by the identifier and information concerning availability of telephone service, e.g., Direct Line (DL), Local Call (LC-384-2341). Toll free call, dial (TF 800-852-7036 or TF 1-800-227-7160). Long Distance (LD 020-426-8800 or LD 1-202-555-1212) etc. The airport NOTAM file identifier will be shown as "NOTAM FILE IAD." Where the FSS is located on the field it will be indicated as "on arpt" following the identifier. Frequencies available will follow. The FSS telephone number will follow along with any significant operational information. FSS's whose name is not the same as the airport on which located will also be listed in the normal alphabetical name listing for the state in which located. Remote Communications Outlet (RCO) providing service to the airport followed by the frequency and name of the Controlling FSS.
FSS's provide information on airport conditions, radio aids and other facilities, and process flight plans. Local Airport Advisory Service is provided on the CTAF by FSS's located at non-tower airports or airports where the tower is not in operation.
(See AIM. Par. 157/158 Traffic Advisory Practices at airports where a tower is not in operation or AC 90 - 42C.)
Aviation weather briefing service is provided by FSS specialists. Flight and weather briefing services are also available by calling the telephone numbers listed.
Remote Communications Outlet (RCO)—An unmanned air/ground communicatons facility, remotely controlled and providing UHF or VHF communications capability to extend the service range of an FSS.
Civil Communications Frequencies—Civil communications frequencies used in the FSS air/ground system are now operated simplex on 122.0, 122.2, 122.3, 122.4, 122.6, 123.6; emergency 121.5; plus receive-only on 122.05, 122.1, 122.15, and 123.6.
  a. 122.0 is assigned as the Enroute Flight Advisory Service channel at selected FSS's.
  b. 122.2 is assigned to most FSS's as a common enroute simplex service.
  c. 123.6 is assigned as the airport advisory channel at non-tower FSS locations, however, it is still in commission at some FSS's collocated with towers to provide part time Local Airport Advisory Service.
  d. 122.1 is the primary receive-only frequency at VOR's. 122.05, 122.15 and 123.6 are assigned at selected VOR's meeting certain criteria.
  e. Some FSS's are assigned 50 kHz channels for simplex operation in the 122-123 MHz band (e.g. 122.35). Pilots using the FSS A/G system should refer to this directory or appropriate charts to determine frequencies available at the FSS or remoted facility through which they wish to communicate.
Part time FSS hours of operation are shown in remarks under facility name.

  Emergency frequency 121.5 is available at all Flight Service Stations, Towers, Approach Control and RADAR facilities, unless indicated as not available.
Frequencies published followed by the letter "T" or "R", indicate that the facility will only transmit or receive respectively on that frequency. All radio aids to navigation frequencies are transmit only.

Legend 7. – Airport/Facility Directory.

8   **DIRECTORY LEGEND**

TERMINAL SERVICES

CTAF—A program designed to get all veicles and aircraft at uncontrolled airports on a common frequency.

ATIS—A continous broadcast of recorded non-control information in selected areas of high activity.

UNICOM—A non-government air/ground radio communications facility utilized to provide general airport advisory service.

APP CON—Approach Control. The symbol Ⓡ indicates radar approach control.

TOWER—Control Tower

GND CON—Ground Control

DEP CON—Departure Control. The symbol Ⓡ indicates radar departure control.

CLNC DEL—Clearance Delivery.

PRE TAXI CLNC—Pre taxi clearance.

VFR ADVSY SVC—VFR Advisory Service. Service provided by Non-Radar Approach Control.
 Advisory Service for VFR aircraft (upon a workload basis) ctc APP CON.

TOWER, APP CON and DEP CON RADIO CALL will be the same as the airport name unless indicated otherwise.

㉒ AIRSPACE

CLASS B—Radar Sequencing and Separation Service for all aircraft in CLASS B airspace

TRSA— Radar Sequencing and Separation Service for participating VFR Aircraft within a Terminal Radar Service Area

Class C, D, and E airspace described in this publication is that airspace usually consisting of a 5 NM radius core surface area that begins at the surface and extends upward to an altitude above the airport elevation (charted in MSL for Class C and Class D).

When CLASS C airspace defaults to CLASS E, the core surface area becomes CLASS E. This will be formatted as: **AIRSPACE: CLASS C** svc "times" ctc **APP CON** other times CLASS E.

When Class C airspace defaults to Class G, the core surface area becomes Class G up to but not including the overlying controlled airspace. There are Class E airspace areas beginning at either 700' or 1200' AGL used to transition to/from the terminal or enroute environment. This will be formatted as: **AIRSPACE: CLASS C** svc "times" ctc **APP CON** other times CLASS G. CLASS E 700' (or 1200') AGL & abv.

NOTE: AIRSPACE SVC EFF "TIMES" INCLUDE ALL ASSOCIATED EXTENSIONS. Arrival extensions for instrument approach procedures become part of the primary core surface area. These extensions may be either Class D or Class E airspace and are effective concurrent with the times of the primary core surface area.

(See CLASS AIRSPACE in the Aeronautical Information Manual for further details)

㉓ RADIO AIDS TO NAVIGATION

The Airport Facility Directory lists by facility name all Radio Aids to Navigation, except Military TACANS, that appear on National Ocean Service Visual or IFR Aeronautical Charts and those upon which the FAA has approved an Instrument Approach Procedure. All VOR, VORTAC ILS and MLS equipment in the National Airspace System has an automatic monitoring and shutdown feature in the event of malfunction. Unmonitored, as used in this publication for any navigational aid, means that FSS or tower personnel cannot observe the malfunction or shutdown signal. The NAVAID NOTAM file identifier will be shown as "NOTAM FILE IAD" and will be listed on the Radio Aids to Navigation line. When two or more NAVAIDS are listed and the NOTAM file identifier is different than shown on the Radio Aids to Navigation line, then it will be shown with the NAVAID listing. NOTAM file identifiers for ILS's and their components (e.g., NDB (LOM) are the same as the identifiers for the associated airports and are not repeated. Hazardous Inflight Weather Advisory Service (HIWAS) will be shown where this service is broadcast over selected VOR's.

NAVAID information is tabulated as indicated in the following sample:

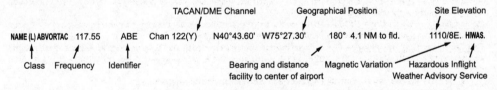

VOR unusable 020°-060° beyond 26 NM below 3500'

Restriction within the normal altitude/range of the navigational aid (See primary alphabetical listing for restrictions on VORTAC and VOR/DME).

    NOTE: Those DME channel numbers with a (Y) suffix require TACAN to be placed in the "Y" mode to receive distance information.

HIWAS—Hazardous Inflight Weather Advisory Service is a continuous broadcast of inflight weather advisories including summarized SIGMETs, convective SIGMETs, AIRMETs and urgent PIREPs. HIWAS is presently broadcast over selected VOR's and will be implemented throughout the conterminous U.S.

ASR/PAR—Indicates that Surveillance (ASR) or Precision (PAR) radar instrument approach minimums are published in the U.S. Terminal Procedures. Only part-time hours of operation will be shown.

Legend 8. – Airport/Facility Directory.

DIRECTORY LEGEND

9

RADIO CLASS DESIGNATIONS

VOR/DME/TACAN Standard Servicie Volume (SSV) Classifications

SSV Class	Altitudes	Distance (NM)
(T) Terminal	1000' to 12,000'	25
(L) Low Altitude	1000' to 18,000'	40
(H) High Altitude	1000' to 14,500'	40
	14,500' to 18,000'	100
	18,000' to 45,000'	130
	45,000' to 60,000'	100

NOTE: Additionally, (H) facilities provide (L) and (T) service volume and (L) facilities provide (T) service. Altitudes are with respect to the station's site elevation. Coverage is not available in a cone of airspace directly above the facility.

The term VOR is, operationally, a general term covering the VHF omnidirectional bearing type of facility without regard to the fact that the power, the frequency protected service volume, the equipment configuration, and operational requirements may vary between facilities at different locations.

AB	Automatic Weather Broadcast
DF	Direction Finding Service.
DME	UHF standard (TACAN compatible) distance measuring equipment.
DME(Y)	UHF standard (TACAN compatible) distance measuring equipment that require TACAN to be placed in the "Y" mode to receive DME.
H	Non-directional radio beacon (homing), power 50 watts to less than 2,000 watts (50 NM at all altitudes).
HH	Non-directional radio beacon (homing), power 2,000 watts or more (75 NM at all altitudes).
H-SAB	Non-directional radio beacons providing automatic transcribed weather service.
ILS	Instrument Landing System (voice, where available, on localizer channel).
ISMLS	Interim Standard Microwave Landing System.
LDA	Localizer Directional Aid.
LMM	Compass locator station when installed at middle marker site (15 NM at all altitudes).
LOM	Compass locator station when installed at outer marker site (15 NM at all altitudes).
MH	Non-directional radio beacon (homing), power less than 50 watts (25 NM at all altitudes).
MLS	Microwave Landing System.
S	Simultaneous range homing signal and/or voice.
SABH	Non-directional radio beacon not authorized for IFR or ATC. Provides automatic weather broadcasts.
SDF	Simplified Direction Facility.
TACAN	UHF navigational facililty-omnidirectional course and distance information.
VOR	VHF navigational facililty-omnidirectional course only.
VOR/DME	Collocated VOR navigational facility and UHF standard distance measuring equipment.
VORTAC	Collocated VOR and TACAN navigational facilities.
W	Without voice on radio facility frequency.
Z	VHF station location marker at a LF facility.

Legend 9. – Airport/Facility Directory.

9.7 *NOTICES TO AIRMEN PUBLICATION (NTAP)*

1. The Notices to Airmen (NOTAM) system disseminates time-critical aeronautical information that either is of a temporary nature or is not sufficiently known in advance to permit publication on aeronautical charts or in other operational publications.

    a. NOTAM information is aeronautical information that could affect your decision to make a flight.

2. NOTAMs are grouped into four types:

    a. **NOTAM (D)** includes information such as airport or primary runway closures; changes in the status of navigational aids, ILSs, and radar service availability; and other information essential to planned en route, terminal, or landing operations. Also included is information on airport taxiways, aprons, ramp areas, and associated lighting.

    b. **FDC NOTAMs** are issued by the Flight Data Center and contain regulatory information such as amendments to published instrument approach charts and other current aeronautical charts.

    c. **Pointer NOTAMs** reduce total NOTAM volume by pointing to other NOTAM (D) and FDC NOTAMs rather than duplicating potentially unnecessary information for an airport or NAVAID. They allow pilots to reference NOTAMs that might not be listed under a given airport or NAVAID identifier.

    d. **Military NOTAMs** reference military airports and NAVAIDs and are rarely of any interest to civilian pilots.

3. The *Notices to Airmen Publication (NTAP)* is issued every 28 days and is an integral part of the NOTAM system. Once a NOTAM is published in the *NTAP*, the NOTAM is not provided during pilot weather briefings unless specifically requested.

    a. The *NTAP* contains (D) NOTAMs that are expected to remain in effect for an extended period and FDC NOTAMs that are current at the time of publication.

QUESTIONS AND ANSWER EXPLANATIONS

All of the private pilot knowledge test questions chosen by the FAA for release as well as additional questions selected by Gleim relating to the material in the previous outlines are reproduced on the following pages. These questions have been organized into the same subunits as the outlines. To the immediate right of each question are the correct answer and answer explanation. You should cover these answers and answer explanations while responding to the questions. Refer to the general discussion in the Introduction on how to take the FAA pilot knowledge test.

Remember that the questions from the FAA pilot knowledge test bank have been reordered by topic and organized into a meaningful sequence. Also, the first line of the answer explanation gives the citation of the authoritative source for the answer.

QUESTIONS

9.1 Longitude and Latitude

1. Which statement about longitude and latitude is true?

A. Lines of longitude are parallel to the Equator.

B. Lines of longitude cross the Equator at right angles.

C. The 0° line of latitude passes through Greenwich, England.

Answer (B) is correct. *(PHAK Chap 15)*
DISCUSSION: Lines of longitude are drawn from the north pole to the south pole and cross the equator at right angles. They indicate the number of degrees east and west of the 0° line of longitude, which passes through Greenwich, England.
Answer (A) is incorrect. Lines of latitude, not longitude, are parallel to the equator. Answer (C) is incorrect. The 0° line of longitude, not latitude, passes through Greenwich, England.

2. (Refer to Figure 21 on page 284.) (Refer to area 3.) Determine the approximate latitude and longitude of Currituck County Airport.

    A.   36°24'N – 76°01'W.

    B.   36°48'N – 76°01'W.

    C.   47°24'N – 75°58'W.

Answer (A) is correct. *(PHAK Chap 15)*
    DISCUSSION: On Fig. 21, find the Currituck County Airport, which is northeast of area 3. Note that the airport symbol is just to the west of 76° longitude (find 76° just north of Virginia Beach and at the bottom of the chart). There are 60 min. between the 76°W and 77°W lines of longitude, with each tick mark depicting 1 min. The airport is one tick to the west of the 76° line, or 76°01'W.
    The latitude is below the 30-min. latitude line across the center of the chart. See the numbered latitude lines at the top (37°) and bottom (36°) of the chart. Since each tick mark represents 1 min. of latitude, and the airport is approximately six ticks south of the 36°30'N latitude, the airport is at 36°24'N latitude. Thus, Currituck County Airport is located at approximately 36°24'N – 76°01'W.
    Answer (B) is incorrect. Currituck County Airport is south of the 36°30'N (not 37°00'N) line of latitude. Answer (C) is incorrect. Currituck County Airport is west (not east) of the 76°W line of longitude and 47°24'N is 11°N of the airport.

3. (Refer to Figure 23 on page 286.) (Refer to area 3.) Determine the approximate latitude and longitude of Shoshone County Airport.

    A.   47°02'N – 116°11'W.

    B.   47°33'N – 116°11'W.

    C.   47°32'N – 116°41'W.

Answer (B) is correct. *(PHAK Chap 15)*
    DISCUSSION: See Fig. 23, just below 3. Shoshone County Airport is just west of the 116° line of longitude (find the 116° line in the 8,000 MSL northwest of Shoshone). There are 60 min. between the 116° line and the 117° line. These are depicted in 1-min. ticks. Shoshone is 11 ticks or 11 min. past the 116° line.
    Note that the 48° line of latitude is labeled. Find the 48° line just northeast of the 116° line. The latitude and longitude lines are presented each 30 min. Since lines of latitude are also divided into 1 min. ticks the airport is three ticks above the 47°30' line or 47°33'N. The correct latitude and longitude is thus 47°33'N – 116°11'W.
    Answer (A) is incorrect. Shoshone Airport is just north of the 47°30' line of latitude (not the 47°00' line). Answer (C) is incorrect. Shoshone Airport is 11 ticks past the 116°00' line of longitude (not the 116°30' line).

4. (Refer to Figure 27 on page 290.) (Refer to area 2.) What is the approximate latitude and longitude of Cooperstown Airport?

    A.   47°25'N – 98°06'W.

    B.   47°25'N – 99°54'W.

    C.   47°55'N – 98°06'W.

Answer (A) is correct. *(PHAK Chap 15)*
    DISCUSSION: First locate the Cooperstown Airport on Fig. 27. It is just above 2, middle right of chart. Note that it is to the left (west) of the 98° line of longitude. The line of longitude on the left side of the chart is 99°. Thus, the longitude is a little bit more than 98°W, but not near 99°W.
    With respect to latitude, note that Cooperstown Airport is just below a line of latitude that is not marked in terms of degrees. However, the next line of latitude below is 47° (see the left side of the chart, northwest of Jamestown Airport). As with longitude, there are two lines of latitude for every degree of latitude; i.e., each line is 30 min. Thus, latitude of the Cooperstown Airport is almost 47°30'N, but not quite. Accordingly, Cooperstown Airport's latitude is 47°25'N and longitude is 98°06'W.
    Answer (B) is incorrect. Cooperstown is just west of the 98° line of longitude (not just east of 99°). Answer (C) is incorrect. Cooperstown is just south of the 47°30' line of latitude (not the 48°00' line).

5. (Refer to Figure 22 on page 285.) (Refer to area 2.) Which airport is located at approximately 47°39'30"N latitude and 100°53'00"W longitude?

    A.   Linrud.

    B.   Crooked Lake.

    C.   Johnson.

Answer (B) is correct. *(PHAK Chap 15)*
    DISCUSSION: On Fig. 22, you are asked to locate an airport at 47°39'30"N latitude and 100°53'W longitude. Note that the 101°W longitude line runs down the middle of the page. Accordingly, the airport you are seeking is 7 min. to the east of that line.
    Each crossline is 1 min. on the latitude and longitude lines. The 48°N latitude line is approximately two-thirds of the way up the chart. The 47°30'N latitude line is about one-fourth of the way up. One-third up from 47°30'N to 48°N latitude would be 47°39'N. At this spot is Crooked Lake Airport.
    Answer (A) is incorrect. Linrud is north of the 48°N latitude line. Answer (C) is incorrect. Both Johnson airports are south of 47°30'N latitude line.

6. (Refer to Figure 22 on page 285.) (Refer to area 3.) Which airport is located at approximately 47°21'N latitude and 101°01'W longitude?

    A. Underwood.

    B. Evenson.

    C. Washburn.

Answer (C) is correct. *(PHAK Chap 15)*
    DISCUSSION: See Fig. 22. Find the 48° line of latitude (2/3 up the figure). Start at the 47°30' line of latitude (the line below the 48° line) and count down nine ticks to the 47°21'N mark and draw a horizontal line on the chart. Next find the 101° line of longitude and go left one tick and draw a vertical line. The closest airport is Washburn.
    Answer (A) is incorrect. Underwood is a city (not an airport) northwest of Washburn by about 1 in. Answer (B) is incorrect. Evenson is north of the 47°36' latitude line.

9.2 Airspace and Altitudes

7. Which is true concerning the blue and magenta colors used to depict airports on Sectional Aeronautical Charts?

    A. Airports with control towers underlying Class A, B, and C airspace are shown in blue; Class D and E airspace are magenta.

    B. Airports with control towers underlying Class C, D, and E airspace are shown in magenta.

    C. Airports with control towers underlying Class B, C, D, and E airspace are shown in blue.

Answer (C) is correct. *(ACL)*
    DISCUSSION: On sectional charts, airports with control towers underlying Class B, C, D, E, or G airspace are shown in blue. Airports with no control towers are shown in magenta.
    Answer (A) is incorrect. There are no airports in Class A airspace. Airports with control towers are shown in blue, and all others are in magenta. Answer (B) is incorrect. Airports with control towers are shown in blue, not magenta.

8. (Refer to Figure 21 on page 284.) (Refer to area 1.) The NALF Fentress (NFE) Airport is in what type of airspace?

    A. Class C.

    B. Class E.

    C. Class G.

Answer (B) is correct. *(ACL)*
    DISCUSSION: The NALF Fentress airport (NFE) is surrounded by a dashed magenta line, indicating Class E airspace from the surface.
    Answer (A) is incorrect. Class C airspace is surrounded by a solid magenta line. The line surrounding NALF Fentress airport (NFE) is dashed magenta. Answer (C) is incorrect. The dashed magenta line surrounding NALF Fentress airport (NFE) indicates Class E begins at the surface. A shaded magenta line would be required to indicate Class G airspace from the surface up to 700 ft. AGL.

9. (Refer to Figure 27 on page 290.) (Refer to area 6.) The airspace overlying and within 5 miles of Barnes County Airport is

    A. Class D airspace from the surface to the floor of the overlying Class E airspace.

    B. Class E airspace from the surface to 1,200 feet MSL.

    C. Class G airspace from the surface to 700 feet AGL.

Answer (C) is correct. *(ACL)*
    DISCUSSION: The requirement is the type of airspace overlying and within 5 SM from Barnes County Airport (Fig. 27). Note at 6 that Barnes County Airport is in the lower right and is surrounded by a shaded magenta (reddish) band, which means the floor of the controlled airspace is 700 ft. Thus, Class G airspace extends from the surface to 700 ft. AGL.
    Answer (A) is incorrect. Class D airspace requires a control tower. The Barnes County Airport does not have a control tower, since the airport identifier is magenta, not blue. Answer (B) is incorrect. An airport located in Class E airspace would be marked by magenta dashed lines, such as the ones surrounding Jamestown Airport to the left. Barnes has no such lines.

10. (Refer to Figure 27 on page 290.) (Refer to area 2.) What hazards to aircraft may exist in areas such as Devils Lake East MOA?

    A. Unusual, often invisible, hazards to aircraft such as artillery firing, aerial gunnery, or guided missiles.

    B. Military training activities that necessitate acrobatic or abrupt flight maneuvers.

    C. High volume of pilot training or an unusual type of aerial activity.

Answer (B) is correct. *(AIM Para 3-4-5)*
    DISCUSSION: Military Operations Areas (MOAs), such as Devils Lake East in Fig. 27 consist of defined lateral and vertical limits that are designated for the purpose of separating military training activities from IFR traffic. Most training activities necessitate acrobatic or abrupt flight maneuvers. Therefore, the likelihood of a collision is increased inside an MOA. VFR traffic is permitted, but extra vigilance should be exercised in seeing and avoiding military aircraft.
    Answer (A) is incorrect. Unusual, often invisible, hazards to aircraft, such as artillery firing, aerial gunnery, or guided missiles, are characteristic of restricted areas, not MOAs. Answer (C) is incorrect. A high volume of pilot training or an unusual type of aerial activity is characteristic of alert areas, not MOAs.

11. (Refer to Figure 27 on page 290.) (Refer to area 1.) Identify the airspace over Lowe Airport.

A. Class G airspace -- surface up to but not including 1,200 feet AGL; Class E airspace -- 1,200 feet AGL up to but not including 18,000 feet MSL.

B. Class G airspace -- surface up to but not including 18,000 feet MSL.

C. Class G airspace -- surface up to but not including 700 feet MSL; Class E airspace -- 700 feet to 14,500 feet MSL.

Answer (A) is correct. *(ACL)*
DISCUSSION: Lowe Airport is located 2 inches left of 1 on Fig. 27. In the lower left-hand corner of the figure, there is a portion of a blue shaded ring. According to the chart legend, this indicates that Class E airspace begins at 1,200 ft. AGL within this ring. Therefore, the Class G airspace would extend from the surface to 1,200 ft. AGL, followed by Class E airspace up to, but not including, 18,000 ft. MSL.
Answer (B) is incorrect. The Class G airspace above Lowe Airport ends at 1,200 ft. AGL (the beginning of Class E airspace), not 18,000 ft. MSL. Answer (C) is incorrect. Class G airspace above Lowe Airport extends to 1,200 ft. AGL, indicated by the shaded blue line in the bottom left corner of the chart excerpt. Class G airspace up to 700 ft. AGL (not MSL) would be indicated by magenta shading surrounding Lowe Airport. Additionally, Class E airspace above Lowe Airport extends to 18,000 ft. MSL, not 14,500 ft. MSL.

12. (Refer to Figure 27 on page 290.) (Refer to area 2.) The visibility and cloud clearance requirements to operate VFR during daylight hours over the town of Cooperstown between 1,200 feet AGL and 10,000 feet MSL are

A. 1 mile and clear of clouds.

B. 1 mile and 1,000 feet above, 500 feet below, and 2,000 feet horizontally from clouds.

C. 3 miles and 1,000 feet above, 500 feet below, and 2,000 feet horizontally from clouds.

Answer (C) is correct. *(FAR 91.155)*
DISCUSSION: The airspace over the town of Cooperstown (Fig. 27, north of 2) is Class G airspace up to 700 ft. AGL, and Class E airspace from 700 ft. AGL up to, but not including, 18,000 ft. MSL (indicated by the magenta shading). Therefore, the visibility and cloud clearance requirements for daylight VFR operation over the town of Cooperstown between 1,200 ft. AGL and 10,000 ft. MSL are 3 miles and 1,000 ft. above, 500 ft. below, and 2,000 ft. horizontally.
Answer (A) is incorrect. One mile and clear of clouds are the visibility and cloud clearance requirements for daylight VFR operation over the town of Cooperstown up to, but not above, 700 ft. AGL (i.e., the visibility and cloud clearance requirements for Class G airspace below 1,200 ft. AGL). Answer (B) is incorrect. One mile and 1,000 ft. above, 500 ft. below, and 2,000 ft. horizontally from clouds are the visibility and cloud clearance requirements for daylight VFR operations at or above 1,200 ft. AGL, but below 10,000 ft. MSL, in Class G airspace. The airspace above Cooperstown in Class E above 700 ft. AGL.

13. With certain exceptions, Class E airspace extends upward from either 700 feet or 1,200 feet AGL to, but does not include,

A. 10,000 feet MSL.

B. 14,500 feet MSL.

C. 18,000 feet MSL.

Answer (C) is correct. *(AIM Para 3-2-6)*
DISCUSSION: Beginning at either 700 ft. AGL or 1,200 ft. AGL, Class E airspace extends up to, but not including, the base of the overlying controlled airspace. With the exception of Class B and Class C airspace, Class E airspace extends up to, but not including, 18,000 ft. MSL, i.e., the floor of Class A airspace.
Answer (A) is incorrect. This is the base of increased VFR visibility and cloud distance requirements and the Mode C requirement. Answer (B) is incorrect. Class G, not Class E, airspace may extend from the surface up to, but not including, 14,500 ft. MSL.

14. (Refer to Figure 27 on page 290.) (Refer to area 3.) When flying over Arrowwood National Wildlife Refuge, a pilot should fly no lower than

A. 2,000 feet AGL.

B. 2,500 feet AGL.

C. 3,000 feet AGL.

Answer (A) is correct. *(AIM Para 7-4-6)*
DISCUSSION: See Fig. 27, which is about 2 in. to the left and slightly below 3. All aircraft are requested to maintain a minimum altitude of 2,000 ft. above the surface of a national wildlife refuge except if forced to land by emergency, landing at a designated site, or on official government business.
Answer (B) is incorrect. This has no significance to wildlife refuges. Answer (C) is incorrect. This has no significance to wildlife refuges.

15. Pilots flying over a national wildlife refuge are requested to fly no lower than

A. 1,000 feet AGL.

B. 2,000 feet AGL.

C. 3,000 feet AGL.

Answer (B) is correct. *(AIM Para 7-4-6)*
DISCUSSION: The Fish and Wildlife Service requests that pilots maintain a minimum altitude of 2,000 ft. above the terrain of national wildlife refuge areas.
Answer (A) is incorrect. This is the required distance above obstructions over congested areas. Answer (C) is incorrect. This has no significance to wildlife refuges.

16. What action should a pilot take when operating under VFR in a Military Operations Area (MOA)?

    A. Obtain a clearance from the controlling agency prior to entering the MOA.

    B. Operate only on the airways that transverse the MOA.

    C. Exercise extreme caution when military activity is being conducted.

Answer (C) is correct. *(AIM Para 3-4-5)*
    DISCUSSION: Military operations areas consist of airspace established for separating military training activities from IFR traffic. VFR traffic should exercise extreme caution when flying within an MOA. Information regarding MOA activity can be obtained from flight service stations (FSSs) within 100 mi. of the MOA.
    Answer (A) is incorrect. A clearance is not required to enter an MOA. Answer (B) is incorrect. VFR flights may fly anywhere in the MOA.

17. (Refer to Figure 21 on page 284.) (Refer to area 4.) What hazards to aircraft may exist in restricted areas such as R-5302B?

    A. Unusual, often invisible, hazards such as aerial gunnery or guided missiles.

    B. High volume of pilot training or an unusual type of aerial activity.

    C. Military training activities that necessitate acrobatic or abrupt flight maneuvers.

Answer (A) is correct. *(AIM Para 3-4-3)*
    DISCUSSION: (See Fig. 21.) Restricted areas denote the existence of unusual, often invisible, hazards to aircraft such as military firing, aerial gunnery, or guided missiles.
    Answer (B) is incorrect. A high volume of pilot training or an unusual type of aerial activity describes an alert, not a warning, area. Answer (C) is incorrect. Military training activities that necessitate acrobatic or abrupt flight maneuvers are characteristic of MOAs, not restricted areas.

18. Flight through a restricted area should not be accomplished unless the pilot has

    A. filed a IFR flight plan.

    B. received prior authorization from the controlling agency.

    C. received prior permission from the commanding officer of the nearest military base.

Answer (B) is correct. *(AIM Para 3-4-3)*
    DISCUSSION: Before an aircraft penetrates a restricted area, authorization must be obtained from the controlling agency. Information pertaining to the agency controlling the restricted area may be found at the bottom of the En Route Chart appropriate to navigation.
    Answer (A) is incorrect. The restriction is to all flight, not just flights without an IFR flight plan. Answer (C) is incorrect. The commanding officer is not necessarily in charge (i.e., controlling agency) of nearby restricted areas.

19. (Refer to Figure 21 on page 284.) (Refer to area 1.) What minimum radio equipment is required to land and take off at Norfolk International?

    A. Mode C transponder and omnireceiver.

    B. Mode C transponder and two-way radio.

    C. Mode C transponder, omnireceiver, and DME.

Answer (B) is correct. *(AIM Para 3-2-4)*
    DISCUSSION: The minimum equipment to land and take off at Norfolk International (Fig. 21) is a Mode C transponder and a two-way radio. Norfolk International is located within Class C airspace. Unless otherwise authorized, a pilot must establish and maintain radio communication with ATC prior to and while operating in the Class C airspace area. Mode C transponders are also required in and above all Class C airspace areas.
    Answer (A) is incorrect. An omnireceiver (VOR) is not required in Class C airspace. Answer (C) is incorrect. Neither an omnireceiver (VOR) nor a DME is required in Class C airspace.

20. (Refer to Figure 21 on page 284.) (Refer to area 2.) The elevation of the Chesapeake Regional Airport is

    A. 20 feet.

    B. 36 feet.

    C. 360 feet.

Answer (A) is correct. *(ACL)*
    DISCUSSION: The requirement is the elevation of the Chesapeake Regional Airport (Fig. 21). East of 2, note that the second line of the airport identifier for Chesapeake Regional reads "20 L 55 123.05." The first number, in bold type, is the altitude of the airport above MSL. It is followed by the L for lighted runway(s), 55 for the length of the longest runway (5,500 ft.), and the CTAF frequency (123.05).
    Answer (B) is incorrect. This is not listed as the elevation of anything near Chesapeake Regional Airport. Answer (C) is incorrect. This is the height above ground of the group obstructions approximately 6 NM southeast of Chesapeake Regional Airport, not the elevation of the airport.

21. (Refer to Figure 23 on page 286.) (Refer to area 1.) The visibility and cloud clearance requirements to operate VFR during daylight hours over Sandpoint Airport at 1,200 feet AGL are

A. 1 mile and clear of clouds.

B. 1 mile and 1,000 feet above, 500 feet below, and 2,000 feet horizontally from each cloud.

C. 3 miles and 1,000 feet above, 500 feet below, and 2,000 feet horizontally from each cloud.

Answer (C) is correct. *(FAR 91.155)*
DISCUSSION: The airspace around Sandpoint Airport is Class G airspace from the surface to 700 ft. AGL and Class E airspace from 700 ft. AGL up to, but not including, 18,000 ft. MSL (indicated by the magenta shading). Therefore, 1,200 ft. AGL is within Class E airspace. The VFR visibility and cloud clearance requirements for operations in Class E airspace below 10,000 ft. MSL are 3 miles and a distance of 1,000 ft. above, 500 ft. below, and 2,000 ft. horizontally from each cloud.
Answer (A) is incorrect. One mile and clear of clouds are the visibility and cloud clearance requirements for VFR operations in Class G, not Class E, airspace at or below 1,200 ft. AGL. Answer (B) is incorrect. One mile and 1,000 ft. above, 500 ft. below, and 2,000 ft. horizontally are the visibility and cloud clearance requirements for VFR operations in Class G, not Class E, airspace at more than, not at, 1,200 ft. AGL but less than 10,000 ft. MSL.

22. (Refer to Figure 23 on page 286.) (Refer to area 3.) The vertical limits of that portion of Class E airspace designated as a Federal Airway over Magee Airport are

A. 1,200 feet AGL to 17,999 feet MSL.

B. 700 feet MSL to 12,500 feet MSL.

C. 7,500 feet MSL to 17,999 feet MSL.

Answer (A) is correct. *(ACL)*
DISCUSSION: Magee Airport on Fig. 23 is northwest of 3. The question asks for the vertical limits of the Class E airspace over the airport. Class E airspace areas extend upwards but do not include 18,000 ft. MSL (base of Class A airspace). The floor of a Class E airspace designated as an airway is 1,200 ft. AGL, unless otherwise indicated.
Answer (B) is incorrect. This airway begins at 1,200 ft. AGL and extends upward to 17,999 ft. MSL, not 12,500 ft. MSL. Answer (C) is incorrect. Class E airspace designated as a Federal Airway begins at 1,200 ft. AGL, not 7,500 ft. MSL, unless otherwise indicated.

23. Information concerning parachute jumping sites may be found in the

A. NOTAM's.

B. Airport/Facility Directory.

C. Graphic Notices and Supplemental Data.

Answer (B) is correct. *(A/FD)*
DISCUSSION: Information concerning parachute jump sites may be found in the *Airport/Facility Directory*.
Answer (A) is incorrect. NOTAMs are only issued for special situations, not routine jump sites. Answer (C) is incorrect. Graphic Notices and Supplemental Data are no longer published.

24. (Refer to Figure 23 on page 286 and Legend 1 on page 283.) (Refer to area 2.) For information about the parachute jumping and glider operations at Silverwood Airport, refer to

A. notes on the border of the chart.

B. the Airport/Facility Directory.

C. the Notices to Airmen (NOTAM) publication.

Answer (B) is correct. *(ACL)*
DISCUSSION: The miniature parachute near the Silverwood Airport (at 2 on Fig. 23) indicates a parachute jumping area. In Legend 1, the symbol for a parachute jumping area instructs you to see the *Airport/Facility Directory (A/FD)* for more information. The *A/FD* will also have information on the glider operations at Silverwood Airport.
Answer (A) is incorrect. The sectional chart legend identifies symbols only. Answer (C) is incorrect. NOTAMs are issued only for hazards to flight.

25. (Refer to Figure 24 on page 287.) (Refer to area 3.) What is the height of the lighted obstacle approximately 6 nautical miles southwest of Savannah International?

A. 1,500 feet MSL.

B. 1,531 feet AGL.

C. 1,549 feet MSL.

Answer (C) is correct. *(ACL)*
DISCUSSION: On Fig. 24, find the lighted obstacle noted by its proximity to Savannah International by being outside the surface area of the Class C airspace, which has a 5-NM radius. It is indicated by the obstacle symbol with arrows or lightning flashes extending from the tip. According to the numbers to the northeast of the symbol, the height of the obstacle is 1,549 ft. MSL or 1,534 ft. AGL.
Answer (A) is incorrect. The unlighted tower 8 NM, not 6 NM, southwest of the airport has a height of 1,500 ft. MSL. Answer (B) is incorrect. An unlighted tower 9 NM, not 6 NM, southwest of the airport has a height of 1,531 ft. AGL.

26. (Refer to Figure 24 on page 287.) (Refer to area 3.) What is the floor of the Savannah Class C airspace at the shelf area (outer circle)?

A. 1,200 feet AGL.

B. 1,300 feet MSL.

C. 1,700 feet MSL.

Answer (B) is correct. *(ACL)*
DISCUSSION: Class C airspace consists of a surface area and a shelf area. The floor of the shelf area is 1,200 ft. above the airport elevation. The Savannah Class C airspace (Fig. 24, area 3) is depicted by solid magenta circles. For each circle there is a number over a number or SFC. The numbers are in hundreds of feet MSL. The bottom number represents the floor of the airspace. Thus, the floor of the shelf area of the Class C airspace is 1,300 ft. MSL (41/13).
Answer (A) is incorrect. The floor of the outer circle of Class C airspace does not vary with the ground elevation. The FAA specifies a fixed MSL altitude, rounded to the nearest 100 ft., which is about 1,200 ft. above the airport elevation. Answer (C) is incorrect. This is the maximum elevation figure (MEF) of the quadrant encompassing Savannah Class C airspace, not the floor of the shelf area.

27. (Refer to Figure 24 on page 287.) (Refer to area 3.) The top of the group obstruction approximately 11 nautical miles from the Savannah VORTAC on the 340° radial is

A. 455 feet MSL.

B. 400 feet AGL.

C. 432 feet MSL.

Answer (A) is correct. *(ACL)*
DISCUSSION: To determine the height of the lighted stack, first find it on Fig. 24. Locate the compass rose and look along the 340° radial, knowing that the compass rose has a 10 NM radius. Just outside the compass rose is a group obstruction (stacks). Its height is 455 ft. MSL; AGL height is not shown.
Answer (B) is incorrect. This is the height of an obstruction to the northeast of the group obstruction. Answer (C) is incorrect. This is the height of a group obstruction on the 320°, not 340°, radial.

28. (Refer to Figure 25 on page 288.) (Refer to area 1.) What minimum altitude is necessary to vertically clear the obstacle on the northeast side of Airpark East Airport by 500 feet?

A. 1,010 feet MSL.

B. 1,273 feet MSL.

C. 1,283 feet MSL.

Answer (B) is correct. *(ACL and FAR 91.119)*
DISCUSSION: Find Airpark East, which is near 1 in Fig. 25. Remember to locate the actual airport symbol, not just the name of the airport. It is the third of three airports in a southwesterly line from the 1. The elevation of the top of the obstacle on the northeast side of the airport is marked in bold as 773 ft. MSL. Minimum altitude to clear the 773-ft. obstacle by 500 ft. is 1,273 ft. MSL.
Answer (A) is incorrect. The airport elevation, not the obstacle, is 510 ft. Answer (C) is incorrect. The AGL altitude of a tower 1 in. west of Caddo Mills Airport appears as 283.

29. (Refer to Figure 25 on page 288.) (Refer to area 2.) What minimum altitude is necessary to vertically clear the obstacle on the southeast side of Winnsboro Airport by 500 feet?

A. 823 feet MSL.

B. 1,013 feet MSL.

C. 1,403 feet MSL.

Answer (C) is correct. *(ACL)*
DISCUSSION: The first step is to find the obstacle on the southeast side of Winnsboro Airport on Fig. 25, near 2. The elevation numbers to the right of the obstruction symbol indicate that its top is 903 ft. MSL or a height of 323 ft. AGL. Thus, the clearance altitude is 1,403 ft. MSL (903 ft. MSL + 500 ft. of clearance).
Answer (A) is incorrect. Since the obstacle height is 323 ft. AGL (number in parentheses), the minimum altitude to clear the obstacle by 500 ft. is 823 ft. AGL, not 823 ft. MSL. Answer (B) is incorrect. This is 500 ft. above the airport elevation (513 ft. MSL), not 500 ft. above the top of the obstacle height of 903 ft.

30. (Refer to Figure 26 on page 289.) At which airports is fixed-wing Special VFR not authorized?

A. Fort Worth Meacham and Fort Worth Spinks.

B. Dallas-Fort Worth International and Dallas Love Field.

C. Addison and Redbird.

Answer (B) is correct. *(ACL)*
DISCUSSION: The first (top) line of the airport data for Dallas-Ft. Worth Int'l. and Dallas Love Field (Fig. 26, areas 5 and 6) indicates NO SVFR, which means no special VFR permitted for a fixed-wing aircraft.
Answer (A) is incorrect. Ft. Worth Meacham permits special VFR operations since it is not indicated otherwise. Ft. Worth Spinks is a non-tower airport; thus ATC does not grant or deny special VFR clearances. Answer (C) is incorrect. Addison and Redbird permit special VFR operations since it is not indicated otherwise.

31. (Refer to Figure 26 on page 289.) (Refer to area 7.) The airspace overlying Mc Kinney (TKI) is controlled from the surface to

A. 700 feet AGL.

B. 2,900 feet MSL.

C. 2,500 feet MSL.

Answer (B) is correct.   *(ACL)*
DISCUSSION: The airspace overlying Mc Kinney airport (TKI) (Fig. 26, northeast of 7) is Class D airspace as denoted by the segmented blue lines. The upper limit is depicted in a broken box in hundreds of feet MSL to the left of the airport symbol. The box contains the number "29," meaning that the vertical limit of the Class D airspace is 2,900 ft. MSL.
Answer (A) is incorrect. This is normally the vertical limit of uncontrolled, not controlled, airspace in the vicinity of non-towered airports with an authorized instrument approach. Answer (C) is incorrect. The height of 2,500 ft. AGL, not MSL, is normally the upper limit of Class D airspace. This is not the case here, where the upper limit is somewhat lower, at about 2,300 ft. AGL [2,900 ft. MSL – 586 ft. AGL (field elevation) = 2,314 ft. AGL].

32. (Refer to Figure 26 on page 289.) (Refer to area 4.) The airspace directly overlying Fort Worth Meacham is

A. Class B airspace to 10,000 feet MSL.

B. Class C airspace to 5,000 feet MSL.

C. Class D airspace to 3,200 feet MSL.

Answer (C) is correct.   *(ACL)*
DISCUSSION: The airspace overlying Fort Worth Meacham (Fig. 26, southeast of 4) is Class D airspace as denoted by the segmented blue lines. The upper limit is depicted in a broken box in hundreds of feet MSL northeast of the airport. Thus, the Class D airspace extends from the surface to 3,200 ft. MSL.
Answer (A) is incorrect. Class D, not Class B, airspace extends from the surface of Ft. Worth Meacham. Class B airspace overlies the airport from 4,000 ft. MSL to 10,000 ft. MSL. Answer (B) is incorrect. Class D, not Class C, airspace directly overlies Ft. Worth Meacham from the surface to 3,200 ft. MSL, not 5,000 ft. MSL.

33. (Refer to Figure 26 on page 289.) (Refer to area 2.) The floor of Class B airspace at Addison Airport is

A. at the surface.

B. 3,000 feet MSL.

C. 3,100 feet MSL.

Answer (B) is correct.   *(ACL, FAR 71.9)*
DISCUSSION: Addison Airport (Fig. 26, area 2) has a segmented blue circle around it depicting Class D airspace. Addison Airport also underlies Class B airspace as depicted by solid blue lines. The altitudes of the Class B airspace are shown as $\frac{110}{30}$ to the east of the airport. The bottom number denotes the floor of the Class B airspace to be 3,000 ft. MSL.
Answer (A) is incorrect. The floor of Class D, not Class B, airspace is at the surface. Answer (C) is incorrect. This is not a defined limit of any airspace over Addison airport.

34. What minimum radio equipment is required for operation within Class C airspace?

A. Two-way radio communications equipment and a 4096-code transponder.

B. Two-way radio communications equipment, a 4096-code transponder, and DME.

C. Two-way radio communications equipment, a 4096-code transponder, and an encoding altimeter.

Answer (C) is correct.   *(AIM Para 3-2-4)*
DISCUSSION: To operate within Class C airspace, the aircraft must have

1. Two-way radio communications equipment,
2. A 4096-code transponder, and
3. An encoding altimeter.

Answer (A) is incorrect. An encoding altimeter (Mode C) is required in Class C airspace. Answer (B) is incorrect. DME is not required in Class C airspace.

35. What minimum radio equipment is required for VFR operation within Class B airspace?

A. Two-way radio communications equipment and a 4096-code transponder.

B. Two-way radio communications equipment, a 4096-code transponder, and an encoding altimeter.

C. Two-way radio communications equipment, a 4096-code transponder, an encoding altimeter, and a VOR or TACAN receiver.

Answer (B) is correct.   *(AIM Para 3-2-3)*
DISCUSSION: To operate within Class B airspace, the aircraft must have

1. Two-way radio communications equipment,
2. A 4096-code transponder, and
3. An encoding altimeter.

Answer (A) is incorrect. An encoding altimeter (Mode C) is also required in Class B airspace. Answer (C) is incorrect. A VOR or TACAN receiver is required for IFR, not VFR, operation within Class B airspace.

36. (Refer to Figure 26 on page 289.) (Refer to area 8.) What minimum altitude is required to fly over the Cedar Hill TV towers in the congested area south of NAS Dallas?

A. 2,555 feet MSL.

B. 3,449 feet MSL.

C. 3,349 feet MSL.

Answer (B) is correct. *(FAR 91.119)*
DISCUSSION: The Cedar Hill TV towers (Fig. 26, west of 8) have an elevation of 2,449 ft. MSL. The minimum safe altitude over a congested area is 1,000 ft. above the highest obstacle within a horizontal radius of 2,000 ft. of the aircraft. Thus, to vertically clear the towers, the minimum altitude is 3,449 ft. MSL (2,449 + 1,000).
Answer (A) is incorrect. The figure of 2,555 ft. AGL, not 2,555 ft. MSL, is the minimum height to fly over the shortest, not the tallest, of the obstructions in the group. Answer (C) is incorrect. This is only 900 ft., not 1,000 ft., above the tallest structure.

37. (Refer to Figure 26 on page 289.) (Refer to area 4.) The floor of Class B airspace overlying Hicks Airport (T67) north-northwest of Fort Worth Meacham Field is

A. at the surface.

B. 3,200 feet MSL.

C. 4,000 feet MSL.

Answer (C) is correct. *(ACL)*
DISCUSSION: Hicks Airport (T67) on Fig. 26 is northeast of 4. Class B airspace is depicted by a solid blue line, as shown just west of the airport. Follow the blue line toward the bottom of the chart until you find a number over a number in blue, $\frac{110}{40}$.
The bottom number denotes the floor of the Class B airspace as 4,000 ft. MSL.
Answer (A) is incorrect. The floor of the Class B airspace would be at the surface if SFC, not 40, was below the 100, as depicted just south of Dallas-Ft. Worth International Airport. Answer (B) is incorrect. This is the upper limit of the Class D airspace for the Ft. Worth/Meacham Airport, not the floor of Class B airspace overlying Hicks Airport.

38. (Refer to Figure 22 on page 285.) The terrain elevation of the light tan area between Minot (area 1) and Audubon Lake (area 2) varies from

A. sea level to 2,000 feet MSL.

B. 2,000 feet to 2,500 feet MSL.

C. 2,000 feet to 2,700 feet MSL.

Answer (B) is correct. *(ACL)*
DISCUSSION: The requirement is the terrain elevation in the tan area between 1 and 2 in Fig. 22. The tan area indicates terrain between 2,000 ft. and 3,000 ft. The elevation contours on sectionals vary by 500 ft. increments. The 2,000 ft. contour line is located where the color changes from light green to light tan. Since there is no other contour line in the light tan area, the terrain elevation is between 2,000 ft. and 2,500 ft. MSL. Also, Poleschook Airport (halfway between 1 and 2) indicates an elevation above MSL of 2,245.
Answer (A) is incorrect. The light tan area indicates terrain elevation from 2,000 ft. to 3,000 ft. MSL, not from sea level to 2,000 ft. MSL. Answer (C) is incorrect. Elevation contours vary by 500 ft., not 700 ft.

39. (Refer to Figure 26 on page 289.) (Refer to area 5.) The navigation facility at Dallas-Ft. Worth International (DFW) is a

A. VOR.

B. VORTAC.

C. VOR/DME.

Answer (C) is correct. *(ACL)*
DISCUSSION: On Fig. 26, DFW is located at the center of the chart and the navigation facility is 1 NM south of the right set of parallel runways. The symbol is a hexagon with a dot in the center within a square. This is the symbol for a VOR/DME navigation facility.
Answer (A) is incorrect. A VOR facility symbol is a hexagon with a dot in the center, but is not located within a square. Answer (B) is incorrect. A VORTAC symbol is a hexagon with a dot in the center and a small rectangle attached to three of the six sides. The Ranger VORTAC is depicted approximately 7 NM to the west of DFW airport.

40. (Refer to Figure 22 on page 285.) (Refer to area 3.) What type military flight operations should a pilot expect along IR 644?

A. IFR training flights above 1,500 feet AGL at speeds in excess of 250 knots.

B. VFR training flights above 1,500 feet AGL at speeds less than 250 knots.

C. Instrument training flights below 1,500 feet AGL at speeds in excess of 150 knots.

Answer (A) is correct. *(AIM Para 3-5-2)*
DISCUSSION: In Fig. 22, IR 644 is below area 3. Military training flights are established to promote proficiency of military pilots in the interest of national defense. Military flight routes below 1,500 ft. are charted with four-digit numbers; those above 1,500 ft. have three-digit numbers. IR means the flights are made in accordance with IFR. (VR would mean they use VFR.) Thus, IR 644, a three-digit number, is above 1,500 ft., and flights will be flown under IFR rules.
Answer (B) is incorrect. VFR flights are coded VR (not IR), and the speeds are in excess of (not less than) 250 kt. Answer (C) is incorrect. Military training flights below 1,500 ft. AGL have four-digit (not three-digit) identifier numbers, and the airspeed is in excess of 250 kt. (not 150 kt.).

9.3   Identifying Landmarks

41.   (Refer to Figure 21 on page 284.) (Refer to area 5.) The CAUTION box denotes what hazard to aircraft?

    A.   Unmarked blimp hangars at 300 feet MSL.

    B.   Unmarked balloon on cable to 3,000 feet AGL.

    C.   Unmarked balloon on cable to 3,000 feet MSL.

Answer (C) is correct.   *(ACL)*
    DISCUSSION: On Fig. 21, northwest of 5, find "CAUTION: UNMARKED BALLOON ON CABLE TO 3,000 MSL." This is self-explanatory.
    Answer (A) is incorrect. The box clearly says that there is an unmarked balloon, not blimp hangars, to 3,000 ft. MSL, not 300 ft. MSL. Answer (B) is incorrect. The balloon extends to 3,000 ft. MSL, not AGL.

42.   (Refer to Figure 21 on page 284.) (Refer to area 2.) The flag symbol at Lake Drummond represents a

    A.   compulsory reporting point for Norfolk Class C airspace.

    B.   compulsory reporting point for Hampton Roads Airport.

    C.   visual checkpoint used to identify position for initial callup to Norfolk Approach Control.

Answer (C) is correct.   *(ACL)*
    DISCUSSION: The magenta (reddish) flag (Fig. 21, north of 2) at Lake Drummond signifies that the lake is a visual check-point that can be used to identify the position for initial callup to the Norfolk approach control.
    Answer (A) is incorrect. Compulsory reporting points are on IFR, not sectional, charts. They are used on IFR flights. Answer (B) is incorrect. Compulsory reporting points are on IFR, not sectional, charts. They are used on IFR flights.

43.   (Refer to Figure 22 on page 285.) Which public use airports depicted are indicated as having fuel?

    A.   Minot Int'l (area 1) and Mercer County Regional Airport (area 3).

    B.   Minot Int'l (area 1) and Garrison (area 2).

    C.   Mercer County Regional Airport (area 3) and Garrison (area 2).

Answer (A) is correct.   *(ACL)*
    DISCUSSION: On Fig. 22, the requirement is to identify the airports having fuel available. Airports having fuel available are designated by small squares extending from the top, bottom, and both sides of the airport symbol. Only Minot (area 1) and Mercer County Regional Airport (area 3) have such symbols.
    Answer (B) is incorrect. Garrison (2 inches left of 2) does not indicate that fuel is available. Answer (C) is incorrect. Garrison (2 inches left of 2) does not indicate that fuel is available.

44.   (Refer to Figure 24 on page 287.) The flag symbols at Statesboro Bullock County Airport, Claxton-Evans County Airport, and Ridgeland Airport are

    A.   outer boundaries of Savannah Class C airspace.

    B.   airports with special traffic patterns.

    C.   visual checkpoints to identify position for initial callup prior to entering Savannah Class C airspace.

Answer (C) is correct.   *(ACL)*
    DISCUSSION: On Fig. 24, note the flag symbols at Claxton-Evans County Airport (1 in. to the left of 2), at Statesboro Bullock County Airport (2 in. above 2), and at Ridgeland Airport (2 in. above 3). These airports are visual checkpoints to identify position for initial callup prior to entering the Savannah Class C airspace.
    Answer (A) is incorrect. They do not indicate outer boundaries of the Class C airspace. The flags are outside the Class C airspace area, the boundaries of which are marked by solid magenta lines. Answer (B) is incorrect. Airports with special traffic patterns are noted in the *Airport/Facility Directory* and also by markings at the airport around the wind sock or tetrahedron.

9.4   Radio Frequencies

45.   (Refer to Figure 27 on page 290.) (Refer to area 2.) What is the recommended communication procedure when inbound to land at Cooperstown Airport?

    A.   Broadcast intentions when 10 miles out on the CTAF/MULTICOM frequency, 122.9 MHz.

    B.   Contact UNICOM when 10 miles out on 122.8 MHz.

    C.   Circle the airport in a left turn prior to entering traffic.

Answer (A) is correct.   *(AIM Para 4-1-9)*
    DISCUSSION: Find Cooperstown Airport, which is at the top of Fig. 27, just north of 2. You should broadcast your intentions when 10 NM out on the CTAF/MULTICOM frequency, 122.9 MHz.
    Answer (B) is incorrect. There is no UNICOM indicated at Cooperstown, and the CTAF is 122.9, not 122.8. Answer (C) is incorrect. A left turn is not a communication procedure.

46. (Refer to Figure 27 on page 290.) (Refer to area 4.) The CTAF/UNICOM frequency at Jamestown Airport is

A. 122.0 MHz.

B. 123.0 MHz.

C. 123.6 MHz.

Answer (B) is correct. *(ACL)*
DISCUSSION: The UNICOM frequency is printed in bold italics in the airport identifier. At Jamestown it is 123.0 MHz. The C next to it indicates it as the CTAF.
Answer (A) is incorrect. This is the Flight Watch frequency, not UNICOM. Answer (C) is incorrect. This is an FSS frequency, not UNICOM.

47. (Refer to Figure 27 on page 290.) (Refer to area 6.) What is the CTAF/UNICOM frequency at Barnes County Airport?

A. 122.0 MHz.

B. 122.8 MHz.

C. 123.6 MHz.

Answer (B) is correct. *(ACL)*
DISCUSSION: In Fig. 27, Barnes County Airport is 1 in. below 6. The CTAF at Barnes County Airport is marked as the UNICOM frequency for the airport, i.e., 122.8.
Answer (A) is incorrect. This is Flight Watch. Answer (C) is incorrect. This is an FSS frequency.

48. (Refer to Figure 21 on page 284.) (Refer to area 3.) What is the recommended communications procedure for a landing at Currituck County Airport?

A. Transmit intentions on 122.9 MHz when 10 miles out and give position reports in the traffic pattern.

B. Contact Elizabeth City FSS for airport advisory service.

C. Contact New Bern FSS for area traffic information.

Answer (A) is correct. *(AIM Para 4-1-9)*
DISCUSSION: Find the symbol for Currituck County Airport, 1/2 in. northeast of 3 in Fig. 21. Incoming flights should use MULTICOM, 122.9, as the CTAF, because it is marked with a C. The recommended procedure is to report 10 NM out and then give position reports in the airport traffic pattern.
Answer (B) is incorrect. There is no Elizabeth City FSS. Elizabeth City is serviced by the Raleigh FSS, as indicated by "Raleigh" just below the identifier box for Elizabeth City VOR. Answer (C) is incorrect. The controlling FSS is Raleigh, not New Bern, and Raleigh FSS does not monitor 122.9, which is marked as the CTAF at Currituck County Airport.

49. As standard operating practice, all inbound traffic to an airport without a control tower should continuously monitor the appropriate facility from a distance of

A. 25 miles.

B. 20 miles.

C. 10 miles.

Answer (C) is correct. *(AIM Para 4-1-9)*
DISCUSSION: As a standard operating practice, pilots of inbound traffic to an airport without a control tower should continuously monitor and communicate, as appropriate, on the designated Common Traffic Advisory Frequency (CTAF) from 10 mi. to landing.
Answer (A) is incorrect. All inbound traffic to an airport without a control tower should continuously monitor the CTAF from a distance of 10 mi., not 25 mi. Answer (B) is incorrect. All inbound traffic to an airport without a control tower should continuously monitor the CTAF from a distance of 10 mi., not 20 mi.

50. (Refer to Figure 22 on page 285.) On what frequency can a pilot receive Hazardous Inflight Weather Advisory Service (HIWAS) in the vicinity of area 1?

A. 117.1 MHz.

B. 118.0 MHz.

C. 122.0 MHz.

Answer (A) is correct. *(ACL)*
DISCUSSION: On Fig. 22, 1 is on the upper left and the Minot VORTAC information box is 1 in. below 1. Availability of Hazardous Inflight Weather Advisory Service (HIWAS) will be indicated by a circle which contains an "H," found in the upper right corner of a navigation frequency box. Note that the Minot VORTAC information box has such a symbol. Accordingly, a HIWAS can be obtained on the VOR frequency of 117.1.
Answer (B) is incorrect. "Ch 118" in the Minot VORTAC information box refers to the TACAN channel (the military equivalent of VOR/DME). Answer (C) is incorrect. This is the universal frequency for Flight Watch.

51. (Refer to Figure 22 on page 285.) (Refer to area 2.) The CTAF/MULTICOM frequency for Garrison Airport is

A. 122.8 MHz.

B. 122.9 MHz.

C. 123.0 MHz.

Answer (B) is correct. *(ACL)*
DISCUSSION: The CTAF for Garrison Municipal Airport (2 inches left of 2 in Fig. 22) is 122.9, because that frequency is marked with a C.
Answer (A) is incorrect. There is no indication of 122.8 at Garrison. Answer (C) is incorrect. There is no indication of 123.0 at Garrison.

52. (Refer to Figure 32 below, and Figure 23 on page 286.) (Refer to area 2 in Figure 23.) What is the correct UNICOM frequency to be used at Coeur D'Alene to request fuel?

A. 135.075 MHz.

B. 122.1/108.8 MHz.

C. 122.8 MHz.

Answer (C) is correct. *(ACL)*
   DISCUSSION: The correct frequency to request fuel at the Coeur D'Alene Airport is the UNICOM frequency 122.8. It is given in Fig. 23, after "L74" in the airport information on the sectional chart. Radio frequencies are also given in Fig. 32, the *Airport/Facility Directory (A/FD)*, under "Communications."
   Answer (A) is incorrect. This is the AWOS frequency for Coeur D'Alene Airport. Answer (B) is incorrect. The COE VOR/DME frequency, not the UNICOM, is 108.8, and 122.1 is not a frequency associated with Coeur D'Alene Airport.

53. (Refer to Figure 32 below, and Figure 23 on page 286.) (Refer to area 2 in Figure 23.) At Coeur D'Alene, which frequency should be used as a Common Traffic Advisory Frequency (CTAF) to monitor airport traffic?

A. 122.05 MHz.

B. 135.075 MHz.

C. 122.8 MHz.

Answer (C) is correct. *(A/FD)*
   DISCUSSION: Fig. 32 is the *A/FD* excerpt for Coeur D'Alene Air Terminal. Look for the section titled **Communications**. On that same line, it states that the CTAF (and UNICOM) frequency is 122.8. The CTAF can also be found in the airport information on the sectional chart.
   Answer (A) is incorrect. This is the remote communication outlet (RCO) frequency to contact Boise FSS in the vicinity of Coeur D'Alene, not the CTAF. Answer (B) is incorrect. This is the AWOS frequency, not the CTAF.

54. (Refer to Figure 32 below, and Figure 23 on page 286.) (Refer to area 2 in Figure 23.) At Coeur D'Alene, which frequency should be used as a Common Traffic Advisory Frequency (CTAF) to self-announce position and intentions?

A. 122.05 MHz.

B. 122.1/108.8 MHz.

C. 122.8 MHz.

Answer (C) is correct. *(A/FD)*
   DISCUSSION: Fig. 32 is the *A/FD* excerpt for Coeur D'Alene Air Terminal. Look for the section titled **Communications**. On that same line, it states the CTAF (and UNICOM) frequency is 122.8.
   Answer (A) is incorrect. This is the remote communications outlet (RCO) frequency to contact Boise FSS in the vicinity of Coeur D'Alene, not the CTAF. Answer (B) is incorrect. The COE VOR/DME frequency, not the CTAF, is 108.8.

---

18   **IDAHO**

COEUR D'ALENE AIR TERMINAL (COE)   9 NW   UTC--8(--7DT)   N47°46.46' W116°49.17'   **GREAT FALLS**
    2318   B   S4   **FUEL** 80, 100, JET A   OX 1, 2   **H--1B, L--9A**
    RWY 05–23: H7400X140 (ASPH--GRVD)   S–57, D–95, DT–165   HIRL   0.7%up NE   **IAP**
        RWY 05: MALSR.   RWY 23: REIL. VASI(V4L)—GA 3.0° TCH 39'.
    RWY 01–19: H5400X75 (ASPH)   S–50, D–83, DT–150   MIRL
        RWY 01: REIL. Rgt. tfc.
    AIRPORT REMARKS: Attended Mon-Fri 1500-0100Z‡. Rwy 05--23 potential standing water and/or ice on center 3000'
        of rwy. Arpt conditions avbl on UNICOM. Rwy 19 is designated calm wind rwy. ACTIVATE MIRL Rwy 01–19, HIRL
        Rwy 05–23 and MALSR Rwy 05—CTAF. REIL Rwy 23 opr only when HIRL on high ints.
    WEATHER DATA SOURCES: AWOS–3 135.075 (208) 772- 8215.
    COMMUNICATIONS: CTAF/UNICOM 122.8
        BOISE FSS (BOI) TF 1-800-WX-BRIEF. NOTAM FILE COE.
    Ⓡ RCO 122.05 (BOISE FSS)
    Ⓡ SPOKANE APP/DEP CON 132.1
    RADIO AIDS TO NAVIGATION: NOTAM FILE COE.
        (T) VORW/DME 108.8   COE   Chan 25   N47°46.42' W116°49.24'   at fld. 2290/19E.
            DME portion unusable 280°–350° byd 15 NM blo 11000' 220°–240° byd 15 NM.
        LEENY NDB (LOM) 347   CO   N47°44.57' W116°57.66'   053° 6.0 NM to fld.
        ILS 110.7   I-COE   Rwy 05   LOM LEENY NDB. ILS localizer/glide slope unmonitored.

Figure 32. – Airport/Facility Directory Excerpt.

55. (Refer to Figure 26 on page 289.) (Refer to area 3.) If Redbird Tower is not in operation, which frequency should be used as a Common Traffic Advisory Frequency (CTAF) to monitor airport traffic?

A. 120.3 MHz.

B. 122.95 MHz.

C. 126.35 MHz.

Answer (A) is correct. *(ACL)*
    DISCUSSION: In Fig. 26, find the Redbird Airport just above 3. When the Redbird tower is not in operation, the CTAF is 120.3 because that frequency is marked with a C.
    Answer (B) is incorrect. This is the UNICOM frequency. Answer (C) is incorrect. This is the ATIS frequency.

56. (Refer to Figure 26 on page 289.) (Refer to area 2.) The control tower frequency for Addison Airport is

A. 122.95 MHz.

B. 126.0 MHz.

C. 133.4 MHz.

Answer (B) is correct. *(ACL)*
    DISCUSSION: Addison Airport (Fig. 26, area 2) control tower frequency is given as the first item in the second line of the airport data to the right of the airport symbol. The control tower (CT) frequency is 126.0 MHz.
    Answer (A) is incorrect. This is the UNICOM, not control tower, frequency for Addison Airport. Answer (C) is incorrect. This is the ATIS, not control tower, frequency for Addison Airport.

9.5 FAA Advisory Circulars

57. FAA advisory circulars (some free, others at cost) are available to all pilots and are obtained by

A. distribution from the nearest FAA district office.

B. ordering those desired from the Government Printing Office.

C. subscribing to the Federal Register.

Answer (B) is correct. *(AC 00-2.15)*
    DISCUSSION: FAA Advisory Circulars are issued with the purpose of informing the public of nonregulatory material of interest. Free advisory circulars can be ordered from the FAA, while those at cost can be ordered from the Government Printing Office.
    Answer (A) is incorrect. FAA offices have their own copies but none for distribution to the public. Answer (C) is incorrect. The *Federal Register* contains Notices of Proposed Rulemaking (NPRM) and final rules. It is a federal government publication.

58. FAA advisory circulars containing subject matter specifically related to Air Traffic Control and General Operations are issued under which subject number?

A. 60.

B. 70.

C. 90.

Answer (C) is correct. *(AC 00-2.15)*
    DISCUSSION: FAA advisory circulars are numbered based on the numbering system used in the FARs:

    60 -- Airmen
    70 -- Airspace
    90 -- Air Traffic Control and General Operation

    Answer (A) is incorrect. This refers to Airmen, not Air Traffic Control. Answer (B) is incorrect. This refers to Airspace, not Air Traffic Control.

59. FAA advisory circulars containing subject matter specifically related to Airmen are issued under which subject number?

A. 60.

B. 70.

C. 90.

Answer (A) is correct. *(AC 00-2.15)*
    DISCUSSION: FAA advisory circulars are numbered based on the numbering system used in the FARs:

    60 -- Airmen
    70 -- Airspace
    90 -- Air Traffic Control and General Operation

    Answer (B) is incorrect. This relates to Airspace, not Airmen. Answer (C) is incorrect. This relates to Air Traffic Control and General Operation (not Airmen).

60. FAA advisory circulars containing subject matter specifically related to Airspace are issued under which subject number?

A. 60.

B. 70.

C. 90.

Answer (B) is correct. *(AC 00-2.15)*
    DISCUSSION: FAA advisory circulars are numbered based on the numbering system used in the FARs:

    60 -- Airmen
    70 -- Airspace
    90 -- Air Traffic Control and General Operation

    Answer (A) is incorrect. This relates to Airmen, not Airspace. Answer (C) is incorrect. This relates to Air Traffic Control and General Operation, not Airspace.

9.6 *Airport/Facility Directory*

61. (Refer to Figure 53 on page 281.) When approaching Lincoln Municipal from the west at noon for the purpose of landing, initial communications should be with

A. Lincoln Approach Control on 124.0 MHz.

B. Minneapolis Center on 128.75 MHz.

C. Lincoln Tower on 118.5 MHz.

Answer (A) is correct. *(A/FD)*
   DISCUSSION: Fig. 53 contains the *A/FD* excerpt for Lincoln Municipal. Locate the section titled Airspace and note that Lincoln Municipal is located in Class C airspace. The Class C airspace is in effect from 0530-0030 local time (1130-0630Z). You should contact approach control (app con) during that time before entering. Move up three lines to App/Dep Con and note that aircraft arriving from the west of Lincoln (i.e., 170° – 349°) at noon should initially contact Lincoln Approach Control on 124.0.
   Answer (B) is incorrect. You would contact Minneapolis Center for basic radar services (i.e., flight following, assistance, etc.) between 0030 and 0530 local time, not at noon. Answer (C) is incorrect. When approaching Lincoln Municipal at noon, your initial contact should be with approach control, not the tower.

62. (Refer to Figure 53 on page 281.) Traffic patterns in effect at Lincoln Municipal are

A. to the right on Runway 17L and Runway 35L; to the left on Runway 17R and Runway 35R.

B. to the left on Runway 17L and Runway 35L; to the right on Runway 17R and Runway 35R.

C. to the right on Runways 14 - 32.

Answer (B) is correct. *(A/FD)*
   DISCUSSION: Fig. 53 contains the *A/FD* excerpt for Lincoln Municipal. For this question, you need to locate the runway end data elements, i.e., Rwy 17R, Rwy 35L, Rwy 14, Rwy 32, Rwy 17L, and Rwy 35R. Traffic patterns are to the left unless right traffic is noted by the contraction Rgt tfc. The only runways with right traffic are Rwy 17R and Rwy 35R.
   Answer (A) is incorrect. Traffic patterns are to the left, not right, for Rwy 17L and Rwy 35L. Traffic patterns are to the right, not left, on Rwy 17R and Rwy 35R. Answer (C) is incorrect. The traffic pattern for Rwy 14 and Rwy 32 is to the left, not right.

63. (Refer to Figure 53 on page 281.) Which type radar service is provided to VFR aircraft at Lincoln Municipal?

A. Sequencing to the primary Class C airport and standard separation.

B. Sequencing to the primary Class C airport and conflict resolution so that radar targets do not touch, or 1,000 feet vertical separation.

C. Sequencing to the primary Class C airport, traffic advisories, conflict resolution, and safety alerts.

Answer (C) is correct. *(A/FD and AIM Para 4-1-17)*
   DISCUSSION: Fig. 53 contains the *A/FD* excerpt for Lincoln Municipal. Locate the section titled Airspace to determine that Lincoln Municipal is located in Class C airspace. Once communications and radar contact are established, VFR aircraft are provided the following services:

1. Sequencing to the primary airport
2. Approved separation between IFR and VFR aircraft
3. Basic radar services, i.e., safety alerts, limited vectoring, and traffic advisories

   The FAA should change "conflict resolution" to "limited vectoring" in the future.
   Answer (A) is incorrect. In addition to sequencing to the primary Class C airport and standard separation, Class C radar service also includes basic radar services, i.e., traffic advisories and safety alerts. Answer (B) is incorrect. One radar service provided to VFR aircraft in Class C airspace provides for traffic advisories and conflict resolution so that radar targets do not touch, or 500 ft., not 1,000 ft., vertical separation.

64. (Refer to Figure 53 on page 281.) Where is Loup City Municipal located with relation to the city?

A. Northeast approximately 3 miles.

B. Northwest approximately 1 mile.

C. East approximately 10 miles.

Answer (B) is correct. *(A/FD)*
   DISCUSSION: Fig. 53 contains the *A/FD* excerpt for Loup City Municipal. On the first line, the third item listed, 1 NW, means that Loup City Municipal is located approximately 1 NM northwest of the city.
   Answer (A) is incorrect. (NE03) is the airport identifier, not an indication that the airport is 3 NM northeast of the city. Answer (C) is incorrect. The airport is approximately 1 NM northwest, not 10 NM east, of the city.

180 **NEBRASKA**

LINCOLN MUNI (LNK) 4 NW UTC–6(–5DT) N40°51.05' W96°45.55'   OMAHA
  1218 B S4 FUEL 100LL. JET A TPA—2218(1000) ARFF Index B   H–1E, 3F, 4F, L –11B
  RWY 17R–35L: **H12901X200** (ASPH–CONC–GRVD) S–100. D–200. DT–400 HIRL   IAP
    RWY 17R: MALSR. VASI(V4L)—GA 3.0° TCH 55'. Rgt tfc. 0.4% down.
    RWY 35L: MALSR. VASI(V4L)—GA 3.0° TCH 55'.
  RWY 14–32: H8620X150 (ASPH–CONC–GRVD) S–80. D–170. DT–280 MIRL
    RWY 14: REIL. VASI(V4L)—GA 3.0° TCH 48'.
    RWY 32: VASI(V4L)—GA 3.0° TCH 53'. Thld dsplcd 431'. Pole. 0.3% up.
  RWY 17L–35R: H5400X100 (ASPH–CONC–AFSC) S–49. D–60 HIRL 0.8% up N
    RWY 17L: PAPI(P4L)—GA 3.0° TCH 33'.   RWY 35R: PAPI(P4L)—GA 3.0° TCH 40'. Pole. Rgt tfc.
  AIRPORT REMARKS: Attended continuously. Birds in vicinity of arpt. Twy D clsd between taxiways S and H indef. For
    MALSR Rwy 17R and Rwy 35L ctc twr. When twr clsd MALSR Rwy 17R and Rwy 35L preset on med ints. and REIL
    Rwy 14 left on when wind favor. NOTE: See Land and Hold Short Operations Section.
  WEATHER DATA SOURCES: ASOS (402) 474–9214. LLWAS
  COMMUNICATIONS: CTAF 118.5   ATIS 118.05   UNICOM 122.95
    COLUMBUS FSS (OLU) TF 1–800–WX–BRIEF. NOTAM FILE LNK.
    RCO 122.65 (COLUMBUS FSS)
  Ⓡ APP/DEP CON 124.0 (170°–349°) 124.8 (350°–169°) (1130–0630Z‡)
  Ⓡ MINNEAPOLIS CENTER APP/DEP CON 128.75 (0630–1130Z‡)
    TOWER 118.5 125.7 (1130–0630Z‡) GND CON 121.9 CLNC DEL 120.7
  AIRSPACE: CLASS C svc 1130–0630Z‡ ctc APP CON other times CLASS E.
  RADIO AIDS TO NAVIGATION: NOTAM FILE LNK. VHF/DF ctc FSS.
    (H) VORTACW 116.1   LNK   Chan 108 N40°55.43' W 96°44.52'   181° 4.5 NM to fld. 1370/9E
    POTTS NDB (MHW/LOM) 385   LN N40°44.83' W 96°45.75'   355° 6.2 NM to fld. Unmonitored when twr clsd.
    ILS 111.1   I–OCZ   Rwy 17R.   MM and OM unmonitored.
    ILS 109.9   I–LNK   Rwy 35L.   LOM POTTS NDB. MM unmonitored. LOM unmonitored when twr clsd.
  COMM/NAVAID REMARKS: Emerg frequency 121.5 not available at tower.

LOUP CITY MUNI (NEØ3) 1 NW UTC –6(–5DT) N41°17.42' W 98°59.44'   OMAHA
  2070 B FUEL 100LL   L-11B
  RWY 15–33: H3200X50 (ASPH) S–8 LIRL
    RWY 33: Trees.
  RWY 04–22: 2100X100 (TURF)
    RWY 04: Tree.   RWY 22: Road.
  AIRPORT REMARKS: Unattended. For svc call 308–745–0328/1244/0664
  COMMUNICATIONS: CTAF 122.9
    COLUMBUS FSS (OLU) TF 1–800–WX–BRIEF. NOTAM FILE OLU.
  RADIO AIDS TO NAVIGATION: NOTAM FILE OLU.
    WOLBACH (H) VORTAC 114.8   OBH   Chan 95 N41°22.54' W 98°21.22'   253° 29.3 NM to fld. 2010/7E.

MARTIN FLD (See SO SIOUX CITY)

MC COOK MUNI (MCK) 2E UTC –6(–5DT) N40°12.36' W 100°35.51'   OMAHA
  2579 B S4 FUEL 100LL. JET A ARFF Index Ltd.   H–20. L–11A
  RWY 12–30: H5999X100 (CONC) S–30. D–38 MIRL 0.6% up NW   IAP
    RWY 12: MALS. VASI(V4L)—GA 3.0° TCH 33'. Tree.   RWY 30: REIL. VASI(V4L)—GA 3.0° TCH 42'.
  RWY 03–21: H3999X75 (CONC) S–30. D–38 MIRL
    RWY 03: VASI(V2L)—GA 3.0° TCH 26'. Rgt tfc.   RWY 21: VASI(V2L)—GA 3.0° TCH 26'.
  RWY 17–35: 1350X200 (TURF)
  AIRPORT REMARKS: Attended daylight hours. Parachute Jumping. Deer on and in vicinity of arpt. Numerous
    waterfowl/migratory birds invof arpt. Arpt closed to air carrier operations with more than 30 passengers except
    24 hour PPR, call arpt manager 308–345–2022. Avoid McCook State (abandoned) arpt 7miles NW on the MCK
    VOR/DME 313° radial at 8.3 DME. ACTIVATE VASI Rwys 12 and 30 and MALS Rwy 12 —CTAF.
  COMMUNICATIONS: CTAF/UNICOM 122.8
    COLUMBUS FSS (OLU) TF 1–800–WX–BRIEF. NOTAM FILE MCK.
    RCO 122.6 (COLUMBUS FSS)
    DENVER CENTER APP/DEP CON 132.7
  AIRSPACE: CLASS E svc effective 1100–0500Z‡ except holidays other times CLASS G.
  RADIO AIDS TO NAVIGATION: NOTAM FILE MCK.
    (H) VORW/DME 115.3   MCK   Chan 100 N40°12.23' W 100°35.65'   at fld. 2570/8E.

Figure 53. – Sectional Chart Excerpt.

65. (Refer to Figure 53 on page 281.) What is the recommended communications procedure for landing at Lincoln Municipal during the hours when the tower is not in operation?

A. Monitor airport traffic and announce your position and intentions on 118.5 MHz.

B. Contact UNICOM on 122.95 MHz for traffic advisories.

C. Monitor ATIS for airport conditions, then announce your position on 122.95 MHz.

Answer (A) is correct. *(A/FD)*
   DISCUSSION: When the Lincoln Municipal tower is closed, you should monitor airport traffic and announce your position and intentions on the CTAF. Fig. 53 contains the *A/FD* excerpt for Lincoln Municipal. Locate the section titled Communications and note that on that same line the CTAF frequency is 118.5.
   Answer (B) is incorrect. When the tower is not in operation, you should monitor other traffic and announce your position and intentions on the specified CTAF. At Lincoln Municipal, the CTAF is the tower frequency of 118.5, not the UNICOM frequency of 122.95. Answer (C) is incorrect. When the tower is not in operation, you should monitor other traffic and announce your position and intentions on the specified CTAF. At Lincoln Municipal, the CTAF is the tower frequency of 118.5, not the UNICOM frequency of 122.95.

9.7 *Notices to Airmen Publication (NTAP)*

66. What information is contained in the *Notices to Airmen Publication (NTAP)*?

A. Current NOTAM (D) and FDC NOTAMs.

B. Military NOTAMs only.

C. Current NOTAM (D), FDC NOTAMs, and military NOTAMs.

Answer (A) is correct. *(AIM Para 5-1-3)*
   DISCUSSION: The *NTAP* contains (D) NOTAMs that are expected to remain in effect for an extended period and FDC NOTAMs that are current at the time of publication.
   Answer (B) is incorrect. Military NOTAMs are not published in the *NTAP*. Answer (C) is incorrect. While current NOTAM (D) and FDC NOTAMs are published in the *NTAP*, military NOTAMs are not.

67. When NOTAMs are published in the *Notices to Airmen Publication (NTAP)*, they are

A. Still a part of a standard weather briefing.

B. Only available in a standard weather briefing if the pilot requests published NOTAMs.

C. Canceled and are no longer valid.

Answer (B) is correct. *(AIM Para 5-1-3)*
   DISCUSSION: Once a NOTAM is published in the *NTAP*, the NOTAM is not provided during pilot weather briefings unless specifically requested.
   Answer (A) is incorrect. Published NOTAMs are only available in a pilot weather briefing if the pilot makes a specific request for them. NOTAMs that have not been published are a part of a standard weather briefing. Answer (C) is incorrect. A published NOTAM remains in effect until its expiration date or until an additional NOTAM is issued to cancel it.

END OF STUDY UNIT

SECTIONAL AERONAUTICAL CHART
SCALE 1:500,000

LEGEND

Airports having **Control Towers** are shown in **Blue**, all others in **Magenta**. Consult Airport/Facility Directory (AFD) for details involving airport lighting, navigation aids, and services. For additional symbol information refer to the Chart User's Guide.

AIRPORTS

Other than hard-surfaced runways

⚓ Seaplane Base

Hard-surfaced runways 1500 ft. to 8069 ft. in length

Hard-surfaced runways greater than 8069 ft. or some multiple runways less than 8069 ft.

Open dot within hard-surfaced runway configuration indicates approximate VOR, VOR-DME, or VORTAC location

All recognizable hard-surfaced runways, including those closed, are shown for visual identification. Airports may be public or private.

Ⓡ Private ("Pvt") – Non-public use having emergency or landmark value.

ADDITIONAL AIRPORT INFORMATION

Military – Other than hard-surfaced. All military airports are identified by abbreviations AFB, NAS, AAF, etc. For complete airport information consult DOD FLIP.

Ⓕ Ultralight Flight Park Selected

Ⓤ Unverified

⊗ Abandoned – paved, having landmark value, 3000 ft. or greater

Ⓗ Heliport Selected

Services–fuel available and field tended during normal working hours depicted by use of ticks around basic airport symbol. (Normal working hours are Mon thru Fri 10:00 A.M. to 4:00 P.M. local time.) Consult A/FD for service availability at airports with surfaced runways greater than 8069 ft.
☆ Rotating airport beacon in operation Sunset to Sunrise.

RADIO AIDS TO NAVIGATION AND COMMUNICATION BOXES

⊙ VHF OMNI RANGE (VOR)

⊙ VORTAC

⊙ VOR-DME

Non-Directional Radiobeacon (NDB)

NDB-DME

⊙ Other facilities, i.e., Commercial Broadcast Stations, FSS Outlets-RCO, etc.

AIRPORT DATA

Box indicates F.A.R. 93 Special Air Traffic Rules & Airport Traffic Patterns

R – 118.3* Ⓛ ATIS 123.8

FSS NO SVFR NAME (NAM) UNICOM Location Identifier

Airport Surveillance Radar CT – 118.3 F.A.R. 91

VFR Advsy 125.0

285 L 72 122.95 Airport of Entry

FSS – Flight Service Station
NO SVFR – Fixed-wing special VFR flight is prohibited.
CT – 118.3 – Control Tower (CT) – primary frequency
NFCT – Non-Federal Control Tower
* – Star indicates operation part-time (see tower frequencies tabulation for hours of operation).
Ⓛ – Indicates Common Traffic Advisory Frequencies (CTAF).
ATIS 123.8 – Automatic Terminal Information Service
ASOS/AWOS 135.42 – Automated Surface Weather Observing Systems. NDBs broadcasting ASOS/AWOS data may not be located at the airport.
UNICOM – Aeronautical advisory station
VFR Advsy – VFR Advisory Service shown where ATIS not available and frequency is other than primary CT frequency
285 – Elevation in feet
Ⓛ – Lighting in operation Sunset to Sunrise
*Ⓛ – Lighting limitations exist, refer to Airport/Facility Directory.
72 – Length of longest runway in hundreds of feet; usable length may be less.
When facility or information is lacking, the respective character is replaced by a dash. All lighting codes refer to runway lights. Lighted runway may not be the longest or lighted full length. All times are local.

122.1R 122.6 123.6
362 *116.8 OAK OAKDALE

Underline indicates no voice on this freq

* – Operates less than continuous or On-Request.
Ⓣ – TWEB
R – Receive only

122.1R

CHICAGO CHI
Heavy line box indicates Flight Service Station (FSS). Freqs. 121.5, 122.2, 243.0, and 255.4 (Canada – 121.5, 126.7 and 243.0) are normally available at all FSSs and are not shown above boxes. All other freqs. are shown.
For Local Airport Advisory use FSS freq. 123.6.

MIAMI
FSS providing voice communication

Frequencies above thin line box are remoted to NAVAID site. Other frequencies at FSS providing voice communication may be available as determined by altitude and terrain. Consult Airport/Facility Directory for complete information.

CONTOUR INTERVAL 500 feet
— 500 —

HIGHEST TERRAIN elevation is 3818 feet located at 34°52'N – 101°59'W

Spot elevation .... •4254
Approximate elevation .... x 3200
Doubtful locations are indicated by omission of the point locator (dot or "x")

TOPOGRAPHIC INFORMATION

Roads
Road Markers
Railroad
Bridges And Viaducts
Power Transmission Lines
Aerial Cable

⊡ Landmark feature – stadium, factory, school, golf course, etc.
Outdoor Theatre
Lookout Tower P-17 (Site Number) 618 (Elevation Base of Tower)
◆ CG Coast Guard Station
Race Track
Tank–water, oil or gas
○ Oil Well ⊙ Water Well
✕ Mines And Quarries
Mountain Pass
11823 (Elevation of Pass)

Rocks
Pier
Dams
Perennial Lake
Non-Perennial Lake

AIRPORT TRAFFIC SERVICE AND AIRSPACE INFORMATION

Only the controlled and reserved airspace effective below 18,000 ft. MSL are shown on this chart. All times are local.

Class B Airspace

Class C Airspace (Mode C) See F.A.R. 91.215/AIM.)

Class D Airspace

[40] Ceiling of Class D Airspace in hundreds of feet. (A minus ceiling value indicates surface up to but not including that value.)

Class E (sfc) Airspace

Class E Airspace with floor 700 ft. above surface

2400 MSL Class E Airspace with floor 1200 ft. or greater above surface that abuts Class G Airspace.

4500 MSL Differentials floors of Class E Airspace greater than 700 ft. above surface

Class E Airspace low altitude Federal Airways are indicated by center line.

Intersection – Arrows are directed towards facilities which establish intersection.

132° V-69 169

Total mileage between NAVAIDs on direct Airways.

Prohibited, Restricted, Warning and Alert Areas Canadian Advisory and Restricted Areas

MOA – Military Operations Area

Special Airport Traffic Areas (See F.A.R. Part 93 for details.)

MODE C (See F.A.R. 91.215/AIM.)

National Security Area

Terminal Radar Service Area (TRSA)

1R211 MTR – Military Training Routes

OBSTRUCTIONS

1000 ft. and higher AGL

below 1000 ft. AGL

Group Obstruction

Obstruction with high-intensity lights May operate part-time

2049 Elevation of the top (1149) above mean sea level
Height above ground
UC Under construction or reported; position and elevation unverified

NOTICE: Guy wires may extend outward from structures.

MISCELLANEOUS

— 1° E Isogonic Line (1995 VALUE)

Ⓤ Ultralight Activity
Ⓗ Hang Glider Activity
Ⓖ Glider Operations
Fl ★ Flashing Light
● Marine Light

NAME (Magenta, Blue, or Black) Visual Check Point
Parachute Jumping Area (See Airport/Facility Directory.)

MILITARY TRAINING ROUTES (MTRs)

All IR and VR MTRs are shown, and may extend from the surface upwards. Only the route centerline, direction of flight along the route and the route designator are depicted – route widths and altitudes are not shown.

Since these routes are subject to change every 56 days, and the charts are reissued every 6 months, you are cautioned and advised to contact the nearest FSS for route dimensions and current status for those routes affecting your flight.

Routes with a change in the alignment of the charted route centerline will be indicated in the Aeronautical Chart Bulletin of the Airport/Facility Directory.

Military Pilots refer to Area Planning AP/1B Military Training Route North and South America for current routes.

—ATTENTION—
THIS CHART CONTAINS MAXIMUM ELEVATION FIGURES (MEF). The Maximum Elevation Figures shown in quadrangles bounded by ticked lines of latitude and longitude are represented in THOUSANDS and HUNDREDS of feet.above mean sea level. The MEF is based on information available concerning the highest known feature in each quadrangle, including terrain and obstructions (trees, towers, antennas, etc.).

12⁵

Example: 12,500 feet ......

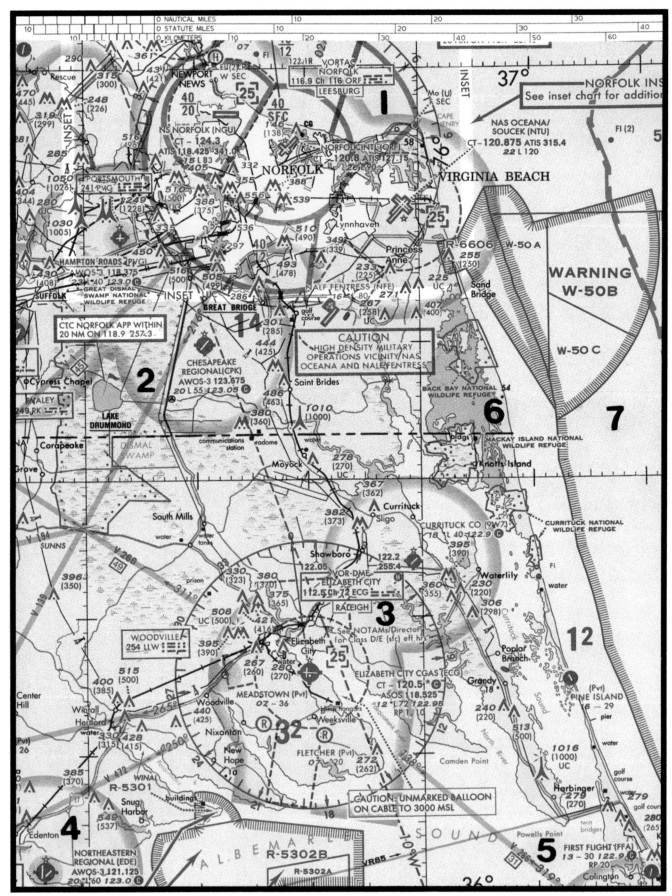

Figure 21. – Sectional Chart Excerpt.

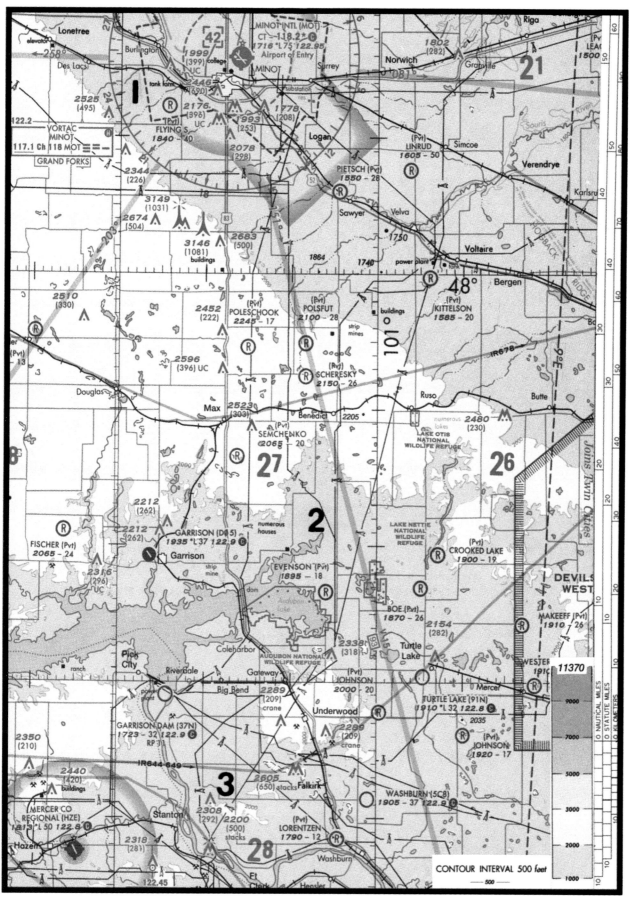

Figure 22. – Sectional Chart Excerpt.

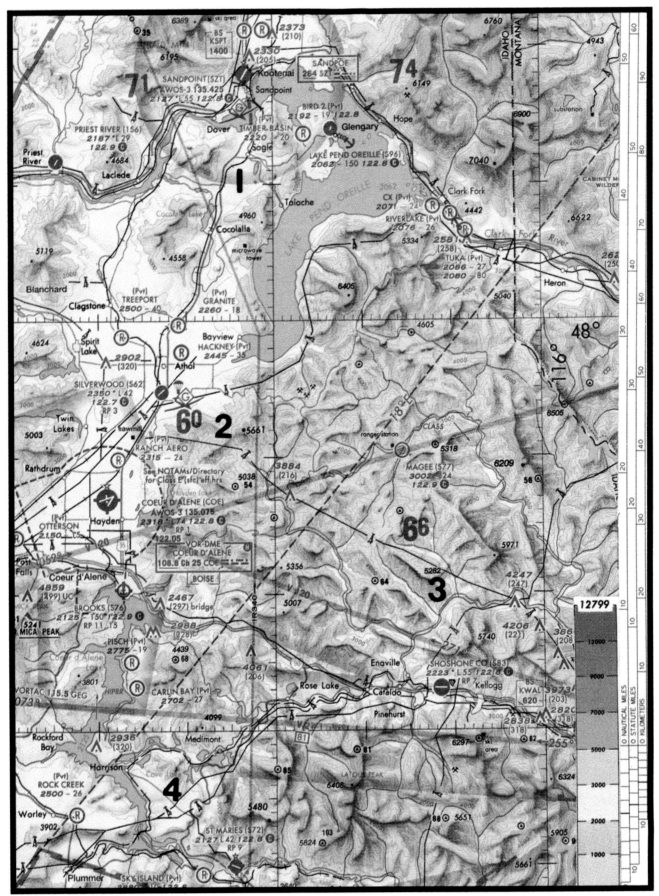

Figure 23. – Sectional Chart Excerpt.

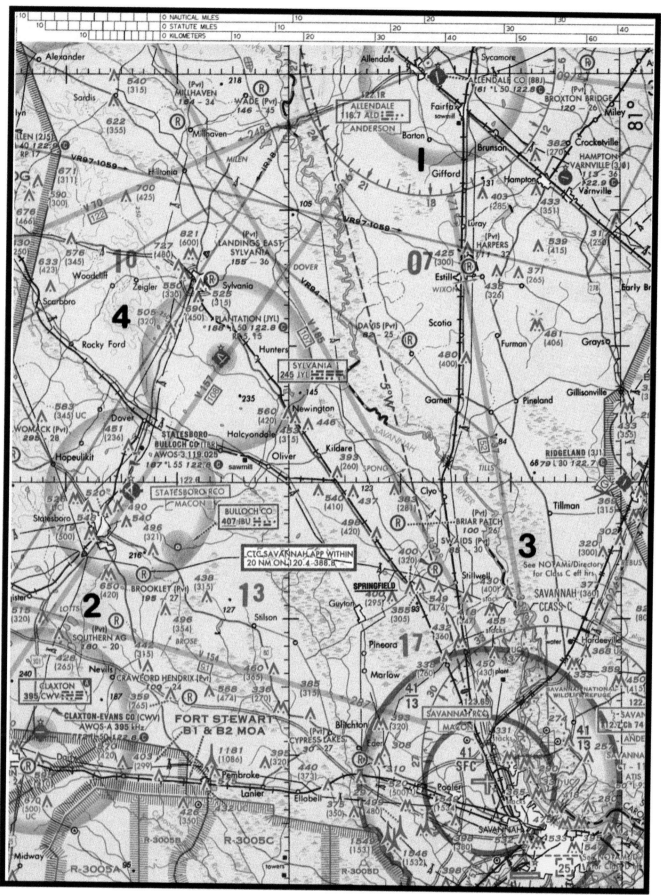

Figure 24. – Sectional Chart Excerpt.

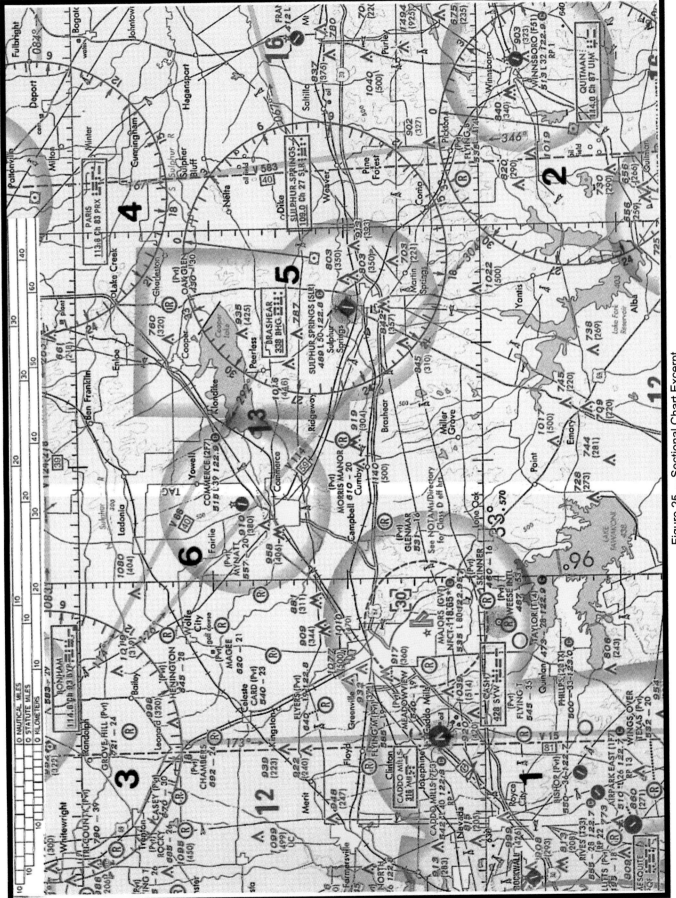

Figure 25. – Sectional Chart Excerpt.

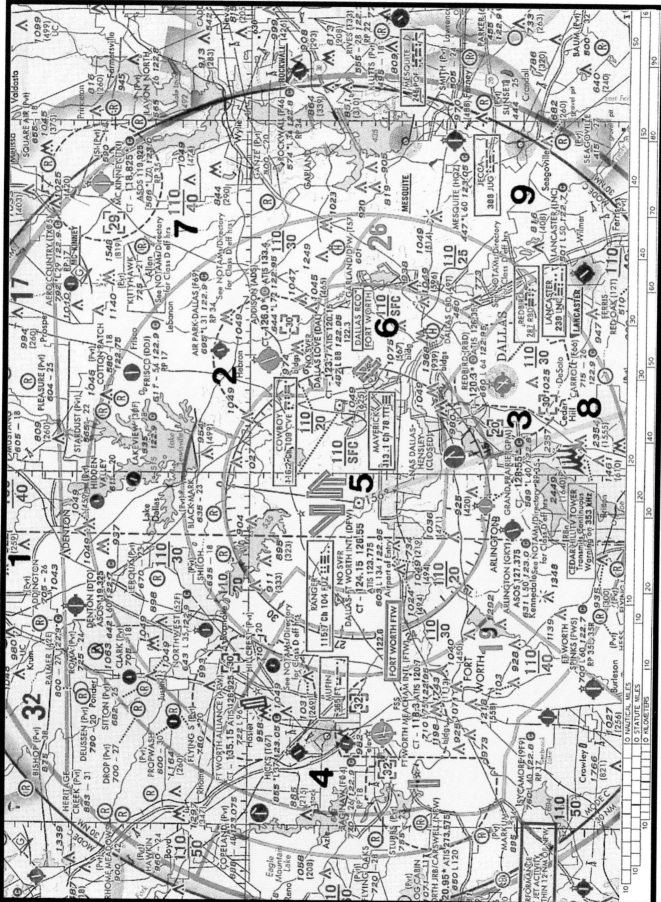

Figure 26. – Sectional Chart Excerpt.

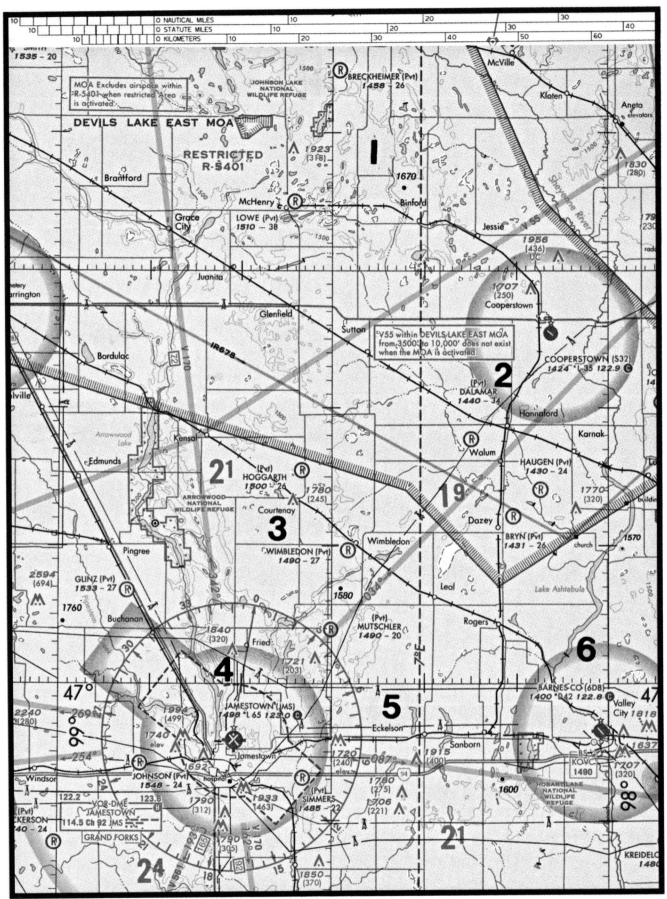

Figure 27. – Sectional Chart Excerpt.

Figure 60. – Sectional Chart Excerpt.

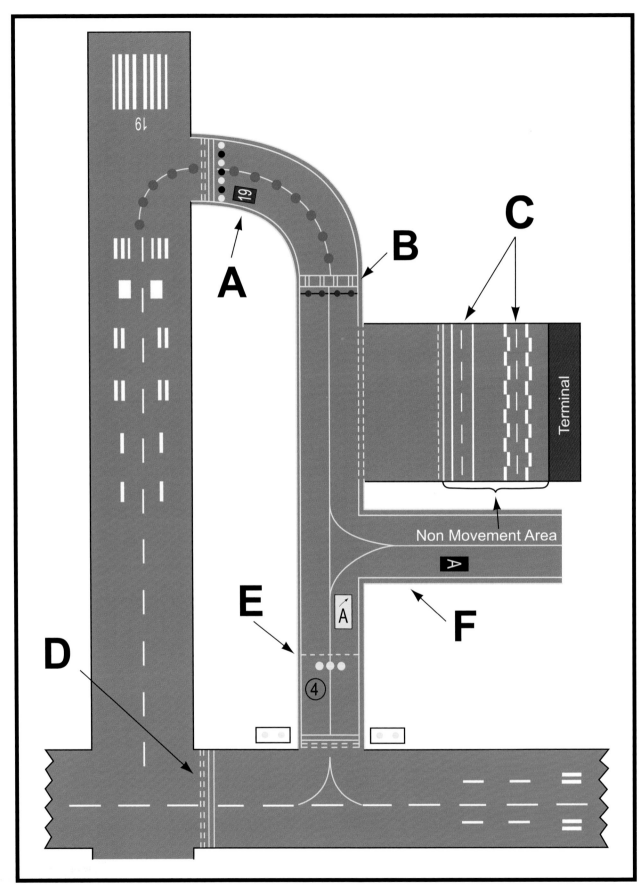

Figure 65. – Airport Markings.

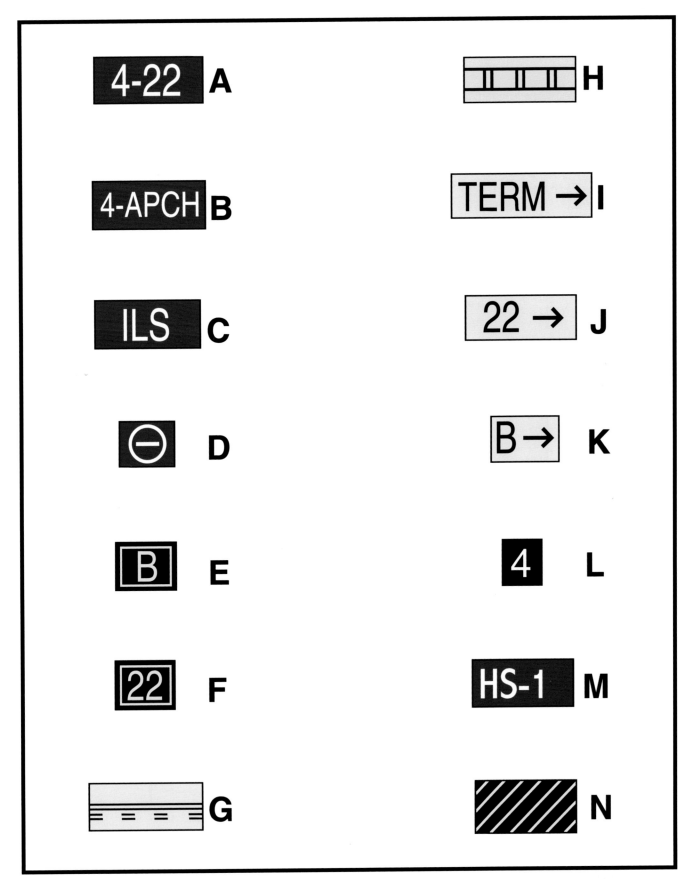

Figure 66. – Airport Signs.

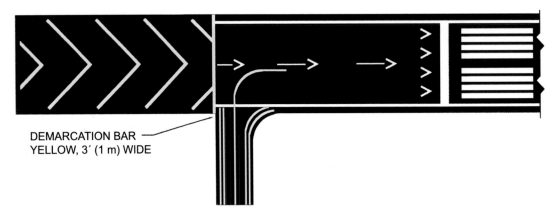

DEMARCATION BAR
YELLOW, 3′ (1 m) WIDE

Figure 68. – Yellow Demarcation Bar.

Figure 4. – Airspeed Indicator.

STUDY UNIT TEN
NAVIGATION SYSTEMS

(4 pages of outline)

This study unit contains outlines of major concepts tested, sample test questions and answers regarding navigation systems, and an explanation of each answer. The table of contents above lists each subunit within this study unit, the number of questions pertaining to that particular subunit, and the pages on which the outlines and questions begin, respectively.

Many of the questions in this study unit ask about the sectional charts, which appear as Legend 1 and Figures 21 through 27. To produce them in color economically, we have put them together on pages 283 through 290. As you will need to turn frequently to these eight pages, mark them with "dog ears" (fold their corners) or a paper clip.

CAUTION: Recall that the **sole purpose** of this book is to expedite your passing the FAA pilot knowledge test for the private pilot certificate. Accordingly, all extraneous material (i.e., topics or regulations not directly tested on the FAA pilot knowledge test) is omitted, even though much more information and knowledge are necessary to fly safely. This additional material is presented in *Pilot Handbook* and *Private Pilot Flight Maneuvers and Practical Test Prep*, available from Gleim Publications, Inc. See the order form on page 371.

10.1 VOR TEST FACILITY (VOT)

1. VOR Test Facilities (VOTs) are available on a specific frequency at certain airports. The facility permits you to check the accuracy of your VOR receiver while you are on the ground.

   a. The airborne use of a VOT is permitted; however, its use is strictly limited to those areas and altitudes specifically authorized by the *Airport/Facilities Directory*.

   b. In each *Airport/Facilities Directory*, there is a section, listed by state, of VOT ground locations and airborne checkpoints.

      1) Frequencies, identifiers, distances, and descriptions, if appropriate, are given to determine where these tests can be conducted.

2. Tune the navigation radio to the specified VOT frequency, and center the course deviation indicator.

   a. The OBS should read either 0° or 180°, regardless of your position at the airport.
   b. If 0°, the TO/FROM indicator should indicate FROM.
   c. If 180°, the TO/FROM indicator should indicate TO.
   d. Accuracy of the VOR should be ±4° for ground checks or ±6° for airborne checks.

10.2 DETERMINING POSITION

1. Several FAA exam questions require you to identify your position based on the intersection of given radials of two VORs.

    a. To locate a position based on VOR radials, draw the radials on your chart or on the plastic overlay during the FAA knowledge test.

    b. Remember that radials are from the VOR, or leaving the VOR.

    c. Make sure you have located the correct radial on the compass rose before drawing your line.

    d. Recheck yourself by counting in 10° or 5° intervals from each of the closest 30° intervals that are numbered and marked with an arrow.

2. Other FAA exam questions require you to identify your position based upon the indications of a single VOR.

    a. You must compare the OBS setting and the TO/FROM indicator with the aircraft heading. To indicate correctly, the OBS (top) setting must correspond roughly with the aircraft heading [e.g., 180° OBS (top) setting, 180° aircraft heading].

    b. The TO/FROM indicator must correspond to the aircraft's flight path in relation to the VOR. Flying TO a VOR with a FROM indication and flying FROM a VOR with a TO indication will result in reverse sensing.

    c. When flying directly from a station, the heading and the radial being flown will correspond (i.e., 360° heading FROM will be the 360° radial).

    d. When flying directly TO a station, the heading flown and the radial being flown will be reciprocals (i.e., 180° heading TO will be on the 360° radial).

    e. With regard to CDI deflection, you must pretend your airplane has the same heading as the OBS setting. A left deviation means you are right of course, and a right deviation means you are left of course.

        1) If your heading and the OBS setting are not roughly the same, the CDI **will not** indicate correctly.

    f. If no TO or FROM flag indication appears, the aircraft is in the area of ambiguity, i.e., 90° away from the radial dialed up on the OBS. To know the side of the station on which the aircraft is located, consult the CDI. The needle points toward the station.

10.3 AUTOMATIC DIRECTION FINDER (ADF)

1. The ADF indicator always has its needle pointing toward the NDB station (nondirectional beacon, also known as a radio beacon).

    a. If the NDB is directly in front of the airplane, the needle will point straight up.

    b. If the NDB is directly off the right wing, i.e., 3 o'clock, the needle will point directly to the right.

    c. If the NDB is directly behind the aircraft, the needle will point straight down, etc.

    d. The figure on the next page illustrates the terms that are used with the ADF.

2. Relative bearing (RB) to the station is the number of degrees you would have to turn to the right to fly directly to the NDB. On a fixed card ADF, the

    a. Relative bearing TO the station is shown by the head of the needle.

        1) In the figure below, the RB to the station is 220°.

    b. Relative bearing FROM is given by the tail of the needle.

        1) In the figure below, the RB from the station is 40° (220° – 180°).

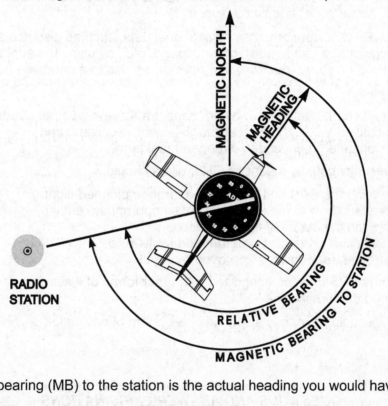

3. Magnetic bearing (MB) to the station is the actual heading you would have to fly to the station.

    a. If you turn right from your present heading to fly to the station, you are adding the number of degrees of turn to your heading.

    b. Thus, magnetic heading + relative bearing = magnetic bearing to the station, or MH + RB = MB (TO).

        1) For MB (FROM), subtract or add 180°.

        2) EXAMPLE: If the airplane shown above has an MH of 40° and an RB of 220°, the MB (TO) is 260° (40° + 220°). The MB (FROM) is 80° (260° – 180°).

4. A fixed card ADF always shows 0° at the top.

    a. Thus, RB may be read directly from the card, and MB must be calculated using the above formula.

    b. If the MB is given, the MH may be calculated as follows: MB – RB = MH.

5. A movable card ADF always shows magnetic heading (MH) at the top.

    a. Thus, MB (TO) may be read directly from the card under the head of the needle.

    b. MB (FROM) is indicated by the tail of the needle.

    c. RB may be calculated as follows: MB – MH = RB.

6. When working ADF problems, it is often helpful to draw the information given (as illustrated above) to provide a picture of the airplane's position relative to the NDB station.

10.4 GLOBAL POSITIONING SYSTEM (GPS)

1.   The Global Positioning System (GPS) is a satellite-based radio navigation system.

2.   GPS is composed of a constellation of 24 satellites.

    a.   The GPS constellation is designed so that at least five satellites are always observable by a user anywhere on earth.

3.   The GPS receiver needs at least four satellites to yield a three-dimensional position (latitude, longitude, and altitude) and time solution.

    a.   The GPS receiver computes navigational data, such as distance and bearing to a waypoint (e.g., an airport), groundspeed, etc., by using the airplane's known latitude/longitude (position) and referencing this to a database built into the receiver. .

4.   GPS receivers

    a.   Contain chart databases, moving maps, traffic, and weather overlays;
    b.   Can contain VOR/DAME/localizer/glideslope receivers; and
    c.   Can compute groundspeed, time, and fuel burn.

5.   To effectively navigate by means of GPS, pilots should

    a.   Determine the GPS unit is approved for their planned flight,
    b.   Understand how to make and cancel all appropriate entries,
    c.   Determine the status of the databases,
    d.   Program and review the programmed route, and
    e.   Ensure the track flown is approved by ATC.

6.   Navigating by GPS must be integrated with other forms of electronic navigation as well as pilotage and dead reckoning.

    a.   Pilots should never rely solely on one system of navigation.

QUESTIONS AND ANSWER EXPLANATIONS

All of the private pilot knowledge test questions chosen by the FAA for release as well as additional questions selected by Gleim relating to the material in the previous outlines are reproduced on the following pages. These questions have been organized into the same subunits as the outlines. To the immediate right of each question are the correct answer and answer explanation. You should cover these answers and answer explanations while responding to the questions. Refer to the general discussion in the Introduction on how to take the FAA pilot knowledge test.

Remember that the questions from the FAA pilot knowledge test bank have been reordered by topic and organized into a meaningful sequence. Also, the first line of the answer explanation gives the citation of the authoritative source for the answer.

QUESTIONS

10.1 VOR Test Facility (VOT)

1. When the course deviation indicator (CDI) needle is centered during an omnireceiver check using a VOR test signal (VOT), the omnibearing selector (OBS) and the TO/FROM indicator should read

    A. 180° FROM, only if the pilot is due north of the VOT.

    B. 0° TO or 180° FROM, regardless of the pilot's position from the VOT.

    C. 0° FROM or 180° TO, regardless of the pilot's position from the VOT.

Answer (C) is correct. *(AIM Para 1-1-4)*
    DISCUSSION: A VOT transmits a 360° (0°) radial in all directions. With the CDI centered, the OBS should indicate 0° with the TO-FROM indicator showing FROM, or 180° TO, regardless of your position from the VOT. A good way to remember the VOT rule is to associate it with the Cessna 182, i.e., 180° TO.
    Answer (A) is incorrect. With the OBS set at 180°, CDI centered, you should have a TO (not FROM) indication, regardless of your position from the VOT. Answer (B) is incorrect. The VOT transmits a 360° radial in all directions; thus, with the CDI centered and the OBS on 0°, you should have a FROM (not TO) indication and 180° TO (not FROM).

2. Where can locations for VOR test facilities be found?

    A. *Aeronautical Information Manual.*

    B. Sectional charts.

    C. *Airport/Facilities Directory.*

Answer (C) is correct. *(AIM Para 1-1-4)*
    DISCUSSION: In each *Airport/Facilities Directory*, there is a section, listed by state, of VOT ground locations and airborne checkpoints.
    Answer (A) is incorrect. The *Aeronautical Information Manual* does not list locations of VOR test facilities. Answer (B) is incorrect. Sectional charts, while showing locations of VORs, do not include locations of VOR test facilities.

3. What should the airborne accuracy of a VOR be?

    A. ±4°

    B. ±5°

    C. ±6°

Answer (C) is correct. *(AIM Para 1-1-4)*
    DISCUSSION: The accuracy of the VOR should be ±4° for ground checks or ±6° for airborne checks.
    Answer (A) is incorrect. The accuracy of an airborne VOR check should be ±6°, not ±4°. Answer (B) is incorrect. The accuracy of an airborne VOR check should be ±6°, not ±5°.

10.2 Determining Position

The sectional chart legend and Figures 21 through 27 appear on pages 283 through 290.

4. (Refer to Figure 21 on page 284.) What is your approximate position on low altitude airway Victor 1, southwest of Norfolk (area 1), if the VOR receiver indicates you are on the 340° radial of Elizabeth City VOR (area 3)?

    A. 15 nautical miles from Norfolk VORTAC.

    B. 18 nautical miles from Norfolk VORTAC.

    C. 23 nautical miles from Norfolk VORTAC.

Answer (B) is correct. *(PHAK Chap 15)*
    DISCUSSION: First find V1 extending SW on the 233° radial from Norfolk VORTAC on Fig. 21. The V1 label appears just above 2. Then, draw along the 340° radial from Elizabeth City VOR (southwest of 3). If you are confused about where the exact VOR is (center of compass rose), draw a line through the entire compass rose so your line coincides with both your radial (here 340°) and its reciprocal (here 160°). Note that the intersection with V1 is 18 NM from the Norfolk VORTAC.
    Answer (A) is incorrect. The position of 15 NM from Norfolk would be on the 345° radial. Answer (C) is incorrect. The position of 23 NM from Norfolk would be on the 330° radial.

5. (Refer to Figure 24 on page 287.) What is the approximate position of the aircraft if the VOR receivers indicate the 320° radial of Savannah VORTAC (area 3) and the 184° radial of Allendale VOR (area 1)?

    A. Town of Guyton.

    B. Town of Springfield.

    C. 3 miles east of Marlow.

Answer (B) is correct. *(PHAK Chap 15)*
    DISCUSSION: To locate a position based on VOR radials, draw the radials on your map or on the plastic overlay during the FAA pilot knowledge test. Remember that radials are from the VOR, or leaving the VOR. On Fig. 24, the 320° radial from Savannah extends northwest, and the 184° radial from Allendale extends south. They intersect over the town of Springfield.
    Answer (A) is incorrect. Guyton is on the 308° radial (not the 320° radial of Savannah VORTAC) and the 188° radial (not the 184° radial of Allendale VOR). Answer (C) is incorrect. The position of 3 NM east of Marlow is on the 300° radial, not the 320° radial of Savannah VORTAC and the 184° radial of Allendale VOR.

6. (Refer to Figure 29 on page 301.) (Refer to illustration 8.) The VOR receiver has the indications shown. What radial is the aircraft crossing?

A. 030°.

B. 210°.

C. 300°.

Answer (A) is correct. *(PHAK Chap 15)*
 DISCUSSION: The OBS is set on 210° with the needle centered. The important factor is the TO indication showing. You are thus crossing the 210° inbound bearing, but with a TO indication it is the 030° radial. If it was a FROM indication, it would be the 210° radial.
 Answer (B) is incorrect. If you were crossing the 210° radial, you would have a FROM (not TO) indication. Answer (C) is incorrect. The 3 at the bottom of the dial means 030° (not 300°).

7. (Refer to Figure 29 on page 301.) (Refer to illustration 1.) The VOR receiver has the indications shown. What is the aircraft's position relative to the station?

A. North.

B. East.

C. South.

Answer (C) is correct. *(PHAK Chap 15)*
 DISCUSSION: The OBS is set to 030°. If the needle were centered, the airplane would be southwest of the station. The CDI is deflected full scale left so you are right of course. You are thus south of the VORTAC.
 Answer (A) is incorrect. To be north would require a right CDI deflection and a FROM indication. Answer (B) is incorrect. To be east would require a left CDI deflection and a FROM indication.

8. (Refer to Figure 29 on page 301.) (Refer to illustration 3.) The VOR receiver has the indications shown. What is the aircraft's position relative to the station?

A. East.

B. Southeast.

C. West.

Answer (B) is correct. *(PHAK Chap 15)*
 DISCUSSION: With no TO or FROM indications showing on VOR 3, Fig. 29, you must be flying in the zone of ambiguity from the VOR, which is perpendicular to the OBS setting, i.e, on the 120° or 300° radials. Since you have a left deflection, you would be on the 120° radial, or southeast of the VOR.
 Answer (A) is incorrect. If you were east, you would have a FROM indication. Answer (C) is incorrect. The 120° radial is southeast, not west.

9. (Refer to Figure 27 on page 290 and Figure 29 on page 301.) The VOR is tuned to Jamestown VOR (area 4 in Figure 27), and the aircraft is positioned over Cooperstown Airport (area 2 in Figure 27). Which VOR indication is correct?

A. 9

B. 2

C. 6

Answer (C) is correct. *(PHAK Chap 15)*
 DISCUSSION: Cooperstown Airport (northeast of 2 in Fig. 27) is located on the 028° radial of the Jamestown VOR (south of 4). With a centered needle, you could have an OBS setting of 028° and a FROM indication or an OBS setting of 208° and a TO indication. VOR 6 fits the aircraft's location over Cooperstown Airport. You have a FROM indication with an OBS setting of 030° and a half-scale deflection of the CDI to the right (because Cooperstown Airport is north of your selected course). You are thus on approximately the 028° radial.
 Answer (A) is incorrect. VOR 9 shows the aircraft's location as southwest of the Jamestown VOR, not over Cooperstown Airport. Answer (B) is incorrect. VOR 2 shows the aircraft's location as somewhere on the 030° radial, which would place it slightly south of, not over, Cooperstown Airport.

10. (Refer to Figure 27 on page 290 and Figure 29 on page 301.) The VOR is tuned to Jamestown VOR (area 4 in Figure 27), and the aircraft is positioned over Cooperstown Airport (area 2 in Figure 27). Which VOR indication is correct?

A. 1

B. 6

C. 4

Answer (B) is correct. *(PHAK Chap 15)*
 DISCUSSION: Cooperstown Airport (northeast of 2 in Fig. 27) is located on the 028° radial of the Jamestown VOR (south of 4). With a centered needle, you could have an OBS setting of 028° and a FROM indication or an OBS setting of 208° and a TO indication. VOR 6 fits the aircraft's location over Cooperstown Airport. You have a FROM indication with an OBS setting of 030° and a half-scale deflection of the CDI to the right (because Cooperstown Airport is north of your selected course). You are thus on approximately the 028° radial.
 Answer (A) is incorrect. VOR 1 shows the aircraft's location as somewhere northwest of the 030° radial, which would place it west of, not over, Cooperstown Airport. Answer (C) is incorrect. VOR 4 shows the aircraft's location as somewhere southeast on the 030° radial, which would place it slightly south of, not over, Cooperstown Airport.

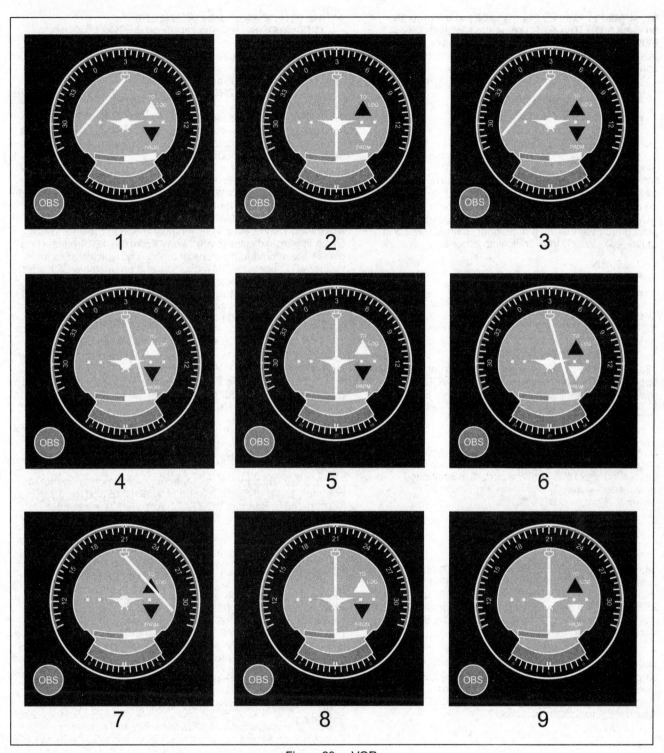

Figure 29. – VOR.

11. (Refer to Figure 21 on page 284 and Figure 29 on page 301.) The VOR is tuned to Elizabeth City VOR (area 3 in Figure 21), and the aircraft is positioned over Shawboro. Which VOR indication is correct?

A. 5

B. 9

C. 2

Answer (C) is correct. *(PHAK Chap 15)*
   DISCUSSION: See Fig. 21, northeast of 3 along the compass rose.
   Shawboro is northeast of the Elizabeth City VOR on the 030° radial. To be over it, the needle should be centered with either an OBS setting of 210° and a TO indication or with an OBS setting of 030° and a FROM indication. VOR 2 matches the latter description.
   Answer (A) is incorrect. VOR 5 indicates that the aircraft is southwest, not northeast, of Elizabeth City VOR. Answer (B) is incorrect. VOR 9 indicates that the aircraft is southwest, not northeast, of Elizabeth City VOR.

12. (Refer to Figure 25 on page 288 and Figure 29 on page 301.) The VOR is tuned to Bonham VORTAC (area 3 in Figure 25) and the aircraft is positioned over the town of Sulphur Springs (area 5 in Figure 25). Which VOR indication is correct?

A. 1

B. 8

C. 7

Answer (C) is correct. *(PHAK Chap 15)*
   DISCUSSION: The town of Sulphur Springs (south-southwest of 5) is on the 120° radial of Bonham VORTAC. Illustration 7 shows the VOR receiver tuned to the 210° radial, which is perpendicular to (90° away from) the 120° radial. This places the aircraft in the zone of ambiguity, which results in neither a TO nor a FROM indication and an unstable CDI, which can be deflected left or right.
   Answer (A) is incorrect. With indication 1, the aircraft would have to be west of Sulphur Springs. Answer (B) is incorrect. It shows the aircraft on the 030° radial, which is well to the north of Sulphur Springs.

13. (Refer to Figure 25 on page 288.) What is the approximate position of the aircraft if the VOR receivers indicate the 245° radial of Sulphur Springs VOR-DME (area 5) and the 140° radial of Bonham VORTAC (area 3)?

A. Glenmar Airport.

B. Meadowview Airport.

C. Majors Airport.

Answer (A) is correct. *(PHAK Chap 15)*
   DISCUSSION: To locate a position based on VOR radials, draw the radials on your map or on the plastic overlay during the FAA knowledge test. Remember that radials are from the VOR, or leaving the VOR.
   On Fig. 25, the 245° radial from Sulphur Springs VOR-DME extends southwest, and the 140° radial from Bonham VORTAC extends southeast. They intersect about 1 mi. east of Glenmar Airport.
   Answer (B) is incorrect. Meadowview Airport is on the 246°, not 245°, radial of Sulphur Springs VOR-DME and the 163°, not 140°, radial of Bonham VORTAC. Answer (C) is incorrect. Majors Airport is on the 157°, not 140°, radial of Bonham VORTAC.

14. (Refer to Figure 26 on page 289.) (Refer to area 5.) The VOR is tuned to the Dallas/Fort Worth VORTAC. The omnibearing selector (OBS) is set on 253°, with a TO indication, and a right course deviation indicator (CDI) deflection. What is the aircraft's position from the VORTAC?

A. East-northeast.

B. North-northeast.

C. West-southwest.

Answer (A) is correct. *(PHAK Chap 15)*
   DISCUSSION: It is not necessary to refer to Fig. 26 to solve this problem. Write the word VOR on a piece of paper. Now draw a line through it, representing the 253° radial and its reciprocal. Now imagine you are flying along this line on a heading of 253°. With a TO indication and a right CDI deflection, you are northeast of the VOR but south of the course.
   NOTE: The FAA previously changed the figure to which this question refers without changing the question. Figure 26 depicts the Dallas-Ft. Worth VOR/DME, not a VORTAC.
   Answer (B) is incorrect. You are south, not north, of the course. Answer (C) is incorrect. You have a TO, not FROM, indication.

*Page
Intentionally
Left Blank*

10.3 Automatic Direction Finder (ADF)

15. (Refer to Figure 30 on page 305.) (Refer to illustration 1.) Determine the magnetic bearing TO the station.

   A. 030°.

   B. 180°.

   C. 210°.

Answer (C) is correct. *(PHAK Chap 15)*
   DISCUSSION: Fig. 30 shows movable card ADFs. In these, the airplane's magnetic heading is always on top and the needle always indicates the magnetic bearing TO the station. Thus, the magnetic bearing TO the station in ADF 1 is 210°.
   Answer (A) is incorrect. The head (not the tail) indicates the magnetic bearing TO the station. Answer (B) is incorrect. The magnetic bearing TO the station is 210° (not 180°).

16. (Refer to Figure 30 on page 305.) (Refer to illustration 1.) What outbound bearing is the aircraft crossing?

   A. 030°.

   B. 150°.

   C. 180°.

Answer (A) is correct. *(PHAK Chap 15)*
   DISCUSSION: The outbound (magnetic) bearing is the bearing FROM the station, which is represented by the tail of the needle in a movable card ADF. The airplane in ADF 1 is crossing the 030° outbound bearing (radial) since the tail of the needle is pointing to 030°.
   Answer (B) is incorrect. This is the reciprocal of the magnetic heading (not bearing). Answer (C) is incorrect. The needle does not point to 180°.

17. (Refer to Figure 30 on page 305.) (Refer to illustration 2.) What magnetic bearing should the pilot use to fly TO the station?

   A. 010°.

   B. 145°.

   C. 190°.

Answer (C) is correct. *(PHAK Chap 15)*
   DISCUSSION: Fig. 30, illustration 2, is a movable card ADF. This ADF displays the airplane's magnetic heading at the top, and the needle always points to the magnetic bearing TO the station. Thus, the magnetic bearing TO the station in ADF 2 is 190°.
   Answer (A) is incorrect. This is the bearing FROM the station. Answer (B) is incorrect. This might be an appropriate heading to intercept the 190° bearing TO the station.

18. (Refer to Figure 30 on page 305.) (Refer to illustration 2.) Determine the approximate heading to intercept the 180° bearing TO the station.

   A. 040°.

   B. 160°.

   C. 220°.

Answer (C) is correct. *(PHAK Chap 15)*
   DISCUSSION: A 180° bearing to the station would put us directly north of the station, assuming no wind. Currently, we are northeast of the station (190° bearing), proceeding in a northwest direction (magnetic heading of 315°). If we want to intercept the 180° bearing to the station, we should turn to the southwest, or 220°.
   Answer (A) is incorrect. This heading would take us in the opposite direction of the station. Answer (B) is incorrect. We are already east of the station, and 160° would put us farther east.

19. (Refer to Figure 30 on page 305.) (Refer to illustration 3.) What is the magnetic bearing FROM the station?

   A. 025°.

   B. 115°.

   C. 295°.

Answer (B) is correct. *(PHAK Chap 15)*
   DISCUSSION: The tail of the needle of an ADF indicates the magnetic bearing FROM the station on a movable card ADF. ADF 3 shows a magnetic bearing of 115° FROM.
   Answer (A) is incorrect. This is the relative bearing TO (not the magnetic bearing FROM) the station. Answer (C) is incorrect. This is the magnetic bearing TO (not FROM) the station.

20. (Refer to Figure 30 on page 305.) (Refer to illustration 1.) What is the relative bearing TO the station?

   A. 030°.

   B. 210°.

   C. 240°.

Answer (C) is correct. *(PHAK Chap 15)*
   DISCUSSION: The relative bearing is measured clockwise from the nose of the airplane to the head of the needle. From ADF 1, the magnetic heading (MH) is 330°, and the magnetic bearing (MB) TO the station is 210°. Use the following standard formula to solve for the relative bearing (RB) TO the station:

$$MH + RB = MB \ (TO)$$
$$330° + RB = 210°$$
$$RB = -120° \ (210° - 330°)$$

Since it is less than 0°, add 360° to determine the RB of 240° (−120° + 360°).
   Answer (A) is incorrect. This is the magnetic bearing FROM the station. Answer (B) is incorrect. This is the magnetic bearing TO the station.

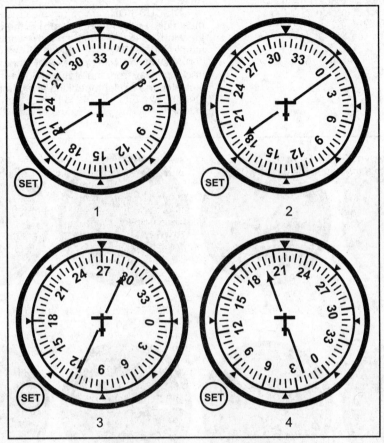

Figure 30. – ADF (Movable Card).

21. (Refer to Figure 30 above.) (Refer to illustration 2.) What is the relative bearing TO the station?

A. 190°.

B. 235°.

C. 315°.

Answer (B) is correct. *(PHAK Chap 15)*
DISCUSSION: The relative bearing is measured clockwise from the nose of the airplane to the head of the needle. Use the following standard formula from ADF 2 in the formula to determine the RB:

$$MH + RB = MB \text{ (TO)}$$
$$315° + RB = 190°$$
$$RB = -125° \ (190° - 315°)$$

Since it is less than 0°, add 360° to determine the RB of 235° (−125° + 360°) TO the station.
Answer (A) is incorrect. This is the magnetic (not relative) bearing TO the station. Answer (C) is incorrect. This is the magnetic heading of the aircraft.

22. (Refer to Figure 30 above.) (Refer to illustration 4.) What is the relative bearing TO the station?

A. 020°.

B. 060°.

C. 340°.

Answer (C) is correct. *(PHAK Chap 15)*
DISCUSSION: The relative bearing (RB) is measured clockwise from the nose of the airplane to the head of the needle. Use the information from ADF 4 in the formula to determine the RB:

$$MH + RB = MB \text{ (TO)}$$
$$220° + RB = 200°$$
$$RB = -20° \ (200° - 220°)$$

Since it is less than 0°, add 360° to determine the RB of 340° (−20° + 360°) TO the station.
Answer (A) is incorrect. This is the magnetic bearing FROM the station. Answer (B) is incorrect. This has no relevance to this problem.

23. (Refer to Figure 30 on page 305.) Which ADF indication represents the aircraft tracking TO the station with a right crosswind?

   A. 1

   B. 2

   C. 4

Answer (C) is correct. *(PHAK Chap 15)*
   DISCUSSION: If you have a crosswind from the right, you must adjust your heading (crab) to the right to compensate for the wind. In that case, the needle would point to the left of the nose, as in ADF 4.
   Answer (A) is incorrect. ADF 1 indicates the airplane tracking away from the station. Answer (B) is incorrect. ADF 2 indicates the airplane tracking away from the station.

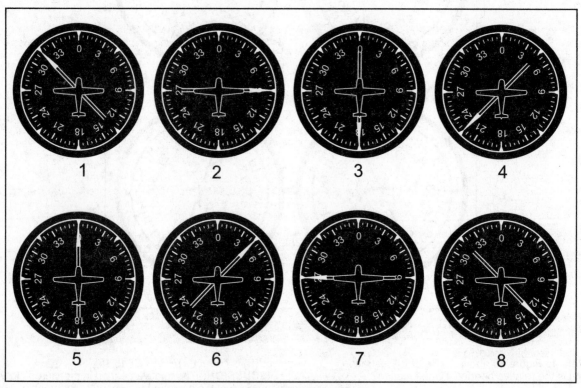

Figure 31. – ADF (Fixed Card).

24. (Refer to Figure 31 above.) (Refer to illustration 3.) The relative bearing TO the station is

   A. 090°.

   B. 180°.

   C. 270°.

Answer (B) is correct. *(PHAK Chap 15)*
   DISCUSSION: The relative bearing (RB) is measured clockwise from the nose of the airplane to the head of the needle. Since this is a fixed card ADF, the needle points to the relative bearing TO the station. ADF 3 in Fig. 31 shows a relative bearing of 180°.
   Answer (A) is incorrect. The needle does not point to the right wing. Answer (C) is incorrect. The needle does not point to the left wing.

25. (Refer to Figure 31 above.) (Refer to illustration 1.) The relative bearing TO the station is

   A. 045°.

   B. 180°.

   C. 315°.

Answer (C) is correct. *(PHAK Chap 15)*
   DISCUSSION: On a fixed card ADF, the needle points to the relative bearing TO the station. ADF 1 in Fig. 31 shows a relative bearing of 315°.
   Answer (A) is incorrect. The needle does not point 45° to the right of the nose. Answer (B) is incorrect. The needle does not point to the tail.

26. (Refer to Figure 31 above.) (Refer to illustration 2.) The relative bearing TO the station is

   A. 090°.

   B. 180°.

   C. 270°.

Answer (A) is correct. *(PHAK Chap 15)*
   DISCUSSION: On a fixed card ADF, the needle points to the relative bearing TO the station. ADF 2 in Fig. 31 shows a relative bearing of 090° TO the station.
   Answer (B) is incorrect. The needle does not point to the tail. Answer (C) is incorrect. The head (not the tail) of the needle indicates relative bearing.

27. (Refer to Figure 31 on page 306.) (Refer to illustration 4.) On a magnetic heading of 320°, the magnetic bearing TO the station is

    A. 005°.

    B. 185°.

    C. 225°.

Answer (B) is correct. *(PHAK Chap 15)*
    DISCUSSION: The magnetic bearing TO the station is required. Use the standard ADF formula.

$$MH + RB = MB \text{ (TO)}$$
$$320° + 225° = MB \text{ (TO)}$$
$$545° = MB \text{ (TO)}$$

    Since it is greater than 360°, subtract 360° to determine that the MB (TO) is 185° (545° − 360°).
    Answer (A) is incorrect. This is not a relevant direction in this problem. Answer (C) is incorrect. This is the relative bearing.

28. (Refer to Figure 31 on page 306.) (Refer to illustration 5.) On a magnetic heading of 035°, the magnetic bearing TO the station is

    A. 035°.

    B. 180°.

    C. 215°.

Answer (A) is correct. *(PHAK Chap 15)*
    DISCUSSION: The magnetic bearing TO the station is required. Use the standard ADF formula.

$$MH + RB = MB \text{ (TO)}$$
$$035° + 0° = MB \text{ (TO)}$$
$$035° = MB \text{ (TO)}$$

    Answer (B) is incorrect. This is the relative bearing FROM the station. Answer (C) is incorrect. This is the magnetic bearing FROM the station.

29. (Refer to Figure 31 on page 306.) (Refer to illustration 6.) On a magnetic heading of 120°, the magnetic bearing TO the station is

    A. 045°.

    B. 165°.

    C. 270°.

Answer (B) is correct. *(PHAK Chap 15)*
    DISCUSSION: The magnetic bearing TO the station is required. Use the standard ADF formula.

$$MH + RB = MB \text{ (TO)}$$
$$120° + 045° = MB \text{ (TO)}$$
$$165° = MB \text{ (TO)}$$

    Answer (A) is incorrect. This is the relative bearing TO the station. Answer (C) is incorrect. This is not a relevant heading in this problem.

30. (Refer to Figure 31 on page 306.) (Refer to illustration 6.) If the magnetic bearing TO the station is 240°, the magnetic heading is

    A. 045°.

    B. 105°.

    C. 195°.

Answer (C) is correct. *(PHAK Chap 15)*
    DISCUSSION: The magnetic heading is required. Use the standard ADF formula.

$$MH + RB = MB \text{ (TO)}$$
$$MH + 045° = 240°$$
$$MH = 240° − 045°$$
$$MH = 195°$$

    Answer (A) is incorrect. This is the relative bearing TO the station. Answer (B) is incorrect. This is not a relevant heading in this problem.

31. (Refer to Figure 31 on page 306.) (Refer to illustration 8.) If the magnetic bearing TO the station is 135°, the magnetic heading is

    A. 135°.

    B. 270°.

    C. 360°.

Answer (C) is correct. *(PHAK Chap 15)*
    DISCUSSION: The magnetic heading is required. Use the standard ADF formula.

$$MH + RB = MB \text{ (TO)}$$
$$MH + 135° = 135°$$
$$MH = 0° = (360°)$$

    Answer (A) is incorrect. This is both the relative and magnetic bearing TO the station. Answer (B) is incorrect. This is not a relevant heading in this problem.

32. (Refer to Figure 31 on page 306.) (Refer to illustration 7.) If the magnetic bearing TO the station is 030°, the magnetic heading is

    A. 060°.

    B. 120°.

    C. 270°.

Answer (B) is correct. *(PHAK Chap 15)*
    DISCUSSION: The magnetic heading is required. Use the standard ADF formula.

$$MH + RB = MB \text{ (TO)}$$
$$MH + 270° = 030°$$
$$MH = 030° − 270°$$
$$MH = −240° \text{ (add 360°)}$$
$$MH = 120°$$

    Answer (A) is incorrect. This is not a relevant heading in this problem. Answer (C) is incorrect. This is the relative bearing TO the station.

10.4 Global Positioning System (GPS)

33. How many satellites make up the Global Positioning System (GPS)?

A. 25

B. 22

C. 24

Answer (C) is correct. *(AIM Para 1-1-19)*
DISCUSSION: The Global Positioning System (GPS) is composed of a constellation of 24 satellites that broadcast signals decoded by a receiver in order to determine a three-dimensional position.
Answer (A) is incorrect. The GPS is composed of 24, not 25, satellites. Answer (B) is incorrect. The GPS is composed of 24, not 22, satellites.

34. What is the minimum number of Global Positioning System (GPS) satellites that are observable by a user anywhere on earth?

A. 6

B. 5

C. 4

Answer (B) is correct. *(AIM Para 1-1-19)*
DISCUSSION: The Global Positioning System is composed of 24 satellites, at least five of which are observable at any given time anywhere on earth.
Answer (A) is incorrect. At least five, not six, satellites are visible anywhere on earth. Answer (C) is incorrect. At least five, not four, satellites are visible anywhere on earth.

35. How many Global Positioning System (GPS) satellites are required to yield a three-dimensional position (latitude, longitude, and altitude) and time solution?

A. 5

B. 6

C. 4

Answer (C) is correct. *(AIM Para 1-1-19)*
DISCUSSION: GPS satellites broadcast radio signals that are decoded by a receiver in order to triangulate a three-dimensional position by calculating distances based on the amount of time it takes the radio signals to reach the receiver. At least four GPS satellites are required to yield a three-dimensional position (latitude, longitude, and altitude) and time solution.
Answer (A) is incorrect. Four, not five, satellites are required for a three-dimensional position and time solution. Answer (B) is incorrect. Four, not six, satellites are required for a three-dimensional position and time solution.

36. The Global Positioning System is

A. ground based.

B. satellite based.

C. antenna based.

Answer (B) is correct. *(PHAK Chap 15)*
DISCUSSION: The Global Positioning System (GPS) is a satellite-based radio navigation system.
Answer (A) is incorrect. GPS is satellite based, not ground based. Answer (C) is incorrect. GPS is satellite based, not antenna based.

37. Which of the following is a true statement concerning the Global Positioning System?

A. Advances in technology make it possible to rely completely on GPS units.

B. GPS databases and paper navigational charts are updated at the same time.

C. Navigating by GPS must be integrated with other forms of navigation.

Answer (C) is correct. *(AAH Chap 3)*
DISCUSSION: Navigating by GPS must be integrated with other forms of electronic navigation as well as pilotage and dead reckoning.
Answer (A) is incorrect. There is always the possibility for equipment failure, so pilots should never rely solely on one system of navigation. Answer (B) is incorrect. GPS databases and paper navigational charts are not necessarily updated on the same schedule.

END OF STUDY UNIT

STUDY UNIT ELEVEN
CROSS-COUNTRY FLIGHT PLANNING

(7 pages of outline)

This study unit contains outlines of major concepts tested, sample test questions and answers regarding cross-country flight planning, and an explanation of each answer. The table of contents above lists each subunit within this study unit, the number of questions pertaining to that particular subunit, and the pages on which the outlines and questions begin, respectively.

This book assumes that you are familiar with the standard flight computer. Full discussion with examples can be found in *Pilot Handbook*, Study Unit 9.

Many of the questions in this study unit ask about the sectional charts, which appear as Legend 1 and Figures 21 through 27. To produce them in color economically, we have put them together on pages 283 through 290. As you will need to turn frequently to these eight pages, mark them with "dog ears" (fold their corners) or paper clips.

CAUTION: Recall that the **sole purpose** of this book is to expedite passing the FAA pilot knowledge test for the private pilot certificate. Accordingly, all extraneous material (i.e., topics or regulations not directly tested) is omitted, even though much more information and knowledge are necessary to fly safely. This additional material is presented in *Pilot Handbook* and *Private Pilot Flight Maneuvers and Practical Test Prep*, available from Gleim Publications, Inc. See the order form on page 371.

11.1 VFR FLIGHT PLAN

1. A VFR flight plan is a form (see Figure 52 on page 317) that contains 17 blocks of information. Only the following blocks are tested on the private pilot knowledge test:

    a. Block 7: "Cruising Altitude." Use only your initial requested altitude on your VFR flight plan.

    b. Block 9: "Destination (Name of airport and city)" should include the airport or place at which you plan to make your last landing for this flight.

        1) Unless you plan a stopover of more than 1 hr. elsewhere en route

    c. Block 12: "Fuel on Board" (in hours and minutes) requires the amount of usable fuel in the airplane at the time of departure, expressed in hours of flying time.

2. You should close your flight plan with the nearest FSS, or, if one is not available, you may request any ATC facility to relay your cancelation to the FSS.

    a. Control towers (and ground control) do not automatically close VFR or DVFR flight plans since they do not know if a particular VFR aircraft is on a flight plan.

11.2 PREFLIGHT INSPECTION

1. During the preflight inspection, the pilot in command is responsible for determining that the airplane is safe for flight.

2. The owner or operator is responsible for maintaining the airplane in an airworthy condition.

3. For the first flight of the day, the preflight inspection should be accomplished by a thorough and systematic means recommended by the manufacturer.

11.3 MISCELLANEOUS AIRSPEED QUESTIONS

1. When turbulence is encountered, the airplane's airspeed should be reduced to maneuvering speed (V_A).

    a. The pilot should attempt to maintain a level flight attitude.

    b. Constant altitude and constant airspeed are usually impossible and result in additional control pressure, which adds stress to the airplane.

2. In the event of a power failure after becoming airborne, the most important thing to do is to immediately establish and maintain the best glide airspeed.

    a. Do not maintain altitude at the expense of airspeed or a stall/spin will result.

    b. The maximum gliding distance of an aircraft is obtained when the total drag on the aircraft is minimized. The best glide speed is calculated by determining when the aircraft produces the least amount of total drag. This occurs at a point where all of the sources of drag add up to a minimum total drag -- any airspeed above or below this point will result in an **increase** in drag.

    c. A constant gliding speed should be maintained because variations in gliding speed will nullify your ability to accurately determine gliding distance and choose a landing spot.

3. Approaches and landings at night should be the same as in daylight (i.e., at same airspeeds and altitudes).

11.4 TAXIING TECHNIQUE

1. When taxiing in strong quartering headwinds, the aileron should be up on the side from which the wind is blowing.

    a. The elevator should be in the neutral position for tricycle-geared airplanes.
    b. The elevator should be in the up position for tailwheel airplanes.

2. When taxiing during strong quartering tailwinds, the aileron should be down on the side from which the wind is blowing.

    a. The elevator should be in the down position (for both tricycle and tailwheel airplanes).

3. When taxiing high-wing, nosewheel-equipped airplanes, the most critical wind condition is a quartering tailwind.

11.5 MAGNETIC COURSE

1. To determine the magnetic course (MC) from one airport to another, correct the true course (TC) only for magnetic variation; i.e., make no allowance for wind correction angle.

    a. Determine the TC by placing the straight edge of a navigational plotter or protractor along the route, with the hole in the plotter on the intersection of the route and a meridian, or line of longitude (the vertical line with little crosslines).

        1) The TC is measured by the numbers on the protractor portion of the plotter (semi-circle) at the meridian.

2) Note that up to four numbers (90° apart) are provided on the plotter. You must determine which is the direction of the flight, using a common sense approximation of your direction.

b. Alternatively, you can use a line of latitude (horizontal line with little crosslines) if your course is in a north or south direction.

1) This is why there are four numbers on the plotter. You may be using either a meridian or line of latitude to measure your course and be going in either direction along the course line.

c. Determine the MC by adjusting the TC for magnetic variation (angle between true north and magnetic north).

1) On sectional charts, a long dashed line provides the number of degrees of magnetic variation. The variation is either east or west and is signified by "E" or "W," e.g., 3°E or 5°W.

2) If the variation is east, subtract; if west, add (memory aid: east is least and west is best). This is from TC to MC.

2. If your course is to or from a VOR, use the compass rose to determine the MC; i.e., no adjustment is needed from TC to MC.

a. Compass roses have about a 3-in. diameter on sectional charts.

b. Every 30° is labeled, as well as marked with an arrow inside the rose pointing out.

c. Use the reciprocal to radials when flying toward the VOR.

1) EXAMPLE: If your course is toward an airport on the 180° radial rather than from the airport, your MC is 360°, not 180°.

11.6 MAGNETIC HEADING

1. To compute magnetic heading (MH) from MC, you must adjust for the wind.

a. To compute wind effect, use the wind side of the computer.

1) Align the magnetic wind direction on the inner scale under the true index (top of the computer).

a) Wind direction is normally given in true, not magnetic, direction. Thus, you must first adjust the wind direction from true to magnetic.

b) Find the magnetic variation on the navigation chart. As with course corrections, add westerly variation and subtract easterly variation.

c) See the discussion of magnetic variation under item 1.c. in Subunit 11.5, "Magnetic Course," above.

d) EXAMPLE: Given wind of 330° and a 20°E variation, the magnetic direction of the wind is 310° (330° − 20°).

2) Slide the grid through the computer until the grommet (the hole in the center) is on the 100-kt. wind line. Measure up the vertical line the amount of wind speed in knots and put a pencil mark on the plastic.

3) Rotate the inner scale so the MC lies under the true index.

4) Slide the grid so that your pencil dot is superimposed over the true airspeed (TAS). The location of the grommet will indicate the groundspeed. This is needed for time en route calculations.

5) The pencil mark will indicate the wind correction angle (WCA). If to the left, it is a negative wind correction. If to the right, it is a positive wind correction.

b. MH is found by adjusting the MC for the wind correction.

1) Add the number of degrees the pencil mark is to the right of the centerline or subtract the number of degrees to the left.

2.   Authors' Note: We suggest converting TC to MC and winds to magnetic before using the wind side of your computer. Then you do not have to convert your final answer from the wind side to magnetic.

    a.   Many courses are to or from VORs, and that allows you to use the VOR compass rose to determine MC directly. That is, you can skip the plotter routine for TC.

3.   **Alternative: E6-B Computer Approach to Magnetic Heading**

    a.   On the previous page, before computing the WCA and groundspeed (GS) on the wind side of the E6-B computer, we converted

        1)   TC to MC
        2)   Winds (true) to winds (magnetic)

    b.   The ALTERNATIVE METHOD suggested on your E6-B is to use TC and winds (true) on the wind side of your E6-B to compute your true heading (TH) and then convert TH to MH. The advantages of this method are

        1)   You convert only one true direction to magnetic, not two.
        2)   This is the way it has always been taught.

    c.   We suggest converting TC and wind (true) to magnetic because pilot activities are in terms of magnetic headings, courses, runways, radials, bearings, and final approach courses. When flying, you should always think magnetic, not true.

        1)   While doing magnetic heading problems, you should visualize yourself flying each problem; e.g., a 270° course with a 300° wind will require a right (270°+) heading and have a headwind component.

4.   On your E6-B, you will find directions such as

    a.   Set Wind Direction opposite True Index.
    b.   Mark Wind Dot up from Grommet.
    c.   Place TC under True Index.
    d.   Slide TAS under Wind Dot.
    e.   Read GS under Grommet.
    f.   Read WCA under Wind Dot.
    g.   Complete the problem by use of the formulas.

5.   The formulas given are

    TH = TC ± WCA (wind correction angle)
    MH = TH ± magnetic variation (E–, W+)
    CH = MH ± compass deviation

6.   We understand this ALTERNATIVE approach is widely used. It was developed prior to the VOR system, compass roses, airways, etc., which are all identified in magnetic direction, **not** true direction. Thus you may use it for textbook exercises, but when flying, think magnetic. (Please use page 373 to comment favorably or unfavorably on our emphasis on magnetic. Thank you.)

11.7 COMPASS HEADING

1.   If a question asks for a compass heading (CH), the MH is converted to a CH by adding or subtracting deviation (installation error), which is indicated on a compass correction card usually mounted on the compass.

    a.   Compass correction cards are needed because the metal, electric motors, and other instruments in each airplane affect the compass causing compass deviation.

    b.   Compass correction cards usually indicate corrections for every 30°.

        1)   For each 30°, you are given the corresponding MH and the heading you should follow, i.e., the CH.

2) The difference between these two headings is the amount to add or subtract.

3) EXAMPLE: Using the compass correction card below, your CH would be 085° if your MH were due east, and 332° if your MH were 330°.

4) If your MH does not coincide with a heading on the correction card, use the nearest one, i.e., interpolate.

FOR (MAGNETIC)	N	30	60	E	120	150
STEER (COMPASS)	0	28	57	85	117	148
FOR (MAGNETIC)	S	210	240	W	300	330
STEER (COMPASS)	180	212	243	274	303	332

Typical Compass Correction Card

11.8 TIME EN ROUTE

1. A number of questions require you to determine the time of arrival at some specified point on a sectional chart given the times that two other points on the chart were crossed. You must

    a. Compute your groundspeed based upon the distance already traveled (i.e., between the two given points) in the given time.

    b. Measure the additional distance to go.

    c. Compute the time required to travel to the next point.

2. First, measure the distances (a) already gone and (b) remaining to go with a navigational plotter or a ruler.

    a. Remember that the scale at the bottom of sectional charts is 1:500,000.

    b. Because the questions give data in NM instead of SM (e.g., wind speed is given in knots), use NM.

3. To compute speed, place the distance already gone on the outer scale of the flight computer adjacent to the number of minutes it took on the inner scale.

    a. Read the number on the outer scale adjacent to the solid triangular pointer (i.e., at 60 min.). This is your groundspeed in knots.

    b. For numbers less than 10 on either scale, add a zero.

    c. EXAMPLE: If you travel 3 NM in 3 min., you are going 60 kt. Place 30 on the outer scale adjacent to 30 on the inner scale. Then the outer scale shows 60 kt. above the solid triangular pointer.

4. To compute the time required to fly to the next point, start with the speed on the outer scale adjacent to 60 min. on the inner scale (just as you had it in the preceding step). Find the remaining NM to go on the outer scale. The adjacent number on the inner scale is the number of minutes to go.

    a. EXAMPLE: Place 12 (for 120 kt.) on the outer scale over the solid triangular pointer (i.e., at 60 min.). Look along the outer scale and find 4 (for 40 NM). It should be directly above 2 (for 20 min.) on the inner scale.

5. Another type of time en route question has you compute the magnetic course, heading, and groundspeed.

    a. Determine the groundspeed as explained previously in this subunit.

    b. Recall that groundspeed appears under the grommet when you slide the grid on the wind side of your flight computer such that your pencil mark is on the TAS arc.

    c. Once you determine groundspeed, put it over 60 min. on the inner scale. Find the distance to go on the outer scale, and read the time en route on the inner scale.

11.9 TIME ZONE CORRECTIONS

1.   To correct for time zones, remember that there is a 1-hr. difference between each time zone, i.e., from the Eastern Time Zone to the Central Time Zone, from the Central Time Zone to the Mountain Time Zone, and from the Mountain Time Zone to the Pacific Time Zone.

   a.   Subtract 1 hr. for each time zone when traveling east to west, and add 1 hr. for each time zone when traveling west to east.

   b.   Additionally, there may be daylight saving time (in the summer) or standard time in effect.

2.   The number of hours to adjust to or from Coordinated Universal Time (UTC) are 4-5-6-7 in the summer and 5-6-7-8 in the winter (for the four zones from east to west).

   a.   EXAMPLE: To compute UTC, add 4 hr. to Eastern Daylight Saving Time and 5 hr. to Eastern Standard Time.

      1)   Add 5 and 6 hr. respectively to Central Time.
      2)   Add 6 and 7 hr. respectively to Mountain Time.
      3)   Add 7 and 8 hr. respectively to Pacific Time.

   b.   Remember, there is always a longer lag in standard than in daylight saving time.

3.   For questions requiring the time of arrival at a destination airport, you should

   a.   First add the hours en route to the time of departure and determine the time of arrival based on the time zone of departure.

   b.   Then adjust the time to the time zone requested, i.e., UTC or time zone of arrival.

   c.   Alternatively, convert the departure time to UTC, add hours en route, and convert to local time.

11.10 FUNDAMENTALS OF FLIGHT

1.   The four fundamentals involved in maneuvering an aircraft are

   a.   Straight-and-level,
   b.   Turns,
   c.   Climbs, and
   d.   Descents.

11.11 RECTANGULAR COURSE

1.   When beginning a rectangular course, the determining factor in deciding the distance from the field boundary at which an aircraft should be flown is the steepness of the bank desired in the turns.

2.   The same techniques of a rectangular course apply when flying an airport traffic pattern.

   a.   On the turn from downwind to base, one goes from a steep to a medium bank.
   b.   On the turn from base to final, one goes from a medium to a shallow bank.
   c.   On a turn from upwind to crosswind, one goes from a shallow to a medium bank.
   d.   On a turn from crosswind to downwind, one goes from a medium to a steep bank.

3.   The corners that require less than a 90° turn in a rectangular course are

   a.   The turn to final and
   b.   The turn to crosswind.

4.   The corners that require more than a 90° turn in a rectangular course are

   a.   The turn to downwind and
   b.   The turn to base.

5. To properly compensate for a crosswind during straight-and-level cruising flight, the pilot should establish a proper heading into the wind by coordinated use of the controls.

11.12 S-TURNS ACROSS A ROAD

1. Groundspeed will be equal when the headwind or tailwind components are the same, e.g., direct crosswind, downwind just out of crosswind and just into crosswind, and the same for upwind.

2. The angle of bank will be steepest when flying in a tailwind.

3. In S-turns, you must be crabbed into the wind the most when you have a full crosswind component.

4. In S-turns, a consistently smaller half-circle will be made on the upwind side of the road when the bank is increased too rapidly during the early part of the turn.

11.13 LANDINGS

1. Under normal conditions, a proper crosswind landing on a runway requires that, at the moment of touchdown, the direction of motion of the aircraft and its longitudinal axis be parallel to the runway.

    a. This minimizes the side load placed on the landing gear during touchdown.

QUESTIONS AND ANSWER EXPLANATIONS

All of the private pilot knowledge test questions chosen by the FAA for release as well as additional questions selected by Gleim relating to the material in the previous outlines are reproduced on the following pages. These questions have been organized into the same subunits as the outlines. To the immediate right of each question are the correct answer and answer explanation. You should cover these answers and answer explanations while responding to the questions. Refer to the general discussion in the Introduction on how to take the FAA pilot knowledge test.

Remember that the questions from the FAA pilot knowledge test bank have been reordered by topic and organized into a meaningful sequence. Also, the first line of the answer explanation gives the citation of the authoritative source for the answer.

QUESTIONS

11.1 VFR Flight Plan

1. How should a VFR flight plan be closed at the completion of the flight at a controlled airport?

A. The tower will automatically close the flight plan when the aircraft turns off the runway.

B. The pilot must close the flight plan with the nearest FSS or other FAA facility upon landing.

C. The tower will relay the instructions to the nearest FSS when the aircraft contacts the tower for landing.

Answer (B) is correct. *(AIM Para 5-1-14)*
    DISCUSSION: A pilot is responsible for ensuring that the VFR or DVFR flight plan is canceled (FAR 91.153). You should close your flight plan with the nearest FSS or, if one is not available, you may request any ATC facility to relay your cancelation to the FSS.
    Answer (A) is incorrect. The tower will automatically close an IFR (not VFR) flight plan. Answer (C) is incorrect. The tower will relay to the nearest FSS only if requested.

2. (Refer to Figure 52 on page 317.) If more than one cruising altitude is intended, which should be entered in block 7 of the flight plan?

A. Initial cruising altitude.

B. Highest cruising altitude.

C. Lowest cruising altitude.

Answer (A) is correct.   *(AIM Para 5-1-4)*
DISCUSSION: Use only your initial requested altitude on your VFR flight plan to assist briefers in providing weather and wind information.
Answer (B) is incorrect. The initial, not highest, altitude should be filed on your VFR flight plan. Answer (C) is incorrect. The initial, not lowest, altitude should be filed on your VFR flight plan.

3. (Refer to Figure 52 on page 317.) What information should be entered in block 9 for a VFR day flight?

A. The name of the airport of first intended landing.

B. The name of destination airport if no stopover for more than 1 hour is anticipated.

C. The name of the airport where the aircraft is based.

Answer (B) is correct.   *(AIM Para 5-1-4)*
DISCUSSION: In Block 9 of the flight plan form in Fig. 52, enter the name of the airport of last intended landing for that flight, as long as no stopover exceeds 1 hr.
Answer (A) is incorrect. The first intended landing, i.e., the end of the first leg of the flight, is included in the route of flight (Block 8). Answer (C) is incorrect. The name of the airport where the airplane is based is entered in Block 14.

4. (Refer to Figure 52 on page 317.) What information should be entered into block 9 for a VFR day flight?

A. The destination airport identifier code and name of the FBO where the airplane will be parked.

B. The destination airport identifier code and city name.

C. The destination city and state.

Answer (B) is correct.   *(AIM Para 5-1-4)*
DISCUSSION: In Block 9 of the flight plan form in Fig. 52, enter the identifier or name of the airport of last intended landing for that flight, as long as no stopover exceeds 1 hr.
Answer (A) is incorrect. The name of the FBO is not required. Answer (C) is incorrect. The destination city may be different from the destination airport's name.

5. (Refer to Figure 52 on page 317.) What information should be entered in block 12 for a VFR day flight?

A. The actual time enroute expressed in hours and minutes.

B. The estimated time in enroute expressed in hours and minutes.

C. The total amount of usable fuel onboard expressed in hours and minutes.

Answer (C) is correct.   *(AIM Para 5-1-4)*
DISCUSSION: Block 12 of the flight plan requires the amount of usable fuel in the airplane at the time of departure. It should be expressed in hours and minutes of flying time.
Answer (A) is incorrect. Block 12 requires the amount of fuel on board expressed in time. Answer (B) is incorrect. Block 12 requires the amount of fuel on board expressed in time.

6. (Refer to Figure 52 on page 317.) What information should be entered in block 12 for a VFR day flight?

A. The estimated time en route plus 30 minutes.

B. The estimated time en route plus 45 minutes.

C. The amount of usable fuel on board expressed in time.

Answer (C) is correct.   *(AIM Para 5-1-4)*
DISCUSSION: Block 12 of the flight plan requires the amount of usable fuel in the airplane at the time of departure. It should be expressed in hours and minutes of flying time.
Answer (A) is incorrect. It states the VFR fuel requirement for day flight. Answer (B) is incorrect. It states the VFR fuel requirement for night flight.

7. (Refer to Figure 52 on page 317.) What information should be entered in block 7 for a VFR day flight?

A. The altitude assigned by the FSS.

B. The initial altitude assigned by ATC.

C. The appropriate VFR cruising altitude.

Answer (C) is correct.   *(AIM Para 5-1-4)*
DISCUSSION: Block 7 should contain the appropriate VFR altitude so the weather briefer can provide you with a more accurate weather advisory.
Answer (A) is incorrect. FSS does not assign altitudes. Answer (B) is incorrect. You will have not yet been assigned an altitude by ATC.

								Form Approved: OMB No. 2120-0026

U.S. DEPARTMENT OF TRANSPORTATION FEDERAL AVIATION ADMINISTRATION

FLIGHT PLAN

(FAA USE ONLY)   ☐ PILOT BRIEFING   ☐ VNR   TIME STARTED   SPECIALIST INITIALS

☐ STOPOVER

1 TYPE	2 AIRCRAFT IDENTIFICATION	3 AIRCRAFT TYPE/ SPECIAL EQUIPMENT	4 TRUE AIRSPEED	5 DEPARTURE POINT	6 DEPARTURE TIME		7 CRUISING ALTITUDE
VFR					PROPOSED (Z)	ACTUAL (Z)	
IFR							
DVFR			KTS				

8 ROUTE OF FLIGHT

9 DESTINATION (Name of airport and city)	10 EST. TIME ROUTE		11 REMARKS
	HOURS	MINUTES	

12 FUEL ON BOARD		13 ALTERNATE AIRPORT(S)	14 PILOT'S NAME, ADDRESS & TELEPHONE NUMBER & AIRCRAFT HOME BASE	15 NUMBER ABOARD
HOURS	MINUTES			
			17 DESTINATION CONTACT/TELEPHONE (OPTIONAL)	

16 COLOR OF AIRCRAFT	CIVIL AIRCRAFT PILOTS. FAR Part 91 requires you file an IFR flight plan to operate under instrument flight rules in controlled airspace. Failure to file could result in a civil penalty not to exceed $1,000 for each violaton (Section 901 of the Federal Aviation Act of 1958, as amended). Filing of a VFR flight plan is recommended as a good operating practice. See also Part 99 for requirements concerning DVFR flight plans.

FAA Form 7233-1 (8-82)   **CLOSE VFR FLIGHT PLAN WITH** _____ **FSS ON ARRIVAL**

Figure 52. – Flight Plan Form.

11.2 Preflight Inspection

8. During the preflight inspection who is responsible for determining the aircraft as safe for flight?

A. The pilot in command.

B. The owner or operator.

C. The certificated mechanic who performed the annual inspection.

Answer (A) is correct. *(FAR 91.7)*
DISCUSSION: During the preflight inspection, the pilot in command is responsible for determining whether the airplane is in condition for safe flight.
Answer (B) is incorrect. The owner or operator is responsible for maintaining the airplane in an airworthy condition, not for determining whether the airplane is safe for flight during the preflight inspection. Answer (C) is incorrect. The pilot in command, not the mechanic who performed the annual inspection, is responsible for determining whether the airplane is safe for flight.

9. Who is primarily responsible for maintaining an aircraft in airworthy condition?

A. Pilot-in-command.

B. Owner or operator.

C. Mechanic.

Answer (B) is correct. *(PHAK Chap 8)*
DISCUSSION: The owner or operator of an airplane is primarily responsible for maintaining an airplane in an airworthy condition, including compliance with all applicable Airworthiness Directives (ADs).
Answer (A) is incorrect. The pilot in command is responsible for determining that the airplane is in airworthy condition, not for maintaining the airplane. Answer (C) is incorrect. The owner or operator, not a mechanic, is responsible for maintaining an airplane in an airworthy condition.

10. How should an aircraft preflight inspection be accomplished for the first flight of the day?

A. Quick walk around with a check of gas and oil.

B. Thorough and systematic means recommended by the manufacturer.

C. Any sequence as determined by the pilot-in-command.

Answer (B) is correct. *(PHAK Chap 8)*
DISCUSSION: For the first flight of the day, the preflight inspection should be accomplished by a thorough and systematic means recommended by the manufacturer.
Answer (A) is incorrect. A quick walk around with a check of gas and oil may be adequate if it is not the first flight of the day in that airplane. Answer (C) is incorrect. A preflight inspection should be done in the sequence recommended by the manufacturer in the POH, not in any sequence determined by the pilot in command.

11.3 Miscellaneous Airspeed Questions

11. Upon encountering severe turbulence, which flight condition should the pilot attempt to maintain?

A. Constant altitude and airspeed.

B. Constant angle of attack.

C. Level flight attitude.

Answer (C) is correct. *(AC 00-24B)*
DISCUSSION: Attempting to hold altitude and airspeed in severe turbulence can lead to overstressing the airplane. Rather, you should set power to what normally will maintain V_A and simply attempt to maintain a level flight attitude.
Answer (A) is incorrect. Maintaining a constant altitude will require additional control movements, adding stress to the airplane. Answer (B) is incorrect. In severe turbulence, the angle of attack will fluctuate due to the wind shears and wind shifts that cause the turbulence.

12. The most important rule to remember in the event of a power failure after becoming airborne is to

A. immediately establish the proper gliding attitude and airspeed.

B. quickly check the fuel supply for possible fuel exhaustion.

C. determine the wind direction to plan for the forced landing.

Answer (A) is correct. *(AFH Chap 8)*
DISCUSSION: In the event of a power failure after becoming airborne, the most important rule to remember is to maintain best glide airspeed. This will usually require a pitch attitude slightly higher than level flight. Invariably, with a power failure, one returns to ground, but emphasis should be put on a controlled return rather than a crash return. Many pilots attempt to maintain altitude at the expense of airspeed, resulting in a stall or stall/spin.
Answer (B) is incorrect. Checking the fuel supply should only be done after a glide has been established and a landing site has been selected. Answer (C) is incorrect. Landing into the wind may not be possible, depending upon altitude and field availability.

13. When executing an emergency approach to land in a single-engine airplane, it is important to maintain a constant glide speed because variations in glide speed

A. increase the chances of shock cooling the engine.

B. assure the proper descent angle is maintained until entering the flare.

C. nullify all attempts at accuracy in judgment of gliding distance and landing spot.

Answer (C) is correct. *(AFH Chap 8)*
DISCUSSION: A constant gliding speed should be maintained because variations of gliding speed nullify all attempts at accuracy in judgment of gliding distance and the landing spot.
Answer (A) is incorrect. Shock cooling the engine can occur when you significantly increase the speed beyond the best glide speed. Answer (B) is incorrect. A constant glide speed may not guarantee a certain descent angle. The angle of descent will be based on many environmental factors. While this statement is potentially valid, it is not the best answer option available for this question.

14. VFR approaches to land at night should be accomplished

A. at a higher airspeed.

B. with a steeper descent.

C. the same as during daytime.

Answer (C) is correct. *(AFH Chap 10)*
DISCUSSION: Every effort should be made to execute approaches and landings at night in the same manner as they are made in the day. Inexperienced pilots often have a tendency to make approaches and landings at night with excessive airspeed.
Answer (A) is incorrect. Approaching at a higher airspeed could result in floating into unseen obstacles at the far end of the runway. Answer (B) is incorrect. A steeper descent is not necessary. You should use the visual glide slope indicators at night whenever they are available.

11.4   Taxiing Technique

15.   (Refer to Figure 9 below.) (Refer to area A.) How should the flight controls be held while taxiing a tricycle-gear equipped airplane into a left quartering headwind?

A.   Left aileron up, elevator neutral.

B.   Left aileron down, elevator neutral.

C.   Left aileron up, elevator down.

Answer (A) is correct. *(AFH Chap 2)*
   DISCUSSION: Given a left quartering headwind, the left aileron should be kept up to spoil the excess lift on the left wing that the crosswind is creating. The elevator should be neutral to keep from putting too much or too little weight on the nosewheel.
   Answer (B) is incorrect. Lowering the left aileron will increase the lift on the left wing. Answer (C) is incorrect. It describes the control setting for a right tailwind in a tailwheel airplane.

16.   (Refer to Figure 9 below.) (Refer to area C.) How should the flight controls be held while taxiing a tricycle-gear equipped airplane with a left quartering tailwind?

A.   Left aileron up, elevator neutral.

B.   Left aileron down, elevator down.

C.   Left aileron up, elevator down.

Answer (B) is correct. *(AFH Chap 2)*
   DISCUSSION: With a left quartering tailwind, the left aileron should be down so the wind does not get under the left wing and flip the airplane over. Also, the elevator should be down, i.e., controls forward, so the wind does not get under the tail and blow the airplane tail over front.
   Answer (A) is incorrect. It describes the control setting for a left headwind. Answer (C) is incorrect. It describes the control setting for a right tailwind.

17.   (Refer to Figure 9 below.) (Refer to area B.) How should the flight controls be held while taxiing a tailwheel airplane into a right quartering headwind?

A.   Right aileron up, elevator up.

B.   Right aileron down, elevator neutral.

C.   Right aileron up, elevator down.

Answer (A) is correct. *(AFH Chap 2)*
   DISCUSSION: When there is a right quartering headwind, the right aileron should be up to spoil the excess lift on the right wing that the crosswind is creating. The elevator should be up to keep weight on the tailwheel to help maintain maneuverability.
   Answer (B) is incorrect. The elevator should be up (not neutral) and the right aileron up (not down) when taxiing a tailwheel airplane in a right quartering headwind. Answer (C) is incorrect. The elevator should be up (not down) when taxiing in a right quartering headwind.

18.   (Refer to Figure 9 below.) (Refer to area C.) How should the flight controls be held while taxiing a tailwheel airplane with a left quartering tailwind?

A.   Left aileron up, elevator neutral.

B.   Left aileron down, elevator neutral.

C.   Left aileron down, elevator down.

Answer (C) is correct. *(AFH Chap 2)*
   DISCUSSION: When there is a left quartering tailwind, the left aileron should be held down so the wind does not get under the left wing and flip the airplane over. Also, the elevator should be down, i.e., controls forward, so the wind does not get under the tail and blow the airplane tail over front.
   Answer (A) is incorrect. The left aileron should be down (not up) and the elevator down (not neutral). Answer (B) is incorrect. The elevator should be down when taxiing with a tailwind.

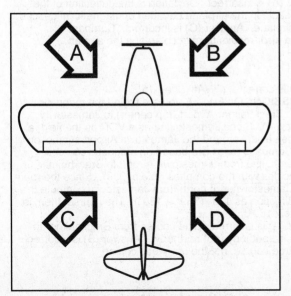

Figure 9. – Control Position for Taxi.

19. When taxiing with strong quartering tailwinds, which aileron positions should be used?

A. Aileron down on the downwind side.

B. Ailerons neutral.

C. Aileron down on the side from which the wind is blowing.

Answer (C) is correct. *(AFH Chap 2)*
   DISCUSSION: When there is a strong quartering tailwind, the aileron should be down on the side from which the wind is blowing (when taxiing away from the wind, turn away from the wind) to help keep the wind from getting under that wing and flipping the airplane over.
   Answer (A) is incorrect. The aileron should be down on the upwind (not downwind) side. Answer (B) is incorrect. The aileron positions help control the airplane while taxiing in windy conditions.

20. Which aileron positions should a pilot generally use when taxiing in strong quartering headwinds?

A. Aileron up on the side from which the wind is blowing.

B. Aileron down on the side from which the wind is blowing.

C. Ailerons neutral.

Answer (A) is correct. *(AFH Chap 2)*
   DISCUSSION: When there is a strong quartering headwind, the aileron should be up on the side from which the wind is blowing to help keep the wind from getting under that wing and blowing the aircraft over. (When taxiing into the wind, turn into the wind.)
   Answer (B) is incorrect. The aileron should be up (not down) on the side from which the wind is blowing (i.e., upwind). Answer (C) is incorrect. The aileron positions help control the airplane while taxiing in windy conditions.

21. Which wind condition would be most critical when taxiing a nosewheel equipped high-wing airplane?

A. Quartering tailwind.

B. Direct crosswind.

C. Quartering headwind.

Answer (A) is correct. *(AFH Chap 2)*
   DISCUSSION: The most critical wind condition when taxiing a nosewheel-equipped high-wing airplane is a quartering tailwind, which can flip a high-wing airplane over on its top. This should be prevented by holding the elevator in the down position, i.e., controls forward, and the aileron down on the side from which the wind is coming.
   Answer (B) is incorrect. A direct crosswind will probably not flip an airplane over. However, it may weathervane the airplane into the wind. Answer (C) is incorrect. A headwind is aerodynamically the condition an airplane is designed for, i.e., wind from the front.

11.5 Magnetic Course

The sectional chart legend and Figures 21 through 27 appear on pages 283 through 290.

22. The angular difference between true north and magnetic north is

A. magnetic deviation.

B. magnetic variation.

C. compass acceleration error.

Answer (B) is correct. *(PHAK Chap 15)*
   DISCUSSION: The angular difference between true and magnetic north is referred to as magnetic variation.
   Answer (A) is incorrect. Deviation is the deflection of the compass needle in the airplane because of magnetic influences within the airplane. Answer (C) is incorrect. Compass acceleration error results from accelerating the aircraft.

23. (Refer to Figure 27 on page 290.) Determine the magnetic course from Breckheimer (Pvt) Airport (area 1) to Jamestown Airport (area 4).

A. 360°.

B. 188°.

C. 180°.

Answer (C) is correct. *(PHAK Chap 15)*
   DISCUSSION: On Fig. 27, you are to find the magnetic course from Breckheimer Airport (top center) to Jamestown Airport (below 4). Since Jamestown has a VOR on the field, a compass rose exists around the Jamestown Airport symbol on the chart. Compass roses are based on magnetic courses. Thus, a straight line from Jamestown Airport to Breckheimer Airport coincides with the compass rose at 359°. Since the route is south to Jamestown, not north from Jamestown, compute the reciprocal direction as 179° (359° − 180°). The course, then, is approximately 180°.
   Answer (A) is incorrect. The course from Breckheimer to Jamestown is southerly (not northerly). Answer (B) is incorrect. This is the true course, not the magnetic course.

24. (Refer to Figure 21 on page 284.) Determine the magnetic course from First Flight Airport (area 5) to Hampton Roads Airport (area 2).

    A. 141°.

    B. 321°.

    C. 331°.

Answer (C) is correct. *(PHAK Chap 15)*
    DISCUSSION: You are to find the magnetic course from First Flight Airport (lower right corner) to Hampton Roads Airport (above 2 on Fig. 21). True course is the degrees clockwise from true north. Determine the true course by placing the straight edge of your plotter along the given route with the grommet at the intersection of your route and a meridian (the north/south line with crosslines). Here, TC is 321°. To convert this to a magnetic course, add the 10° westerly variation (indicated by the slanted dashed line across the upper right of the sectional), and find the magnetic course of 331°. Remember to subtract easterly variation and add westerly variation.
    Answer (A) is incorrect. This is the true, not magnetic, course for a flight from Hampton Roads Airport to First Flight Airport, not for a flight from First Flight to Hampton Roads. Answer (B) is incorrect. This is the true, not magnetic, course.

25. (Refer to Figure 22 on page 285.) What course should be selected on the omnibearing selector (OBS) to make a direct flight from Mercer County Regional Airport (area 3) to the Minot VORTAC (area 1) with a TO indication?

    A. 359°.

    B. 179°.

    C. 001°.

Answer (A) is correct. *(PHAK Chap 15)*
    DISCUSSION: Use Fig. 22 to find the course (omnibearing selector with a "TO" indication) from Mercer County Regional Airport (lower left corner) to the Minot VORTAC (right of 1). Note the compass rose (based on magnetic courses) that indicates the Minot VORTAC. A straight line from Mercer to Minot Airport coincides the compass rose at 179°. Since the route is north TO Minot, not south from Minot, compute the reciprocal direction as 359° (179° + 180°).
    Answer (B) is incorrect. This is the radial on which a direct flight from Mercer County Regional Airport to the Minot VORTAC would be flown. If 179° is selected on the OBS, it will result in a FROM indication and reverse sensing. Answer (C) is incorrect. This would be the proper OBS setting for a flight originating 5 NM west of Mercer County Regional Airport, rather than directly from it.

26. (Refer to Figure 24 on page 287.) On what course should the VOR receiver (OBS) be set to navigate direct from Hampton Varnville Airport (area 1) to Savannah VORTAC (area 3)?

    A. 003°.

    B. 183°.

    C. 200°.

Answer (B) is correct. *(PHAK Chap 15)*
    DISCUSSION: You are to find the OBS course setting from Hampton Varnville Airport (right of 1) to Savannah VORTAC (below 3 on Fig. 24). Since compass roses are based on magnetic courses, you can find that a straight line from Hampton Varnville Airport to Savannah VORTAC coincides the Savannah VORTAC compass rose at 003°. Since the route is south to (not north from) Savannah, compute the reciprocal direction as 183° magnetic (003° + 180°). To use the VOR properly when flying to a VOR station, the course you select with the OBS should be the reciprocal of the radial you will be tracking. If this is not done, reverse sensing occurs.
    Answer (A) is incorrect. This would be the course north from, not south to, Savannah. Answer (C) is incorrect. This would be the course from Ridgeland Airport, not Hampton Varnville Airport, to Savannah VORTAC.

27. (Refer to Figure 25 on page 288.) Determine the magnetic course from Airpark East Airport (area 1) to Winnsboro Airport (area 2). Magnetic variation is 6°30'E.

    A. 075°.

    B. 082°.

    C. 091°.

Answer (A) is correct. *(PHAK Chap 15)*
    DISCUSSION: To find the magnetic course from Airpark East Airport (lower left of chart) to Winnsboro Airport (right of 2 on Fig. 25), you must find true course and correct it for magnetic variation. Determine the true course by placing the straight edge of your plotter along the given route such that the grommet (center hole) is on a meridian (the north/south line with crosslines). True course of 82° is the number of degrees clockwise from true north. It is read on the protractor portion of your plotter at the intersection of the meridian. To convert this to a magnetic course, subtract the 6°30'E (or round up to 7°E) easterly variation and find that the magnetic course is 075°. Remember to subtract easterly variation and add westerly variation.
    Answer (B) is incorrect. This is the true, not magnetic, course. Answer (C) is incorrect. You must subtract, not add, an easterly variation.

28. (Refer to Figure 25 on page 288.) On what course should the VOR receiver (OBS) be set in order to navigate direct from Majors Airport (area 1) to Quitman VORTAC (area 2)?

    A. 101°.

    B. 108°.

    C. 281°.

Answer (A) is correct. *(PHAK Chap 15)*
    DISCUSSION: You are to find the radial to navigate direct from Majors Airport (less than 2 in. north and east of 1) to Quitman VORTAC (southeast of 2 on Fig. 25). A compass rose, based on magnetic course, exists around the Quitman VORTAC. A straight line from Majors Airport to Quitman VORTAC coincides with this compass rose at 281°. Since the route is east to (not west from) Quitman, compute the reciprocal direction as 101° magnetic (281° − 180°).
    Answer (B) is incorrect. This is the true, not magnetic, course from Majors to Quitman VORTAC. A VORTAC always uses magnetic direction. Answer (C) is incorrect. This is the course west from, not east to, Quitman VORTAC.

11.6 Magnetic Heading

29. (Refer to Figure 22 on page 285.) Determine the magnetic heading for a flight from Mercer County Regional Airport (area 3) to Minot International (area 1). The wind is from 330° at 25 knots, the true airspeed is 100 knots, and the magnetic variation is 10°E.

    A. 002°.

    B. 012°.

    C. 352°.

Answer (C) is correct. *(PHAK Chap 15)*
    DISCUSSION: On Fig. 22, begin by computing the true course (TC) from Mercer Co. Reg. (lower left corner) to Minot Int'l. (upper left center) by drawing a line between the two airports. Next, determine the TC by placing the grommet on the plotter at the intersection of the course line and a meridian (vertical line with cross-hatchings) and the top of the plotter aligned with the course line. Note the 012° TC on the edge of the protractor.
    Next, subtract the 10° east magnetic variation from the TC to obtain a magnetic course (MC) of 002°. Since the wind is given true, subtract the 10° east magnetic variation to obtain a magnetic wind direction of 320° (330° − 10°).
    Now use the wind side of your computer to plot the wind direction and velocity. Place the magnetic wind direction of 320° on the inner scale on the true index. Mark 25 kt. up from the grommet with a pencil. Turn the inner scale to the magnetic course of 002°. Slide the grid up until the pencil mark lies over the line for true airspeed (TAS) of 100 kt. Correct for the 10° left wind angle by subtracting from the magnetic course of 002° to obtain a magnetic heading of 352°. This is intuitively correct because, given the magnetic course of 002° and a northwesterly wind, you must turn to the left (crab into the wind) to correct for it.
    Answer (A) is incorrect. This is the magnetic course, not heading; i.e., you must still correct for wind drift. Answer (B) is incorrect. This is the true course, not magnetic heading.

30. (Refer to Figure 23 on page 286.) What is the magnetic heading for a flight from Priest River Airport (area 1) to Shoshone County Airport (area 3)? The wind is from 030° at 12 knots and the true airspeed is 95 knots.

    A. 118°.

    B. 143°.

    C. 136°.

Answer (A) is correct. *(PHAK Chap 15)*
    DISCUSSION: On Fig. 23, begin by computing the true course from Priest River Airport (upper left corner) to Shoshone County Airport (just below 3) by laying a flight plotter between the two airports. The grommet should coincide with the meridian (vertical line with cross-hatchings). Note the 143° true course on the edge of the protractor.
    Next, find the magnetic variation that is given by the dashed line marked 18°E, slanting in a northeasterly fashion just south of Carlin Bay private airport. Subtract the 18°E variation from TC to obtain a magnetic course of 125°. Since the wind is given true, reduce the true wind direction of 30° by the magnetic variation of 18°E to a magnetic wind direction of 12°.
    Now use the wind side of your computer. Turning the inner circle to 12° under the true index, mark 12 kt. above the grommet. Set the magnetic course of 125° under the true index. Slide the grid so the pencil mark is on 95 kt. TAS. Note that the pencil mark is 7° left of the center line, requiring you to adjust the magnetic course to a 118° magnetic heading (125° − 7°). Subtract left, add right. That is, if you are on an easterly flight and the wind is from the north, you will want to correct to the left.
    Answer (B) is incorrect. This is the true course, not the magnetic heading. Answer (C) is incorrect. This would be the magnetic heading if the wind was from 215° at 19 kt., not 030°, at 12 kt.

31. (Refer to Figure 23 on page 286.) Determine the magnetic heading for a flight from St. Maries Airport (area 4) to Priest River Airport (area 1). The wind is from 340° at 10 knots and the true airspeed is 90 knots.

    A. 327°.

    B. 320°.

    C. 345°.

Answer (A) is correct. *(PHAK Chap 15)*
DISCUSSION:

1. This flight is from St. Maries (just below 4) to Priest River (upper left corner) on Fig. 23.
2. TC is 346°.
3. MC = 346° − 18°E variation = 328°.
4. Wind magnetic = 340° − 18° = 322°.
5. Mark 10 kt. up when 322° under true index.
6. Put MC 328° under true index.
7. Slide grid so pencil mark is on 90 kt. TAS.
8. Note that the pencil mark is 1° left.
9. Subtract 1° from 328° MC for 327° MH.

Answer (B) is incorrect. This would be the magnetic heading if the wind was from 300° at 14 kt., not 340° at 10 kt. Answer (C) is incorrect. This is the approximate true course, not magnetic heading.

32. (Refer to Figure 23 on page 286.) Determine the magnetic heading for a flight from Sandpoint Airport (area 1) to St. Maries Airport (area 4). The wind is from 215° at 25 knots and the true airspeed is 125 knots.

    A. 349°.

    B. 169°.

    C. 187°.

Answer (B) is correct. *(PHAK Chap 15)*
DISCUSSION:

1. This flight is from Sandpoint Airport (above 1), to St. Maries (below 4) on Fig. 23.
2. TC = 181°.
3. MC = 181° − 18°E variation = 163°.
4. Wind magnetic = 215° − 18° = 197°.
5. Mark up 25 kt. with 197° under true index.
6. Put MC 163° under true index.
7. Slide grid so pencil mark is on 125 kt. TAS.
8. Note that the pencil mark is 6° right.
9. Add 6° to 163° MC for 169° MH.

Answer (A) is incorrect. This would be the magnetic heading for a flight from St. Maries Airport to Sandpoint Airport, not from Sandpoint to St. Maries, with the wind from 145°, not 215°, at 25 kt. Answer (C) is incorrect. This is the true heading, not the magnetic heading. Correction for 18° of easterly magnetic variation has not been applied.

33. (Refer to Figure 26 on page 289.) Determine the magnetic heading for a flight from Fort Worth Meacham (area 4) to Denton Muni (area 1). The wind is from 330° at 25 knots, the true airspeed is 110 knots, and the magnetic variation is 7°E.

    A. 003°.

    B. 017°.

    C. 023°.

Answer (A) is correct. *(PHAK Chap 15)*
DISCUSSION:

1. This flight is from Fort Worth Meacham (southeast of 4) to Denton Muni (southwest of 1) on Fig. 26.
2. TC = 021°.
3. MC = 021° − 7°E variation = 014°.
4. Wind magnetic = 330° − 7°E variation = 323°.
5. Mark up 25 kt. with 323° under true index.
6. Put MC 014° under true index.
7. Slide grid so pencil mark is on 110 kt. TAS.
8. Note that the pencil mark is 11° left.
9. Subtract 11° from 014° MC for 003° MH.

Answer (B) is incorrect. You must subtract (not add) an easterly variation. Answer (C) is incorrect. You must subtract (not add) a left wind correction.

34. (Refer to Figure 25 on page 288.) Determine the magnetic heading for a flight from Majors Airport (area 1) to Winnsboro Airport (area 2). The wind is from 340° at 12 knots, the true airspeed is 36 knots, and the magnetic variation is 6° 30'E.

A. 089°.

B. 096°.

C. 101°.

Answer (A) is correct.   *(PHAK Chap 15)*
DISCUSSION: On Fig. 25, begin by computing the true course (TC) from Majors Airport (area 1) to Winnsboro Airport (area 2) by drawing a line between the two airports. Next, determine the TC by placing the grommet on the plotter at the intersection on the course line and a meridian (vertical line with cross-hatchings) and the top of the plotter aligned with the course line. Note the TC of 101° TC on the edge of the protractor. Next, subtract the 6° east magnetic variation from the TC to obtain a magnetic course (MC) of 95°. Since the wind is given true, subtract the 6° magnetic variation to obtain a magnetic wind direction of 334° (340° – 6°). Now use the wind side of your computer to plot the wind direction and velocity. Place the magnetic wind direction of 334° on the inner scale on the true index. Mark 12 kt. up from the grommet with a pencil. Turn the inner scale to the magnetic course of 95°. Slide the grid up until the pencil mark lies over the line for true airspeed (TAS) of 36 kts. Correct for the 6° left wind angle by subtracting from the magnetic course of 095° to obtain a magnetic heading of 089°. This is intuitively correct because, given the magnetic course of 095° and a northwesterly wind, you must turn to the left (crab into the wind) to correct for it.
    Answer (B) is incorrect. This is the heading you would get if you added magnetic variation instead of subtracting it from the true course and wind direction. Answer (C) is incorrect. Correcting to the right would result in a magnetic heading of 101°.

35. (Refer to Figure 24 on page 287.) Determine the magnetic heading for a flight from Allendale County Airport (area 1) to Claxton-Evans County Airport (area 2). The wind is from 090° at 16 knots and the true airspeed is 90 knots.

A. 230°.

B. 212°.

C. 208°.

Answer (C) is correct.   *(PHAK Chap 15)*
DISCUSSION:

1. This flight is from Allendale County (above 1) to Claxton-Evans County Airport (left of 2) on Fig. 24. Variation is shown on Fig. 24 as 5°W.
2. TC = 212°.
3. MC = 212° TC + 5°W variation = 217°.
4. Wind magnetic = 090° + 5°W variation = 095°.
5. Mark up 16 kt. with 095° under true index.
6. Place MC 217° under true index.
7. Move wind mark to 90 kt. TAS arc.
8. Note that the pencil mark is 9° left.
9. Subtract 9° from 217° MC for 208° MH.

Answer (A) is incorrect. This would be the approximate magnetic heading if the wind was out of 330° at 23 kt., not 090° at 16 kt. Answer (B) is incorrect. This is the true heading, not the magnetic heading.

36. If a true heading of 135° results in a ground track of 130° and a true airspeed of 135 knots results in a groundspeed of 140 knots, the wind would be from

A. 019° and 12 knots.

B. 200° and 13 knots.

C. 246° and 13 knots.

Answer (C) is correct.   *(FI Comp)*
DISCUSSION: To estimate your wind given true heading and a ground track, place the groundspeed under the grommet (140 kt.) with the ground track of 130° under the true index. Then find the true airspeed on the true airspeed arc of 135 kt., and put a pencil mark for a 5° right deviation (135° – 130° = 5°). Place the pencil mark on the centerline under the true index and note a wind from 246° under the true index. The pencil mark is now on 153 kt., which is about 13 kt. up from the grommet (153 – 140).
    Answer (A) is incorrect. These would be your approximate wind and velocity for a 5° left wind correction, not right. Answer (B) is incorrect. These would be your approximate wind and velocity when at a true airspeed of 140 kt., not 135 kt., and a groundspeed of 135 kt., not 140 kt.

37. When converting from true course to magnetic heading, a pilot should

    A. subtract easterly variation and right wind correction angle.

    B. add westerly variation and subtract left wind correction angle.

    C. subtract westerly variation and add right wind correction angle.

Answer (B) is correct. *(PHAK Chap 15)*
    DISCUSSION: When converting true course to magnetic heading, you should remember two rules. With magnetic variation, east variation is subtracted and west variation is added. With wind corrections, left correction is subtracted and right correction is added.
    Answer (A) is incorrect. Right wind correction is added, not subtracted. Answer (C) is incorrect. Westerly variation is added, not subtracted.

11.7 Compass Heading

38. (Refer to Figure 59 below, and Figure 24 on page 287.) Determine the compass heading for a flight from Claxton-Evans County Airport (area 2) to Hampton Varnville Airport (area 1). The wind is from 280° at 8 knots, and the true airspeed is 85 knots.

    A. 033°.

    B. 042°.

    C. 038°.

Answer (B) is correct. *(PHAK Chap 15)*
    DISCUSSION:

1. This flight is from Claxton-Evans (left of 2) to Hampton Varnville (right of 1) on Fig. 24.
2. TC = 045°.
3. MC = 045° TC + 5°W variation = 050°.
4. Wind magnetic = 280° + 5°W variation = 285°.
5. Mark up 8 kt. with 285° under true index.
6. Place MC 050° under true index.
7. Move wind mark to 85 kt. TAS arc.
8. Note that the pencil mark is 5° left.
9. Subtract 5° from 050° MC for 045° MH.
10. Subtract 3° compass variation (obtained from Fig. 59) from 045° to find the compass heading of 042°.

    Answer (A) is incorrect. This would be the approximate compass heading if the wind were out of 295° at 22 kt., not 280° at 8 kt. Answer (C) is incorrect. This would be the approximate compass heading if the wind were out of 295° at 12 kt.

For	N	30	60	E	120	150
Steer	0	27	56	85	116	148
For	S	210	240	W	300	330
Steer	181	214	244	274	303	332

Figure 59. – Compass Card.

11.8 Time En Route

39. (Refer to Figure 21 on page 284.) En route to First Flight Airport (area 5), your flight passes over Hampton Roads Airport (area 2) at 1456 and then over Chesapeake Municipal at 1501. At what time should your flight arrive at First Flight?

    A. 1516.

    B. 1521.

    C. 1526.

Answer (C) is correct. *(PHAK Chap 15)*
    DISCUSSION: The distance between Hampton Roads Airport (about 2 in. north of 2) and Chesapeake Municipal (northeast of 2 on Fig. 21) is 10 NM. It took 5 min. (1501 – 1456) to go 10 NM, so the airplane is traveling at 2 NM per minute. The distance from Chesapeake Municipal to First Flight (right of 5) is 50 NM. At 2 NM per minute, it will take 25 min. Twenty-five minutes added to the time you passed Chesapeake Municipal (1501) is 1526.
    Note: There is a discrepancy between this question and the figure. "Chesapeake Municipal" is labeled "Chesapeake Regional" on the chart.
    Answer (A) is incorrect. At 2 NM per min., it will take 25 min., not 15 min., to reach first flight. Answer (B) is incorrect. The 25 min. must be added to 1501, not 1456.

40. (Refer to Figure 22 on page 285.) What is the estimated time en route from Mercer County Regional Airport (area 3) to Minot International (area 1)? The wind is from 330° at 25 knots and the true airspeed is 100 knots. Add 3-1/2 minutes for departure and climb-out.

   A. 44 minutes.

   B. 48 minutes.

   C. 52 minutes.

Answer (B) is correct. *(PHAK Chap 15)*
   DISCUSSION: The requirement is time en route and not magnetic heading, so there is no need to convert TC to MC.
   Using Fig. 22, the time en route from Mercer Co. Reg. Airport (lower left corner) to Minot (right of 1) is determined by measuring the distance (58 NM measured with a plotter), determining the time based on groundspeed, and adding 3.5 min. for takeoff and climb. The TC is 012° as measured with a plotter. The wind is from 330° at 25 kt.
   On the wind side of your flight computer, place the wind direction 330° under the true index and mark 25 kt. up. Rotate TC of 012° under the true index. Slide the grid so the pencil mark is on the arc for TAS of 100 kt. Read 80 kt. groundspeed under the grommet.
   Turn to the calculator side and place the groundspeed of 80 kt. on the outer scale over 60 min. Find 58 NM on outer scale and note 44.5 min. on the inner scale. Add 3.5 min. to 44.5 min. for climb for en route time of 48 min.
   Answer (A) is incorrect. You must add 3.5 min. for departure and climbout. Answer (C) is incorrect. The time en route is 48 min. (not 52 min.).

41. (Refer to Figure 23 on page 286.) Determine the estimated time en route for a flight from Priest River Airport (area 1) to Shoshone County Airport (area 3). The wind is from 030 at 12 knots and the true airspeed is 95 knots. Add 2 minutes for climb-out.

   A. 29 minutes.

   B. 27 minutes.

   C. 31 minutes.

Answer (C) is correct. *(PHAK Chap 15)*
   DISCUSSION: The requirement is time en route and not magnetic heading, so there is no need to convert TC to MC.

1. To find the en route time from Priest River Airport (area 1) to Shoshone County Airport (area 3) use Fig. 23.
2. Measure the distance with plotter to be 48 NM.
3. TC = 143°.
4. Mark up 12 kt. with 030° under true index.
5. Put TC of 143° under true index.
6. Slide the grid so the pencil mark is on TAS of 95 kt.
7. Read the groundspeed of 99 kt. under the grommet.
8. On the calculator side, place 99 kt. on the outer scale over 60 min.
9. Read 29 min. on the inner scale below 48 NM on the outer scale.
10. Add 2 min for climb-out and the en route time is 31 min.

   Answer (A) is incorrect. Twenty-nine minutes would be the approximate time en route if you forgot to add 2 min. for climb-out. Answer (B) is incorrect. You must add, not subtract, the 2 min. for climb-out.

42. (Refer to Figure 23 on page 286.) What is the estimated time en route from Sandpoint Airport (area 1) to St. Maries Airport (area 4)? The wind is from 215° at 25 knots, and the true airspeed is 125 knots.

   A. 38 minutes.

   B. 30 minutes.

   C. 34 minutes.

Answer (C) is correct. *(PHAK Chap 15)*
   DISCUSSION: The requirement is time en route and not magnetic heading, so there is no need to convert TC to MC.

1. You are to find the en route time from Sandpoint Airport (north of 1) to St. Maries Airport (southeast of 4) on Fig. 23.
2. Measure the distance with plotter to be 59 NM.
3. TC = 181°.
4. Mark up 25 kt. with 215° under true index.
5. Put TC of 181° under true index.
6. Slide the grid so the pencil mark is on TAS of 125 kt.
7. Read the groundspeed of 104 kt. under the grommet.
8. On the calculator side, place 104 kt. on the outer scale over 60 min.
9. Find 59 NM on the outer scale and read 34 min. on the inner scale.

   Answer (A) is incorrect. To make the trip in 38 min. would require a groundspeed of 93 kt., not 104 kt. Answer (B) is incorrect. To make the trip in 30 min. would require a groundspeed of 118 kt., not 104 kt.

43. (Refer to Figure 23 on page 286.) What is the estimated time en route for a flight from St. Maries Airport (area 4) to Priest River Airport (area 1)? The wind is from 300° at 14 knots and the true airspeed is 90 knots. Add 3 minutes for climb-out.

A. 38 minutes.

B. 43 minutes.

C. 48 minutes.

Answer (B) is correct. *(PHAK Chap 15)*
DISCUSSION: The requirement is time en route and not magnetic heading, so there is no need to convert TC to MC.

1. Time en route from St. Maries Airport (southeast of 4) to Priest River Airport (upper left corner) on Fig. 23.
2. Measure the distance with plotter to be 54 NM.
3. TC = 346°.
4. Mark up 14 kt. with 300° under true index.
5. Put TC of 346° under true index.
6. Slide the grid so the pencil mark is on TAS of 90 kt.
7. Read the groundspeed of 80 kt. under the grommet.
8. On the calculator side, place 80 kt. on the outer scale over 60 min.
9. Find 54 NM on the outer scale and read 40 min. on the inner scale.
10. Add 3 min for climb-out to get time en route of 43 min.

Answer (A) is incorrect. To make the trip in 38 min. would require a groundspeed of 92 kt. (not 80 kt.). Answer (C) is incorrect. To make the trip in 48 min. would require a groundspeed of 72 kt. (not 80 kt.).

44. (Refer to Figure 24 on page 287.) While en route on Victor 185, a flight crosses the 248° radial of Allendale VOR at 0953 and then crosses the 216° radial of Allendale VOR at 1000. What is the estimated time of arrival at Savannah VORTAC?

A. 1023.

B. 1028.

C. 1036.

Answer (B) is correct. *(PHAK Chap 15)*
DISCUSSION: The first step is to find the three points involved. V185 runs southeast from the top left of Fig. 24. The first intersection (V70 and V185) is about 1 in. from the top of the chart. The second intersection (V157 and V185) is about 1-1/2 in. farther along V185. The Savannah VORTAC is about 6 in. farther down V185.

Use the sectional scale 1:500,000. From the first intersection (V70 and V185), it is about 10 NM to the intersection of V185 and V157. From there it is 40 NM to Savannah VORTAC.

On your flight computer, place the 7 min. the first leg took (1000 – 0953) on the inner scale under 10 NM on the outer scale. Then find 40 NM on the outer scale. Read 28 min. on the inner scale, which is the time en route from the V185 and V157 intersection to the Savannah VORTAC. Arrival time over Savannah VORTAC is therefore 1028.

Answer (A) is incorrect. You must add 28 min. to 1000 to obtain the correct ETA of 1028. Answer (C) is incorrect. You must add 28 min. to 1000 to obtain the correct ETA of 1028.

45. (Refer to Figure 24 on page 287.) What is the estimated time en route for a flight from Allendale County Airport (area 1) to Claxton-Evans County Airport (area 2)? The wind is from 100° at 18 knots and the true airspeed is 115 knots. Add 2 minutes for climb-out.

A. 33 minutes.

B. 27 minutes.

C. 30 minutes.

Answer (C) is correct. *(PHAK Chap 15)*
DISCUSSION: The requirement is time en route and not magnetic heading, so there is no need to convert TC to MC.

1. To find the en route time from Allendale County (northeast of 1) to Claxton-Evans (southeast of 2), use Fig. 24.
2. Measure the distance with plotter to be 55 NM.
3. TC = 212°.
4. Mark up 18 kt. with 100° under true index.
5. Put TC of 212° under true index.
6. Slide the grid so the pencil mark is on TAS of 115 kt.
7. Read the groundspeed of 120 kt. under the grommet.
8. On the calculator side, place 120 kt. on the outer scale over 60 min.
9. Read 28 min. on the inner scale below 55 NM on the outer scale.
10. Add 2 min for climb-out and the en route time is 30 min.

Answer (A) is incorrect. The groundspeed is 120 kt., not 105 kt. Answer (B) is incorrect. The groundspeed is 120 kt., not 130 kt.

46. (Refer to Figure 24 on page 287.) What is the estimated time en route for a flight from Claxton-Evans County Airport (area 2) to Hampton Varnville Airport (area 1)? The wind is from 290° at 18 knots and the true airspeed is 85 knots. Add 2 minutes for climb-out.

    A. 35 minutes.

    B. 39 minutes.

    C. 44 minutes.

Answer (B) is correct. *(PHAK Chap 15)*
    DISCUSSION: The distance en route from Claxton-Evans (southwest of 2) to Hampton Varnville (east of 1 on Fig. 24) is approximately 57 NM. Also use your plotter to determine that the TC is 45°. The requirement is time en route and not magnetic heading, so there is no need to convert TC to MC.
    Using the wind side of your computer, turn your true index to the wind direction of 290° and mark 18 kt. above the grommet with your pencil. Then turn the inner scale so that the true index is above the TC of 45°. Place the pencil mark on the TAS of 85 kt. and note the groundspeed of 91 kt. Turn your flight computer over and set the speed of 91 kt. above the 60-min. index on the inner scale. Then find the distance of 57 NM on the outer scale to determine a time en route of 37 min. Add 2 min. for climb-out, and the en route time is 39 min.
    Answer (A) is incorrect. You must add (not subtract) 2 min. for climb-out to the time en route. Answer (C) is incorrect. The groundspeed is 91 kt. (not 81 kt.).

47. (Refer to Figure 25 on page 288.) Estimate the time en route from Majors Airport (area 1) to Winnsboro Airport (area 2). The wind is from 340° at 12 knots and the true airspeed is 36 knots.

    A. 55 minutes.

    B. 59 minutes.

    C. 63 minutes.

Answer (B) is correct. *(PHAK Chap 15)*
    DISCUSSION: The distance between Majors Airport and Winnsboro Airport is 41 NM. The approximate MH heading to Winnsboro Airport is 100 degrees. The wind produces a tailwind of approximately 6 knots. Therefore, the aircraft will travel at 42 knots (36 knots + 6 knots). Accordingly, 41 NM/42 NM/hr = .976 hr.
    Answer (A) is incorrect. The wind's effect is a tailwind of 6 knots, not 9 knots. Answer (C) is incorrect. The wind's effect is a tailwind of 6 knots, not 3 knots.

48. (Refer to Figure 26 on page 289.) What is the estimated time en route for a flight from Denton Muni (area 1) to Addison (area 2)? The wind is from 200° at 20 knots, the true airspeed is 110 knots, and the magnetic variation is 7° east.

    A. 13 minutes.

    B. 16 minutes.

    C. 19 minutes.

Answer (A) is correct. *(PHAK Chap 15)*
    DISCUSSION: The requirement is time en route and not magnetic heading, so there is no need to convert TC to MC.

1. To find the en route time from Denton Muni (southwest of 1) to Addison (southwest of 2), use Fig. 26.
2. Measure the distance with plotter to be 23 NM.
3. TC = 128°.
4. Mark up 20 kt. with 200° under true index.
5. Put TC of 128° under true index.
6. Slide the grid so the pencil mark is on TAS of 110 kt.
7. Read the groundspeed of 102 kt. under the grommet.
8. On the calculator side, place 102 kt. on the outer scale over 60 min.
9. Read 13 min. on the inner scale below 23 NM on the outer scale.

    Answer (B) is incorrect. The groundspeed is 102 kt. (not 86 kt.). Answer (C) is incorrect. The groundspeed is 102 kt. (not 73 kt.).

49. (Refer to Figure 26 on page 289.) Estimate the time en route from Addison (area 2) to Redbird (area 3). The wind is from 300° at 15 knots, the true airspeed is 120 knots, and the magnetic variation is 7° east.

    A. 8 minutes.

    B. 11 minutes.

    C. 14 minutes.

Answer (A) is correct. *(PHAK Chap 15)*
    DISCUSSION: The requirement is time en route and not magnetic heading, so there is no need to convert TC to MC.

1. To find the en route time from Addison (southwest of 2) to Redbird (above 3), use Fig. 26.
2. Measure the distance with plotter to be 18 NM.
3. TC = 186°.
4. Mark up 15 kt. with 300° under true index.
5. Put TC of 186° under true index.
6. Slide the grid so the pencil mark is on TAS of 120 kt.
7. Read the groundspeed of 125 kt. under the grommet.
8. On the calculator side, place 125 kt. on the outer scale over 60 min.
9. Read 8.5 min. on the inner scale below 18 NM on the outer scale.

    Answer (B) is incorrect. The groundspeed is 125 kt. (not 98 kt.). Answer (C) is incorrect. The groundspeed is 125 kt. (not 77 kt.).

50. How far will an aircraft travel in 2-1/2 minutes with a groundspeed of 98 knots?

    A. 2.45 NM.

    B. 3.35 NM.

    C. 4.08 NM.

Answer (C) is correct. *(FI Comp)*
    DISCUSSION: To determine the distance traveled in 2-1/2 min. at 98 kt., note that 98 kt. is 1.6 NM/min. (98 ÷ 60 = 1.6). Thus, in 2-1/2 min., you will have traveled a total of 4.08 NM (1.6 × 2.5 = 4.08). Alternatively, put 98 on the outer scale of your flight computer over the index on the inner scale. Find 2.5 min. on the inner scale, above which is 4.1 NM.
    Answer (A) is incorrect. For 2.45 NM to be true, you would need a groundspeed of approximately 59 kt. Answer (B) is incorrect. For 3.35 NM to be true, you would need a groundspeed of approximately 80 kt.

51. How far will an aircraft travel in 7.5 minutes with a ground speed of 114 knots?

    A. 14.25 NM.

    B. 15.00 NM.

    C. 14.50 NM.

Answer (A) is correct. *(FI Comp)*
    DISCUSSION: To determine the distance traveled in 7-1/2 minutes at 114 kt., first determine the distance traveled per minute (114 ÷ 60 = 1.9). In 1 minute, the aircraft travels 1.9 NM. Thus, in 7.5 minutes, the plane will have traveled 14.25 NM (1.9 × 7.5 = 14.25). Alternatively, put 114 on the outer scale of your flight computer over the index on the inner scale. Find 7.5 minutes on the inner scale, above which is 14.25 miles.
    Answer (B) is incorrect. The airplane would require a groundspeed of 120 kt. to travel 15.00 NM in 7.5 minutes. Answer (C) is incorrect. The airplane would require a groundspeed of 116 kt. to travel 14.50 NM in 7.5 minutes.

52. On a cross-country flight, point A is crossed at 1500 hours and the plan is to reach point B at 1530 hours. Use the following information to determine the indicated airspeed required to reach point B on schedule.

Distance between A and B.................. 70 NM
Forecast wind....................................... 310° at 15 kts
Pressure altitude...................................8,000 ft
Ambient temperature .... −10°C
True course..270°

The required indicated airspeed would be approximately

    A. 126 knots.

    B. 137 knots.

    C. 152 knots.

Answer (B) is correct. *(FI Comp)*
    DISCUSSION: First determine the required groundspeed to reach point B at 1530 by placing 70 NM on the outer scale over 30 min. on the inner scale to determine a groundspeed of 140 kt. On the wind side of the computer, put the wind direction of 310° under the true index and put a pencil mark 15 kt. up from the grommet. Next, turn the inner scale so the 270° true course is under the true index and put the grommet over the groundspeed. Note that to obtain the 140-kt. groundspeed, you need a 152-kt. true airspeed. Next, on the computer side, put the air temperature of −10°C over 8,000 ft. altitude. Then find the true airspeed of 152 kt. on the outer scale, which lies over approximately 137 kt. indicated airspeed on the inner scale.
    Answer (A) is incorrect. This would be your indicated airspeed if you had a true airspeed of 140 kt. Note that 140 kt. is your groundspeed, not true airspeed. Answer (C) is incorrect. This is your required true airspeed, not indicated airspeed.

11.9 Time Zone Corrections

53. (Refer to Figure 28 on page 331.) An aircraft departs an airport in the eastern daylight time zone at 0945 EDT for a 2-hour flight to an airport located in the central daylight time zone. The landing should be at what coordinated universal time?

    A. 1345Z.

    B. 1445Z.

    C. 1545Z.

Answer (C) is correct. *(Figure 28)*
    DISCUSSION: First convert the departure time to coordinated universal time (Z) by using the time conversion table in Fig. 28. To convert from eastern daylight time (EDT), add 4 hr. to get 1345Z (0945 + 4 hr). A 2-hr. flight would have you arriving at your destination airport at 1545Z.
    Answer (A) is incorrect. This is the departure time. Answer (B) is incorrect. You would arrive at an airport at 1445Z if the flight were 1 (not 2) hr.

54. (Refer to Figure 28 on page 331.) An aircraft departs an airport in the central standard time zone at 0930 CST for a 2-hour flight to an airport located in the mountain standard time zone. The landing should be at what time?

    A. 0930 MST.

    B. 1030 MST.

    C. 1130 MST.

Answer (B) is correct. *(Figure 28)*
    DISCUSSION: Flying from the Central Standard Time Zone to the Mountain Standard Time Zone results in a 1-hr. gain due to time zone changes. A 2-hr. flight leaving at 0930 CST will arrive in the Mountain Standard Time Zone at 1130 CST, which is 1030 MST.
    Answer (A) is incorrect. The aircraft departed at 0930 CST (not MST). Answer (C) is incorrect. A landing at 1130 MST would be correct for a 3-hr. (not 2-hr.) flight departing from the CST zone at 0930 CST to the MST zone.

55. (Refer to Figure 28 on page 331.) An aircraft departs an airport in the central standard time zone at 0845 CST for a 2-hour flight to an airport located in the mountain standard time zone. The landing should be at what coordinated universal time?

    A. 1345Z.

    B. 1445Z.

    C. 1645Z.

Answer (C) is correct. *(Figure 28)*
    DISCUSSION: First convert the departure time to coordinated universal time (Z) by using the time conversion table in Fig. 28. To convert from CST to Z, you must add 6 hr., thus 0845 CST is 1445Z (0845 + 6 hr.). A 2-hr. flight would make the estimated landing time at 1645Z (1445 + 2 hr.).
    Answer (A) is incorrect. This is the departure time at 0845 CDT, not CST. Answer (B) is incorrect. This is the departure (not landing) time.

56. (Refer to Figure 28 on page 331.) An aircraft departs an airport in the mountain standard time zone at 1615 MST for a 2-hour 15-minute flight to an airport located in the Pacific standard time zone. The estimated time of arrival at the destination airport should be

    A. 1630 PST.

    B. 1730 PST.

    C. 1830 PST.

Answer (B) is correct. *(Figure 28)*
    DISCUSSION: Departing the Mountain Standard Time Zone at 1615 MST for a 2-hr. 15-min. flight would result in arrival in the Pacific Standard Time Zone at 1830 MST. Because there is a 1-hr. difference between Mountain Standard Time and Pacific Standard Time, 1 hr. must be subtracted from the 1830 MST arrival to determine the 1730 PST arrival.
    Answer (A) is incorrect. An arrival time of 1630 PST would be for a 1-hr. 15-min. (not a 2-hr. 15-min.) flight. Answer (C) is incorrect. The time of 1830 MST (not PST) is the estimated time of arrival at the destination airport.

57. (Refer to Figure 28 on page 331.) An aircraft departs an airport in the Pacific standard time zone at 1030 PST for a 4-hour flight to an airport located in the central standard time zone. The landing should be at what coordinated universal time?

    A. 2030Z.

    B. 2130Z.

    C. 2230Z.

Answer (C) is correct. *(Figure 28)*
    DISCUSSION: First, convert the departure time to coordinated universal time (Z) by using the time conversion table in Fig. 28. To convert from PST to Z, you must add 8 hr.; thus, 1030 PST is 1830Z (1030 + 8 hr.). A 4-hr. flight would make the proposed landing time at 2230Z (1830 + 4 hr.).
    Answer (A) is incorrect. This is for a flight of 2 (not 4) hr. Answer (B) is incorrect. This is the proposed landing time if the departure time were 1030 PDT, not PST.

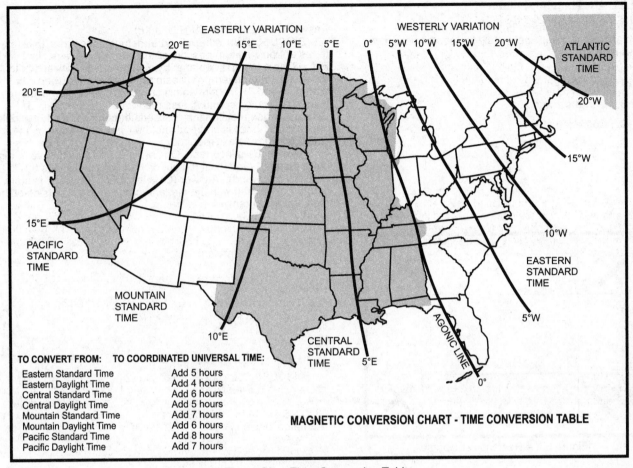

EASTERLY VARIATION   WESTERLY VARIATION

20°E 15°E 10°E 5°E 0° 5°W 10°W 15°W 20°W   ATLANTIC STANDARD TIME

20°E

20°W

15°W

15°E

PACIFIC STANDARD TIME

10°W

MOUNTAIN STANDARD TIME

EASTERN STANDARD TIME

10°E

5°W

CENTRAL STANDARD TIME   5°E

AGONIC LINE

0°

TO CONVERT FROM:	TO COORDINATED UNIVERSAL TIME:
Eastern Standard Time	Add 5 hours
Eastern Daylight Time	Add 4 hours
Central Standard Time	Add 6 hours
Central Daylight Time	Add 5 hours
Mountain Standard Time	Add 7 hours
Mountain Daylight Time	Add 6 hours
Pacific Standard Time	Add 8 hours
Pacific Daylight Time	Add 7 hours

MAGNETIC CONVERSION CHART - TIME CONVERSION TABLE

Figure 28. – Time Conversion Table.

58. (Refer to Figure 28 above.) An aircraft departs an airport in the mountain standard time zone at 1515 MST for a 2-hour 30-minute flight to an airport located in the Pacific standard time zone. What is the estimated time of arrival at the destination airport?

A. 1645 PST.

B. 1745 PST.

C. 1845 PST.

Answer (A) is correct. *(Figure 28)*
DISCUSSION: Departing the Mountain Standard Time (MST) Zone at 1515 MST for a 2-hr. 30-min. flight would result in arrival in the Pacific Standard Time (PST) Zone at 1745 MST. Because there is a 1-hr. difference between MST and PST, 1 hr. must be subtracted from the 1745 MST arrival to determine the 1645 PST estimated time of arrival at the destination airport.
Answer (B) is incorrect. The estimated time of arrival at the destination airport is 1745 MST, not PST. Answer (C) is incorrect. This would be the estimated arrival time for a 3-hr. 30-min. (not 2-hr. 30-min.) flight from the MST zone.

11.10 Fundamentals of Flight

59. Select the four flight fundamentals involved in maneuvering an aircraft.

A. Aircraft power, pitch, bank, and trim.

B. Starting, taxiing, takeoff, and landing.

C. Straight-and-level flight, turns, climbs, and descents.

Answer (C) is correct. *(AFH Chap 3)*
DISCUSSION: Maneuvering an airplane is generally divided into four flight fundamentals: straight-and-level flight, turns, climbs, and descents. All controlled flight consists of one or a combination of more than one of these basic maneuvers.
Answer (A) is incorrect. It lists variable factors necessary for the performance of the flight fundamentals. Answer (B) is incorrect. It lists a combination of basic maneuvers, not flight fundamentals.

11.11 Rectangular Course

60. (Refer to Figure 63 below.) In flying the rectangular course, when would the aircraft be turned less than 90°?

    A. Corners 1 and 4.

    B. Corners 1 and 2.

    C. Corners 2 and 4.

Answer (A) is correct. *(AFH Chap 6)*
    DISCUSSION: When doing a rectangular course, think in terms of traffic pattern descriptions of the various legs. In Fig. 63, note that the airplane is going counterclockwise about the rectangular pattern. While on the base leg (between corners 3 and 4), the airplane is crabbed to the inside of the course. Thus, on corner 4, less than a 90° turn is required. Similarly, when the airplane proceeds through corner 1, it should roll out such that it is crabbed into the wind, and, again, a less-than-90° angle is required.
    Answer (B) is incorrect. On corner 2, you would have to turn more than 90° (you must roll out of your crab angle plus 90° to be heading downwind). Answer (C) is incorrect. Corner 2 is more than 90° (you start with the airplane crabbed to the outside of the rectangular course).

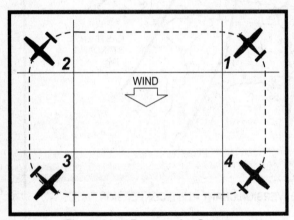

Figure 63. – Rectangular Course

61. (Refer to Figure 63 above.) In flying the rectangular course, when should the aircraft bank vary from a steep bank to a medium bank?

    A. Corner 1.

    B. Corner 3.

    C. Corner 2 and 3.

Answer (B) is correct. *(AFH Chap 6)*
    DISCUSSION: When flying a rectangular course, imagine that the course is a traffic pattern at an airport. On the downwind leg, the wind is a tailwind and results in an increased groundspeed. Accordingly, the turn on the next leg requires a fast roll-in with a steep bank. When the tailwind component diminishes, the bank angle is reduced.
    Answer (A) is incorrect. Corner 1 requires a turn that varies from shallow to medium. Answer (C) is incorrect. Corner 2 requires a turn that varies from medium to steep.

11.12 S-Turns across a Road

62. (Refer to Figure 67 below.) While practicing S-turns, a consistently smaller half-circle is made on one side of the road than on the other, and this turn is not completed before crossing the road or reference line. This would most likely occur in turn

A. 1-2-3 because the bank is decreased too rapidly during the latter part of the turn.

B. 4-5-6 because the bank is increased too rapidly during the early part of the turn.

C. 4-5-6 because the bank is increased too slowly during the latter part of the turn.

Answer (B) is correct. *(AFH Chap 6)*
DISCUSSION: Note that the wind in Fig. 67 is coming up from the bottom rather than from the top as in Fig. 63. The consistently smaller half-circle is made when on the upwind side of the road, i.e., 4-5-6. The initial bank is increased too rapidly, resulting in a smaller half-circle. Then an attempt is made to widen the turn out in the latter stages. Thus, the recrossing of the road is done at less than a 90° angle.
Answer (A) is incorrect. Decreasing the bank too rapidly in the latter stages of 1, 2, and 3 on the downwind side of the road increases, not decreases, the size of that half-circle. Answer (C) is incorrect. Increasing the bank too slowly at the latter stages of 4, 5, and 6 would make the half-circle larger, not smaller.

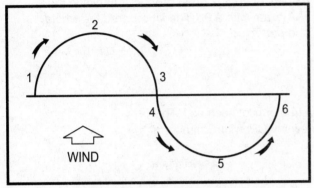

Figure 67. – S-Turn Diagram

11.13 Landings

63. To minimize the side loads placed on the landing gear during touchdown, the pilot should keep the

A. direction of motion of the aircraft parallel to the runway.

B. longitudinal axis of the aircraft parallel to the direction of its motion.

C. downwind wing lowered sufficiently to eliminate the tendency for the aircraft to drift.

Answer (B) is correct. *(AFH Chap 8)*
DISCUSSION: At touchdown when landing, the longitudinal axis of the airplane should be parallel to the direction of its motion, i.e., no side loads to stress the landing gear.
Answer (A) is incorrect. It is important that the longitudinal axis also parallels the runway to avoid side loads on the landing gear. Answer (C) is incorrect. The upwind wing, not the downwind wing, needs to be lowered.

END OF STUDY UNIT

Success stories!

Just for the record I previously used only Gleim study guides for my instrument, commercial, and instructor tests, with first time high passing scores on all of them! Thanks again! You can quote me and my belief that the Gleim Study Guides are the best in the industry.

- Jay Rickmeyer

I want to thank you and all the Gleim team for such a great product. Thanks to the Gleim Private Pilot kit I passed the FAA exam with a 95. The kit covered the whole material. It was easy, simple, clear, and portable.

- Guerline Mejia

It was a learning experience and a great review for an instructor/pilot such as I. This is not my first renewal from Gleim and I will continue to renew my CFI with Gleim for a long ... long... time!

- Les Hirahara

These courses are extremely easy to comprehend and assimilate into practical use.

- Christopher Butera

An excellent review which will allow me to present this information in a manner that a student will absorb more easily.

- Rock Rockcastle, CFII/MEI

Excellent course for CFI renewal. It includes thoughtful and well presented information for the CFI community. All questions and comments about various facts and or presentations are immediately replied to via email. Well done Gleim! I'll be back!

- J.A. Reinhardt

Once again the experience of using Gleim for any recurrent training or training for a new certificate is the best. Can't be beat!

- Karin Case

GLEIM
KNOWLEDGE
TRANSFER
SYSTEMS®

APPENDIX A
PRIVATE PILOT PRACTICE TEST

The following 60 questions have been randomly selected from the airplane questions in our private pilot test bank. You will be referred to figures (charts, tables, etc.) throughout this book. Be careful not to consult the answers or answer explanations when you look for and at the figures. Topical coverage in this practice test is similar to that of the FAA pilot knowledge test. Use the correct answer listing on page 340 to grade your practice test.

NOTE: Our **FAA Test Prep Online** provides unlimited Study and Test Sessions for your personal use. See the discussion on pages 17 through 19 in the Introduction of this book.

1. (Refer to Figure 8 on page 176.) What is the effect of a temperature increase from 25 to 50° F on the density altitude if the pressure altitude remains at 5,000 feet?

A — 1,200-foot increase.
B — 1,400-foot increase.
C — 1,650-foot increase.

2. (Refer to Figure 41 on page 178.) Determine the total distance required for takeoff to clear a 50-foot obstacle.

OAT . . . . Std
Pressure altitude . . . . 4,000 ft
Takeoff weight . . . . 2,800 lb
Headwind component . . . . Calm

A — 1,500 feet.
B — 1,750 feet.
C — 2,000 feet.

3. (Refer to Figure 25 on page 288.) Determine the magnetic course from Airpark East Airport (area 1) to Winnsboro Airport (area 2). Magnetic variation is 6°30'E.

A — 075°.
B — 082°.
C — 091°.

4. (Refer to Figure 24 on page 287.) What is the estimated time en route for a flight from Claxton-Evans County Airport (area 2) to Hampton Varnville Airport (area 1)? The wind is from 290° at 18 knots and the true airspeed is 85 knots. Add 2 minutes for climb-out.

A — 35 minutes.
B — 39 minutes.
C — 44 minutes.

5. (Refer to Figure 37 on page 182.) With a reported wind of south at 20 knots, which runway is appropriate for an airplane with a 13-knot maximum crosswind component?

A — Runway 10.
B — Runway 14.
C — Runway 24.

6. If an aircraft is loaded 90 pounds over maximum certificated gross weight and fuel (gasoline) is drained to bring the aircraft weight within limits, how much fuel should be drained?

A — 10 gallons.
B — 12 gallons.
C — 15 gallons.

7. (Refer to Figure 33 on page 194 and Figure 34 on page 195.) Which action can adjust the airplane's weight to maximum gross weight and the CG within limits for takeoff?

Front seat occupants . . . . 425 lb
Rear seat occupants . . . . 300 lb
Fuel, main tanks . . . . 44 gal

A — Drain 12 gallons of fuel.
B — Drain 9 gallons of fuel.
C — Transfer 12 gallons of fuel from the main tanks to the auxiliary tanks.

8. What is true altitude?

A — The vertical distance of the aircraft above sea level.
B — The vertical distance of the aircraft above the surface.
C — The height above the standard datum plane.

9. (Refer to Figure 26 on page 289.) (Refer to area 4.) The airspace directly overlying Fort Worth Meacham is

A — Class B airspace to 10,000 feet MSL.
B — Class C airspace to 5,000 feet MSL.
C — Class D airspace to 3,200 feet MSL.

10. (Refer to Figure 12 on page 236.) Which of the reporting stations have VFR weather?

A — All.
B — KINK, KBOI, and KJFK.
C — KINK, KBOI, and KLAX.

11. (Refer to Figure 14 on page 238.) If the terrain elevation is 1,295 feet MSL, what is the height above ground level of the base of the ceiling?

A — 505 feet AGL.
B — 1,295 feet AGL.
C — 6,586 feet AGL.

12. (Refer to Figure 21 on page 284.) (Refer to area 3.) Determine the approximate latitude and longitude of Currituck County Airport.

A — 36°24'N – 76°01'W.
B — 36°48'N – 76°01'W.
C — 47°24'N – 75°58'W.

13. (Refer to Figure 24 on page 287.) (Refer to area 3.) What is the height of the lighted obstacle approximately 6 nautical miles southwest of Savannah International?

A — 1,500 feet MSL.
B — 1,531 feet AGL.
C — 1,549 feet MSL.

14. (Refer to Figure 26 on page 289.) (Refer to area 2.) The floor of Class B airspace at Addison Airport is

A — at the surface.
B — 3,000 feet MSL.
C — 3,100 feet MSL.

15. (Refer to Figure 27 on page 290.) (Refer to area 2.) What is the recommended communication procedure when inbound to land at Cooperstown Airport?

A — Broadcast intentions when 10 miles out on the CTAF/MULTICOM frequency, 122.9 MHz.
B — Contact UNICOM when 10 miles out on 122.8 MHz.
C — Circle the airport in a left turn prior to entering traffic.

16. (Refer to Figure 26 on page 289.) (Refer to area 2.) The control tower frequency for Addison Airport is

A — 122.95 MHz.
B — 126.0 MHz.
C — 133.4 MHz.

17. (Refer to Figure 15 on page 243 and Figure 18 on page 244.) The only cloud type forecast in TAF reports is

A — Nimbostratus.
B — Cumulonimbus.
C — Scattered cumulus.

18. (Refer to Figure 4 on page 49 and in color on page 294.) What is the full flap operating range for the airplane?

A — 60 to 100 MPH.
B — 60 to 208 MPH.
C — 65 to 165 MPH.

19. (Refer to Figure 29 on page 301.) (Refer to illustration 3.) The VOR receiver has the indications shown. What is the aircraft's position relative to the station?

A — East.
B — Southeast.
C — West.

20. (Refer to Figure 31 on page 306.) (Refer to illustration 1.) The relative bearing TO the station is

A — 045°.
B — 180°.
C — 315°.

21. (Refer to Figure 25 on page 288.) What is the approximate position of the aircraft if the VOR receivers indicate the 245° radial of Sulphur Springs VOR-DME (area 5) and the 140° radial of Bonham VORTAC (area 3)?

A — Glenmar Airport.
B — Meadowview Airport.
C — Majors Airport.

22. (Refer to Figure 9 on page 319.) (Refer to area C.) How should the flight controls be held while taxiing a tailwheel airplane with a left quartering tailwind?

A — Left aileron up, elevator neutral.
B — Left aileron down, elevator neutral.
C — Left aileron down, elevator down.

23. Detonation occurs in a reciprocating aircraft engine when

A — the spark plugs are fouled or shorted out or the wiring is defective.
B — hot spots in the combustion chamber ignite the fuel/air mixture in advance of normal ignition.
C — the unburned charge in the cylinders explodes instead of burning normally.

24. During a night flight, you observe a steady red light and a flashing red light ahead and at the same altitude. What is the general direction of movement of the other aircraft?

A — The other aircraft is crossing to the left.
B — The other aircraft is crossing to the right.
C — The other aircraft is approaching head-on.

25. (Refer to Figure 48 on page 87.) Illustration A indicates that the aircraft is

A — below the glide slope.
B — on the glide slope.
C — above the glide slope.

26. Prior to entering an Airport Advisory Area, a pilot should

A — monitor ATIS for weather and traffic advisories.
B — contact approach control for vectors to the traffic pattern.
C — contact the local FSS for airport and traffic advisories.

27. What ATC facility should the pilot contact to receive a special VFR departure clearance in Class D airspace?

A — Automated Flight Service Station.
B — Air Traffic Control Tower.
C — Air Route Traffic Control Center.

28. During operations within controlled airspace at altitudes of less than 1,200 feet AGL, the minimum horizontal distance from clouds requirement for VFR flight is

A — 1,000 feet.
B — 1,500 feet.
C — 2,000 feet.

29. Which condition would cause the altimeter to indicate a lower altitude than true altitude?

A — Air temperature lower than standard.
B — Atmospheric pressure lower than standard.
C — Air temperature warmer than standard.

30. Which condition is most favorable to the development of carburetor icing?

A — Any temperature below freezing and a relative humidity of less than 50 percent.
B — Temperature between 32°F and 50°F and low humidity.
C — Temperature between 20°F and 70°F and high humidity.

31. Prior to starting each maneuver, pilots should

A — check altitude, airspeed, and heading indications.
B — visually scan the entire area for collision avoidance.
C — announce their intentions on the nearest CTAF.

32. (Refer to Figure 26 on page 289.) Determine the magnetic heading for a flight from Fort Worth Meacham (area 4) to Denton Muni (area 1). The wind is from 330° at 25 knots, the true airspeed is 110 knots, and the magnetic variation is 7°E.

A — 003°.
B — 017°.
C — 023°.

33. What determines the longitudinal stability of an airplane?

A — The location of the CG with respect to the center of lift.
B — The effectiveness of the horizontal stabilizer, rudder, and rudder trim tab.
C — The relationship of thrust and lift to weight and drag.

34. In the Northern Hemisphere, the magnetic compass will normally indicate a turn toward the south when

A — a left turn is entered from an east heading.
B — a right turn is entered from a west heading.
C — the aircraft is decelerated while on a west heading.

35. VFR approaches to land at night should be accomplished

A — at a higher airspeed.
B — with a steeper descent.
C — the same as during daytime.

36. Each pilot of an aircraft approaching to land on a runway served by a visual approach slope indicator (VASI) shall

A — maintain a 3° glide to the runway.
B — maintain an altitude at or above the glide slope.
C — stay high until the runway can be reached in a power-off landing.

37. When are the four forces that act on an airplane in equilibrium?

A — During unaccelerated flight.
B — When the aircraft is accelerating.
C — When the aircraft is at rest on the ground.

38. What is an important airspeed limitation that is not color coded on airspeed indicators?

A — Never-exceed speed.
B — Maximum structural cruising speed.
C — Maneuvering speed.

39. Which in-flight advisory would contain information on severe icing not associated with thunderstorms?

A — Convective SIGMET.
B — SIGMET.
C — AIRMET.

40. The presence of ice pellets at the surface is evidence that there

A — are thunderstorms in the area.
B — has been cold frontal passage.
C — is a temperature inversion with freezing rain at a higher altitude.

41. Which statement best defines hypoxia?

A — A state of oxygen deficiency in the body.
B — An abnormal increase in the volume of air breathed.
C — A condition of gas bubble formation around the joints or muscles.

42.   If the pitot tube and outside static vents become clogged, which instruments would be affected?

A — The altimeter, airspeed indicator, and turn-and-slip indicator.

B — The altimeter, airspeed indicator, and vertical speed indicator.

C — The altimeter, attitude indicator, and turn-and-slip indicator.

43.   What action can a pilot take to aid in cooling an engine that is overheating during a climb?

A — Reduce rate of climb and increase airspeed.

B — Reduce climb speed and increase RPM.

C — Increase climb speed and increase RPM.

44.   What information is provided by the Radar Summary Chart that is not shown on other weather charts?

A — Lines and cells of hazardous thunderstorms.

B — Ceilings and precipitation between reporting stations.

C — Types of clouds between reporting stations.

45.   Which incident requires an immediate notification be made to the nearest NTSB field office?

A — An overdue aircraft that is believed to be involved in an accident.

B — An in-flight radio communications failure.

C — An in-flight generator or alternator failure.

46.   With respect to the certification of airmen, which is a category of aircraft?

A — Gyroplane, helicopter, airship, free balloon.

B — Airplane, rotorcraft, glider, lighter-than-air.

C — Single-engine land and sea, multiengine land and sea.

47.   A 100-hour inspection was due at 3302.5 hours. The 100-hour inspection was actually done at 3309.5 hours. When is the next 100-hour inspection due?

A — 3312.5 hours.

B — 3402.5 hours.

C — 3409.5 hours.

48.   Prior to takeoff, the altimeter should be set to which altitude or altimeter setting?

A — The current local altimeter setting, if available, or the departure airport elevation.

B — The corrected density altitude of the departure airport.

C — The corrected pressure altitude for the departure airport.

49.   When must a current pilot certificate be in the pilot's personal possession or readily accessible in the aircraft?

A — When acting as a crew chief during launch and recovery.

B — Only when passengers are carried.

C — Anytime when acting as pilot in command or as a required crewmember.

50.   An airplane and an airship are converging. If the airship is left of the airplane's position, which aircraft has the right-of-way?

A — The airship.

B — The airplane.

C — Each pilot should alter course to the right.

51.   Unless each occupant is provided with supplemental oxygen, no person may operate a civil aircraft of U.S. registry above a maximum cabin pressure altitude of

A — 12,500 feet MSL.

B — 14,000 feet MSL.

C — 15,000 feet MSL.

52.   The three takeoffs and landings that are required to act as pilot in command at night must be done during the time period from

A — sunset to sunrise.

B — 1 hour after sunset to 1 hour before sunrise.

C — the end of evening civil twilight to the beginning of morning civil twilight.

53.   In regard to privileges and limitations, a private pilot may

A — act as pilot in command of an aircraft carrying a passenger for compensation if the flight is in connection with a business or employment.

B — not pay less than the pro rata share of the operating expenses of a flight with passengers provided the expenses involve only fuel, oil, airport expenditures, or rental fees.

C — not be paid in any manner for the operating expenses of a flight.

54.   Under what condition, if any, may a pilot allow a person who is obviously under the influence of drugs to be carried aboard an aircraft?

A — In an emergency or if the person is a medical patient under proper care.

B — Only if the person does not have access to the cockpit or pilot's compartment.

C — Under no condition.

55. What is one purpose of wing flaps?

A — To enable the pilot to make steeper approaches to a landing without increasing the airspeed.

B — To relieve the pilot of maintaining continuous pressure on the controls.

C — To decrease wing area to vary the lift.

56. Thunderstorms reach their greatest intensity during the

A — mature stage.

B — downdraft stage.

C — cumulus stage.

57. A stable air mass is most likely to have which characteristic?

A — Showery precipitation.

B — Turbulent air.

C — Poor surface visibility.

58. Every physical process of weather is accompanied by, or is the result of, a

A — movement of air.

B — pressure differential.

C — heat exchange.

59. Convective circulation patterns associated with sea breezes are caused by

A — warm, dense air moving inland from over the water.

B — water absorbing and radiating heat faster than the land.

C — cool, dense air moving inland from over the water.

60. Where does wind shear occur?

A — Only at higher altitudes.

B — Only at lower altitudes.

C — At all altitudes, in all directions.

PRACTICE TEST LIST OF ANSWERS

Listed below are the answers to the practice test. To the immediate right of each answer is the page number on which the question, as well as correct and incorrect answer explanations, can be found.

Q. #	Answer	Page	Q. #	Answer	Page	Q. #	Answer	Page	Q. #	Answer	Page
1.	C	177	16.	B	279	31.	B	91	46.	B	120
2.	B	179	17.	B	243	32.	A	323	47.	B	157
3.	A	321	18.	A	48	33.	A	32	48.	A	144
4.	B	328	19.	B	300	34.	C	45	49.	C	124
5.	B	183	20.	C	306	35.	C	318	50.	A	141
6.	C	187	21.	A	302	36.	B	88	51.	C	154
7.	B	197	22.	C	319	37.	A	28	52.	B	129
8.	A	51	23.	C	61	38.	C	46	53.	B	131
9.	C	274	24.	A	90	39.	B	252	54.	A	139
10.	C	236	25.	B	87	40.	C	217	55.	A	26
11.	A	238	26.	C	94	41.	A	202	56.	A	215
12.	A	268	27.	B	151	42.	B	46	57.	C	222
13.	C	272	28.	C	150	43.	A	57	58.	C	213
14.	B	274	29.	C	53	44.	A	246	59.	C	214
15.	A	276	30.	C	59	45.	A	160	60.	C	218

APPENDIX B
INTERPOLATION

The following is a tutorial based on information that has appeared in the FAA's *Pilot's Handbook of Aeronautical Knowledge*. Interpolation is required in questions found in the following two subunits:

Study Unit 5 - "Airplane Performance and Weight and Balance"
Subunit 5.2, "Density Altitude Computations" (pages 165, 175)
Subunit 5.4, "Cruise Power Settings" (pages 166, 180)

A. To interpolate means to compute intermediate values between a series of given values.

1. In many instances when performance is critical, an accurate determination of the performance values is the only acceptable means to enhance safe flight.

2. Guessing to determine these values should be avoided.

B. Interpolation is simple to perform if the method is understood. The following are examples of how to interpolate, or accurately determine the intermediate values, between a series of given values.

C. The numbers in column A range from 10 to 30, and the numbers in column B range from 50 to 100. Determine the intermediate numerical value in column B that would correspond with an intermediate value of 20 placed in column A.

A	B
10	50
20	X = Unknown
30	100

1. It can be visualized that 20 is halfway between 10 and 30; therefore, the corresponding value of the unknown number in column B would be halfway between 50 and 100, or 75.

D. Many interpolation problems are more difficult to visualize than the preceding example; therefore, a systematic method must be used to determine the required intermediate value. The following describes one method that can be used.

1. The numbers in column A range from 10 to 30 with intermediate values of 15, 20, and 25. Determine the intermediate numerical value in column B that would correspond with 15 in column A.

A	B
10	50
15	
20	
25	
30	100

2. First, in column A, determine the relationship of 15 to the range between 10 and 30 as follows:

$$\frac{15 - 10}{30 - 10} = \frac{5}{20} \text{ or } 1/4$$

a. It should be noted that 15 is 1/4 of the range between 10 and 30.

3.   Now determine 1/4 of the range of column B between 50 and 100 as follows:

$$100 - 50 = 50$$
$$1/4 \text{ of } 50 = 12.5$$

a.   The answer 12.5 represents the number of units, but to arrive at the correct value, 12.5 must be added to the lower number in column B as follows:

$$50 + 12.5 = 62.5$$

4.   The interpolation has been completed and 62.5 is the actual value which is 1/4 of the range of column B.

E.   Another method of interpolation is shown below:

1.   Using the same numbers as in the previous example, a proportion problem based on the relationship of the number can be set up.

Proportion: $\dfrac{5}{20} = \dfrac{X}{50}$

$$20X = 250$$
$$X = 12.5$$

a.   The answer, 12.5, must be added to 50 to arrive at the actual value of 62.5.

F.   The following example illustrates the use of interpolation applied to a problem dealing with one aspect of airplane performance:

Temperature (°F)	Takeoff Distance (ft.)
70	1,173
80	1,356

1.   If a distance of 1,173 feet is required for takeoff when the temperature is 70°F and 1,356 feet is required at 80°F, what distance is required when the temperature is 75°F? The solution to the problem can be determined as follows:

$$\frac{5}{10} = \frac{X}{183}$$
$$10X = 915$$
$$X = 91.5$$

a.   The answer, 91.5, must be added to 1,173 to arrive at the actual value of 1,264.5 ft.

APPENDIX C
FAA FIGURES WITHOUT ASSOCIATED QUESTIONS

The FAA has not released any questions associated with Figure 64, which is included in the FAA's *Computer Testing Supplement for Recreational Pilot and Private Pilot*. You may encounter one or two questions that reference this figure.

Figure 64 (on the next page) is an *Airport/Facility Directory* excerpt for Toledo. Your experience answering questions in Study Units 9 and 11 will prepare you for any pretest questions concerning Figure 64.

OHIO

TOLEDO

METCALF FLD   (TDZ)   6 SE   UTC-5(-4DT)   N41°33.89'   W83°28.94'   DETROIT
   622   B   S4   FUEL 1OOLL, JET A   OX1, 3   H-3H, L-23C
   RWY 14-32: H5830X100 (ASPH)   S-63, D-85   MIRL   IAP
     RWY 14: REIL. Tower.   **RWY 32:** VASI(V4L)---GA 3.0° TCH 43'.
     Road.
   RWY 04-22: H3664X150 (ASPH)   S-63, D-85   MIRL
     RWY 04: REIL. VASI(V4L)---GA 3.0° TCH 45'.   Trees.
     RWY 22: REIL. VASI(V4R)---GA 3.O° TCH 39'.   Thld dsplcd 90'.   Tree.
   RUNWAY DECLARED DISTANCE INFORMATION
     RWY 14:   TORA 4600   TODA 4600   ASDA 5242   LDA 4680
     RWY 32:   TORA 5268   TODA 5268   ASDA 5268   LDA 4680
   AIRPORT REMARKS: Attended Mon-Fri all hrs: Sat-Sun 1200-0100Z‡.
     Parallel twy Rwy 04-22 and Rwy 14-32 25' wide. Seagulls on and
     invof arpt. Ldg fee. ACTIVATE REILs Rwy 04 and Rwy 22--CTAF.
     Rwy 32 VASI OTS indef. REIL Rwy 14 OTS Indef.
   WEATHER DATA SOURCES: ASOS 119.275 (419) 838-5034.
   COMMUNICATIONS: CTAF/UNICOM 123.05
     CLEVELAND FSS (CLE) TF 1-800-WX-BRIEF. NOTAM FILE TDZ.
   Ⓡ **TOLEDO APP/DEP CON** 134.35   **CLNC DEL** 125.6 OTS indef.
   RADIO AIDS TO NAVIGATION: NOTAM FILE CLE.
     WATERVILLE (L) VOR/DWE 113.1 VWV Chan 78   N41°27.09'
     W83°38.32'   048° 9.8 NM to fld. 660/2W.

TOLEDO EXPRESS   (TOL)   10 W   UTC-5(-4DT)   N41°35.21' W83°48.47'   DETROIT
   684   B   S4   FUEL 100LL.JET A   OX3   LRA   ARFF Index B   H-3H, L-23C
   Rwy 07-25: H10600X150 (ASPH-GRVD)   S-100. D-174,DT-300, DDT-550   HIRL   CL   IAP
     Rwy 07: ALSF2. TDZL. Tree. Arresting device.
     Rwy 25: MALSR. VASI(V4L)---GA 3.O° TCH 51'. Tree. Arresting device. 0.3% up.
   RWY 16-34: H5599X150 (ASPH-GRVD)   S-100, D-174, DT-300   MIRL
     RWY 16: REIL. Trees.   **RWY 34:** REIL VASI(V4L)---GA 3.0° TCH 35'. Trees.
   AIRPORT REMARKS: Attended continuously. Fuel and svc avbl 1300-0500Z‡. Birds and deer on and invof arpt.
     Customs: Sat-Sun req must be made prior to 2200Z‡ on Fri. phone 419-259-6424. Twy C restricted to B-727
     acft or smaller. Rwy 34 REIL OTS indef. NOTE: See Land and Hold Short Operations Section.
   WEATHER DATA SOURCES: ASOS (419) 865-8351.
   COMMUNICATIONS: ATIS 118.75   **UNICOM** 122.95
     CLEVELAND FSS (CLE) TF 1-800-WX-BRIEF. NOTAM FILE TOL.
   Ⓡ **APP/DEP CON** 126.1 (180°-359°) 134.35 (360°-179°) 123.975
     TOWER 118.1   **GND CON** 121.9   **CLNC DEL** 121.75
   AIRSPACE: CLASS C svc continuous ctc APP CON
   RADIO AIDS TO NAVIGATION: NOTAM FILE CLE.
     WATERVILLE (L) VOR/DME 113.1 VWV Chan 78   N41°27.09' W83°38.32'   319° 11.1 NM to fld. 660/2W.
     TOPHR NDB (LOM) 219 TO N41°33.21' W83°55.27'   074° 5.5 NM to fld. Unmonitored. NOTAM FILE TOL.
     ILS 109.7.   I-TOL   Rwy 07.   LOM TOPHR NDB.
     ILS 108.7   I-BQE   Rwy25.
     ASR

SEAGATE HELISTOP   (6T2)   00 N   UTC-5(-4DT)   N41°39.25' W83°31.88'   DETROIT
   650
   HELIPAD H1: 50X50 (CONC)
   HELIPORT REMARKS: Unattended. Ldg fee. ACTIVATE orange perimeter lgts----CTAF. Helipad H1 NSTD 1-box (2 VASIS).
     For heliport access to street phone 419-247-2172: 2 days in advance. Helipad H1 not marked with "H."
     Helipad H1 perimeter lgts.
   COMMUNICATIONS: CTAF/UNICOM 123.05
     CLEVELAND FSS (FDY) TF 800-WX-BRIEF. NOTAM FILE CLE.

MICHIGAN

ADRIAN

LENAWEE CO   (ADG)   3 SW   UTC-5(-4DT)   N41°52.17'   W84°04.49'   DETROIT
   798   B   S4   FUEL 1OOLL, JET A   L-23C
   RWY 05-23: H3994X75 (ASPH)   S-20   MIRL   IAP
     RWY 05: REIL. VASI(V4L)---GA 3.0 TCH 40'. Road.   **RWY 23:** PAPI(P4R)---GA 3.2° TCH 30'. Tree.
   RWY 11-29: 2400X270 (TURF)
     RWY 11: Trees.   **RWY 29:** Trees.
   AIRPORT REMARKS: Attended 1300Z‡-dusk. Arpt unattended major holidays except by prior arrangement; call arpt
     manager 517-263-0045.   Rwy 11-29 CLOSED Dec-Apr and when snow covered. Snow removal Rwy 05-23 only.
     Extensive glider ops weekends. Rgt tfc Rwy 05 for glider ops. Perimeters twy marked with reflectors. Taxi on
     hard surfaces only during spring thaw and wet conditions. Rwy 11-29 marked with 3' yellow cones. MIRL Rwy
     05-23 preset low inst; to increase ints and ACTIVATE REIL Rwy 05; VASI Rwy 05 and PAPI Rwy 23-CTAF.
   WEATHER DATA SOURCES: ASOS 118.375 (517) 265-9089.
   COMMUNICATIONS: CTAF/UNICOM 122.8
     LANSING FSS (LAN) TF 1-800-WX-BRIEF. NOTAM FILE ADG.
   Ⓡ **TOLEDO APP/DEP CON** 126.1
   RADIO AIDS TO NAVIGATION: NOTAM FILE JXN.
     JACKSON (L) VORW/DME 109.6 JXN Chan 33   N42°15.55'   W84°27.52'   149° 29 NM to fld. 1000/5W.
     ADRIAN NDB (MHW) 278 ADG   N41°52.20'   W84°04.66'   at fld. NOTAM FILE ADG. Unmonitored.

Figure 64. – Airport/Facility Directory Excerpt.

FAA LISTING OF
LEARNING STATEMENT CODES

Reprinted below and on the next two pages are the FAA's learning statement codes for the questions presented in this book. These are the codes that will appear on your Airman Computer Test Report. See the example on page 13. Your test report will list the learning statement code of each question answered incorrectly.

When you receive your Airman Computer Test Report, you can trace the learning statement codes listed on it to these pages to find out which topics you had difficulty with. You should discuss your test results with your CFI.

Additionally, you should trace the learning statement codes on your Airman Computer Test Report to our cross-reference listing of questions beginning on page 349. Determine which Gleim subunits you need to review.

Effective September 30, 2007, all knowledge test grade reports ceased to offer subject matter knowledge codes. The FAA has introduced learning statements that have taken the place of the previous knowledge code system. These statements are designed to represent the knowledge test topic areas in clear verbal terms and encourage applicants to study the entire area of identified weakness instead of merely studying a specific question area.

The following learning statements are all used in this test bank. The FAA will periodically revise the existing learning codes and add new ones. Any changes to FAA learning statement codes will be available through the Gleim online updates system at www.gleim.com/updates.

To determine the knowledge area in which a particular question was incorrectly answered, compare the learning statement code(s) on your Airman Computer Test Report to the listing that follows. The total number of test items missed may differ from the number of learning statement codes shown on your test report because you may have missed more than one question in a certain knowledge area.

Code	Description
PLT003	Calculate aircraft performance - center of gravity
PLT005	Calculate aircraft performance - density altitude
PLT008	Calculate aircraft performance - landing
PLT011	Calculate aircraft performance - takeoff
PLT012	Calculate aircraft performance - time/speed/distance/course/fuel/wind
PLT013	Calculate crosswind / headwind components
PLT014	Calculate distance / bearing from/to a station
PLT015	Calculate flight performance / planning - range
PLT019	Calculate pressure altitude
PLT021	Calculate weight and balance
PLT023	Define altitude - absolute / true / indicated / density / pressure
PLT025	Define Bernoulli's principle
PLT026	Define ceiling
PLT039	Interpret airport landing indicator
PLT040	Interpret airspace classes - charts / diagrams
PLT041	Interpret altimeter - readings / settings
PLT044	Interpret ATC communications / instructions / terminology
PLT059	Interpret information on a METAR / SPECI report
PLT061	Interpret information on a PIREP
PLT063	Interpret information on a Radar Summary Chart
PLT064	Interpret information on a Sectional Chart
PLT068	Interpret information on a Significant Weather Prognostic Chart
PLT071	Interpret information on a Surface Analysis Chart
PLT072	Interpret information on a Terminal Aerodrome Forecast (TAF)
PLT075	Interpret information on a Weather Depiction Chart
PLT076	Interpret information on a Winds and Temperatures Aloft Forecast (FB)
PLT077	Interpret information on an Airport Diagram
PLT078	Interpret information in an Airport Facility Directory (AFD)
PLT081	Interpret information on an Aviation Area Forecast (FA)
PLT088	Interpret speed indicator readings
PLT090	Interpret VOR - charts / indications / CDI / ADF / NAV
PLT091	Interpret VOR / ADF / NDB / CDI / RMI - illustrations / indications / procedures
PLT092	Interpret weight and balance - diagram
PLT097	Recall aeromedical factors - effects of carbon monoxide poisoning
PLT099	Recall aeromedical factors - scanning procedures
PLT101	Recall aeronautical charts - pilotage
PLT103	Recall Aeronautical Decision Making (ADM) - hazardous attitudes
PLT104	Recall Aeronautical Decision Making (ADM) - human factors / CRM
PLT112	Recall aircraft controls - proper use / techniques
PLT115	Recall aircraft engine - detonation/backfiring/ after firing, cause/characteristics
PLT116	Recall aircraft general knowledge / publications / AIM / navigational aids
PLT118	Recall aircraft instruments - gyroscopic
PLT119	Recall aircraft lighting - anti-collision / landing / navigation

PLT121   Recall aircraft loading - computations
PLT123   Recall aircraft performance - airspeed
PLT124   Recall aircraft performance - atmospheric effects
PLT125   Recall aircraft performance - climb / descent
PLT127   Recall aircraft performance - density altitude
PLT128   Recall aircraft performance - effects of icing
PLT131   Recall aircraft performance - ground effect
PLT132   Recall aircraft performance - instrument markings / airspeed / definitions / indications
PLT133   Recall aircraft performance - normal climb / descent rates
PLT134   Recall aircraft performance - takeoff
PLT136   Recall aircraft systems - anti-icing / deicing
PLT140   Recall airport operations - LAHSO
PLT141   Recall airport operations - markings / signs / lighting
PLT147   Recall airport operations - visual glideslope indicators
PLT150   Recall airport traffic patterns - entry procedures
PLT161   Recall airspace classes - limits / requirements / restrictions / airspeeds / equipment
PLT162   Recall airspace requirements - operations
PLT163   Recall airspace requirements - visibility / cloud clearance
PLT165   Recall altimeter - effect of temperature changes
PLT166   Recall altimeter - settings / setting procedures
PLT167   Recall altimeters - characteristics / accuracy
PLT168   Recall angle of attack - characteristics / forces / principles
PLT170   Recall approach / landing / taxiing techniques
PLT171   Recall ATC - reporting
PLT172   Recall ATC - system / services
PLT173   Recall atmospheric conditions - measurements / pressure / stability
PLT187   Recall basic instrument flying - turn coordinator / turn and slip indicator
PLT189   Recall carburetor - effects of carburetor heat / heat control
PLT190   Recall carburetor ice - factors affecting / causing
PLT191   Recall carburetors - types / components / operating principles / characteristics
PLT192   Recall clouds - types / formation / resulting weather
PLT194   Recall collision avoidance - scanning techniques
PLT196   Recall communications - ATIS broadcasts
PLT200   Recall dead reckoning - calculations / charts
PLT201   Recall departure procedures - ODP / SID
PLT204   Recall effective communication - basic elements
PLT206   Recall effects of temperature - density altitude / icing
PLT207   Recall electrical system - components / operating principles / characteristics / static bonding and shielding
PLT208   Recall emergency conditions / procedures
PLT213   Recall flight characteristics - longitudinal stability / instability
PLT215   Recall flight instruments - magnetic compass
PLT219   Recall flight operations - maneuvers
PLT220   Recall flight operations - night and high altitude operations
PLT221   Recall flight operations - takeoff / landing maneuvers
PLT222   Recall flight operations - takeoff procedures
PLT225   Recall flight plan - requirements
PLT226   Recall fog - types / formation / resulting weather
PLT235   Recall forces acting on aircraft - aerodynamics
PLT236   Recall forces acting on aircraft - airfoil / center of pressure / mean camber line
PLT242   Recall forces acting on aircraft - lift / drag / thrust / weight / stall / limitations

PLT243   Recall forces acting on aircraft - propeller / torque
PLT245   Recall forces acting on aircraft - stalls / spins
PLT248   Recall forces acting on aircraft - turns
PLT249   Recall fuel - air mixture
PLT250   Recall fuel - types / characteristics / contamination / fueling / defueling / precautions
PLT251   Recall fuel characteristics / contaminants / additives
PLT253   Recall fuel system - components / operating principles / characteristics / leaks
PLT258   Recall ground reference maneuvers - ground track diagram
PLT263   Recall hazardous weather - fog / icing / turbulence / visibility restriction
PLT271   Recall human factors (ADM) - judgment
PLT274   Recall icing - formation / characteristics
PLT278   Recall indicating systems - airspeed / angle of attack / attitude / heading / manifold pressure / synchro / EGT
PLT279   Recall Inertial/Doppler Navigation System principles / regulations / requirements / limitations
PLT281   Recall information in an Airport Facility Directory
PLT284   Recall information on a Forecast Winds and Temperatures Aloft (FD)
PLT289   Recall information on a Weather Depiction Chart
PLT290   Recall information on AIRMETS / SIGMETS
PLT291   Recall information on an Aviation Area Forecast (FA)
PLT301   Recall inversion layer - characteristics
PLT309   Recall load factor - angle of bank
PLT311   Recall load factor - effect of airspeed
PLT312   Recall load factor - maneuvering / stall speed
PLT316   Recall meteorology - severe weather watch (WW)
PLT320   Recall navigation - true north / magnetic north
PLT322   Recall navigation - VOR / NAV system
PLT323   Recall NOTAMS - classes / information / distribution
PLT324   Recall oil system - types / components / functions / oil specifications
PLT328   Recall performance planning - aircraft loading
PLT330   Recall physiological factors - cause / effects of hypoxia
PLT332   Recall physiological factors - hyperventilation
PLT334   Recall physiological factors - spatial disorientation
PLT335   Recall pilotage - calculations
PLT337   Recall pitot-static system - components / operating principles / characteristics
PLT342   Recall powerplant - controlling engine temperature
PLT343   Recall powerplant - operating principles / operational characteristics / inspecting
PLT345   Recall pressure altitude
PLT346   Recall primary / secondary flight controls - types / purpose / functionality / operation
PLT350   Recall propeller operations - constant / variable speed
PLT351   Recall propeller system - types / components / operating principles / characteristics
PLT353   Recall Radar Summary Chart
PLT354   Recall radio - GPS / RNAV / RAIM
PLT362   Recall radio - VHF / Direction Finding
PLT363   Recall radio - VOR / VOT
PLT366   Recall regulations - accident / incident reporting and preserving wreckage
PLT369   Recall regulations - aerobatic flight requirements

PLT370 Recall regulations - Air Traffic Control authorization / clearances

PLT371 Recall regulations - Aircraft Category / Class

PLT372 Recall regulations - aircraft inspection / records / expiration

PLT373 Recall regulations - aircraft operating limitations

PLT374 Recall regulations - aircraft owner / operator responsibilities

PLT375 Recall regulations - aircraft return to service

PLT376 Recall regulations - airspace special use / TFRS

PLT377 Recall regulations - airworthiness certificates / requirements / responsibilities

PLT378 Recall regulations - Airworthiness Directives

PLT381 Recall regulations - altimeter settings

PLT383 Recall regulations - basic flight rules

PLT384 Recall regulations - briefing of passengers

PLT387 Recall regulations - change of address

PLT393 Recall regulations - controlled / restricted airspace - requirements

PLT395 Recall regulations - definitions

PLT399 Recall regulations - display / inspection of licenses and certificates

PLT400 Recall regulations - documents to be carried on aircraft during flight

PLT401 Recall regulations - dropping / aerial application / towing restrictions

PLT402 Recall regulations - ELT requirements

PLT403 Recall regulations - emergency deviation from regulations

PLT405 Recall regulations - equipment / instrument / certificate requirements

PLT407 Recall regulations - experience / training requirements

PLT411 Recall regulations - flight instructor limitations / qualifications

PLT413 Recall regulations - fuel requirements

PLT414 Recall regulations - general right-of-way rules

PLT425 Recall regulations - maintenance reports / records / entries

PLT426 Recall regulations - maintenance requirements

PLT427 Recall regulations - medical certificate requirements / validity

PLT430 Recall regulations - minimum safe / flight altitude

PLT431 Recall regulations - operating near other aircraft

PLT435 Recall regulations - operational procedures for an uncontrolled airport

PLT438 Recall regulations - oxygen requirements

PLT439 Recall regulations - persons authorized to perform maintenance

PLT440 Recall regulations - Pilot / Crew duties and responsibilities

PLT442 Recall regulations - pilot currency requirements

PLT443 Recall regulations - pilot qualifications / privileges / responsibilities / crew complement

PLT444 Recall regulations - pilot-in-command authority / responsibility

PLT445 Recall regulations - preflight requirements

PLT446 Recall regulations - preventative maintenance

PLT447 Recall regulations - privileges / limitations of medical certificates

PLT448 Recall regulations - privileges / limitations of pilot certificates

PLT449 Recall regulations - proficiency check requirements

PLT451 Recall regulations - ratings issued / experience requirements / limitations

PLT455 Recall regulations - requirements of a flight plan release

PLT457 Recall regulations - student pilot endorsements / other endorsements

PLT462 Recall regulations - use of microphone / megaphone / interphone / public address system

PLT463 Recall regulations - alcohol or drugs

PLT464 Recall regulations - use of safety belts / harnesses (crew member)

PLT465 Recall regulations - use of seats / safety belts / harnesses (passenger)

PLT466 Recall regulations - V speeds

PLT467 Recall regulations - visual flight rules and limitations

PLT468 Recall regulations - Visual Meteorological Conditions (VMC)

PLT473 Recall secondary flight controls - types / purpose / functionality

PLT475 Recall squall lines - formation / characteristics / resulting weather

PLT477 Recall stalls - characteristics / factors / recovery / precautions

PLT478 Recall starter / ignition system - types / components / operating principles / characteristics

PLT479 Recall starter system - starting procedures

PLT485 Recall taxiing / crosswind / techniques

PLT486 Recall taxiing / takeoff - techniques / procedures

PLT492 Recall temperature - effects on weather formations

PLT493 Recall the dynamics of frost / ice / snow formation on an aircraft

PLT494 Recall thermals - types / characteristics / formation / locating / maneuvering / corrective actions

PLT495 Recall thunderstorms - types / characteristics / formation / hazards / precipitation static

PLT497 Recall transponder - codes / operations / usage

PLT502 Recall universal signals - hand / light / visual

PLT506 Recall V speeds - maneuvering / flap extended / gear extended / V1, V2, r, ne, mo, mc, mg, etc.

PLT507 Recall VOR - indications / VOR / VOT / CDI

PLT508 Recall VOR/altimeter/transponder checks - identification / tuning / identifying / logging

PLT509 Recall wake turbulence - characteristics / avoidance techniques

PLT511 Recall weather associated with frontal activity / air masses

PLT512 Recall weather conditions - temperature / moisture / dewpoint

PLT514 Recall weather reporting systems - briefings / forecasts / reports / AWOS / ASOS

PLT515 Recall weather services - EFAS / TIBS / TPC / WFO / AFSS / HIWAS

PLT516 Recall winds - types / characteristics

PLT518 Recall windshear - characteristics / hazards / power management

PLT520 Calculate density altitude

CROSS-REFERENCES TO
THE FAA LEARNING STATEMENT CODES

Pages 349 through 356 contain a listing of all of the questions from our private pilot knowledge test bank. Non-airplane questions are excluded. The questions are in FAA Learning Statement Code (LSC) sequence. To the right of each LSC, we present our study unit/question number and our answer. For example, note that in one instance below, PLT003 is cross-referenced to 1-34, which represents our Study Unit 1, question 34; the correct answer is B.

Pages 345 through 347 contain a complete listing of all the FAA Learning Statement Codes associated with all of the private pilot questions presented in this book. Use this list to identify the specific topic associated to each Learning Statement Code.

The first line of each of our answer explanations in Study Units 1 through 11 contains

1. The correct answer and
2. A reference for the answer explanation, e.g., *AFH Chap 1*. If this reference is not useful, use the following chart to identify the learning statement code to determine the specific reference appropriate for the question.

FAA Learning Code	Gleim SU/ Q. No.	Gleim Answer	FAA Learning Code	Gleim SU/ Q. No.	Gleim Answer	FAA Learning Code	Gleim SU/ Q. No.	Gleim Answer
PLT003	1–34	B	PLT012	11–23	C	PLT012	11–56	B
PLT003	5–50	C	PLT012	11–25	A	PLT012	11–57	C
PLT003	5–51	A	PLT012	11–27	A	PLT012	11–58	A
PLT005	5–16	C	PLT012	11–30	A	PLT013	5–28	A
PLT005	5–18	C	PLT012	11–31	A	PLT013	5–29	B
PLT008	5–34	B	PLT012	11–32	B	PLT013	5–30	C
PLT008	5–35	B	PLT012	11–35	C	PLT013	5–31	C
PLT008	5–37	C	PLT012	11–39	C	PLT013	5–32	C
PLT008	5–38	B	PLT012	11–40	B	PLT013	5–33	B
PLT008	5–39	B	PLT012	11–41	C	PLT014	10–23	C
PLT008	5–40	C	PLT012	11–42	C	PLT015	5–26	C
PLT008	5–41	B	PLT012	11–43	B	PLT015	5–27	C
PLT008	5–42	B	PLT012	11–44	B	PLT019	5–13	A
PLT008	5–43	B	PLT012	11–45	C	PLT019	5–15	A
PLT008	5–44	A	PLT012	11–46	B	PLT019	5–17	B
PLT011	5–9	C	PLT012	11–47	B	PLT021	5–46	C
PLT011	5–19	A	PLT012	11–48	A	PLT021	5–47	C
PLT011	5–20	B	PLT012	11–49	A	PLT021	5–48	B
PLT011	5–21	B	PLT012	11–50	C	PLT021	5–56	B
PLT011	5–22	A	PLT012	11–51	A	PLT021	5–57	B
PLT012	5–23	B	PLT012	11–52	B	PLT021	5–58	A
PLT012	5–24	B	PLT012	11–53	C	PLT021	5–60	B
PLT012	5–25	B	PLT012	11–54	B	PLT021	5–61	B
PLT012	9–38	B	PLT012	11–55	C	PLT021	5–62	A

FAA Learning Code	Gleim SU/ Q. No.	Gleim Answer	FAA Learning Code	Gleim SU/ Q. No.	Gleim Answer	FAA Learning Code	Gleim SU/ Q. No.	Gleim Answer
PLT021	5–63	A	PLT064	9–1	B	PLT068	8–62	B
PLT023	2–29	B	PLT064	9–2	A	PLT068	8–63	A
PLT023	2–30	A	PLT064	9–3	B	PLT068	8–64	C
PLT023	2–31	B	PLT064	9–4	A	PLT068	8–65	A
PLT023	2–33	B	PLT064	9–5	B	PLT071	8–39	B
PLT023	2–34	B	PLT064	9–6	C	PLT072	8–30	A
PLT023	2–35	B	PLT064	9–8	B	PLT072	8–31	A
PLT023	2–42	A	PLT064	9–10	B	PLT072	8–32	A
PLT025	1–13	C	PLT064	9–12	C	PLT072	8–33	C
PLT026	8–11	B	PLT064	9–14	A	PLT072	8–34	B
PLT039	3–28	A	PLT064	9–15	B	PLT072	8–35	B
PLT039	3–29	C	PLT064	9–17	A	PLT072	8–36	B
PLT039	3–30	A	PLT064	9–20	A	PLT072	8–37	B
PLT039	3–31	C	PLT064	9–25	C	PLT075	8–42	C
PLT039	3–34	B	PLT064	9–27	A	PLT075	8–43	A
PLT040	3–72	C	PLT064	9–28	B	PLT076	8–56	A
PLT040	9–7	C	PLT064	9–29	C	PLT076	8–57	C
PLT040	9–9	C	PLT064	9–30	B	PLT076	8–58	B
PLT040	9–11	A	PLT064	9–31	B	PLT076	8–59	C
PLT040	9–26	B	PLT064	9–33	B	PLT076	8–60	B
PLT040	9–32	C	PLT064	9–36	B	PLT077	3–4	B
PLT041	2–25	C	PLT064	9–37	C	PLT077	3–5	B
PLT041	2–26	C	PLT064	9–39	C	PLT077	3–6	A
PLT041	2–27	A	PLT064	9–40	A	PLT077	3–7	C
PLT041	2–28	B	PLT064	9–41	C	PLT077	3–32	A
PLT041	2–36	C	PLT064	9–42	C	PLT078	9–61	A
PLT044	3–68	A	PLT064	9–43	A	PLT078	9–62	B
PLT044	3–76	A	PLT064	9–44	C	PLT078	9–63	C
PLT059	8–12	C	PLT064	9–45	A	PLT078	9–64	B
PLT059	8–13	A	PLT064	9–46	B	PLT078	9–65	A
PLT059	8–14	A	PLT064	9–47	B	PLT081	8–26	B
PLT059	8–15	B	PLT064	9–48	A	PLT081	8–27	C
PLT059	8–16	B	PLT064	9–50	A	PLT081	8–28	A
PLT061	8–17	A	PLT064	9–51	B	PLT081	8–29	A
PLT061	8–18	C	PLT064	9–52	C	PLT088	2–14	C
PLT061	8–19	A	PLT064	9–53	C	PLT088	2–16	C
PLT061	8–20	C	PLT064	9–54	C	PLT088	2–17	C
PLT061	8–21	C	PLT064	9–55	A	PLT088	2–18	A
PLT063	8–44	C	PLT064	9–56	B	PLT088	2–19	C
PLT063	8–46	B	PLT064	11–24	C	PLT090	10–4	B
PLT063	8–47	A	PLT064	11–26	B	PLT090	10–6	A
PLT063	8–48	C	PLT068	8–61	B	PLT090	10–14	A

FAA Learning Code	Gleim SU/ Q. No.	Gleim Answer	FAA Learning Code	Gleim SU/ Q. No.	Gleim Answer	FAA Learning Code	Gleim SU/ Q. No.	Gleim Answer
PLT091	10–7	C	PLT103	6–28	A	PLT132	2–20	C
PLT091	10–8	B	PLT103	6–29	A	PLT132	2–21	B
PLT091	10–9	C	PLT103	6–30	C	PLT132	2–22	C
PLT091	10–10	B	PLT103	6–31	C	PLT132	2–23	C
PLT091	10–11	C	PLT103	6–32	C	PLT132	2–24	B
PLT091	10–12	C	PLT103	6–33	B	PLT132	2–45	C
PLT091	10–15	C	PLT103	6–34	C	PLT133	4–12	C
PLT091	10–16	A	PLT103	6–35	C	PLT134	1–23	A
PLT091	10–17	C	PLT104	6–36	C	PLT136	2–62	A
PLT091	10–18	C	PLT112	11–15	A	PLT140	3–114	C
PLT091	10–19	B	PLT112	11–16	B	PLT140	3–115	A
PLT091	10–20	C	PLT112	11–17	A	PLT140	3–117	A
PLT091	10–21	B	PLT112	11–18	C	PLT140	3–118	A
PLT091	10–22	C	PLT115	2–74	C	PLT140	3–119	B
PLT091	10–24	B	PLT115	2–75	A	PLT141	3–1	C
PLT091	10–25	C	PLT115	2–76	B	PLT141	3–2	C
PLT091	10–26	A	PLT116	9–57	B	PLT141	3–8	B
PLT091	10–27	B	PLT116	9–58	C	PLT141	3–9	C
PLT091	10–28	A	PLT116	9–59	A	PLT141	3–10	C
PLT091	10–29	B	PLT116	9–60	B	PLT141	3–11	C
PLT091	10–30	C	PLT116	10–36	B	PLT141	3–12	C
PLT091	10–31	C	PLT118	2–47	C	PLT141	3–13	A
PLT091	10–32	B	PLT119	3–57	A	PLT141	3–14	A
PLT092	5–49	C	PLT119	3–58	A	PLT141	3–15	A
PLT092	5–52	B	PLT119	3–59	C	PLT141	3–16	B
PLT092	5–53	B	PLT119	3–67	B	PLT141	3–17	B
PLT092	5–54	B	PLT119	4–161	C	PLT141	3–20	A
PLT092	5–55	A	PLT121	5–59	C	PLT141	3–21	B
PLT097	6–20	B	PLT123	4–13	A	PLT141	3–22	A
PLT097	6–21	A	PLT124	5–2	B	PLT141	3–23	B
PLT097	6–22	C	PLT124	5–12	A	PLT141	3–24	B
PLT097	6–23	A	PLT124	5–14	C	PLT141	3–25	B
PLT099	3–61	C	PLT125	3–64	A	PLT141	3–33	C
PLT099	6–16	C	PLT127	5–5	B	PLT141	3–104	A
PLT099	6–17	C	PLT127	5–6	C	PLT147	3–36	B
PLT099	6–18	B	PLT127	5–7	C	PLT147	3–37	C
PLT099	6–19	A	PLT127	5–8	A	PLT147	3–38	A
PLT101	10–13	A	PLT128	1–24	C	PLT147	3–39	C
PLT101	11–29	C	PLT131	1–26	A	PLT147	3–40	B
PLT103	6–25	A	PLT131	1–27	A	PLT147	3–41	B
PLT103	6–26	A	PLT131	1–28	B	PLT147	3–42	B
PLT103	6–27	A	PLT131	1–29	B	PLT147	3–43	B

FAA Learning Code	Gleim SU/ Q. No.	Gleim Answer	FAA Learning Code	Gleim SU/ Q. No.	Gleim Answer	FAA Learning Code	Gleim SU/ Q. No.	Gleim Answer
PLT147	3–45	B	PLT163	4–134	A	PLT192	7–38	B
PLT147	3–47	B	PLT163	4–135	C	PLT192	7–39	B
PLT150	3–79	C	PLT163	4–137	A	PLT192	7–40	B
PLT150	3–109	A	PLT163	4–139	B	PLT192	7–41	B
PLT161	3–3	B	PLT163	4–140	B	PLT192	7–42	C
PLT161	3–73	B	PLT163	4–141	C	PLT192	7–46	A
PLT161	3–75	C	PLT163	4–142	B	PLT192	7–47	C
PLT161	3–77	C	PLT163	4–143	A	PLT194	2–48	B
PLT161	3–81	C	PLT163	4–144	A	PLT194	2–49	C
PLT161	3–82	B	PLT163	4–145	C	PLT194	3–60	B
PLT161	3–83	C	PLT163	4–146	B	PLT194	3–62	C
PLT161	3–84	C	PLT165	2–43	A	PLT194	3–63	B
PLT161	3–85	A	PLT165	7–2	A	PLT194	3–66	B
PLT161	3–89	B	PLT166	2–37	C	PLT194	3–99	B
PLT161	4–65	C	PLT166	2–44	C	PLT194	3–100	A
PLT161	4–66	C	PLT167	2–38	C	PLT194	3–101	C
PLT161	4–80	A	PLT167	2–39	C	PLT194	3–102	C
PLT161	4–81	B	PLT167	2–40	B	PLT194	6–15	A
PLT161	4–107	B	PLT167	2–41	C	PLT196	3–70	C
PLT161	4–109	A	PLT167	4–151	A	PLT196	3–71	C
PLT161	4–126	B	PLT168	1–15	A	PLT200	11–33	A
PLT161	4–127	A	PLT168	1–16	A	PLT200	11–34	A
PLT161	4–129	A	PLT168	1–17	B	PLT200	11–36	C
PLT161	4–147	B	PLT168	1–18	B	PLT200	11–37	B
PLT161	4–150	C	PLT168	1–19	C	PLT200	11–38	B
PLT161	4–164	A	PLT170	11–63	B	PLT201	3–27	C
PLT161	9–13	C	PLT171	3–98	B	PLT204	3–87	C
PLT161	9–19	B	PLT172	3–88	A	PLT204	3–95	A
PLT161	9–21	C	PLT173	7–45	A	PLT204	3–96	A
PLT161	9–22	A	PLT173	7–52	B	PLT204	3–97	C
PLT161	9–35	B	PLT173	7–53	A	PLT206	5–3	B
PLT162	3–74	C	PLT187	2–46	A	PLT207	2–86	A
PLT162	3–78	C	PLT189	2–68	A	PLT207	2–87	B
PLT163	4–67	A	PLT189	2–69	B	PLT207	2–88	C
PLT163	4–69	B	PLT189	2–70	B	PLT207	2–89	C
PLT163	4–70	B	PLT190	2–63	C	PLT208	5–36	B
PLT163	4–71	C	PLT190	2–64	A	PLT208	11–12	A
PLT163	4–72	B	PLT190	2–65	C	PLT208	11–13	C
PLT163	4–77	B	PLT190	2–67	C	PLT213	1–31	B
PLT163	4–78	B	PLT191	2–66	B	PLT213	1–32	A
PLT163	4–132	B	PLT192	7–22	C	PLT215	2–1	C
PLT163	4–133	C	PLT192	7–23	B	PLT215	2–2	A

FAA Learning Code	Gleim SU/ Q. No.	Gleim Answer	FAA Learning Code	Gleim SU/ Q. No.	Gleim Answer	FAA Learning Code	Gleim SU/ Q. No.	Gleim Answer
PLT215	2–3	C	PLT258	11–62	B	PLT324	2–50	A
PLT215	2–4	B	PLT263	7–24	A	PLT328	1–36	A
PLT215	2–5	B	PLT263	11–11	C	PLT328	5–45	A
PLT215	2–6	B	PLT271	6–24	C	PLT330	6–1	A
PLT215	2–7	C	PLT271	7–18	C	PLT330	6–2	C
PLT215	2–8	C	PLT274	7–19	C	PLT330	6–3	A
PLT215	2–9	B	PLT274	7–20	C	PLT332	6–4	B
PLT219	11–60	A	PLT274	8–23	A	PLT332	6–5	A
PLT220	11–14	C	PLT278	2–15	C	PLT332	6–6	A
PLT221	3–35	B	PLT279	10–37	C	PLT332	6–7	A
PLT221	3–44	B	PLT281	3–112	B	PLT332	6–8	A
PLT221	3–46	B	PLT281	9–23	B	PLT334	6–9	B
PLT222	3–80	C	PLT281	9–24	B	PLT334	6–10	B
PLT225	11–2	A	PLT284	8–54	B	PLT334	6–11	A
PLT225	11–3	B	PLT284	8–55	C	PLT334	6–12	A
PLT225	11–4	B	PLT289	8–38	A	PLT334	6–13	A
PLT226	7–34	C	PLT289	8–40	B	PLT334	6–14	B
PLT226	7–35	B	PLT289	8–41	A	PLT335	10–5	B
PLT226	7–36	A	PLT290	8–66	C	PLT337	2–10	C
PLT226	7–37	C	PLT290	8–67	B	PLT337	2–11	C
PLT235	1–11	A	PLT290	8–68	B	PLT337	2–12	B
PLT235	1–12	A	PLT290	8–69	A	PLT337	2–13	C
PLT236	1–33	C	PLT290	8–70	C	PLT342	2–51	B
PLT242	1–10	A	PLT291	8–24	C	PLT342	2–52	C
PLT242	1–30	A	PLT291	8–25	C	PLT342	2–53	C
PLT243	1–37	A	PLT301	7–21	C	PLT342	2–55	A
PLT243	1–38	B	PLT301	7–55	A	PLT342	2–56	A
PLT243	1–39	B	PLT301	7–56	C	PLT343	2–54	C
PLT245	1–14	A	PLT301	7–57	A	PLT343	2–85	B
PLT245	1–21	C	PLT309	1–41	B	PLT345	2–32	C
PLT245	1–22	A	PLT309	1–43	C	PLT345	5–1	A
PLT248	11–59	C	PLT309	1–44	C	PLT346	1–3	A
PLT249	2–72	B	PLT309	1–45	B	PLT346	1–4	A
PLT249	2–73	A	PLT311	1–40	B	PLT346	1–5	B
PLT250	2–77	C	PLT312	1–42	A	PLT346	1–6	A
PLT250	2–79	A	PLT316	8–45	B	PLT346	1–7	B
PLT250	2–81	C	PLT320	11–22	B	PLT346	1–8	A
PLT250	2–83	A	PLT322	11–28	A	PLT346	1–9	A
PLT251	2–80	C	PLT323	3–18	C	PLT350	2–57	A
PLT253	2–71	A	PLT323	3–19	C	PLT351	1–35	B
PLT253	2–82	B	PLT323	9–66	A	PLT351	2–58	B
PLT258	11–61	B	PLT323	9–67	B	PLT351	2–59	B

FAA Learning Code	Gleim SU/ Q. No.	Gleim Answer	FAA Learning Code	Gleim SU/ Q. No.	Gleim Answer	FAA Learning Code	Gleim SU/ Q. No.	Gleim Answer
PLT351	5–4	B	PLT375	4–180	B	PLT414	4–99	A
PLT353	8–49	A	PLT376	4–148	B	PLT414	4–100	C
PLT353	8–50	B	PLT377	4–14	C	PLT414	4–101	A
PLT354	10–33	C	PLT377	4–190	C	PLT414	4–102	C
PLT354	10–34	B	PLT378	4–16	C	PLT414	4–103	B
PLT354	10–35	C	PLT381	4–115	A	PLT414	4–104	B
PLT362	3–113	A	PLT381	4–116	B	PLT414	4–105	B
PLT363	10–1	C	PLT381	4–117	B	PLT414	4–106	B
PLT366	4–193	A	PLT383	4–108	B	PLT425	4–189	A
PLT366	4–194	B	PLT383	4–174	B	PLT426	4–15	B
PLT366	4–195	A	PLT384	4–191	A	PLT426	4–183	A
PLT366	4–196	C	PLT387	4–47	A	PLT426	4–185	C
PLT366	4–197	B	PLT393	4–110	A	PLT427	4–29	C
PLT366	4–198	C	PLT393	4–124	A	PLT427	4–30	C
PLT366	4–199	C	PLT393	4–125	C	PLT427	4–31	C
PLT369	4–166	B	PLT393	4–128	B	PLT427	4–32	B
PLT369	4–167	A	PLT393	9–16	C	PLT430	4–111	A
PLT369	4–168	B	PLT393	9–18	B	PLT430	4–112	C
PLT369	4–169	B	PLT393	9–34	C	PLT430	4–113	B
PLT369	4–170	B	PLT395	4–3	C	PLT430	4–114	A
PLT370	3–86	C	PLT395	4–34	C	PLT431	4–98	C
PLT370	4–6	C	PLT399	4–22	C	PLT435	9–49	C
PLT370	4–122	B	PLT399	4–23	B	PLT438	4–162	C
PLT371	4–1	B	PLT399	4–24	C	PLT438	4–163	C
PLT371	4–2	B	PLT399	4–25	C	PLT439	4–18	B
PLT371	4–4	A	PLT400	4–156	C	PLT440	4–93	A
PLT371	4–5	A	PLT401	4–75	C	PLT440	4–118	C
PLT372	4–181	B	PLT401	4–86	B	PLT442	4–38	C
PLT372	4–182	C	PLT402	3–110	B	PLT442	4–39	A
PLT372	4–184	B	PLT402	3–111	C	PLT442	4–40	C
PLT372	4–187	B	PLT402	4–157	B	PLT442	4–42	C
PLT372	4–188	C	PLT402	4–158	C	PLT442	4–44	B
PLT373	4–84	B	PLT402	4–159	B	PLT442	4–45	A
PLT373	4–85	B	PLT402	4–160	A	PLT442	4–46	C
PLT373	4–173	B	PLT403	4–119	A	PLT443	4–76	C
PLT374	4–175	B	PLT405	4–171	C	PLT444	3–65	B
PLT374	4–176	A	PLT405	4–172	A	PLT444	3–116	A
PLT374	4–177	A	PLT407	4–48	B	PLT444	4–82	B
PLT374	11–9	B	PLT407	4–49	C	PLT444	4–83	B
PLT375	4–17	B	PLT411	4–41	A	PLT444	4–91	C
PLT375	4–178	A	PLT413	4–130	C	PLT444	4–120	B
PLT375	4–179	B	PLT413	4–131	B	PLT444	4–121	C

FAA Learning Code	Gleim SU/ Q. No.	Gleim Answer	FAA Learning Code	Gleim SU/ Q. No.	Gleim Answer	FAA Learning Code	Gleim SU/ Q. No.	Gleim Answer
PLT444	4–123	B	PLT463	4–88	A	PLT497	3–92	A
PLT444	4–192	B	PLT463	4–89	A	PLT497	3–93	A
PLT444	11–8	A	PLT464	4–94	C	PLT497	3–94	A
PLT445	4–90	C	PLT465	4–95	B	PLT497	4–79	A
PLT445	4–92	B	PLT465	4–96	B	PLT497	4–165	A
PLT445	8–5	C	PLT465	4–97	A	PLT502	3–69	B
PLT445	8–6	B	PLT466	4–7	A	PLT502	3–103	B
PLT445	11–10	B	PLT466	4–9	C	PLT502	3–105	C
PLT446	4–19	C	PLT466	4–11	A	PLT502	3–106	B
PLT446	4–20	A	PLT467	4–61	B	PLT502	3–107	A
PLT446	4–21	B	PLT467	4–138	B	PLT502	3–108	B
PLT447	4–28	B	PLT467	4–149	B	PLT506	4–8	A
PLT448	4–33	C	PLT467	4–152	C	PLT506	4–10	A
PLT448	4–50	B	PLT467	4–153	C	PLT507	10–2	C
PLT448	4–51	B	PLT467	4–154	B	PLT507	10–3	C
PLT448	4–52	B	PLT467	4–155	B	PLT508	4–186	C
PLT448	4–53	C	PLT468	4–136	B	PLT509	3–48	C
PLT448	4–54	A	PLT473	1–1	A	PLT509	3–49	A
PLT448	4–55	C	PLT473	1–2	C	PLT509	3–50	C
PLT448	4–56	B	PLT475	7–10	B	PLT509	3–51	C
PLT448	4–57	A	PLT477	1–20	C	PLT509	3–52	C
PLT448	4–58	B	PLT478	2–60	A	PLT509	3–53	B
PLT448	4–59	C	PLT478	2–61	B	PLT509	3–54	B
PLT448	4–60	B	PLT478	2–78	B	PLT509	3–55	A
PLT448	4–62	C	PLT479	2–84	A	PLT509	3–56	A
PLT448	4–63	C	PLT485	11–20	A	PLT511	7–6	C
PLT448	4–64	A	PLT485	11–21	A	PLT511	7–7	A
PLT448	4–68	A	PLT486	11–19	C	PLT511	7–8	A
PLT448	4–73	C	PLT492	7–54	A	PLT511	7–48	A
PLT448	4–74	C	PLT493	1–25	A	PLT511	7–49	C
PLT449	4–37	C	PLT493	7–32	B	PLT511	7–50	C
PLT451	4–35	B	PLT494	7–5	C	PLT511	7–51	A
PLT455	11–1	B	PLT495	7–9	B	PLT512	7–1	C
PLT455	11–5	C	PLT495	7–11	A	PLT512	7–28	C
PLT455	11–6	C	PLT495	7–12	B	PLT512	7–29	C
PLT455	11–7	C	PLT495	7–13	A	PLT512	7–30	B
PLT457	4–36	B	PLT495	7–14	B	PLT512	7–31	A
PLT457	4–43	C	PLT495	7–15	B	PLT512	7–33	A
PLT462	3–26	C	PLT495	7–16	A	PLT512	7–43	B
PLT463	4–26	A	PLT495	7–17	A	PLT512	7–44	C
PLT463	4–27	B	PLT497	3–90	C	PLT514	8–1	C
PLT463	4–87	C	PLT497	3–91	B	PLT514	8–2	C

FAA Learning Code	Gleim SU/ Q. No.	Gleim Answer
PLT514	8–3	C
PLT514	8–4	A
PLT514	8–7	A
PLT514	8–8	C
PLT514	8–9	A
PLT514	8–10	A
PLT514	8–22	A
PLT515	8–51	C
PLT515	8–52	A
PLT515	8–53	A
PLT516	7–3	B
PLT516	7–4	C
PLT518	7–25	C
PLT518	7–26	C
PLT518	7–27	B
PLT520	5–10	A
PLT520	5–11	C

AUTHORS' RECOMMENDATIONS

The Experimental Aircraft Association, Inc., is a very successful and effective nonprofit organization that represents and serves those of us interested in flying, in general, and in sport aviation, in particular. We personally invite you to enjoy becoming a member. Visit their website at www.eaa.org.

Types of EAA Memberships:

$40 -   Individual (includes subscription to *EAA Sport Aviation* magazine)
$50 -   Family (extends all benefits to member's spouse and children under 18, except for an additional EAA magazine subscription)
Free -   Student (for those age 18 or under who have completed the EAA Young Eagles program)
$1,295 - Lifetime

Write: EAA Aviation Center   *Call:* (920) 426-4800
3000 Poberezny Rd.   (800) JOIN-EAA
Oshkosh, Wisconsin 54902   *Email:* membership@eaa.org

The annual EAA Oshkosh AirVenture is an unbelievable aviation spectacular with over 10,000 airplanes at one airport and virtually everything aviation-oriented you can imagine! Plan to spend at least 1 day (not everything can be seen in a day) in Oshkosh (100 miles northwest of Milwaukee). Visit the AirVenture website at www.airventure.org.

Convention dates: 2014 -- July 28 through August 3
2015 -- July 27 through August 2

The annual Sun 'n Fun EAA Fly-In is also highly recommended. It is held at the Lakeland, FL (KLAL), airport (between Orlando and Tampa). Visit the Sun 'n Fun website at www.sun-n-fun.org.

Convention dates: 2014 -- April 1 through April 6
2015 -- April 21 through April 26
2016 -- April 5 through April 10

AIRCRAFT OWNERS AND PILOTS ASSOCIATION

AOPA is the largest, most influential aviation association in the world, with more than 415,000 members--two thirds of all pilots in the United States. AOPA's most important contribution to the world's most accessible, safest, least expensive, friendliest, easiest-to-use general aviation environment is their lobbying on our behalf at the federal, state, and local levels. AOPA also provides legal services, advice, and other assistance to the aviation community.

We recommend that you become an AOPA member, which costs only $45 annually. To join, call 1-800-USA-AOPA or visit the AOPA website at www.aopa.org.

LET'S GO FLYING!

The Aircraft Owners and Pilots Association (AOPA) hosts an informational web page on getting started in aviation. "Let's Go Flying!" contains information for those still dreaming about flying, those who are ready to begin, and those who are already making the journey.

The goal of this program is to encourage people to experience their dreams of flying through an introductory flight. Interested individuals can order a FREE copy of *Let's Go Flying: Your Invitation to Fly*, which explains how amazing it is to be a pilot. Other resources are available, such as a flight school finder, a guide on what to expect throughout training, an explanation of pilot certification options, a FREE monthly flight training newsletter, and much more. To learn more, visit www.aopa.org/letsgoflying.

358

INSTRUCTOR CERTIFICATION FORM
PRIVATE PILOT KNOWLEDGE TEST

Name: _____

     I certify that I have reviewed the above individual's preparation for the FAA Private Pilot—Airplane knowledge test [covering the topics specified in 14 CFR 61.105(b)(1) through (13)] using the *Private Pilot FAA Knowledge Test* book, software, and/or online course by Irvin N. Gleim and find him/her competent to pass the knowledge test.

_____ _____ _____ _____ _____

       Signed   Date   Name   CFI Number   Expiration Date

* *

INSTRUCTOR CERTIFICATION FORM
RECREATIONAL PILOT KNOWLEDGE TEST

Name: _____

     I certify that I have reviewed the above individual's preparation for the FAA Recreational Pilot—Airplane knowledge test [covering the topics specified in 14 CFR 61.97(b)(1) through (12)] using the *Recreational Pilot FAA Knowledge Test* book, software, and/or online course by Irvin N. Gleim and find him/her competent to pass the knowledge test.

_____ _____ _____ _____ _____

       Signed   Date   Name   CFI Number   Expiration Date

ABBREVIATIONS AND ACRONYMS IN
PRIVATE PILOT FAA KNOWLEDGE TEST

A/FD	*Airport/Facility Directory*		METAR	aviation routine weather report
AAH	*Advanced Avionics Handbook*		MH	magnetic heading
AC	Advisory Circular		MOA	Military Operations Area
ACL	Aeronautical Chart Legend		MSL	mean sea level
AD	Airworthiness Directive		MTR	Military Training Route
ADF	automatic direction finder		MVFR	marginal VFR
AFH	*Airplane Flying Handbook*		NDB	nondirectional radio beacon
AFSS	Automated Flight Service Station		NFCT	nonfederal control tower
AGL	above ground level		NM	nautical mile
AIM	*Aeronautical Information Manual*		NOTAM	notice to airmen
AIRMET	Airmen's Meteorological Information		NPRM	Notice of Proposed Rulemaking
AME	aviation medical examiner		NTSB	National Transportation Safety Board Regulations
ANDS	accelerate north, decelerate south			
AOE	airport of entry		OAT	outside air temperature
ARTS	Automated Radar Terminal System		OBS	omnibearing selector
ASEL	airplane single-engine land		PAPI	precision approach path indicator
ATA	actual time of arrival		PCL	pilot-controlled lighting
ATC	Air Traffic Control		*PHAK*	*Pilot's Handbook of Aeronautical Knowledge*
ATIS	Automatic Terminal Information Service			
AvW	*Aviation Weather*		PIC	pilot in command
AWBH	*Aircraft Weight and Balance Handbook*		PIREP	Pilot Weather Report
AWS	*Aviation Weather Services*		RB	relative bearing
CDI	course deviation indicator		SFC	surface
CDT	central daylight time		SIGMET	Significant Meteorological Information
CFI	Certificated Flight Instructor		SM	statute mile
CG	center of gravity		STC	supplemental type certificate
CH	compass heading		SVFR	special VFR
CT	control tower		TACAN	Tactical Air Navigation
CTAF	Common Traffic Advisory Frequency		TAF	terminal aerodrome forecast
DME	distance measuring equipment		TAS	true airspeed
DT	daylight time		TC	true course
DUAT	Direct User Access Terminal		TH	true heading
EFAS	En Route Flight Advisory System		TWEB	Transcribed Weather Broadcast
ELT	emergency locator transmitter		UHF	ultra high frequency
ETA	estimated time of arrival		UTC	Coordinated Universal Time
ETD	estimated time of departure		V_A	maneuvering speed
FA	area forecast		VASI	visual approach slope indicator
FAA	Federal Aviation Administration		V_{FE}	maximum flap extended speed
FAR	Federal Aviation Regulations		VFR	visual flight rules
FBO	Fixed-Base Operator		VHF	very high frequency
FCC	Federal Communications Commission		VHF/DF	VHF direction finder
FD	winds and temperatures aloft forecast		V_{LE}	maximum landing gear extended speed
Fl Comp	Flight Computer		V_{NE}	never-exceed speed
FL	flight level		V_{NO}	maximum structural cruising speed
FSDO	Flight Standards District Office		VOR	VHF omnidirectional range
FSS	Flight Service Station		VORTAC	Collocated VOR and TACAN
GPH	gallons per hour		VOT	VOR test facility
Hg	mercury		VR	visual route
HP	horsepower		V_{S0}	stalling speed or the minimum steady flight speed in the landing configuration
IAS	indicated airspeed			
ICAO	International Civil Aviation Organization		V_{S1}	stalling speed or the minimum steady flight speed obtained in a specific configuration
IFH	*Instrument Flying Handbook*			
IFR	instrument flight rules		V_X	speed for best angle of climb
IR	instrument route		V_Y	speed for best rate of climb
ISA	International Standard Atmosphere		WCA	wind correction angle
LLWAS	lowlevel wind-shear alert system		Z	Zulu or UTC time
mb	millibar			
MB	magnetic bearing			
MC	magnetic course			
MEF	maximum elevation figure			

INDEX OF LEGENDS AND FIGURES

INDEX

GLEIM® Pilot Training Kits with Online Ground School

Sport	$199.95 _____
Private	$249.95 _____
Instrument	$249.95 _____
Commercial	$174.95 _____
Instrument/Commercial	$341.95 _____
Sport Pilot Flight Instructor	$174.95 _____
Flight/Ground Instructor	$174.95 _____
ATP	$189.95 _____

Also Available:

Flight Engineer Online Ground School	$99.95 _____
Flight Engineer Test Prep Online	$64.95 _____

Shipping (nonrefundable): **$20 per kit** $ _____
(Alaska and Hawaii please call for shipping price)
Add applicable sales tax for shipments within the state of Florida. $ _____
For orders outside the United States, please visit our website at
www.gleim.com/aviation/products.php to place your order.   TOTAL $ _____

Reference Materials and Other Accessories Available by Contacting Gleim.

TOLL FREE: 800.874.5346 ext. 471
WEBSITE: gleim.com

LOCAL: 352.375.0772 ext. 471
FAX: 352.375.6940
EMAIL: aviationteam@gleim.com

Gleim Publications, Inc.
P.O. Box 12848
Gainesville, FL 32604

NAME (please print) _____

ADDRESS_____ Apt. _____
                  (street address required for UPS)

_____

CITY _____ STATE _____ ZIP _____

_____ MC/VISA/DISC _____ Check/M.O.   Daytime Telephone (_____) ____ - _____

Credit Card # _____ - _____ - _____ - _____

Exp. _____ / _____   Signature _____
        Mo./Yr.

Email Address _____

1. We process and ship orders daily, within one business day over 98.8% of the time. Call by 3:00 pm for same day service.

2. Gleim Publications, Inc. guarantees the immediate refund of all resalable texts, unopened and un-downloaded Test Prep Software, and unopened and un-downloaded audios returned within 30 days of purchase. Accounting and Academic Test Prep Online and other online courses may be canceled within 30 days of purchase if no more than the first study unit or lesson has been accessed. In addition, Online CPE courses may be canceled within 30 days of adding the course to your Personal Transcript if the outline has not yet been accessed. Accounting Exam Rehearsals and Practice Exams may be canceled within 30 days of purchase if they have not been started. Aviation Test Prep Online may be canceled within 30 days of purchase if no more than the first study unit has been accessed. Other Aviation online courses may be canceled within 30 days of purchase if no more than two study units have been accessed. This policy applies only to products that are purchased directly from Gleim Publications, Inc. No refunds will be provided on opened or downloaded Test Prep Software or audios, partial returns of package sets, or shipping and handling charges. Any freight charges incurred for returned or refused packages will be the purchaser's responsibility.

3. Please PHOTOCOPY this order form for others.

4. No CODs. Orders from individuals must be prepaid   Prices subject to change without notice. 06/13

If you see topics covered on your FAA knowledge test that are not contained in this book, please email us at aviation@gleim.com or mail us this form to report your experience and help us fine-tune our test preparation materials.

This form can be mailed to **Irvin N. Gleim • c/o Gleim Publications, Inc. • P.O. Box 12848 • University Station • Gainesville, Florida • 32604.** Please include your name and address so we can properly thank you for your interest.

1. _____

2. _____

3. _____

4. _____

5. _____

6. _____

7. _____

8. _____

9. _____

10. _____

11. _____

12. _____

13. _____

14. _____

15. _____

16. _____

17. _____

18. _____

Name: _____

Address: _____

City/State/Zip: _____

Telephone: Home: _____ Work: _____ Fax: _____

Email: _____